Deep Learning and Computer Vision in Remote Sensing

Deep Learning and Computer Vision in Remote Sensing

Editors

Fahimeh Farahnakian
Jukka Heikkonen
Pouya Jafarzadeh

MDPI • Basel • Beijing • Wuhan • Barcelona • Belgrade • Manchester • Tokyo • Cluj • Tianjin

Editors
Fahimeh Farahnakian
University of Turku
Turku
Finland

Jukka Heikkonen
University of Turku
Turku
Finland

Pouya Jafarzadeh
University of Turku
Turku
Finland

Editorial Office
MDPI
St. Alban-Anlage 66
4052 Basel, Switzerland

This is a reprint of articles from the Special Issue published online in the open access journal *Remote Sensing* (ISSN 2072-4292) (available at: https://www.mdpi.com/journal/remotesensing/special_issues/deep_learning_computer_vision_remote_sensing).

For citation purposes, cite each article independently as indicated on the article page online and as indicated below:

LastName, A.A.; LastName, B.B.; LastName, C.C. Article Title. *Journal Name* **Year**, *Volume Number*, Page Range.

ISBN 978-3-0365-6368-8 (Hbk)
ISBN 978-3-0365-6369-5 (PDF)

Contents

About the Editors

Fahimeh Farahnakian

Fahimeh Farahnakian is currently an adjunct professor (docent) in the Algorithms and Computational Intelligence Research Lab, Department of Future Technologies, University of Turku, Finland. Her research interests include the theory and algorithms of machine learning, computer vision and data analysis methods, and their applications in various different fields. She has published +30 articles in journal and conference proceedings. She is a member of the IEEE and has also served on the program committees of numerous scientific conferences.

Jukka Heikkonen

Jukka Heikkonen is a full professor and head of the Algorithms and Computational Intelligence Research Lab, University of Turku, Finland. His research focuses on data analytics, machine learning, and autonomous systems. He has worked at top-level research laboratories and Centers of Excellence in Finland and international organizations (the European Commission and Japan) and has led many international and national research projects. He has authored more than 150 peer-reviewed scientific articles. He has served as an organizing/program committee member in numerous conferences and has acted as a guest editor in five Special Issues of scientific journals.

Pouya Jafarzadeh

Pouya Jafarzadeh received an MS degree in Technological Competence Management from the University of Applied Silence, Turku, Finland. He is currently working toward a PhD degree in the Algorithms and Computational Intelligence Research Lab, University of Turku, Finland. His research interests include artificial intelligence, machine learning, deep learning, computer vision, and data analysis. He is a frequent reviewer for research journals.

remote sensing

MDPI

Article

Convolutional Neural Network Algorithms for Semantic Segmentation of Volcanic Ash Plumes Using Visible Camera Imagery

José Francisco Guerrero Tello [1,*], Mauro Coltelli [1], Maria Marsella [2], Angela Celauro [2] and José Antonio Palenzuela Baena [2]

[1] Istituto Nazionale di Geofisica e Vulcanologia, Osservatorio Etneo, Piazza Roma 2, 95125 Catania, Italy
[2] Department of Civil, Building and Environmental Engineering, Sapienza University of Rome, Via Eudossiana 18, 00184 Roma, Italy
* Correspondence: francisco.guerrero@ingv.it

Abstract: In the last decade, video surveillance cameras have experienced a great technological advance, making capturing and processing of digital images and videos more reliable in many fields of application. Hence, video-camera-based systems appear as one of the techniques most widely used in the world for monitoring volcanoes, providing a low cost and handy tool in emergency phases, although the processing of large data volumes from continuous acquisition still represents a challenge. To make these systems more effective in cases of emergency, each pixel of the acquired images must be assigned to class labels to categorise them and to locate and segment the observable eruptive activity. This paper is focused on the detection and segmentation of volcanic ash plumes using convolutional neural networks. Two well-established architectures, the segNet and the U-Net, have been used for the processing of in situ images to validate their usability in the field of volcanology. The dataset fed into the two CNN models was acquired from in situ visible video cameras from a ground-based network (Etna_NETVIS) located on Mount Etna (Italy) during the eruptive episode of 24th December 2018, when 560 images were captured from three different stations: CATANIA-CUAD, BRONTE, and Mt. CAGLIATO. In the preprocessing phase, data labelling for computer vision was used, adding one meaningful and informative label to provide eruptive context and the appropriate input for the training of the machine-learning neural network. Methods presented in this work offer a generalised toolset for volcano monitoring to detect, segment, and track ash plume emissions. The automatic detection of plumes helps to significantly reduce the storage of useless data, starting to register and save eruptive events at the time of unrest when a volcano leaves the rest status, and the semantic segmentation allows volcanic plumes to be tracked automatically and allows geometric parameters to be calculated.

Keywords: ANN; automatic classification; risk mitigation; machine learning

Citation: Guerrero Tello, J.F.; Coltelli, M.; Marsella, M.; Celauro, A.; Palenzuela Baena, J.A. Convolutional Neural Network Algorithms for Semantic Segmentation of Volcanic Ash Plumes Using Visible Camera Imagery. *Remote Sens.* **2022**, *14*, 4477. https://doi.org/10.3390/rs14184477

Academic Editors: Jukka Heikkonen, Fahimeh Farahnakian and Pouya Jafarzadeh

Received: 4 July 2022
Accepted: 29 August 2022
Published: 8 September 2022

1. Introduction

Volcano monitoring is composed of a set of techniques that enable the measurement of different parameters (geochemical, seismic, thermal, deformational, etc.) [1]. Keeping these parameters under surveillance is essential for risk mitigation and guarantees security to the population. These parameters allow us to know the state of internal and external activity of a volcano and to know if there are changes in the behaviour of the volcano that can lead to an eruption or to understand if there are changes during an eruptive event. Although seismic and geodetic instruments permit quasi-real-time monitoring, video cameras are also currently a standard and necessary tool for effective volcano observation [2,3].

Explosive volcanic eruptions eject a big quantity of pyroclastic products into the atmosphere. In these events, continuous surveillance is mandatory to avoid significant damage in rural and metropolitan areas [4] that may disrupt the surface and air traffic [5], and even may cause negative impacts on human health [6]. In 1985, the eruption of "Nevado del Ruiz" volcano in Colombia ejected more than 35 tons of pyroclastic flow that reached 30 km in height. This eruption melted the ice and created four lahars that descended through the slopes of the volcano and destroyed a whole town called "Armero" located 50 km from the volcano, with a loss of 24.800 lives [7]. To counteract further disasters, it is fundamental to create new methodologies and instruments based on innovation for risk mitigation. Video cameras have proven suitable for tracking those pyroclastic products in many volcanoes in the world, whether with visible (0.4–0.7 μm) or near-infrared (~1 μm) wavelength. Both sensors are suitable to collect and analyse information at a long distance.

Video cameras installed on volcanoes often experience limited performance in relation to crisis episodes. They are programmed to capture images in a specific time range (i.e., one capture per minute, one capture every two minutes, etc.); those settings lead to the storage of unnecessary data that need to be deleted manually by an operator with time-consuming tasks. On the other hand, video cameras do not have an internal software to deeply analyse images in real time. This work is carried out after downloading by applying different computer vision techniques to calibrate the sensor [8] and extract relevant information by edge-detection algorithms and GIS-based methods, such as contours detections and statistics classification, such as PCA [9]. All these kinds of postprocessing procedures involve semi-automatics and time-consuming tasks.

These limitations can be faced through machine-learning techniques for computing vision. In the last decade, technological innovation has increased dramatically in the world of artificial intelligence (AI) and machine learning (ML) in parallel to video cameras [10]. The convolutional neural networks (CNN) became popular because they outperformed any other network architecture on computer vision [11]. Specifically, the architecture U-Net is nowadays being routinely and successfully used in image processing, reaching an accuracy similar to or even higher than other existing ANN, for example, of the FCN type [12–14], providing multiple applications where pattern recognition and feature extraction play an essential role. CNNs have been applied to find solutions to mitigate risk in different environmental fields, such as for the detection and segmentation of smoke and forest fires [15,16], flood detection [17], and to find solutions regarding global warming, for example, through monitoring of the ice of the poles [18,19]. CNNs have been applied in several studies in the field of volcanology for earthquake detection and classification [20,21], for the classification of volcanic ash particles [22], and to validate their capability for real-time monitoring of the persistent explosive activity of Stromboli volcano [23], for video data characterisation [2], detection of volcanic unrest [24], and volcanic eruption detection using satellite images [25–27]. Thus, the importance of applying architectures based on CNN could be an alternative to improve the results obtained in the different scientific works performed till now.

This research aims to create algorithms that help solve computer vision problems based on deep learning for the detection and segmentation of the volcanic plume, providing an effective tool for emergency management to risk management practitioners. The concept of this tool focuses on a neural network which is fed with data from the 24th to 27th December 2018 eruptive event. The eruption that began at noon was preceded by 130 earthquake tremors, the two strongest of which measured 4.0 and 3.9 on the Richter scale. From this eruptive event, 560 images were collected and then preprocessed and split into 80% training and 20% validation. The training dataset was used in the training of two very consolidated models: the SegNet Deep Convolutional Encoder-Decoder and U-net architectures. In this groundwork phase, more consolidated models were sought to have a large comparative pool and to substantiate their use in the volcanological field. As a result, a trained model is generated to automatically detect the beginning of an eruptive activity and tracking the entire eruptive episode. Automatic detection of the volcanic plume supports volcanic monitoring to store useful information enabling real-time tracking of the plume and the extraction of concerning geometric parameters. By developing a comprehensive and reliable approach, it is possible to extend it to many other explosive volcanoes. The current results encourage a broader research objective that will be oriented towards the creation of more advanced neural networks [2], deepening the real-time monitoring for observing precursors, such as change in degassing state.

2. Geological Settings

Mt. Etna is a basaltic volcano located in Sicily in the middle of Gela-Catania foredeep, at the front of the Hyblean Foreland [28] (Figure 1). This volcano is one of the most active in the world with its nearly continuous eruptions and lava flow emissions and, with its dimensions, it represents a major potential risk to the community inhabiting its surroundings.

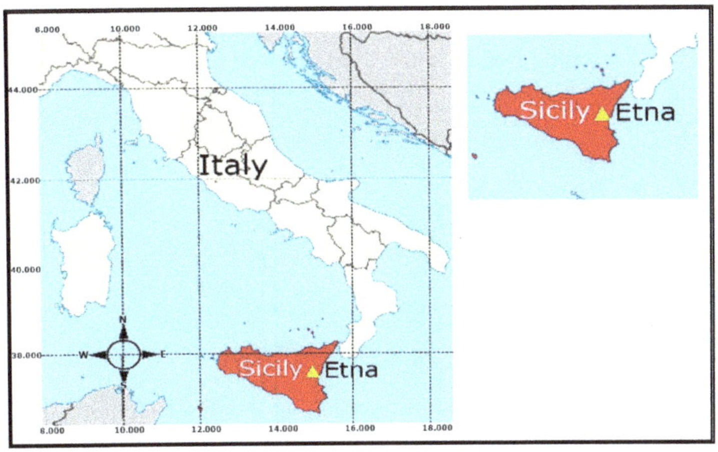

Figure 1. Location of Etna volcano.

The geological map, updated in 2011 [29] at the scale of 1:50,000, is a dataset of the Etna eruptions that occurred throughout its history (Figure 2, from [29], with modifications). This information is fundamental for land management and emergency planning.

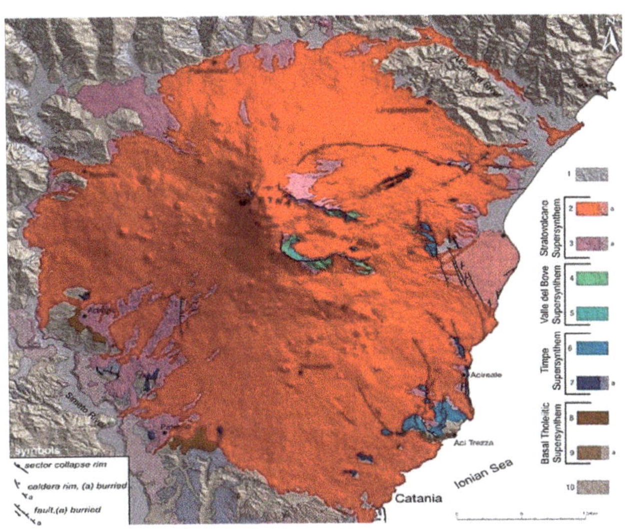

Figure 2. Geological map of Mt. Etna.

3. Etna_NETVIS Network

Mt. Etna has become one of the better monitored volcanoes in the world by using several instrumental networks. One of them is the permanent terrestrial Network of Thermal and Visible Sensors of Mount Etna, which comprises thermal and visible cameras located at different sites on the southern and eastern flanks of Etna. The network, initially composed of CANON VC-C4R visible (V) and FLIR A40 Thermal (T) cameras installed in Etna Cuad (ECV), Etna Milo (EMV), Etna Montagnola (EMOV and EMOT), and Etna Nicolosi (ENV and ENT), has been recently upgraded (since 2011) by adding high-resolution (H) sensors (VIVOTEK IP8172 and FLIR A320) at the Etna Mt. Cagliato (EMCT and EMCH), Etna Montagnola (EMOH), and Etna Bronte (EBVH) sites [3]. Visible spectrum video cameras used in this work and examples of field of view (FOV), Bronte, Catania, and Mt. Cagliato are shown in Figure 3. These surveillance cameras do not allow 3D model extraction due to poor overlap, unfavourable baseline, and low image resolution. Despite this, simulation of the camera network geometry and sensor configuration have been carried out in a previous project (MEDSUV project [3]) and will be adopted as a reference for future implementation of the Etna Network.

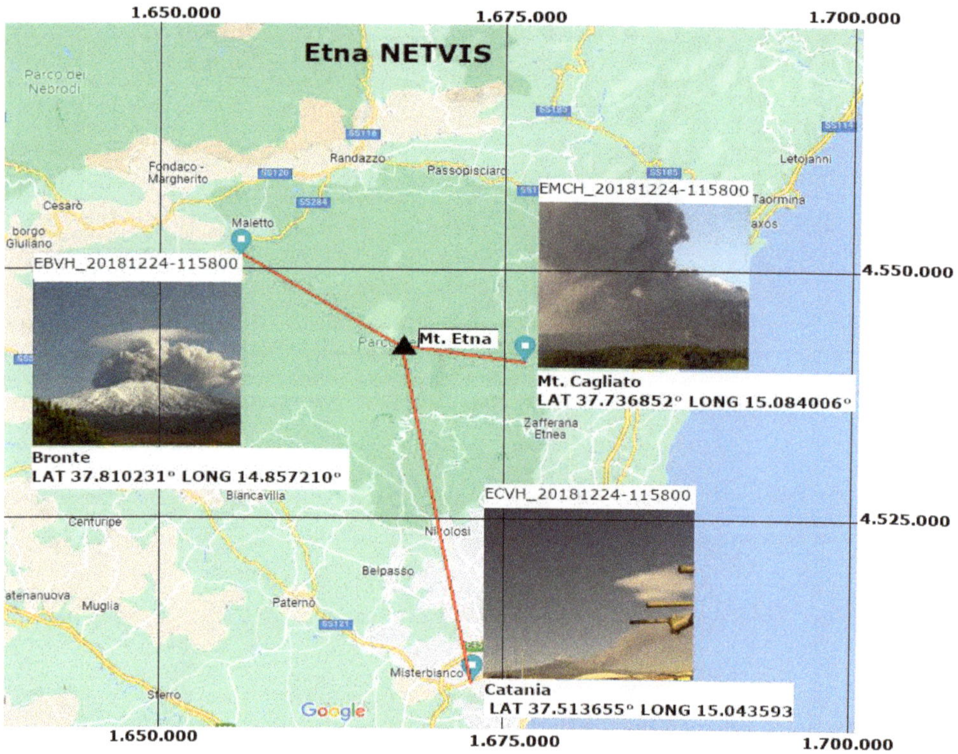

Figure 3. Etna_Netvis surveillance network.

The technical specifications of Etna_NETVIS network cameras used in this work, such as pixel resolution, linear distance to the vent, and horizontal and vertical field of view (HFOV and VFOV), are described in Table 1.

Table 1. Characteristics of the ETNA NETVIS cameras.

ETNA NETVIS					
Station Name	**Resolution Pixel**	**Distance to the Vent**	**Image Captured per Minute**	**Model**	**Angular FOV (deg)**
BRONTE	760 × 1040	13.78 km	1	VIVOTEK	33_~93_ (horizontal), 24_~68_ (vertical)
CATANIA	2560 × 1920	27 km	1		
MONTE CAGLIATO	2560 × 1920	8 km	2	VIVOTEK	33_~93_ (horizontal), 24_~68_ (vertical)

4. Materials and Methods

4.1. Materials: Data Preparation

The paradigm used for this work was a supervised learning based on a set of samples consisting of a pair of data; input variables (x) and output labelled variables (y). Data labelling is the crucial part of the data preprocessing in the workflow to build a neural network model, which requires large volumes of high-quality training data. The processes for creating label data are expensive, complicated, and time-consuming. Many open-source libraries, such as MNIST by Keras, offer a full dataset ready to use, but it covers neither all types of objects nor labelled data for volcanic ash plume shapes. Thus, the 560 images collected were manually labelled using an open-source image editor "GIMP" to delineate the boundaries of volcanic plums and generate the ground truth mask (Figure 4). The samples were split into two sets: training and validation in a proportion of 80% and 20%, respectively. As this research deals with a binary classification problem, the neural network is contextualised within volcanic plume shapes by assigning pixel level. Thus, pixels that are inside the ash column contour are assigned values of 255 or, otherwise, 0. Inputs with large integer values could collapse the bias value or slow down the learning process, so, to avoid this effect, pixels were normalised between 0 and 1 by applying Equation (1):

$$x' = \frac{(x - x_{min})}{(x_{max} - x_{min})} \tag{1}$$

where x is the pixel to normalize, x_{min} is the minimum value of pixels of the image, and x_{max} is the maximum value pixel of the image. To keep size consistency across the dataset while reducing memory consumption, images were resized to (768px × 768px) by applying bilinear interpolation.

Finally, to improve the robustness of the inputs, the training data were augmented through a technique called "data augmentation". It was applied with the Keras library "ImageDataGenerator" class that artificially expands the size of the dataset, creating some perturbating in our images as horizontal flips, zoom, random noise, and rotations (Figure 5). Data augmentation avoids overfitting in the training stage.

4.2. Methods: ANN and UNET

The perceptron, core concept of deep learning and convolutional neural network introduced by Rosenblatt [30], in brief, consists of a single-layer neural network whose base algorithms are the threshold function and the gradient descent [31]. The latter method is the most popular algorithm that performs parametrisation and optimisation of the parameters in the artificial neural network (ANN), by means of labelled samples and process iterations for the prediction of accurate outputs [31].

The optimisation minimises the loss function (or cost function), represented by the cross-entropy as a measure of the difference between the actual and predicted classes. Finally, the learning rate is an important parameter, used in the following sections to control the time of the algorithm and the network parameter training at every iteration, which is crucial to reach the expected results of the refined model. These parameters are here briefly introduced, leaving the theoretical digression to dedicated sources [30,31].

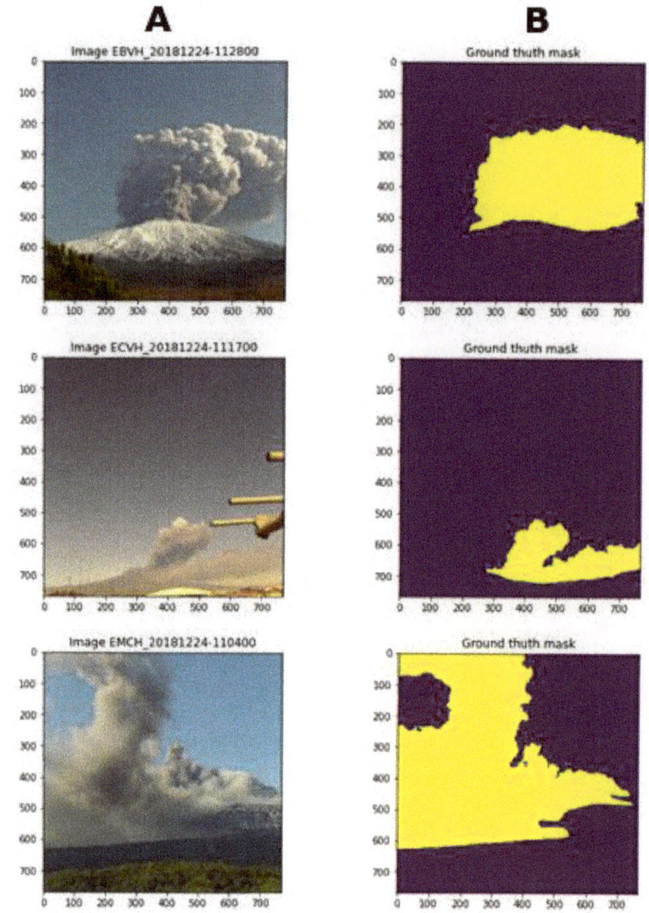

Figure 4. Examples of variable pairs (in (**A**) the real images are shown and (**B**) represents the ground truth mask).

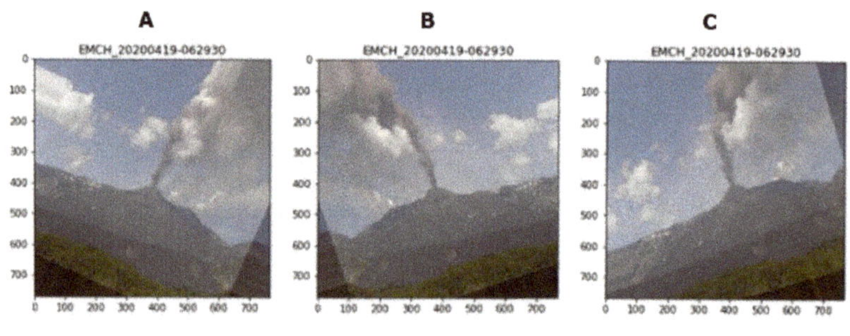

Data augmentation (Vertical flips)

Figure 5. Example of data augmentation with vertical and horizontal flips ((**A**) is a vertical right flipped image of 60 inclination degrees, (**B**) is a horizontal and vertical flipped and (**C**) is a horizontal and vertical flipped with distortion).

Convolutional Neural Network Architectures

Segmentation is a fundamental task for image analysis. Semantic segmentation describes the process of associating each pixel in an image with a class label. Segmenting images of volcanic plumes is a complicated task, different from segmenting other objects, such as people, cars, roads, buildings, and other entities that are well differentiated from their background. Those types of objects are considered homogeneous and regular in form and radiometry, but a volcanic plume can have very different physical properties [32], such as shapes, colour, and density. In deep learning, CNN appears as a class of ANN based on the shared-weight architecture of the convolution kernels [11] and proved very efficient for pattern recognition, feature extraction for applications in computer vision analysis and image recognition [33], classification [34], and segmentation [35]. This is useful to solve problems as faced in this paper. Thus, this paper presents developed models based on specific CNN architectures.

Different algorithms were implemented to develop a tool able to segment a volcanic ash plume from in situ images, creating two models based on architectures of SegNet [36] and U-Net [37]. Those trained models were carried out using Tensorflow GPU version 2.12 [38], Python 3.6 language, and Keras 2.9 [39], all of these based on open-source libraries and built on Tensorflow framework. Keras appears here as the core language for ANN programming, as it contains numerous implementations of commonly used neural network building blocks, such as layers, activation functions, optimizers, metrics, and tools, to preprocess images.

The U-net (Figure 6) is a CNN architecture for the segmentation of images, developed by Olaf Ronneberger et al. [37] and used for medical scope, but now applied in several other fields [40–43]. It is built upon the symmetric fully convolutional network and is made up of two parts. The down-sampling network (encoder) reduces dimensionality of the features while losing spatial information; instead, the up-sampling network (decoder) enables the up-sampling of an input feature map to a desired output feature map using some learnable parameters based on transposed convolutions. Thus, it is an end-to-end fully convolutional network (FCN) that makes it possible to accept images of any size.

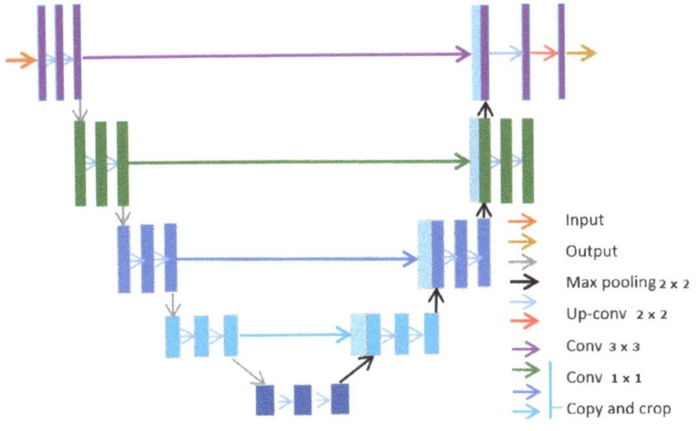

Figure 6. U-net architecture.

On the other hand, the SegNet architecture [36] FCN is based on decoupled encoder–decoder, where the encoder network is based on convolutional layers, while the decoder is based on up-samples. The architecture of this model is shown in Figure 7. It is a symmetric network where each layer of encoder has a corresponding layer in the decoder.

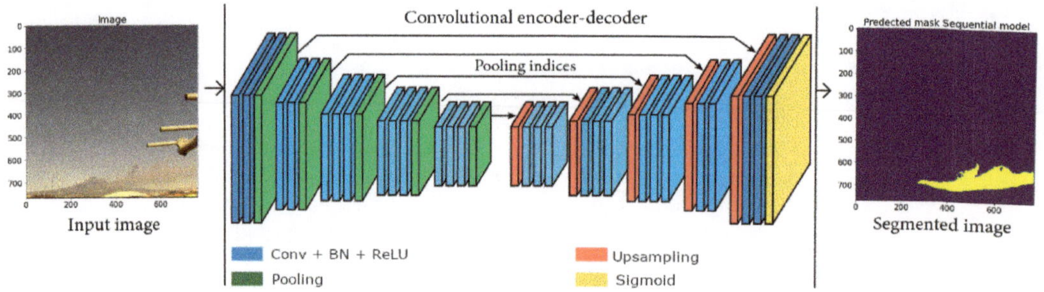

Figure 7. SegNet architecture.

Loss functions are used to optimize the model during training stage, aiming at minimising the loss function (error). The lower the value of loss function, the better the model. Cross-entropy loss is the most important loss function to face classification problems. The problem tackled in this work is a single classification problem and the loss function applied was a binary cross-entropy (Equation (2)):

$$Loss = -\frac{1}{N}\sum_{i=1}^{N} y_i * log y_i' + (1 - y_i) * log(1 - y_i') \quad (2)$$

where y_i' is the i-th scalar value in the model output, y_i is the corresponding target value, and **N** is the number of scalar values in the model output.

A deep learning model is highly dependent on hyperparameters, and hyperparameter optimisation is essential to reach good results. In this work, a CNN based on U-net architecture was built, capable of segmenting volcanic plumes from visible cameras. The values assigned to model parameters are shown in Table 2.

Table 2. Hyperparameters required for the training phase for both CNN architectures.

Hyperparameters Required for Training	
Learning Rate	0.0001
Batch_Size	4
Compile networks	
Optimiser	adam
Loss	binary_crossentropy
Metrics	Accuracy; iou_score
Fit Generator	
Step_per_epoch	112
Validation_steps	28
epochs	100

The encoder and encoder networks contain five layers with the configuration shown in Table 3.

Table 3. Convolutional layers description for U-Net architecture.

Input Layer	**A 2D Image with Shape (768, 768, 3)**						
			Encoder Network				
Convolutional Layer	**Filters**	**Kernel Size**	**Pooling Layer**	**Activations**	**Kernel Initialiser**	**Stride**	**Dropout**
Conv1	16	3 × 3	yes	ReLU	he_normal	1 × 1	No
Conv2	32	3 × 3	yes	ReLU	he_normal	1 × 1	No
Conv3	64	3 × 3	yes	ReLU	he_normal	1 × 1	No
Conv4	128	3 × 3	yes	ReLU	he_normal	1 × 1	No
Conv5	256	3 × 3	yes	ReLU	he_normal	1 × 1	No
Bottle neck	512	3 × 3	No	ReLU	he_normal		0.5
			Decoder Network				
Convolutional Layer	**Filters**	**Kernel Size**	**Concatenate Layer**	**Up-Sampling**	**Activations**	**Kernel Initializer**	**Stride**
Conv6	256	3 × 3	Conv5-Conv6	yes	ReLU	he_normal	1 × 1
Conv7	128	3 × 3	Conv4-Conv7	yes	ReLU	he_normal	1 × 1
Conv8	64	3 × 3	Conv3-Conv8	yes	ReLU	he_normal	1 × 1
Conv9	32	3 × 3	Conv2-Conv9	yes	ReLU	he_normal	1 × 1
Conv10	16	3 × 3	Conv1-Conv10	yes	ReLU	he_normal	1 × 1
Output layer	1	1 × 1	No	No	Sigmoid	he_normal	
Total trainable params	7.775.877						

The encoder and encoder networks contain five layers with the configuration shown in Table 4.

Table 4. Convolutional layers description for SegNet architecture.

Input Layer	**A 2D Image with Shape (768, 768, 3)**					
			Encoder Network			
Convolutional Layer	**Filters**	**Kernel Size**	**Pooling Layer**	**Activations**	**Stride**	**Dropout**
Conv1	16	3 × 3	yes	ReLU	1 × 1	No
Conv2	32	3 × 3	yes	ReLU	1 × 1	No
Conv3	64	3 × 3	yes	ReLU	1 × 1	No
Conv4	128	3 × 3	yes	ReLU	1 × 1	0.5
Conv5	256	3 × 3	yes	ReLU	1 × 1	0.5
Bottle neck	512	3 × 3	No	ReLU		0.5
			Decoder Network			
Convolutional Layer	**Filters**	**Kernel Size**	**Up-Sampling**	**Activations**	**Stride**	**Dropout**
Conv6	256	3 × 3	yes	ReLU	1 × 1	No
Conv7	128	3 × 3	yes	ReLU	1 × 1	No
Conv8	64	3 × 3	yes	ReLU	1 × 1	No
Conv9	32	3 × 3	yes	ReLU	1 × 1	No
Conv10	16	3 × 3	yes	ReLU	1 × 1	No
Output layer	1	1 × 1	No	Sigmoid		No
Total trainable params	11.005.841					

In order to show the models built and the difference in the architecture used in this work, Keras provides a function to create a plot of the neural network graph that can make more complex models easier to understand, as is shown in Figure 8.

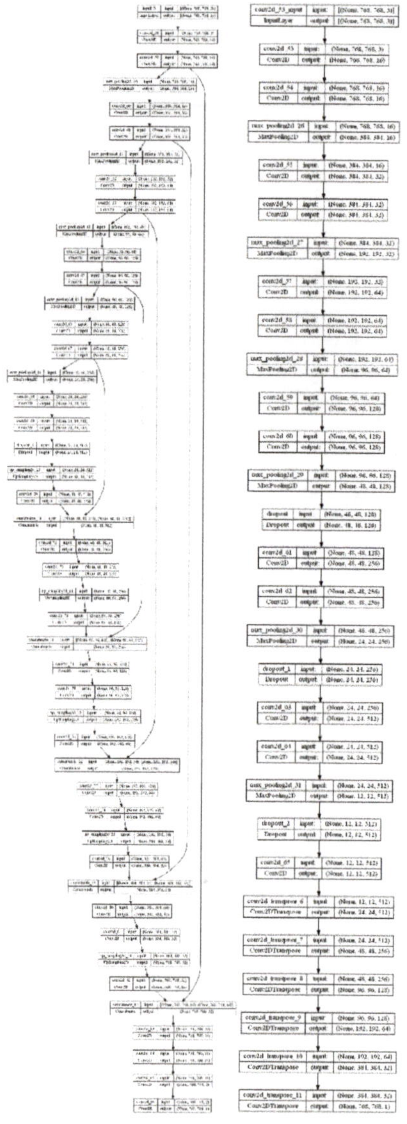

Figure 8. Left sketch of the U-net model with Deepest 4, right sketch of the SegNet model (the images are available with higher resolution at the links in [44,45]).

4.3. Evaluation of the Proposed Model

Various evaluation metrics are used to calculate the performance of the model. The evaluation metrics used in this research are explained below:

Accuracy score: it is the ratio of number of correct pixel predictions to the total number of input samples (Equation (3)).

$$Accuracy = TP/TNP \quad (3)$$

where **TP** is the number of true positives and **NPT** is the total number of predictions.

Jaccard index is the Intersection over Union (Equation (4)), where the perfect intersection has a minimum value equal to zero.

$$L(A,B) = 1 - (A \cap B/A \cup B) \tag{4}$$

where: $(\boldsymbol{A} \cap \boldsymbol{B}/\boldsymbol{A} \cup \boldsymbol{B})$ is the predicted masks overlap coefficient with the real masks between the union of that masks.

Validation curves: the trend of a learning curve can be used to evaluate the behaviour of a model and, in turn, it suggests the type of configuration changes that may be made to improve learning performance [46]. On these curve plots, both the training error (blue line) and the validation error (orange line) of the model are shown. By visually analysing both of these errors, it is possible to diagnose if the model is suffering from high bias or high variance. There are three common trends in learning curves: underfitting (high bias, low variance), overfitting (low bias, high variance) and best fitting (Figure 9).

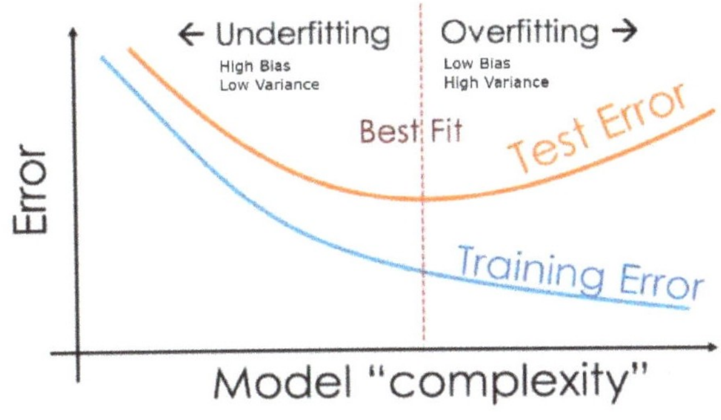

Figure 9. Underfitting, overfitting, and best fit example.

Figure 10 shows a trend graph of the cross-entropy loss of both architectures (Y axis) over number of epochs (X axis) for the training (blue) and validation (orange) datasets. For the U-Net architecture, the plot shows that the training process of our model converges well and that the plot of training loss decreases to a point of stability. Moreover, the plot of validation loss decreases to a point of stability and has a small gap with the training loss. On the other hand, for the SegNet architecture, the plot shows that the training process of our model converged well until epoch 30, then showed an increase in variance, taking to a possible overfitting. This means that the model pays a lot of attention to training data and does not generalise on the data that it has not seen before. As a result, the SegNet model performs very well on training data but has more error rates than U-net model on test data.

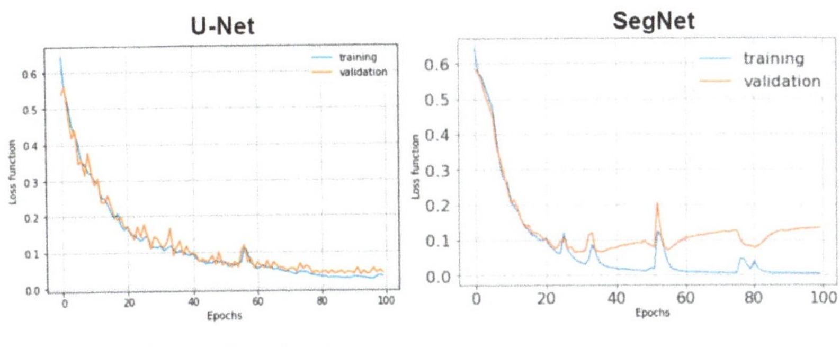

Figure 10. Trend curve of loss function.

The loss function for U-Net architecture for the training dataset is 0.026 and validation 0.316 and, for SegNet, for the training dataset is 0.018, while for the validation dataset is 0.142.

Figure 11 shows a trend graph of the accuracy metric (Y axis) over the number of epochs (X axis) for the training (blue) and validation (orange) datasets. In the Epoch 100, the accuracy value reached for the U-Net architecture training dataset is 98.35% and validation dataset is 98.28; while, for SegNet, the accuracy value for the training dataset is 98.15% and validation dataset is 97.56.

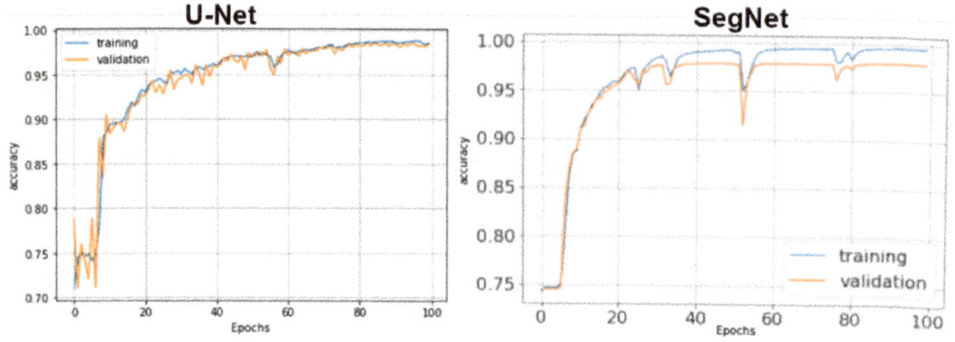

Figure 11. Trend curve of accuracy metric of training and validation dataset.

IoU (Intersection over Union) or Jaccard index is the most commonly used metric to evaluate models of semantic segmentation. It is a straightforward metric but extremely effective (metric ranges from 0 to 1, where 1 is the perfect IoU). Thus, in order to quantify the results, for both architectures, the IoUs were calculated using the validation dataset with 112 images with a step of 28 per epoch that represent 20% of the whole dataset. An average of IoU of 0.9013 was obtained for U-Net architecture and, for SegNet, an average value of IoU of 0.88 (Figure 12).

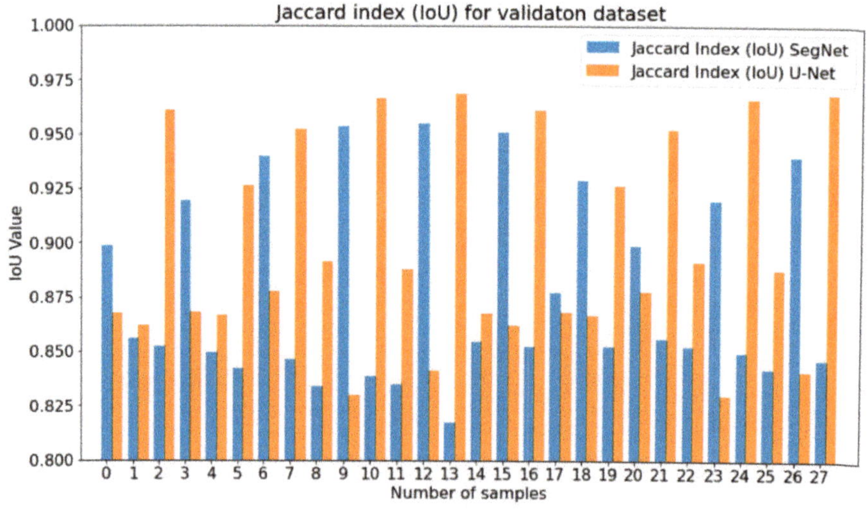

Figure 12. Jaccard index percentage for validation dataset of Unet (orange colour) and SegNet (blue colour) architectures.

In Figure 13, the predicted mask results of three samples of the validation dataset are shown, where (a) is the image, (b) is the ground truth mask (mask made by hand), (c) is the predicted mask by SegNet model, and (d) is the predicted mask by U-Net model.

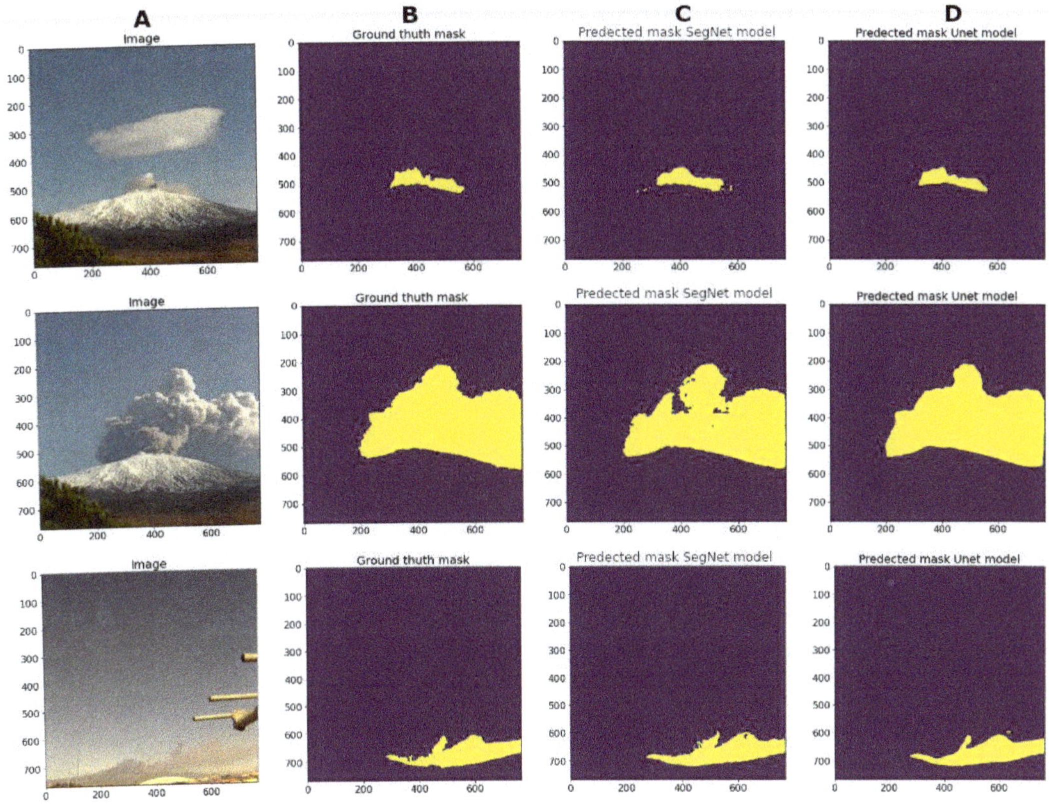

Figure 13. Original image (**A**), ground truth mask (**B**), predicted mask by SegNet (**C**), predicted mask by U-net (**D**).

Once the model was completely trained and after verifying training and validation metrics, in order to evaluate how the models performed, a test dataset (data not previously used in training and validation) was used. The samples of the data used provide an unbiased evaluation as the test dataset is the crucial standard to evaluate the model, it is well curated, and it contains carefully sampled data that cover several classes that the trained model will deal with when used in the real world, for example, images non acquired from Etna_NETVIS Network, eruptions in cloudy time, and images from other volcanoes different from Mt. Etna.

Figure 14 shows examples of photographs of different eruptive events, of which two were taken by local citizens during the Etna eruption; the one following belongs to photos of the Monte Cagliato Etna station, the fourth shows the summit crater on a cloudy day, and a last one photo was taken by local people during an eruptive event of the Galeras volcano in Colombia, where the column reached 6 km in height.

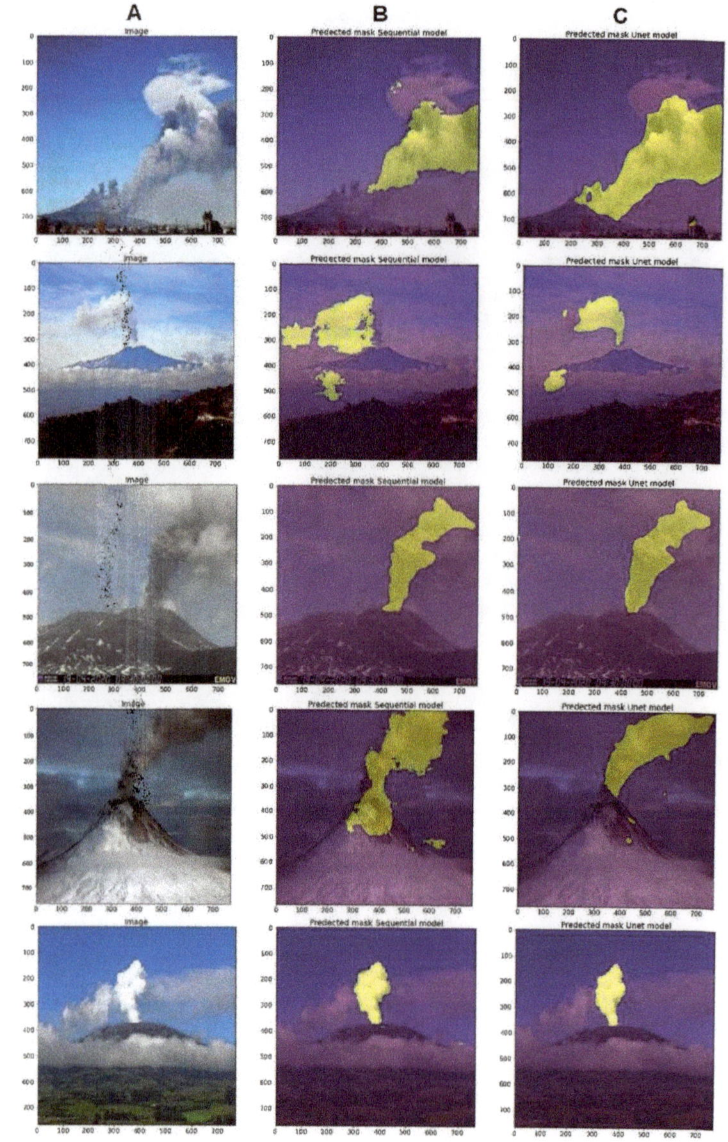

Figure 14. Semantic segmentation of results from test dataset: original image (**A**), predicted mask by SegNet (**B**), predicted mask by U-net (**C**).

5. Discussion and Concluding Remarks

In this paper, we proposed a new innovative approach based on AI for volcanic monitoring focused on the use of visible high-resolution images coming from a surveillance network of Mount Etna (Etna NETVIS). Considering that optical RGB channels and the wavelength of in situ images carry enough information, the primary aim was using all these data to solve problems related to the characterisation and monitoring of ash plumes during an explosive eruption. For this, a deep convolutional neural network was built to extract ash plume shapes automatically.

Before reaching the final results, we had to face several challenges, as the amount of data was limited; in fact, the accuracy of a neural network largely depends on the quality, quantity, and contextual meaning of training data. Even though our amount of data was limited (560 images), not enough for a model of machine learning, we hypothesised that there could have been a possible overfitting; therefore, to avoid this problem, we artificially increased the amount of data by generating new ones from the existing dataset through "data-augmentation" technique. The use of supervised learning paradigm applied in this work required that the data collected were labelled, and these preprocessing and data labelling tasks were other challenges faced in this work, which took 60% of the whole time of the full project.

In order to assess the performance of our trained deep CNN models, firstly, we measured our model error through metrics combination in a learning curve (training loss and validation loss over time). The training loss indicates how well the model is fitting the training data, while the validation loss indicates how well the model fits new data. Loss measured in the U-Net model error was of 0.026 for the training dataset and 0.0316 for the validation dataset. Secondly, we measured in the learning curve with an accuracy of 0.9835 for the training dataset and 98.28 for the validation dataset, evidencing that our model performance increased over time, which means that the model improved with experience. To reach the optimal fitting during our training, a regularisation named "early stopping" was applied to block our training when detecting an increase in the loss function value, thus avoiding the overfitting. To determine the robustness of our preliminary results, we computed the Jaccard similarity coefficient [47] to measure the similarity and diversity of sample sets. The average (IoU) value obtained from 20% of our validation dataset was equal to 91.3% of similarity. On the other hand, loss measured in SegNet model error was of 0.018 for the training dataset and 0.142 for the validation dataset. In the learning curve, an accuracy of 0.9815 was reached for the training dataset and 97.56 for the validation dataset. These results are interpreted as an increasing model performance over time but giving greater importance to the training data, which means an increase in the value of the variance, leading to possible errors in the segmentation of new data. It should be noted that the SegNet model obtained good results but always lower than those of the U-Net architecture.

The developed method is currently tested for analysis of visible images. As a future work, this method can also be integrated with images acquired from satellite sensors when the terrestrial cameras are out of coverage range. Extensive testing will be performed by exploiting the data of the open-source and on-demand platforms to validate their suitability for different types of explosive volcanoes. Moreover, this is a semi-automatic tool because the data need to be downloaded from a server storage and loaded into the deep NN. Concerning this, the creation of an internal software into the cameras is planned, which can collect and automatically analyse them by deep CNN; this will improve the performance by allowing real-time monitoring and having at disposal a powerful tool in times of emergency.

Predictably, deep learning will become one of the most transformative technologies for volcano monitoring applications. We found that deep CNN architecture was useful for the identification and classification of ash plumes by using visible images. Further studies should concentrate on the effectiveness of deep CNN architectures with large high-quality datasets obtained from remote sensing monitoring networks [25,48].

Concerning the aim of the research in the current phase, the method has been, so far, developed for plume monitoring purposes, such as detection and measurement of ash clouds emitted by large explosive eruptions, focusing on the capability of measuring the height of the plume, as the most relevant parameter to understand the magnitude of the explosion, and not yet for observing eruption precursors. By extending the procedure to process large time series of images, additional parameters can be extracted, such as elevation increase rate and temporal evolution, which can significantly contribute to set up a low-cost monitoring tool to help mitigate volcanic hazards. Furthermore, additional precious information usable as precursor indices can be derived from the monitoring of the degassing state of volcanoes. As is already noticeable in Figure 14, the algorithm allowed the distinction of a lenticular meteorological cloud from volcanic water vapor emission, excluding it from the eruption ash plume. These water vapour clouds can give important indications about changes in a volcano's degassing, considered as eruption precursors, so their discerning may be profitable for the mitigation of risks in volcanic context. However, the data used in this research are still insufficient and inadequate to detect other parameters as indicators of dew point or humidity. The important difference is that a large eruption plume is recognizable from the meteorological clouds in the background. Conversely, the degassing plume is subject to the physical condition of the atmosphere.

The results shown in this work demonstrated that this innovative approach based on deep learning is capable of detecting and segmenting volcanic ash plume and can be a powerful tool for volcano monitoring; also, the proposed method can be widely used by volcano observatories, since the trained model can be installed on standard computers where they can analyse images acquired by either own surveillance cams or from other sources through internet, as long as visibility allows, enhancing the observatory capacity in volcano monitoring.

Author Contributions: J.F.G.T. developed the neural network and performed the analysis in this chapter under the supervision of M.M. and M.C. as principal tutors; J.F.G.T. prepared the original draft; J.A.P.B., A.C., M.M. and M.C. contributed to the writing, review, and editing of the manuscript. All authors have read and agreed to the published version of the manuscript.

Funding: This research was conducted during a PhD course, with a studentship by CEIBA Colombia foundation (https://ceiba.org.co/ (accessed on 1 August 2022)), the APC was funded by Istituto Nazionale di Geofisica e Vulcanologia (INGV).

Data Availability Statement: Etna eruption 24-12-2018 dataset is curated by INGV Osservatorio Etneo Catania and is available on request (https://www.ingv.it (accessed on 1 August 2022)). Requests to access these datasets should be directed to https://www.ingv (accessed on 1 August 2022). Data presented in this study are available upon request from the corresponding author. The data is not publicly available due to source for security policy is not possible to access to data from external.

Acknowledgments: Dataset was obtained from INGV; The neural network was training in laboratory of Department of Civil, Building and Environmental Engineering of Sapienza of Roma university. We thank INGV for financial support for publishing this paper. We thank reviewers for their comments on an earlier version of the paper.

Conflicts of Interest: The authors declare no conflict of interest.

References

1. Moran, S.C.; Freymueller, J.T.; La Husen, R.G.; McGee, K.A.; Poland, M.P.; Power, J.A.; Schmidt, D.A.; Schneider, D.J.; Stephens, G.; Werner, C.A.; et al. *Instrumentation Recommendations for Volcano Monitoring at U.S. Volcanoes under the National Volcano Early Warning System*; U.S. G. S.: Scientific Investigations Report; U.S. Geological Survey: Liston, VA, USA, 2008; pp. 1–47. [CrossRef]
2. Witsil, A.J.C.; Johnson, J.B. Volcano video data characterized and classified using computer vision and machine learning algorithms. *GSF* **2020**, *11*, 1789–1803. [CrossRef]
3. Coltelli, M.; D'Aranno, P.J.V.; De Bonis, R.; Guerrero Tello, J.F.; Marsella, M.; Nardinocchi, C.; Pecora, E.; Proietti, C.; Scifoni, S.; Scutti, M.; et al. The use of surveillance cameras for the rapid mapping of lava flows: An application to Mount Etna Volcano. *Remote Sens.* **2017**, *9*, 192. [CrossRef]
4. Wilson, G.; Wilson, T.; Deligne, N.I.; Cole, J. Volcanic hazard impacts to critical infrastructure: A review. *J. Volcanol. Geotherm. Res.* **2014**, *286*, 148–182. [CrossRef]
5. Bursik, M.I.; Kobs, S.E.; Burns, A.; Braitseva, O.A.; Bazanova, L.I.; Melekestsev, I.V.; Kurbatov, A.; Pieri, D.C. Volcanic plumes and wind: Jetstream interaction examples and implications for air traffic. *J. Volcanol. Geotherm. Res.* **2009**, *186*, 60–67. [CrossRef]
6. Barsotti, S.; Andronico, D.; Neri, A.; Del Carlo, P.; Baxter, P.J.; Aspinall, W.P.; Hincks, T. Quantitative assessment of volcanic ash hazards for health and infrastructure at Mt. Etna (Italy) by numerical simulation. *J. Volcanol. Geotherm. Res.* **2010**, *192*, 85–96. [CrossRef]
7. Voight, B. The 1985 Nevado del Ruiz volcano catastrophe: Anatomy and retrospection. *J. Volcanol. Geotherm. Res.* **1990**, *42*, 151–188. [CrossRef]
8. Scollo, S.; Prestifilippo, M.; Pecora, E.; Corradini, S.; Merucci, L.; Spata, G.; Coltelli, M. Eruption Column Height Estimation: The 2011–2013 Etna lava fountains. *Ann. Geophys.* **2014**, *57*, S0214.

9. Li, C.; Dai, Y.; Zhao, J.; Zhou, S.; Yin, J.; Xue, D. Remote Sensing Monitoring of Volcanic Ash Clouds Based on PCA Method. *Acta Geophys.* **2015**, *63*, 432–450. [CrossRef]
10. Alzubaidi, L.; Zhang, J.; Humaidi, A.J.; AlDujaili, A.; Duan, Y.; AlShamma, O.; Santamaría, J.; Fadhel, M.A.; AlAmidie, M.; Farhan, L. Review of deep learning: Concepts, CNN architectures, challenges, applications, future directions. *J. Big Data* **2001**, *8*, 53. [CrossRef]
11. Zhang, W.; Itoh, K.; Tanida, J.; Ichioka, Y. Parallel distributed processing model with local space-invariant interconnections and its optical architecture. *Appl. Opt.* **1990**, *29*, 4790–4797. [CrossRef] [PubMed]
12. Öztürk, O.; Saritürk, B.; Seker, D.Z. Comparison of Fully Convolutional Networks (FCN) and U-Net for Road Segmentation from High Resolution Imageries. *Int. J. Geoinform.* **2020**, *7*, 272–279. [CrossRef]
13. Ran, S.; Ding, J.; Liu, B.; Ge, X.; Ma, G. Multi-U-Net: Residual Module under Multisensory Field and Attention Mechanism Based Optimized U-Net for VHR Image Semantic Segmentation. *Sensors* **2021**, *21*, 1794. [CrossRef] [PubMed]
14. John, D.; Zhang, C. An attention-based U-Net for detecting deforestation within satellite sensor imagery. *Int. J. Appl. Earth Obs. Geoinf.* **2022**, *107*, 102685. [CrossRef]
15. Ghali, R.; Akhloufi, M.A.; Jmal, M.; Souidene Mseddi, W.; Attia, R. Wildfire Segmentation Using Deep Vision Transformers. *Remote Sens.* **2021**, *13*, 3527. [CrossRef]
16. Frizzi, S.; Bouchouicha, M.; Ginoux, J.M.; Moreau, E.; Sayadi, M. Convolutional neural network for smoke and fire semantic segmentation. *IET Image Process* **2021**, *15*, 634–647. [CrossRef]
17. Jain, P.; Schoen-Phelan, B.; Ross, R. Automatic flood detection in Sentinel-2 imagesusing deep convolutional neural networks. In *SAC '20: Proceedings of the 35th Annual ACM Symposium on Applied Computing*; Association for Computing Machinery: New York, NY, USA, 2020; pp. 617–623.
18. Khaleghian, S.; Ullah, H.; Kræmer, T.; Hughes, N.; Eltoft, T.; Marinoni, A. Sea Ice Classification of SAR Imagery Based on Convolution Neural Networks. *Remote Sens.* **2021**, *13*, 1734. [CrossRef]
19. Zhang, C.; Chen, X.; Ji, S. Semantic image segmentation for sea ice parameters recognition using deep convolutional neural networks. *Int. J. Appl. Earth Obs. Geoinf.* **2022**, *112*, 102885. [CrossRef]
20. Perol, T.; Gharbi, M.; Denolle, M. Convolutional neural network for earthquake detection and location. *Sci. Adv.* **2018**, *4*, e1700578. [CrossRef]
21. Manley, G.; Mather, T.; Pyle, D.; Clifton, D. A deep active learning approach to the automatic classification of volcano-seismic events. *Front. Earth Sci.* **2022**, *10*, 7926. [CrossRef]
22. Shoji, D.; Noguchi, R.; Otsuki, S. Classification of volcanic ash particles using a convolutional neural network and probability. *Sci. Rep.* **2018**, *8*, 8111. [CrossRef]
23. Bertucco, L.; Coltelli, M.; Nunnari, G.; Occhipinti, L. Cellular neural networks for real-time monitoring of volcanic activity. *Comput. Geosci.* **1999**, *25*, 101–117. [CrossRef]
24. Gaddes, M.E.; Hooper, A.; Bagnardi, M. Using machine learning to automatically detect volcanic unrest in a time series of interferograms. *J. Geophys. Res. Solid Earth* **2019**, *124*, 12304–12322. [CrossRef]
25. Del Rosso, M.P.; Sebastianelli, A.; Spiller, D.; Mathieu, P.P.; Ullo, S.L. On-board volcanic eruption detection through CNNs and Satellite Multispectral Imagery. *Remote Sens.* **2021**, *13*, 3479. [CrossRef]
26. Efremenko, D.S.; Loyola R., D.G.; Hedelt, P.; Robert, J.D.; Spurr, R.J.D. Volcanic SO2 plume height retrieval from UV sensors using a full-physics inverse learning machine algorithm. *Int. J. Remote Sens.* **2017**, *1*, 1–27. [CrossRef]
27. Corradino, C.; Ganci, G.; Cappello, A.; Bilotta, G.; Hérault, A.; Del Negro, C. Mapping Recent Lava Flows at Mount Etna Using Multispectral Sentinel-2 Images and Machine Learning Techniques. *Remote Sens.* **2019**, *11*, 1916. [CrossRef]
28. Lentini, F.; Carbone, S. Geologia della Sicilia—Geology of Sicily III-Il dominio orogenic -The orogenic domain. *Mem. Descr. Carta Geol. Ital.* **2014**, *95*, 7–414.
29. Branca, S.; Coltelli, M.; Groppelli, G.; Lentini, F. Geological map of Etna volcano, 1:50,000 scale. *Italian J. Geosci.* **2011**, *130*, 265–291. [CrossRef]
30. Rosenblatt, F. The perceptron: A probabilistic model for information storage and organization in the brain. *Psychol. Rev.* **1958**, *65*, 386–408. [CrossRef]
31. Eli Berdesky's Webstite. Understanding Gradient Descent. Available online: https://eli.thegreenplace.net/2016/understanding-gradient-descent/ (accessed on 1 April 2021).
32. Aizawa, K.; Cimarelli, C.; Alatorre-Ibargüengoitia, M.A.; Yokoo, A.; Dingwell, D.B.; Iguchi, M. Physical properties of volcanic lightning: Constraints from magnetotelluric and video observations at Sakurajima volcano, Japan. *EPSL* **2016**, *444*, 45–55. [CrossRef]
33. Hijazi, S.; Kumar, R.; Rowen, C. Using Convolutional Neural Networks for Image Recognition. Cadence Design Systems Inc. Available online: https://ip.cadence.com/uploads/901/cnn_wp-pdf (accessed on 1 April 2021).
34. Wang, J.; Yang, Y.; Mao, J.; Huang, Z.; Huang, C.; Xu, W. Cnn-rnn: A unified framework for multi-label image classification. In Proceedings of the 2016 IEEE Conference on Computer Vision and Pattern Recognition, Las Vegas, NV, USA; 2016; pp. 2285–2294.
35. Sultana, F.; Sufian, A.; Dutta, P. Evolution of Image Segmentation using Deep Convolutional Neural Network: A Survey. *Knowl.-Based Syst.* **2020**, *201*, 106062. [CrossRef]

36. Badrinarayanan, V.; Kendall, A.; Cipolla, R. SegNet: A Deep Convolutional Encoder-Decoder Architecture for Image Segmentation. *IEEE Trans. Pattern Anal. Mach. Intell.* **2017**, *39*, 2481–2495. [CrossRef] [PubMed]
37. Ronneberger, O.; Fischer, P.; Brox, T. U-Net: Convolutional networks for biomedical image segmentation. In *Medical Image Computing and Computer-Assisted Intervention 2015*; Navab, N., Hornegger, J., Wells, W.M., Frangi, A.F., Eds.; Springer International Publishing: Cham, Switzerland, 2015; pp. 234–241.
38. TensorFlow. Available online: https://www.tensorflow.org/ (accessed on 1 August 2022).
39. Wikipedia–Keras. Available online: https://en.wikipedia.org/wiki/Keras (accessed on 1 August 2022).
40. Pugliatti, M.; Maestrini, M.; Di Lizia, P.; Topputo, F. Onboard Small-Body semantic segmentation based on morphological features with U-Net. In Proceedings of the 31st AAS/AIAA Space Flight Mechanics Meeting, Charlotte, NC, USA, 31 January–4 February 2021; pp. 1–20.
41. Gonzales, C.; Sakla, W. Semantic Segmentation of Clouds in Satellite Imagery Using Deep Pre-trained U-Nets. In Proceedings of the 2019 IEEE Applied Imagery Pattern Recognition Workshop (AIPR), Washington, DC, USA, 15–17 October 2019; pp. 1–7. [CrossRef]
42. Tapasvi, B.; Udaya Kumar, N.; Gnanamanoharan, E. A Survey on Semantic Segmentation using Deep Learning Techniques. *Int. J. Eng. Res. Technol.* **2021**, *9*, 50–56.
43. Leichter, A.; Almeev, R.R.; Wittich, D.; Beckmann, P.; Rottensteiner, F.; Holtz, F.; Sester, M. Automated segmentation of olivine phenocrysts in a volcanic rock thin section using a fully convolutional neural network. *Front. Earth Sci.* **2022**, *10*, 740638. [CrossRef]
44. Github–Semantic-Segmentation-Ash- Plumes-U-net. Available online: https://github.com/jfranciscoguerrero/semantic-segmentation-ash-plumes-U-Net/blob/main/fig10_%20Sketch%20of%20the%20U-Net%20model%20with%20deepest%204.png (accessed on 30 June 2022).
45. Github-Semantic-Segmentation-Ash-Plumes-U-Net. Available online: https://github.com/jfranciscoguerrero/semantic-segmentation-ash-plumes-U-Net/blob/main/model_SegNet_volcanic.png (accessed on 2 August 2022).
46. Ghojogh, B.; Crowley, M. The theory behind overfitting, cross validation, regularization, bagging, and boosting: Tutorial. *arXiv* **2019**; arXiv:1905.12787.
47. da Fontoura Costa, L. Further generalization of the Jaccard Index. *arXiv* **2021**, arXiv:2110.09619.
48. Carniel, R.; Guzmán, S.R. Machine Learning in Volcanology: A Review. In *Updates in Volcanology-Transdisciplinary Nature of Volcano Science*; Károly, N., Ed.; IntechOpen: London, UK, 2020. [CrossRef]

remote sensing

Article

Mutual Guidance Meets Supervised Contrastive Learning: Vehicle Detection in Remote Sensing Images

Hoàng-Ân Lê [1,*], Heng Zhang [2], Minh-Tan Pham [1] and Sébastien Lefèvre [1]

1 Institut de Recherche en Informatique et Systèmes Aléatoires (IRISA), Université Bretagne Sud, UMR 6074, F-56000 Vannes, France; minh-tan.pham@irisa.fr (M.-T.P.); sebastien.lefevre@irisa.fr (S.L.)
2 Institut de Recherche en Informatique et Systèmes Aléatoires (IRISA), Université Rennes 1, F-35000 Rennes, France; heng.zhang@irisa.fr
* Correspondence: hoang-an.le@irisa.fr

Abstract: Vehicle detection is an important but challenging problem in Earth observation due to the intricately small sizes and varied appearances of the objects of interest. In this paper, we use these issues to our advantage by considering them results of latent image augmentation. In particular, we propose using supervised contrastive loss in combination with a mutual guidance matching process to helps learn stronger object representations and tackles the misalignment of localization and classification in object detection. Extensive experiments are performed to understand the combination of the two strategies and show the benefits for vehicle detection on aerial and satellite images, achieving performance on par with state-of-the-art methods designed for small and very small object detection. As the proposed method is domain-agnostic, it might also be used for visual representation learning in generic computer vision problems.

Keywords: contrastive learning; mutual guidance; spatial misalignment; vehicle detection

Citation: Lê, H.-Â.; Zhang, H.; Pham, M.-T; Lefèvre, S. Mutual Guidance Meets Supervised Contrastive Learning: Vehicle Detection in Remote Sensing Images. *Remote Sens.* **2022**, *14*, 3689. https://doi.org/10.3390/rs14153689

Academic Editors: Jukka Heikkonen, Fahimeh Farahnakian and Pouya Jafarzadeh

Received: 31 May 2022
Accepted: 27 July 2022
Published: 1 August 2022

1. Introduction

Object detection consists of two tasks: localization and classification. As they are different in nature [1] yet contribute toward the overall detection performance, deep architectures usually have two distinct prediction heads, which share the same features extracted from an input. The separated branches, despite the shared parameters, have shown inefficiency as classification scores might not well reflect proper localization [2,3], while the intersection-over-union (IOU) scores of anchor boxes might miss the semantic information [4].

The misalignment of localization and classification may be aggravated depending on the domain of application. Vehicle detection is a challenging but important problem in Earth observation. It is instrumental for traffic surveillance and management [5], road safety [6], traffic modeling [7], and urban planning [8] due to large coverage from aerial viewpoints [9]. The intrinsic challenges include, but are not limited to, the small and diverse sizes of vehicles, inter-class similarity, illumination variation, and background complexity [10,11].

A simple method to combine the localization and classification score to mutually guide the training process, recently introduced by Zhang et al. [4], has shown effectiveness in alleviating the task misalignment problem on generic computer vision datasets MS-COCO [12] and PASCAL-VOC [13]. Its ability to cope with the intricacies of remote sensing vehicle detection yet remains unexplored.

In this paper, we propose a framework inspired by the mutual guidance idea [4] for vehicle detection from remote sensing images (Figure 1). The idea is that the intersection-over-union (IOU) of an anchor box should contribute toward the predicted category and vice versa; the learned semantic information could help in providing more fitting bounding boxes.

Figure 1. Vehicle detection from the VEDAI's aerial images performed by the proposed contrastive mutual guidance loss. Class labels include car (1), truck (2), pickup (3), tractor (4), camping (5), boat (6), van (7), other (8).

To improve the semantic understanding and overcome the varied object sizes and appearances, we also propose a loss module based on the contrastive learning notion [14,15]: for each detected object, the other objects of the same class are pulled closer in the embedding space, while those of different classes are pushed away. The underlying intuition is that the features of the same-class objects should be close together in the latent space, and by explicitly imposing this, the network is forced to learn representations that better underline intra-class characteristics.

Contrastive learning is a discriminative approach to visual representation learning, which has proven effective for pre-training networks before transferring to an actual downstream task [16–20]. The well-known SimCLR framework [16] proposes applying image augmentation to create an image's positive counterpart, eliminating the need for manual annotations for pretext tasks, hence self-supervision. Our hypothesis is that different objects of the same class from aerial points of view could be considered as a result of compositions of multiple augmentation operations, such as cropping, scaling, re-coloring, adding noises, etc., which, as shown by SimCLR, should be beneficial for representation learning (Figure 2). Thus, by pulling together same-class objects and pushing away the others, the network could learn to overcome the environmental diversity and better recognize the objects of interest.

As we rely on ground truth labels to form positive and negative contrastive pairs, the proposed contrastive loss could be seen as being inspired by supervised contrastive learning [17], but applied here to object detection. The differences are that the contrastive pairs are drawn from object-instance level, not image level, and that contrastive loss is employed as an auxiliary loss in combination with the mutually guided detection loss.

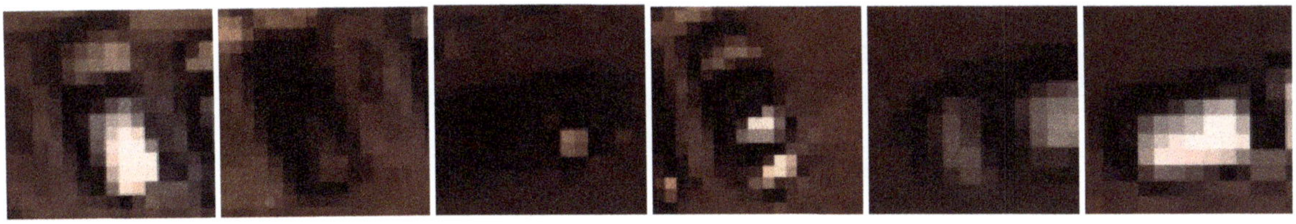

Figure 2. Different objects of the same class, "car", from an aerial point of view could be considered as passing through various compositions of image augmentation, such as cropping, rotation, re-coloring, noise adding, etc.

The contributions of the paper are fourfold, i.e.,

- applying the mutual guidance idea to a remote sensing context;
- formulating supervised contrastive learning as an auxiliary loss in a detection problem, which, to the best of our knowledge, is the first approach using supervised contrastive learning for object detection, especially in the context of Earth observation;
- improving existing detection networks for vehicle detection by combining mutual guidance and contrastive learning, termed contrastive mutual guidance or CMG;
- providing new state-of-the-art results on benchmarked datasets including VEDAI (aerial images) [21] and xView (satellite images) [22].

2. Related Work

2.1. Vehicle Detection in Remote Sensing

Deep-learning-based vehicle detection from aerial and satellite images has been an active research topic in remote sensing for Earth observation within the last decade due to intrinsically challenging natures such as intricately small vehicle sizes, various types and orientations, heterogeneous backgrounds, etc. General approaches include adapting state-of-the-art detectors from the computer vision community to apply to Earth observation context [11,23,24]. Similar to the general object detection task [25], most of the proposed methods could be divided into one-stage and two-stage approaches and are generally based on anchor box prediction. Famous anchor-based detector families such as Faster-RCNN, SSD, and YOLO have been widely exploited in remote sensing object detection, including vehicles. In [26,27], the authors proposed to modify and improve the Faster-RCNN detector for vehicle detection from aerial remote sensing images. Multi-scaled feature fusion and data augmentation techniques such as oversampling or homography transformation have proven to help two-stage detectors to provide better object proposals.

In [28,29], YOLOv3 and YOLOv4 were modified and adapted to tackle small vehicle detection from both Unmanned Aerial Vehicle (UAV) and satellite images with the objective of providing a real-time operational context. In the proposed YOLO-fine [28] and YOLO-RTUAV [29] models, the authors attempted to remove unnecessary network layers from the backbones of YOLOv3 and YOLOv4-tiny, respectively, while adding some others to focus on small object searching. In [23], the Tiramisu segmentation model as well as the YOLOv3 detector were experimented and compared for their capacity to detect very small vehicles from 50-cm Pleiades satellite images. The authors finally proposed a late fusion technique to obtain the combined benefits from both models. In [30], the authors focused on the detection of dense construction vehicles from UAV images using an orientation-aware feature fusion based on the one-stage SSD models.

As the use of anchor boxes introduces many hyper-parameters and design choices, such as the number of boxes, sizes, and aspect ratios [9], some recent works have also investigated anchor-free detection frameworks with feature enhancement or multi-scaled dense path feature aggregation to better characterize vehicle features in latent spaces [9,31,32]. We refer interested readers to these studies for more details about anchor-free methods. As anchor-free networks usually require various extra constraints on the loss functions, well-established anchor-based approaches remain popular in the computer vision community for their stability. Therefore, within the scope of this paper, we base our work on anchor-based approaches.

2.2. Misalignment in Object Detection

Object detection involves two tasks: classification and localization. Apparently, precise detection results require high-quality joint predictions of both tasks. Most object detection models regard these two tasks as independent ones and ignore their potential interactions, leading to the misalignment between classification and localization tasks. Indeed, detection results with correct classification but imprecise localization or with precise localization but wrong classification will both reduce the overall precision, and should be prevented.

The authors of IoU-Net [2] were the first to study this task-wise misalignment problem. Their solution is to use an additional prediction head to estimate the localization confidence (i.e., the intersection-over-union (IoU) between the regressed box and the true box), and then aggregate this localization confidence into the final classification score. In this way, the classification prediction contains information from the localization prediction, and the misalignment is greatly alleviated.

Along this direction, the authors of Double-Head RCNN [1] propose to apply different network architectures for classification and localization networks. Specifically, they find the fully connected layers more suitable for the classification task, and the convolutional layers more suitable for the localization task.

TSD [3] further proposes to use disentangled proposals for classification and localization predictions. To achieve the best performance of both tasks, two dedicated region of interest (RoI) proposals are estimated for classification and localization tasks, respectively, and the final detection result comes from the combination of both proposals.

The recently proposed MutualGuidance [4] addresses the misalignment problem from the perspective of label assignment. It introduces an adaptive matching strategy between anchor boxes and true objects, where the labels for one task are assigned according to the prediction quality on the other task, and vice versa. Compared to the aforementioned methods, the main advantage of MutualGuidance is that its improvement only involves the loss computation, while the architecture of the detection network remains unchanged, so it can be generalized to different detection models and application cases. These features motivate us to rely on this method in our study, and to explore its potential in Earth observation.

2.3. Contrastive Learning

Contrastive learning has been predominantly employed to transfer representations learned from a pretext task, usually without provided labels, to a different actual task, by finetuning using accompanied annotations [14–16,18–20,33]. The pretext tasks involving mostly feature vectors in embedding space are usually trained with metric distance learning such as N-pair loss [34] or triplet [35].

Depending on the downstream tasks, the corresponding pretexts are chosen accordingly. Chen et al. [16] propose a simple framework, called SimCLR, exploiting image augmentation to pretrain a network using the temperature-scaled N-pair loss and demonstrate an improvement in classifying images. An image paired with the augmented version and used against its pairing with other images in a mini-batch for optimization helps in learning decent visual representations. The representations can be further improved when they participate in the contrastive loss by non-linear transformed proxy. This notion is employed in our paper as the projection head.

Contrastive learning trained on image-level tasks, i.e., a single feature vector per image, however, is shown to be sub-optimal for downstream tasks requiring instance-level or dense pixel-level prediction, such as detection [18] or segmentation [20], 3respectively. The reasons are attributed to the missing of dedicated properties such as spatial sensitivity, translation, and scale invariance. Consequently, different pretext schemes are proposed to effectively pretrain a network conforming to particular downstream tasks, including but not limited to DenseCL [36], SoCo [18], DetCo [19], and PixPro [20]. The common feature of these methods is the use of explicit image augmentation to generate positive pairs, following SimCLR's proposal, for pretraining networks. In our method, we acquire the augmentation principles yet consider the aerial views of different same-class objects as their augmented versions; hence, no extra views are generated during training. Moreover, the contrastive loss is not used as pretext but as auxiliary loss to improve the semantic information in the mutual guidance process.

In contrast to most works that apply contrastive learning in a self-supervised context, Khosla et al. [17] leverage label information and formulate the batch contrastive approach in the supervised setting by pulling together features of the same class and pushing apart those from different classes in the embedding space. They also unify the contrastive loss function to be used for either self-supervised or supervised learning while consistently outperforming cross-entropy on image classification. The contrastive loss employed in our paper could be considered as being inspired by the same work but repurposed for a detection problem.

3. Method

In this paper, we follow the generic one-stage architecture for anchor-based object detection comprising a backbone network for feature extraction and 2 output heads for localization and classification. The overview of our framework is shown in Figure 3. For illustration purposes, a 2-image batch size, single spatial resolution features, and 6 anchor boxes are shown, yet the idea is seamlessly applicable to larger batch sizes with different numbers of anchor boxes, and multi-scaled feature extraction such as FPN [37].

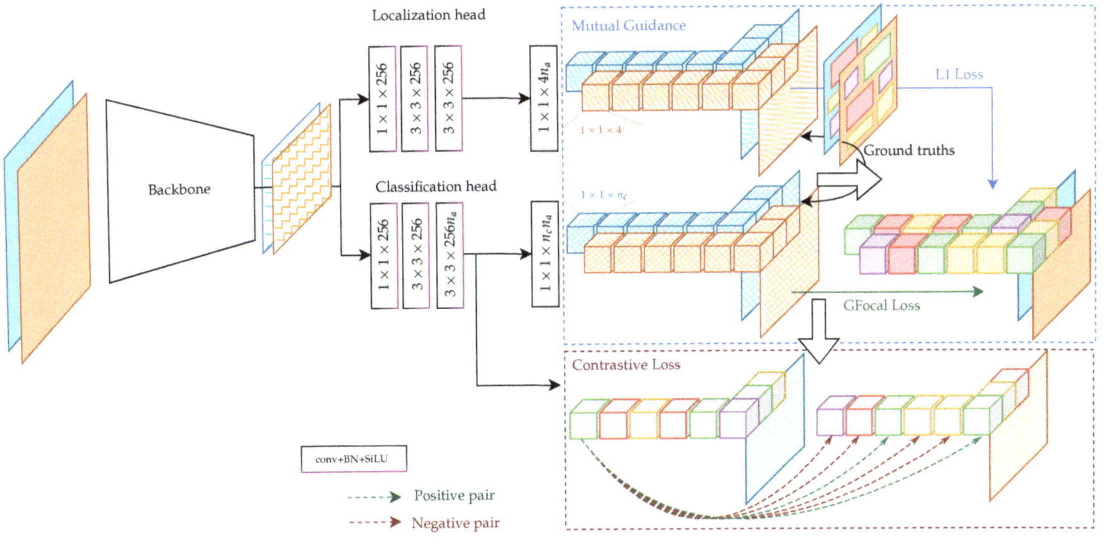

Figure 3. An overview of our framework: the backbone network encodes a batching input before passing the extracted features to the localization and classification heads, which predict 4-tuple bounding box values and n_c-class confidence scores for each anchor box. The mutual guidance module re-ranks the anchor boxes based on semantic information from the classification branch and improves the confidence score with localization information. The ground truth categories of the anchor boxes are used to supervise the contrastive loss. The pipeline is illustrated with a batch size of 2 and the number of anchor boxes $n_a = 6$.

The 2 output heads have the same network architecture: two parallel branches with two 3×3 convolution layers, followed by one 1×1 convolution layer for localization and classification predictions. The former classifies each anchor box into foreground (positive) or background (negative), while the latter refines anchor boxes via bounding-box regression to better suit target boxes. Instead of optimizing the 2 head networks independently, mutual guidance [4] introduces a task-based bidirectional supervision strategy to align the model predictions of localization and classification tasks.

3.1. Generation of Detection Targets

A general supervised object detection provides, for each input image, a list of ground truth bounding boxes $B \in \mathbb{R}^{n_B \times 4}$ accompanied by a list of labels $L \in \mathbb{R}^{n_B}$, where n_B is the number of ground truth boxes annotated for the image. Each box is represented by a 4-tuple (l, t, w, h) (in MS-COCO [12] format) or (x_c, y_c, w, h) (in YOLO [38] format), where (l, t) and (x_c, y_c) are the (x, y) coordinates of a box's top-left corner and center, respectively, and w, h are the box's width and height. The ground truth boxes are arbitrary and unordered and thus usually adapted into targets of a different form that is more compatible for optimization in a deep network. The process is called *matching*.

The idea is to define a list of fixed-size boxes called *anchors*, $A \in \mathbb{R}^{n_A \times 4}$, for each vector in a CNN output feature map, where n_A is the total number of predefined anchors per image. For a 512×512 input image with $n_a = 6$ predefined anchor sizes per vector, a 3-level FPN-based feature extraction network with output scale of $(8, 16, 32)$ can produce up to

$$\left(\frac{512}{8} \times \frac{512}{8} + \frac{512}{16} \times \frac{512}{16} + \frac{512}{32} \times \frac{512}{32}\right) \times 6 = 32{,}256 \tag{1}$$

anchors. As the anchors are defined at every vector in an output feature map, they are directly compatible with loss calculation and thus are used as targets for optimization.

Conventional matching. Depending on how similar each anchor is to the real ground truth boxes, it is marked as a positive (i.e., object) or negative target (i.e., background). The most common similarity metric is the Jaccard index [39], which measures the ratio of the overlapping area of 2 boxes (an anchor and a ground truth box) over their area of union, as shown in Equation (2).

$$\mathcal{J}(X, Y) = \frac{X \cap Y}{X \cup Y}. \tag{2}$$

Specifically, the matrix M containing the Jaccard indices between all pairs of ground truth and anchor boxes is computed. We define the Jaccard index over the Cartesian product of two sets of boxes as the Jaccard indices of all the pairs of boxes in the sets as follows:

$$\mathcal{J}(\mathcal{X} \times \mathcal{Y}) = \{\mathcal{J}(X, Y) | X \in \mathcal{X} \text{ and } Y \in \mathcal{Y}\}. \tag{3}$$

Thus, $M = \mathcal{J}(B \times A)$. An anchor is matched to a ground truth box if (1) this anchor is the closest that the ground truth box can have (among all anchors) or (2) this ground truth box is the closest that the anchor can have (among all other ground truths). A threshold can be applied to further filter out the matched anchors with low intersection-over-union scores. Subsequently, each anchor is associated with, at most, 1 ground truth box, i.e., *positive target*, or none, i.e., background or *negative target*. Some of the positive targets can be marked as *ignored* and do not contribute to the optimization process. The concrete algorithm is shown in Algorithm 1.

Mutual matching. Mutual guidance [4] formulates the process of label assignment in a mutual supervision manner. In particular, it constrains anchors that are well localized to be well classified (localize to classify), and those well classified to be well localized (classify to localize).

Localize to classify. The target anchor box corresponding to a feature vector that well localizes an object must be covering semantically important parts of the underlying object; therefore, it should be prioritized as a target for classification. A step-by-step procedure is shown in Algorithm 2. To this end, the Jaccard matrices between all ground truth and predicted boxes are computed, i.e., $\hat{M} = \mathcal{J}(B \times \hat{B})$ (see Algorithm 2, Line 1). The top-K anchors per ground truth box are shortlisted as positive classification targets, while the rest are considered negative targets. Concretely, we keep the Jaccard score of the best ground truth box (if any) for each anchor and zero out the other ground truth boxes, i.e., a column in the Jaccard matrix now has at most a single non-zero entry (Line 3–5). Then, each ground box will have all anchors besides the K with the highest score removed (Line 6–7). The remaining ground truth box per anchor is associated with it. We also use their Jaccard scores as soft-label targets for the loss function by replacing 1s in one-hot vectors with the corresponding scores. The loss is shown in Section 3.2.

Algorithm 1 Generating targets with common matching

Input: list of ground truth boxes $B \in \mathbb{R}^{n_B \times 4}$, and corresponding labels $L \in \mathbb{R}^{n_B}$,
list of anchors $A \in \mathbb{R}^{n_A \times 4}$,
negative and positive threshold θ_n, θ_p, where $\theta_n \leq \theta_p$
Output: list of target boxes $\tilde{B} \in \mathbb{R}^{n_A \times 4}$,
and corresponding target labels $\tilde{L} \in \mathbb{R}^{n_A}$ for each anchor

1: $M \leftarrow \mathcal{J}(B \times A)$ # $M \in \mathbb{R}^{n_B \times n_A}$
2: $\tilde{L} \leftarrow \begin{bmatrix} 0 & 0 & \cdots & 0 \end{bmatrix}$
3: $\tilde{B} \leftarrow A$ # the target boxes are the anchor boxes
4: **for each** column index c of M **do**
5: $\quad iou \leftarrow \max(M_{*c})$ # Processing condition 2
6: $\quad i \leftarrow \text{argmax}(M_{*c})$
7: $\quad$ **if** $iou \geq \theta$
8: $\quad\quad \tilde{L}_c \leftarrow L_i$
9: $\quad\quad \tilde{B}_{c*} \leftarrow B_i$
10: $\quad$ **else if** $iou < \theta_n$
11: $\quad\quad \tilde{L}_c \leftarrow -1$
12: **for each** row index r of M **do**
13: $\quad iou \leftarrow \max(M_{r*})$ # Overwritten with condition 1
14: $\quad i \leftarrow \text{argmax}(M_{r*})$
15: $\quad$ **if** $iou \geq \theta$
16: $\quad\quad \tilde{L}_i \leftarrow L_r$
17: $\quad\quad \tilde{B}_i \leftarrow B_r$
18: $\quad$ **else if** $iou < \theta_n$
19: $\quad\quad \tilde{L}_i \leftarrow -1$

Algorithm 2 Generating classification targets from predicted localization

Input: list of ground truth boxes $B \in \mathbb{R}^{n_B \times 4}$, and corresponding labels $L \in \mathbb{R}^{n_B}$,
list of anchors $A \in \mathbb{R}^{n_A \times 4}$,
list of predicted boxes $\hat{B} \in \mathbb{R}^{n_A \times 4}$
Output: list of target labels for all anchors $\tilde{L} \in \mathbb{R}^{n_A}$

1: $\hat{M} \leftarrow \mathcal{J}(B \times \hat{B})$ # $\hat{M} \in \mathbb{R}^{n_B \times n_A}$
2: $\tilde{L} \leftarrow \begin{bmatrix} 0 & 0 & \cdots & 0 \end{bmatrix}$
3: **for each** column index c of $\hat{M}$ **do**
4: $\quad i \leftarrow \text{argmax}(\hat{M}_{*c})$
5: $\quad \hat{M}_{kc} \leftarrow 0, \quad \forall\, k \neq i$
6: **for each** row index r of $\hat{M}$ **do**
7: $\quad \hat{M}_{rk} \leftarrow 0, \quad \forall\, k \notin \text{topk}(\hat{M}_{r*})$
8: **for each** column index c of $\hat{M}$ **do**
9: $\quad i \leftarrow \text{argmax}(\hat{M}_{*c})$
10: $\quad \tilde{L}_c \leftarrow L_i$

Classify to localize. Likewise, a feature vector at the output layer that induces correct classification indicates the notable location and shape of the corresponding target anchor box. As such, the anchor should be prioritized for bounding box regression. To this end, the Jaccard similarity between a ground truth and anchor box is scaled by the confidence score of the anchor's corresponding feature vector for the given ground truth box. Concretely, a curated list $\tilde{C} \in \mathbb{R}^{n_B \times n_A}$ of confidence scores for the class of each given ground truth box is obtained from the all-class input scores $\hat{C} \in \mathbb{R}^{n_A \times n_C}$, as shown in Algorithm 3 on Line 2–4, where n_C is the number of classes in the classification task. The Jaccard similarity between a ground truth and anchor box M (similar to conventional detection matching) is scaled by the corresponding confidence score and clamped to the range $[0, 1]$ (Line 5, where $\odot$ indicates the Hadamard product). The rest of the algorithm proceeds as shown in the previous algorithm with the updated similarity matrix $\tilde{M}$ in lieu of the predicted similarity matrix $\hat{M}$.

Algorithm 3 Generating localization targets from predicted class labels

Input: list of ground truth boxes $B \in \mathbb{R}^{n_B \times 4}$, and corresponding labels $L \in \mathbb{R}^{n_B}$,
list of anchors $A \in \mathbb{R}^{n_A \times 4}$,
list of confidence scores for all classes $\hat{C} \in \mathbb{R}^{n_A \times n_C}$,
Output: list of target box specifications for all anchors $\tilde{B} \in \mathbb{R}^{n_A \times 4}$
1: $M \leftarrow \mathcal{J}(B \times A)$ # $M \in \mathbb{R}^{n_B \times n_A}$
2: **for each** row index r of M **do**
3: $\quad l \leftarrow L_{r*}$
4: $\quad \tilde{C}_{r*} \leftarrow \exp\left(\frac{\hat{C}_{l*}}{\sigma}\right)$ # $\tilde{C} \in \mathbb{R}^{n_B \times n_A}$
5: $\tilde{M} \leftarrow \max(0,\ \min(1,\ M \odot \tilde{C}))$
6: $\tilde{L} \leftarrow [0 \quad 0 \quad \cdots \quad 0]$
7: **for each** column index c of $\tilde{M}$ **do**
8: $\quad i \leftarrow \operatorname{argmax}(\tilde{M}_{*c})$
9: $\quad \tilde{M}_{kc} \leftarrow 0, \quad \forall\, k \neq i$
10: **for each** row index r of $\tilde{M}$ **do**
11: $\quad \tilde{M}_{rk} \leftarrow 0, \quad \forall\, k \notin \operatorname{topk}(\tilde{M}_{r*})$
12: **for each** column index c of $\tilde{M}$ **do**
13: $\quad i \leftarrow \operatorname{argmax}(\tilde{M}_{*c})$
14: $\quad \tilde{B}_{c*} \leftarrow B_{i*}$

3.2. Losses

Classification loss. For classification, we adopt the Generalized Focal Loss [40] with soft target given by the Jaccard scores of predicted localization and ground truth boxes. The loss is given by Equation (4):

$$\mathcal{L}_{\text{class}}(\hat{y}, \tilde{y}) = -|\tilde{y} - \hat{y}|^2 \sum_{i}^{n_C} \tilde{y}_i \log \hat{y}_i, \tag{4}$$

where $\tilde{y} \in \mathbb{R}^{n_C}$ is the one-hot target label given by $\tilde{C}$, softened by the predicted Jaccard scores, and $\hat{y} \in \mathbb{R}^{n_C}$ is the anchor's confidence score.

Localization loss. We employ the balanced L1 loss [41], derived from the conventional smooth L1 loss, for the localization task to promote the crucial regression gradients from accurate samples (inliers) by separating inliers from outliers, and we clip the large gradients produced by outliers with a maximum value of β. This is expected to rebalance the involved samples and tasks, thus achieving a more balanced training within classification, overall localization, and accurate localization. We first define the balanced loss $L_b(x)$ as follows:

$$L_b(x) = \begin{cases} \frac{\alpha}{b}(b|x| + 1)\ln\left(b\frac{|x|}{\beta} + 1\right) - \alpha|x|, & \text{if } |x| < \beta \\ \gamma|x| + \frac{\gamma}{b} - \alpha * \beta, & \text{otherwise,} \end{cases} \tag{5}$$

where $\alpha = 0.5, \beta = 0.11, \gamma = 1.5$, and b is constant such that

$$\alpha \ln(b + 1) = \gamma. \tag{6}$$

The localization loss using balanced L1 loss is defined as $L_{\text{loc}} = L_b(pred - target)$.

Contrastive Loss. The mutual guidance process assigns to each anchor box a confidence score $s_i \in [0,1]$ from the prediction of the feature vector associated with it, and a category label $c_i > 0$ if the anchor box is deemed to be an object target or $c_i = 0$ if background target. Let $\mathcal{B}_k^{\phi} = \{i \neq k : c_i = \phi\}$ be the index set of all anchor boxes other than k, whose labels follow the condition ϕ and $\mathbf{z}$ be a feature vector at the before-last layer in the classification branch (Figure 3). Following SupCo [17], we experiment with two versions of the loss function, $\mathcal{L}_{\text{out}}$, with summation being outside of the logarithm, and $\mathcal{L}_{\text{in}}$ inside, whose equations are given as follows:

$$\mathcal{L}_{\text{in}} = \frac{-1}{|\mathcal{B}|} \sum_{i \in \mathcal{B}} \log \left(\frac{1}{|\mathcal{B}_i^{c_i}|} \frac{\sum_{j \in \mathcal{B}_i^{c_i}} \delta(\mathbf{z}_i, \mathbf{z}_j)}{\sum_{k \in \mathcal{B}_i} \delta(\mathbf{z}_i, \mathbf{z}_k)} \right), \tag{7}$$

$$\mathcal{L}_{\text{out}} = \frac{-1}{|\mathcal{B}|} \sum_{i \in \mathcal{B}} \frac{1}{|\mathcal{B}_i^{c_i}|} \sum_{j \in \mathcal{B}_i^{c_i}} \log \frac{\delta(\mathbf{z}_i, \mathbf{z}_j)}{\sum_{k \in \mathcal{B}_i} \delta(\mathbf{z}_i, \mathbf{z}_k)}, \tag{8}$$

where $\delta(\mathbf{v}_1, \mathbf{v}_2) = \exp\left(\frac{1}{\tau} \frac{\mathbf{v}_1 \cdot \mathbf{v}_2}{\|\mathbf{v}_1\| \|\mathbf{v}_2\|}\right)$ is the temperature-scaled similarity function. In this paper we choose $\tau = 1$.

4. Experiments

4.1. Setup

In this section, the proposed modules are analyzed and tested using the YOLOX small (-s) and medium (-m) backbones, which are adopted exactly from the YOLOv5 backbone and its scaling rules, as well as the YOLOv3 backbone (DarkNet53+SPP bottleneck) due to its simplicity and broad compatibility, and hence popularity, in various applied domains. More detailed descriptions can be referred to in the YOLOX paper [42]. We also perform an ablation study to analyze the effects of different components and a comparative study with state-of-the-art detectors including EfficientDet [43], YOLOv3 [38], YOLO-fine [28] YOLOv4, and Scaled-YOLOv4 [44].

For fair comparison, the input image size is fixed to 512 × 512 pixels for all experiments.

Dataset. We use the VEDAI aerial image dataset [21] and xView satellite image dataset [22] to conduct our experiments. For VEDAI, there exist two RGB versions with 12.5-cm and 25-cm spatial resolutions. We name them as VEDAI12 and VEDAI25, respectively, in our experimental results. The original data contain 3757 vehicles of 9 different classes, including *car, truck, pickup, tractor, camper, ship, van, plane*, and *others*. As done by the authors in [28], we merge class *plane* into class *others* since there are only a few *plane* instances. Next, the images from the xView dataset were collected from the WorldView-3 satellite at 30-cm spatial resolution. We followed the setup in [28] to gather 19 vehicle classes into a single *vehicle* class. The dataset contains a total number of around 35,000 vehicles. It should be noted that our intention to benchmark these two datasets is based on their complementary characteristics. The VEDAI dataset contains aerial images with multiple classes of vehicles from different types of backgrounds (urban, rural, desert, forest, etc.). Moreover, the numbers of images and objects are quite limited (e.g., 1200 and 3757, respectively). Meanwhile, the xView dataset involves satellite images of lower resolution, with a single merged class of very small vehicle sizes. It also contains more images and objects (e.g., 7400 and 35,000, respectively).

Metric. We report per-class average precision (AP) and their mean values (mAP) following the PASCAL VOC [13] metric. An intersection-over-union (IOU) threshold computed by the Jaccard index [39] is used for identifying positive boxes during evaluation. IOU values vary between 0 (no overlapping) and 1 (tight overlapping). Within the context of vehicle detection in remote sensing images, we follow [28] to set a small threshold, i.e., testing threshold is set to 0.1 unless stated otherwise.

To be more informative, we also show the widely used precision–recall (PR) curves in later experiments. The recall and precision are computed by Equations (9) and (10), respectively.

$$\text{Recall} = \frac{\text{number of correct detections}}{\text{number of existing objects}} = \frac{TP}{TP + FN} \quad (9)$$

$$\text{Precision} = \frac{\text{number of correct detections}}{\text{number of detected objects}} = \frac{TP}{TP + FP}, \quad (10)$$

where TP, FP, and FN denote true positive, false positive, and false negative, respectively.

The PR curve plots the precision values, which usually decrease, at each recall rate. Higher recall rates correspond to lower testing confidence thresholds, thus indicating a higher likelihood of false positives and a lower precision rate. On the other hand, lower recall rates mean stricter testing thresholds and a reduced likelihood of false positives, thus resulting in better precision. The visualization of the precision–recall curve gives a global vision of the compromise between precision and recall.

4.2. Mutual Guidance

In this section, we show the impact of mutual guidance on the remote sensing data by applying it directly for vehicle detection, apart from the other modules. The baseline is the same backbone with a generic setup, as used in [4]. As they use focal loss [45] in their setup, we include the mutual guidance with the same loss for a fair comparison.

The results in Table 1 show the improvement when switching from the IOU-based scheme to mutual guidance. The impact is diminished with YOLOX-m as was already efficient to begin with. The use of GFocal loss shows even further improvement for both architectures.

Table 1. Mutual guidance for different backbone architectures on VEDAI25 dataset. The best performance per column is shown in boldface.

Matching Strategy	Loss	YOLOX-s	YOLOX-m	YOLOv3
IOU-Based	Focal	70.20	74.30	70.78
Mutual Guidance	Focal	71.48	74.47	74.13
Mutual Guidance	GFocal	**73.04**	**79.82**	**74.88**

4.3. Contrastive Loss

Similar to the previous subsection, here, we aim to test the ability of contrastive loss in the context of vehicle detection. To this end, the contrastive loss is used together with the detection losses using the IOU-based matching strategy. Following [17], we also test the two possibilities of loss function, namely $\mathcal{L}_{\text{in}}$ (Equation (7)) and $\mathcal{L}_{\text{out}}$ (Equation (8)). The results are shown in Table 2.

Table 2. YOLOX-s performance on VEDAI25 with different contrastive loss functions.

Matching Strategy	Loss	YOLOX-s	YOLOX-m	YOLOv3
IOU-Based	Focal	70.20	74.30	70.78
	GFocal $+\mathcal{L}_{\text{in}}$	71.53	**79.89**	**75.53**
	GFocal $+\mathcal{L}_{\text{out}}$	**74.20**	77.81	74.41

The contrastive loss seems to have the reverse effect of mutual guidance on the two YOLOX backbones. The additional auxiliary loss does not improve the performance of YOLOX-s as highly as YOLOX-m, and, for the case of the outside loss, it even has negative impacts. This shows that YOLOX-m does not suffer from the misalignment problem as much as YOLOX-s does; thus, it can benefit more from the improvement in visual representation brought about by the contrastive loss.

4.4. Mutual Guidance Meets Contrastive Learning

The results of YOLOX with the mutual guidance strategy and contrastive learning are shown in Table 3. Contrastive loss shows great benefit to the network when the misalignment between localization and classification is alleviated by mutual guidance. The improvement seems balanced between both backbones. Although the inside contrastive loss seems to dominate over the outside one in the previous experiment, it becomes inferior when the semantic information from the classification branch and projection head is properly utilized in the localization process, conforming to the finding from [17]. The combination of mutual guidance and outside contrastive loss is coined contrastive mutual guidance, or CMG.

Table 3. Performance of YOLOX backbones on VEDAI25 when training with mutual guidance (MG) and contrastive loss.

Matching Strategy	Loss	YOLOX-s	YOLOX-m	YOLOv3
Mutual Guidance	GFocal	73.04	79.82	74.88
	GFocal $+\mathcal{L}_{\text{in}}$	75.57	80.95	76.26
	GFocal $+\mathcal{L}_{\text{out}}$	**76.67**	**81.57**	**77.41**

Multiple datasets. We further show the results on different datasets with different resolutions in Table 4 and the corresponding precision-recall curve in Figure 4.

Table 4. Performance of YOLOX-s vanilla with mutual guidance (MG) and contrastive mutual guidance (CMG) on the 3 datasets. The contrastive mutual guidance strategy consistently outperforms other configurations, showing its benefit.

Configuration	VEDAI12	VEDAI25	xView30
vanilla	78.68	70.20	79.96
+MG	79.70	73.04	83.49
+CMG	**81.25**	**76.67**	**83.67**

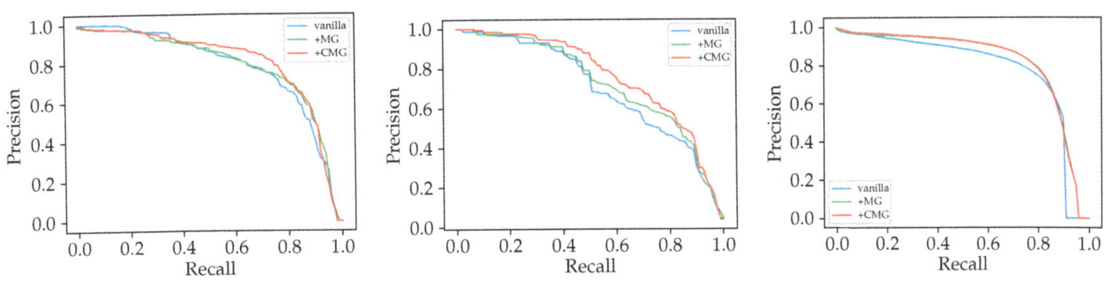

Figure 4. Precision–recall curve of YOLOX-s on 3 datasets, from left to right: VEDAI12, VEDAI25, and xView30. The methods with +CMG gain improvement over the others at around recall level of 0.5 for the VEDAI datasets and both +MG and +CMG outperform the vanilla method on the xView dataset.

The methods with +CMG gain an improvement over the others at around a recall level of 0.5 for the VEDAI datasets and both +MG and +CMG outperform the vanilla method on the xView dataset.

Some qualitative results on the VEDAI25 and xView datasets can be found in Figures 5 and 6, respectively. Several objects are missing in the second and third columns, while the CMG strategy (last column) is able to recognize objects of complex shape and appearance.

Comparison to the state-of-the-art. In Table 5, we compare our method with several state-of-the-art methods on the three datasets. Our YOLOX backbone with the CMG strategy outperforms others on the VEDAI datasets and is on par with YOLO-fine on xView. From the qualitative results in Figures 7 and 8, respectively, for the VEDAI and xView, it can be seen that although the xView dataset contains extremely small objects, our method, without deliberate operations for tiny object detection, can approach the state-of-the-art method specifically designed for small vehicle detection [28]. A breakdown of performance for each class of VEDAI is shown in Table 6.

Table 5. Performance of different YOLOX backbones with CMG compared to the state-of-the-art methods. Our method outperforms or is on par with the methods designed for tiny object recognition.

Architecture	VEDAI12	VEDAI25	xView30
EfficientDet	74.01	51.36	82.45
YOLOv3	73.11	62.09	78.93
YOLO-fine	76.00	68.18	84.14
YOLOv4	79.93	73.14	79.19
Scaled-YOLOv4	78.57	72.78	81.39
YOLOX-s+CMG (ours)	81.25	76.67	83.67
YOLOX-m+CMG (ours)	**83.07**	**81.57**	**84.79**
YOLOv3+CMG (ours)	78.09	77.41	83.54

Table 6. Per-class performance of YOLOX backbones with CMG on VEDAI25 dataset. Our method outperforms the state-of-the-art for all classes.

Model	Car	Truck	Pickup	Tractor	Camping	Boat	Van	Other	mAP
EfficientDet	69.08	61.20	65.74	47.18	69.08	33.65	16.55	36.67	51.36
YOLOv3	75.22	73.53	65.69	57.02	59.27	47.20	71.55	47.20	62.09
YOLOv3-tiny	64.11	41.21	48.38	30.04	42.37	24.64	68.25	40.77	44.97
YOLOv3-spp	79.03	68.57	72.30	61.67	63.41	44.26	60.68	42.43	61.57
YOLO-fine	76.77	63.45	74.35	**78.12**	64.74	70.04	77.91	45.04	68.18
YOLOv4	87.50	80.47	78.63	65.80	81.07	75.92	66.56	49.16	73.14
Scaled-YOLOv4	86.78	79.37	81.54	73.83	71.58	76.53	63.90	48.70	72.78
YOLOX-s+CMG (ours)	88.92	85.92	79.66	77.16	81.21	65.22	64.90	**70.33**	76.67
YOLOX-m+CMG (ours)	91.26	85.34	84.91	76.22	85.03	**78.68**	**82.02**	69.08	**81.57**
YOLOv3 +CMG (ours)	**92.20**	**85.98**	**87.34**	77.27	**85.56**	53.74	73.94	64.13	77.41

Two failure cases are shown in the last columns of Figures 7 and 8. We can see that our method has difficulty in recognizing the "other" class (VEDAI), which comprises various object types, and might wrongly detect objects of extreme resemblance (xView).

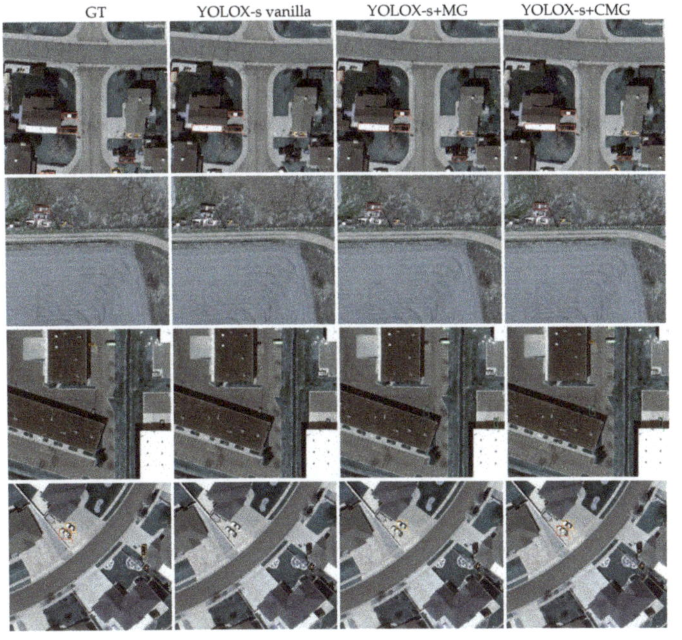

Figure 5. Qualitative results of YOLOX-s on VEDAI25. The contrastive mutual guidance helps to recognize intricate objects. The number and color of each box correspond to one of the classes, i.e., (1) car, (2) truck, (3) pickup, (4) tractor, (5) camper, (6) ship, (7) van, and (8) plane.

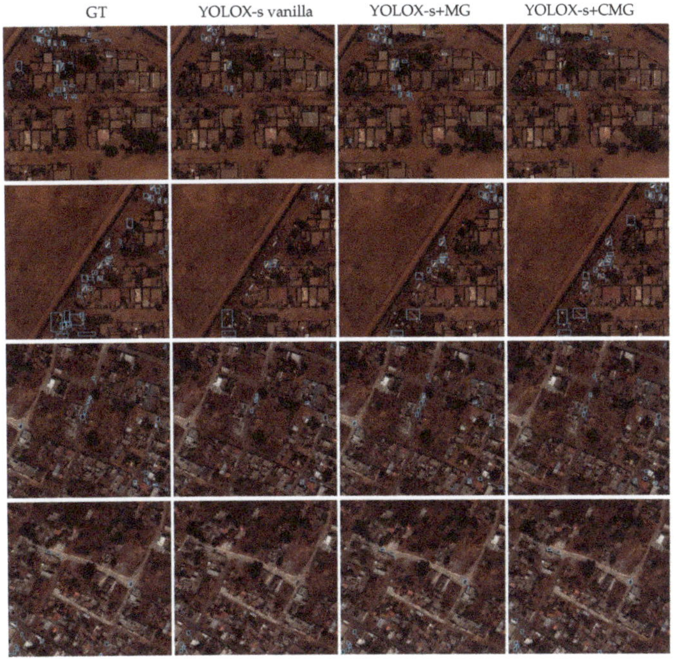

Figure 6. Qualitative results of YOLOX-s on xView. The contrastive mutual guidance helps to recognize intricate objects. The number and color of each box indicate the vehicle class.

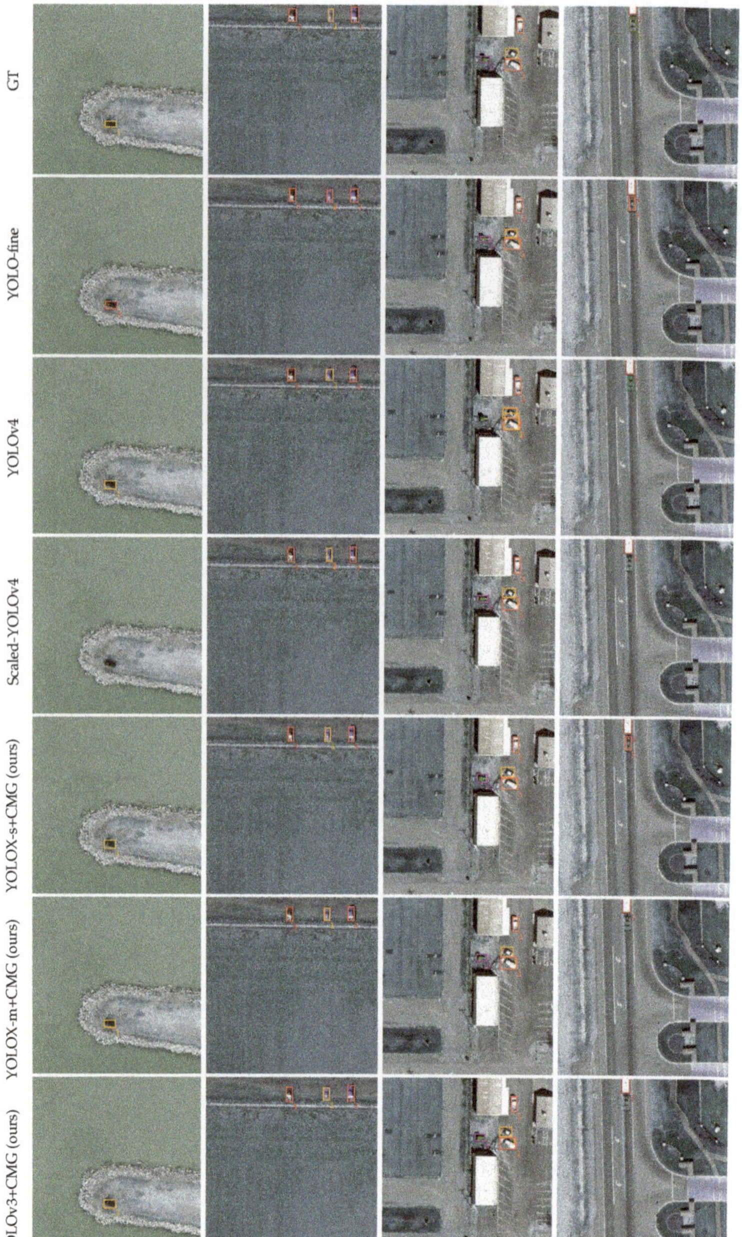

Figure 7. Qualitative results of our methods and state-of-the-art methods on VEDAI25. The number and color of each box correspond to one of the classes, i.e. (1) car, (2) truck, (3) pickup, (4) tractor, (5) camper, (6) ship, (7) van, and (8) plane. The last column shows a failure case. Our method has difficulties in recognizing the "other" class, which comprises various object types.

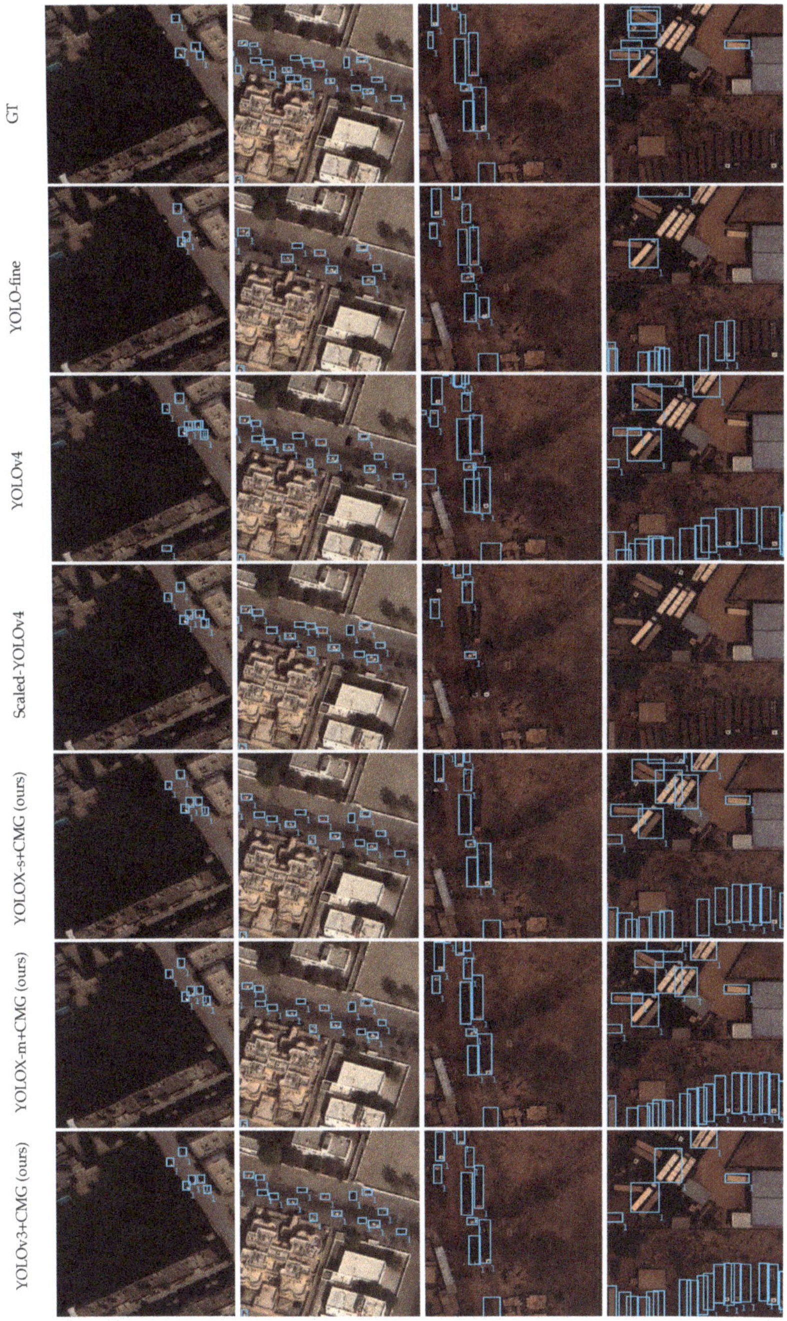

Figure 8. Qualitative results of our methods and state-of-the-art methods on xView. The number and color of each box indicates the vehicle class. The last column shows a failure case. Our method could recognize objects of various shapes and would wrongly detect objects of extreme resemblance (although this might have been because of the faulty annotations).

5. Discussion

Although supervised contrastive loss has been shown to be able to replace cross-entropy for classification problems [17], in this paper, contrastive loss is applied as an auxiliary loss besides the main localization and classification losses. This is because only a small number of anchors are involved in the contrastive process due to the large number of anchors, especially negative anchors.

However, contrastive loss shows weakness when the annotations are noisy, such as those of the xView dataset. Several boxes are missing for (what appear to be) legitimate objects, as shown in Figure 9.

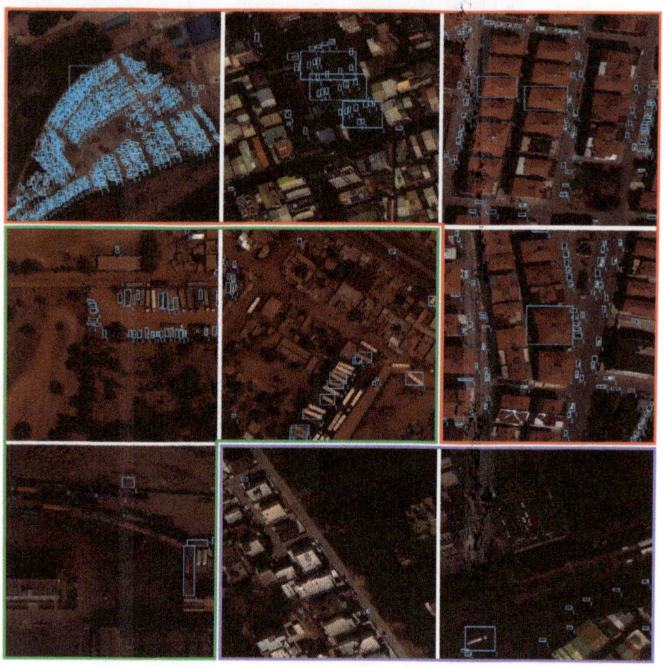

Figure 9. Examples of faulty annotations in the xView dataset: non-vehicle annotation (red border), missing annotations of container trucks (green border), and cars (blue border). The number and color of each box indicates the vehicle class.

It is shown from the experimental results that inward contrastive loss is not always inferior to its outward counterpart, as shown in [17]. We speculate that this could be due to the auxiliary role of contrastive loss in the detection problem and/or the characteristics of small objects in remote sensing images.

6. Conclusions

This paper presents a combination of a mutual guidance matching strategy and supervised contrastive loss for the vehicle detection problem. The mutual guidance helps in better connecting the localization and classification branches of a detection network, while contrastive loss improves the visual representation, which provides better semantic information. The vehicle detection task is generally complicated due to the varied object sizes and similar appearances from the aerial point of view. This, however, provides an opportunity for contrastive learning, as it can be regarded as image augmentation, which has been shown to be beneficial for learning visual representations. Although the paper is presented in a remote sensing context, we believe that this idea could be expanded to generic computer vision applications.

Author Contributions: Conceptualization, H.-Â.L. and S.L.; methodology, H.-Â.L., H.Z. and M.-T.P.; software, H.-Â.L. and H.Z.; validation, H.-Â.L. and M.-T.P.; formal analysis, H.-Â.L.; investigation, H.-Â.L.; writing—original draft preparation, H.-Â.L., H.Z. and M.-T.P.; writing—review and editing, H.-Â.L., M.-T.P. and S.L.; visualization, H.-Â.L.; supervision, M.-T.P. and S.L.; project administration, M.-T.P. and S.L.; funding acquisition, S.L. All authors have read and agreed to the published version of the manuscript.

Funding: This work was supported by the SAD 2021-ROMMEO project (ID 21007759).

Data Availability Statement: The VEDAI and xView datasets are publicly available. Source code and dataset will be available at https://lhoangan.github.io/CMG_vehicle/.

Conflicts of Interest: The authors declare no conflict of interest.

References

1. Wu, Y.; Chen, Y.; Yuan, L.; Liu, Z.; Wang, L.; Li, H.; Fu, Y. Rethinking Classification and Localization for Object Detection. In Proceedings of the IEEE/CVF Conference on Computer Vision and Pattern Recognition (CVPR), Seattle, WA, USA, 13–19 June 2020.
2. Jiang, B.; Luo, R.; Mao, J.; Xiao, T.; Jiang, Y. Acquisition of Localization Confidence for Accurate Object Detection. In Proceedings of the European Conference on Computer Vision (ECCV), Munich, Germany, 8–14 September 2018.
3. Song, G.; Liu, Y.; Wang, X. Revisiting the Sibling Head in Object Detector. In Proceedings of the IEEE/CVF Conference on Computer Vision and Pattern Recognition (CVPR), Seattle, WA, USA, 13–19 June 2020.
4. Zhang, H.; Fromont, E.; Lefevre, S.; Avignon, B. Localize to Classify and Classify to Localize: Mutual Guidance in Object Detection. In Proceedings of the Asian Conference on Computer Vision (ACCV), Online, 30 November–4 December 2020.
5. Kaack, L.H.; Chen, G.H.; Morgan, M.G. Truck Traffic Monitoring with Satellite Images. In Proceedings of the ACM SIGCAS Conference on Computing and Sustainable Societies, Accra, Ghana, 3–5 July 2019.
6. Arora, N.; Kumar, Y.; Karkra, R.; Kumar, M. Automatic vehicle detection system in different environment conditions using fast R-CNN. *Multimed. Tools Appl.* **2022**, *81*, 18715–18735. [CrossRef]
7. Zhou, H.; Creighton, D.; Wei, L.; Gao, D.Y.; Nahavandi, S. Video Driven Traffic Modelling. In Proceedings of the IEEE/ASME International Conference on Advanced Intelligent Mechatronics, Wollongong, NSW, Australia, 9–12 July 2013.
8. Kamenetsky, D.; Sherrah, J. Aerial Car Detection and Urban Understanding. In Proceedings of the International Conference on Digital Image Computing: Techniques and Applications (DICTA), Adelaide, SA, Australia, 23–25 November 2015.
9. Shi, F.; Zhang, T.; Zhang, T. Orientation-Aware Vehicle Detection in Aerial Images via an Anchor-Free Object Detection Approach. *IEEE Trans. Geosci. Remote. Sens.* **2021**, *59*, 5221–5233. [CrossRef]
10. Zheng, K.; Wei, M.; Sun, G.; Anas, B.; Li, Y. Using Vehicle Synthesis Generative Adversarial Networks to Improve Vehicle Detection in Remote Sensing Images. *ISPRS Int. J. -Geo-Inf.* **2019**, *8*, 390. [CrossRef]
11. Bouguettaya, A.; Zarzour, H.; Kechida, A.; Taberkit, A.M. Vehicle Detection From UAV Imagery With Deep Learning: A Review. *IEEE Trans. Neural Netw. Learn. Syst.* **2021**. [CrossRef] [PubMed]
12. Lin, T.Y.; Maire, M.; Belongie, S.; Hays, J.; Perona, P.; Ramanan, D.; Dollar, P.; Zitnick, C.L. Microsoft COCO: Common Objects in Context. In Proceedings of the European Conference on Computer Vision (ECCV), Zurich, Switzerland, 6–12 September 2014.
13. Everingham, M.; Van Gool, L.; Williams, C.K.I.; Winn, J.; Zisserman, A. The Pascal Visual Object Classes (VOC) Challenge. *Int. J. Comput. Vis.* **2010**, *88*, 303–338. [CrossRef]
14. Bachman, P.; Hjelm, R.D.; Buchwalter, W. Learning Representations by Maximizing Mutual Information across Views. In Proceedings of the 33rd International Conference on Neural Information Processing Systems, Vancouver, BC, Canada, 8–14 December 2019.
15. Dosovitskiy, A.; Springenberg, J.T.; Riedmiller, M.; Brox, T. Discriminative Unsupervised Feature Learning with Convolutional Neural Networks. In Proceedings of the Advances in Neural Information Processing Systems, Montreal, QC, Canada, 8–13 December 2014.
16. Chen, T.; Kornblith, S.; Norouzi, M.; Hinton, G. A Simple Framework for Contrastive Learning of Visual Representations. In Proceedings of the ICML, 2020, Machine Learning Research, Vienna, Austria, 13–18 July 2020.
17. Khosla, P.; Teterwak, P.; Wang, C.; Sarna, A.; Tian, Y.; Isola, P.; Maschinot, A.; Liu, C.; Krishnan, D. Supervised Contrastive Learning. In Proceedings of the Advances in Neural Information Processing Systems, Virtual, 6–12 December 2020.
18. Wei, F.; Gao, Y.; Wu, Z.; Hu, H.; Lin, S. Aligning Pretraining for Detection via Object-Level Contrastive Learning. In Proceedings of the Advances in Neural Information Processing Systems, Virtual, 6–14 December 2021.
19. Xie, E.; Ding, J.; Wang, W.; Zhan, X.; Xu, H.; Sun, P.; Li, Z.; Luo, P. DetCo: Unsupervised Contrastive Learning for Object Detection. In Proceedings of the IEEE/CVF International Conference on Computer Vision (ICCV), Montreal, QC, Canada, 10–17 October 2021.
20. Xie, Z.; Lin, Y.; Zhang, Z.; Cao, Y.; Lin, S.; Hu, H. Propagate Yourself: Exploring Pixel-Level Consistency for Unsupervised Visual Representation Learning. In Proceedings of the IEEE/CVF Conference on Computer Vision and Pattern Recognition (CVPR), Nashville, TN, USA, 20–25 June 2021.
21. Razakarivony, S.; Jurie, F. Vehicle detection in aerial imagery: A small target detection benchmark. *J. Vis. Commun. Image Represent.* **2016**, *34*, 187–203. [CrossRef]

22. Lam, D.; Kuzma, R.; McGee, K.; Dooley, S.; Laielli, M.; Klaric, M.; Bulatov, Y.; McCord, B. xView: Objects in Context in Overhead Imagery. *arXiv* **2018**, arXiv:1802.07856.
23. Froidevaux, A.; Julier, A.; Lifschitz, A.; Pham, M.T.; Dambreville, R.; Lefèvre, S.; Lassalle, P. Vehicle detection and counting from VHR satellite images: Efforts and open issues. In Proceedings of the IGARSS 2020-2020 IEEE International Geoscience and Remote Sensing Symposium, Waikoloa, HI, USA, 26 September–2 October 2020.
24. Srivastava, S.; Narayan, S.; Mittal, S. A survey of deep learning techniques for vehicle detection from UAV images. *J. Syst. Archit.* **2021**, *117*, 102152. [CrossRef]
25. Zou, Z.; Shi, Z.; Guo, Y.; Ye, J. Object detection in 20 years: A survey. *arXiv* **2019**, arXiv:1905.05055.
26. Ji, H.; Gao, Z.; Mei, T.; Li, Y. Improved faster R-CNN with multiscale feature fusion and homography augmentation for vehicle detection in remote sensing images. *IEEE Geosci. Remote. Sens. Lett.* **2019**, *16*, 1761–1765. [CrossRef]
27. Mo, N.; Yan, L. Improved faster RCNN based on feature amplification and oversampling data augmentation for oriented vehicle detection in aerial images. *Remote. Sens.* **2020**, *12*, 2558. [CrossRef]
28. Pham, M.T.; Courtrai, L.; Friguet, C.; Lefèvre, S.; Baussard, A. YOLO-Fine: One-Stage Detector of Small Objects Under Various Backgrounds in Remote Sensing Images. *Remote. Sens.* **2020**, *12*, 2501. [CrossRef]
29. Koay, H.V.; Chuah, J.H.; Chow, C.O.; Chang, Y.L.; Yong, K.K. YOLO-RTUAV: Towards Real-Time Vehicle Detection through Aerial Images with Low-Cost Edge Devices. *Remote Sens.* **2021**, *13*, 4196. [CrossRef]
30. Guo, Y.; Xu, Y.; Li, S. Dense construction vehicle detection based on orientation-aware feature fusion convolutional neural network. *Autom. Constr.* **2020**, *112*, 103124. [CrossRef]
31. Yang, J.; Xie, X.; Shi, G.; Yang, W. A feature-enhanced anchor-free network for UAV vehicle detection. *Remote. Sens.* **2020**, *12*, 2729. [CrossRef]
32. Li, Y.; Pei, X.; Huang, Q.; Jiao, L.; Shang, R.; Marturi, N. Anchor-free single stage detector in remote sensing images based on multiscale dense path aggregation feature pyramid network. *IEEE Access* **2020**, *8*, 63121–63133. [CrossRef]
33. Tseng, W.H.; Lê, H.Â.; Boulch, A.; Lefèvre, S.; Tiede, D. CroCo: Cross-Modal Contrastive Learning for Localization of Earth Observation Data. In Proceedings of the ISPRS Annals of the Photogrammetry, Remote Sensing and Spatial Information Sciences, Nice, France, 6–11 June 2022.
34. Sohn, K. Improved Deep Metric Learning with Multi-class N-pair Loss Objective. In Proceedings of the Advances in Neural Information Processing Systems, Barcelona, Spain, 5–10 December 2016.
35. Weinberger, K.Q.; Blitzer, J.; Saul, L. Distance Metric Learning for Large Margin Nearest Neighbor Classification. In Proceedings of the Advances in Neural Information Processing Systems, Vancouver, BC, Canada, 5–8 December 2005.
36. Wang, X.; Zhang, R.; Shen, C.; Kong, T.; Li, L. Dense Contrastive Learning for Self-Supervised Visual Pre-Training. In Proceedings of the IEEE/CVF Conference on Computer Vision and Pattern Recognition (CVPR), Nashville, TN, USA, 20–25 June 2021.
37. Lin, T.Y.; Dollar, P.; Girshick, R.; He, K.; Hariharan, B.; Belongie, S. Feature Pyramid Networks for Object Detection. In Proceedings of the IEEE Conference on Computer Vision and Pattern Recognition (CVPR), Honolulu, HI, USA, 21–26 July 2017.
38. Redmon, J.; Farhadi, A. Yolov3: An incremental improvement. *arXiv* **2018**, arXiv:1804.02767.
39. Jaccard, P. The distribution of the Flora in the Alpine Zone. 1. *New Phytol.* **1912**, *11*, 37–50. [CrossRef]
40. Li, X.; Wang, W.; Wu, L.; Chen, S.; Hu, X.; Li, J.; Tang, J.; Yang, J. Generalized Focal Loss: Learning Qualified and Distributed Bounding Boxes for Dense Object Detection. In Proceedings of the Advances in Neural Information Processing Systems, Online, 6–12 December 2020.
41. Pang, J.; Chen, K.; Shi, J.; Feng, H.; Ouyang, W.; Lin, D. Libra R-CNN: Towards Balanced Learning for Object Detection. In Proceedings of the IEEE Conference on Computer Vision and Pattern Recognition, Long Beach, CA, USA, 16–20 June 2019.
42. Ge, Z.; Liu, S.; Wang, F.; Li, Z.; Sun, J. YOLOX: Exceeding YOLO Series in 2021. *arXiv* **2021**, arXiv:2107.08430.
43. Tan, M.; Pang, R.; Le, Q.V. Efficientdet: Scalable and efficient object detection. In Proceedings of the IEEE/CVF Conference on Computer Vision and Pattern Recognition, Seattle, WA, USA, 13–19 June 2020.
44. Wang, C.Y.; Bochkovskiy, A.; Liao, H.Y.M. Scaled-yolov4: Scaling cross stage partial network. In Proceedings of the IEEE/CVF Conference on Computer Vision and Pattern Recognition, Nashville, TN, USA, 20–25 June 2021.
45. Lin, T.Y.; Goyal, P.; Girshick, R.; He, K.; Dollar, P. Focal Loss for Dense Object Detection. In Proceedings of the IEEE International Conference on Computer Vision (ICCV), Venice, Italy, 22–29 October 2017.

remote sensing

MDPI

Article

Detection of River Plastic Using UAV Sensor Data and Deep Learning

Nisha Maharjan [1,*], Hiroyuki Miyazaki [1,2], Bipun Man Pati [3], Matthew N. Dailey [1], Sangam Shrestha [4] and Tai Nakamura [1]

1. Department of Information and Communication Technologies, School of Engineering and Technology, Asian Institute of Technology, Pathum Thani 12120, Thailand; miyazaki@ait.asia (H.M.); mdailey@ait.asia (M.N.D.); nakamura-tai@ait.asia (T.N.)
2. Center for Spatial Information Science, The University of Tokyo, Chiba 277-8568, Japan
3. Faculty of Pharmaceutical Sciences, Chulalongkorn University, 254 Phayathai Road, Patumwan District, Bangkok 10330, Thailand; bipun.m@chula.ac.th
4. Department of Civil and Infrastructure Engineering, School of Engineering and Technology, Asian Institute of Technology, Pathum Thani 12120, Thailand; sangam@ait.asia

* Correspondence: nisha.maharjan065@gmail.com

Abstract: Plastic pollution is a critical global issue. Increases in plastic consumption have triggered increased production, which in turn has led to increased plastic disposal. In situ observation of plastic litter is tedious and cumbersome, especially in rural areas and around transboundary rivers. We therefore propose automatic mapping of plastic in rivers using unmanned aerial vehicles (UAVs) and deep learning (DL) models that require modest compute resources. We evaluate the method at two different sites: the Houay Mak Hiao River, a tributary of the Mekong River in Vientiane, Laos, and Khlong Nueng canal in Talad Thai, Khlong Luang, Pathum Thani, Thailand. Detection models in the You Only Look Once (YOLO) family are evaluated in terms of runtime resources and mean average Precision (mAP) at an Intersection over Union (IoU) threshold of 0.5. YOLOv5s is found to be the most effective model, with low computational cost and a very high mAP of 0.81 without transfer learning for the Houay Mak Hiao dataset. The performance of all models is improved by transfer learning from Talad Thai to Houay Mak Hiao. Pre-trained YOLOv4 with transfer learning obtains the overall highest accuracy, with a 3.0% increase in mAP to 0.83, compared to the marginal increase of 2% in mAP for pre-trained YOLOv5s. YOLOv3, when trained from scratch, shows the greatest benefit from transfer learning, with an increase in mAP from 0.59 to 0.81 after transfer learning from Talad Thai to Houay Mak Hiao. The pre-trained YOLOv5s model using the Houay Mak Hiao dataset is found to provide the best tradeoff between accuracy and computational complexity, requiring model resources yet providing reliable plastic detection with or without transfer learning. Various stakeholders in the effort to monitor and reduce plastic waste in our waterways can utilize the resulting deep learning approach irrespective of location.

Keywords: deep learning; transfer learning; plastic; UAVs

Citation: Maharjan, N.; Miyazaki, H.; Pati, B.M.; Dailey, M.N.; Shrestha, S.; Nakamura, T. Detection of River Plastic Using UAV Sensor Data and Deep Learning. *Remote Sens.* **2022**, *14*, 3049. https://doi.org/10.3390/rs14133049

Academic Editors: Jukka Heikkonen, Fahimeh Farahnakian and Pouya Jafarzadeh

Received: 9 May 2022
Accepted: 22 June 2022
Published: 25 June 2022

1. Introduction

Plastic is used extensively in households and industry. Plastic takes hundreds of years to degrade, so it affects both the terrestrial and marine ecosystems. Marine litter has been recognized as a serious global environmental issue since the rise of the plastic industry in the mid-1950s [1]. Hence, the need for research into plastic management solutions is self-evident [2]. The UN Environment Programme (UNEP) estimates that 15% of marine litter floats on the sea's surface, 15% remains in the water column, and 70% rests on the seabed. Up to 80% of the plastic in the ocean is from land-based sources and reaches the ocean via rivers [3]. Nevertheless, riverine plastics are understudied compared to marine plastics [4]. The earliest research on riverine plastic began in the 2010s, with a study on a sample of waterways in Europe and North America, particularly the Los Angeles area [5] and the Seine [6].

Current government regulations do not adequately address marine litter and plastics. There is also a gap in regional frameworks addressing the issue of plastic litter. Establishing proper waste collection systems and changing peoples' perceptions are two major hurdles to plastic litter prevention, and both goals remain a distant dream in southeast Asian countries. Thoroughly surveying plastic litter distribution in rural areas manually is time-consuming and complex, so automatic mapping of plastic litter using unmanned aerial vehicles (UAVs) is a better option, especially in inaccessible locations.

UAVs (abbreviations used throughout the paper are listed in "Abbreviations" in alphabetical order) are relatively low-cost and can operate at low-altitudes with minimal risk. They provide images with high resolution and high image acquisition frequency [7]. UAV-based real-time data collection of imagery is important for surveillance, mapping, and disaster monitoring [8,9]. UAVs are widely used for data collection, object detection, and tracking [10]. UAVs can be categorized as low- or high-altitude platforms [11] and can be roughly categorized into three classes: small, medium, and large, according to their maximum altitude and range. The maximum altitude for small drones is usually below 300 m; the maximum altitude for large drones is normally above 5500 m. Altitudes vary within these ranges for medium size UAVs. Regarding maximum range, small UAVs can typically cover less than 3 km, while medium UAVs can cover 150–250 km, and large ones can cover even larger distances. High-altitude UAVs can image large areas quickly, while low attitude UAVs can capture more detailed features in smaller fields of view. High-altitude UAV scans can be used as a preliminary to reduce the overhead involved in finding the correct areas for more detailed surveys. Once a high-altitude survey is completed, the plastic in a river can be precisely detected and catalogued based on a follow-up low-altitude UAV survey. Since UAVs at such low-altitudes can provide centimeter-level or better pixel resolution with high accuracy [12], they open the door for ordinary individuals to collect and analyze high-quality imagery through automatic methods irrespective of whether satellite or aerial imagery is available from formal sources. Given a specific camera selected and mounted on a UAV, an appropriate flight altitude should be determined to obtain a suitable ground sampling distance (GSD) for measuring sizes of items captured in the images and for efficiently covering the target area. The GSD is the size of the projection of one pixel on the ground and is a function of the focal length of the camera, flight altitude, and physical dimensions of sensor's pixels. The GSD places a lower limit on the precision achievable for points on the ground [13]. In addition, flight altitude, camera properties determine the resolution of the images captured. Though we obtain good resolution with a 4K camera at 30 m, other researchers [13–15] conducted flights at ranges of 6–10 m for better image resolution. UAVs flying at a low-altitude provide high-resolution data, which are useful in detecting plastic, metal, and other litter in rivers. The focal length also affects image quality and plays a vital role in obtaining accurate annotations and precise plastic detection [16]. Simple color-based approaches to categorization of litter in UAV images [17] are less dependent on flight altitude and GSD than object detectors, which typically require high resolution images captured at lower altitudes.

UAVs have already been used in monitoring marine macro-litter (2.5 cm to 50 cm) in remote islands [18–20], which suggests that low-cost UAVs are suitable for low-altitude, high-resolution surveys (from 6 m to 30 m). Estimates of plastic litter in global surface waters are available [2], but we are far from having a global inventory of litter along shores due to the low efficiency and limited extent of surveys along shores thus far [21]. However, UAV images have been found effective for analyzing the spatial distribution of plastic litter cross-shore and long-shore, as well as for measuring the sizes of detected items using semi-automated image processing techniques [22]. Moreover, UAV applications were found to be effective for monitoring coastal morphology, the extent of morphological changes, and interaction of marine litter dynamics on the beach [23].

Floating litter surveys conducted by UAVs at altitudes of 20 m and 120 m have been found to be more accurate than beach litter surveys at altitudes of 20 m and 40 m [24]. The authors attribute this to seawater being a more homogeneous background than sand. Floating litter surveys, however, have the risk of losing the UAV while it is flying over the sea, and beach litter surveys are less affected by environmental challenges. According to Martin et al. [20], manual screening of UAV images of beaches taken from a height of ten meters was 39 times faster and 62% more accurate than the standard ground-based visual census method. Researchers also pointed out that training citizen scientists to annotate plastic litter datasets acquired through UAVs is effective [25,26]. However, machine learning-based automatic mapping combined with manual screening was found to be even faster and more cost-effective [19,20].

Since rigorous interpretation of aerial images from UAVs by humans is time-consuming, error-prone, and costly, modern deep learning (DL) methods using convolutional neural networks (CNNs) are a preferable alternative [27]. DL is already well established in remote sensing analysis of satellite images. UAV technology integrated with deep learning techniques is now widely used for disaster monitoring in real time, yielding post-disaster identification of changes with very higher accuracy [28,29]. DL has emerged as an extremely effective technique in modern computer vision due to its ability to handle a variety of conditions, such as scale transformations, changes in background, occlusion, clutter, and low resolution, partly due to model capacity and partly due to the use of extensive image augmentation during training [30]. DL has proven superior to traditional machine learning techniques in many fields of computer vision, especially object detection, which involves precise localization and identification of objects in an image [17,31]. Classification, segmentation, and object detection in multispectral ortho imagery through CNNs has been successful [32]. In UAV mapping applications involving detection of objects, changes in viewing angles and illumination introduce complications, but CNNs nevertheless extract useful distinguishable features. CNNs are very effective for per-pixel image classification.

Although deep learning methods have been shown to provide accurate and fast detection of marine litter [33], little research integrating UAVs and deep learning has been conducted in the context of monitoring plastics on beaches and rivers. Once a model has been trained, processing UAV images for detection of plastics with the model is straightforward. However, deep learning methods require a great deal of computing resources for offline training and online inference, as models are required to perform well across various conditions, increasing their complexity. Furthermore, training of modern object detection models requires a great deal of manual labor to label data, as the data preparation requires accurate bounding boxes in addition to class labels, making the data engineering more intensive than that required for classification models. To minimize these costs, plastic monitoring application should analyze georeferenced UAV patch images ensuring appropriate image quality and little redundancy. To determine whether a given training dataset is sufficiently representative for the plastic detection in similar georeferenced patch images after model development, we advocate evaluation of the method at multiple locations.

It is time consuming to train a deep neural network for detection from scratch. It can be more effective to fine-tune an existing pre-trained model on a new task without defining and training a new network, gathering millions of images, or having an especially powerful GPU. Using a pre-trained network as a beginning point rather than starting from scratch (called transfer learning) can help accelerate learning of features in new datasets with small amounts of training data while avoiding overfitting. This approach is therefore potentially particularly useful for detection of plastic in a modest-scale dataset. OverFeat [34], the winner of the localization task in the ILSVRC2013 competition, used transfer learning. Google DeepMind uses transfer learning to build deep Q-network agents that use pixels from 210 $\times$ 160 color video at 60 Hz and the game score as input and learn new games across different environments with the same algorithms and minimal knowledge. This model was the first artificial agent to learn a wide variety of challenging tasks without task-specific engineering [35]. Nearly every object detection method in use today makes use of transfer learning from the ImageNet and COCO datasets. The use of transfer learning provides the following advantages [36]:

1. higher baseline performance;
2. less time to develop the model;
3. better final performance.

We therefore investigated the performance of pretrained and tabula rasa object detection models for plastic detection using data acquired from a Mekong river tributary, the Houay Mak Hiao (HMH) river in Vientiane, Laos, as well as a canal in the Bangkok area, Khlong Nueng in Talad Thai (TT), Khlong Luang, Pathum Thani, Thailand. We explored how a model trained on one location performs in a different location in terms of compute resources, accuracy, and time.

This paper makes three main contributions to the state of the art in riverine plastic monitoring:

1. We examine the performance of object detection models in the You Only Look Once (YOLO) family for plastic detection in ortho imagery acquired by low-altitude UAVs.
2. We examine the transferability of the knowledge encapsulated in a detection model from one location to another.
3. We contribute a new dataset comprising images with annotations for the public to use to develop and evaluate riverine plastic monitoring systems.

We believe that this research will provide practitioners with tools to save computing resources and manual labor costs in the process of developing deep learning models for plastic detection in rivers. The techniques introduced here should scale up to various types of landscapes all over the world.

2. Materials and Methods

In this section, we describe the study area for the research and the materials and methods adopted to perform experiments on the task of plastic detection from UAV imagery in two locations through deep learning.

2.1. Study Area

We gathered data at two locations, viz., Khlong Nueng Canal, Talad Thai, Pathum Thani (TT), Thailand and Houay Mak Hiao river in Vientiane, Laos (HMH) as in Figure 1. HMH is in a sub-basin of the Mekong River basin with a land area of 436.91 km^2, located in Vientiane, the capital city of Laos as in Figure 2. The study area was at coordinates 17.95°N 102.91°E. This river contributes pollutant to the Mekong River basin. TT is in Khlong Luang district, Thailand with coordinates 14.08°N 100.62°E, as shown in Figure 3. The study areas were selected based on their contribution to pollution downstream and the ease and safety of accessibility for data collection considering UAV survey zone restriction in Laos and Thailand. As no study of individual plastic object detection in these areas has yet been performed, they were found to be ideal for evaluating plastic monitoring methods.

2.2. Materials

UAV surveys 30 m above the terrain were carried out at Houay Mak Hiao river (HMH) in Vientiane, Laos and Khlong Nueng Canal (TT) in Talad Thai, Pathum Thani, Thailand with a DJI Phantom 4 with a 4K resolution camera resulting in a ground sampling distance of 0.82 cm to assess the plastic monitoring methods for these waterways.

The computing resources comprised two environments: (1) Anaconda with Jupyter running on a personal computer with an Intel®Core™ i7-10750H CPU @2.60 GHz, 16 GB RAM, and NVIDIA GeForce RTX 2060 GPU with 6 GB GPU RAM, and (2) Google Colaboratory Pro. The personal computer was used for YOLOv3 and YOLOv5, and Google Colaboratory Pro was used for YOLOv2 and YOLOv4.

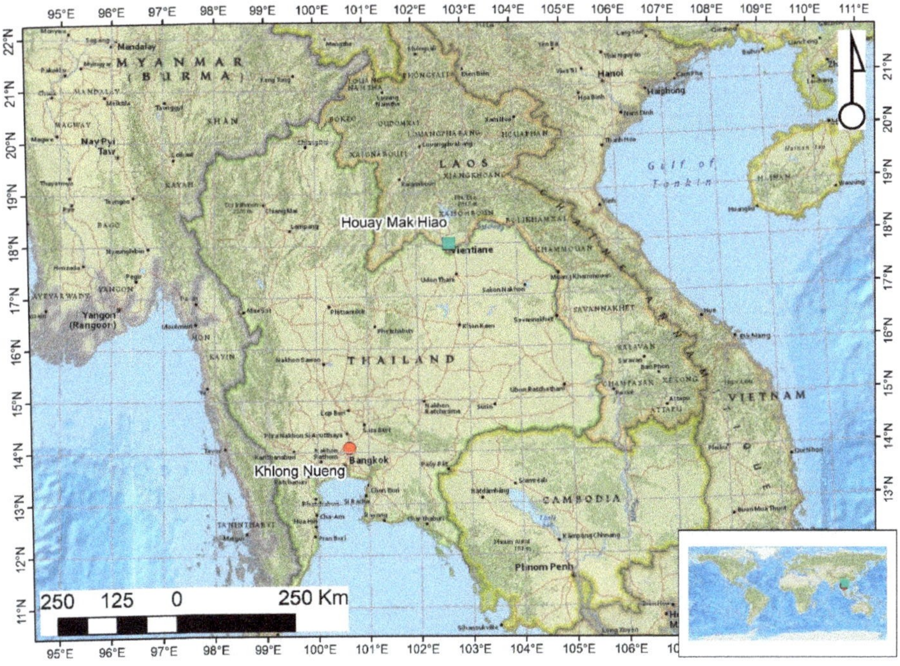

Figure 1. Location of study sites (Background map: OpenStreetMap, 2021).

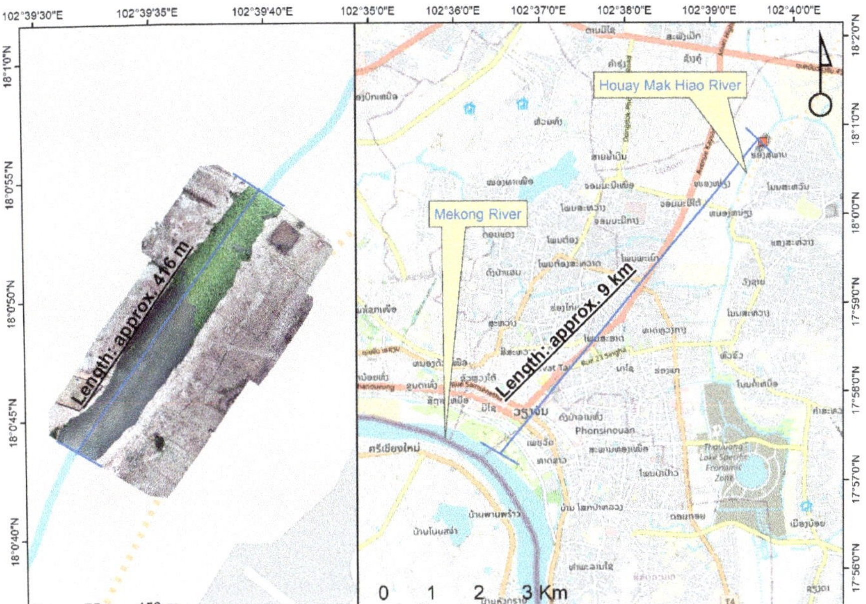

Figure 2. Study area showing Houay Mak Hiao River, Vientiane, Laos. (Background map: OpenStreetMap, 2021).

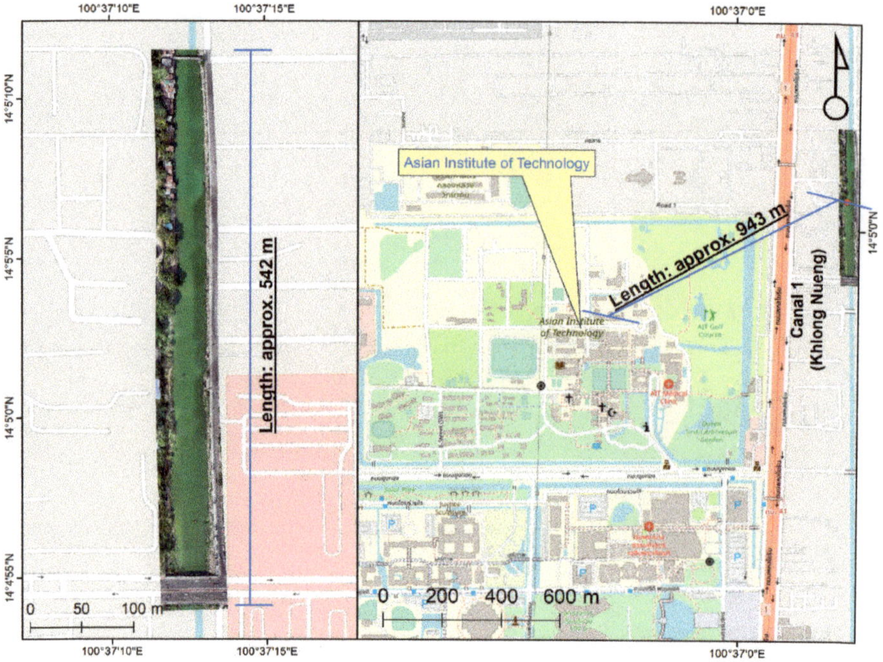

Figure 3. Study area showing Khlong Nueng, Talad Thai, Pathum Thani, Thailand (Background map: OpenStreetMap, 2021).

2.3. Methodology

In this section, the proposed methodology for detection of plastic in rivers is discussed, along with the various deep learning model architectures used in the experiments. We aim to assess model performance in the task of identifying plastic in rivers using georeferenced ortho-imagery and deep learning approaches utilizing minimal computing resources, as shown in Figure 4.

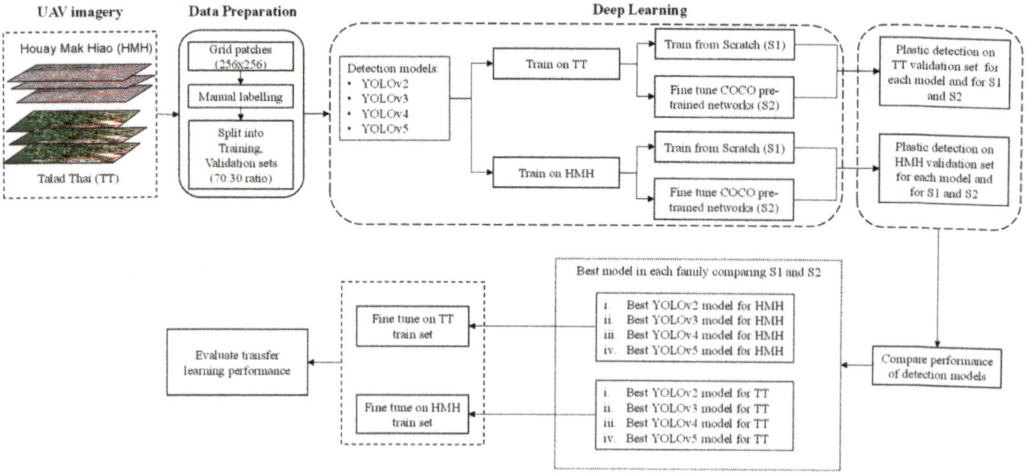

Figure 4. Methodological framework for assessment of performance of deep learning architectures for plastic detection.

2.3.1. Deep Learning Models for Object Detection

CNNs can locate multiple objects in an image, effectively separating foreground from background [37]. We thus evaluate various CNN-based object detection models on riverine plastic detection. Object detection has two main functions: to find regions of interest and to classify those regions. Regions of interest can be obtained in two ways, by region proposal methods or direct regression. Region proposal methods involve two stages, the first of which involves finding regions of interest through color contrast and superpixel straddling, and the second of which involves classifying the resulting proposals with CNNs. The direct regression method, on the other hand, is a one step-method in which region proposals and object detection are carried out in a single step. Single-step models tend to find it difficult to locate small objects in an image due to a limited number of possible bounding boxes at fine levels of detail. YOLO is the most popular single-stage detector. It carries out both the bounding box identification and object classification tasks in a single pass of the network. R-CNN is a representative of two-stage detectors. Some of the older detection models use a full CNN classifier such as VGG-16 or ResNet as the classifier while most modern detectors such as YOLO use a CNN classifier as a backbone for feature extraction followed by a small "head" for classification.

Early versions of YOLO had better performance in both speed and accuracy than extant models such as MobileNetSSDv2 and Faster R-CNN. YOLO makes use of a single CNN to detect objects by processing the entire image at once without creating region proposals. It predicts a detection tensor directly based on a small set of possible bounding boxes. Features at the deeper layers used for the final detection have receptive fields spanning the entire image, making it less likely to predict false positives in background regions. YOLO models output bounding box coordinates, confidence scores, and object class scores directly with an image as input. The confidence scores signify the probability that a predicted box contains an object. YOLO is fast, running at 45 FPS in real-time, and Fast YOLO is faster at 155 FPS [38]. The original YOLO architecture predicts just two bounding boxes per grid cell [39]. The total of 98 bounding boxes per image is small compared to the 2000 boxes predicted by Selective Search. Though most of the early detection frameworks depended on heavy feature extractors such as VGG-16, which uses 30.69 billion floating operations in a single pass for a single image of 224 $\times$ 224 resolution, YOLO used the more lightweight GoogLeNet architecture, with only 8.52 billion operations [40], albeit with lower accuracy as a backbone than VGG-16. YOLO has no localization error and hence is less likely to predict false positives in the background [41].

YOLOv2 was introduced to improve the speed-accuracy trade-offs in YOLO. The custom GoogLeNet [42] network was replaced by DarkNet19, and batch normalization [43] was introduced. The fully connected layers in GoogLeNet were also removed, and anchor boxes with aspect ratios learned through k-means were introduced along with multiscale training. Despite these improvements, YOLOv2 has low recall [38], so YOLOv3 was subsequently introduced with further improvements. YOLOv3 is tuned for small objects with multi-scale features [44]. YOLOv3 is much more complicated than the previous model, and the speed and accuracy can be varied by changing model size. YOLOv3 provides good average precision (AP) at an Intersection over Union (IoU) threshold of 0.5, but the AP decreases at higher IoU levels because YOLOv3 does not predict ground truth bounding box boundaries very accurately. YOLOv3-SPP (spatial pyramid pooling) adds a SPP module, which uses the concept of the spatial feature pyramid, realizing both local and global features. This solves the issue of image distortion caused by cropping and zooming the image area and repeated feature extraction by the CNN. The smaller version of YOLOv3, called Tiny YOLOv3, is designed for mobile machine learning and low-powered computing devices such as the Internet of Things (IoT) devices and shows better performance in terms of speed accordingly [45]. The size of the Tiny YOLOv3 CNN is about 20% that of YOLOv3, and it runs several times faster, making it usable for real-time detection on small devices. From YOLOv2 to YOLOv3, the computational complexity in terms of GFLOPs (billion floating-point operations), which mostly depends on the number and types of layers used in the network, increases from 30 to 140, with an increase in mAP from 21% to 33%. The added complexity, however, means it cannot be considered a light-weight model [44].

YOLOv4 and YOLOv5 were developed to increase the speed of YOLOv3 while keeping high accuracy. YOLOv3 was known not to perform well on images with multiple features or on small objects. Among other improvements, YOLOv4 uses the Darknet53 backbone augmented with cross-stage partial blocks (CSPDarknet53), improving over YOLOv3 using only 66% of the parameters of YOLOv3, accounting for its fast speed and accuracy [46]. The YOLOv5 model pushes this further, with a size of only 27 megabytes (MB), compared to the 244 MB of YOLOv4. YOLOv5 models pre-trained on MS COCO achieve mAPs from 36.8% (YOLOv5s) to 50.1% (YOLOv5x). YOLOv5 and YOLOv4 have similar network architectures; both use CSPDarknet53 as the backbone, and both use a path aggregation network (PANet) and SPP in the neck and YOLOv3 head layers. YOLOv5's reference implementation is based on the PyTorch framework for training rather than the Darknet C++ library of YOLOv4. This makes YOLOv5 more convenient to train on a custom dataset to build a real time object detection model.

Yao et al. [47] consider the fact that UAVs normally capture images of objects with high interclass similarity and intraclass diversity. Under these conditions, anchor-free detectors using point features are simple and fast but have unsatisfactory performance due to losing semantic information about objects resulting from their arbitrary orientations. The authors' solution uses a stacked rotation convolution module and a class-specific semantic enhancement module to extract points with representations that are more class-specific, increasing mAP by 2.4%. Future work could compare YOLO-type detectors with improved point feature-based detectors such as R^2 IPoints. However, it is difficult to detect small objects with dense arrangements using this detector due to the sensitiveness of IoU to the deviation of the position of small objects.

The use of transformer neural networks [48] has led a new direction in computer vision. Transformers use stacked self-attention layers to handle sequence-to-sequence tasks without recursion, and transformers have recently been applied to vision tasks such as object detection. The vision Transformer (ViT) was the first high accuracy transformer for image classification [49]. However, ViT can only use small-sized images as input, which results in loss of information. The detection transformer (DETR) [50] performs object detection and segmentation. DETR matches the performance of highly optimized Faster R-CNN on the COCO dataset [51]. The Swin transformer [52] has been proposed as a backbone for computer vision. Swin stands for shifted window which is a general-purpose backbone for computer vision. Swin is a hierarchical transformer that limits the self-attention computation to non-overlapping local windows and allows cross-window connection through shifted window to address the issue of a large variation in scale and resolution of images, leading to relatively good efficiency on general hardware, running in time linear in the image size. The Swin transformer achieves current state-of-the-art performance on the COCO object detection task (58.7 box AP and 51.1 mask AP on COCO test-dev) and ADE20K semantic segmentation (53.5 mIoU on ADE20Kval).

CNNs have a natural inductive bias for image processing problems, such as translation equivariance and contrast adaptivity, but the transformer lacks these properties, resulting in requirements for much larger datasets or stronger data enhancement [53] to achieve the best performance. Since our goal is to perform well on moderate-sized datasets using modest compute resources, we do not consider transformers at this time.

2.3.2. Selection of Object Detection Models

Various object detection models have been used in research related to plastic litter detection. Majchrowska et al. [54] use EfficientDet-D2 to localize litter and EfficientNet-B2 to classify waste into seven categories. The researchers obtained 75% classification accuracy and 70% mean average precision.

Córdova et al. [55] conducted a comparative study on state-of-the-art approaches for object detection using the PlastOPol and TACO datasets and found that YOLOv5-based detectors perform well in litter detection. On the PlastOPol dataset, YOLO-v5x obtains a best AP@0.5 of 84.9, and YOLO-v5s obtains best AP@0.5 of 79.9. On the TACO dataset, YOLO-v5x obtains a best AP@0.5 of 63.3, and YOLO-v5s obtains a best AP@0.5 of 54.7 for YOLO-v5s. YOLO-v5s was found to be 4.87, 5.31, 6.05, and 13.38 times faster than RetinaNet, Faster R-CNN, Mask R-CNN, and EfficientDet-d5, respectively.

Kraft et al. [56] use calibrated onboard cameras with GNSS and GPS to capture images and use YOLOv3, YOLOv4, and EfficientDet for object detection [57]. They find that YOLOv4 and EfficientDet-d3 show the highest mean average precision (mAP) for trash detection. Kumar et al. [58] analyze the efficiency of YOLOv3 and YOLOv3-tiny in separating waste into bio-degradable and non-biodegradable types. Their research shows that YOLOv3 has better predictive performance than YOLOv3-tiny, with accuracies of 85.29% and 26.47%, respectively. This research used 6437 images drawn from six classes (cardboard, paper, glass, plastic, metal, and organic waste) and found that YOLOv3-tiny needs four times less computation time than YOLOv3, demonstrating a wide speed-accuracy tradeoff.

Fulton et al. [59] evaluate the performance of object detection algorithms (YOLOv2, Tiny-YOLO, Faster R-CNN with Inception v2, and Single Shot MultiBox Detector (SSD) with MobileNetV2 for underwater trash detection and removal of trash using autonomous underwater vehicles. (AUVs). The models detect three classes of objects in the J-EDI (JAMSTEC E-Library of Deep-Sea Images) dataset, i.e., plastic, remotely operated vehicles (ROVs), and a "bio" class (plants, fish, detritus, etc.). All the above-mentioned models are fine-tuned from their pre-trained states. The authors' transfer learning method for the YOLO model only updates weights in the last three layers. The authors find that the YOLOv2 models have good speed, but YOLOv2 and tiny-YOLO have low mAP. They also find that transfer learning increases accuracy for the bio-class to a level sufficient for deployment in real time scenarios.

Tata et al. [60] describe the DeepPlastic project for marine debris detection in the epipelagic layer of the ocean. This project includes the development of the DeepTrash dataset comprising annotated data captured from videos of marine plastic using off-the-shelf cameras (GoPro Hero 9) in three study sites in California (South Lake Tahoe, Bodega Bay, and San Francisco Bay) and also incorporating the J-EDI dataset to represent marine plastics in different locations. The research used low-cost GPUs and the deep learning architectures YOLOv4-tiny, Faster R-CNN, SSD, and YOLOv5s for detection with the aim to build a real-time monitoring system. The YOLOv5s model achieved a mAP of 85%, which is higher than that of the YOLOv4-tiny model (84%). These models outperformed a model for detection of deep-sea and riverine plastic by the University of Minnesota [59], which had mAPs of 82.3% using YOLOv2 and 83.3% using Faster R-CNN. The authors therefore selected YOLOv4-tiny and YOLOv5s, which have good accuracy and sufficiently high inference speeds for real-time object detection. Since there are several models with different speed-accuracy tradeoffs in the YOLOv5 group of detectors, various YOLOv5 models have been used in research related to the detection of plastic [61]. This family of object detection models offers flexibility in terms of architecture and can be adjusted for the best performance in different tasks. From YOLOv5s to YOLOv5l, the number of parameters, depth, and width increases steadily resulting in higher model complexity but better accuracy. We use the YOLO family of algorithms for plastic detection in the river in this research due to its good performance in terms of speed and accuracy of detection in real-world environments with limited computing resources and data. We trained different pre-trained YOLOv2 models (YOLOv2, YOLOv2-tiny), YOLOv3 models (YOLOv3, YOLOv3-tiny, and YOLOv3-spp), YOLOv4 models (YOLOv4, YOLOv4-tiny), and YOLOv5 models (YOLOv5s, YOLOv5m, YOLOv5l, YOLOv5x) to perform plastic detection in UAV images. In addition, fine-tuning the pre-trained models, we also trained each of the aforementioned models from scratch to determine which approach performs best with limited time and capacity. As previously discussed, YOLOv5s was previously found to perform best for plastic detection in the epipelagic layer of the ocean, with a mAP of 0.851 [60], so we use a similar methodology to evaluate performance of plastic detection models for rivers using various YOLO architectures according to mAP at different IoUs.

2.3.3. Transfer Learning

Training deep CNNs from scratch is difficult, as they need a large amount of training data and labeling expertise. Transfer learning can speed up model development compared to training from scratch by fine-tuning some or all of the layers of a pretrained network to perform well on a new dataset [62]. Transfer learning reduces training time, as the model does not need to be trained for many iterations to give good performance. There are two methods of transfer learning, feature extraction and fine-tuning. Feature extraction uses knowledge of features learned on one model to extract meaningful features from a new dataset. In this transfer learning setup, weights of the feature extraction portion of the pre-trained network are not updated during training on the new dataset. Instead, some of the deepest layers are unfrozen, and the model is trained with a low learning rate for both the new classifier layer and the previously existing deepest layers of the base model. Transfer learning via fine-tuning, on the other hand, allows all the layers or some of the layers of the base model to be unfrozen, and model is retrained end-to-end, again with a very low learning rate. The outcome is to fine-tune the weights of the pre-trained network to extract high-order features more appropriate for the specific new task.

2.3.4. Performance Assessment of Transfer Learning

In addition, in-sample test performance, we also assess each model's capacity for knowledge transfer to another location. Deep learning models learn features representative of their training datasets. Early layers tend to learn general features, while later layers tend to learn features that are high level and more specific to the training dataset. We perform transfer learning on models pre-trained on one location, fine-tuning them by either (1) freezing weights of all the initial layers of the network of the pre-trained models and then changing the weights of the last two layers of the respective network, allowing them to learn features from data of the new location, or (2) fine-tuning all parameters in every layer. The best weights for the best model for plastic detection at one location are used as a basis for training at the other location. The same performance metrics are computed for each of the transferred models to find the best approach to transfer learning about the plastic detection task to a new location at low computing cost with minimal compute time.

The following basic steps are required to perform the comparison of deep learning techniques.

a. Data preparation: Prepare the data set in the appropriate format (e.g., DarkNet format for YOLOv4-tiny and PyTorch format for YOLOv5s) and then split it into training and validation sets.
b. Input: Prepare images and label files for training and validation dataset along with the pre-trained weights and configuration file for training.
c. Output: Save trained model to a file containing optimized weights.

(A) Training models from pre-trained networks (S1):

To train neural networks for plastic detection beginning with pre-trained networks, we perform the following steps.

i. Load pre-trained weights (optimized for the COCO dataset) into the model.
ii. Freeze the initial N_1 layers and unfreeze the last N_2 layers of the model.
iii. Select a hyperparameter configuration from Table 1.
iv. Train the model and stop training when average loss stops decreasing.
v. Record final average loss.
vi. Repeat steps iii–v for all combinations of hyperparameters.
vii. Select the model with hyperparameters that achieve the lowest average loss.

Table 1. Selection of hyperparameters.

Parameters	Value
Batch size *	16, 32, 64 and 128
Learning rate	0.01 to 0.001
No. of filters in YOLO layers	18 **

* YOLOv5 requires a batch size 4 for all experiments due to limited GPU memory; ** Replace number of filters $(80 + 5) \cdot 3$ for COCO with $(1 + 5) \cdot 3$ in the convolutional layer before each *YOLO* layer.

(B) Training from scratch (S2):

The following steps are undertaken to carry out model training from scratch. The steps are the same as for pre-trained networks (S1) with modifications to step (ii) as follows:

i. Load the pre-trained weights (trained on COCO dataset).
ii. Unfreeze all layers and initialize weights to random values from Gaussian distributions having mean zero and standard deviation $\sqrt{(2/n)}$, where n denotes unit's fan in (number of input units). This initialization controls the initial output and improves convergence empirically [63].
iii. Select a subset of hyperparameters from Table 1.
iv. Train the model and stop training when average loss stops decreasing.
v. Record average loss.
vi. Repeat steps iii–v for all combinations of hyperparameters.
vii. Select the model with hyperparameters that achieve the lowest average loss.

(C) Transfer learning:

To evaluate transfer of learning from one location to another, the following steps are carried out.

i. Collect best weights for each model and each type of training at one location.
ii. Load the best weights for one location and one model.
iii. Freeze initial N_1 layers and fine-tune the last N_2 layers.
iv. Select a subset of hyperparameters from Table 1.
v. Train the model in a new location and stop training when average loss stops decreasing.
vi. Calculate average loss.
vii. Repeat steps iv–vi for all combinations of hyperparameters, for all models.

2.3.5. Performance Indicators

We evaluate the performance of detection models using the performance metrics described in this section.

(A) Mean Average Precision (mAP):

It is unrealistic to expect perfect matches between the ground truth and predicted bounding boxes due to variations in labeling and quantization. The area under a precision versus recall curve gives the average precision for a specific class for the set of predictions of a model. The average of this value, calculated over all classes and multiple IoU thresholds, is called mAP. mAP measures the performance of an object detector based on the IoU between the predicted and ground truth bounding boxes across all classes in the dataset. The Jaccard similarity or IoU is a measure of how well a predicted bounding box fits a ground truth bounding box for an object, defined by

$$\text{IoU} = \frac{\text{Area of Overlap}}{\text{Area of Union}}. \tag{1}$$

The numerator is the area of the intersection of the predicted and ground-truth bounding boxes, while the denominator is the total area covered by the union of the predicted and ground truth bounding boxes. IoU ranges from 0 to 1. Closer rectangles give higher IoU values. If the IoU threshold is 0.5, and a predicted bounding box has an IoU with a ground-truth bounding box of more than 0.5, the prediction is considered a true positive

(TP). If a predicted bounding box has IoUs less than 0.5 for all ground-truth bounding boxes, it is considered a false positive (FP). IoU is well suited to unbalanced datasets [64]. We use an IoU threshold of 0.5.

mAP is a widely used metric and the benchmark for comparing models on the COCO data set. AP gives information about the accuracy of a detector's predicted bounding boxes (precision) and the proportion of relevant objects found (recall). Precision is the number of the correctly identified objects of a specific class in class, divided by the total number of objects of that class in an image set.

$$\text{Precision} = \frac{\text{TP}}{\text{TP} + \text{FP}} \quad (2)$$

In the equation, TP and FP are the total number of true positives and false positives.

The recall is the number of correctly detected objects divided by the total number of objects in the dataset. It signifies how well the ground truth objects are detected.

$$\text{Recall} = \frac{\text{TP}}{\text{TP} + \text{FN}} \quad (3)$$

FN is the number of false negatives. A false negative is a ground truth bounding box with insufficient overlap with any predicted bounding box [65]. Perfect detection is a precision of 1 at all recall levels [66]. There is usually a tradeoff between precision and recall; precision decreases as recall increases and vice-versa. AP averages the model's precision over several levels of recall.

(B) F1-Score:

F1 is a measure of a model's accuracy on a dataset at a specific confidence level and IoU threshold. It is the harmonic mean of the model's precision and recall [67]. It ranges from 0 to 1. A F1-score of 1 indicates perfect precision and recall. The maximum F1 score refers to the best harmonic mean of precision and recall obtained from a search over confidence score thresholds for the test set.

$$\text{F1} - \text{Score} = \frac{2 \cdot \text{Precision} \cdot \text{Recall}}{\text{Precision} + \text{Recall}} \quad (4)$$

3. Results

3.1. Dataset Preparation

The image dataset comprised tiled ortho-images cropped to a size of 256 × 256 pixels corresponding to 2 m × 2 m patches of terrain. We annotated 500 tiles for each river using the YoloLabel tool [68] to record the bounding box for each identifiable piece of plastic in each image. Sample images from Laos (HMH) and Talad Thai (TT) datasets are shown in Figure 5.

Manual labeling of plastic in the image is a work-intensive task. However, labelers have done their best to identify only plastic though there will be some unavoidable errors in the labeling due to difficulty in perceiving the material [69]. Plastic litter is the bulk of the litter in the marine environment and the greatest threat to marine ecosystems. Marine plastic is the biggest concern for the world, most of the marine plastic comes from rivers, etc.

The images were randomly assigned to training and validation sets in a ratio of 70:30 for preparing object detection models using different versions of YOLO. The objects in the HMH dataset tended to be brighter and more distinct-shaped than in the TT dataset, in which the objects were darker, occluded with sand, and mostly trapped among vegetation. Variations in datasets should result in learning of better features and more robust predictions. In most cases, only a small portion of each image contains plastic. Most deep learning methods do not generalize well across different locations [70]. The datasets represent only floating plastic and plastic visible on riverbanks. Submerged plastic was not considered. Similar analysis of the training data representative of plastic has been conducted in the context of automatic mapping of plastic using a video camera and deep learning in five locations of Indonesia [71].

Figure 5. Sample images from datasets used for training deep learning models for plastic detection in rivers. (**a**) HMH in Laos with co-ordinates (887,503.069 m, 1,995,416.74 m); (887,501.986 m, 1,995,416.537 m); and (887,501.418 m, 1,995,417.692 m) (**b**) TT in Thailand with co-ordinates 674,902.457 m, 1,557,870.257 m); (674,903.403 m, 1,557,860.135 m); and (674,925.317 m, 1,557,850.965 m) under WGS_1984_UTM_Zone_47N.

3.2. Experimental Parameter Sets

The individual experiments we carried out to assess the performance of plastic detection with various models in the YOLO family for two locations are tabulated in Table 2. The parameters that are considered for YOLOv3 and YOLOv5 families are batch size 4, epoch 100, and batch size 16 for YOLOv2 and YOLOv4 families with a learning rate 0.001. Mostly, the batch size is adjusted according to the GPU memory with possible allowed high value to simulate model [72]. The models are set up to train on HMH and TT datasets separately from pre-trained networks and from scratch with various YOLO models. Transfer learning from one location HMH (Laos) to another location TT (Thailand), and vice-versa is performed taking the best weights from the best model in each YOLO family to transfer the knowledge to different locations through fine-tuning.

We evaluate the experimental results through the calculation of mAP, computational complexity in terms of GFLOPs, and F1-score. We also calculate the total volume of plastic in terms of estimated surface area covered by plastic objects, using the pixel size in cm and each bounding box's size. We also analyze the smallest and largest plastics that can be detected by the best model. We report the results in this section.

Table 2. Plastic detection experiment details using Houay Mak Hiao river (HMH) and Khlong Nueng Canal (TT) datasets.

Experiment	Training Dataset	Testing Dataset	Training Method	Models (YOLO Family)
I	HMH	TT	Scratch	YOLOv2 YOLOv2-tiny YOLOv3 YOLOv3-tiny YOLOv3-spp YOLOv4 YOLOv4-tiny YOLOv5s YOLOv5m YOLOv5l YOLOv5x
II			Using pre-trained model	
III	TT	HMH	Scratch	
IV			Using pre-trained model	
V	HMH	TT	Fine-tuning	YOLOv5s, YOLOv4, YOLOv3-spp, and YOLOv2 trained in II
VI	TT	HMH	Fine-tuning	YOLOv5s, YOLOv4, YOLOv3-spp, and YOLOv2 trained in IV
VII	Plastic volume estimation using pre-trained YOLOv5s in terms of surface area			

3.3. Experiments I, II, III, and IV: Plastic Detection in UAV Imagery

Plastic detection results without transfer learning given in Tables 3 and 4 are for the HMH and TT datasets, respectively.

The performance of YOLOv2-tiny is clearly worse than that of YOLOv2, YOLOv3, and YOLOv3-tiny as small objects tend to be ignored by YOLOv2. This is likely due to the lack of multi-scale feature maps in YOLOv2 [73]. Previous research [59] found that YOLOv2 provides mAP 47.9 with average IoU 54.7 in the plastic detection compared to 0.809 at IoU 0.5 for YOLOv4 pre-trained here. YOLOv3-tiny scratch has the best inference time of 0.004 s when there is no detection in the HMH dataset.

In our research, the F1 is highest with a value of 0.78 for pre-trained YOLOv4, YOLOv5s, and YOLOv5l for HMH, while the highest F1 is 0.78 and 0.61 for the TT, for pre-trained YOLOv4 and YOLOv5s. Overall, pre-trained YOLOv5s is small, requiring 13.6 MB for weights on disk, and has lower computational complexity than other models, requiring only 16.3 GFLOPs compared to YOLOv4's 244.2 MB model size and 59.563 GFLOPs. Moreover, YOLOv5s takes less time to train than the other models. It exhibits fast inference speed and produces real-time results. Because YOLOv5 is implemented in PyTorch, while YOLOv4 requires the Darknet environment, it is slightly easier to test and deploy in the field, though we note that both Darknet models and PyTorch models can be converted to ONNX and deployed easily. With all of these considerations in mind, we conclude that YOLOv5s is better than YOLOv4 for plastic detection in rivers.

Table 3. Experiment I and II results. Detection Performance on HMH dataset.

Model	Training Time (h)	Inference Time per Image (s)	Model Size (MB)	Computational Complexity (GFLOPs)	mAP @ 0.5 IoU for Validation Dataset	Map @ 0.5 IoU for Testing Dataset	Highest F1 Score	Computing Platform
Pre-trained YOLOv2	0.359	4.74	192.9	29.338	0.723	0.442	0.66	Google Colab
YOLOv2 scratch	0.367	4.84	192.9	29.338	0.581	0.259	0.6	
Pre-trained YOLOv2-tiny	0.166	3.53	42.1	5.344	0.467	0.293	0.38	
YOLOv2-tiny scratch	0.23	3.52	42.1	5.344	0.348	0.286	0.44	
Pre-trained YOLOv3 tiny	0.082	0.01	16.5	12.9	0.714	0.366	0.7	Intel®Core™ i7-10750H CPU @2.60 GHz, 16 GB RAM, and GPU as NVIDIA GeForce RTX 2060
YOLOv3-tiny scratch	0.082	0.004	16.5	12.9	0.555	0.336	0.58	
Pre-trained YOLOv3	0.259	0.018	117	154.9	0.735	0.396	0.72	
YOLOv3 scratch	0.258	0.017	117	154.9	0.479	0.311	0.54	
Pre-trained YOLOv3-spp	0.266	0.017	119	155.7	0.787	0.402	0.75	
YOLOv3-spp scratch	0.279	0.014	119	155.7	0.59	0.265	0.57	
Pre-trained YOLOv4	1.884	6.85	244.2	59.563	0.809	0.463	0.78	Google Colab
YOLOv4 scratch	1.961	5.54	244.2	59.563	0.766	0.373	0.74	
Pre-trained YOLOv4-tiny	0.899	2.92	22.4	6.787	0.758	0.418	0.76	
YOLOv4-tiny scratch	0.968	2.72	22.4	6.787	0.732	0.355	0.73	
Pre-trained YOLOv5s	0.146	0.019	13.6	16.3	0.810	0.424	0.78	Intel®Core™ i7-10750H CPU @2.60 GHz, 16 GB RAM, and GPU as NVIDIA GeForce RTX 2060
YOLOv5s scratch	0.149	0.017	13.6	16.3	0.740	0.272	0.67	
Pre-trained YOLOv5m	0.195	0.041	40.4	50.3	0.787	0.434	0.77	
YOLOv5m scratch	0.197	0.04	40.4	50.3	0.695	0.331	0.70	
Pre-trained YOLOv5l	0.265	0.027	89.3	114.1	0.810	0.422	0.78	
YOLOv5l scratch	0.262	0.032	89.3	114.1	0.669	0.176	0.67	
Pre-trained YOLOv5x	0.402	0.036	166	217.1	0.781	0.367	0.76	
YOLOv5x scratch	0.399	0.042	166	217.1	0.710	0.316	0.69	

Table 4. Experiment III and IV results. Detection Performance on Talad Thai dataset.

Model	Training Time (h)	Inference Time per Image (s)	mAP@ 0.5 IoU for Validation Dataset	mAP @ 0.5 IoU for Testing Dataset	Highest F1 Score	Computing Platform
Pre-trained YOLOv2	0.649	4.74	0.499	0.452	0.52	Google Colab
YOLOv2 scratch	0.648	4.94	0.368	0.327	0.44	
Pre-trained YOLOv2-tiny	0.162	3.53	0.328	0.256	0.33	
YOLOv2-tiny scratch	0.174	3.43	0.302	0.220	0.32	
Pre-trained YOLOv3-tiny	0.087	0.007	0.495	0.483	0.53	Intel®Core™ i7-10750H CPU @2.60 GHz, 16 GB RAM, and GPU as NVIDIA GeForce RTX 2060
YOLOv3-tiny scratch	0.088	0.007	0.409	0.562	0.47	
Pre-trained YOLOv3	0.282	0.017	0.571	0.743	0.59	
YOLOv3 scratch	0.286	0.016	0.359	0.358	0.43	
Pre-trained YOLOv3-spp	0.285	0.016	0.570	0.748	0.60	
YOLOv3-spp scratch	0.28	0.016	0.390	0.511	0.41	
Pre-trained YOLOv4	1.86	4.54	0.608	0.553	0.78	Google Colab
YOLOv4 scratch	1.89	4.63	0.544	0.524	0.75	
Pre-trained YOLOv4-tiny	0.949	2.85	0.609	0.568	0.59	
YOLOv4-tiny scratch	0.44	3.33	0.560	0.434	0.54	
Pre-trained YOLOv5s	0.146	0.029	0.610	0.767	0.61	Intel®Core™ i7-10750H CPU @2.60 GHz, 16 GB RAM, and GPU as NVIDIA GeForce RTX 2060
YOLOv5s scratch	0.155	0.025	0.530	0.622	0.59	
Pre-trained YOLOv5m	0.22	0.036	0.562	0.761	0.57	
YOLOv5m scratch	0.221	0.036	0.426	0.494	0.49	
Pre-trained YOLOv5l	0.273	0.026	0.579	0.767	0.60	
YOLOv5l scratch	0.283	0.027	0.442	0.529	0.49	
Pre-trained YOLOv5x	0.41	0.035	0.575	0.779	0.57	
YOLOv5x scratch	0.393	0.035	0.363	0.456	0.45	

3.4. Experiment V and VI: Transfer Learning from One Location to Another

The results of the transfer learning experiments are shown in Table 5.

Table 5. Experiment V and VI results. Performance comparison between models trained from scratch, without transfer learning, and with transfer learning by location based on mAP.

YOLO Family	Best Model (Pre-Trained)	Evaluation Dataset	Mean Average Precision (mAP)			
			Training from Scratch	Pretraining on COCO; No Transfer Learning	Transfer from	Pretraining on COCO + Transfer
YOLOv5	YOLOv5s	HMH	0.74	0.81	TT	0.83
		TT	0.53	0.61	HMH	0.62
YOLOv4	YOLOv4	HMH	0.76	0.80	TT	0.83
		TT	0.54	0.60	HMH	0.61
YOLOv3	YOLOv3-spp	HMH	0.59	0.79	TT	0.81
		TT	0.39	0.57	HMH	0.59
YOLOv2	YOLOv2	HMH	0.58	0.72	TT	0.77
		TT	0.37	0.49	HMH	0.51

Transfer learning with fine-tuning is only marginally better than transfer learning without fine-tuning, but both are substantially better than training from scratch. Though mAP on HMH for YOLOv4 and YOLOv5s transfer without fine-tuning is similar (0.81), with fine-tuning, YOLOv4 shows a 3% increase in mAP compared to 1% for YOLOv5s. The number of ground truth objects in HMH is 592 compared to 796 for TT so we see that the model of TT transfers better than HMH with a 2.7% increase in mAP by YOLOv3-spp to 0.81 in compared to training from scratch but still, it is less than by mAP obtained by transfer learning using pre-trained YOLOv4 and YOLOv5s. The YOLOv3-spp model is large (119MB) and has high computational complexity (155.7 GFLOPs) compared to YOLOv5s (13.6 MB and 16.3 GFLOPs). YOLOv4 and YOLOv5 are also faster than YOLOv3. Hence, considering model simplicity, speed, and accuracy, the pre-trained YOLOv5s model for HMH is good for detection with or without transfer learning.

3.5. Experiment VII: Estimation of Plastic Volume in Different Detection Cases

Experiments I-VI lead to the conclusion that the pre-trained YOLOv5s is the best in terms of mAP, inference time, and detection resources. The minimum and maximum size of detected plastic objects are measured using the surface area covered by the detected bounding box using the best pre-trained YOLOv5s model are shown in Figure 6. The smallest and largest ground truth bounding box areas are approximately 26 cm^2 and 4422 cm^2 for HMH, while they are 30 cm^2 and 3336 cm^2 for TT, respectively.

The smallest size of plastic detected is approximately 47 cm^2 in HMH, while the largest size of plastic detected is approximately 7329 cm^2, in TT. The applicable size range for detected plastic depends not only on the models but also on the GSD. The GSD, in turn, depends on the flight altitude and geometric properties of the camera (focal length and sensor size) [74]. Here, we used a single camera for capturing images at both locations, so higher spatial resolution images captured at lower altitudes using the same high-resolution camera could improve the detection of the smaller plastic objects.

Smallest size: 47 cm² Largest size: 3855 cm²

(**a**)

Smallest size: 48 cm² Largest size: 3234 cm²

(**b**)

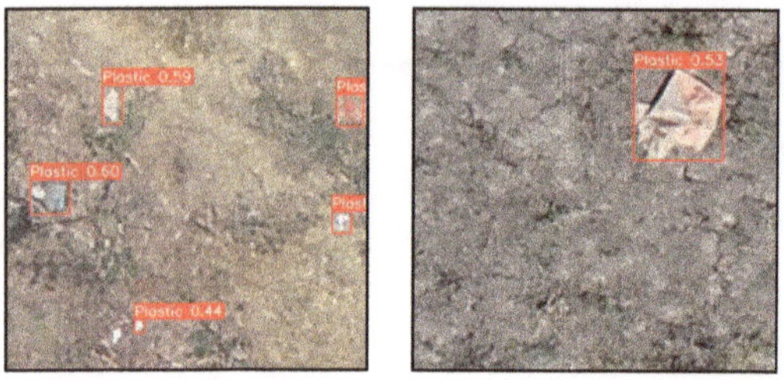

Smallest size: 150.61 cm² Largest size: 2796 cm²

(**c**)

Smallest size: 48 cm² Largest size: 7329 cm²

(**d**)

Figure 6. Experiment VII results. Smallest and largest plastics detected. (**a**) HMH. (**b**) TT. (**c**) Transfer from TT to HMH. (**d**) Transfer from HMH to TT. For reference, the actual dimensions of a 600 mL bottle of water are 23×5 cm = 75 cm^2.

4. Discussion

In this section, we discuss the detection results, examining specific examples of detection using the best pre-trained YOLOv5s model. We also discuss the performance of the model under transfer to a new location.

We find that bright plastics are well detected by the Houay Mak Hiao (HMH) models, while darker and rougher plastics are better detected by the Talad Thai (TT) models. Neither model detects soil-covered or very bright plastic well. This result is sensible, as the HMH data include varied types of rigid plastic objects that are bright and irregular, while the TT data include objects that are more irregular and darker in appearance. Under both transfer and direct training, we find that the TT dataset is more difficult than HMH. The TT dataset has a wider variety of plastic in terms of shape, color, and size.

4.1. Analysis of Sample Plastic Detection Cases with/without Transfer Learning from HMH to TT

First, we consider transfer learning from HMH to TT. Figure 7 shows some of the good results obtained by a model trained on HMH then fine-tuned on TT. The HMH model was originally trained on brighter and rigid objects; hence, the brighter rigid objects in the TT dataset are well detected. However, plastic filled with sand and soil or affected by shadow are ignored.

Figure 8 shows some of the weak results for the HMH model fine-tuned on TT. Amorphous plastic is detected with high confidence by the TT model but with lower confidence by the HMH model fine-tuned on TT. The HMH model appears biased toward rigid and bright objects.

(**a**)

(**b**)

Figure 7. The HMH model fine-tuned on TT performs well in some cases. (**a**) TT model result on TT. (**b**) HMH model results on TT with fine-tuning. (Note: bar-like objects are galvanized stainless steel roof sheets).

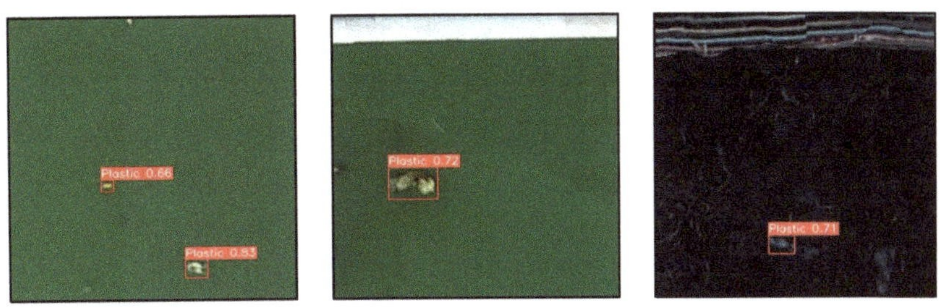

(**a**)

(**b**)

Figure 8. Fine-tuning the HMH model on TT is weak in some cases. (**a**) TT model result on TT. (**b**) HMH model results on TT with fine-tuning. Transfer learning confidence scores are lower. (Note: bar-like objects are galvanized stainless steel roof sheets).

Figure 9 shows some cases in which no plastic is detected by either the TT model or the HMH model after fine-tuning on TT. The plastic is very bright and looks like water or sticks. Apart from the brightness, it is known that the turbidity or cloudiness of the water also affects detection in shallow water, making plastic detection difficult [75]. Shadows and reflections also make detection difficult [19]. Hence, image capture should be performed under optimal weather conditions from a nadir viewing angle [76]. Unavoidable remaining shadows in the image can be rectified through statistical analysis or by applying filters such as gamma correction [77]. In addition, the flight height of the UAV, temperature, and wind speed need to be considered to minimize the effects of atmospheric condition on the images.

Figure 9. Both the TT model and the HMH model transferred to TT fail in some cases. Neither model detected any plastic in these images from TT.

4.2. Analysis of Sample Plastic Detection Cases with/without Transfer Learning from TT to HMH

Next, we consider transfer learning from TT to HMH. Figure 10 shows good results obtained by training on TT then transferring to HMH with fine-tuning. The TT model was originally trained on the amorphous dark objects typical of the TT dataset; hence, these types of objects in the HMH dataset are well detected, showing that model does retain some positive bias from the initial training set.

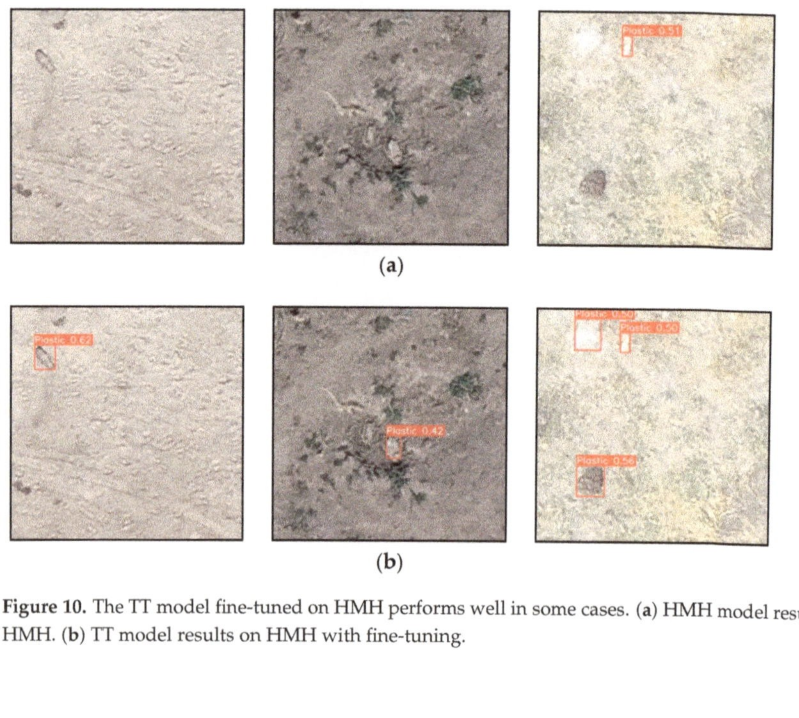

(a)

(b)

Figure 10. The TT model fine-tuned on HMH performs well in some cases. (**a**) HMH model result on HMH. (**b**) TT model results on HMH with fine-tuning.

Figure 11 shows weak results for the TT model fine-tuned on HMH. Rigid, bright, and colored objects are well detected with high confidence by the HMH model but with lower confidence by the TT model fine-tuned on HMH, as the TT data are biased toward dark irregular objects.

(**a**)

(**b**)

Figure 11. Fine-tuning the TT model on HMH is weak or fails in some cases. (**a**) HMH model result on HMH. (**b**) TT model results on HMH with fine-tuning.

Figure 12 shows some cases in which no plastic is detected by either the HMH model or the model using transfer learning from TT to HMH. Neither model detected objects that are soil-like or bright objects floating in the water. Transparent plastic partially floating on the water surface is particularly difficult to identify, as it is affected by the light transmitted through and reflected by the plastic [72].

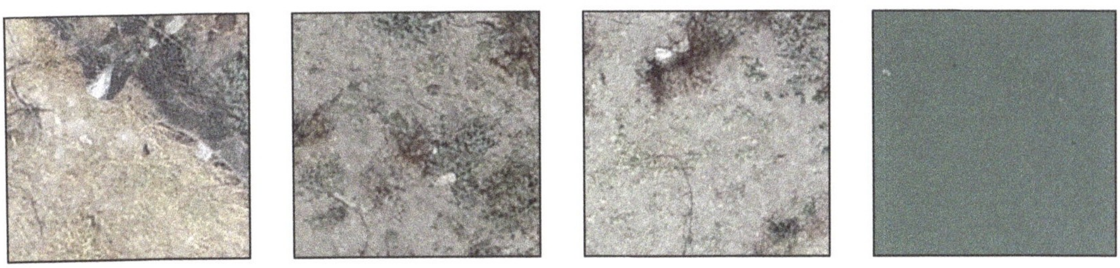

Figure 12. Both the HMH model and the TT model with transfer learning fail in some cases. Neither model detected any plastic in these images.

4.3. Analysis of Performance of YOLO Models for Detection

Models generally improve in accuracy over time as new techniques are introduced, but it is important to evaluate the various models' effectiveness in terms of computational complexity and operational considerations as well as in terms of accuracy. In our experiments, the mAP measurements of the best pre-trained models are higher than those of the best scratch-trained models at the same number of training epochs. The mAP results from the pre-trained YOLOv4 and YOLOv5s models are similar, with values of 0.809 and

0.81 in HMH, respectively, and 0.608 and 0.610 in TT, respectively. This result is consistent with the results of research by the Roboflow team on a custom trained blood cell detection model [78]. A custom dataset of 364 images with three classes (red blood cells, white blood cells, and platelets) was used in their research. The researchers found that YOLOv4 and YOLOv5s had similar performance, with 0.91 mAP @ 0.5 IoU for red blood cells and white blood cells.

According to our method, the pre-trained YOLOv5s model outperforms other YOLO algorithms regardless of the study area. However, the plastic in the HMH dataset appears to be easier to detect than in the TT dataset. Training the pre-trained YOLOv5s model on the HMH or TT dataset gives the best result that dataset in terms of speed, accuracy, and compute resources. We also find that transfer learning improves mAP. Transfer learning from HMH to TT with fine-tuning performs better than training on TT only in the case of bright objects, while TT to HMH works better for dark objects. Pre-trained YOLOv4 and YOLOv5s on TT before fine-tuning on HMH shows high mAP. In other work [78], YOLOv5s has been found to be as accurate as YOLOv4 on small datasets, while YOLOv4 can make better use of large datasets. YOLOv5s has good generalization, while YOLOv4 has more accurate localization. However, YOLOv5s is 88% smaller than YOLOv4 and easier to deploy than YOLOv4, as the YOLOv5 implementation is based on PyTorch, making it easier to deploy in production.

Multiple kinds of research on plastic detection in UAV images using deep learning algorithms have found that plastic can be detected using deep learning techniques [72,76,79], but choosing appropriate models is important. Research with different versions of YOLO on object detection [80,81] have found that YOLOv3 is less capable than YOLOv4 and YOLOV5, perhaps because YOLOv3 uses DarkNet53, which has low resolution for small objects [44]. YOLOv4 extends YOLOv3 with the "bag of freebies" and "bag of specials," that substantially increase accuracy [46]. Research applying YOLOv5s and YOLOv4-tiny models in the epipelagic layer in the ocean [60] found that YOLOv5s performed the best, with high mAP and F1 scores. They found that the VGG19 architecture obtained the best prediction, with an overall accuracy of 77.60% and F1 score of 77.42% [25]. The F1 score of 77.6% is a big improvement over previous research [20] on automatic detection of litter using Faster R–CNN, which obtained an F1 score which found an F-score of 44.2 $\pm$ 2.0%. Consistent with these results, our research shows that YOLOv5s is a fast, efficient, and robust model for real time plastic detection. YOLOv5 uses a Focus structure with CSP-Darknet53 to increase speed and accuracy [81]. Compared to DarkNet53, this structure utilizes less CUDA memory during both forward and backward propagation. YOLOv5 also integrates an anchor box selection process that automatically selects the best anchor boxes for training [82]. Overall, we find that the lightweight YOLOv5s is the most user-friendly model and framework for implementing real-world plastic detection.

4.4. Challenges in Plastic Detection and Future Opportunities for Improvement

There are several challenges involved in detecting plastic in rivers. The reflectance properties of water and other objects influences plastic detection. Previous research [83] found that floating debris caught in river plumes can be identified as plastic when images are analyzed by the floating debris index (FDI) and spectral signatures. Clear water is efficient in absorbing light in the near infrared (NIR) spectrum, while floating plastic and weeds reflect NIR. These spectral properties make floating plastic more visible depending on the spectrum used. Seaweed absorbs shortwave infrared (SWIR) light at 1610 nm more than seawater or plastic, but SWIR absorption has high variation due to atmospheric correction. Timber has peak reflection in the NIR band and is also reflects strongly in the red and SWIR ranges. These properties would help distinguish plastic litter from other materials more effectively if hyperspectral sensors were adopted.

It is sometimes difficult to detect plastic in RGB images due to their limited spectral range and precision [84]. A UAV with a RGB camera may be accurate enough for larger objects but will depends on the objects having distinctive color and weather condition being good for the best performance [85]. UAVs with multispectral or hyperspectral sensors can achieve centimeter-level or decimeter-level resolution while flying at an altitude of several hundred meters and have great potential for monitoring of plastic debris [86]. Though multi-spectral and hyperspectral remote sensing is still in its early stages, it has long-term and global potential for monitoring plastic litter, due to the broader wavelength range and differing absorption and reflectance properties of different materials at different wavelengths. Multispectral sensors can also improve litter categorization. Research by Gonçalves et al. [87] used multispectral orthophotos to categorize litter types and materials applying the sample angle mapping (SAM) technique considering five multispectral bands (B, R, G, RedEdge, and NIR) providing a F1 score of 0.64. However, dunes, grass, and partly buried items were challenges for the litter detection process obtaining a low number of false positives (FP) was crucial to outputting reliable litter distribution estimates.

According to research by Guffogg et al. [88], spectral feature analysis enables detection of synthetic material at a sub-pixel. The minimum surface cover required to detect plastic on a sandy surface was found to be merely 2–8% for different polymer types. The use of spectral features in the near and shortwave infrared (SWIR) regions of the electromagnetic spectrum (800–2500 nm) that characterize plastic polymers can deal with the challenges that occurred due to variable plastic size and shape. Spectral absorption features at 1215 nm and 1732 nm proved useful for detecting plastic in a complex natural environment in Indian Ocean, whereas RGB video and imagery can be complicated by variable light and the color of plastic. Other research [89] has used SWIR spectral features to find large plastics and found that airborne hyperspectral sensors can be used to detect floating plastics covering only 5% of a pixel. However, plastic detection can be affected by the presence of wood or spume, and spectral feature analysis is susceptible to plastic transparency [90].

The characteristics of plastic litter in a river also affect detection quality. Plastic litter does not have a definite shape, size, or thickness in every river. In a study of some beaches of Maldives, more than 87% of litter objects larger than 5 cm were visible in images captured with a UAV at 10 m altitude with a 12.4 MP camera [19]. However, on beaches and in rivers, small plastic objects cause confusion, especially in crowded images [55], while larger plastic items are easily identified, as they span a greater number of pixels and are distinct from surrounding objects. Some plastics can be easily identified through color, but color fades with time, and plastic structure can also degrade in response to exposure to natural elements. Some plastics are flexible, with no distinct edges, and are easily occluded by water and sand. In addition, some transparent objects that look like plastic can be easily misclassified as plastic. Watergrass and strong sunlight reflections interfere with riverine plastic monitoring, as do natural wood debris and algae [91–93]. Different types of vegetation have unique roles in trapping different litter categories, and this phenomenon can increase the difficulty of plastic litter detection [22]. However, including such images in the training set does improve the robustness of the trained model. We therefore include such data in the training sets in this research. Shadows also disrupt the quality of visual information and can impair detectors [94]. It is also difficult to collect a large amount of training data in a short period of time in real environments.

The UAV platform and the performance of its sensors are also important for obtaining good image quality with low observation time. High-performance sensors operated at high-altitudes can cover a broader area more quickly than a low-performance sensor at low-altitudes [95]. The wide coverage area achievable with UAV mapping provides more detailed information on the distribution of plastic in a given area than other survey methods [96]. In future work, the use of hyperspectral sensors [95,97] should be explored, as plastic reflects various wavelengths differently than other objects and materials. Imaging conditions such as brightness, camera properties, and camera height affect the quality of the image. It is also difficult to obtain high quality marine plastic litter monitoring data under different wind speeds and river velocities. Such operating conditions can affect plastic detection accuracy by 39% to 75% [98]. Detection of plastics is easier when the study area has a homogenous substrate on the riverbank.

In summary, plastic detection and monitoring is highly dependent on plastic characteristics and imaging conditions. The global orthomap could be combined with the grid-wise plastic litter detections over the whole study region to create detailed litter maps that would guide stakeholders in effective management of plastic litter.

5. Conclusions

In this paper, we have examined the performance of object detection models in the YOLO family for plastic detection in rivers using UAV imagery with reasonable computing resources. Pre-trained deep learning YOLO models transfer well to plastic detection in terms of precision and speed of training. YOLOv5s is small size with low computational complexity and fast inference speeds, while YOLOv4 is better at localization. Transfer learning with fine-tuning using YOLOv5s improves plastic detection. Hence, we find the pre-trained YOLOv5s model most useful for plastic detection in rivers in UAV imagery.

We make the following main observations from the experiments.

1. Our experiments provide insight into the spatial resolution needed by UAV imaging and computational capacity required for deep learning of YOLO models for precise plastic detection.
2. Transfer learning from one location to another with fine-tuning improves performance.
3. Detection ability depends on a variety of features of the objects imaged including the type of plastic, as well as its brightness, shape, size, and color.
4. The datasets used in this research can be used as references for detection of plastic in other regions as well.

This research introduces a simple to use and efficient model for effective plastic detection and examines the applicability of transfer learning based on the nature of the available plastic samples acquired during a limited period of time. The study should provide plastic management authorities with the means to perform automated plastic monitoring in rivers in inaccessible areas of rivers using deep learning techniques. Furthermore, the research was carried out over limited river stretches during a specific limited period of time. Hence, a UAV survey with wide coverage area and longer flight time may add more prominent data, which would in turn enhance the performance of the detection of plastic.

Author Contributions: N.M., H.M., T.N. and B.M.P. conceived the research. N.M., H.M. and B.M.P. contributed to data management, methodology, experiments, interpretations of the result, and drafting the manuscript. H.M. and B.M.P. supervised the research. H.M. arranged the funding for the research. M.N.D. provided ideas in shaping an improved version of the research and manuscript. T.N. and S.S. contributed ideas and suggestions to the research. All authors have read and agreed to the published version of the manuscript.

Funding: This research is a part of the doctoral of engineering study in the Asian Institute of Technology, Thailand, supported by the Japanese Government Scholarship (August 2017). We would like to express sincere gratitude to Japan Society for the Promotion of Science (JSPS) for providing grant for this research as Grant-in-Aid for Scientific Research (B): 20H01483 through The University of Tokyo, Japan. In addition, we would like to thank GLODAL, Inc. Japan for providing technical assistance and private grant as financial assistance to accomplish this research.

Data Availability Statement: The plastic dataset with images and annotations has been uploaded to: https://github.com/Nisha484/Nisha/tree/main/Datagithub (accessed on 8 May 2022).

Acknowledgments: The authors would like to express sincere thanks to The Government of Japan. The authors would like to acknowledge Kakuko Nagatani-Yoshida, Regional Coordinator for Chemicals, Waste and Air Quality, United Nations Environment Programme, Regional Office for Asia and the Pacific (UNEP/ROAP) for providing an opportunity for data collection. In addition, we would like to express sincere thanks to Kavinda Gunasekara and Dan Tran of Geoinformatics Center (GIC) for their kind support and ideas in data collection. We would like to thank Chathumal Madhuranga and Rajitha Athukorala, Research Associates of GIC, for their kind cooperation in data collection and management. Lastly, we would like to thank Anil Aryal from University of Yamanashi, Japan for assisting in overall research.

Conflicts of Interest: The authors declare no conflict of interest.

Abbreviations

The abbreviations including in the text are reported alphabetically.

AP	Average Precision
AUVs	Autonomous Underwater Vehicles
CNNs	Convolutional Neural Networks
COCO	Microsoft Common Objects in Context
CSM	Class-specific Semantic enhancement Module
CSP	Cross Stage Partial
DETR	Detection Transformer
DL	Deep Learning
FDI	Floating Debris Index
FN	False Negative
FP	False Positive
FPS	Floating Point Systems
GFLOPs	One billion Floating-point Operations Per Second
GNSS	Global Navigation Satellite System
GPS	Global Positioning System
GPU	Graphics Processing Unit
GSD	Ground Sampling Distance
HMH	Houay Mak Hiao
ILSVRC2013	ImageNet Large Scale Visual Recognition Challenge 2013
IoU	Intersection over Union
J-EDI	JAMSTEC E-Library of Deep-sea Images
mAP	Mean Average Precision
NIR	Near Infrared
PANet	Path Aggregation Network
R-CNN	Region-Based Convolutional Neural Networks
RNN	Recurrent Neural Network
R^2 IPoints	Rotation-Insensitive Points
ROVs	Remotely Operated Vehicles
SAM	Sample Angle Mapping
SPP	Spatial Pyramid Pooling
SRM	Stacked rotation convolution module
SSD	Single Shot Detector
SWIR	Short-wave Infrared
TP	True Positive
TT	Talad Thai
TACO	Trash Annotations in Context Dataset
UAVs	Unmanned Aerial Vehicles
UNEP	United Nations Environment Programme
VGG-16	Visual Geometry Group-16
YOLO	You Only Look Once

References

1. Kershaw, P. *Marine Plastic Debris and Microplastics–Global Lessons and Research to Inspire Action and Guide Policy Change*; United Nations Environment Programme: Nairobi, Kenya, 2016.
2. Lebreton, L.C.M.; van der Zwet, J.; Damsteeg, J.W.; Slat, B.; Andrady, A.; Reisser, J. River plastic emissions to the world's oceans. *Nat. Commun.* **2017**, *8*, 15611. [CrossRef] [PubMed]
3. Jambeck, J.R.; Geyer, R.; Wilcox, C.; Siegler, T.R.; Perryman, M.; Andrady, A.; Naray, R. Plastic waste inputs from land into the ocean. *Science* **2015**, *347*, 768–771. [CrossRef] [PubMed]
4. Blettler, M.C.M.; Abrial, E.; Khan, F.R.; Sivri, N.; Espinola, L.A. Freshwater plastic pollution: Recognizing research biases and identifying knowledge gaps. *Water Res.* **2018**, *143*, 416–424. [CrossRef] [PubMed]
5. Moore, C.J.; Lattin, G.L.; Zellers, A.F. Este artigo está disponível em. *J. Integr. Coast. Zone Manag.* **2011**, *11*, 65–73.

6. Gasperi, J.; Dris, R.; Bonin, T.; Rocher, V.; Tassin, B. Assessment of floating plastic debris in surface water along the seine river. *Environ. Pollut.* **2014**, *195*, 163–166. [CrossRef] [PubMed]
7. Yao, X.; Wang, N.; Liu, Y.; Cheng, T.; Tian, Y.; Chen, Q.; Zhu, Y. Estimation of wheat LAI at middle to high levels using unmanned aerial vehicle narrowband multispectral imagery. *Remote Sens.* **2017**, *9*, 1304. [CrossRef]
8. Papakonstantinou, A.; Kavroudakis, D.; Kourtzellis, Y.; Chtenellis, M.; Kopsachilis, V.; Topouzelis, K.; Vaitis, M. Mapping cultural heritage in coastal areas with UAS: The case study of Lesvos Island. *Heritage* **2019**, *2*, 1404–1422. [CrossRef]
9. Watts, A.C.; Ambrosia, V.G.; Hinkley, E.A. Unmanned aircraft systems in remote sensing and scientific research: Classification and considerations of use. *Remote Sens.* **2012**, *4*, 1671–1692. [CrossRef]
10. Shakhatreh, H.; Sawalmeh, A.; Al-Fuqaha, A.; Dou, Z.; Almaita, E.; Khalil, I.; Othman, N.S.; Khreishah, A.; Guizani, M. Unmanned aerial vehicles: A survey on civil applications and key research challenges. *IEEE Access* **2018**, *7*, 48572–48634. [CrossRef]
11. Reynaud, L.; Rasheed, T. Deployable aerial communication networks: Challenges for futuristic applications. In Proceedings of the 9th ACM Symposium on Performance Evaluation of Wireless Ad Hoc, Sensor, and Ubiquitous Networks, Paphos, Cyprus, 24–25 October 2012.
12. Colomina, I.; Molina, P. Unmanned aerial systems for photogrammetry and remote sensing: A review. *ISPRS J. Photogramm. Remote Sens.* **2014**, *92*, 79–97. [CrossRef]
13. Mugnai, F.; Longinotti, P.; Vezzosi, F.; Tucci, G. Performing low-altitude photogrammetric surveys, a comparative analysis of user-grade unmanned aircraft systems. *Appl. Geomat.* **2022**, *14*, 211–223. [CrossRef]
14. Martin, C.; Zhang, Q.; Zhai, D.; Zhang, X.; Duarte, C.M. Enabling a large-scale assessment of litter along Saudi Arabian Red Sea shores by combining drones and machine learning. *Environ. Pollut.* **2021**, *277*, 116730. [CrossRef]
15. Merlino, S.; Paterni, M.; Berton, A.; Massetti, L. Unmanned aerial vehicles for debris survey in coastal areas: Long-term monitoring programme to study spatial and temporal accumulation of the dynamics of beached marine litter. *Remote Sens.* **2020**, *12*, 1260. [CrossRef]
16. Andriolo, U.; Gonçalves, G.; Rangel-Buitrago, N.; Paterni, M.; Bessa, F.; Gonçalves, L.M.S.; Sobral, P.; Bini, M.; Duarte, D.; Fontán-Bouzas, Á.; et al. Drones for litter mapping: An inter-operator concordance test in marking beached items on aerial images. *Mar. Pollut. Bull.* **2021**, *169*, 112542. [CrossRef] [PubMed]
17. Pinto, L.; Andriolo, U.; Gonçalves, G. Detecting stranded macro-litter categories on drone orthophoto by a multi-class neural network. *Mar. Pollut. Bull.* **2021**, *169*, 112594. [CrossRef]
18. Deidun, A.; Gauci, A.; Lagorio, S.; Galgani, F. Optimising beached litter monitoring protocols through aerial imagery. *Mar. Pollut. Bull.* **2018**, *131*, 212–217. [CrossRef]
19. Fallati, L.; Polidori, A.; Salvatore, C.; Saponari, L.; Savini, A.; Galli, P. Anthropogenic marine debris assessment with unmanned aerial vehicle imagery and deep learning: A case study along the beaches of the Republic of Maldives. *Sci. Total Environ.* **2019**, *693*, 133581. [CrossRef]
20. Martin, C.; Parkes, S.; Zhang, Q.; Zhang, X.; McCabe, M.F.; Duarte, C.M. Use of unmanned aerial vehicles for efficient beach litter monitoring. *Mar. Pollut. Bull.* **2018**, *131*, 662–673. [CrossRef]
21. Nelms, S.E.; Coombes, C.; Foster, L.C.; Galloway, T.S.; Godley, B.J.; Lindeque, P.K.; Witt, M.J. Marine anthropogenic litter on british beaches: A 10-year nationwide assessment using citizen science data. *Sci. Total Environ.* **2017**, *579*, 1399–1409. [CrossRef]
22. Andriolo, U.; Gonçalves, G.; Sobral, P.; Bessa, F. Spatial and size distribution of macro-litter on coastal dunes from drone images: A case study on the Atlantic Coast. *Mar. Pollut. Bull.* **2021**, *169*, 112490. [CrossRef]
23. Andriolo, U.; Gonçalves, G.; Sobral, P.; Fontán-Bouzas, Á.; Bessa, F. Beach-dune morphodynamics and marine macro-litter abundance: An integrated approach with unmanned aerial system. *Sci. Total Environ.* **2020**, *749*, 432–439. [CrossRef] [PubMed]
24. Andriolo, U.; Garcia-Garin, O.; Vighi, M.; Borrell, A.; Gonçalves, G. Beached and floating litter surveys by unmanned aerial vehicles: Operational analogies and differences. *Remote Sens.* **2022**, *14*, 1336. [CrossRef]
25. Papakonstantinou, A.; Batsaris, M.; Spondylidis, S.; Topouzelis, K. A citizen science unmanned aerial system data acquisition protocol and deep learning techniques for the automatic detection and mapping of marine litter concentrations in the coastal zone. *Drones* **2021**, *5*, 6. [CrossRef]
26. Merlino, S.; Paterni, M.; Locritani, M.; Andriolo, U.; Gonçalves, G.; Massetti, L. Citizen science for marine litter detection and classification on unmanned aerial vehicle images. *Water* **2021**, *13*, 3349. [CrossRef]
27. Ham, S.; Oh, Y.; Choi, K.; Lee, I. Semantic segmentation and unregistered building detection from UAV images using a deconvolutional network. In Proceedings of the International Archives of the Photogrammetry, Remote Sensing and Spatial Information Sciences—ISPRS Archives; International Society for Photogrammetry and Remote Sensing, Niece, France, 30 May 2018; Volume 42, pp. 419–424.
28. Kamilaris, A.; Prenafeta-Boldú, F.X. Disaster Monitoring using unmanned aerial vehicles and deep learning. *arXiv* **2018**, arXiv:1807.11805.
29. Zeggada, A.; Benbraika, S.; Melgani, F.; Mokhtari, Z. Multilabel conditional random field classification for UAV images. *IEEE Geosci. Remote Sens. Lett.* **2018**, *15*, 399–403. [CrossRef]
30. Zhao, Z.; Zheng, P.; Xu, S.; Wu, X. Object detection with deep learning: A review. *IEEE Trans. Neural Netw. Learn. Syst.* **2019**, *30*, 3212–3232. [CrossRef]
31. Viola, P.; Jones, M.J. Robust Real-Time Object Detection; 2001. In Proceedings of the Workshop on Statistical and Computational Theories of Vision, Cambridge Research Laboratory, Cambridge, MA, USA, 25 February 2001; Volume 266, p. 56.

32. Längkvist, M.; Kiselev, A.; Alirezaie, M.; Loutfi, A. Classification and segmentation of satellite orthoimagery using convolutional neural networks. *Remote Sens.* **2016**, *8*, 329. [CrossRef]
33. LeCun, Y.; Bengio, Y.; Hinton, G. Deep learning. *Nature* **2015**, *521*, 436–444. [CrossRef]
34. Sermanet, P.; Eigen, D.; Zhang, X.; Mathieu, M.; Fergus, R.; LeCun, Y. OverFeat: Integrated recognition, localization and detection using convolutional networks. *arXiv* **2013**, arXiv:1312.6229.
35. Mnih, V.; Kavukcuoglu, K.; Silver, D.; Rusu, A.A.; Veness, J.; Bellemare, M.G.; Graves, A.; Riedmiller, M.; Fidjeland, A.K.; Ostrovski, G.; et al. Human-level control through deep reinforcement learning. *Nature* **2015**, *518*, 529–533. [CrossRef] [PubMed]
36. Maitra, D.S.; Bhattacharya, U.; Parui, S.K. CNN based common approach to handwritten character recognition of multiple scripts. In Proceedings of the International Conference on Document Analysis and Recognition, ICDAR; IEEE Computer Society, Tunis, Tunisia, 23–26 August 2015; Volume 2015, pp. 1021–1025.
37. Girshick, R.; Donahue, J.; Darrell, T.; Malik, J. Region-based convolutional networks for accurate object detection and segmentation. *IEEE Trans. Pattern Anal. Mach. Intell.* **2016**, *38*, 142–158. [CrossRef] [PubMed]
38. Redmon, J.; Farhadi, A. YOLO9000: Better, faster, stronger. In Proceedings of the IEEE Conference on Computer Vision and Pattern Recognition, Honolulu, HI, USA, 21–26 July 2017; pp. 7263–7271.
39. Redmon, J.; Divvala, S.; Girshick, R.; Farhadi, A. You only look once: Unified, real-time object detection. In Proceedings of the IEEE Conference on Computer Vision and Pattern Recognition, Las Vegas, NV, USA, 27–30 June 2016; pp. 779–788.
40. Lin, M.; Chen, Q.; Yan, S. Network in network. *arXiv* **2013**, arXiv:1312.4400.b.
41. Sarkar, P.; Gupta, M.A. Object Recognition with Text and Vocal Representation. *Int. J. Eng. Res. Appl.* **2020**, *10*, 63–77.
42. Szegedy, C.; Liu, W.; Jia, Y.; Sermanet, P.; Reed, S.; Anguelov, D.; Erhan, D.; Vanhoucke, V.; Rabinovich, A. Going Deeper with Convolutions. *arXiv* **2014**, arXiv:1409.4842.
43. He, K.; Zhang, X.; Ren, S.; Sun, J. Deep residual learning for image recognition. In Proceedings of the IEEE Conference on Computer Vision and Pattern Recognition, Las Vegas, NV, USA, 27–30 June 2016; pp. 770–778.
44. Redmon, J.; Farhadi, A. YOLOv3: An incremental improvement. *arXiv* **2018**, arXiv:1804.02767.
45. Salimi, I.; Bayu Dewantara, B.S.; Wibowo, I.K. Visual-based trash detection and classification system for smart trash bin robot. In Proceedings of the 2018 International Electronics Symposium on Knowledge Creation and Intelligent Computing (IES-KCIC), Bali, Indonesia, 29–30 October 2018; pp. 378–383.
46. Bochkovskiy, A.; Wang, C.-Y.; Liao, H.-Y.M. YOLOv4: Optimal speed and accuracy of object detection. *arXiv* **2020**, arXiv:2004.10934.
47. Yao, X.; Shen, H.; Feng, X.; Cheng, G.; Han, J. R^2 IPoints: Pursuing rotation-insensitive point representation for aerial object detection. *IEEE Trans. Geosci. Remote Sens.* **2022**, *60*, 5623512. [CrossRef]
48. Vaswani, A.; Brain, G.; Shazeer, N.; Parmar, N.; Uszkoreit, J.; Jones, L.; Gomez, A.N.; Kaiser, Ł.; Polosukhin, I. Attention is all you need. *Adv. Neural Inf. Processing Syst.* **2017**, *30*, 6000–6010.
49. Bazi, Y.; Bashmal, L.; al Rahhal, M.M.; al Dayil, R.; al Ajlan, N. Vision Transformers for Remote Sensing Image Classification. *Remote Sens.* **2021**, *13*, 516. [CrossRef]
50. Zhu, X.; Su, W.; Lu, L.; Li, B.; Wang, X.; Dai, J. Deformable DETR: Deformable transformers for end-to-end object detection. *arXiv* **2020**, arXiv:2010.04159.
51. Carion, N.; Massa, F.; Synnaeve, G.; Usunier, N.; Kirillov, A.; Zagoruyko, S. End-to-end object detection with transformers. In Proceedings of the European Conference on Computer Vision, Glasgow, UK, 23–28 August 2020; Springer: Berlin/Heidelberg, Germany, 2020; pp. 213–229.
52. Liu, Z.; Lin, Y.; Cao, Y.; Hu, H.; Wei, Y.; Zhang, Z.; Lin, S.; Guo, B. Swin transformer: Hierarchical vision transformer using shifted windows. In Proceedings of the IEEE/CVF International Conference on Computer Vision, Montreal, QC, Canada, 11 October 2021.
53. Touvron, H.; Cord, M.; Douze, M.; Massa, F.; Sablayrolles, A.; Jégou, H. Training data-efficient image transformers & distillation through attention. *arXiv* **2021**, arXiv:2012.12877.
54. Majchrowska, S.; Mikołajczyk, A.; Ferlin, M.; Klawikowska, Z.; Plantykow, M.A.; Kwasigroch, A.; Majek, K. Deep learning-based waste detection in natural and urban environments. *Waste Manag.* **2022**, *138*, 274–284. [CrossRef]
55. Córdova, M.; Pinto, A.; Hellevik, C.C.; Alaliyat, S.A.A.; Hameed, I.A.; Pedrini, H.; da Torres, R.S. Litter detection with deep learning: A comparative study. *Sensors* **2022**, *22*, 548. [CrossRef]
56. Kraft, M.; Piechocki, M.; Ptak, B.; Walas, K. Autonomous, onboard vision-based trash and litter detection in low altitude aerial images collected by an unmanned aerial vehicle. *Remote Sens.* **2021**, *13*, 965. [CrossRef]
57. Tan, M.; Pang, R.; Le, Q.V. EfficientDet: Scalable and efficient object detection. In Proceedings of the IEEE Computer Society Conference on Computer Vision and Pattern Recognition, Seattle, DC, USA, 13–19 June 2020; pp. 10778–10787. [CrossRef]
58. Kumar, S.; Yadav, D.; Gupta, H.; Verma, O.P.; Ansari, I.A.; Ahn, C.W. A Novel Yolov3 algorithm-based deep learning approach for waste segregation: Towards smart waste management. *Electronics* **2021**, 14. [CrossRef]
59. Fulton, M.; Hong, J.; Islam, M.J.; Sattar, J. Robotic detection of marine litter using deep visual detection models. *arXiv* **2018**, arXiv:1804.01079.
60. Tata, G.; Royer, S.-J.; Poirion, O.; Lowe, J. A robotic approach towards quantifying epipelagic bound plastic using deep visual models. *arXiv* **2021**, arXiv:2105.01882.

61. Luo, W.; Han, W.; Fu, P.; Wang, H.; Zhao, Y.; Liu, K.; Liu, Y.; Zhao, Z.; Zhu, M.; Xu, R.; et al. A water surface contaminants monitoring method based on airborne depth reasoning. *Processes* **2022**, *10*, 131. [CrossRef]
62. Pati, B.M.; Kaneko, M.; Taparugssanagorn, A. A deep convolutional neural network based transfer learning method for non-cooperative spectrum sensing. *IEEE Access* **2020**, *8*, 164529–164545. [CrossRef]
63. Huang, Z.; Pan, Z.; Lei, B. Transfer learning with deep convolutional neural network for SAR target classification with limited labeled data. *Remote Sens.* **2017**, *9*, 907. [CrossRef]
64. Long, J.; Shelhamer, E.; Darrell, T. Fully convolutional networks for semantic segmentation. In Proceedings of the IEEE Conference on Computer Vision and Pattern Recognition, Boston, MA, USA, 7–12 June 2015; pp. 3431–3440. [CrossRef]
65. Li, L.; Zhang, S.; Wu, J. Efficient object detection framework and hardware architecture for remote sensing images. *Remote Sens.* **2019**, *11*, 2376. [CrossRef]
66. Boutell, M.R.; Luo, J.; Shen, X.; Brown, C.M. Learning multi-label scene classification. *Pattern Recognit.* **2004**, *37*, 1757–1771. [CrossRef]
67. Russell, S.; Norvig, P. *Artificial Intelligence: A Modern Approach*, 3rd ed.; Pearson Education, Inc.: Upper Saddle River, NJ, USA, 2009.
68. Kwon, Y. Yolo_Label: GUI for Marking Bounded Boxes of Objects in Images for Training Neural Network Yolo v3 and v2. Available online: https://github.com/developer0hye/Yolo_Label.git (accessed on 24 December 2021).
69. Huang, K.; Lei, H.; Jiao, Z.; Zhong, Z. Recycling waste classification using vision transformer on portable device. *Sustainability* **2021**, *13*, 1572. [CrossRef]
70. Devries, T.; Misra, I.; Wang, C.; van der Maaten, L. Does object recognition work for everyone. *arXiv* **2019**, arXiv:1906.02659. [CrossRef]
71. van Lieshout, C.; van Oeveren, K.; van Emmerik, T.; Postma, E. Automated River plastic monitoring using deep learning and cameras. *Earth Space Sci.* **2020**, *7*, e2019EA000960. [CrossRef]
72. Jakovljevic, G.; Govedarica, M.; Alvarez-Taboada, F. A deep learning model for automatic plastic mapping using unmanned aerial vehicle (UAV) data. *Remote Sens.* **2020**, *12*, 1515. [CrossRef]
73. Lin, F.; Hou, T.; Jin, Q.; You, A. Improved yolo based detection algorithm for floating debris in waterway. *Entropy* **2021**, *23*, 1111. [CrossRef]
74. Colica, E.; D'Amico, S.; Iannucci, R.; Martino, S.; Gauci, A.; Galone, L.; Galea, P.; Paciello, A. Using unmanned aerial vehicle photogrammetry for digital geological surveys: Case study of Selmun promontory, northern of Malta. *Environ. Earth Sci.* **2021**, *80*, 12538. [CrossRef]
75. Lu, H.; Li, Y.; Xu, X.; He, L.; Li, Y.; Dansereau, D.; Serikawa, S. underwater image descattering and quality assessment. In Proceedings of the 2016 IEEE International Conference on Image Processing (ICIP), Phoenix, AZ, USA, 25–28 September 2016; pp. 1998–2002.
76. Wolf, M.; van den Berg, K.; Garaba, S.P.; Gnann, N.; Sattler, K.; Stahl, F.; Zielinski, O. Machine learning for aquatic plastic litter detection, classification and quantification (APLASTIC-Q). *Environ. Res. Lett.* **2020**, *15*, 094075. [CrossRef]
77. Silva, G.F.; Carneiro, G.B.; Doth, R.; Amaral, L.A.; de Azevedo, D.F.G. Near real-time shadow detection and removal in aerial motion imagery application. *ISPRS J. Photogramm. Remote Sens.* **2018**, *140*, 104–121. [CrossRef]
78. Nelson, J.; Solawetz, J. Responding to the Controversy about YOLOv5. Available online: https://blog.roboflow.com/yolov4-versus-yolov5/ (accessed on 30 July 2020).
79. Garcia-Garin, O.; Monleón-Getino, T.; López-Brosa, P.; Borrell, A.; Aguilar, A.; Borja-Robalino, R.; Cardona, L.; Vighi, M. Automatic detection and quantification of floating marine macro-litter in aerial images: Introducing a novel deep learning approach connected to a web application in R. *Environ. Pollut.* **2021**, *273*, 116490. [CrossRef] [PubMed]
80. Ge, Z.; Liu, S.; Wang, F.; Li, Z.; Sun, J. YOLOX: Exceeding YOLO series in 2021 V100 batch 1 latency (Ms) YOLOX-L YOLOv5-L YOLOX-DarkNet53 YOLOv5-Darknet53 EfficientDet5 COCO AP (%) number of parameters (M) figure 1: Speed-accuracy trade-off of accurate models (Top) and size-accuracy curve of lite models on mobile devices (Bottom) for YOLOX and other state-of-the-art object detectors. *arXiv* **2021**, arXiv:2107.08430.
81. Nepal, U.; Eslamiat, H. Comparing YOLOv3, YOLOv4 and YOLOv5 for autonomous landing spot detection in faulty UAVs. *Sensors* **2022**, *22*, 464. [CrossRef]
82. Glenn, J. Ultralytics/Yolov5. Available online: https://github.com/ultralytics/yolov5/releases (accessed on 5 April 2022).
83. Biermann, L.; Clewley, D.; Martinez-Vicente, V.; Topouzelis, K. Finding plastic patches in coastal waters using optical satellite data. *Sci. Rep.* **2020**, *10*, 5364. [CrossRef]
84. Gonçalves, G.; Andriolo, U.; Gonçalves, L.; Sobral, P.; Bessa, F. Quantifying marine macro litter abundance on a sandy beach using unmanned aerial systems and object-oriented machine learning methods. *Remote Sens.* **2020**, *12*, 2599. [CrossRef]
85. Escobar-Sánchez, G.; Haseler, M.; Oppelt, N.; Schernewski, G. Efficiency of aerial drones for macrolitter monitoring on Baltic Sea Beaches. *Front. Environ. Sci.* **2021**, *8*, 237. [CrossRef]
86. Cao, H.; Gu, X.; Sun, Y.; Gao, H.; Tao, Z.; Shi, S. Comparing, validating and improving the performance of reflectance obtention method for UAV-remote sensing. *Int. J. Appl. Earth Obs. Geoinf.* **2021**, *102*, 102391. [CrossRef]
87. Gonçalves, G.; Andriolo, U. Operational use of multispectral images for macro-litter mapping and categorization by unmanned aerial vehicle. *Mar. Pollut. Bull.* **2022**, *176*, 113431. [CrossRef]

88. Guffogg, J.A.; Blades, S.M.; Soto-Berelov, M.; Bellman, C.J.; Skidmore, A.K.; Jones, S.D. Quantifying marine plastic debris in a beach environment using spectral analysis. *Remote Sens.* **2021**, *13*, 4548. [CrossRef]
89. Garaba, S.P.; Aitken, J.; Slat, B.; Dierssen, H.M.; Lebreton, L.; Zielinski, O.; Reisser, J. Sensing ocean plastics with an airborne hyperspectral shortwave infrared imager. *Environ. Sci. Technol.* **2018**, *52*, 11699–11707. [CrossRef] [PubMed]
90. Goddijn-Murphy, L.; Dufaur, J. Proof of concept for a model of light reflectance of plastics floating on natural waters. *Mar. Pollut. Bull.* **2018**, *135*, 1145–1157. [CrossRef] [PubMed]
91. Taddia, Y.; Corbau, C.; Buoninsegni, J.; Simeoni, U.; Pellegrinelli, A. UAV approach for detecting plastic marine debris on the beach: A case study in the Po River Delta (Italy). *Drones* **2021**, *5*, 140. [CrossRef]
92. Gonçalves, G.; Andriolo, U.; Pinto, L.; Bessa, F. Mapping marine litter using UAS on a beach-dune system: A multidisciplinary approach. *Sci. Total Environ.* **2020**, *706*, 135742. [CrossRef] [PubMed]
93. Geraeds, M.; van Emmerik, T.; de Vries, R.; bin Ab Razak, M.S. Riverine plastic litter monitoring using unmanned aerial vehicles (UAVs). *Remote Sens.* **2019**, *11*, 2045. [CrossRef]
94. Makarau, A.; Richter, R.; Muller, R.; Reinartz, P. Adaptive shadow detection using a blackbody radiator model. *IEEE Trans. Geosci. Remote Sens.* **2011**, *49*, 2049–2059. [CrossRef]
95. Balsi, M.; Moroni, M.; Chiarabini, V.; Tanda, G. High-resolution aerial detection of marine plastic litter by hyperspectral sensing. *Remote Sens.* **2021**, *13*, 1557. [CrossRef]
96. Andriolo, U.; Gonçalves, G.; Bessa, F.; Sobral, P. Mapping marine litter on coastal dunes with unmanned aerial systems: A showcase on the Atlantic Coast. *Sci. Total Environ.* **2020**, *736*, 139632. [CrossRef]
97. Topouzelis, K.; Papakonstantinou, A.; Garaba, S.P. Detection of floating plastics from satellite and unmanned aerial systems (plastic litter project 2018). *Int. J. Appl. Earth Obs. Geoinf.* **2019**, *79*, 175–183. [CrossRef]
98. Lo, H.S.; Wong, L.C.; Kwok, S.H.; Lee, Y.K.; Po, B.H.K.; Wong, C.Y.; Tam, N.F.Y.; Cheung, S.G. Field test of beach litter assessment by commercial aerial drone. *Mar. Pollut. Bull.* **2020**, *151*, 110823. [CrossRef]

remote sensing

MDPI

Article

Point RCNN: An Angle-Free Framework for Rotated Object Detection

Qiang Zhou [1,†,*] and Chaohui Yu [2,†]

1 Alibaba Group, Hangzhou 311121, China
2 Alibaba Group, Beijing 100102, China; huakun.ych@alibaba-inc.com
* Correspondence: jianchong.zq@alibaba-inc.com; Tel.: +86-1307-369-6312
† These authors contributed equally to this work.

Abstract: Rotated object detection in aerial images is still challenging due to arbitrary orientations, large scale and aspect ratio variations, and extreme density of objects. Existing state-of-the-art rotated object detection methods mainly rely on angle-based detectors. However, angle-based detectors can easily suffer from a long-standing boundary problem. To tackle this problem, we propose a purely angle-free framework for rotated object detection, called Point RCNN. Point RCNN is a two-stage detector including both PointRPN and PointReg which are angle-free. Given an input aerial image, first, the backbone-FPN extracts hierarchical features, then, the PointRPN module generates an accurate rotated region of interests (RRoIs) by converting the learned representative points of each rotated object using the MinAreaRect function of OpenCV. Motivated by RepPoints, we designed a coarse-to-fine process to regress and refine the representative points for more accurate RRoIs. Next, based on the learned RRoIs of PointRPN, the PointReg module learns to regress and refine the corner points of each RRoI to perform more accurate rotated object detection. Finally, the final rotated bounding box of each rotated object can be attained based on the learned four corner points. In addition, aerial images are often severely unbalanced in categories, and existing rotated object detection methods almost ignore this problem. To tackle the severely unbalanced dataset problem, we propose a balanced dataset strategy. We experimentally verified that re-sampling the images of the rare categories can stabilize the training procedure and further improve the detection performance. Specifically, the performance was improved from 80.37 mAP to 80.71 mAP in DOTA-v1.0. Without unnecessary elaboration, our Point RCNN method achieved new state-of-the-art detection performance on multiple large-scale aerial image datasets, including DOTA-v1.0, DOTA-v1.5, HRSC2016, and UCAS-AOD. Specifically, in DOTA-v1.0, our Point RCNN achieved better detection performance of 80.71 mAP. In DOTA-v1.5, Point RCNN achieved 79.31 mAP, which significantly improved the performance by 2.86 mAP (from ReDet's 76.45 to our 79.31). In HRSC2016 and UCAS-AOD, our Point RCNN achieved higher performance of 90.53 mAP and 90.04 mAP, respectively.

Keywords: rotated object detection; angle-based detector; angle-free framework; rotated region of interests (RRoIs); representative points

Citation: Qiang, Z.; Chaohui, Y. Point RCNN: An Angle-Free Framework for Rotated Object Detection. *Remote Sens.* **2022**, *14*, 2605. https://doi.org/10.3390/rs14112605

Academic Editors: Fahimeh Farahnakian, Jukka Heikkonen and Pouya Jafarzadeh

Received: 30 March 2022
Accepted: 27 May 2022
Published: 29 May 2022

1. Introduction

Object detection has been a fundamental task in computer vision and has progressed dramatically in the past few years using deep learning. It aims to predict a set of bounding boxes and the corresponding categories in an image. Modern object detection methods of natural images can be categorized into two main categories: two-stage detectors, exemplified by Faster RCNN [1] and Mask RCNN [2], and one-stage detectors, such as YOLO [3], SSD [4], and RetinaNet [5].

Although object detection has achieved significant progress in natural images, it still remains challenging for rotated object detection in aerial images, due to the arbitrary orientations, large scale and aspect ratio variations, and extreme density of objects [6]. Rotated object detection in aerial images aims to predict a set of oriented bounding boxes (OBBs) and the corresponding classes in an aerial image, which serves an important role in many applications, e.g., urban management, emergency rescue, precise agriculture, automatic monitoring, and geographic information system (GIS) updating [7,8]. Among these applications, antenna systems are very important for object detection, and many excellent examples [9–11] have been proposed.

Modern rotated object detectors can be divided into two categories in terms of the representation of OBB: angle-based detectors and angle-free detectors.

In angle-based detectors, an OBB of a rotated object is usually represented as a five-parameter vector (x, y, w, h, θ). Most existing state-of-the-art methods are angle-based detectors relying on two-stage RCNN frameworks [12–16]. Generally, these methods use an RPN to generate horizontal or rotated region of interests (RoIs), then a designed RoI pooling operator is used to extract features from these RoIs. Finally, an RCNN head is used to predict the OBB and the corresponding classes. Compared to two-stage detectors, one-stage angle-based detectors [17–21] directly regress the OBB and classify them based on dense anchors for efficiency. However, angle-based detectors usually introduce a long-standing boundary discontinuity problem [22,23] due to the periodicity of the angle and the exchange of edges. Moreover, the unit between (x, y, w, h) and angle θ of the five-parameter representation is not consistent. These obstacles can cause the training to be unstable and limit the performance.

In contrast to angle-based detectors, angle-free detectors usually represent a rotated object as an eight-parameter OBB $(x_1, y_1, x_2, y_2, x_3, y_3, x_4, y_4)$, which denotes the four corner points of a rotated object. Modern angle-free detectors [24–27] directly perform quadrilateral regression, which is more straightforward than the angle-based representation. Unfortunately, although abandoning angle regression and the parameter unit is consistent, the performance of existing angle-free detectors is still relatively limited.

How to design a more straightforward and effective framework to alleviate the boundary discontinuity problem is the key to the success of rotated object detectors.

However, all the above methods use predefined (rotated) anchor boxes, whether angle-based or using angle-free methods. Compared to anchor boxes, representation points can provide more precise object localization, including shape and pose. Thus, the features extracted from the representative points may be less influenced by background content or uninformative foreground areas that contain little semantic information. In this paper, based on the learning of representative points, we propose a purely angle-free framework for rotated object detection in aerial images, called Point RCNN, which can alleviate the boundary discontinuity problem and attain state-of-the-art performance. Our Point RCNN is a two-stage detector and mainly consists of an RPN (PointRPN) and an RCNN head (PointReg), which are both angle-free. PointRPN serves as an RPN network. Given an input feature map, first, PointRPN learns a set of representative points for each feature point in a coarse-to-fine manner. Then, a rotated RoI (RRoI) is generated through the `MinAreaRect` function of OpenCV [28]. Finally, serving as an angle-free RCNN head, PointReg applies a rotate RoI Align [13,15] operator to extract RRoI features, and then refines and classifies the eight-parameter OBB of the corner points. In addition, the existing methods almost ignore the category imbalance in aerial images, and we propose to resample images of rare categories to stabilize convergence during training.

The main contributions of this paper are summarized as follows:

- We propose Point RCNN, a purely angle-free framework for rotated object detection in aerial images. Without introducing angle prediction, Point RCNN is able to address the boundary discontinuity problem.
- We propose PointRPN as an RPN network, which aims to learn a set of representative points for each object of interest, and can provide better detection recall for rotated objects in aerial images.

- We propose PointReg as an RCNN head, which can responsively regress and refine the four corners of the rotated proposals generated by PointRPN.
- Aerial images are usually long-tail distributed. We further propose to resample images of rare categories to stabilize training and improve the overall performance.
- Compared with state-of-the-art methods, extensive experiments demonstrate that our Point RCNN framework attains higher detection performance on multiple large-scale datasets and achieves new state-of-the-art performance.

2. Materials and Methods

2.1. Related Work

2.1.1. Horizontal Object Detection

In the past decade, object detection has become an important computer vision task and has received considerable attention from academia and industry. Traditional methods use hand-crafted features (e.g., HoG, SIFT) to solve detection as classification on a set of candidate bounding boxes. With the development of deep convolutional neural networks (CNN), modern horizontal object detection methods can be mainly categorized into three types: two-stage detectors, one-stage detectors, and recent end-to-end detectors.

One line of research focuses on two-stage detectors [2,29–33], which first generate a sparse set of regions of interests (RoIs) with a region proposal network (RPN), and then perform classification and bounding box regression. While two-stage detectors still attract much attention, another line of research focuses on developing efficient one-stage detectors due to their much simpler and cleaner design [3–5,34–37], in which SSD [4] and YOLO [3] are the fundamental methods that use a set of pre-defined anchor boxes to predict object category and anchor box offsets. Note that anchors were first proposed in the RPN module of faster RCNN to generate proposals. Recently, more studies [38,39] that use bounding boxes for object detection have been reported. In addition, effort has been spent on designing anchor-free detectors [35,40]. FCOS [35] and Foveabox [40] use the center region of targets as positive samples. In addition, FCOS introduces the so-called centerness score to make non-maximum suppression (NMS) more accurate. The authors of [41] propose an adaptive training sample selection (ATSS) scheme to automatically define positive and negative training samples. PAA [42] involves a probabilistic anchor assignment strategy, leading to easier training compared to heuristic IoU hard-label assignment strategies. In addition to improve the assignment strategy of FCOS, efforts has been devoted to the detection features [43] and loss functions [44] to further boost anchor-free detector performance.

Very recently, several studies have proposed end-to-end frameworks for horizontal object detection by removing NMS from the pipeline. DETR [45] introduces a transformer-based attention mechanism to object detection. Essentially the sequence-to-sequence learning task in [46] was solved in parallel by a self-attention-based transformer rather than RNN. Deformable DETR [47] accelerates the training convergence of DETR by proposing to only attend to a small set of key sampling points. DeFCN [48] adopts a one-to-one matching strategy to enable end-to-end object detection based on a fully convolutional network with competitive performance. PSS [49] involves a compact and plug-in PSS head to eliminate heuristic NMS and achieve better performance.

2.1.2. Rotated Object Detection

With the development of deep-learning technology, rotated object detection in aerial images has achieved great success in the past few years, especially with the release of the largest aerial image dataset DOTA [6], which has become a standard benchmark and has significantly boosted the development of rotated object detectors. In terms of the representation of the oriented bounding box (OBB), modern rotated object detectors can be mainly divided into two categories: angle-based detectors and angle-free detectors. As depicted in Figure 1, we show the main differences between angle-based detectors and angle-free detectors. Figure 1a shows the learning targets (x, y, w, h, θ) of angle-based detectors, where (x, y) denote the coordinates of the center points, (w, h) denote the shorter and longer edges of the rotated bounding box, and θ denotes the angle between the longer edge and the horizontal axis. Figure 1b shows the learning targets $(x_1, y_1, x_2, y_2, x_3, y_3, x_4, y_4)$ of angle-free detectors, which represent the coordinates of four corner points of a rotated bounding box. Compared to angle-based detectors, angle-free detectors are more efficient since they are more straightforward and can alleviate the boundary discontinuity problem without introducing angle prediction.

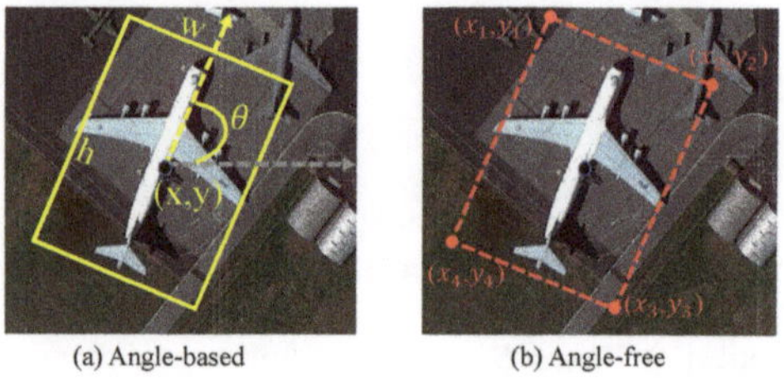

Figure 1. Comparison of angle-based and angle-free detectors.

Angle-based detectors: Figure 2 illustrates three different methods for generating RRoIs: (a) and (b) denote two classical and mainstream RRoI generating methods. As shown in Figure 2a, one line of early research in the generation of RRoIs is the rotated region proposal network (rotated RPN) [17,18], which sets 54 anchors with different scales, angles, and aspect rations (three scales × six angles × three ratios) on each location to cover oriented objects. With the help of densely rotated anchors, the detection recall performance is thus improved. However, the introduction of massive rotated anchors increases the computational complexity and memory consumption, which limits the application of these methods. To tackle this issue, as shown in Figure 2b, the RoI transformer [13] proposes that RRoIs learn from horizontal RoIs by transforming default horizontal RoIs into RRoIs. The RoI transformer avoids introducing abundant anchors; however, it involves RPN, RoI alignment and regression, which are also complex processes. R^2CNN [12] proposes the detection of the horizontal and rotated bounding box simultaneously with multi-task learning. SCRDet [14] enhances features with an attention module and proposes an IoU-smooth L_1 loss to alleviate the loss discontinuity issue. SCRDet++ [23] extends SCRDet with image-level and instance-level de-noising modules to enhance the detection of small and cluttered objects. CSL [19] reformulates angle prediction from regression to classification to alleviate the discontinuous boundary problem. GWD [50] and KLD [51] propose a more efficient loss function for OBB regression. R3Det [20] proposes a refined single-stage rotation detector for fast and accurate object detection using a progressive regression approach from coarse to fine granularity. Constraint loss [52] proposes a decoupling modulation mechanism to overcome the problem of sudden changes in loss. S^2A-Net [21] proposes a single-shot alignment network to realize full feature alignment and alleviates the inconsistency between regression and classification. Recently, ReDet [15] has proposed the use of a rotation-equivariant network to encode rotation equivariance explicitly and presents a rotation-invariant RoI aligned to extract rotation-invariant features. The oriented RCNN [16] proposes a two-stage detector that consists of an oriented RPN for generating the RRoI and an oriented RCNN for refining the RRoI. Both ReDet and the oriented RCNN provide promising accuracy.

However, the boundary problem in the angle regression learning still causes training to be unstable and limits the performance. While angle-based detectors still find many applications, angle-free methods are receiving more and more attention from the research community.

Angle-free detectors: Textboxes++ [53] directly predict arbitrarily oriented word bounding boxes via a regression model by quadrilateral representation. ICN [24] proposes to directly estimate the four vertices of a quadrilateral to regress an oriented object based on an image pyramid and feature pyramid. RSDet [25] and gliding vertex [26] achieve more accurate rotated object detection via directly quadrilateral regression prediction. LR-

TSDet [8] proposes an effective tiny ship detector for low-resolution remote-sensing images based on horizontal bounding box regression. TPR-R2CNN [54] proposes an improved R2CNN based on a double-detection head structure and a three-point regression method. Recently, BBAVectors [27] have extended the horizontal keypoint-based object detector to an oriented object detection task. CFA [55] proposes a convex-hull feature adaptation approach for configuring convolutional features. Compared to angle-based methods, angle-free detectors are more straightforward and can alleviate the boundary problem to a large extent. However, the performance of current angle-free oriented object detectors is still relatively limited.

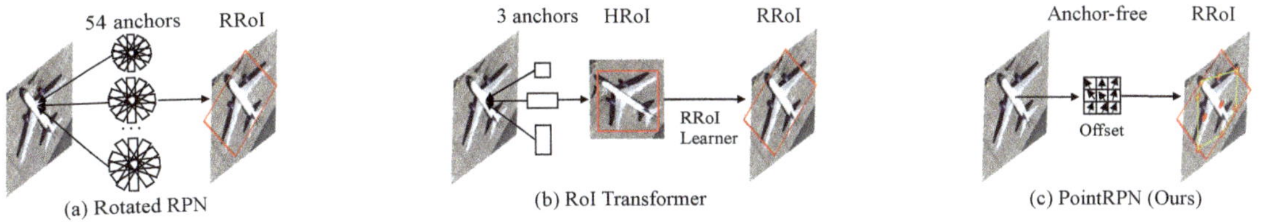

Figure 2. Comparison of different methods for generating rotated RoI (RRoI). (**a**) Rotated RPN places multiple rotated anchors with different angles, scales, and aspect ratios. (**b**) RoI transformer proposes an RRoI learner to model the RRoI from the horizontal RoI (HRoI) for each feature point based on 3 anchors. (**c**) Our proposed PointRPN generates accurate RRoI in an anchor-free and angle-free manner.

In this paper, we propose an effective angle-free framework for rotated object detection, called Point RCNN, which mainly consists of an RPN network (PointRPN) and an RCNN head (PointReg). Compared to the methods of Figure 2a,b, our proposed PointRPN generates accurate RRoIs in an anchor-free and angle-free manner. Specifically, PointRPN directly learns a set of implicit representative points for each rotated object. Based on these points, RRoIs can be easily attained with the `MinAreaRect` function of OpenCV. Without introducing anchors and angle regression, PointRPN becomes more efficient and accurate.

2.2. Methods

The overall structure of our Point RCNN is depicted in Figure 3. We start by revisiting the boundary discontinuity problem of angle-based detectors. Then, we describe the overall pipeline of Point RCNN. Finally, we elaborate the PointRPN and PointReg modules, and propose a balanced dataset strategy to rebalance the long-tailed datasets during training.

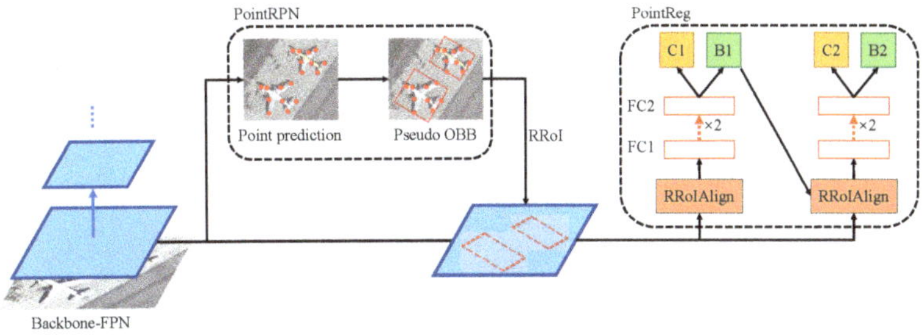

Figure 3. The overall pipeline of the proposed angle-free Point RCNN framework for rotated object detection. Point RCNN mainly consists of two modules: PointRPN for generating rotated proposals, and PointReg for refining for more accurate detection. "RRoI" denotes rotated RoI, "FC" denotes fully-connected layer, "C" and "B" represent the predicted category and rotated box coordinates of each RRoI, respectively.

2.2.1. Boundary Discontinuity Problem

The boundary discontinuity problem [22,23] is a long-standing problem that has existed in angle-based detectors. Taking the commonly used five-parameter OBB representation (x, y, w, h, θ) as an example, where (x, y) represent the center coordinates, (w, h) represent the shorter and longer edges of the bounding box, and θ represents the angle between the longer edge and the horizontal axis. As shown in Figure 4, when the target box is approximately square, a slight variation in edge length may cause w and h to swap, leading to a substantial variation in $\pi/2$ in angle θ.

This boundary discontinuity issue in angle prediction will confuse the optimization of the network and limit the detection performance.

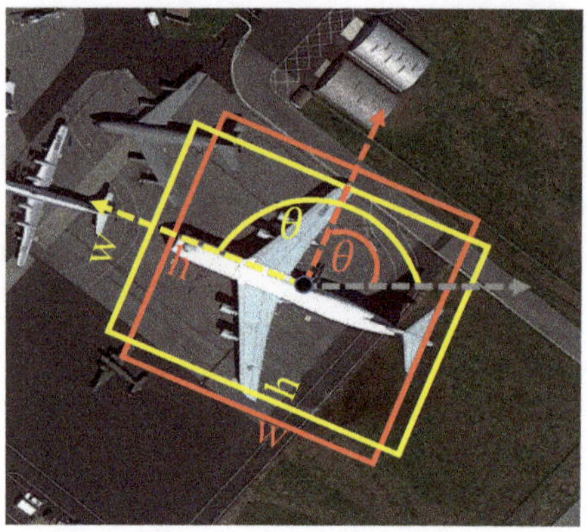

Figure 4. Boundary discontinuity problem of angle prediction. The red and yellow bounding boxes indicate two different targets. Although the two square-like targets have slightly different edge (w and h) lengths, there is a huge gap between the angle target θ.

2.2.2. Overview

To tackle the boundary problem in angle regression, in this paper, we propose a straightforward and efficient angle-free framework for rotated object detection. Instead of predicting the angle, as many previous angle-based two-stage methods do [13,15,16], our proposed Point RCNN reformulates the oriented bounding box (OBB) task as learning the representative points of the object in the RPN phase and modeling the corner points in the RCNN refine phase, which are both totally angle-free. Figure 5 shows the entire detection process, from the representative point learning to the final refined four corners of the oriented object.

(a) (b) (c)

Figure 5. Illustration of the detection process of the Point RCNN framework. (**a**) denotes the predicted representative points with the PointRPN module. (**b**) denotes the conversion from the representative points to the rotated proposals. (**c**) denotes the refinement process of the corner points with the PointReg module.

The overall pipeline of Point RCNN is shown in Figure 3. During training, Backbone-FPN first extracts pyramid feature maps given an input image. Then, PointRPN performs representative points regression and generates a pseudo-OBB for the rotated RoI (RRoI). Finally, for each RRoI, PointReg regresses and refines the corner points and classifies them for final detection results. Furthermore, we propose to resample images of rare categories to stabilize training and further improve the overall performance.

The overall training objective is described as:

$$\mathcal{L} = \mathcal{L}_{PointRPN} + \mathcal{L}_{PointReg}, \quad (1)$$

where $\mathcal{L}_{PointRPN}$ denotes the losses in PointRPN, and $\mathcal{L}_{PointReg}$ denotes the losses in PointReg. We will describe them in detail in the following sections.

2.2.3. PointRPN

Existing rotated object detection methods generate rotated proposals indirectly by transforming the outputs of RPN [1] and suffer from the boundary discontinuity problem caused by angle prediction. For example, Refs. [13,15] use an RoI transformer to convert horizontal proposals to rotated proposals with an additional angle prediction task. Unlike these methods, in this paper, we propose to directly predict the rotated proposals with representative point learning. The learning of points is more flexible, and the distribution of points can reflect the angle and size of the rotated object. The boundary discontinuity problem can thus be alleviated without angle regression.

Representative Points Prediction: Inspired by RepPoints [37] and CFA [55], we propose PointRPN to predict the representative points in the RPN stage. The predicted points can effectively represent the rotating box and can be easily converted to rotated proposals in subsequent RCNN stages.

As shown in Figure 6, PointRPN learns a set of representative points for each feature point. In order to make the features adapt more effectively to the representative points learning, we adopt a coarse-to-fine prediction approach. In this way, the features are refined with deformable convolutional networks (DCN) [56] and predicted offsets in the initial stage. For each feature point, the predicted representative points of the two stages are as follows:

$$\begin{aligned} \mathcal{R}^{init} &= \{(x_i^0 + \Delta x_i^0, y_i^0 + \Delta y_i^0)\}_{i=1}^{K}, \\ \mathcal{R}^{refine} &= \{(x_i^1 + \Delta x_i^1, y_i^1 + \Delta y_i^1)\}_{i=1}^{K}, \end{aligned} \quad (2)$$

where K denotes the number of predicted representative points and we set $K = 9$ by default. $\{(x_i^0, y_i^0)\}_{i=1}^{K}$ denotes the initial location, $\{(\Delta x_i^0, \Delta y_i^0)\}_{i=1}^{K}$ denote the learned offsets in the initial stage, and $\{(\Delta x_i^1, \Delta y_i^1)\}_{i=1}^{K}$ denote the learned offsets in the refine stage.

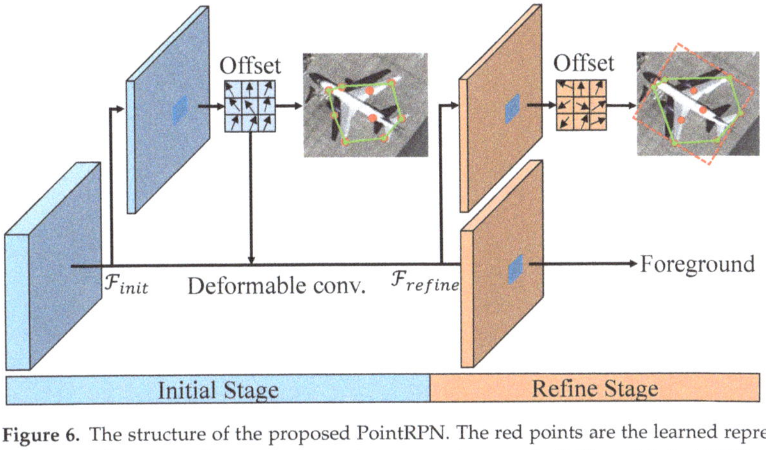

Figure 6. The structure of the proposed PointRPN. The red points are the learned representative points, and the green polygon represents the converted convex-hull. The red dotted OBB is converted from the representative points with the `MinAreaRect` function of OpenCV [28] for generating RRoI.

Label Assignment: PointRPN predicts representative points for each feature point in the initial and refine stages. This section will describe how we determine the positive samples among all feature points for these two stages.

For the initial stage (see the initial stage in Figure 6), we project each ground-truth box to the corresponding feature level l_i according to its area, and then select the feature point closest to its center as the positive sample. The rule used for projecting the ground-truth box b_i^* to the corresponding feature level is defined as:

$$l_i = \log_2\left(\sqrt{\frac{w_i h_i}{s}}\right), \tag{3}$$

where s is a hyper-parameter and is set to 16 by default. w_i and h_i are the width and height of the ground-truth box b_i^*. The calculated l_i will be further limited to the range of $[3,7]$, since we make predictions for the five feature levels of (P_3, P_4, P_5, P_6, P_7). It is beneficial to optimize the overall detector by placing objects with different scales into different feature levels.

For the refine stage (see the refine stage in Figure 6), considering that the initial stage can already provide coarse prediction, we use the predicted representative points from the initial stage to help determine the positive samples for refined results. To be specific, for each feature point with its corresponding prediction $\mathcal{R}^{init}$, if the maximum convex-hull GIoU (defined in Equation (6)) between $\mathcal{R}^{init}$ and ground-truth boxes exceeds the threshold τ, we select this feature point as a positive sample. We set $\tau = 0.1$ in all our experiments.

Optimization: The optimization of the proposed PointRPN is driven by classification loss and rotated object localization loss. The learning objective is formulated as follows:

$$\mathcal{L}_{PointRPN} = \lambda_1 \, {}^{+}\mathcal{L}_{loc}^{init} + \lambda_2 \, \mathcal{L}_{cls}^{refine} + \lambda_3 \, {}^{+}\mathcal{L}_{loc}^{refine}, \tag{4}$$

where λ_1, λ_2, and λ_3 are the trade-off parameters and are set to 0.5, 1.0, and 1.0 by default, respectively. ${}^{+}\mathcal{L}_{loc}^{init}$ denotes the localization loss of the initial stage. $\mathcal{L}_{cls}^{refine}$ and ${}^{+}\mathcal{L}_{loc}^{refine}$ denote the classification loss and localization loss of the refine stage. Note that the classification loss is only calculated in the refine stage, and the two localization losses are only calculated for the positive samples.

In the initial stage, the localization loss is calculated between the convex-hulls converted from the learned points $\mathcal{R}^{init}$ and the ground-truth OBBs, respectively. We use convex-hull GIoU loss [55] to calculate the localization loss:

$${}^{+}\mathcal{L}_{loc}^{init} = \frac{1}{N_{pos}^{0}} \sum_i \Big(1 - \mathrm{CIoU}(\Gamma(\mathcal{R}_i^{init}), \Gamma(b_i^*))\Big), \tag{5}$$

where N_{pos}^{0} indicates the number of positive samples of the initial stage. b_i^* is the matched ground-truth OBB. CIoU represents the convex-hull GIoU between the two convex-hulls $\Gamma(\mathcal{R}_i^{init})$ and $\Gamma(b_i^*)$, which are differential and can be calculated as follows:

$$\mathrm{CIoU}(\Gamma(\mathcal{R}_i^{init}), \Gamma(b_i^*)) = \frac{|\Gamma(\mathcal{R}_i^{init}) \cap \Gamma(b_i^*)|}{|\Gamma(\mathcal{R}_i^{init}) \cup \Gamma(b_i^*)|} - \frac{|\mathcal{P}_i \setminus (\Gamma(\mathcal{R}_i^{init}) \cup \Gamma(b_i^*))|}{\mathcal{P}_i}, \tag{6}$$

where the first term denotes the convex-hull IoU, and $\mathcal{P}_i$ denotes the smallest enclosing convex object area of $\Gamma(\mathcal{R}_i^{init})$ and $\Gamma(b_i^*)$. $\Gamma(\cdot)$ denotes Jarvis's march algorithm [57] used to calculate the convex-hull from points.

The learning of the refine stage, which is responsible for outputting more accurate rotated proposals, is driven by both classification loss and localization loss. $\mathcal{L}_{cls}^{refine}$ is a standard focal loss [5], which can be calculated as:

$$\mathcal{L}_{cls}^{refine} = \frac{1}{N_{pos}^{1}} \sum_i \mathrm{FL}(p_i, c_i^*), \tag{7}$$

$$\mathrm{FL}(p_i, c_i^*) = \begin{cases} -\alpha(1-p_i)^\gamma \log(p_i), & \text{if } c_i^* > 0; \\ -(1-\alpha)p_i^\gamma \log(1-p_i), & \text{otherwise,} \end{cases} \quad (8)$$

where N_{pos}^1 denotes the number of positive samples in the refine stage, p_i and c_i^* are the classification output and the assigned ground-truth category, respectively. α and γ are hyper-parameters and are set to 0.25 and 2.0 by default. The localization loss $\mathcal{L}_{loc}^{refine}$ is similar to Equation (5) and can be formulated as:

$$\mathcal{L}_{loc}^{refine} = \frac{1}{N_{pos}^1} \sum_i \Big(1 - \mathrm{CIoU}\big(\Gamma(\mathcal{R}_i^{refine}), \Gamma(b_i^*)\big)\Big). \quad (9)$$

With the refined representative points, the pseudo-OBB (see red-dotted OBB in Figure 6) is converted using the `MinAreaRect` function of OpenCV [28], which is then used for generating the RRoI for PointReg.

2.2.4. PointReg

Corner Points Refine: The rotated proposals generated by PointRPN already provide a reasonable estimate for the target rotated objects. To avoid the problems caused by angle regression and to further improve the detection performance, we refine the four corners of the rotated proposals in the RCNN stage. As shown in Figure 7, with the rotated proposals as input, we use an RRoI feature extractor [13,15] to extract the RRoI features. Then, given the RRoI features, two consecutive fully connected and ReLU layers are used to encode the RRoI features. Finally, two fully connected layers are responsible for predicting the class probability P and refined corners $\mathcal{C}$ of the corresponding rotated object. The refined corner points can be represented as follows:

$$\mathcal{C} = \{(x_i + \Delta x_i, y_i + \Delta y_i)\}_{i=1}^4, \quad (10)$$

where $\{(x_i, y_i)\}_{i=1}^4$ denotes the four corner coordinates of the input rotated proposals, and we denote the corresponding four predicted corner offsets as $\{(\Delta x_i, \Delta y_i)\}_{i=1}^4$.

In PointReg, instead of directly performing angle prediction, we refine the four corners of the input rotated proposals. There are three advantages of adopting corner points refinement: (1) it can alleviate the boundary discontinuity problem caused by angle prediction; (2) the parameter units are consistent among the eight parameters $\{(x_i, y_i)\}_{i=1}^4$; and (3) it is possible to improve the localization accuracy using a coarse-to-fine approach.

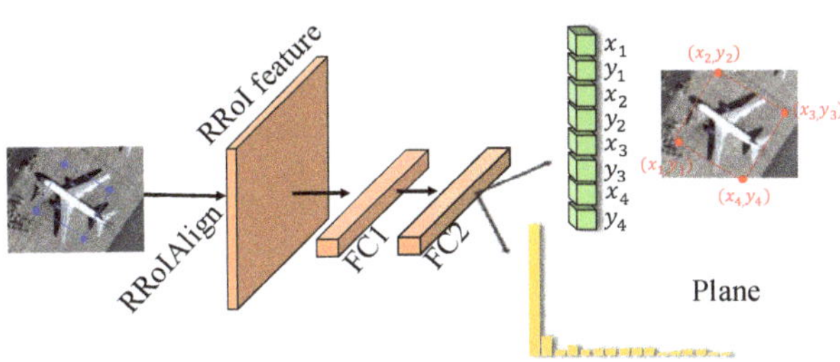

Figure 7. The diagram of the proposed PointReg. For simplicity, we only show the first stage of PointReg. The blue and red points represent the four corner points of the input RRoI and the refined results, respectively.

We can easily extend PointReg to a cascade structure for better performance. As shown in Figure 3, in the cascade structure, the refined rotated proposals of the previous stage are used as the input of the current stage.

Optimization: The learning of PointReg is driven by the classification loss and the rotated object localization loss:

$$\mathcal{L}_{PointReg} = \mu_1 \mathcal{L}_{cls} + \mu_2 \, {}^{+}\mathcal{L}_{loc}, \tag{11}$$

where μ_1 and μ_2 are the trade-off coefficients and are both set to 1.0 by default. $\mathcal{L}_{cls}$ indicates the classification loss, which is a standard cross-entropy loss:

$$\mathcal{L}_{cls} = -\frac{1}{N}\sum_{i}\sum_{c=0}^{C} Y_{i\to c}\log(P_i), \tag{12}$$

where N denotes the number of training samples in PointReg, C is the number of categories excluding the background, P_i is the predicted classification probability of the i_{th} RRoI. $Y_{i\to c} = 1$ if the ground-truth class of the i_{th} RRoI is c; otherwise it is 0. ${}^{+}\mathcal{L}_{loc}$ represents the localization loss between the refined corners and the corners of the ground-truth OBB. We use L_1 loss to optimize the corner points refinement learning which can be calculated as:

$${}^{+}\mathcal{L}_{loc} = \frac{1}{N}\sum_{i}|\mathcal{C}_i - \vartheta(b_i^*)|, \tag{13}$$

where we let $\mathcal{C}_i (= \{(x_j, y_j)\}_{j=1}^{4})$ denote the refined corners for the i_{th} rotated proposal, let $b_i^* (= \{(x_j^*, y_j^*)\}_{j=1}^{4})$ denote the corners of the matched ground-truth OBB. $\vartheta(b_i^*)$ denotes the permutation of four corners of b_i^* with the smallest L_1 loss $|\mathcal{C}_i - \vartheta(b_i^*)|$, which can alleviate the sudden loss change issue in angle-free detectors. Note that ${}^{+}\mathcal{L}_{loc}$ is only calculated for positive training samples.

2.2.5. Balanced Dataset Strategy

The extremely non-uniform object densities of aerial images usually make the dataset long-tailed, which may cause the training process to be unstable and limit the detection performance. For instance, DOTA-v1.0 [6] contains 52, 516 ship instances but only 678 ground-track field instances [7]. To alleviate this issue, in this section, we propose a balanced dataset strategy. Specifically, we resample the images of rare categories, which was inspired by [58]. More concretely, first, for each category $c \in C$, we compute the fraction of images F_c that contains this category. Then, we compute the category-level repeat factor for each category:

$$r_c = \max(1.0, \sqrt{\beta_{thr}/F_c}), \tag{14}$$

where β_{thr} is a threshold which indicates that there will not be oversampling if "$F_c > \beta_{thr}$". Next, we compute the image-level repeat factor r_I for each image I:

$$r_I = \max_{c\in C_I}(r_c), \tag{15}$$

where C_I denotes the categories contained in image I. Finally, we can resample the images according to the image-level repeat factor. In other words, those images that contain long-tailed categories will have a greater chance of being resampled during training.

3. Results

In this section, we describe the dataset, evaluation protocol, implementation details, and demonstrate an overall evaluation and describe detailed ablation studies of the proposed method.

3.1. Datasets

To evaluate the effectiveness of our proposed Point RCNN framework, we performed experiments on four popular large-scale oriented object detection datasets: DOTA-v1.0 [6], DOTA-v1.5, HRSC2016 [59], and UCAS-AOD [60], which are widely used for rotated object detection. The statistic information comparison of these datasets is depicted in Table 1.

DOTA [6] is a large-scale and challenging aerial image dataset for oriented object detection with three released versions: DOTA-v1.0, DOTA-v1.5 and DOTA-v2.0. To compare the performance with the state-of-the-art methods, we performed experiments on DOTA-v1.0 and DOTA-v1.5. DOTA-v1.0 contains 2806 images ranging in size from 800 × 800 to 4000 × 4000, and contains 188, 282 instances in 15 categories, abbreviated as: Bridge (BR), Harbor (HA), Ship (SH), Plane (PL), Helicopter (HC), Small vehicle (SV), Large vehicle (LV), Baseball diamond (BD), Ground track field (GTF), Tennis court (TC), Basketball court (BC), Soccer-ball field (SBF), Roundabout (RA), Swimming pool (SP), and Storage tank (ST). The dataset is divided into a training set, validation set, and test set, which account for one half, one sixth, and one third of the total dataset, respectively. DOTA-v1.5 is an updated version of DOTA-v1.0. It has the same images as DOTA-v1.0 but contains 402, 089 instances. DOTA-v1.5 has revised and updated the annotation of objects, where many small object instances about or below 10 pixels that were missed in DOTA-v1.0 have been additionally annotated. This is a more challenging dataset, which introduces a new category Container Crane (CC) and more small instances. For a fair comparison, we used the training set and validation set for training, and the test set was used to verify the performance of our model. The performances were obtained by submitting the prediction results to DOTA's evaluation server. The official evaluation protocol of the DOTA dataset in terms of the mAP was used.

Table 1. The statistic information comparison of the datasets. OBB denotes the oriented bounding box.

Dataset	Source	Annotation	Categories	Instances	Images	Year
UCAS-AOD [60]	Google Earth	OBB	2	14,596	1510	2015
HRSC2016 [59]	Google Earth	OBB	1	2976	1061	2016
DOTA-v1.0 [6]	multi source	OBB	14	188,282	2806	2018
DOTA-v1.5	multi source	OBB	15	402,089	2806	2019

HRSC2016 [59] is another popular dataset for oriented object detection. The images of this dataset were mainly collected from two scenarios, including ships on the sea and ships close to the shore. The dataset contains 1061 aerial images with size ranges from 300 × 300 to 1500 × 900, with most larger than 1000 × 600. There are more than 25 types of ships with large varieties in scale, position, rotation, shape, and appearance. This dataset can be divided into a training set, validation set and test set. There are 436 images, 181 images, and 444 images in the training set, validation set and test set, respectively. For a fair comparison, we used both the training and validation sets for training. The standard evaluation protocol of HRSC2016 dataset in terms of mAP was used.

UCAS-AOD [60] is another dataset for small oriented object detection with two categories (car and plane), which contains 1510 aerial images with 510 car images and 1000 airplane images. There are 14,596 instances in total, and the image size is approximately 659 × 1280. For a fair comparison, equivalent to the UCAS-AOD-benchmark (https://github.com/ming71/UCAS-AOD-benchmark, accessed on 29 March 2022), we also divided the dataset into 755 images for training, 302 images for validation, and 453 images for testing with a ratio of 5:2:3. The standard evaluation protocol of the UCAS-AOD dataset in terms of mAP was used.

3.2. Implementation Details

We implemented Point RCNN using the MMDetection tool-box [61]. We followed ReDet [15] to use ReResNet with ReFPN as our backbone (ReR50-ReFPN), which has shown the ability to extract rotation-equivariant features. We also verified with the more generalized transformer backbone (Swin-Tiny) to show the generalization and scalability of our Point RCNN framework.

For the DOTA dataset, following previous methods [13,15,21], we cropped the images to 1024 × 1024 with 824 pixels as a stride and we also resized the image to three scales $\{0,5,1.0,1.5\}$ to prepare multi-scale data. Random horizontal flipping and random rotation ($[-45°, 45°]$) were adopted as the data augmentation for multi-scale training. For the HRSC2016 dataset, as in the previous method [15], we resized all the images to (800, 512), and we used random horizontal flipping as the data augmentation method during training.

For the UCAS-AOD dataset, following the UCAS-AOD-benchmark, we resized all the images to (800, 800) and only used the training set for training. We also used random horizontal flipping, HSV augment and random rotation as the data augmentation approach during training. Unless otherwise specified, we trained all the models with 19 epochs for DOTA, 36 epochs for HRSC2016, and 36 epochs for UCAS-AOD. Specifically, we trained all the models using the AdamW [62] optimizer with $\beta_1 = 0.9$ and $\beta_2 = 0.999$. The initial learning rate was set to 0.0002 with warming up for 500 iterations, with the learning rate decaying by a factor of 10 at each decay step. The weight decay was set to 0.05, and the mini-batch size was set to 16 (two images per GPU). We conducted the experiments on a server with 8 Tesla-V100 GPUs. The code will be released.

3.3. Main Results

We compared our Point RCNN framework with other state-of-the-art methods for four datasets: DOTA-v1.0, DOTA-v1.5, HRSC2016, and UCAS-AOD. As shown in Tables ??–5, without unnecessary elaboration, our Point RCNN demonstrated superior performance compared to state-of-the-art methods.

Results on DOTA-v1.0: As reported in Table ??, we first evaluated our method on the DOTA-v1.0 dataset and compared it with the popular and the state-of-the-art rotated object detection methods. We obtained the overall detection performance by submitting our results to the official DOTA-v1.0 evaluation server. In this comparison experiment, we compared many classic and impressive methods [13,19,21,23,26,27,55,63–65] and some state-of-the-art methods, e.g., Oriented RCNN [16] and ReDet [15].

As shown in Table ??, our Point RCNN method achieved new, state-of-the-art, detection performance against the comparison methods. More specifically, with the ReR50-ReFPN backbone, our Point RCNN improved the detection performance by 0.61 mAP against ReDet (form 80.10 to 80.71). Compared with Oriented RCNN, Point RCNN improved the performance by about 0.2 mAP. We observed that, with the proposed balanced dataset strategy, our Point RCNN was able to improve the performance by 0.34 mAP (from 80.37 to 80.71), which confirms that the extremely non-uniform rotated object densities of aerial images do limit detection performance.

In addition, we also evaluated our Point RCNN with the more generalized transformer backbone Swin-Tiny [66] (Swin-T). The Swin transformer [66] is a new vision transformer, which has been used as a general backbone of computer vision in recent years. Our proposed Point RCNN was able to further improve the performance by 0.61% (from 80.71 to 81.32), indicating that Point RCNN is scalable to general backbone networks.

Results on DOTA-v1.5: As reported in Table 3, we then evaluated our method on the DOTA-v1.5 dataset, which is a more challenging dataset, since it contains more categories and more small object instances. We obtained the overall detection performance by submitting our results to the official DOTA-v1.5 evaluation server. In this experiment, we compared some traditional strong two-stage oriented object detectors, e.g., Faster RCNN OBB (FR-O) [6], Mask RCNN [2], the Hybrid Task Cascade (HTC) [67] and state-of-the-art methods, including Oriented RCNN [16] and ReDet [15].

As shown in Table 3, our Point RCNN method achieved the new state-of-the-art detection performance on DOTA-v1.5 against the comparison methods. More specifically, our Point RCNN improved the detection performance by 2.51 mAP against ReDet (form 76.80 to 79.31), which represents a significant improvement for oriented object detection. Compared with Oriented RCNN, Point RCNN also significantly improved the performance by 2.86 mAP (from 76.45 to 79.31). We also observed that, with our proposed balanced dataset strategy, Point RCNN was able to further improve the performance by 0.57 mAP based on a high performance baseline (from 78.74 to 79.31). We also evaluated Point RCNN with the Swin-Tiny [66] (Swin-T) backbone. With the more generalized transformer backbone, our proposed Point RCNN was able to further improve the performance by **0.83** mAP (from 79.31 to 80.14), indicating that Point RCNN is scalable to general backbone networks and more challenging aerial image datasets.

Table 2. Performance comparisons on the DOTA-v1.0 test set (AP (%) for each category and overall mAP (%)). * denotes multi-scale training and testing, *† denotes the results of using our balanced dataset strategy. "R50" denotes ResNet-50, "R101" denotes ResNet-101, "R152" denotes ResNet-152, "H104" denotes Hourglass-104, "ReR50" denotes ReResNet-50, "Swin-T" denotes Swin Transformer Tiny.

Method	Backbone	PL	BD	BR	GTF	SV	LV	SH	TC
RoI Trans. * [13]	R101-FPN	88.64	78.52	43.44	75.92	68.81	73.68	83.59	90.74
O^2-DNet * [63]	H104	89.30	83.30	50.10	72.10	71.10	75.60	78.70	90.90
DRN * [64]	H104	89.71	82.34	47.22	64.10	76.22	74.43	85.84	90.57
Gliding Vertex * [26]	R101-FPN	89.64	85.00	52.26	77.34	73.01	73.14	86.82	90.74
BBAVectors * [27]	R101	88.63	84.06	52.13	69.56	78.26	80.40	88.06	90.87
CenterMap * [65]	R101-FPN	89.83	84.41	54.60	70.25	77.66	78.32	87.19	90.66
CSL * [19]	R152-FPN	90.25	85.53	54.64	75.31	70.44	73.51	77.62	90.84
SCRDet++ * [23]	R152-FPN	88.68	85.22	54.70	73.71	71.92	84.14	79.39	90.82
CFA * [55]	R-152	89.08	83.20	54.37	66.87	81.23	80.96	87.17	90.21
S^2A-Net * [21]	R50-FPN	88.89	83.60	57.74	81.95	79.94	83.19	89.11	90.78
ReDet * [15]	ReR50-ReFPN	88.81	82.48	60.83	80.82	78.34	86.06	88.31	90.87
Oriented RCNN * [16]	R101-FPN	90.26	84.74	62.01	80.42	79.04	85.07	88.52	90.85
Point RCNN * (Ours)	ReR50-ReFPN	82.99	85.73	61.16	79.98	77.82	85.90	88.94	90.89
Point RCNN *† (Ours)	ReR50-ReFPN	86.21	86.44	60.30	80.12	76.45	86.17	88.58	90.84
Point RCNN *† (Ours)	Swin-T-FPN	86.59	85.72	61.64	81.08	81.01	86.49	88.84	90.83
		BC	ST	SBF	RA	HA	SP	HC	mAP
RoI Trans. * [13]	R101-FPN	77.27	81.46	58.39	53.54	62.83	58.93	47.67	69.56
O^2-DNet * [63]	H104	79.90	82.90	60.20	60.00	64.60	68.90	65.70	72.80
DRN * [64]	H104	86.18	84.89	57.65	61.93	69.30	69.63	58.48	73.23
Gliding Vertex * [26]	R101-FPN	79.02	86.81	59.55	70.91	72.94	70.86	57.32	75.02
BBAVectors * [27]	R101	87.23	86.39	56.11	65.62	67.10	72.08	63.96	75.36
CenterMap * [65]	R101-FPN	84.89	85.27	56.46	69.23	74.13	71.56	66.06	76.03
CSL * [19]	R152-FPN	86.15	86.69	69.60	68.04	73.83	71.10	68.93	76.17
SCRDet++ * [23]	R152-FPN	87.04	86.02	67.90	60.86	74.52	70.76	72.66	76.56
CFA * [55]	R-152	84.32	86.09	52.34	69.94	75.52	80.76	67.96	76.67
S^2A-Net * [21]	R50-FPN	84.87	87.81	70.30	68.25	78.30	77.01	69.58	79.42
ReDet * [15]	ReR50-ReFPN	88.77	87.03	68.65	66.90	79.26	79.71	74.67	80.10
Oriented RCNN * [16]	R101-FPN	87.24	87.96	72.26	70.03	82.93	78.46	68.05	80.52
Point RCNN * (Ours)	ReR50-ReFPN	88.89	88.16	71.84	68.21	79.03	80.32	75.71	80.37
Point RCNN *† (Ours)	ReR50-ReFPN	88.58	88.44	73.03	70.10	79.26	79.02	77.15	80.71
Point RCNN *† (Ours)	Swin-T-FPN	87.22	88.23	68.85	71.48	82.09	83.60	76.08	81.32

Table 3. Performance comparisons on DOTA-v1.5 test set (AP (%) for each category and overall mAP (%)). * denotes multi-scale training and testing, *† denotes the results of using balanced dataset strategy. Note that the results of Faster RCNN OBB (FR-O) [6], RetinaNet OBB (RetinaNet-O) [5], Mask RCNN [2] and Hybrid Task Cascade (HTC) [67] are excerpted from ReDet [15]. The results of Oriented RCNN* and ReDet* with Swin-T-FPN backbone are our re-implementations based on their released official code. "R50" denotes ResNet-50, "R101" denotes ResNet-101, "ReR50" denotes ReResNet-50, "Swin-T" denotes Swin Transformer Tiny.

Method	**Backbone**	**PL**	**BD**	**BR**	**GTF**	**SV**	**LV**	**SH**	**TC**
RetinaNet-O [5]	R50-FPN	71.43	77.64	42.12	64.65	44.53	56.79	73.31	90.84
FR-O [6]	R50-FPN	71.89	74.47	44.45	59.87	51.28	68.98	79.37	90.78
Mask RCNN [2]	R50-FPN	76.84	73.51	49.90	57.80	51.31	71.34	79.75	90.46
HTC [67]	R50-FPN	77.80	73.67	51.40	63.99	51.54	73.31	80.31	90.48
OWSR * [68]	R101-FPN	-	-	-	-	-	-	-	-
Oriented RCNN * [16]	R101-FPN	87.20	84.67	60.13	80.79	67.51	81.63	89.74	90.88
ReDet * [15]	ReR50-ReFPN	88.51	86.45	61.23	81.20	67.60	83.65	90.00	90.86
ReDet * [15]	Swin-T-FPN	80.90	85.13	60.61	80.83	67.07	83.32	89.80	90.79
Point RCNN * (Ours)	ReR50-ReFPN	83.40	86.59	60.76	80.25	79.92	83.37	90.04	90.86
Point RCNN *† (Ours)	ReR50-ReFPN	83.12	86.55	60.84	82.43	80.60	83.39	90.01	90.88
Point RCNN * (Ours)	Swin-T-FPN	83.88	85.22	60.76	79.40	81.64	83.48	89.98	90.75
Point RCNN *† (Ours)	Swin-T-FPN	86.93	85.79	59.52	80.42	81.91	81.92	89.95	90.35

		BC	ST	SBF	RA	HA	SP	HC	CC	mAP
RetinaNet-O [5]	R50-FPN	76.02	59.96	46.95	69.24	59.65	64.52	48.06	0.83	59.16
FR-O [6]	R50-FPN	77.38	67.50	47.75	69.72	61.22	65.28	60.47	1.54	62.00
Mask RCNN [2]	R50-FPN	74.21	66.07	46.21	70.61	63.07	64.46	57.81	9.42	62.67
HTC [67]	R50-FPN	75.12	67.34	48.51	70.63	64.84	64.48	55.87	5.15	63.40
OWSR * [68]	R101-FPN	-	-	-	-	-	-	-	-	74.90
Oriented RCNN * [16]	R101-FPN	82.21	78.51	70.98	78.63	79.46	75.40	75.71	39.69	76.45
ReDet * [15]	ReR50-ReFPN	84.30	75.33	71.49	72.06	78.32	74.73	76.10	46.98	76.80
ReDet * [15]	Swin-T-FPN	86.04	78.69	75.35	77.38	78.48	75.41	79.51	61.95	78.20
Point RCNN * (Ours)	ReR50-ReFPN	87.45	84.50	72.79	77.32	78.29	77.48	78.92	47.97	78.74
Point RCNN *† (Ours)	ReR50-ReFPN	87.25	84.60	73.49	78.51	78.75	78.41	76.12	54.12	79.31
Point RCNN * (Ours)	Swin-T-FPN	87.00	84.65	70.70	77.87	78.32	79.50	74.35	63.80	79.46
Point RCNN *† (Ours)	Swin-T-FPN	85.72	85.84	68.57	76.35	78.79	81.24	78.64	69.23	80.14

Results on HRSC2016: We also verified our Point RCNN method on the HRSC2016 dataset, which contains many ship objects with arbitrary orientations. In this experiment, we compared our proposed Point RCNN method with some classic methods, e.g., RRPN [17], RoI-Trans. ref. [13], R^3Det [20], and S^2A-Net [21], and the state-of-the-art methods, Oriented RCNN [16] and ReDet [15]. Some methods were evaluated under the VOC2007 metric, while others were compared under the VOC2012 metric. To make a comprehensive comparison, we report the results for both metrics.

We report the experimental results in Table 4. We can observe that our Point RCNN method attained a new state-of-the-art performance under both the VOC2007 and VOC2012 metrics. Specifically, under the VOC2007 metric, our Point RCNN achieved 90.53 mAP, which exceeded the results for the comparison methods. It is worth noting that the Point RCNN significantly improved the performance by 0.90 and 0.93 mAP against ReDet and Oriented RCNN under the VOC2012 metric, respectively.

Table 4. Performance comparisons for the HRSC2016 test set. mAP_{07} and mAP_{12} indicate that the results were evaluated under VOC2007 and VOC2012 metrics (%), respectively. We report both results for fair comparison. "R50" denotes ResNet-50, "R101" denotes ResNet-101, "R152" denotes ResNet-152, "H34" denotes Hourglass-34, "ReR50" denotes ReResNet-50.

Method	Backbone	mAP_{07} (%)	mAP_{12} (%)
RC2 [69]	VGG16	75.70	-
RRPN [17]	R101	79.08	85.64
R^2PN [18]	VGG16	79.60	-
RRD [70]	VGG16	84.30	-
RoI-Trans. [13]	R101-FPN	86.20	-
Gliding Vertex [26]	R101-FPN	88.20	-
R^3Det [20]	R101-FPN	89.26	-
DRN [64]	H34	-	92.7
CenterMap [65]	R50-FPN	-	92.8
CSL [19]	R152-FPN	89.62	-
S^2A-Net [21]	R101-FPN	90.17	95.01
ReDet [15]	ReR50-ReFPN	90.46	97.63
Orient RCNN [16]	R101-FPN	90.50	97.60
Point RCNN (Ours)	ReR50-ReFPN	90.53	98.53

Results on UCAS-AOD: The UCAS-AOD dataset consists of a large number of small rotated objects, which are often overwhelmed by complex scenes in aerial images. We evaluated our proposed Point RCNN method on UCAS-AOD and report the comparison results in Table 5. For a fair comparison, we report the results under the VOC2007 metric. As shown in Table 5, our proposed method achieved the best performance of **90.04** mAP_{07}, in which a value of 89.60 was obtained for the car detection, and a value of 90.48 was obtained for the airplane detection.

Table 5. Performance comparisons for the UCAS-AOD test set (AP (%) for each category and overall mAP (%)). All models were evaluated via the VOC2007 metric (%).

Method	Backbone	Car	Airplane	mAP_{07} (%)
R-Yolov3 [71]	Darknet53	74.63	89.52	82.08
R-RetinaNet [5]	ResNet50	84.64	90.51	87.57
RoI-Trans. [13]	ResNet50	88.02	90.02	89.02
DAL [72]	ResNet50	89.25	90.49	89.87
S^2A-Net [21]	ResNet50	89.56	90.42	89.99
Point RCNN (Ours)	ReR50-ReFPN	89.60	90.48	90.04

3.4. *Ablation Study*

In this section, we report an ablation study of our proposed Point RCNN framework. If not specified, all the models were trained only on the training and validation set with a scale of 1.0 for simplicity, and were tested using multi-scale testing. The metric mAP was evaluated on the DOTA-v1.5 test set and obtained by submitting prediction results to DOTA-v1.5's evaluation server. In the following sections, we mainly elaborate the effectiveness of our angle-free Point RCNN framework, including PointRPN, PoingReg, and the balanced dataset strategy.

3.4.1. Analysis of PointRPN

To analysis the efficiency of the proposed PointRPN, which serves as an RPN network, we evaluated the detection recall of PointRPN on the validation set of DOTA-v1.5. For simplicity, we trained the models on the training set with scale 1.0 and evaluated the detection recall on the validation set with scale 1.0 as well. The positive intersection over union (IoU) threshold was set to 0.5. We selected the top-300, top-1000, and top-

2000 proposals to calculate their recall values, respectively. The experimental results are reported in Table 6. We found that when the number of proposals reached 2000, as for the settings of many state-of-the-art methods [15,16], our PointRPN was able to attain 90.00% detection recall. When the number of proposals changed from top-2000 to top-1000, the detection recall value only dropped by 0.17%. Even if there were only top-300 proposals, our PointRPN was still able to achieve 85.93% detection recall. The high detection recall observed demonstrates that our angle-free PointRPN can alleviate the boundary discontinuity problem caused by angle prediction and effectively detect more oriented objects with arbitrary orientations in aerial images.

Table 6. Comparison of the detection recall results by varying the number of proposals of each image patch. The metric recall is evaluated on the DOTA-v1.5 validation set. Recall_{300}, Recall_{1000}, and Recall_{2000} represent the detection recall of the top-300, top-1000, and top-2000 proposals, respectively.

Method	Recall_{300} (%)	Recall_{1000} (%)	Recall_{2000} (%)
PointRPN	85.93	89.83	90.00

We also performed visualization analysis of PointRPN. As shown in Figure 8, we visualized some examples of the learned representative points of our PointRPN on the DOTA-v1.0 test set. The visualization results demonstrated that the proposed PointRPN was able to automatically learn the extreme points and the semantic key points of the rotated objects with arbitrary orientations, the large scale and aspect ratio variations, and the extreme non-uniform object densities.

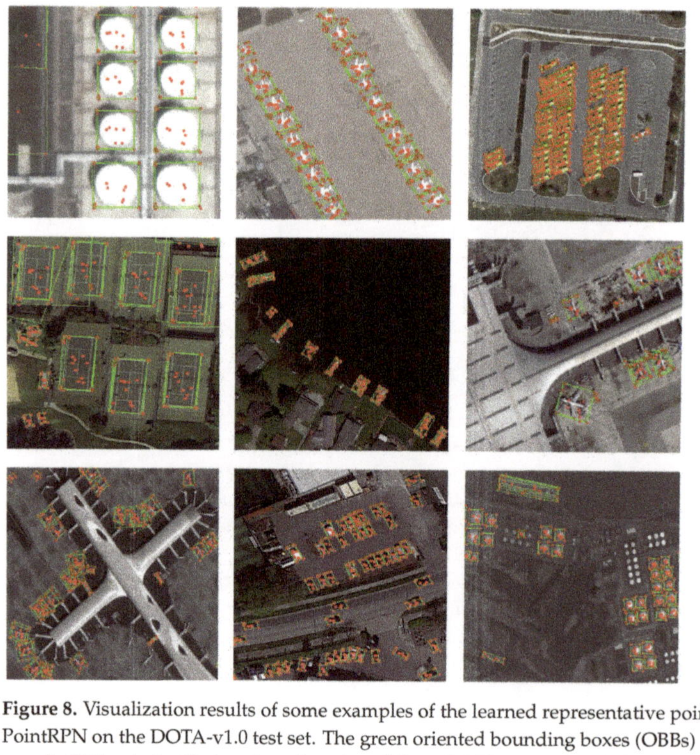

Figure 8. Visualization results of some examples of the learned representative points (red points) of PointRPN on the DOTA-v1.0 test set. The green oriented bounding boxes (OBBs) are the converted pseudo-OBBs via the `MinAreaRect` function of OpenCV. The score threshold was set to 0.001 without using NMS.

3.4.2. Effectiveness of PointReg

In this section, we provide an analysis of the effectiveness of the proposed PointReg module. We evaluated different OBB regression types of our PointReg and the results are reported in Table 7; compared to the five-parameter (x, y, w, h, θ) representation, the eight-parameter (also called corner points) $(x_1, y_1, x_2, y_2, x_3, y_3, x_4, y_4)$ regression type achieved higher detection performance (77.60 vs. 77.25) for oriented objects. In other words, our angle-free PointReg was shown to be capable of alleviating the boundary discontinuity problem caused by angle prediction and to achieve higher performance.

Table 7. Analysis of the effectiveness of OBB regression type of PointReg. The metric mAP was evaluated for the DOTA-v1.5 test set.

Regression Type of PointReg	mAP (%)
(x, y, w, h, θ)	77.25
$(x_1, y_1, x_2, y_2, x_3, y_3, x_4, y_4)$	77.60

3.4.3. Analysis of Balanced Dataset Strategy

In this section, an analysis of the impact of the oversampling threshold β_{thr} of the proposed balanced dataset strategy is provided. As shown in Table 8, we achieved the best detection accuracy of 77.60 mAP at $\beta_{thr} = 0.3$. Therefore, we set $\beta_{thr} = 0.3$ in all other experiments, unless otherwise stated.

Table 8. Comparison of detection accuracy by varying the oversampling threshold β_{thr}. The metric mAP was evaluated on the DOTA-v1.5 test set.

Oversampling Threshold (β_{thr})	mAP (%)
0	73.52
0.1	76.49
0.2	77.44
0.3	77.60
0.4	77.48

3.4.4. Factor-by-Factor Experiment

To explore the effectiveness of each module of our proposed Point RCNN framework, we conducted a factor-by-factor experiment on the proposed PointRPN, PointReg and balanced dataset strategy. The results are depicted in Table 9. Each component had a positive effect, and all components were combined to obtain the best performance.

Table 9. Factor-by-factor ablation experiments. The detection performance was evaluated on the test set of DOTA-v1.5 dataset.

Method	PointRPN	Balanced Dataset Strategy	PointReg	mAP (%)
Baseline				71.36
Point RCNN	✓			74.17
		✓		74.22
	✓	✓		77.25
	✓	✓	✓	77.60

3.4.5. Visualization Analysis

We visualized some detection results for rotated objects for the DOTA-v1.0 test set. Figure 8 depicts some examples of the learned representative points of our PointRPN, which indicates that PointRPN was capable of learning the representative points of the rotated object. Specifically, PointRPN was able to automatically learn the extreme points, e.g., the corner points of the rotated objects, and the semantic key points, e.g., the meaningful area of the rotated object.

Based on the reasonable prediction of high detection recall for the target rotated objects of PointRPN, our PointReg was able to continuously optimize and refine the corner points of the rotated objects. Some quantitative results for the DOTA-v1.0 test set are shown in Figure 9; the red points represent the corner points of the rotated objects learned by PointReg and the colored OBBs converted by the `MinAreaRect` function of OpenCV denote the final detection results. We also provide a visualization of the detection results for the UCAS-AOD and HRSC2016 datasets in Figures 10 and 11, respectively. The visualization results demonstrate the remarkable efficiency of our proposed angle-free Point RCNN framework for rotated object detection.

Figure 9. Visualization of the detection results of Point RCNN for the DOTA-v1.0 test set. The score threshold was set to 0.01. Each color represents a category. The red points and colored OBBs are the predicted corner points and the converted OBBs of PointReg.

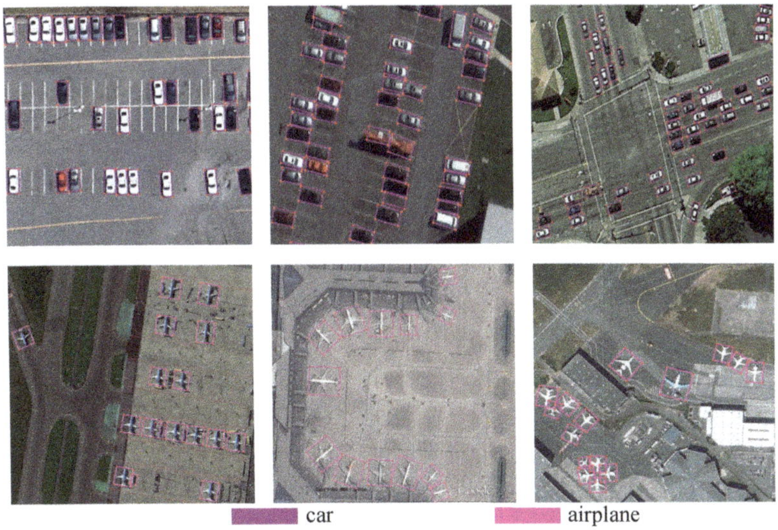

Figure 10. Visualization of the detection results of Point RCNN for the UCAS-AOD test set. The score threshold was set to 0.01. The red points and colored OBBs are the predicted corner points and the converted OBBs of PointReg.

Figure 11. Visualization of the detection results of Point RCNN for the HRSC2016 test set. The score threshold was set to 0.01. The red points and colored OBBs are the predicted corner points and the converted OBBs of PointReg.

4. Discussion

Although the experiments undertaken substantiate the superiority of our proposed Point RCNN framework over state-of-the-art methods, our method did not perform well enough in some categories, e.g., PL (Plane) in the DOTA dataset, which requires further exploration. In addition, as with existing oriented object detectors, our Point RCNN also needs to use rotate non-maximum suppression (NMS) to remove duplicate results, which may mistakenly remove the true positive (TP) predictions and thus limit the final performance. Transformer-based methods [45] may provide potential solutions, which will be pursued in future work.

5. Conclusions

In this study, we revisited rotated object detection and proposed a purely angle-free framework for rotated object detection, named Point RCNN, which mainly consists of a PointRPN for generating accurate RRoIs, and a PointReg for refining corner points based on the generated RRoIs. In addition, we proposed a balanced dataset strategy to overcome the long-tailed distribution of different object classes in aerial images. Compared to existing rotated object detection methods, which mainly rely on angle prediction and suffer from the boundary discontinuity problem, our proposed Point RCNN framework is purely angle-free and can alleviate the boundary problem without introducing angle prediction. Extensive experiments on multiple large-scale benchmarks demonstrated the significant superiority of our proposed Point RCNN framework against state-of-the-art methods. Specifically, Point RCNN achieved new state-of-the-art performances of 80.71, 79.31, 98.53, and 90.04 mAPs on DOTA-v1.0, DOTA-v1.5, HRSC2016, and UCAS-AOD datasets, respectively.

Author Contributions: Conceptualization, Q.Z. and C.Y.; methodology, Q.Z.; validation, Q.Z. and C.Y.; formal analysis, Q.Z.; investigation, Q.Z. and C.Y.; resources, C.Y.; data curation, C.Y.; writing—original draft preparation, Q.Z. and C.Y.; writing—review and editing, Q.Z. and C.Y.; visualization, C.Y.; supervision, Q.Z.; project administration, Q.Z.; funding acquisition, Q.Z. All authors have read and agreed to the published version of the manuscript.

Funding: This research received no external funding.

Institutional Review Board Statement: Not applicable.

Informed Consent Statement: Not applicable.

Data Availability Statement: Not applicable.

Conflicts of Interest: The authors declare no conflict of interest.

References

1. Ren, S.; He, K.; Girshick, R.B.; Sun, J. Faster R-CNN: Towards Real-Time Object Detection with Region Proposal Networks. In Proceedings of the NeurIPS, Montreal, ON, Canada, 7–12 December 2015.
2. He, K.; Gkioxari, G.; Dollár, P.; Girshick, R. Mask r-cnn. In Proceedings of the IEEE International Conference on Computer Vision, Venice, Italy, 22–29 October 2017; pp. 2961–2969.
3. Redmon, J.; Divvala, S.; Girshick, R.; Farhadi, A. You Only Look Once: Unified, Real-Time Object Detection. In Proceedings of the CVPR 2016, Las Vegas, NV, USA, 26 June–1 July 2016.
4. Liu, W.; Anguelov, D.; Erhan, D.; Szegedy, C.; Reed, S.; Fu, C.Y.; Berg, A.C. SSD: Single Shot MultiBox Detector. In Proceedings of the ECCV 2016, Amsterdam, The Netherlands, 8–16 October 2016.
5. Lin, T.Y.; Goyal, P.; Girshick, R.B.; He, K.; Dollár, P. Focal Loss for Dense Object Detection. In Proceedings of the IEEE International Conference on Computer Vision 2017, Venice, Italy, 22–29 October 2017.
6. Xia, G.S.; Bai, X.; Ding, J.; Zhu, Z.; Belongie, S.; Luo, J.; Datcu, M.; Pelillo, M.; Zhang, L. DOTA: A large-scale dataset for object detection in aerial images. In Proceedings of the IEEE Conference on Computer Vision and Pattern Recognition, Salt Lake City, UT, USA, 18–22 June 2018; pp. 3974–3983.
7. Ding, J.; Xue, N.; Xia, G.S.; Bai, X.; Yang, W.; Yang, M.Y.; Belongie, S.; Luo, J.; Datcu, M.; Pelillo, M.; et al. Object detection in aerial images: A large-scale benchmark and challenges. *arXiv* **2021**, arXiv:2102.12219.
8. Wu, J.; Pan, Z.; Lei, B.; Hu, Y. LR-TSDet: Towards Tiny Ship Detection in Low-Resolution Remote Sensing Images. *Remote Sens.* **2021**, *13*, 3890. [CrossRef]
9. Alibakhshikenari, M.; Virdee, B.S.; Althuwayb, A.A.; Aïssa, S.; See, C.H.; Abd-Alhameed, R.A.; Falcone, F.; Limiti, E. Study on on-chip antenna design based on metamaterial-inspired and substrate-integrated waveguide properties for millimetre-wave and THz integrated-circuit applications. *J. Infrared. Millim. Terahertz Waves* **2021**, *42*, 17–28. [CrossRef]
10. Althuwayb, A.A. On-chip antenna design using the concepts of metamaterial and SIW principles applicable to terahertz integrated circuits operating over 0.6–0.622 THz. *Int. J. Antennas Propag.* **2020**, *2020*, 6653095. [CrossRef]
11. Shirkolaei, M.M.; Jafari, M. A new class of wideband microstrip falcate patch antennas with reconfigurable capability at circular-polarization. *Microw. Opt. Technol. Lett.* **2020**, *62*, 3922–3927. [CrossRef]
12. Jiang, Y.; Zhu, X.; Wang, X.; Yang, S.; Li, W.; Wang, H.; Fu, P.; Luo, Z. R 2 cnn: Rotational region cnn for arbitrarily-oriented scene text detection. In Proceedings of the 2018 24th International Conference on Pattern Recognition, Beijing, China , 20–24 August 2018; pp. 3610–3615.
13. Ding, J.; Xue, N.; Long, Y.; Xia, G.S.; Lu, Q. Learning roi transformer for oriented object detection in aerial images. In Proceedings of the IEEE/CVF Conference on Computer Vision and Pattern Recognition, Long Beach, CA, USA, 16–17 June 2019; pp. 2849–2858.
14. Yang, X.; Yang, J.; Yan, J.; Zhang, Y.; Zhang, T.; Guo, Z.; Sun, X.; Fu, K. Scrdet: Towards more robust detection for small, cluttered and rotated objects. In Proceedings of the IEEE/CVF Conference on Computer Vision and Pattern Recognition, Long Beach, CA, USA, 16–17 June 2019 pp. 8232–8241.
15. Han, J.; Ding, J.; Xue, N.; Xia, G.S. Redet: A rotation-equivariant detector for aerial object detection. In Proceedings of the IEEE/CVF Conference on Computer Vision and Pattern Recognition, Nashville, TN, USA, 20–25 June 2021; pp. 2786–2795.

16. Xie, X.; Cheng, G.; Wang, J.; Yao, X.; Han, J. Oriented R-CNN for Object Detection. *arXiv* **2021**, arXiv:2108.05699.
17. Ma, J.; Shao, W.; Ye, H.; Wang, L.; Wang, H.; Zheng, Y.; Xue, X. Arbitrary-oriented scene text detection via rotation proposals. *IEEE Trans. Multimed.* **2018**, *20*, 3111–3122. [CrossRef]
18. Zhang, Z.; Guo, W.; Zhu, S.; Yu, W. Toward arbitrary-oriented ship detection with rotated region proposal and discrimination networks. *IEEE Geosci. Remote Sens. Lett.* **2018**, *15*, 1745–1749. [CrossRef]
19. Yang, X.; Yan, J. Arbitrary-oriented object detection with circular smooth label. In Proceedings of the European Conference on Computer Vision, Glasgow, UK, 23–28 August 2020; pp. 677–694.
20. Yang, X.; Liu, Q.; Yan, J.; Li, A.; Zhang, Z.; Yu, G. R3det: Refined single-stage detector with feature refinement for rotating object. *arXiv* **2019**, arXiv:1908.05612.
21. Han, J.; Ding, J.; Li, J.; Xia, G.S. Align deep features for oriented object detection. *IEEE Trans. Geosci. Remote Sens.* **2021**, *60*, 1–11. [CrossRef]
22. Yang, X.; Hou, L.; Zhou, Y.; Wang, W.; Yan, J. Dense Label Encoding for Boundary Discontinuity Free Rotation Detection. In Proceedings of the CVPR 2021, Nashville, TN, USA, 20–25 June 2021.
23. Yang, X.; Yan, J.; Yang, X.; Tang, J.; Liao, W.; He, T. Scrdet++: Detecting small, cluttered and rotated objects via instance-level feature denoising and rotation loss smoothing. *arXiv* **2020**, arXiv:2004.13316.
24. Azimi, S.M.; Vig, E.; Bahmanyar, R.; Körner, M.; Reinartz, P. Towards multi-class object detection in unconstrained remote sensing imagery. In Proceedings of the Asian Conference on Computer Vision, Perth, Australia, 2–6 December 2018; pp. 150–165.
25. Qian, W.; Yang, X.; Peng, S.; Guo, Y.; Yan, J. Learning modulated loss for rotated object detection. *arXiv* **2019**, arXiv:1911.08299.
26. Xu, Y.; Fu, M.; Wang, Q.; Wang, Y.; Chen, K.; Xia, G.S.; Bai, X. Gliding vertex on the horizontal bounding box for multi-oriented object detection. *IEEE Trans. Pattern Anal. Mach. Intell.* **2020**, *43*, 1452–1459. [CrossRef] [PubMed]
27. Yi, J.; Wu, P.; Liu, B.; Huang, Q.; Qu, H.; Metaxas, D. Oriented object detection in aerial images with box boundary-aware vectors. In Proceedings of the IEEE/CVF Winter Conference on Applications of Computer Vision, Waikoloa, HI, USA, 3–8 January 2021; pp. 2150–2159.
28. Bradski, G. The OpenCV Library. *Dr. Dobb'S J. Softw. Tools* **2000**, *25*, 120–123.
29. Girshick, R.; Donahue, J.; Darrell, T.; Malik, J. Rich feature hierarchies for accurate object detection and semantic segmentation. In Proceedings of the IEEE Conference on Computer Vision and Pattern Recognition (CVPR), Columbus, OH, USA, 23–28 June 2014; pp. 580–587.
30. Girshick, R. Fast r-cnn. In Proceedings of the IEEE International Conference on Computer Vision (ICCV), Santiago, Chile, 7–13 December 2015; pp. 1440–1448.
31. Lin, T.Y.; Dollár, P.; Girshick, R.; He, K.; Hariharan, B.; Belongie, S. Feature pyramid networks for object detection. In Proceedings of the IEEE Conference on Computer Vision and Pattern Recognition (CVPR) 2017, Honolulu, HI, USA, 21–26 July 2017; pp. 2117–2125.
32. Cai, Z.; Vasconcelos, N. Cascade r-cnn: Delving into high quality object detection. In Proceedings of the IEEE Conference on Computer Vision and Pattern Recognition (CVPR), Salt Lake City, UT, USA, 18–22 June 2018; pp. 6154–6162.
33. Hu, H.; Gu, J.; Zhang, Z.; Dai, J.; Wei, Y. Relation networks for object detection. In Proceedings of the IEEE Conference on Computer Vision and Pattern Recognition (CVPR) 2018, Salt Lake City, UT, USA, 18–22 June 2018; pp. 3588–3597.
34. Law, H.; Deng, J. Cornernet: Detecting objects as paired keypoints. In Proceedings of the European Conference on Computer Vision (ECCV), Munich, Germanym 8–14 September 2018; pp. 734–750.
35. Tian, Z.; Shen, C.; Chen, H.; He, T. Fcos: Fully convolutional one-stage object detection. In Proceedings of the IEEE International Conference on Computer Vision (ICCV), Seoul, Korea, 27 October–2 November 2019; pp. 9627–9636.
36. Duan, K.; Bai, S.; Xie, L.; Qi, H.; Huang, Q.; Tian, Q. Centernet: Keypoint triplets for object detection. In Proceedings of the International Conference on Computer Vision (ICCV), Seoul, Korea, 27 October–2 November 2019; pp. 6569–6578.
37. Yang, Z.; Liu, S.; Hu, H.; Wang, L.; Lin, S. Reppoints: Point set representation for object detection. In Proceedings of the IEEE/CVF International Conference on Computer Vision, Seoul, Korea, 27 October–2 November 2019; pp. 9657–9666.
38. Wang, H.; Zhang, X.; Zhou, L.; Lu, X.; Wang, C. Intersection detection algorithm based on hybrid bounding box for geological modeling with faults. *IEEE Access* **2020**, *8*, 29538–29546. [CrossRef]
39. Premachandra, H.W.H.; Yamada, M.; Premachandra, C.; Kawanaka, H. Low-Computational-Cost Algorithm for Inclination Correction of Independent Handwritten Digits on Microcontrollers. *Electronics* **2022**, *11*, 1073. [CrossRef]
40. Kong, T.; Sun, F.; Liu, H.; Jiang, Y.; Li, L.; Shi, J. FoveaBox: Beyond Anchor-based Object Detector. *IEEE Trans. Image Process.* **2020**, *29*, 7389–7398. [CrossRef]
41. Zhang, S.; Chi, C.; Yao, Y.; Lei, Z.; Li, S.Z. Bridging the gap between anchor-based and anchor-free detection via adaptive training sample selection. In Proceedings of the CVPR 2020, Seattle, WA, USA, 14–19 June 2020; pp. 9759–9768.
42. Kim, K.; Lee, H.S. Probabilistic Anchor Assignment with IoU Prediction for Object Detection. In Proceedings of the ECCV 2020, Glasgow, UK, 23–28 August 2020.
43. Qiu, H.; Ma, Y.; Li, Z.; Liu, S.; Sun, J. BorderDet: Border Feature for Dense Object Detection. In Proceedings of the ECCV 2020, Glasgow, UK, 23–28 August 2020; pp. 549–564.
44. Li, X.; Wang, W.; Wu, L.; Chen, S.; Hu, X.; Li, J.; Tang, J.; Yang, J. Generalized Focal Loss: Learning Qualified and Distributed Bounding Boxes for Dense Object Detection. In Proceedings of the NeurIPS 2020, Online, 6–12 December 2020.

45. Carion, N.; Massa, F.; Synnaeve, G.; Usunier, N.; Kirillov, A.; Zagoruyko, S. End-to-End Object Detection with Transformers. In Proceedings of the ECCV 2020, Glasgow, UK, 23–28 August 2020.
46. Stewart, R.; Andriluka, M.; Ng, A.Y. End-to-end people detection in crowded scenes. In Proceedings of the CVPR 2016, Las Vegas, NV, USA, 27–30 June 2016.
47. Zhu, X.; Su, W.; Lu, L.; Li, B.; Wang, X.; Dai, J. Deformable DETR: Deformable Transformers for End-to-End Object Detection. In Proceedings of the ICLR 2020, Addis Ababa, Ethiopia, 26–30 April 2020.
48. Wang, J.; Song, L.; Li, Z.; Sun, H.; Sun, J.; Zheng, N. End-to-End Object Detection with Fully Convolutional Network. In Proceedings of the CVPR 2021, Online, 19–25 June 2021.
49. Zhou, Q.; Yu, C.; Shen, C.; Wang, Z.; Li, H. Object Detection Made Simpler by Eliminating Heuristic NMS. *arXiv* **2021**, arXiv:2101.11782.
50. Yang, X.; Yan, J.; Ming, Q.; Wang, W.; Zhang, X.; Tian, Q. Rethinking Rotated Object Detection with Gaussian Wasserstein Distance Loss. *arXiv* **2021**, arXiv:2101.11952.2021.
51. Yang, X.; Yang, X.; Yang, J.; Ming, Q.; Wang, W.; Tian, Q.; Yan, J. Learning High-Precision Bounding Box for Rotated Object Detection via Kullback-Leibler Divergence. In Proceedings of the 2021 Annual Conference on Neural Information Processing Systems, Online, 6–14 December 2021.
52. Zhang, L.; Wang, H.; Wang, L.; Pan, C.; Liu, Q.; Wang, X. Constraint Loss for Rotated Object Detection in Remote Sensing Images. *Remote Sens.* **2021**, *13*, 4291. [CrossRef]
53. Liao, M.; Shi, B.; Bai, X. Textboxes++: A single-shot oriented scene text detector. *IEEE Trans. Image Process.* **2018**, *27*, 3676–3690. [CrossRef] [PubMed]
54. Wu, F.; He, J.; Zhou, G.; Li, H.; Liu, Y.; Sui, X. Improved Oriented Object Detection in Remote Sensing Images Based on a Three-Point Regression Method. *Remote Sens.* **2021**, *13*, 4517. [CrossRef]
55. Guo, Z.; Liu, C.; Zhang, X.; Jiao, J.; Ji, X.; Ye, Q. Beyond Bounding-Box: Convex-Hull Feature Adaptation for Oriented and Densely Packed Object Detection. In Proceedings of the CVPR 2021, Online, 19–25 June 2021; pp. 8792–8801.
56. Dai, J.; Qi, H.; Xiong, Y.; Li, Y.; Zhang, G.; Hu, H.; Wei, Y. Deformable convolutional networks. In Proceedings of the IEEE International Conference on Computer Vision 2017, Venice, Italy, 22–29 October 2017; pp. 764–773.
57. Jarvis, R.A. On the identification of the convex hull of a finite set of points in the plane. *Inf. Process. Lett.* **1973**, *2*, 18–21. [CrossRef]
58. Gupta, A.; Dollár, P.; Girshick, R.B. LVIS: A Dataset for Large Vocabulary Instance Segmentation. In Proceedings of the CVPR 2019, Long Beach, CA, USA, 15–25 June 2019.
59. Liu, Z.; Yuan, L.; Weng, L.; Yang, Y. A high resolution optical satellite image dataset for ship recognition and some new baselines. In Proceedings of the 2017 ICPRAM, Porto, Portugal, 24–26 February 2017; Volume 2, pp. 324–331.
60. Zhu, H.; Chen, X.; Dai, W.; Fu, K.; Ye, Q.; Jiao, J. Orientation robust object detection in aerial images using deep convolutional neural network. In Proceedings of the 2015 IEEE International Conference on Image Processing (ICIP), Quebec, QC, Canada, 27–30 September 2015; pp. 3735–3739.
61. Chen, K.; Wang, J.; Pang, J.; Cao, Y.; Xiong, Y.; Li, X.; Sun, S.; Feng, W.; Liu, Z.; Xu, J.; et al. MMDetection: Open MMLab Detection Toolbox and Benchmark. *arXiv* **2019**, arXiv:1906.07155.
62. Kingma, D.P.; Ba, J. Adam: A Method for Stochastic Optimization. In Proceedings of the ICLR 2015, San Diego, CA, USA, 7–9 May 2015.
63. Wei, H.; Zhang, Y.; Chang, Z.; Li, H.; Wang, H.; Sun, X. Oriented objects as pairs of middle lines. *ISPRS J. Photogramm. Remote Sens.* **2020**, *169*, 268–279. [CrossRef]
64. Pan, X.; Ren, Y.; Sheng, K.; Dong, W.; Yuan, H.; Guo, X.; Ma, C.; Xu, C. Dynamic refinement network for oriented and densely packed object detection. In Proceedings of the IEEE/CVF Conference on Computer Vision and Pattern Recognition 2020, Seattle, WA, USA, 13–19 June 2020; pp. 11207–11216.
65. Wang, J.; Yang, W.; Li, H.C.; Zhang, H.; Xia, G.S. Learning center probability map for detecting objects in aerial images. *IEEE Trans. Geosci. Remote Sens.* **2020**, *59*, 4307–4323. [CrossRef]
66. Liu, Z.; Lin, Y.; Cao, Y.; Hu, H.; Wei, Y.; Zhang, Z.; Lin, S.; Guo, B. Swin transformer: Hierarchical vision transformer using shifted windows. *arXiv* **2021**, arXiv:2103.14030.
67. Chen, K.; Pang, J.; Wang, J.; Xiong, Y.; Li, X.; Sun, S.; Feng, W.; Liu, Z.; Shi, J.; Ouyang, W.; et al. Hybrid task cascade for instance segmentation. In Proceedings of the IEEE/CVF Conference on Computer Vision and Pattern Recognition 2019, Long Beach, CA, USA, 15–20 June 2019; pp. 4974–4983.
68. Li, C.; Xu, C.; Cui, Z.; Wang, D.; Jie, Z.; Zhang, T.; Yang, J. Learning object-wise semantic representation for detection in remote sensing imagery. In Proceedings of the IEEE/CVF Conference on Computer Vision and Pattern Recognition 2019, Long Beach, CA, USA, 15–20 June 2019; pp. 20–27.
69. Liu, L.; Pan, Z.; Lei, B. Learning a rotation invariant detector with rotatable bounding box. *arXiv* **2017**, arXiv:1711.09405.
70. Liao, M.; Zhu, Z.; Shi, B.; Xia, G.s.; Bai, X. Rotation-sensitive regression for oriented scene text detection. In Proceedings of the IEEE Conference on Computer Vision and Pattern Recognition 2018, Salt Lake City, UT, USA, 18–23 June 2018; pp. 5909–5918.
71. Redmon, J.; Farhadi, A. Yolov3: An incremental improvement. *arXiv* **2018**, arXiv:1804.02767.
72. Ming, Q.; Zhou, Z.; Miao, L.; Zhang, H.; Li, L. Dynamic anchor learning for arbitrary-oriented object detection. *arXiv* **2020**, arXiv:2012.04150.

remote sensing

MDPI

Article

LSNet: Learned Sampling Network for 3D Object Detection from Point Clouds

Mingming Wang [1], Qingkui Chen [1,2,*] and Zhibing Fu [1]

[1] Department of Systems Science, Business School, University of Shanghai for Science and Technology, Shanghai 200093, China; 171310068@st.usst.edu.cn (M.W.); zbfu@usst.edu.cn (Z.F.)

[2] Department of Computer Science and Engineering, School of Optical-Electrical and Computer Engineering, University of Shanghai for Science and Technology, Shanghai 200093, China

* Correspondence: chenqingkui@usst.edu.cn; Tel.: +86-131-2238-1881

Abstract: The3D object detection of LiDAR point cloud data has generated widespread discussion and implementation in recent years. In this paper, we concentrate on exploring the sampling method of point-based 3D object detection in autonomous driving scenarios, a process which attempts to reduce expenditure by reaching sufficient accuracy using fewer selected points. FPS (farthest point sampling), the most used sampling method, works poorly in small sampling size cases, and, limited by the massive points, some newly proposed sampling methods using deep learning are not suitable for autonomous driving scenarios. To address these issues, we propose the learned sampling network (LSNet), a single-stage 3D object detection network containing an LS module that can sample important points through deep learning. This advanced approach can sample points with a task-specific focus while also being differentiable. Additionally, the LS module is streamlined for computational efficiency and transferability to replace more primitive sampling methods in other point-based networks. To reduce the issue of the high repetition rates of sampled points, a sampling loss algorithm was developed. The LS module was validated with the KITTI dataset and outperformed the other sampling methods, such as FPS and F-FPS (FPS based on feature distance). Finally, LSNet achieves acceptable accuracy with only 128 sampled points and shows promising results when the number of sampled points is small, yielding up to a 60% improvement against competing methods with eight sampled points.

Keywords: 3D object detection; point cloud; sampling; single-stage

Citation: Wang, M.; Chen, Q.; Fu, Z. LSNet: Learned Sampling Network for 3D Object Detection from Point Clouds. *Remote Sens.* **2022**, *14*, 1539. https://doi.org/10.3390/rs14071539

Academic Editors: Fahimeh Farahnakian, Jukka Heikkonen and Pouya Jafarzadeh

Received: 14 February 2022
Accepted: 19 March 2022
Published: 23 March 2022

1. Introduction

Three-dimensional data captured by LiDAR and the RGB-D camera have applications in various fields such as autonomous driving, virtual reality, and robotics. Many deep learning techniques have been applied to point cloud tasks such as point cloud classification, segmentation, completion, and generation. In this paper, we focus on 3D object detection of autonomous driving.

In recent years, 3D object detection of autonomous driving has been a major focus. Refs. [1–4] fuse point clouds and images together to detect 3D objects. In this paper, we focus on the processing of point clouds. With a point cloud captured by LiDAR, the different methodologies to approach this issue can be classified as view-based, point-based, and voxel-based methods. Additionally, some methods utilize the advantages of both the point-based method and voxel-based method to enable both high-quality 3D proposal generation and flexible receptive fields to improve 3D detection performance. With the massive number of raw points in a point cloud, it is not trivial to downsample the point cloud data efficiently and reserve as many meaningful points as possible. With this said, the sampling approaches themselves have received comparatively less attention.

View-based methods project the 3D point cloud data into different 2D views so that mature 2D convolution techniques can be applied to solve the problem efficiently. The downsampling process is reflected in both the pooling process and the step size of convolution. Voxel-based methods view the 3D point cloud space as a cube and divide it into voxels. This means the size of the sampled point subset can be controlled by the length, width, and height of each voxel, while the step size of 3D convolution and 3D pooling can also downsample the data. Additionally, point-based methods take the raw point cloud as input and generate predictions based on each point. This causes the point-based methods to suffer from a heavy computational burden due to the need to process so much data. Ref. [5] addressed this issue and proposed an efficient and lightweight neural architecture for semantic segmentation task of large-scale point clouds. Ref. [6] introduced kernel point convolution to improve the efficiency of feature extraction in point-based methods. Hence, developing an appropriate sampling strategy has become a crucial issue.

In a point-based model, one naive approach is to sample points randomly. The most widely used method is furthest-point-sampling (FPS), which selects a group of points that are farthest apart from each other based on their 3D Euclidean distance. However, there is one sampling approach that first voxelizes the whole 3D space and only preserves one point in each voxel. KPConv [6] used grid subsampling and chose barycenters of the original input points contained in all non-empty grid cells. The 3DSSD [7] utilizes the F-FPS and FS methods. F-FPS samples points based on feature distance instead of Euclidean distance in FPS, while FS is the fusion of D-FPS(FPS) and F-FPS. Crucially, these sampling strategies are non-learned approaches and cannot preserve important points when the sampling size is small, leading to poor performance. Recently there have been a few learned approaches. Ref. [8–10] proposed learning-based methods, but they are limited to simple datasets such as ModelNet40 [11] and are not suitable to autonomous driving scenarios.

In conclusion, the small sampling size can save the cost of both memory and computation. However, existing sampling approaches either perform poorly in small sampling size cases or are not suitable for autonomous driving scenarios. Motivated by these issues, in this paper, we present a novel architecture named LSNet, shown in Figure which contains a learning-based sampling module and works extraordinarily well in low sampling size cases. The sampling process faces two main challenges. The first is how to allow backpropagation and the second is how to avoid excessive time consumption. The learned sampling module of LSNet is a deep learning network that must be kept streamlined to avoid the issue of excessive computation time because if the sampling network is too complex, the resources and time invested would render the downsampling strategy moot. The LS module outputs a one-hot-like sampling matrix and uses matrix multiplication to create the sampling subset of points. Since the sampling process itself is discrete and is not trainable, we instead adjust the grouping method in the SA module and use the τ-based softmax function in the LS module to make it differentiable. Additionally, we add random relaxation to the sampling matrix in the early part of the training with the degree of relaxation decaying to zero along the training step. However, the sampling matrix of the LS module cannot ensure that the sampled points will not be redundant. To solve this issue, a new sampling loss was proposed. Finally, of major importance is that the entire LSNet model is end-to-end trainable.

We evaluate the model on the widely used KITTI [12] dataset. To verify the effectiveness of the LS module, we compare it with random sampling, D-FPS, F-FPS, and FS. The results of these comparisons show that the LS module method outperformed the other methods and it was close to the state-of-the-art 3D detectors with 512 sampled points. Specifically, LSNet with 128 sampled points has relatively little accuracy loss and achieves acceptable accuracy. It is also shown that the fewer the sampling points, the better the improvement. Figure 1 shows the results of different sampling methods with only eight sampled points. Unlike other sampling methods, such as FPS, this learning-based sampling approach utilizes semantically high-level representations, which is reflected in the fact that the points sampled by the LS module are distributed around the target objects. Furthermore, it pays more attention to regions of interest and is less sensitive to outliers.

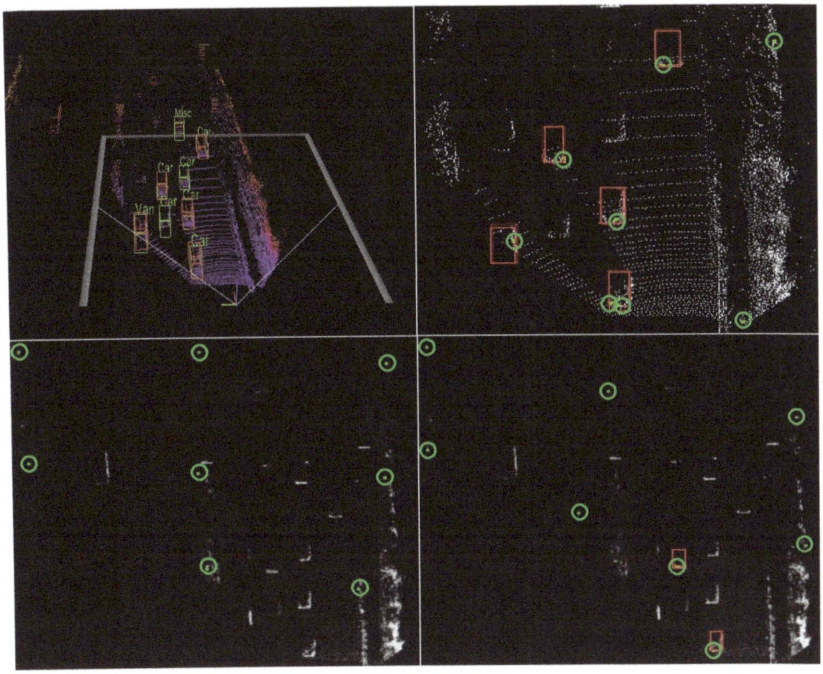

Figure 1. The results of different sampling methods processing the same eight sampled points in the same scene. The **top-left** picture is the 3D object detection results of our model and the green box shows the ground truth, while the red box shows the detection of our model with eight points. The remaining three pictures demonstrate the points before sampling (4096 white points) and the points after sampling (eight green points inside green circle) in the bird's eye view (BEV). **Top-right**: sampling results of the LS module, zoomed in and cropped for better illustration since there are no outliers, unlike the other two pictures. **Bottom-left**: sampling results of D-FPS (FPS). **Bottom-right**: sampling results of F-FPS.

In addition, the proposed LS module can be viewed as a complete standalone module. This means it can be attached to another model to sample points flexibly. Following the results of the end-to-end training is the study of the multi-stage training process. Based on the results from the experiment evaluation, when given a trained task network with limited training time, the number of sampled points can be reduced by half. This can be accomplished quickly with the only cost an affordable loss of accuracy.

To summarize, the key contribution of the proposed model lies in the following four points:

- First, the proposed LSNet, a point-based 3D object detector with a novel sampling approach (LS module), can be trained end-to-end to sample points with consideration for a specific task. The approach nears parity with state-of-the-art 3D detectors when using 512 sampling points while still achieving acceptable performance with only 128 sampling points.
- Second, to enable backpropagation of the sampling process and make it differentiable, the vanilla SA module's grouping method was adjusted and the τ-based softmax function was used to approximate one-hot-encoding while also applying random relaxation to the sampling matrix to boost the performance.
- Third, to address the issue of duplicate sampling, a new sampling loss technique was used. This resulted in a significant increase of unique samples as well as improved accuracy.

- Fourth, the LS module can be flexibly transferred and inserted into other point-based detection models to reduce the number of points needed. Of significant importance is the fact that the multi-stage training method enables the LS module to be easily attached to other trained models, while reducing the necessary number of points with relatively little training time.

2. Related Work

In this section, recent advances in 3D object detection of autonomous driving are reviewed, after which some of the pioneer works related to point cloud sampling methods are examined.

For the purposes of 3D object detection, recent 3D object detection models based on LiDAR point clouds can be roughly categorized into view-based methods, voxel-based methods, point-based methods, and integrated methods.

With the rapid development of computer vision, much effort has been devoted to detecting objects from images. In the service of this effort, representing 3D point clouds as 2D views is helpful as it makes it easy to apply off-the-shelf and mature computer vision skills to the problem. The most used views are front view ([13–15]), bird's eye view ([1,3,16–18]), and range view ([19,20]). However, these methods cannot localize 3D objects accurately due to the loss of information.

In the voxel-based methods ([21–26]), the point clouds are divided into 3D voxels equally to be processed by 3D CNN. Due to the massive amount of empty voxels, 3D sparse convolution [23,27] is introduced for efficient computation. For example, ref. [22] used 3D sparse convolutions through the entire network. VoxelNet ([24]), SECOND ([23]), and PointPillars ([25]) learn the representation of each voxel with the voxel feature encoding (VFE) layer. TANet ([26]) learns a more discriminative and robust representation for each voxel through triple attention (channel-wise, point-wise, and voxel-wise attention). Then, the 3D bounding boxes are computed by a region proposal network based on the learned voxel representation.

Point-based methods are mostly based on the PointNet series [28,29]. The set abstraction operation proposed by PointNet is widely used in point-based approaches [7]. PointRCNN [30] generates 3D proposals directly from the whole point clouds. Qi, Litany, He, and Guibas proposed VoteNet [31], the Hough voting strategy for better object feature grouping. The work in [32] introduces StarNet, a flexible, local point-based object detector. The work in [33] proposed PointGNN, a new object detection approach using a graph neural network on the point cloud.

PV-RCNN [34] takes advantages of both the voxel-based and point-based methods for 3D point-cloud feature learning, leading to improved performance of 3D object detection with manageable memory consumption. The work in [35] combines both voxel-based CNN and point-based shared-MLP for efficient point cloud feature learning.

In relation to point clouds sampling, farthest point sampling (FPS) is widely used in many models ([7,29,31,33]) to handle the downsampling issue inherent in using point clouds. Ref. [36] applied graph-based filters to extract features. Haar-like low/highpass graph filters are used to preserve specific points efficiently, and 3DSSD [7] proposed F-FPS and FS. According to [8], the proposed simplification network, termed S-Net, is the first learned point clouds sampling approach. After this, SampleNet [9] further improved the performance with sampled point clouds to classify and reconstruct the tasks based on it. Ref. [10] used Gumbel subset sampling to replace FPS to improve its accuracy.

3. Methods

3.1. Problem Formulation

Consider a general matrix representation of a point cloud with N points and K attributes,

$$\mathrm{P} = \begin{bmatrix} \mathbf{f}_1 & \mathbf{f}_2 & \dots & \mathbf{f}_K \end{bmatrix} = \begin{bmatrix} \mathbf{p}_1^T \\ \mathbf{p}_2^T \\ \vdots \\ \mathbf{p}_N^T \end{bmatrix} \in \mathbb{R}^{N \times K}, \tag{1}$$

where $\mathbf{f}_i \in \mathbb{R}^N$ denotes the ith attribute and $\mathbf{p}_j \in \mathbb{R}^K$ denotes the jth point. Specifically, the actual number of K varies according to the output feature size of each layer. The attributes contain 3D coordinates and context features. The context features can be the original input features or the extracted features. For instance, the input feature of velodyne LiDAR is the one-dimensional laser reflection intensity, and it is the three-dimensional RGB colors of the RGB-D camera. Additionally, the extracted features come from the neural network layers. To distinguish 3D coordinates from the other attributes, we store them in the first three columns of P and call that submatrix $\mathrm{P}_c \in \mathbb{R}^{N \times 3}$, while storing the rest in the last $K-3$ columns of P and call that submatrix $\mathrm{P}_o \in \mathbb{R}^{N \times (K-3)}$.

The target of the LS module in Figure 2 is to create a sampling matrix,

$$\mathrm{S} = \begin{bmatrix} \mathbf{p}'_1 & \mathbf{p}'_2 & \dots & \mathbf{p}'_{N'} \end{bmatrix} = \begin{bmatrix} \mathbf{p}_1^T \\ \mathbf{p}_2^T \\ \vdots \\ \mathbf{p}_N^T \end{bmatrix} \in \mathbb{R}^{N \times N'}, \tag{2}$$

where $\mathbf{p}'_i \in \mathbb{R}^N$ represents the ith sampled point and $\mathbf{p}_j \in \mathbb{R}^{N'}$ represents the jth point before sampling. N is the original points size and N' is the sampled points size. This matrix is used to select $N'(N' < N)$ points from the original points. Let the sampled point cloud be $\mathrm{P}_{N'} \in \mathbb{R}^{N' \times K}$ and the original point cloud be $\mathrm{P}_N \in \mathbb{R}^{N \times K}$. To achieve this, column $\mathbf{p}'_i$ should be a one-hot vector, defined as

$$\mathbf{p}'_{i,j} = \begin{cases} 1, & j = \text{the index of selected point in N original points;} \\ 0, & \text{otherwise.} \end{cases} \tag{3}$$

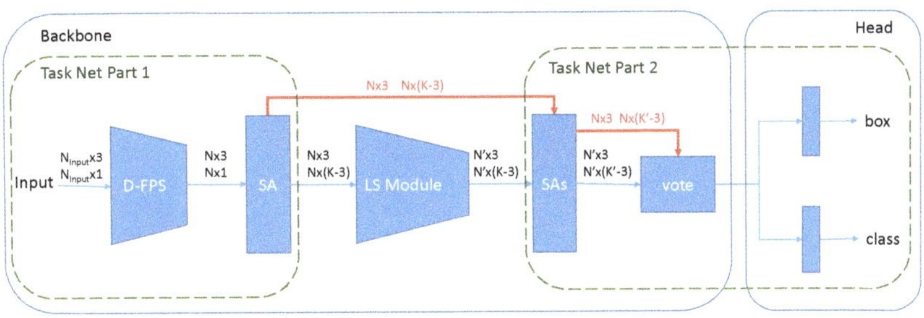

Figure 2. The overall architecture of the proposed LSNet. The input data of each module contain coordinates data ($N \times 3$) and feature data ($N \times K$). The raw coordinates information is kept for point grouping and feature extraction. The blue arrows represent the main data flow of LSNet, while the red arrows demonstrate the data flow in the multi-stage training method when the LS module is skipped. There are two ways to split the entire network for a concise model description in the paper. One is dividing the network into a feature extraction backbone and a detection head. The other is dividing the network into a task network and a sampling network (LS module).

There should be only one original point selected in each column $\mathbf{p}'_i$, defined as

$$\sum_{j=1}^{N} \left(\mathbf{p}'_{i,j} \right) = 1. \tag{4}$$

With the sampling matrix S and original point cloud P_N, we can acquire the new sampling point cloud $P_{N'}$ through matrix multiplication:

$$P_{N'} = S^T \otimes P_N, \qquad P_{N'} \in \mathbb{R}^{N' \times K}; S^T \in \mathbb{R}^{N' \times N}; P_N \in \mathbb{R}^{N \times K}. \tag{5}$$

The invariance properties of the sampling approach are pivotal. Since the intrinsic distribution of 3D points remains the same when we permutate, shift, and rotate a point cloud, the outputs of the sampling strategy are also not expected to be changed. These invariance properties will be analyzed on the coordinate matrix P_c alone because the features of each point (P_o) will not be influenced by them.

Definition 1. *A sampling strategy is permutation-invariant when, given input* $P_N \in \mathbb{R}^{N \times K}$, $\forall$ *permutation matrix* M_p *of size N,*

$$\text{SAMPLE}(M_p \cdot P_N) = \text{SAMPLE}(P_N). \tag{6}$$

Definition 2. *A sampling strategy is shift-invariant when, given input* $P_N \in \mathbb{R}^{N \times K}$, $\forall$ *shift matrix* M_s *of size 3,*

$$\text{SAMPLE}(M_s \cdot P_N) = \text{SAMPLE}(P_N). \tag{7}$$

Definition 3. *A sampling strategy is rotation-invariant when, given input* $P_N \in \mathbb{R}^{N \times K}$, $\forall$ *rotation matrix* M_r *of size 3,*

$$\text{SAMPLE}(M_r \cdot P_N) = \text{SAMPLE}(P_N). \tag{8}$$

The softmax function is also permutation-invariant, which is already proved in [10].

Lemma 1. *Given* $A \in \mathbb{R}^{N \times N}$, $\forall$ *permutation matrix* M_p *of size N,*

$$\text{softmax}(M_p A M_p^T) = M_p \, \text{softmax}(A) M_p^T. \tag{9}$$

3.2. Network Architecture

The entire network structure of LSNet is displayed in Figure 2. It is a point-based, single-stage 3D object detection network with a feature extraction backbone and a detection head. The backbone, similar to many other point-based methods [7,29,31,34], uses the multi-scale set abstraction(SA) proposed by PointNet++ [29] to gather neighborhood information and extract features, making it a PointNet-based model as well. Multiple SA modules were stacked to abstract high-level features and enlarge the receptive field. Inspired by VoteNet [31] and 3DSSD [7], a vote layer was added to improve network performance. For downsampling points, the FPS sampling method, i.e., D-FPS in 3DSSD [7], is used to downsample the raw points roughly, while the LS module is used to further sample the points delicately. In addition, there are two 3D detection heads in the proposed model, one for box regression and the other for classification.

In relation to the LiDAR point cloud, the inputs of the model consist of 3D coordinates and 1D laser reflection intensity, i.e., $P_{input} = [P_c \;\; P_o]$, $P_{input} \in \mathbb{R}^{N \times 4}, P_c \in \mathbb{R}^{N \times 3}$, $P_o \in \mathbb{R}^{N \times 1}$. The predicted object in the KITTI 3D object detection dataset can be represented by a 3D bounding box $(c_x, c_y, c_z, h, w, l, \theta)$, including its center, c_x, c_y, c_z, size, h, w, l, and orientation, θ, which indicates the heading angle around the up-axis.

First, FPS based on 3D Euclidean distance is used to sample a subset of the raw points P_{input}. Then, the vanilla multi-scale SA module is applied to extract the low-level features $P_o \in \mathbb{R}^{N \times (K-3)}$, which will be viewed as the inputs of the LS module with their coordinates. Working from these middle features, the LS module generates the sampling point cloud $P_{N'}$ and $P_{N'} \subset P_N$. After several SA modules and a vote layer, the final features are fed into the detection head to predict the box and class of the object. After this, NMS is applied to remove the redundant boxes. Non-maximum suppression (NMS) is a critical post-processing procedure to suppress redundant bounding boxes based on the order of detection confidence, which is widely used in object detection tasks.

According to PointNet++, the SA module has many 1D-convolution-like layers, which are composed of shared-MLP layers. For each point, the SA module groups the surrounding points within a specific radius and uses shared-MLP to extract the features. The box regression head and the classification head are both fully connected (FC) layers.

3.3. LS Module

The traditional sampling approaches are neither differentiable nor task-agnostic. Therefore, they cannot be trained using the loss method. Since the sampling process is discrete, we need to convert it to a continuous issue to smoothly integrate the sampling operation into a neural network. Ref. [8] proposed S-Net and [9] proposed its variant SampleNet to ameliorate this shortcoming. These sampling strategies have several defects. First, they generate new point coordinates, which are not in the subset of the original points. In addition, they can only be placed at the beginning of the total network and the entire model lacks the ability to be trained end-to-end. Another issue is due to the fact that the sampling network extracts features from coordinate inputs, while the task network also extracts features from the raw inputs. This duplicated effort inevitably results in a level of redundant extraction in regard to low-level features. A final issue is that the sampling network is relatively complex and time-consuming. This problem will become more severe as the number of points grows. A sampling process that requires burdensome levels of computation to function defeats the purpose of its application to the issue. In consideration of these issues, the discussed methods are not suitable for autonomous driving tasks.

To overcome such problems, the LS module was developed. As illustrated in Figure 3, the network architecture of the LS module has only a few layers, which keeps the complexity low. Rather than extracting useful features to create a sampling matrix from a fresh start, these features are instead extracted by the task network part 1 and are shared, and the matrix is the output based on them to improve computational efficiency and to avoid the repeated extraction of the underlying features.

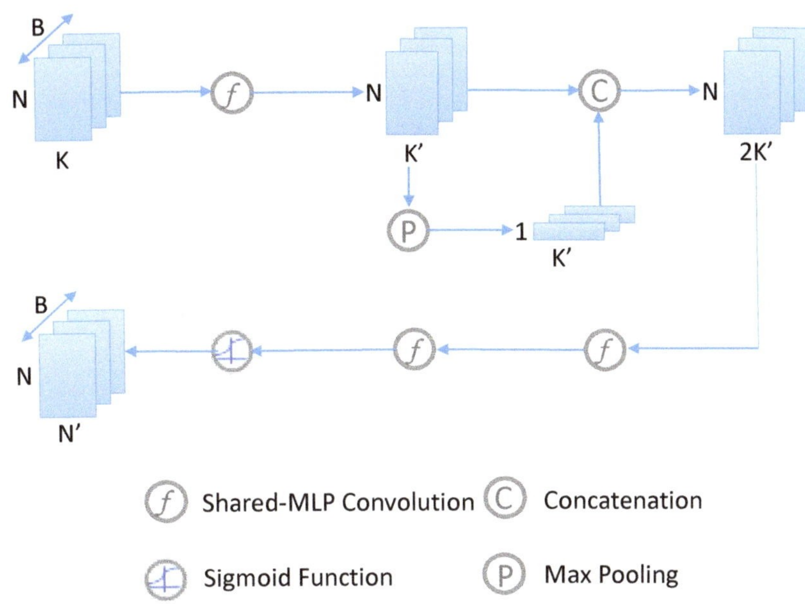

Figure 3. The details of the LS module's network structure, where B is the batch size, N is the points size, and K is the feature size.

The input of the LS module is P_N, which is the subset of the points sampled by FPS with the features extracted by the former SA module. First a shared-MLP convolution layer is applied to obtain the local feature F_{local} of each point,

$$F_{local} = f(P_N|W_1), F_{local} \in \mathbb{R}^{N\times k'}. \tag{10}$$

Function f represents the shared-MLP convolution layer with its weights W. Then, a symmetric feature-wise max pooling operation is used to obtain a global feature vector F_{global},

$$F_{global} = MaxPool(F_{local}) \quad F_{global} \in \mathbb{R}^{1\times k'}. \tag{11}$$

With the global features and the local features, we concatenate them of each point and pass these features to the shared-MLP convolution layers and use the sigmoid function to generate a matrix $\hat{S}$, defined as

$$\hat{S} = f(f(concat(F_{local}, F_{global})|W_2)|W_3), \tag{12}$$

$$\hat{S} = Sigmoid(\hat{S}) \quad \hat{S} \in \mathbb{R}^{N\times N'}. \tag{13}$$

$\hat{S}$ has the same shape as the sampling matrix S. It is the output of the LS module while also being the middle value of S.

To sample data based on P_N, the sampling matrix is further adjusted to S (used in the inference stage) or S′ (used in the training stage). S can be computed as

$$S = \underset{p'_i \in \mathbb{R}^N, i\in[1,N']}{one\_hot\_encoding(argmax(\hat{S}))}, \tag{14}$$

where the *argmax* function and the *one_hot_encoding* function are applied to each column of $\hat{S}$, i.e., $\mathbf{p}'_i$ with the shape of original points size N. Since $\hat{S}$ has N' columns, corresponding to N' sampled points, and each column of S is a one-hot vector, Equation (5) can be used to obtain the final sampled points $P_{N'}$.

However, the *argmax* operation and the *one_hot_encoding* operation are not differentiable, indicating that Equation (14) cannot be used in the training stage to enable backpropagation. Inspired by the Gumbel-softmax trick [10,37,38], softmax is applied to each column of $\hat{S}$ with parameter τ to approximate the *one_hot_encoding* operation. The generated sampling matrix is called S′,

$$S' = \underset{p'_i \in \mathbb{R}^N, i\in[1,N']}{softmax}(\hat{S}/\tau), \tag{15}$$

where parameter $\tau > 0$ is the annealing temperature, as $\tau \to 0^+$, each column in S′ degenerates into a one-hot distribution such as S. When the distribution of each column in S′ does not degenerate to a one-hot distribution, the features of sampled points $P_{o,N'}$ are not the same as before. $P_{o,N'}$ is computed by the matrix multiplication with S′,

$$P_{o,N'} = S'^T \otimes P_{o,N}, P_{o,N'} \in \mathbb{R}^{N'\times(K-3)}; S'^T \in \mathbb{R}^{N'\times N}; P_{o,N} \in \mathbb{R}^{N\times(K-3)}. \tag{16}$$

Nevertheless, it is desirable to keep the coordinates of the sampled points the same as they were previously. So, the *argmax* operation and the *one_hot_encoding* operation are applied to S′ to generate sampling matrix S. Then, the coordinates of the sampled points $P_{c,N'}$ are computed as

$$P_{c,N'} = S^T \otimes P_{c,N}, P_{c,N'} \in \mathbb{R}^{N'\times 3}; S^T \in \mathbb{R}^{N'\times N}; P_{c,N} \in \mathbb{R}^{N\times 3}. \tag{17}$$

Additionally, before Equation (15), a random relaxation trick is employed to further boost the performance of the model, represented as

$$\gamma = r^{\frac{current\_step}{decay\_steps}}, r \in [0,1]; \tag{18}$$

$$\hat{S} = \hat{S} + Random(\gamma), \quad Random(\gamma) \in \mathbb{R}^{N \times N'}, \tag{19}$$

where r is the decay rate and γ is the upper boundary of the random number. Parameter γ is decayed with the training step exponentially and eventually approaches 0 when there is no relaxation.

In actuality, the sampling matrix S' introduces the attention mechanism to the model. Each column of S' indicates the newly generated sampling point's attention on old points. Then, the new features in $P_{o,N'}$ contain the point-wise attention on the old points. Since each column of S is a one-hot distribution, the coordinates of the sampled points $P_{c,N'}$ calculated with S mean its attention is focused on the single old point when it comes to coordinate generation.

In all the above functions, the shared-MLP function f and the $Sigmoid$ function are point-wise operations, while the random relaxation is an element-wise operation. In addition, the $MaxPool$ function operates from the feature dimension and selects the max value of each feature from all points. This means these functions do not change the permutation equivariance of the LS module. Separate from these functions, Lemma 1 shows the permutation invariance of $softmax$. Thus, our proposed sampling method is permutation-invariant (Definition 1).

3.4. SA Module

The set abstraction procedure proposed by Qi et al., PointNet++, which is widely used in many point-based models, can be roughly divided into a sampling layer, grouping layer, and a PointNet layer. To obtain better coverage of the entire point set, PointNet++ uses FPS to select N' grouping center points from N input points in the sampling layer. Based on the coordinates of these center points, the model will gather Q points within a specified radius, contributing to a group set. In relation to the PointNet layer, a mini-PointNet (composed of multiple shared-MLP layers) is used to encode the local region patterns of each group into feature vectors. In this paper, the grouping layer and the PointNet layer are retained in our SA module. The LS module is used instead of FPS to generate a subset of points serving as the grouping center points, while the grouping layer is adjusted to fit our learned sampling model.

As shown in Figure 4, multi-scale grouping is used to group the points of each center point with different scales. Features at different scales are learned by different shared-MLP layers and then concatenated to form a multi-scale feature. If the points sampled by the LS module are viewed as ball centers and perform the ball grouping process on the original dataset N, similar to PointNet++, the entire network cannot be trained through backpropagation since the outputs of the LS module are not passed to the following network explicitly. Two methods have been developed to address this issue. The first method is to ignore the old dataset before sampling and instead use the newly sampled dataset for both the grouping center points and grouping pool. The other possibility is to use the new sampled dataset as grouping center point and replace the points of the old dataset with the new points in their corresponding positions. Using this method, it is possible to concatenate the features of the new sampled points to each group and pass the outputs (new points) of the LS module to the network.

Within each group, the local relative location of each point from the center point is used to replace the absolute location P_c. Importantly, the extracted features P_o will not be affected by shifting or rotating the point cloud. So, it follows that the inputs to the LS module remain the same despite the shift and rotation operations, which also indicates that the proposed sampling method is shift-invariant (Definition 2) and rotation-invariant (Definition 3).

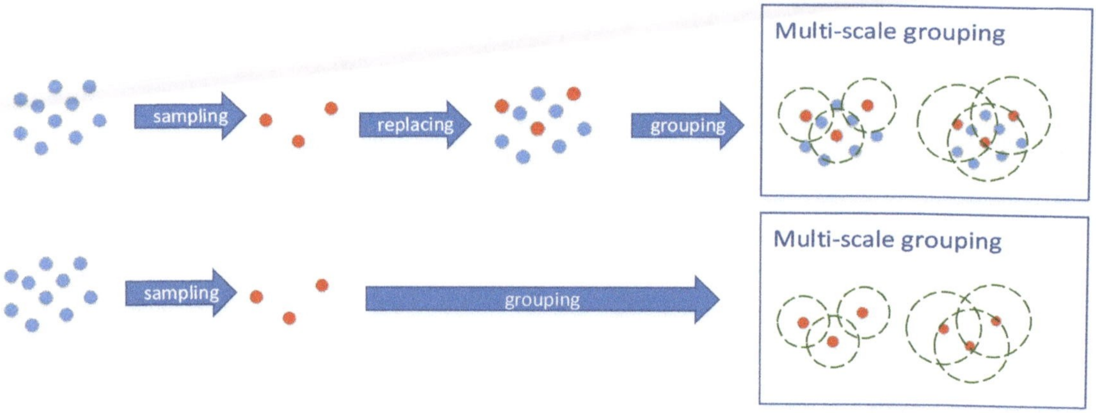

Figure 4. Adjusted multi-scale grouping methods. The red points are sampled by the LS module, while the blue points are old points before sampling. The dotted circle represents a ball of a particular radius. **Top**: Grouping with old points and new points. **Bottom**: Grouping with new points only.

3.5. Loss

Sampling loss. Unlike the D-FPS and F-FPS methods, the point in the sampling subset generated by the LS module is not unique and the high duplicate rate will result in unwanted levels of computational usage while being unable to make full use of a limited sampling size. This problem increases in severity as the sampling decreases in size.

As illustrated in Equation (20), a sampling loss has been presented to reduce the duplicate rate and sample unique points to as great an extent as feasible. We accumulate each row of S', i.e., $\mathbf{p}_j \in \mathbb{R}^{N'}$. $\mathbf{p}_j$ represents the sampling value of each point in the original dataset P_N. The ideal case is that the point in P_N is sampled 0 or 1 time. Since each column in S' can be summarized to 1 and tends to be a one-hot distribution, the accumulation of $\mathbf{p}_j$ should tend to be near 0 or 1 if the point is not sampled more than once. Equation (20) is designed to control this issue. The more the accumulation of $\mathbf{p}_j$ nears 0 or 1, the less the loss.

$$L_{sample} = \frac{1}{N}\sum_{j=1}^{N}\left(\left|\left|\sum_{i=1}^{N'} S'[j,i] - 0.5\right| - 0.5\right|\right) \tag{20}$$

Each row of S' indicates the old point's attention on the newly generated sampling points. If there are many high values in one row, this old point is highly relevant to more than one new point, and the new point's features will be deeply affected by the old point with high attention when each column in S' tends to be a one-hot distribution. That is, these new points tends to be similar to the same old point, which leads to repeated sampling. However, we expect a variety of new sampling points. In a word, we utilized Equation (20) to restrain each old point's attention.

Task loss. In the 3D object detection task, the task loss consists of 3D bounding box regression loss L_r, classification loss L_c, and vote loss L_{vote}. θ_1, θ_2, and θ_3 are the balance weights for these loss terms, respectively.

$$L_{task} = \theta_1 L_r + \theta_2 L_c + \theta_3 L_{vote} \tag{21}$$

Cross-entropy loss is used to calculate classification loss L_r while vote loss related to the vote layer is calculated as VoteNet [31]. Additionally, the regression loss in the model is similar to the regression loss in 3DSSD [7]. The regression loss includes distance regression loss L_{dist}, size regression loss L_{size}, angle regression loss L_{angle}, and corner loss L_{corner}. The smooth-l_1 loss is utilized for L_{dist} and L_{size}, in which the targets are offsets from the candidate points to their corresponding instance centers and sizes of the corresponding instances, respectively. Angle regression loss contains orientation classification loss and residual prediction loss. Corner loss is the distance between the predicted eight corners and assigned ground-truth.

Total loss. The overall loss is composed of sampling loss and task loss with α and β adopted to balance these two losses.

$$L = \alpha L_{sample} + \beta L_{task} \tag{22}$$

3.6. *Training Method*

3.6.1. End-to-End Training

Problem statement: Given a point set $\mathrm{P}_N \in \mathbb{R}^{N\times K}$, a sample size $N' \leq N$, and a untrained task network T, find a subset $\mathrm{P}^*_{N'} \in \mathbb{R}^{N'\times K}$ of N' points and a group of weights W of T that minimize the total objective function L:

$$\mathrm{P}^*_{N'} = \underset{\mathrm{P}_{N'},W}{\arg\min}\, L(T(\mathrm{P}_{N'})|W), \quad \mathrm{P}_{N'} \subseteq \mathrm{P}_N, |\mathrm{P}_{N'}| = N' \leq N. \tag{23}$$

For the end-to-end training method, the task network T and the LS module are trained simultaneously using the total loss L. Compared to the network in the multi-stage training method, the task network part 2 is trained and inferred on the same sampling points distribution. Thus, the entire network is well trained with a certain sampling size.

3.6.2. Multi-Stage Training and Flexibility of the LS Module

Problem statement: Given a point set $\mathrm{P}_N \in \mathbb{R}^{N\times K}$, a sample size $N' \leq N$, and a trained task network T, find a subset $\mathrm{P}^*_{N'} \in \mathbb{R}^{N'\times K}$ of N' points that minimizes the total objective function L:

$$\mathrm{P}^*_{N'} = \underset{\mathrm{P}_{N'}}{\arg\min}\, L(T(\mathrm{P}_{N'})), \quad \mathrm{P}_{N'} \subseteq \mathrm{P}_N, |\mathrm{P}_{N'}| = N' \leq N. \tag{24}$$

Figure 5 shows the flexibility of the LS module and the multi-stage training procedure. The task network part 2 is first trained on sampling points distribution D_N. After this, the task network parts are loaded and fixed to train the sampling network(LS module). Therefore, it is possible to obtain a learned sampling points distribution $D_{N'}$. Subsequently, in the inference stage, the distribution $D_{N'}$ is passed to the task network part 2 for detection. Due to these factors, the task network part 2 is trained and inferred on different sampling points distribution. With the sampled dataset $\mathrm{P}^*_{N'}$ being the best subset of P_N that can make full use of the trained task network, the performance of the network using this method is relatively inferior to the performance of an end-to-end training network because the task network part 2 has not been fully trained with the sampled dataset $\mathrm{P}^*_{N'}$.

In relation to the flexibility of the LS module, the effectiveness of the multi-stage training demonstrates that the LS module can be transferred and adjusted to other point-based models to replace FPS or any other sampling approaches concisely. Even in the case of an already trained task network, point size can still be reduced simply by attaching the LS module to the existing task network and training the LS module solely. This training process can be accomplished quickly because stage 1 is skipped and the LS module is relatively simple and small.

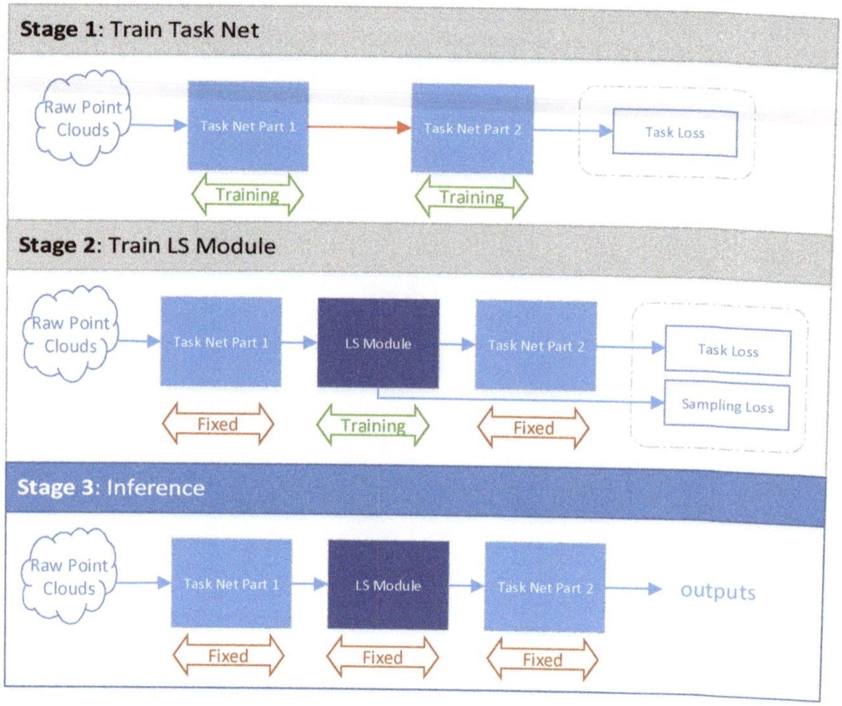

Figure 5. Flexibility and multi-stage training. Illustration of the proposed multi-stage training and inference procedure. In stage 1, the LS module is skipped and the task network is trained on N points data with task loss. In stage 2, we use the trained weights from the former stage and fix the weights of the task network layers, after which the LS module is trained through task loss and sampling loss. The LS module will output N' sampled points. In stage 3, the inference step, the trained LS module is used to sample data and generate the results.

4. Experimental Results

4.1. Setup

Datasets. The KITTI Dataset [12] is one of the most popular dataset for 3D object detection for autonomous driving. All of the experiments for the proposed module are conducted on it. The KITTI dataset collects point cloud data using a 64-scanning-line LiDAR and contains 7481 training samples and 7518 test samples. The training samples are generally divided into the training split (3712 samples) and the val split (3769 samples). Each sample provides both the point cloud and the camera image. Using this approach, only the point cloud is used. Since the dataset only annotates objects that are visible within the image, the point cloud is processed only within the field of view of the image. The KITTI benchmark evaluates the mean average precision (mAP) of three types of objects: car, pedestrian and cyclist. We perform all our experiments on the car objects. Three difficulty levels are involved (easy, moderate, and hard), which depend on the size, occlusion level, and truncation of the 3D objects. For training purposes, samples that do not contain objects of interest are removed.

Data Augmentation. To prevent overfitting, data augmentation is performed on the training data. The point cloud is randomly rotated by yaw $\Delta\theta \sim \mathcal{U}(-\pi/4, +\pi/4)$ and flipped along its x-axis. Each axis is also shifted by Δx, Δy, and Δz (independently drawn from $\mathcal{N}(0, 0.25)$). The mix-up strategy used in SECOND [23] is also used to randomly add foreground instances from other scenes to the current scene. During the translation, it is checked to avoid collisions among boxes, or between background points and boxes.

Network Architecture Details. The network architecture is illustrated in Figure 2. FPS is used to sample 4096 points from the raw input. The LS module will sample points from these 4096 points. There are four multi-scale SA modules in the network with a different shared-MLP structure and a different grouping radius. The shared-MLP layer is a stack of "FC–BN–FC–BN–FC–BN".

Training and Inference Details. All of the experiments are conducted on a single RTX 2080Ti GPU card. The Adam optimizer [39] is used in the training stage with a learning rate of 0.002. The mini-batch size differs according to each sampling size.

Evaluation Metric. Mean average precision (mAP) is utilized as the evaluation metric. For a fair comparison, the official evaluation protocol is followed. Specifically, the IoU threshold is set to 0.7 for cars. As for the unique rate of the sampled points, this is determined by taking the size of the unique points divided by the size of the entire points.

4.2. 3D Object Detection on the KITTI Dataset

LSNet is evaluated along two points. First is submitting the results of the car objects to the KITTI 3D object detection benchmark and the BEV object detection benchmark. Table 1 shows a comparison of the submitted results and the existing literature on the KITTI test dataset. The LSNet-512 (LSNet with 512 sampled points) model is applied to detect the 3D objects of the test dataset, with the results showing that LSNet outperforms other 3D detectors [1–3,23–25] with only 512 points, and the performance of LSNet-512 is similar to 3DSSD [7] on the easy difficulty level and is a little worse than it on the moderate and hard difficulty levels. For a more specific and detailed comparison of the two models, Table 2 compares their precision and speed on the KITTI validation set. LSNet-1024 (LSNet with 1024 sampled points) works better than LSNet-512 and shows competitive accuracy compared to 3DSSD. However, LSNet-1024 runs slower than LSNet-512. Although LSNet-512 sacrifices some accuracy, it runs faster than 3DSSD. To balance the accuracy and speed, we chose LSNet-512 as the final model which was used to generate results on the KITTI test set in Table 1.

Table 1. The mean average precision (mAP) comparison of 3D object detection and bird's eye view(BEV) object detection on the KITTI test set.

Method	Modality	Car-3D (%)			Car-BEV (%)		
		Easy	Moderate	Hard	Easy	Moderate	Hard
MV3D [1]	Image + LiDAR	74.97	63.63	54.00	86.62	78.93	69.80
F-PointNet [2]	Image + LiDAR	82.19	69.79	60.59	91.17	84.67	74.77
AVOD-FPN [3]	Image + LiDAR	83.07	71.76	65.73	90.99	84.82	79.62
VoxelNet [24]	LiDAR	81.97	65.46	62.85	89.60	84.81	78.57
PointPillars [25]	LiDAR	82.58	74.31	68.99	90.07	86.56	82.81
SECOND [23]	LiDAR	83.34	72.55	65.82	89.39	83.77	78.59
3DSSD [7]	LiDAR	88.36	79.57	74.55	92.66	89.02	85.86
LSNet (ours)	LiDAR	86.13	73.55	68.58	92.12	85.89	80.80

Table 2. The mean average precision (mAP) and speed comparison of 3D object detection on the KITTI validation set between 3DSSD and LSNet.

Method	Speed (fps)	Car-3D (%)		
		Easy	Moderate	Hard
3DSSD	10.89	90.87	82.62	79.82
LSNet-512	12.17	89.29	78.36	75.46
LSNet-1024	10.71	91.04	82.15	78.98

Second, Tables 3–5 compare the mAP of different sampling approaches with different sampling sizes. To make a fair comparison, the only change is replacing the LS module with other sampling methods such as random, FPS, F-FPS, and FS sampling, with the rest of the model remaining unchanged. F-FPS and FS are sampling methods raised by 3DSSD [7]. After detailed study about the structure and code of SampleNet, we found that the sampling method of SampleNet [9] is too heavy and not suitable for massive points scenarios such as autonomous driving. Therefore, it is not necessary to conduct experiments on it. The red values between parentheses in these tables are calculated by subtracting the mean of random, FPS, F-FPS, and FS from the value of the LS module. With only eight sampled points, LSNet outperforms other sampling methods significantly with a 60% mAP gain on the easy difficulty level, a 42% mAP gain on the moderate difficulty level, and a 33% mAP gain on the hard difficulty level. Also shown is the fact that when the number of sampling points is decreased, the LS module increasingly outperforms the other approaches. However, once the number of points reaches 512, the differences between these approaches are small. The cause of this behavior is due to the fact that there are already enough points to describe the whole 3D space and the sampling mode does not affect the coverage of key information.

Table 3. Performance comparison on the *easy* difficulty level between different sampling methods on the KITTI validation set. The results are evaluated using the mean average precision (mAP).

Sampled Points	Random (%)	FPS (%)	F-FPS (%)	FS (%)	LS Module (%)
8	4.12	0.18	2.06	0.06	61.28 (+59.67)
16	18.71	1.83	10.87	0.15	66.59 (+58.70)
32	35.95	8.29	46.38	9.09	73.48 (+48.55)
64	51.40	32.61	77.21	45.30	83.57 (+31.94)
128	70.55	64.82	86.63	74.94	88.19 (+13.95)
256	72.81	76.10	89.66	86.10	88.56 (+7.39)
512	78.93	87.75	89.17	85.27	89.29 (+4.01)

Table 4. Performance comparison on the *moderate* difficulty level between different sampling methods on the KITTI validation set. The results are evaluated using the mean average precision (mAP).

Sampled Points	Random (%)	FPS (%)	F-FPS (%)	FS (%)	LS Module (%)
8	3.19	0.18	0.32	0.08	42.49 (+41.54)
16	13.30	2.07	8.57	0.24	46.53 (+40.49)
32	27.11	7.82	35.70	7.56	53.29 (+37.74)
64	39.21	28.60	64.28	34.34	65.18 (+23.57)
128	54.87	54.77	75.87	63.01	72.64 (+10.51)
256	61.20	64.66	79.08	74.72	74.51 (+4.60)
512	66.97	76.79	79.52	76.83	78.36 (+3.33)

Table 5. Performance comparison on the *hard* difficulty level between different sampling methods on the KITTI validation set. The results are evaluated using the mean average precision (mAP).

Sampled Points	Random (%)	FPS (%)	F-FPS (%)	FS (%)	LS Module (%)
8	2.16	0.32	1.31	0.03	33.46 (+32.50)
16	11.76	1.62	7.54	0.28	39.17 (+33.87)
32	24.18	7.49	30.77	6.82	46.28 (+28.97)
64	34.77	26.89	58.65	30.54	58.43 (+20.71)
128	51.20	52.31	71.62	57.05	67.19 (+9.15)
256	57.99	62.84	76.43	70.60	72.63 (+5.67)
512	64.82	73.94	78.83	74.45	75.46 (+2.45)

In Figures 6–8, visual examples of the described behavior are shown that illustrate the advantages of the LS module. Firstly, by comparing these three sampling methods, we can see that our sampling approach generates more points within the region of interest and near the target object, which is the reason why LSNet works extremely well when the sampling size is small. Furthermore, Figures 6 and 7 depict a complex scene with various features and a simple scene with relatively less different features. It is obvious that FPS and F-FPS performed poorly in the complex scene because there is relatively more distraction. In contrast, our sampling approach can still locate the key areas by selecting the corresponding points nearby.

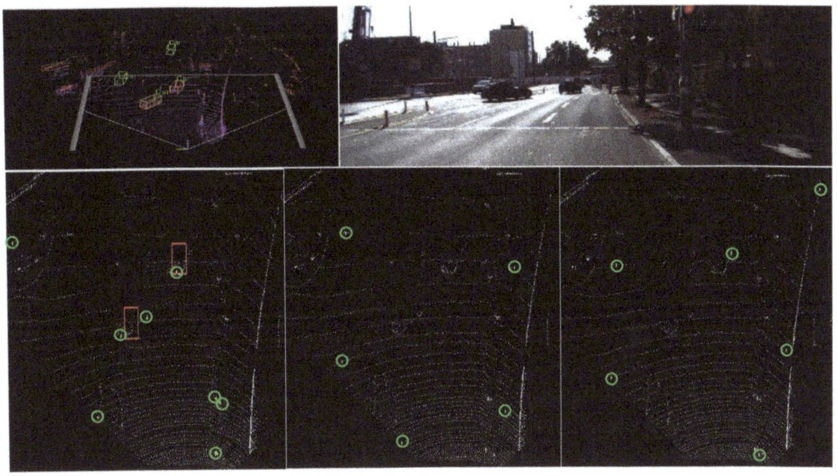

Figure 6. Visualizing the results of LSNet with 16 sampled points and different sampling approaches. The **top-left** frame presents the 3D object detection results, where ground truth and predictions are labeled in red and green, respectively. Moreover, the area surrounded by gray lines is the visible area within the image, which can be also recognized as a region of interest. The **top-right** frame displays the image of the scene. The second line illustrates the sampling results of the LS module, D-FPS, and F-FPS, where the sampled points are displayed in green and the 4096 original points before sampling are displayed in white.

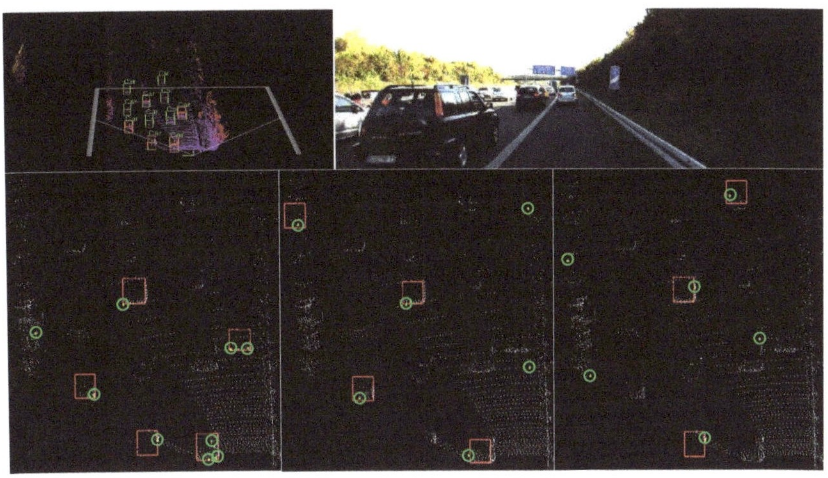

Figure 7. Visualizing the results of LSNet with 16 sampled points and different sampling approaches. This is an easier scene compared to Figure 6 .

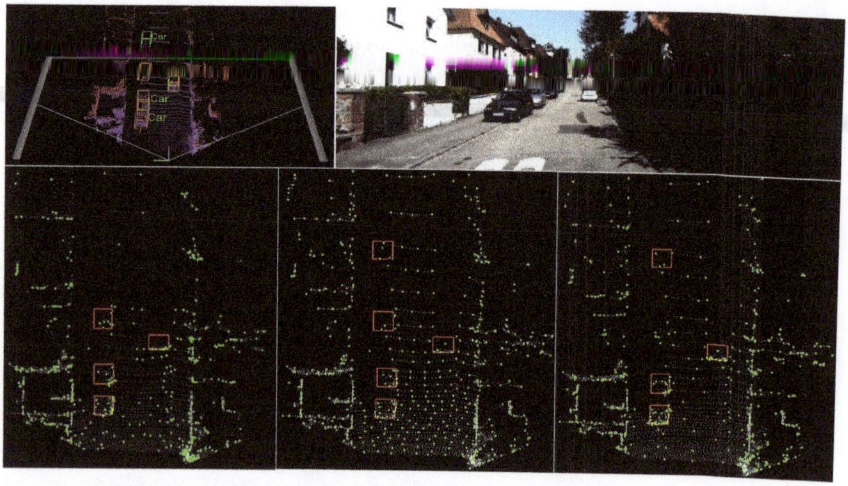

Figure 8. Visualizing the results of LSNet with 512 sampled points and different sampling approaches.

4.3. *Effects of Multi-Stage Training and the Flexibility of the LS Module*

As previously described, only the LS module is replaced, with the rest of the architecture remaining the same. This ease of insertion and the removal of the part of the LS module illustrates how flexibly it can be used. Additionally, Tables 6 and 7 show the results of multi-stage training when starting from an already-trained task network. In this paper, the model trained with the F-FPS sampling approach was used as the task network. Following this, F-FPS was substituted with the LS module. Once the substitution was finished, the weights were set as fixed to solely train the LS module. Finally, Tables 6 and 7 show the results inferred by the new model with the LS module. Using the LS module, the number of sampled points is halved in short order with only a small loss of accuracy. The number of sampled points of the fixed task model is 256 in Table 6 and 512 in Table 7. This illustrates that the greater the difference in the number of sampled points between the original task network and the LS module, the larger the performance degradation. For example, with 128 sampling points, task-network-256 leads task-network-512 by over 17% mAP gain in moderate difficulty. Figure 9 shows that the training time of the LS module is much shorter in comparison to the time required for the end-to-end training. Furthermore, the growth trend of the time cost for training the LS module is more gentle.

Table 6. Multi-stage training on the trained task network with 256 sampled points using the F-FPS sampling method. The first line of the table shows the original performance of the trained model and the results are evaluated by the mean average precision (mAP).

Sampled Points	Easy (%)	Moderate (%)	Hard (%)
256 (task-net)	89.66	79.08	76.43
8	30.19	18.23	14.56
16	42.61	27.13	22.34
32	69.24	50.36	42.28
64	76.47	59.29	50.82
128	84.13	70.08	65.21

Table 7. Multi-stage training on the trained task network with 512 sampled points using the F-FPS sampling method. The first line of the table shows the original performance of the trained model and the results are evaluated by the mean average precision (mAP).

Sampled Points	Easy (%)	Moderate (%)	Hard (%)
512 (task-net)	89.17	79.52	78.83
8	6.80	4.36	3.31
16	21.43	15.33	12.52
32	50.31	32.29	26.48
64	64.28	46.47	39.61
128	79.36	52.86	55.31
256	85.58	70.36	66.53

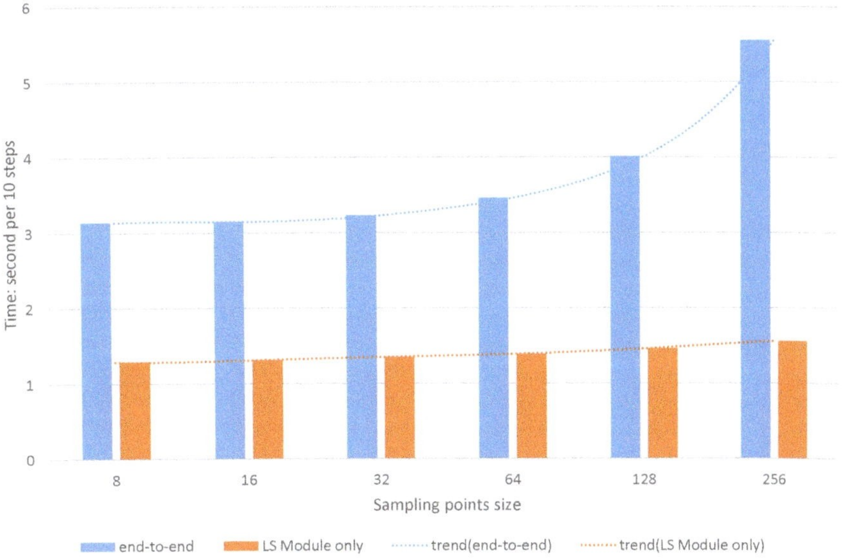

Figure 9. Time comparison between training the entire model end-to-end and training the LS module only with a batch size of eight.

5. Discussion

Ablation Study

In this section, extensive ablation experiments are conducted to analyze the individual components of the proposed approaches. All the models are trained on the training split and evaluated on the validation split for the car class of the KITTI dataset [12].

Effects of the Different Grouping Methods. Table 8 compares the performance between the grouping with the old points and new points together versus the grouping with the new points only. The result is that while there is no difference when the number of points is large, when the number of points is very small, the approach of grouping old and new points together gains higher accuracy. For example, the mAP of eight sampled points for "new points only" is lower than the one for "old + new", which is caused by the relatively smaller information loss of grouping old and new points together.

Table 8. The mAP results for different groupings.

Sampled Points	Old + New (%)			New Points Only (%)		
	Easy	Moderate	Hard	Easy	Moderate	Hard
8	61.24	42.36	33.12	46.49	35.01	29.15
16	66.38	46.09	39.31	69.16	51.61	42.29
32	73.07	53.35	46.21	74.08	57.81	50.26
64	83.47	65.81	58.43	83.13	69.42	63.16
128	88.12	72.42	67.35	87.52	74.16	69.63
256	88.63	74.36	72.61	88.17	74.56	70.47
512	89.36	78.61	75.47	89.35	75.43	72.39

Effects of the Sampling Loss. As shown in Table 9, the proposed sampling loss can boost the unique rate significantly. With our sampling loss, the average unique rate of the points can be stabilized at around 95%. On the contrary, once we remove the sampling loss, the repetition rate climbs to 88% with 512 sampled points and 77% with 256 sampled points. Another issue is that the model performs poorly with a large number of repetition points when it comes to mAP.

Table 9. The effectiveness of sampling loss evaluated by unique rate and mAP results.

Sampled Points	Unique Rate (%)		mAP (%)					
	With SL	Without SL	With SL			Without SL		
			Easy	Moderate	Hard	Easy	Moderate	Hard
8	95	94	46.49	35.01	29.15	47.35	35.12	30.56
16	95	85	69.16	51.61	42.29	63.26	45.63	39.18
32	95	75	74.08	57.81	50.26	72.53	55.45	49.32
64	95	54	83.13	69.42	63.16	70.64	56.36	50.21
128	96	51	87.52	74.16	69.63	79.10	65.12	60.03
256	96	33	88.17	74.56	70.47	80.59	65.51	60.52
512	95	22	89.35	75.43	72.39	76.53	62.59	56.24

Effects of Relaxation. Table 10 confirms that the random relaxation strategy of the sampling matrix yields a higher mAP, i.e., increasing the mAP by an average of 2.96%, 2.94%, and 1.97% on the easy, moderate, and hard difficulty levels, respectively.

Table 10. The effectiveness of random relaxation evaluated by mAP results.

Sampled Points	With Relaxation (%)			Without Relaxation (%)		
	Easy	Moderate	Hard	Easy	Moderate	Hard
8	61.24	42.36	33.12	54.23	36.19	30.09
16	66.38	46.09	39.31	60.63	41.71	35.32
32	73.07	53.35	46.21	70.52	49.63	44.37
64	83.47	65.81	58.43	81.25	63.61	56.47
128	88.12	72.42	67.35	86.21	70.45	65.21
256	88.63	74.36	72.61	88.04	73.54	71.58
512	89.36	78.61	75.47	88.54	77.31	75.65

Speed Analysis of LSNet. All the speed experiments were run on a 2080Ti GPU. Table 11 illustrates the inference speed of the entire network in fps(frames per second). The processing time of each model with different sampling approaches has little variation, which proves that replacing the original sampling strategy in other models with the LS module will not introduce excessive time consumption. Thus, this shows that the LS module is lightweight and can be plugged into other models without encumbering them. Under the consideration of both inference speed and accuracy, LSNet outperforms the other tested methods according to Figure 10. The green gradient background of the table shows the overall performance of the method, and the darker the color, the better the performance. Then, we can see that LSNet gains higher overall performance, especially when it comes to faster inference speed. Furthermore, we add several auxiliary lines (black dashed lines) in Figure 10 to address this superiority. Each auxiliary line indicates the same accuracy, and LSNet runs faster than other methods with the same accuracy. In addition, FPS collapses very quickly at speeds above 15 fps. The inference time of LSNet-256 is 73 ms and the inference time of LSNet-8 is 64 ms.

Table 11. Speed comparison between different sampling methods by checking the fps (frames per second) of the entire model.

Sampled Points	Random	FPS	F-FPS	LS module
8	15.48	15.64	15.10	15.53
16	15.38	15.58	15.09	15.49
32	15.38	15.53	15.04	15.13
64	15.31	15.47	15.00	15.05
128	14.96	14.99	14.93	14.83
256	14.00	14.02	13.73	13.66
512	13.02	12.92	12.51	12.17

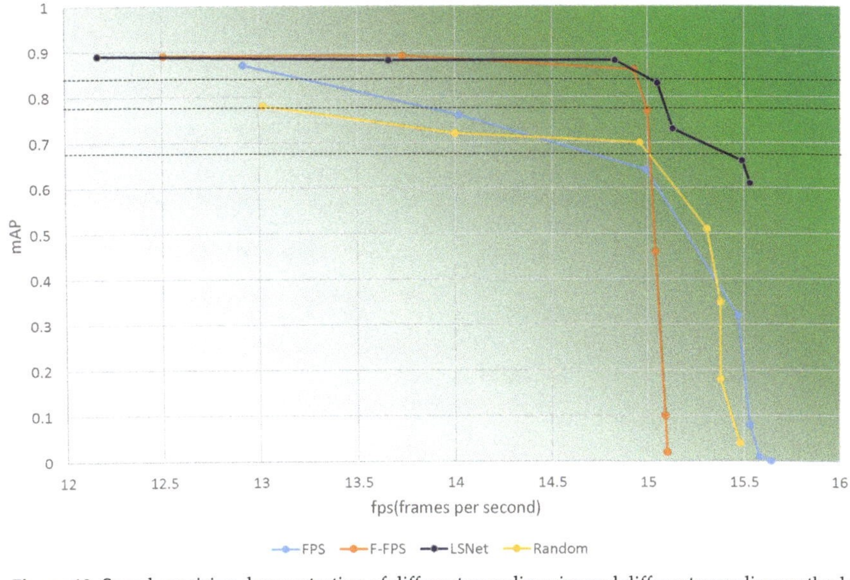

Figure 10. Speed-precision demonstration of different sampling size and different sampling methods.

6. Conclusions

In this paper, LSNet was proposed to solve the 3D object detection task that operates on LiDAR point clouds. Importantly, the LS module, which is a novel deep-learning-based sampling approach that is differentiable and task-related, was presented. Specifically, with 128 sampled points, it attained a computational acceleration at the cost of acceptable accuracy loss. In addition, the random relaxation method was introduced to the sampling matrix. Evaluated on the challenging KITTI dataset, the LS module of LSNet was found to work extremely well when only using a small amount of sampling data in comparison to the D-FPS and F-FPS methods. The proposed sampling loss was proven to be highly effective in ameliorating the issue of sampling duplicates. Finally, it has been shown that, with an already trained point-based task network, the LS module can be attached to the task network flexibly to replace the original sampling method such as FPS.

As the proposed method has been shown to be superior in comparison to other sampling methods for usage in low sampling size cases and complex scenarios, it is therefore particularly appropriate for autonomous driving usage on urban roads. This is due to the increased complexity faced on urban roads in comparison to highway driving. Additionally, if autonomous vehicles, i.e., trucks, are equipped with multiple LiDARs, this would greatly increase the initial amount of raw points in the system, an issue this sampling method is well suited to handling, giving rise to a reduction in the required memory and computational cost. In a similar vein, the large amount of exploration undertaken recently in China on vehicle-to-everything (V2X) scenarios can also benefit from the LS module. As V2X involves multiple sensors containing LiDAR, they inevitably produce more point cloud data than vehicle-only scenarios. Once again, this means that the module's efficiency in dealing with such issues is applicable. These varied use cases show the widespread potential and applicability of the LS module.

The LS module tends to sample more points in dense objects than sparse objects, which results in relatively weak performance in moderate and hard categories. In the future, we will work on sampling points evenly on each object and regard their density. Furthermore, it is expected to keep at least one point, even when the object is badly shaded. In addition, we look forward to achieving better accuracy with less points in the following study.

Author Contributions: Conceptualization, M.W.; Data curation, M.W. and Z.F.; Formal analysis, M.W.; Funding acquisition, Q.C.; Investigation, M.W.; Methodology, M.W.; Project administration, M.W. and Q.C.; Resources, Q.C.; Software, M.W.; Supervision, M.W. and Q.C.; Validation, M.W. and Z.F.; Visualization, M.W.; Writing—original draft, M.W.; Writing—review and editing, M.W., Q.C., and Z.F. All authors have read and agreed to the published version of the manuscript.

Funding: This work was supported by the Shanghai Key Science and Technology Project (19DZ1208903); National Natural Science Foundation of China (Grant Nos. 61572325 and 60970012); Ministry of Education Doctoral Fund of Ph.D. Supervisor of China (Grant No. 20113120110 0 08); Shanghai Key Science and Technology Project in Information Technology Field (Grant Nos. 14511107902 and 16DZ1203603); Shanghai Leading Academic Discipline Project (No. XTKX2012); Shanghai Engineering Research Center Project (Nos. GCZX14014 and C14001).

Data Availability Statement: Data available in a publicly accessible repository that does not issue DOIs. Publicly available datasets were analyzed in this study. This data can be found here: [http://www.cvlibs.net/datasets/kitti/index.php], accessed on 10 March 2022.

Acknowledgments: The authors would like to acknowledge the support from the Flow Computing Laboratory at University of Shanghai for Science and Technology.

Conflicts of Interest: The authors declare no conflict of interest.

References

1. Chen, X.; Ma, H.; Wan, J.; Li, B.; Xia, T. Multi-View 3D Object Detection Network for Autonomous Driving. In Proceedings of the 2017 IEEE Conference on Computer Vision and Pattern Recognition (CVPR), Honolulu, HI, USA, 21–26 July 2017.
2. Qi, C.R.; Liu, W.; Wu, C.; Su, H.; Guibas, L.J. Frustum pointnets for 3d object detection from rgb-d data. In Proceedings of the IEEE Conference on Computer Vision and Pattern Recognition, Salt Lake City, UT, USA, 18–22 June 2018; pp. 918–927.
3. Ku, J.; Mozifian, M.; Lee, J.; Harakeh, A.; Waslander, S. Joint 3D Proposal Generation and Object Detection from View Aggregation. In Proceedings of the 2018 IEEE/RSJ International Conference on Intelligent Robots and Systems (IROS), Madrid, Spain, 1–5 October 2019.
4. Wang, J.; Zhu, M.; Wang, B.; Sun, D.; Wei, H.; Liu, C.; Nie, H. KDA3D: Key-Point Densification and Multi-Attention Guidance for 3D Object Detection. *Remote Sens.* **2020**, *12*, 1895. [CrossRef]
5. Hu, Q.; Yang, B.; Xie, L.; Rosa, S.; Guo, Y.; Wang, Z.; Trigoni, N.; Markham, A. RandLA-Net: Efficient Semantic Segmentation of Large-Scale Point Clouds. In Proceedings of the IEEE Conference on Computer Vision and Pattern Recognition, Seattle, WA, USA, 14–19 June 2020.
6. Thomas, H.; Qi, C.R.; Deschaud, J.E.; Marcotegui, B.; Goulette, F.; Guibas, L.J. KPConv: Flexible and Deformable Convolution for Point Clouds. In Proceedings of the IEEE International Conference on Computer Vision, Seoul, Korea, 27 October–3 November 2019.

7. Yang, Z.; Sun, Y.; Liu, S.; Jia, J. 3dssd: Point-based 3d single stage object detector. In Proceedings of the IEEE/CVF Conference on Computer Vision and Pattern Recognition, Seattle, WA, USA, 14–19 June 2020; pp. 11040–11048.
8. Dovrat, O.; Lang, I.; Avidan, S. Learning to sample. In Proceedings of the IEEE Conference on Computer Vision and Pattern Recognition, Long Beach, CA, USA, 16–20 June 2019; pp. 2760–2769.
9. Lang, I.; Manor, A.; Avidan, S. SampleNet: Differentiable Point Cloud Sampling. In Proceedings of the IEEE/CVF Conference on Computer Vision and Pattern Recognition, Seattle, WA, USA, 14–19 June 2020; pp. 7578–7588.
10. Yang, J.; Zhang, Q.; Ni, B.; Li, L.; Liu, J.; Zhou, M.; Tian, Q. Modeling point clouds with self-attention and gumbel subset sampling. In Proceedings of the IEEE Conference on Computer Vision and Pattern Recognition, Long Beach, CA, USA, 16–20 June 2019; pp. 3323–3332.
11. Wu, Z.; Song, S.; Khosla, A.; Yu, F.; Zhang, L.; Tang, X.; Xiao, J. 3d shapenets: A deep representation for volumetric shapes. In Proceedings of the IEEE Conference on Computer Vision and Pattern Recognition, Boston, MA, USA, 7–12 June 2015; pp. 1912–1920.
12. Geiger, A.; Lenz, P.; Urtasun, R. Are we ready for autonomous driving? the kitti vision benchmark suite. In Proceedings of the 2012 IEEE Conference on Computer Vision and Pattern Recognition, Providence, RI, USA, 16–21 June 2012; pp. 3354–3361.
13. Song, S.; Chandraker, M. Joint SFM and detection cues for monocular 3D localization in road scenes. In Proceedings of the IEEE Conference on Computer Vision and Pattern Recognition, Boston, MA, USA, 7–12 June 2015; pp. 3734–3742.
14. Chen, X.; Kundu, K.; Zhang, Z.; Ma, H.; Fidler, S.; Urtasun, R. Monocular 3d object detection for autonomous driving. In Proceedings of the IEEE Conference on Computer Vision and Pattern Recognition, Las Vegas, NV, USA, 27–30 June 2016; pp. 2147–2156.
15. Mousavian, A.; Anguelov, D.; Flynn, J.; Kosecka, J. 3d bounding box estimation using deep learning and geometry. In Proceedings of the IEEE Conference on Computer Vision and Pattern Recognition, Honolulu, HI, USA, 21–27 June 2017; pp. 7074–7082.
16. Yang, B.; Luo, W.; Urtasun, R. Pixor: Real-time 3d object detection from point clouds. In Proceedings of the IEEE Conference on Computer Vision and Pattern Recognition, Salt Lake City, UT, USA, 18–23 June 2018; pp. 7652–7660.
17. Simony, M.; Milzy, S.; Amendey, K.; Gross, H.M. Complex-YOLO: An Euler-Region-Proposal for Real-time 3D Object Detection on Point Clouds. In Proceedings of the European Conference on Computer Vision (ECCV) Workshops, Munich, Germany, 1–4 September 2018.
18. Yang, B.; Liang, M.; Urtasun, R. Hdnet: Exploiting hd maps for 3d object detection. In Proceedings of the Conference on Robot Learning, Zurich, Switzerland, 29–31 October 2018; pp. 146–155.
19. Li, B.; Zhang, T.; Xia, T. Vehicle detection from 3d LiDAR using fully convolutional network. *arXiv* **2016**, arXiv:1608.07916.
20. Chai, Y.; Sun, P.; Ngiam, J.; Wang, W.; Caine, B.; Vasudevan, V.; Zhang, X.; Anguelov, D. To the Point: Efficient 3D Object Detection in the Range Image With Graph Convolution Kernels. In Proceedings of the IEEE/CVF Conference on Computer Vision and Pattern Recognition (CVPR), Virtual, 19–25 June 2021; pp. 16000–16009.
21. Chen, Y.; Liu, S.; Shen, X.; Jia, J. Fast point r-cnn. In Proceedings of the IEEE International Conference on Computer Vision, Seoul, Korea, 27 October–3 November 2019; pp. 9775–9784.
22. Shi, S.; Wang, Z.; Shi, J.; Wang, X.; Li, H. From points to parts: 3d object detection from point cloud with part-aware and part-aggregation network. *IEEE Trans. Pattern Anal. Mach. Intell.* **2020**, *43*, 2647–2664. [CrossRef] [PubMed]
23. Yan, Y.; Mao, Y.; Li, B. Second: Sparsely embedded convolutional detection. *Sensors* **2018**, *18*, 3337. [CrossRef] [PubMed]
24. Zhou, Y.; Tuzel, O. Voxelnet: End-to-end learning for point cloud based 3d object detection. In Proceedings of the IEEE Conference on Computer Vision and Pattern Recognition, Salt Lake City, UT, USA, 18–23 June 2018; pp. 4490–4499.
25. Lang, A.H.; Vora, S.; Caesar, H.; Zhou, L.; Yang, J.; Beijbom, O. Pointpillars: Fast encoders for object detection from point clouds. In Proceedings of the IEEE Conference on Computer Vision and Pattern Recognition, Seoul, Korea, 27 October–3 November 2019; pp. 12697–12705.
26. Liu, Z.; Zhao, X.; Huang, T.; Hu, R.; Zhou, Y.; Bai, X. TANet: Robust 3D Object Detection from Point Clouds with Triple Attention. In Proceedings of the AAAI Conference on Artificial Intelligence , New York, NY, USA, 7–12 February 2020; pp. 11677–11684.
27. Graham, B.; Engelcke, M.; Van Der Maaten, L. 3d semantic segmentation with submanifold sparse convolutional networks. In Proceedings of the IEEE Conference on Computer Vision and Pattern Recognition, Salt Lake City, UT, USA, 18–23 June 2018; pp. 9224–9232.
28. Qi, C.R.; Su, H.; Mo, K.; Guibas, L.J. Pointnet: Deep learning on point sets for 3d classification and segmentation. In Proceedings of the IEEE Conference on Computer Vision and Pattern Recognition, Honolulu, HI, USA, 21–27 June 2017; pp. 652–660.
29. Qi, C.R.; Yi, L.; Su, H.; Guibas, L.J. Pointnet++: Deep hierarchical feature learning on point sets in a metric space. *Adv. Neural Inf. Process. Syst.* **2017**, *30*, 5099–5108.
30. Shi, S.; Wang, X.; Li, H. Pointrcnn: 3d object proposal generation and detection from point cloud. In Proceedings of the IEEE Conference on Computer Vision and Pattern Recognition, Long Beach, CA, USA, 16–20 June 2019; pp. 770–779.
31. Qi, C.R.; Litany, O.; He, K.; Guibas, L.J. Deep hough voting for 3d object detection in point clouds. In Proceedings of the IEEE International Conference on Computer Vision, Seoul, Korea, 27 October–3 November 2019; pp. 9277–9286.
32. Ngiam, J.; Caine, B.; Han, W.; Yang, B.; Chai, Y.; Sun, P.; Zhou, Y.; Yi, X.; Alsharif, O.; Nguyen, P.; et al. Starnet: Targeted computation for object detection in point clouds. *arXiv* **2019**, arXiv:1908.11069.
33. Shi, W.; Rajkumar, R. Point-gnn: Graph neural network for 3d object detection in a point cloud. In Proceedings of the IEEE/CVF Conference on Computer Vision and Pattern Recognition, Seattle, WA, USA, 14–19 June 2020; pp. 1711–1719.

34. Shi, S.; Guo, C.; Jiang, L.; Wang, Z.; Shi, J.; Wang, X.; Li, H. Pv-rcnn: Point-voxel feature set abstraction for 3d object detection. In Proceedings of the IEEE/CVF Conference on Computer Vision and Pattern Recognition, Seattle, WA, USA, 14–19 June 2020; pp. 10529–10538.
35. Liu, Z.; Tang, H.; Lin, Y.; Han, S. Point-Voxel CNN for efficient 3D deep learning. *Adv. Neural Inf. Process. Syst.* **2019**, *32*, 965–975.
36. Chen, S.; Tian, D.; Feng, C.; Vetro, A.; Kovačević, J. Fast resampling of three-dimensional point clouds via graphs. *IEEE Trans. Signal Process.* **2017**, *66*, 666–681. [CrossRef]
37. Jang, E.; Gu, S.; Poole, B. Categorical reparameterization with gumbel-softmax. *arXiv* **2016**, arXiv:1611.01144.
38. Maddison, C.J.; Mnih, A.; Teh, Y.W. The concrete distribution: A continuous relaxation of discrete random variables. *arXiv* **2016**, arXiv:1611.00712.
39. Kingma, D.P.; Ba, J. Adam: A method for stochastic optimization. *arXiv* **2014**, arXiv:1412.6980.

remote sensing

Article

Oriented Object Detection in Remote Sensing Images with Anchor-Free Oriented Region Proposal Network

Jianxiang Li , Yan Tian *, Yiping Xu and Zili Zhang

School of Electronic Information and Communications, Huazhong University of Science and Technology, Wuhan 430074, China; jianxiang_li@hust.edu.cn (J.L.); xuyiping@hust.edu.cn (Y.X.); zhangzili@hust.edu.cn (Z.Z.)
* Correspondence: tianyan@hust.edu.cn

Abstract: Oriented object detection is a fundamental and challenging task in remote sensing image analysis that has recently drawn much attention. Currently, mainstream oriented object detectors are based on densely placed predefined anchors. However, the high number of anchors aggravates the positive and negative sample imbalance problem, which may lead to duplicate detections or missed detections. To address the problem, this paper proposes a novel anchor-free two-stage oriented object detector. We propose the Anchor-Free Oriented Region Proposal Network (AFO-RPN) to generate high-quality oriented proposals without enormous predefined anchors. To deal with rotation problems, we also propose a new representation of an oriented box based on a polar coordinate system. To solve the severe appearance ambiguity problems faced by anchor-free methods, we use a Criss-Cross Attention Feature Pyramid Network (CCA-FPN) to exploit the contextual information of each pixel and its neighbors in order to enhance the feature representation. Extensive experiments on three public remote sensing benchmarks—DOTA, DIOR-R, and HRSC2016—demonstrate that our method can achieve very promising detection performance, with a mean average precision (mAP) of 80.68%, 67.15%, and 90.45%, respectively, on the benchmarks.

Keywords: remote sensing images; oriented object detection; contextual information; Anchor Free Region Proposal Network; polar representation

Citation: Li, J.; Tian, Y.; Xu, Y.; Zhang, Z. Oriented Object Detection in Remote Sensing Images with Anchor-Free Oriented Region Proposal Network. *Remote Sens.* **2022**, *14*, 1246. https://doi.org/10.3390/rs14051246

Academic Editor: Józef Lisowski

Received: 19 January 2022
Accepted: 1 March 2022
Published: 3 March 2022

1. Introduction

Object detection is a fundamental and challenging task in computer vision. Object detection in remote sensing images (RSIs) [1–9], which recognizes and locates the objects of interest such as vehicles [4,5], ships [6,7], and airplanes [8,9] on the ground, has enabled applications in fields such as traffic planning and land surveying.

Traditional object detection methods [10], like object-based image analysis (OBIA) [11], usually take two steps to accomplish object detection: firstly, extract regions that may contain potential objects, then extract hand-designed features and apply classifiers to obtain the class information. However, their detection performance is unsatisfactory because the handcrafted features have limited representational power with insufficient semantic information.

Benefitting from the rapid development of deep convolutional neural networks (DCNNs) [12] and publicly available large-scale benchmarks, generic object detection [13–19] has made extensive progress in natural scenes, which has also prompted the increased development of object detection in RSIs. Generic object detectors employ an axis-aligned bounding box, also called a horizontal bounding box (HBB), to localize the object in the image. However, detecting objects in RSIs with HBBs remains a challenge. Because RSIs are photographed from a bird's eye view, the objects in RSIs often have large aspect ratios and dense arrangements, as is the case with, for example, ships docked in a harbor. As a result, oriented bounding box (OBB) has recently been adopted to describe the position of the arbitrary-rotated object in RSIs.

Currently, mainstream oriented object detectors [20–23] are based on densely placed predefined anchors. Several early rotation detectors use a horizontal anchor-based Region Proposal Network (RPN) to generate horizontal regions of interest (RoIs), and then design novel network modules to convert the horizontal RoIs into OBBs. For example, Ding et al. [20] build a rotated RoI learner to transform horizontal RoIs into rotated RoIs (RRoIs), and then regress the RRoIs to obtain the final results. However, the horizontal RoI typically contains massive ground pixels and other objects due to the arbitrary orientation and dense distribution of the objects, as shown in Figure 1a. The mismatch between the horizontal anchors and rotation objects causes difficulties in network training and further degrades performance [21].

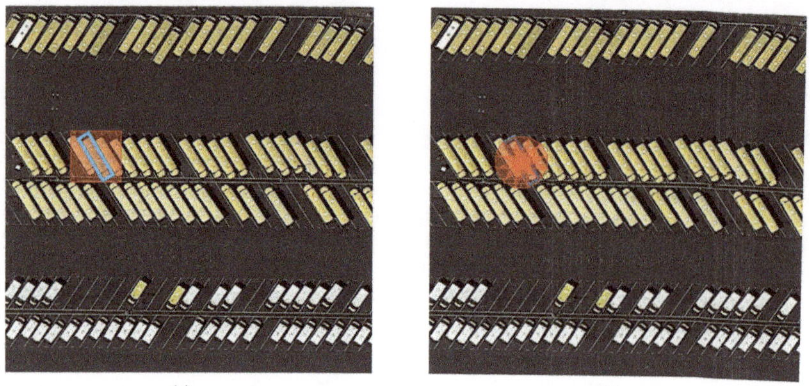

(a) (b)

Figure 1. Disadvantages of anchor-based detectors. The blue rectangle represents the ground truth, and the orange rectangle represents the anchor box. (**a**) The horizontal anchor contains massive ground pixels and other objects. (**b**) RRPN often places too many oriented anchors to ensure a high recall rate.

To address the problem, some detectors use a rotated anchor-based RPN (RRPN) [23] to generate RRoIs. Nevertheless, the Intersection over Union (IoU) is highly sensitive to the angle. To ensure the high recall rate, RRPN places 54 rotated anchors (six orientations, three aspect ratios, and three scales) for each sample point on the feature map, as shown in Figure 1b. However, the high number of anchors increases the computational burden and aggravates the imbalance between positive and negative samples. Moreover, dense anchors may lead to duplicate detections of the same object and missed detections [21] after the non-maximum suppression (NMS).

Owing to the above problems, the use of anchor-free oriented object detectors is increasing. Anchor-free detectors directly locate the objects without manually defined anchors. In particular, keypoint-based methods use several points, such as corners [24], extreme points [25], and the center [26], to represent the positive samples and directly regress the categories and locations of the objects from the features of the keypoints. For example, CenterNet [26] uses one center point to represent the object and directly regresses other properties, such as object size, dimension, and pose, from the features at the center position. Most anchor-free oriented object detectors are inherent from CenterNet for high efficiency and generality, having achieved performance competitive with anchor-based detectors. For example, Pan et al. [27] extend the CenterNet by adding a branch to regress the orientations of the OBBs, and the proposed DRN achieved consistent gains across multiple datasets in comparison with baseline approaches.

However, keypoint-based anchor-free object detectors face severe appearance ambiguity problems with backgrounds or other categories. As shown in Figure 2, the central areas of the objects are similar to the backgrounds, and some objects belonging to different categories even share the same center parts. The main reason for this is that the commonly used fully convolutional networks have insufficient contextual information [28] because of the limited local receptive fields due to fixed DCNN structures. Furthermore, nearly all anchor-free detectors are one-stage detectors, which usually encounter severe misalignment [29] between the axis-aligned convolutional features extracted by the DCNNs and rotational bounding boxes. However, the feature warping module of the two-stage detectors, such as RRoI Pooling [23] or RRoI Align [20], can alleviate this problem.

Figure 2. Appearance ambiguity problems of the keypoint-based anchor-free object detectors. (**a**) The central areas of the objects are similar to the backgrounds. (**b**) Some different categories objects share the same center parts.

Based on the above discussion, we propose a novel two-stage oriented object detector, following the coarse- to fine-detection paradigm. Our method consists of four components: a backbone, a Criss-Cross Attention Feature Pyramid Network (CCA-FPN), an Anchor-Free Oriented Region Proposal Network (AFO-RPN) and oriented RCNN heads.

At the outset, we use the proposed AFO-RPN to generate high quality–oriented proposals without placing excessive fix-shaped anchors on the feature map. To enhance the feature representation of each pixel in the feature map, we adopt CCA-FPN to exploit the contextual information from full image patch. To deal with rotation problems, we propose a new representation of OBB based on polar coordinate system. Finally, we apply an AlignConv to align the features and then use oriented RCNN heads to predict the classification scores and regress the final OBBs. To demonstrate the effectiveness of our method, we conducted extensive experiments on three public RSI oriented object detection datasets—DOTA [30], DIOR-R [31], and HRSC2016 [7].

The contributions of this paper can be summarized as follows: (1) We propose a new anchor-free oriented object detector following the two-stage coarse-to-refined detection paradigm. Specifically, we proposed AFO-RPN to generate high-quality proposals without enormous predefined anchors and a new representation method of OBB in the polar coordinate system, which can better handle the rotation problem; (2) We apply CCA module into FPN to enhance the feature representation of each pixel by capturing the contextual information from the full patch image; and (3) Experimental results on three publicly available datasets show that our method achieves promising results and outperforms previous state-of-the-art methods.

The rest of this paper is organized as follows. Section 2 reviews the related work and explains our method in details. Section 3 compares the propsed method with state-of-the-art methods on different datasets. Section 4 discusses the ablation experiments of the proposed method. Section 5 offers our conclusions.

2. Materials and Methods

2.1. Related Work

2.1.1. Generic Object Detection

With recent advances in deep learning techniques, the performance of DCNN-based generic object detectors has improved significantly. Generic object detectors aim to detect general objects in natural scenes with HBBs to locate objects. The mainstream generic object detection methods can be roughly divided according to the following standards: two- or single-stage object detection, and anchor-free or anchor-based object detection.

Two-stage object detectors, such as Faster RCNN [13] and Mask RCNN [14], first generate RoIs, which can be treated as coarse class-agnostic detection results, and then in the second stage extract the RoI features to perform refined classification and location. Two-stage object detectors can achieve high detection accuracy, but their inference speed is slow. One-stage object detectors, such as YOLO series [15–17], SSD [18], and RetinaNet [19], directly regress the complete detection results through one-step prediction. One-stage detectors are fast and can achieve real-time inference, but they are less accurate than two-stage detectors. The design of anchors has been popularized by Faster R-CNN in its RPN and has become the convention in many modern object detectors.

Although anchor-based detectors currently dominate in the object detection arena, they involve placing a dense set of predefined anchors at each location of the feature map, which dramatically increases the computational cost. As a result, anchor-free detectors [24–26,32–34], which directly locate the object without manually defined anchors, have become popular. For example, CornerNet [24] directly regresses the top-left and bottom-right corner points and then groups them to form the final HBB. ExtremeNet [25] predicts four extreme points (top-most, left-most, right-most, and bottom-most) and one center point, and then groups them into the HBB. CenterNet [26] models an object as one single point and directly regresses the center point of the HBB. Unlike key point-based anchor-free detectors, which treat several key points of the objects as positive samples, pixel-based anchor-free detectors attempt to solve the problem in a per-pixel prediction fashion. RepPoints [32] introduces a set of representative points that adaptively learn to position themselves over an object. Tian et al. [33] regard all the pixels inside the object HBB as positive samples. Motivated by the human eye system, Kong et al. [34] regard the pixels inside the fovea area of the object HBB as the positive samples. Both of them predict four distances to the four sides of HBB from the positive pixels to form the HBB. Anchor-free detection methods are fast in inference and also achieve competitive detection results with anchor-based detection methods.

2.1.2. Oriented Object Detection

Oriented object detection is receiving significant attention in areas such as remote sensing images and natural scene text. Oriented object detectors use OBBs to locate arbitrary-rotated objects other than HBBs because the objects in those scenes usually have large aspect ratios and are densely packed.

Oriented object detectors often use generic object detectors as a baseline and then add specially designed modules to regress OBB from HBB. Based on Faster-RCNN [13], RRPN [23] uses Rotation RPN and Rotation RoI pooling for arbitrary-oriented text detection. The RoI Transformer [20] utilizes a learnable module to transform horizontal RoIs to RRoIs. Xu et al. [22] propose to glide each vertex on the four corresponding sides of HBB to represent OBB, and Ye et al. [35] introduce feature fusion and feature filtration modules to exploit multilevel context information.

Based on RetinaNet [19], ADT-Det [36] uses a feature pyramid transformer that enhances features through feature interaction with multiple scales and layers. S^2A-Net [29] utilizes a feature alignment module for full feature alignment and an oriented detection module to alleviate the inconsistency between classification and regression. R^3Det [37] uses a feature refinement module to re-encode the position information and then reconstruct the entire feature map through pixel-wise interpolation.

Some research has adopted the OBB based on the semantic-segmentation network, such as Mask RCNN [14]. Mask OBB [38] is the first to treat the oriented object detection as an instance segmentation problem. Wang et al. [39] propose a center probability map OBB that gives a better OBB representation by reducing the influence of background pixels inside the OBB and obtaining higher detection performance.

Aside from the above anchor-based detectors, some rotation object detectors use an anchor-free approach. Based on CenterNet [26], Pan et al. [27] propose DRN by adding a branch to regress the orientations of the OBBs, and Shi et al. [40] develop a multi-task learning procedure to weight multi-task loss function during training. Other anchor-free detectors use new OBB representations. Xiao et al. [41] adopt FCOS [33] as the baseline and propose axis learning to detect oriented objects by predicting the axis of the object. Guo et al. [42] propose CFA, which uses RepPoints [32] as its baseline, and construct a convex-hull set for each oriented object.

2.1.3. Contextual Information and Attention Mechanisms

Numerous studies have shown that using contextual information and attention mechanisms can improve the performance of vision tasks such as scene classification, object detection, and instance segmentation.

For example, Wang et al. [43] use a novel locality and structure regularized low-rank representation method to characterize the global and local structures for hyper-spectral image classification task. ARCnet [44] utilizes a novel recurrent attention structure to force the scene classifiers to learn to focus on some critical areas of the very high-resolution RSIs, which often contain complex objects. AGMFA-Net [45] uses an attention-guided multi-layer feature aggregation network to capture more complete semantic regions for more powerful scene representation.

Contextual information aggregation has been widely adopted in semantic segmentation networks. To enhance the ability of the network to distinguish small-scale objects, CFEM [46] uses a context-based feature enhancement module to enhance the discriminant ability to distinguish small objects. HRCNet [47] utilizes a lightweight high-resolution context extraction network to acquire global context information and recognize the boundary being.

The usefulness of contextual information has been verified by many studies [35,48,49] in aerial object detection, especially when object appearances are insufficient due to small size, occlusion, or complicated backgrounds. CADNet [48] incorporates global and local contextual information and has a spatial-and-scale awareness attention module for object detection in RSIs. Wu et al. [49] propose a local context module that establishes the positional relationships between a proposal and its surrounding region pixels to help detect objects. $\mathcal{F}$3-Net [35] uses a feature fusion module that extracts the contextual information at different scales.

Attention mechanisms also show promise in oriented object detection by guiding the processing to more informative and relevant regions. ROSD [50] uses an orientation attention module to enhance the orientation sensitivity for accurate rotated object regression. CFC-Net [51] utilizes polarized attention to construct task-specific critical features. Li et al. [52] use a center-boundary dual attention module to extract attention features on the oriented objects' center and boundary regions. RADet [53] uses a multi-layer attention network focused simultaneously on objects' spatial position and features. SCRDet [54] uses a supervised multi-dimensional attention network consisting of a pixel attention network and channel attention network to suppress the noise and highlight the foreground.

2.1.4. OBB Representation Methods

The two most widely used OBB representation methods are the angle-based five-parameter representation method and the vertex-based eight-parameter representation method. The more commonly used five-parameter representation directly adds an angle parameter θ to HBB representation (x, y, w, h), and the definition of the angle θ is the acute angle determined by the long side of the rectangle and X-axis. The eight-parameter representation directly adopts the four corners of the OBB, e.g., $(x_1, y_1, x_2, y_2, x_3, y_3, x_4, y_4)$.

Although oriented object detectors with either form of OBB representation have demonstrated good performance, the inherent drawbacks of these two representations hinder the further improvement of the detection results [55]. The angular parameters embedded in the five-parameter representation encounter the problem of angular periodicity, leading to difficulty in the learning process. In contrast, the eight-parameter representation requires the exact same points order of ground truth and prediction, which otherwise leads to an unstable training process.

To handle these problems, some detectors have introduced new representations along with the anchor-free model. Axis learning [41] locates objects by predicting their axis and width, the latter of which is vertical to the axis. O^2DNet [56] treats the objects as pairs of middle lines. SAR [57] uses a brand-new representation with a circle-cut horizontal rectangle. Wu et al. [58] propose a novel projection-based method for describing OBB. Yi et al. [59] propose BBAVectors to regress one center point and four middle points of the corresponding sides to form the OBB. X-LineNet [9] uses paired appearance-based intersecting line segments to represent aircraft.

The above representations are all based on cartesian coordinates, and recently, the representation based on polar coordinates has been employed for rotated object detection and instance segmentation. Polar Mask [60], which model instance masks in the polar coordinates as one center and n rays, achieves competitive performance with much simpler and more flexible. Polar coordinates-based representations have been proved helpful in rotation and direction-related problems. Following Polar Mask, some rotated object detectors [61,62] also adopt polar representation and show great potential. PolarDet [61] represents the OBB by multiple angles and shorter-polar diameter ratios. However, the OBB representation of PolarDet needs 13 parameters, and some of them are redundant. In contrast, we propose a similar but more efficient representation method with only seven parameters. P-RSDet [62] regresses three parameters in polar coordinates, which include a polar radius ρ and the first two angles, to form the OBB and put forward a new Polar Ring Area Loss to improve the prediction accuracy.

2.2. *Method*

2.2.1. Overall Architecture

As shown in Figure 3, the proposed detector follows the two-stage detection paradigm, and contains four modules: the backbone for feature extraction, a CCA-FPN for feature representation enhancement with contextual information, an AFO-RPN for RRoI generation, and oriented RCNN heads for the final class and locations of the rotational object. For the backbone, we adopted ResNet [12], which is commonly used in many oriented detectors.

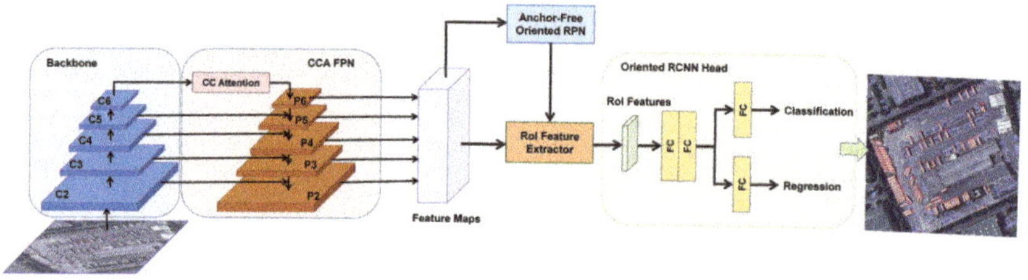

Figure 3. Overall architecture of the proposed method. There are four modules: backbone, Criss-Cross Attention FPN, anchor-free oriented RPN, and oriented RCNN heads.

2.2.2. Criss-Cross Attention FPN

Contextual information has been shown to be helpful in many computer vision tasks, such as scene classification, object detection, and semantic segmentation. In general, contextual information in vision describes the relationship between a pixel and its surrounding pixels.

One of the characteristics of RSIs is that the same category objects are often distributed in a particular region, such as vehicles in a parking lot or ships in a harbor. Another characteristic is that objects are closely related to the scene—for example, airplanes are closely related to an airport, and ships are closely related to the water.

Motivated by the above observations and analysis, we propose a Criss-Cross Attention FPN to fully exploit the contextual information of each pixel and its neighbors, which enhances the feature representation of the objects. Specifically, we embed the cascaded criss-cross attention modules into the FPN to enhance the pixel representations. The criss-cross attention module first used in CC-Net [28] is designed to collect the contextual information in the criss-cross path in order to enhance the pixel representative ability by modeling full-patch image dependencies over local features.

Given a feature map $\mathbf{H} \in \mathbb{R}^{C\times W\times H}$, we first apply three 1×1 convolutional layers on $\mathbf{H}$ to obtain three feature maps: queries map $\mathbf{Q}$, keys map $\mathbf{K}$, and values map $\mathbf{V}$. Note that $\mathbf{Q}$ and $\mathbf{K}$ have the same dimension, where $\{\mathbf{Q}, \mathbf{K}\} \in \mathbb{R}^{C'\times W\times H}$, and $\mathbf{V}$ has the same dimension as $\mathbf{H}$. We set C' less than C for the purpose of dimension reduction.

Next, we obtain a vector $\mathbf{Q_u}$ at each spatial position $\mathbf{u}$ of $\mathbf{Q}$ and the set $\mathbf{\Omega_u}$ in which the vectors are extracted from the same row and column with spatial position $\mathbf{u}$ from keys map $\mathbf{K}$. The correlation vector $\mathbf{D_u}$ is calculated by applying affinity operation on query vector $\mathbf{Q_u}$ and key vector set $\mathbf{\Omega_u}$ as follows:

$$\mathbf{D_u} = \mathbf{Q_u}\mathbf{\Omega_u}^T, \tag{1}$$

where $\mathbf{D_u} \in \mathbb{R}^{W+H-1}$. Next, we calculate the attention vector $\mathbf{A_u}$ by applying softmax function on $\mathbf{D_u}$ over the channel dimension, as follows:

$$\mathbf{A_u} = softmax(\mathbf{D_u}). \tag{2}$$

Then, we obtain the value vector set $\mathbf{\Phi_u}$, in which the value vectors are extracted from the same row and column with position $\mathbf{u}$ of $\mathbf{V}$. The contextual information is collected by an aggregation operation defined as:

$$\mathbf{H}'_\mathbf{u} = \sum_{i=0}^{W+H-1} \mathbf{A}_{i,\mathbf{u}}\mathbf{\Phi}_{i,\mathbf{u}} + \mathbf{H_u}, \tag{3}$$

where $\mathbf{H}' \in \mathbb{R}^{C\times W\times H}$ is the output of criss-cross attention module, which aggregates contextual information together with each pixel. A single criss-cross attention module can only capture contextual information of pixels in horizontal and vertical directions. However, it is not sufficient to focus only on the criss-cross path information for the problem of oriented object detection. To capture the contextual information in other directions, we use two cascaded criss-cross attention modules, following CC-Net [28].

2.2.3. Anchor-Free Oriented Region Proposal Network

As shown in Figure 3, the CCA-FPN produces five levels of feature maps $\{P_2, P_3, P_4, P_5, P_6\}$, where their strides $\{s_2, s_3, s_4, s_5, s_6\}$ are 4, 8, 16, 32, and 64, respectively. The proposed AFO-RPN takes the feature map P_i as input and outputs a set of oriented proposals, as shown in Figure 4. We introduce the polar representation method of OBB and then present the details of AFO-RPN.

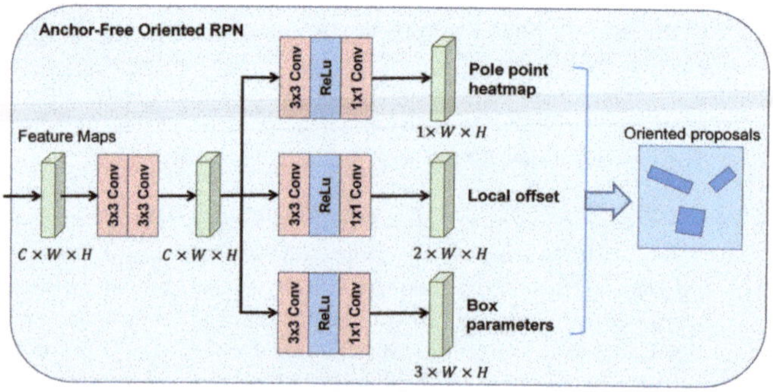

Figure 4. Details of the proposed AFO-RPN.

2.2.4. Polar Representation of OBB

Instead of the commonly used Cartesian-based OBB representation, we use the polar-based OBB representation in this paper, as shown in Figure 5. Specifically, the centroid of each object is used as the origin of the polar coordinates, and we use $(c_x, c_y, \rho, \gamma, \varphi)$ to represent the OBB, where c_x, c_y are the centroids of the OBB, which are also the poles of the polar coordinates. ρ is the radius, which calculates the distance from the centroid to the vertex, and γ is the central angle corresponding to the short side of the OBB. This representation is more robust than the one that uses w and h to represent a rectangular box's long and short sides. The reason is that using w and h to represent the rectangular box is prone to the problem of confusion between w and h when the rectangular box is close to the square [55]. However, by using ρ and γ to represent rectangular, the confusion between w and h can be avoided. φ represents the rotation angle of the OBB, which is defined in the polar coordinate system. We define the beginning angle 0° to coincide with the positive y-axis and increase the angle counterclockwise.

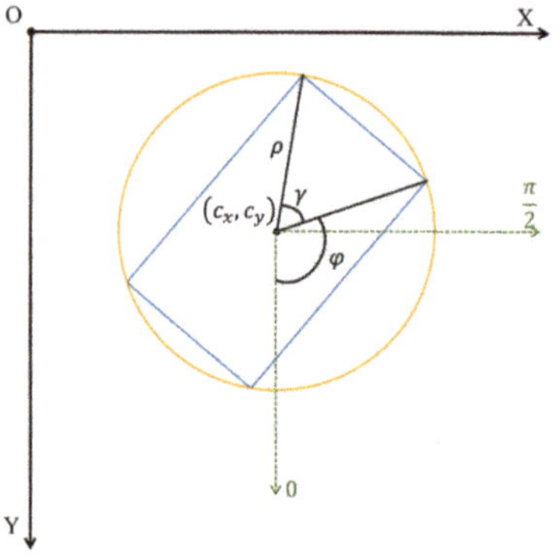

Figure 5. Proposed polar representation of OBB.

2.2.5. Pole Point Regression

Following previous work such as CenterNet [26], we use pole point (center point of the OBB) heatmaps to represent the location and objectness of the objects. Unlike CenterNet, which uses a 2D Gaussian kernel with a diagonal correlation matrix to map the key point to heatmaps, we use the rotated Gaussian kernel with a correlation matrix related to the rotation angle of the ground truth box.

Specifically, for an OBB ground truth (c_x, c_y, w, h, θ), we place a 2D Gaussian distribution $\mathcal{N}(\mathbf{m}, \mathbf{\Sigma})$ to form the ground truth heatmap in the training stage. Here, $\mathbf{m} = \left(\left\lfloor \frac{c_x}{s} \right\rfloor, \left\lfloor \frac{c_y}{s} \right\rfloor \right)$

represents the center of the gaussian distribution mapped into the feature map, where s is the downsampling stride of each feature map. The correlation matrix $\boldsymbol{\Sigma}$ is calculated as:

$$\boldsymbol{\Sigma}^{\frac{1}{2}} = \mathbf{R}(\theta)\mathbf{S}\mathbf{R}^T(\theta), \tag{4}$$

where the rotation matrix $\mathbf{R}(\theta)$ is defined as:

$$\mathbf{R}(\theta) = \begin{bmatrix} \cos\theta & -\sin\theta \\ \sin\theta & \cos\theta \end{bmatrix}. \tag{5}$$

$\mathbf{S} = diag(\sigma_x, \sigma_y)$ is the standard deviation matrix, where $\sigma_x = w \times \sigma_p$, $\sigma_y = h \times \sigma_p$, and σ_p is an object size-adaptive standard deviation [26].

In the training stage, only the peaks of the Gaussians are treated as the positive samples; all the other points are negative. To handle the imbalance between the positive and negative samples, we use a pixel-wise logistic regression with variant focal loss as CenterNet [26]:

$$\mathcal{L}_k = \frac{-1}{N}\sum_{xy}\begin{cases} (1-\hat{Y}_{xy})^{\alpha}\log(\hat{Y}_{xy}), & \text{if } \hat{Y}_{xy} = 1 \\ (1-Y_{xy})^{\beta}\log(\hat{Y}_{xy})^{\alpha}\log(1-\hat{Y}_{xy}), & \text{otherwise} \end{cases}, \tag{6}$$

where $\hat{Y}_{xy}$ and Y_{xy} refer to the ground-truth and the predicted heatmap values, α and β are the hyper-parameters of the focal loss that control the contribution of each point, and N is the number of the objects in the input image.

Furthermore, to compensate for the quantization error caused by the output stride, we additionally predict a local offset map $\mathbf{O} \in \mathbb{R}^{2\times H\times W}$, slightly adjust the center point locations before remapping them to the input resolution, and the offset of the OBB center point is defined as $\mathbf{o} = \left(\frac{c_x}{s} - \lfloor\frac{c_x}{s}\rfloor, \frac{c_y}{s} - \lfloor\frac{c_y}{s}\rfloor\right)$.

The offset is optimized with a smooth L_1 loss [13]:

$$\mathcal{L}_{\mathbf{O}} = \frac{1}{N}\sum_k Smooth_{L_1}(\mathbf{o}_k - \hat{\mathbf{o}}_k), \tag{7}$$

where $\hat{\mathbf{o}}_k$ and $\mathbf{o}_k$ refer to the ground-truth and the predicted local offset of the kth object, respectively. The smooth L_1 loss is defined as:

$$Smooth_{L_1} = \begin{cases} 0.5x^2, & \text{if } |x| < 1 \\ |x| - 0.5, & \text{otherwise} \end{cases}. \tag{8}$$

2.2.6. Box Parameters Regression

The box parameters are defined as $\mathbf{b} = (\rho, \gamma, \varphi)$, where ρ is the radius that calculates the distance from the centroid to the vertex, γ is the central angle corresponding to the short side of the OBB, and φ represents the rotation angle of the OBB, as depicted in Figure 5. We predict the box parameter map $\mathbf{B} \in \mathbb{R}^{3\times W\times H}$ with a smooth L1 loss:

$$\mathcal{L}_{\mathbf{B}} = \frac{1}{N}\sum_k Smooth_{L_1}\left(\mathbf{b}_k - \hat{\mathbf{b}}_k\right), \tag{9}$$

where $\hat{\mathbf{b}}_k$ and $\mathbf{b}_k$ refer to the ground truth and the predicted box parameters of the kth object, respectively.

The overall training loss of AFO-RPN is:

$$\mathcal{L}_{AFO-RPN} = \mathcal{L}_k + \lambda_O\mathcal{L}_{\mathbf{O}} + \lambda_B\mathcal{L}_{\mathbf{B}}, \tag{10}$$

where λ_O and λ_B are the weighted factors to control the contributions of each item, and we set $\lambda_O = 1$ and $\lambda_B = 0.1$ in our experiments.

2.2.7. Oriented RCNN Heads

As shown in Figure 6, the RoI feature extractor takes a group of feature maps {P2, P3, P4, P5, P6} and a set of oriented proposals as input. We use the align conv module to extract a fix-sized RoI feature from the corresponding feature map. The details of the align conv can be referred to S^2A-Net [29]. Then we use two fully connected layers and two sibling fully connected layers to predict the classification scores and regress the final oriented bounding boxes, as shown in Figure 3. The loss of RCNN heads is the same as that in [20]. The RCNN heads loss is given by:

$$\mathcal{L}_{head} = \frac{1}{N_{cls}} \sum_i \mathcal{L}_{cls} + \frac{1}{N_{reg}} \sum_i p_i^\star \mathcal{L}_{reg}, \quad (11)$$

where N_{cls} and N_{reg} are the number of proposals generated by AFO-RPN and the positive proposals in a mini batch, respectively. $p_i^\star$ is an index and when ith proposal is positive, it is 1, otherwise it is 0.

The total loss function of the proposed method follows the multitask learning way, and it is defined as:

$$\mathcal{L}_{total} = \lambda_{AFO-RPN} \mathcal{L}_{AFO-RPN} + \lambda_{head} \mathcal{L}_{head}, \quad (12)$$

where $\lambda_{AFO-RPN}$ and λ_{head} are the weighted factors, and we set $\lambda_{AFO-RPN} = 1$ and $\lambda_{head} = 1$.

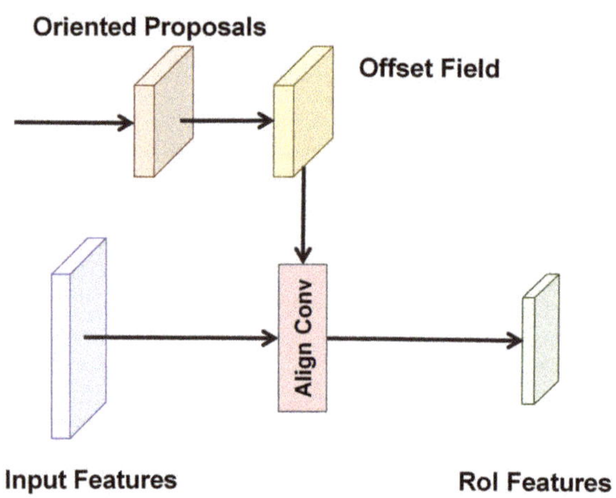

Figure 6. The details of RoI feature extractor module.

3. Results

3.1. Datasets

3.1.1. DOTA

DOTA [30] is one of the largest public aerial image detection datasets. It contains 2806 images ranging from 800 × 800 to 4000 × 4000 pixels and 188,282 instances labeled by arbitrarily oriented quadrilaterals over 15 categories: plane (PL), baseball diamond (BD), bridge (BR), ground track field (GTF), small vehicle (SV), large vehicle (LV), ship (SH), tennis court (TC), basketball court (BC), storage tank (ST), soccer-ball field (SBF), roundabout (RA), harbor (HA), swimming pool (SP), and helicopter (HC). The total dataset is divided into the training set (1411 images), validation set (458 images), and test set (937 images). We used the training set for network training and the validation set for evaluation in the ablation experiments. In a comparison with state-of-the-art object detectors, the training set and validation set were both used for network training, and the corresponding results on the test set were submitted to the official evaluation server at https://captain-whu.github.io/DOTA/evaluation.html (accessed on 27 January 2022). Following [20], we crop the original images into 1024 × 1024 patches with a stride 200 for training and testing. For multi-scale training and testing, we first resize original images at three scales (0.5, 1.0, and 1.5) which are chosen empirically, and then crop them into 1024 × 1024 patches with a stride of 512.

3.1.2. DIOR-R

DIOR-R [31] is a revised dataset of DIOR [1], which is another publicly available arbitrary-oriented object detection dataset in the earth observation community. It contains 23,463 images with a fixed size of 800 × 800 pixels and 192,518 annotated instances, covering a wide range of scenes. The spatial resolutions range from 0.5 m to 30 m. The objects of this dataset belong to 20 categories: airplane (APL), airport (APO), baseball field (BF), basketball court (BC), bridge (BR), chimney (CH), expressway service area (ESA), expressway toll station (ETS), dam (DAM), golf field (GF), ground track field (GTF), harbor (HA), overpass (OP), ship (SH), stadium (STA), storage tank (STO), tennis court (TC), train station (TS), vehicle (VE), and windmill (WM). The dataset is divided into the training (5862 images), validation (5863 images), and test (11,738 images) sets. For a fair comparison with other methods, the proposed detector is trained on the train+val set and evaluated on the test set.

3.1.3. HRSC2016

HRSC2016 [7] is an oriented ship detection dataset that contains 1061 images of rotated ships with large aspect ratios, collected from six famous harbors, including ships on the sea and close in-shore. The images range from 300 × 300 to 1500 × 900 pixels, and the ground sample distances are between 2 m and 0.4 m. The dataset is randomly split into the training set, validation set, and test set, containing 436 images including 1207 instances, 181 images including 541 instances, and 444 images including 1228 instances, respectively. We used both the training and validation sets for training and the test set for evaluation in our experiments. All images were resized to 800 × 1333 without changing the aspect ratio.

3.2. Implementation Details

We used ResNet 101 [12] as the backbone network for comparisons with state-of-the-art methods. Our model was implemented on the mmdetection [20] library. We optimized the model by using the SGD algorithm, and the initial learning rate was set to 0.005. The momentum and weight decay were 0.9 and 0.0001, respectively. The DOTA and DIOR-R datasets were trained by 12 epochs in total, and the learning rate was divided by 10 at eight epochs and 11 epochs, respectively. The HRSC2016 dataset was trained by 36 epochs in total, and the decay steps were 24 and 33 epochs. We used one Nvidia Titan XP GPU for all the experiments.

In this article, we adopt the mean Average Precision (mAP) metric to evaluate the multi-class detection accuracy of all experiments. mAP is the average of AP values for all classes:

$$\text{mAP} = \frac{\sum_{i=1}^{N} \text{AP}_i}{N}, \tag{13}$$

where N is number of classes. The AP metric is measured by the area under the precision-recall curve. The higher the mAP value, the better the performance, and vice versa.

3.3. Comparisons with State-of-the-Art Methods

3.3.1. Results on DOTA

To validate the effectiveness of our method, we compared it with several state-of-the-art methods on the DOTA dataset test set. The results were evaluated by the official DOTA evaluation server. As shown in Table 1, our model achieved a 76.57% mAP, which is higher than many advanced methods. With the multi-scale training and testing strategy, our model achieved an 80.68% mAP. Some detection results are shown in Figure 7.

Table 1. Comparisons with state-of-the-art methods on DOTA dataset test set. * means multi-scale training and testing. **Bold** denotes the best detection results.

Method	Backbone	PL	BD	BR	GTF	SV	LV	SH	TC	BC	ST	SBF	RA	HA	SP	HC	mAP(%)
One-stage																	
DAL [63]	ResNet 101	88.61	79.69	46.27	70.37	65.89	76.10	78.53	90.84	79.98	78.41	58.71	62.02	69.23	71.32	60.65	71.78
ProjBB-R [58]	ResNet 101	88.96	79.32	53.98	70.21	60.67	76.20	89.71	90.22	78.94	76.82	60.49	63.62	73.12	71.43	61.96	73.03
RSDet [64]	ResNet 152	90.2	83.5	53.6	70.1	64.6	79.4	67.3	91.0	**88.3**	82.5	64.1	68.7	62.8	69.5	66.9	73.5
CFC-Net [51]	ResNet 50	89.08	80.41	52.41	70.02	76.28	78.11	87.21	90.89	84.47	85.64	60.51	61.52	67.82	68.02	50.09	73.50
R^3Det [37]	ResNet 101	88.76	83.09	50.91	67.27	76.23	80.39	86.72	90.78	84.68	83.24	61.98	61.35	66.91	70.63	53.94	73.79
SLA [21]	ResNet 50	85.23	83.78	48.89	71.65	76.43	76.80	86.83	90.62	88.17	86.88	49.67	66.13	75.34	72.11	64.88	74.89
RDD [65]	ResNet 101	89.70	84.33	46.35	68.62	73.89	73.19	86.92	90.41	86.46	84.30	64.22	64.95	73.55	72.59	73.31	75.52
Two-stage																	
FR-O [30]	ResNet 101	79.42	77.13	17.7	64.05	35.3	38.02	37.16	89.41	69.64	59.28	50.3	52.91	47.89	47.4	46.3	54.13
RRPN [23]	ResNet 101	88.52	71.20	31.66	59.30	51.85	56.19	57.25	90.81	72.84	67.38	56.69	52.84	53.08	51.94	53.58	61.01
FFA [66]	ResNet 101	81.36	74.30	47.70	70.32	64.89	67.82	69.98	90.76	79.06	78.20	53.64	62.90	67.02	64.17	50.23	68.16
RADet [53]	ResNeXt 101	79.45	76.99	48.05	65.83	65.46	74.40	68.86	89.70	78.14	74.97	49.92	64.63	66.14	71.58	62.16	69.09
RoI Transformer [20]	ResNet 101	88.64	78.52	43.44	75.92	68.81	73.68	83.59	90.74	77.27	81.46	58.39	53.54	62.83	58.93	47.67	69.56
CAD-Net [48]	ResNet 101	87.8	82.4	49.4	73.5	71.1	63.5	76.7	90.9	79.2	73.3	48.4	60.9	62.0	67.0	62.2	69.9
SCR-Det [54]	ResNet 101	89.98	80.65	52.09	68.36	68.36	60.32	72.41	90.85	87.94	86.86	65.02	66.68	66.25	68.24	65.21	72.64
ROSD [50]	ResNet 101	88.88	82.13	52.85	69.76	78.21	77.32	87.08	90.86	86.40	82.66	56.73	65.15	74.43	68.24	63.18	74.92
Gliding Vertex [22]	ResNet 101	89.64	85.00	52.26	77.34	73.01	73.14	86.82	90.74	79.02	86.81	59.55	70.91	72.94	70.86	57.32	75.02
SAR [57]	ResNet 101	89.67	79.78	54.17	68.29	71.70	77.90	84.63	**90.91**	88.22	87.07	60.49	66.95	75.13	70.01	64.29	75.28
Mask-OBB [38]	ResNeXt 101	89.56	85.95	54.21	72.90	76.52	74.16	85.63	89.85	83.81	86.48	54.89	69.64	73.94	69.06	63.32	75.33
APE [67]	ResNet 50	89.96	83.62	53.42	76.03	74.01	77.16	79.45	90.83	87.15	84.51	67.72	60.33	74.61	71.84	65.55	75.75
CenterMap-Net [39]	ResNet 101	89.83	84.41	54.60	70.25	77.66	78.32	87.19	90.66	84.89	85.27	56.46	69.23	74.13	71.56	66.06	76.03
CSL [55]	ResNet 152	90.25	85.53	54.64	75.31	70.44	73.51	77.62	90.84	86.15	86.69	69.60	68.04	73.83	71.10	68.93	76.17
ReDet [68]	ResNet 50	88.79	82.64	53.97	74.00	78.13	84.06	88.04	90.89	87.78	85.75	61.76	60.39	75.96	68.07	63.59	76.25
OPLD [69]	ResNet 101	89.37	85.82	54.10	79.58	75.00	75.13	86.92	90.88	86.42	86.62	62.46	68.41	73.98	68.11	63.69	76.43
HSP [70]	ResNet 101	**90.39**	86.23	56.12	**80.59**	77.52	73.26	83.78	90.80	87.19	85.67	69.08	**72.02**	76.98	72.50	67.96	78.01
Anchor-free																	
CenterNet-O [26]	Hourglass 104	89.02	69.71	37.62	63.42	65.23	63.74	77.28	90.51	79.24	77.93	44.83	54.64	55.93	61.11	45.71	65.04
Axis Learning [41]	ResNet 101	79.53	77.15	38.59	61.15	67.53	70.49	76.30	89.66	79.07	83.53	47.27	61.01	56.28	66.06	36.05	65.98
P-RSDet [62]	ResNet 101	88.58	77.84	50.44	69.29	71.10	75.79	78.66	90.88	80.10	81.71	57.92	63.03	66.30	69.70	63.13	72.30
BBAVectors [59]	ResNet 101	88.35	79.96	50.69	62.18	78.43	78.98	87.94	90.85	83.58	84.35	54.13	60.24	65.22	64.28	55.70	72.32
O^2-Det [56]	Hourglass 104	89.3	83.3	50.1	72.1	71.1	75.6	78.7	90.9	79.9	82.9	60.2	60.0	64.6	68.9	65.7	72.8
PolarDet [61]	ResNet 50	89.73	**87.05**	45.30	63.32	78.44	76.65	87.13	90.79	80.58	85.89	60.97	67.94	68.20	74.63	68.67	75.02
AOPG [31]	ResNet 101	89.14	82.74	51.87	69.28	77.65	82.42	88.08	90.89	86.26	85.13	60.60	66.30	74.05	67.76	58.77	75.39
CBDANet [52]	DLA 34	89.17	85.92	50.28	65.02	77.72	82.32	87.89	90.48	86.47	85.90	66.85	66.48	67.41	71.33	62.89	75.74
CFA [42]	ResNet 152	89.08	83.20	54.37	66.87	**81.23**	80.96	87.17	90.21	84.32	86.09	52.34	69.94	75.52	**80.76**	67.96	76.67
Proposed Method	ResNet 101	89.23	84.50	52.90	76.93	78.51	76.93	87.40	90.89	87.42	84.66	64.40	63.97	75.01	73.39	62.37	76.57
Proposed Method *	ResNet 101	90.20	84.94	**61.04**	79.66	79.73	**84.37**	**88.78**	90.88	86.16	**87.66**	**71.85**	70.40	**81.37**	79.71	**73.51**	**80.68**

Figure 7. Depictions of the detection results on the DOTA dataset test set. We use bounding boxes of different colors to represent different categories.

3.3.2. Results on DIOR-R

DIOR-R is a new oriented object detection dataset, so we retrained and tested several advanced methods to conduct fair performance comparisons. As shown in Table 2, Faster RCNN OBB [30], as the baseline two-stage oriented method, and RetinaNet OBB [19], as the baseline single-stage oriented method, achieved 57.14% and 55.92% mAP, respectively. As the advanced methods, RoI Transformer [20] and Gliding Vertex [22] achieved 65.93% and 61.81% mAP, respectively. AOPG [31], as the baseline method in the DIOR-R dataset, achieved 64.41% mAP. Our model achieved 65.80% mAP with ResNet 50 [12] as the backbone and 67.15% mAP with ResNet 101 [12] as the backbone. The detection results are depicted in Figure 8.

Table 2. Comparisons with state-of-the-art methods on DIOR-R dataset test set. **Bold** denotes the best detection results.

Method	Backbone	APL	APO	BF	BC	BR	CH	DAM	ETS	ESA	GF	GTF	HA	OP	SH	STA	STO	TC	TS	VE	WM	mAP
RetinaNet-O [19]	ResNet 101	64.20	21.97	73.99	86.76	17.57	72.62	72.36	47.22	22.08	77.90	76.60	36.61	30.94	74.97	63.35	49.21	83.44	44.93	37.53	64.18	55.92
FR-O [30]	ResNet 101	61.33	14.73	71.47	86.46	19.86	72.24	59.78	55.98	19.72	77.08	81.47	39.21	33.30	78.78	70.05	61.85	81.31	53.44	39.90	64.81	57.14
Gliding Vertex [22]	ResNet 101	61.58	36.02	71.61	86.87	33.48	72.37	72.85	64.62	25.78	76.03	81.81	42.41	47.25	80.57	69.63	61.98	86.74	58.20	41.87	64.48	61.81
AOPG [31]	ResNet 50	62.39	37.79	71.62	87.63	40.90	72.47	31.08	65.42	77.99	73.20	81.94	42.32	54.45	81.17	72.69	71.31	81.49	60.04	52.38	69.99	64.41
RoI Trans [20]	ResNet 101	61.54	45.46	71.90	87.48	**41.43**	**72.67**	78.67	**67.17**	**38.26**	**81.83**	**83.40**	**48.94**	55.61	81.18	**75.06**	62.63	88.36	**63.09**	47.80	66.10	65.93
Proposed Method	ResNet 50	**68.26**	38.34	77.35	88.10	40.68	72.48	78.90	62.52	30.64	73.51	81.32	45.51	**55.78**	**88.74**	71.24	71.12	**88.60**	59.74	**52.95**	**70.30**	65.80
Proposed Method	ResNet 101	61.65	**47.58**	**77.59**	**88.39**	40.98	72.55	**81.90**	63.76	38.17	79.49	81.82	45.39	54.94	88.67	73.48	**75.75**	87.69	61.69	52.43	69.00	**67.15**

Figure 8. Depictions of the detection results on the DIOR-R dataset test set. We use bounding boxes of different colors to represent different categories.

3.3.3. Results on HRSC2016

The HRSC2016 dataset contains many densely packed ship instances with arbitrary orientation and large aspect ratios. Table 3 shows the results of our comparison of the proposed method with several state-of-the-art methods. Our model achieved 89.96% mAP with ResNet 50 as the backbone and 90.45% mAP with ResNet 101 as the backbone, which shows the effectiveness of dealing with such objects. As shown in Figure 9, our model accurately detects ships in complex remote sensing images.

Table 3. Comparisons with other methods on HRSC2016 dataset test set. **Bold** denotes the best detection results.

Method	Backbone	Image Size	mAP
Axis Learning [41]	ResNet 101	800 × 800	78.15
SLA [21]	ResNet 50	768 × 768	87.14
SAR [57]	ResNet 101	896 × 896	88.11
Gliding Vertex [22]	ResNet 101	-	88.2
OPLD [69]	ResNet 50	1024 × 1333	88.44
BBAVectors [59]	ResNet 101	608 × 608	88.6
DAL [63]	ResNet 101	800 × 800	88.6
ProjBB-R [58]	ResNet 101	800 × 800	89.41
CSL [55]	ResNet 152	-	89.62
CFC-Net [51]	ResNet 101	800 × 800	89.7
ROSD [50]	ResNet 101	1000 × 800	90.08
PolarDet [61]	ResNet 50	800 × 800	90.13
AOPG [31]	ResNet 101	800 × 1333	90.34
ReDet [68]	ResNet 50	800 × 512	90.46
CBDANet [52]	DLA 34	512 × 512	**90.5**
Proposed Method	ResNet 50	800 × 1333	89.96
Proposed Method	ResNet 101	800 × 1333	90.45

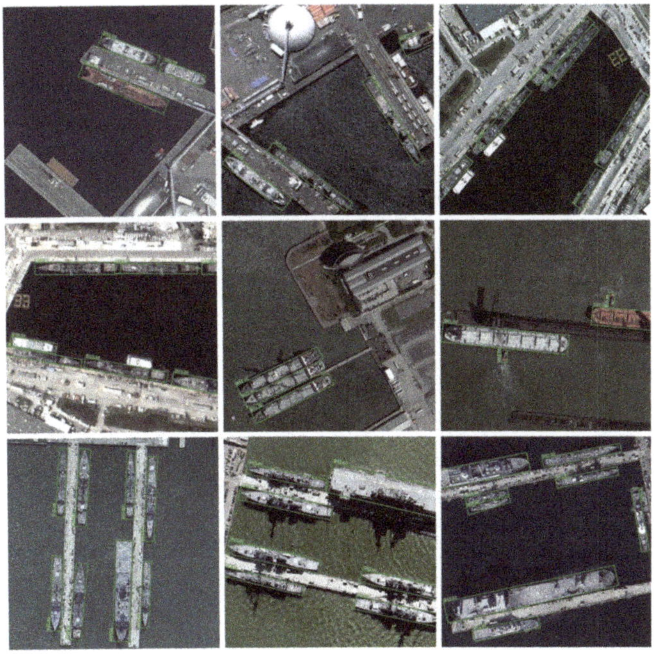

Figure 9. Depictions of the detection results on the HRSC2016 dataset test set.

4. Discussion

4.1. Ablation Study

To verify the effectiveness of the proposed method, we conducted ablation studies on the DOTA dataset test set. We used the RoI Transformer [20] with ResNet 101 [12] as the baseline in the experiments. It can be seen from the first row in Table 4 that the baseline method achieved 69.56% mAP, and from the fourth row that the proposed method with both CCA-FPN and AFO-RPN modules achieved a significant improvement of 7.01% mAP. Some visual comparison examples are shown in Figure 10.

Table 4. Ablation study of proposed modules on DOTA dataset test set.

Method	CCA-FPN	AFO-RPN	PL	BD	BR	GTF	SV	LV	SH	TC	BC	ST	SBF	RA	HA	SP	HC	mAP(%)
Baseline [20]	-	-	88.64	78.52	43.44	75.92	68.81	73.68	83.59	90.74	77.27	81.46	58.39	53.54	62.83	58.93	47.67	69.56
Proposed Method	✓	-	88.59	81.60	52.27	68.19	78.02	73.69	86.64	90.74	82.97	85.12	56.31	65.38	69.66	68.50	56.75	73.63 (+4.07)
	-	✓	88.88	84.06	52.13	69.55	70.96	76.59	79.52	90.87	87.23	86.19	56.14	65.35	66.96	72.08	64.20	74.05 (+4.49)
	✓	✓	89.23	84.50	52.90	76.93	78.51	76.93	87.40	90.89	87.42	84.66	64.40	63.97	75.01	73.39	62.37	76.57 (+7.01)

(a) (b)

Figure 10. Depictions of the detection results on the DOTA dataset test set. (**a**) Baseline [20]. (**b**) Proposed method.

4.1.1. Effect of the Proposed AFO-RPN

The third row of Table 4 shows 4.49% increases in terms of mAP with the AFO-RPN module. The proposed AFO-RPN is designed to generate high quality–oriented proposals without placing excessive fix-shaped anchors on the feature map. The accuracy for hard instance categories such as BD, BR, LV, BC, and HC increased by 5.54%, 8.69%, 2.91%, 9.96%, and 16.53% in terms of mAP, respectively. However, the accuracy for some categories such as GTF, SH, SBF decreased by 6.37%, 4.07%, and 2.25% in terms of mAP. The reason is that AFO-RPN is keypoint-based anchor-free method and it could face severe appearance ambiguity problems with backgrounds or other categories, as shown in Figure 2. The results prove the weakness of the anchor-free method

4.1.2. Effect of the CCA-FPN

The second row of Table 4 shows 4.07% increases in terms of mAP with CCA-FPN module. CCA-FPN is designed to enhance the feature representation of each pixel by capturing the contextual information. The accuracy for some hard instance categories, such as BR, SV, SH, BC, and RA, increased by 8.83%, 9.21%, 3.05%, 5.7%, and 11.84% in terms of mAP, respectively. It can be seen from the last two rows in Table 4, the performances for GTF, SH, SBF increased by 7.38%, 7.88%, 8.26% in terms of mAP, respectively. It shows contextual information is useful to enhance the representation of each point on the feature map.

We also compared the model's parameters (Params) and calculations (FLOPs) of the proposed method with baseline. The sizes of the input image are 800 × 800 pixels. The smaller Params and FLOPs, the higher the efficiency and the shorter inference time of the detector. The second row of Table 5 shows that the proposed method with AFO-RPN module has fewer parameters and low computational complexity. However, the third row of Table 5 shows that the CCA-FPN module brings huge parameters and a high computational burden.

Table 5. Evaluation results with the parameters and computational complexity.

Method	CCA-FPN	AFO-RPN	Params(M)	FLOPs(G)
Baseline [20]	-	-	55.13	148.38
Proposed Method	-	✓	41.73	134.38
	✓	✓	65.66	376.99

4.1.3. Effect of the Proposed Polar Representation of OBB

To explore the impacts of different OBB representation methods, we compared the proposed polar representation method with two commonly used Cartesian system representation methods—angle-based representation (x, y, w, h, θ) and vertex-based representation $(x_1, y_1, x_2, y_2, x_3, y_3, x_4, y_4)$—on the DOTA, DIOR-R, and HRSC2016 datasets. As shown in Table 6, the proposed polar representation method achieved a significant increase over the Cartesian system representation methods in all three datasets.

Table 6. Ablation study of proposed polar representation method of OBB.

Cartesian System	Polar System	DOTA mAP(%)	DIOR-R mAP(%)	HRSC2016 mAP(%)
(x, y, w, h, θ)	-	73.84	64.81	88.12
$(x_1, y_1, x_2, y_2, x_3, y_3, x_4, y_4)$	-	72.58	63.48	84.84
-	$(x, y, \rho, \gamma, \varphi)$	76.57	67.15	90.45

4.2. Limitations

As shown in Table 4, the utilization of the proposed AFO-RPN module improves the performance on many categories but degrades the performance on several categories. To solve this problem, we apply an attention module Criss-Cross Attention into FPN to enhance the feature representation by exploiting the contextual information. The proposed method with both CCA-FPN and AFO-RPN modules achieved a significant improvement while encountering another problem of calculation complexity, as shown in Table 5. This is a problem to be solved in future work.

5. Conclusions

In this paper, we analyzed the drawbacks of the mainstream anchor-based methods and found that both horizontal anchors and oriented anchors will hinder the further improvement of the oriented object detection results. To address this, we propose a two-stage coarse-to-fine oriented detector. The proposed method has the following novel features: (1) the proposed AFO-RPN, which generates high-quality oriented proposals without enormous predefined anchors; (2) the CCA-FPN, which enhances the feature representation of each pixel by capturing the contextual information; and (3) a new representation method of the OBB in the polar coordinates system, which slightly improves the detection performance. Extensive ablation studies have shown the superiority of the proposed modules. We achieved mAPs of 80.68% on the DOTA dataset, 67.15% on the DIOR-R dataset, and 90.45% on the HRSC2016 dataset, demonstrating that our method can achieve promising performance compared with the state-of-the-art methods.

However, despite the good performance, our method increased the parameters and computation cost. We will focus on improving the method and reducing the calculation burden in our future work.

Author Contributions: Conceptualization, J.L. and Y.T.; methodology, J.L.; software, J.L.; validation, J.L., Y.X. and Z.Z.; formal analysis, J.L. and Y.T.; investigation, Y.X. and Z.Z.; resources, Y.T. and Y.X.; data curation, J.L., Y.X. and Z.Z.; writing—original draft preparation, J.L.; writing—review and editing, Y.T.; visualization, J.L. and Z.Z.; supervision, Y.T.; project administration, J.L. All authors have read and agreed to the published version of the manuscript.

Funding: This research received no external funding.

Institutional Review Board Statement: Not applicable.

Informed Consent Statement: Not applicable.

Data Availability Statement: The datasets used in this study are available on request from the corresponding author.

Conflicts of Interest: The authors declare no conflict of interest.

Abbreviations

The following abbreviations are used in this manuscript:

RSI	Remote Sensing Image
DCNN	Deep Convolutional Neural Network
RSI	Remote Sensing Image
HBB	Horizontal Bounding Box
OBB	Oriented Bounding Box
RPN	Region Proposal Network
RoI	Region of Interest
FPN	Feature Pyramid Network
mAP	mean Average Precision
AFO-RPN	Anchor-Free Oriented Region Proposal Network
CCA-FPN	Criss-Cross Attention Feature Pyramid Network

References

1. Li, K.; Wan, G.; Cheng, G.; Meng, L.; Han, J. Object detection in optical remote sensing images: A survey and a new benchmark. *ISPRS J. Photogramm. Remote Sens.* **2020**, *159*, 296–307. [CrossRef]
2. Cheng, G.; Zhou, P.; Han, J. Learning Rotation-Invariant Convolutional Neural Networks for Object Detection in VHR Optical Remote Sensing Images. *IEEE Trans. Geosci. Remote Sens.* **2016**, *54*, 7405–7415. [CrossRef]
3. Han, J.; Zhang, D.; Cheng, G.; Guo, L.; Ren, J. Object Detection in Optical Remote Sensing Images Based on Weakly Supervised Learning and High-Level Feature Learning. *IEEE Trans. Geosci. Remote Sens.* **2015**, *53*, 3325–3337. [CrossRef]
4. Audebert, N.; Le Saux, B.; Lefèvre, S. Segment-before-Detect: Vehicle Detection and Classification through Semantic Segmentation of Aerial Images. *Remote Sens.* **2017**, *9*, 368. [CrossRef]

5. Li, J.; Zhang, Z.; Tian, Y.; Xu, Y.; Wen, Y.; Wang, S. Target-Guided Feature Super-Resolution for Vehicle Detection in Remote Sensing Images. *IEEE Geosci. Remote Sens. Lett.* **2021**, *19*, 1–5. [CrossRef]
6. Zou, Z.; Shi, Z. Ship Detection in Spaceborne Optical Image With SVD Networks. *IEEE Trans. Geosci. Remote Sens.* **2016**, *54*, 5832–5845. [CrossRef]
7. Liu, Z.; Yuan, L.; Weng, L.; Yang, Y. A high resolution optical satellite image dataset for ship recognition and some new baselines. In Proceedings of the 6th International Conference on Pattern Recognition Applications and Methods, Porto, Portugal, 24–26 February 2017; Volume 2, pp. 324–331.
8. Zhou, M.; Zou, Z.; Shi, Z.; Zeng, W.J.; Gui, J. Local Attention Networks for Occluded Airplane Detection in Remote Sensing Images. *IEEE Geosci. Remote Sens. Lett.* **2020**, *17*, 381–385. [CrossRef]
9. Wei, H.; Zhang, Y.; Wang, B.; Yang, Y.; Li, H.; Wang, H. X-LineNet: Detecting Aircraft in Remote Sensing Images by a Pair of Intersecting Line Segments. *IEEE Trans. Geosci. Remote Sens.* **2021**, *59*, 1645–1659. [CrossRef]
10. Cheng, G.; Han, J. A survey on object detection in optical remote sensing images. *ISPRS J. Photogramm. Remote Sens.* **2016**, *117*, 11–28. [CrossRef]
11. Blaschke, T. Object based image analysis for remote sensing. *ISPRS J. Photogramm. Remote Sens.* **2010**, *65*, 2–16. [CrossRef]
12. He, K.; Zhang, X.; Ren, S.; Sun, J. Deep Residual Learning for Image Recognition. In Proceedings of the 2016 IEEE Conference on Computer Vision and Pattern Recognition (CVPR), Las Vegas, NV, USA, 26 June–1 July 2016; pp. 770–778.
13. Ren, S.; He, K.; Girshick, R.; Sun, J. Faster R-CNN: Towards Real-Time Object Detection with Region Proposal Networks. *IEEE Trans. Pattern Anal. Mach. Intell.* **2016**, *39*, 1137–1149. [CrossRef] [PubMed]
14. He, K.; Gkioxari, G.; Dollar, P.; Girshick, R. Mask R-CNN. In Proceedings of the 2017 IEEE International Conference on Computer Vision (ICCV), Venice, Italy, 22–29 October 2017; pp. 2980–2988.
15. Redmon, J.; Divvala, S.; Girshick, R.; Farhadi, A. You Only Look Once: Unified, Real-Time Object Detection. In Proceedings of the 2016 IEEE Conference on Computer Vision and Pattern Recognition (CVPR), Las Vegas, NV, USA, 26 June–1 July 2016; pp. 779–788.
16. Redmon, J.; Farhadi, A. YOLO9000: Better, Faster, Stronger. In Proceedings of the 2017 IEEE/CVF Conference on Computer Vision and Pattern Recognition (CVPR), Honolulu, HI, USA, 21–26 July 2017; pp. 6517–6525.
17. Redmon, J.; Farhadi, A. YOLOv3: An Incremental Improvement. *arXiv* **2018**, arXiv:1804.02767.
18. Liu, W.; Anguelov, D.; Erhan, D.; Szegedy, C.; Reed, S.; Fu, C.Y.; Berg, A.C. SSD: Single Shot MultiBox Detector. In Proceedings of the European Conference on Computer Vision (ECCV), Amsterdam, The Netherlands, 11–14 October 2016; pp. 21–37.
19. Lin, T.Y.; Goyal, P.; Girshick, R.; He, K.; Dollar, P. Focal Loss for Dense Object Detection. In Proceedings of the 2017 IEEE International Conference on Computer Vision (ICCV), Venice, Italy, 22–29 October 2017; pp. 2999–3007.
20. Ding, J.; Xue, N.; Long, Y.; Xia, G.S.; Lu, Q. Learning RoI Transformer for Oriented Object Detection in Aerial Images. In Proceedings of the 2019 IEEE/CVF Conference on Computer Vision and Pattern Recognition (CVPR), Long Beach, CA, USA, 16–20 June 2019; pp. 2844–2853.
21. Ming, Q.; Miao, L.; Zhou, Z.; Song, J.; Yang, X. Sparse Label Assignment for Oriented Object Detection in Aerial Images. *Remote Sens.* **2021**, *13*, 2664. [CrossRef]
22. Xu, Y.; Fu, M.; Wang, Q.; Wang, Y.; Chen, K.; Xia, G.S.; Bai, X. Gliding Vertex on the Horizontal Bounding Box for Multi-Oriented Object Detection. *IEEE Trans. Pattern Anal. Mach. Intell.* **2021**, *43*, 1452–1459. [CrossRef]
23. Ma, J.; Shao, W.; Ye, H.; Wang, L.; Wang, H.; Zheng, Y.; Xue, X. Arbitrary-Oriented Scene Text Detection via Rotation Proposals. *IEEE Trans. Multimedia* **2018**, *20*, 3111–3122. [CrossRef]
24. Law, H.; Deng, J. CornerNet: Detecting Objects as Paired Keypoints. In Proceedings of the European Conference on Computer Vision (ECCV), Munich, Germany, 8–14 September 2018; pp. 734–750.
25. Zhou, X.; Zhuo, J.; Krahenbuhl, P. Bottom-Up Object Detection by Grouping Extreme and Center Points. In Proceedings of the 2019 IEEE/CVF Conference on Computer Vision and Pattern Recognition (CVPR), Long Beach, CA, USA, 16–20 June 2019; pp. 850–859
26. Zhou, X.; Wang, D.; Krähenbühl, P. Objects as Points. *arXiv* **2019**, arXiv:1904.07850.
27. Pan, X.; Ren, Y.; Sheng, K.; Dong, W.; Yuan, H.; Guo, X.; Ma, C.; Xu, C. Dynamic Refinement Network for Oriented and Densely Packed Object Detection. In Proceedings of the 2020 IEEE/CVF Conference on Computer Vision and Pattern Recognition (CVPR), Seattle, WA, USA, 14–19 June 2020; pp. 11204–11213.
28. Huang, Z.; Wang, X.; Huang, L.; Huang, C.; Wei, Y.; Liu, W. CCNet: Criss-Cross Attention for Semantic Segmentation. In Proceedings of the 2019 IEEE/CVF International Conference on Computer Vision (ICCV), Seoul, Korea, 20–26 October 2019; pp. 603–612.
29. Han, J.; Ding, J.; Li, J.; Xia, G.S. Align Deep Features for Oriented Object Detection. *IEEE Trans. Geosci. Remote Sens.* 2021, *accepted*. [CrossRef]
30. Xia, G.S.; Bai, X.; Ding, J.; Zhu, Z.; Belongie, S.; Luo, J.; Datcu, M.; Pelillo, M.; Zhang, L. DOTA: A Large-Scale Dataset for Object Detection in Aerial Images. In Proceedings of the 2018 IEEE/CVF Conference on Computer Vision and Pattern Recognition (CVPR), Salt Lake City, UT, USA, 18–22 June 2018; pp. 3974–3983.
31. Cheng, G.; Wang, J.; Li, K.; Xie, X.; Lang, C.; Yao, Y.; Han, J. Anchor-free Oriented Proposal Generator for Object Detection. *arXiv* **2021**, arXiv:2110.01931.

32. Yang, Z.; Liu, S.; Hu, H.; Wang, L.; Lin, S. RepPoints: Point Set Representation for Object Detection. In Proceedings of the 2019 IEEE/CVF International Conference on Computer Vision (ICCV), Seoul, Korea, 20–26 October 2019; pp. 9656–9665.
33. Tian, Z.; Shen, C.; Chen, H.; He, T. FCOS: Fully Convolutional One-Stage Object Detection. In Proceedings of the 2019 IEEE/CVF International Conference on Computer Vision (ICCV), Seoul, Korea, 20–26 October 2019; pp. 9626–9635.
34. Kong, T.; Sun, F.; Liu, H.; Jiang, Y.; Li, L.; Shi, J. FoveaBox: Beyound Anchor-Based Object Detection. *IEEE Trans. Image Process.* **2020**, *29*, 7389–7398. [CrossRef]
35. Ye, X.; Xiong, F.; Lu, J.; Zhou, J.; Qian, Y. $\mathcal{F}$3-Net: Feature Fusion and Filtration Network for Object Detection in Optical Remote Sensing Images. *Remote Sens.* **2020**, *12*, 4027. [CrossRef]
36. Zheng, Y.; Sun, P.; Zhou, Z.; Xu, W.; Ren, Q. ADT-Det: Adaptive Dynamic Refined Single-Stage Transformer Detector for Arbitrary-Oriented Object Detection in Satellite Optical Imagery. *Remote Sens.* **2021**, *13*, 2623. [CrossRef]
37. Yang, X.; Yan, J.C.; Feng, Z.M.; Hen, T. R3Det: Refined Single-Stage Detector with Feature Refinement for Rotating Object. In Proceedings of the AAAI Conference on Artificial Intelligence, Palo Alto, CA, USA, 2–9 February 2021.
38. Wang, J.; Ding, J.; Guo, H.; Cheng, W.; Pan, T.; Yang, W. Mask OBB: A Semantic Attention-Based Mask Oriented Bounding Box Representation for Multi-Category Object Detection in Aerial Images. *Remote Sens.* **2019**, *11*, 2930. [CrossRef]
39. Wang, J.; Yang, W.; Li, H.C.; Zhang, H.; Xia, G.S. Learning Center Probability Map for Detecting Objects in Aerial Images. *IEEE Trans. Geosci. Remote Sens.* **2021**, *59*, 4307–4323. [CrossRef]
40. Shi, F.; Zhang, T.; Zhang, T. Orientation-Aware Vehicle Detection in Aerial Images via an Anchor-Free Object Detection Approach. *IEEE Trans. Geosci. Remote Sens.* **2021**, *59*, 5221–5233. [CrossRef]
41. Xiao, Z.; Qian, L.; Shao, W.; Tan, X.; Wang, K. Axis Learning for Orientated Objects Detection in Aerial Images. *Remote Sens.* **2020**, *12*, 908. [CrossRef]
42. Guo, Z.; Liu, C.; Zhang, X.; Jiao, J.; Ji, X.; Ye, Q. Beyond Bounding-Box: Convex-Hull Feature Adaptation for Oriented and Densely Packed Object Detection. In Proceedings of the 2021 IEEE/CVF Conference on Computer Vision and Pattern Recognition, Nashville, TN, USA, 19–25 June 2021; pp. 8788–8797.
43. Wang, Q.; He, X.; Li, X. Locality and Structure Regularized Low Rank Representation for Hyperspectral Image Classification. *IEEE Trans. Geosci. Remote Sens.* **2019**, *57*, 911–923. [CrossRef]
44. Wang, Q.; Liu, S.; Chanussot, J.; Li, X. Scene Classification with Recurrent Attention of VHR Remote Sensing Images. *IEEE Trans. Geosci. Remote Sens.* **2019**, *57*, 1155–1167. [CrossRef]
45. Li, M.; Lei, L.; Tang, Y.; Sun, Y.; Kuang, G. An Attention-Guided Multilayer Feature Aggregation Network for Remote Sensing Image Scene Classification. *Remote Sens.* **2021**, *13*, 3113. [CrossRef]
46. Chong, Y.; Chen, X.; Pan, S. Context Union Edge Network for Semantic Segmentation of Small-Scale Objects in Very High Resolution Remote Sensing Images. *IEEE Geosci. Remote Sens. Lett.* **2020**, *19*, 6000305. [CrossRef]
47. Xu, Z.; Zhang, W.; Zhang, T.; Li, J. HRCNet: High-Resolution Context Extraction Network for Semantic Segmentation of Remote Sensing Images. *Remote Sens.* **2020**, *13*, 71. [CrossRef]
48. Zhang, G.; Lu, S.; Zhang, W. CAD-Net: A Context-Aware Detection Network for Objects in Remote Sensing Imagery. *IEEE Trans. Geosci. Remote Sens.* **2019**, *57*, 10015–10024. [CrossRef]
49. Wu, Y.; Zhang, K.; Wang, J.; Wang, Y.; Wang, Q.; Li, Q. CDD-Net: A Context-Driven Detection Network for Multiclass Object Detection. *IEEE Geosci. Remote Sens. Lett.* **2020**, *19*, 8004905. [CrossRef]
50. Zhang, K.; Zeng, Q.; Yu, X. ROSD: Refined Oriented Staged Detector for Object Detection in Aerial Image. *IEEE Access* **2021** *9*, 66560–66569. [CrossRef]
51. Ming, Q.; Miao, L.; Zhou, Z.; Dong, Y. CFC-Net: A Critical Feature Capturing Network for Arbitrary-Oriented Object Detection in Remote-Sensing Images. *IEEE Trans. Geosci. Remote Sens.* **2021**, *60*, 5605814 . [CrossRef]
52. Liu, S.; Zhang, L.; Lu, H.; He, Y. Center-Boundary Dual Attention for Oriented Object Detection in Remote Sensing Images. *IEEE Trans. Geosci. Remote Sens.* **2021**, *60*, 5603914. [CrossRef]
53. Li, Y.; Huang, Q.; Pei, X.; Jiao, L.; Shang, R. RADet: Refine Feature Pyramid Network and Multi-Layer Attention Network for Arbitrary-Oriented Object Detection of Remote Sensing Images. *Remote Sens.* **2020**, *12*, 389. [CrossRef]
54. Yang, X.; Yang, J.; Yan, J.; Zhang, Y.; Zhang, T.; Guo, Z.; Sun, X.; Fu, K. SCRDet: Towards More Robust Detection for Small, Cluttered and Rotated Objects. In Proceedings of the 2019 IEEE/CVF International Conference on Computer Vision (ICCV), Seoul, Korea, 20–26 October 2019; pp. 8231–8240.
55. Yang, X.; Yan, J. Arbitrary-Oriented Object Detection with Circular Smooth Label. In Proceedings of the European Conference on Computer Vision (ECCV), Virtual, 23–28 August 2020; pp. 677–694.
56. Wei, H.; Zhang, Y.; Chang, Z.; Li, H.; Wang, H.; Sun, X. Oriented objects as pairs of middle lines. *ISPRS J. Photogramm. Remote Sens.* **2020**, *169*, 268–279. [CrossRef]
57. Lu, J.; Li, T.; Ma, J.; Li, Z.; Jia, H. SAR: Single-Stage Anchor-Free Rotating Object Detection. *IEEE Access* **2020**, *8*, 205902–205912. [CrossRef]
58. Wu, Q.; Xiang, W.; Tang, R.; Zhu, J. Bounding Box Projection for Regression Uncertainty in Oriented Object Detection. *IEEE Access* **2021**, *9*, 58768–58779. [CrossRef]
59. Yi, J.; Wu, P.; Liu, B.; Huang, Q.; Qu, H.; Metaxas, D. Oriented Object Detection in Aerial Images with Box Boundary-Aware Vectors. In Proceedings of the 2021 IEEE Winter Conference on Applications of Computer Vision (WACV), Waikoloa, HI, USA, 5–9 January 2021; pp. 2149–2158.

60. Xie, E.; Sun, P.; Song, X.; Wang, W.; Liu, X.; Liang, D.; Shen, C.; Luo, P. PolarMask: Single Shot Instance Segmentation with Polar Representation. In Proceedings of the 2020 IEEE/CVF Conference on Computer Vision and Pattern Recognition (CVPR), Seattle, WA, USA, 14–19 June 2020; pp. 12190–12199.
61. Zhao, P.; Qu, Z.; Bu, Y.; Tan, W.; Guan, Q. PolarDet: A fast, more precise detector for rotated target in aerial images. *Int. J. Remote Sens.* **2021**, *42*, 5831–5861. [CrossRef]
62. Zhou, L.; Wei, H.; Li, H.; Zhao, W.; Zhang, Y.; Zhang, Y. Arbitrary-Oriented Object Detection in Remote Sensing Images Based on Polar Coordinates. *IEEE Access* **2020**, *8*, 223373–223384. [CrossRef]
63. Ming, Q.; Zhou, Z.; Miao, L.; Zhang, H.; Li, L. Dynamic Anchor Learning for Arbitrary-Oriented Object Detection. In Proceedings of the AAAI Conference on Artificial Intelligence, Palo Alto, CA, USA, 2–9 February 2021.
64. Qian, W.; Yang, X.; Peng, S.; Yan, J.; Guo, Y. Learning modulated loss for rotated object detection. In Proceedings of the AAAI Conference on Artificial Intelligence, Palo Alto, CA, USA, 2–9 February 2021.
65. Zhong, B.; Ao, K. Single-Stage Rotation-Decoupled Detector for Oriented Object. *Remote Sens.* **2020**, *12*, 3262. [CrossRef]
66. Fu, K.; Chang, Z.; Zhang, Y.; Xu, G.; Zhang, K.; Sun, X. Rotation-aware and multi-scale convolutional neural network for object detection in remote sensing images. *ISPRS J. Photogramm. Remote Sens.* **2020**, *161*, 294–308. [CrossRef]
67. Zhu, Y.; Du, J.; Wu, X. Adaptive Period Embedding for Representing Oriented Objects in Aerial Images. *IEEE Trans. Geosci. Remote Sens.* **2020**, *58*, 7247–7257. [CrossRef]
68. Han, J.; Ding, J.; Xue, N.; Xia, G.S. ReDet: A Rotation-equivariant Detector for Aerial Object Detection. In Proceedings of the 2021 IEEE/CVF Conference on Computer Vision and Pattern Recognition, Nashville, TN, USA, 19–25 June 2021; pp. 2785–2794.
69. Song, Q.; Yang, F.; Yang, L.; Liu, C.; Hu, M.; Xia, L. Learning Point-Guided Localization for Detection in Remote Sensing Images. *IEEE J. Sel. Top. Appl. Earth Obs. Remote Sens.* **2021**, *14*, 1084–1094. [CrossRef]
70. Xu, C.; Li, C.; Cui, Z.; Zhang, T.; Yang, J. Hierarchical Semantic Propagation for Object Detection in Remote Sensing Imagery. *IEEE Trans. Geosci. Remote Sens.* **2020**, *58*, 4353–4364. [CrossRef]

remote sensing MDPI

Article

Multiview Image Matching of Optical Satellite and UAV Based on a Joint Description Neural Network

Chuan Xu [1], Chang Liu [1,*], Hongli Li [2], Zhiwei Ye [1], Haigang Sui [3] and Wei Yang [4]

1 School of Computer Science, Hubei University of Technology, Wuhan 430068, China; 20200064@hbut.edu.cn (C.X.); hgcsyzw@mail.hbut.edu.cn (Z.Y.)
2 School of Computer Science and Engineering, Wuhan Institute of Technology, Wuhan 430070, China; lhl@wit.edu.cn
3 State Key Laboratory of Information Engineering in Surveying, Mapping and Remote Sensing, Wuhan University, Wuhan 430072, China; 00201543@whu.edu.cn
4 School of Information Science and Engineering, Wuchang Shouyi University, Wuhan 430064, China; yangwei403@wsyu.edu.cn
* Correspondence: 101910727@hbut.edu.cn; Tel.: +86-188-7140-7395

Abstract: Matching aerial and satellite optical images with large dip angles is a core technology and is essential for target positioning and dynamic monitoring in sensitive areas. However, due to the long distances and large dip angle observations of the aerial platform, there are significant perspective, radiation, and scale differences between heterologous space-sky images, which seriously affect the accuracy and robustness of feature matching. In this paper, a multiview satellite and unmanned aerial vehicle (UAV) image matching method based on deep learning is proposed to solve this problem. The main innovation of this approach is to propose a joint descriptor consisting of soft descriptions and hard descriptions. Hard descriptions are used as the main description to ensure matching accuracy. Soft descriptions are used not only as auxiliary descriptions but also for the process of network training. Experiments on several problems show that the proposed method ensures matching efficiency and achieves better matching accuracy for multiview satellite and UAV images than other traditional methods. In addition, the matching accuracy of our method in optical satellite and UAV images is within 3 pixels, and can nearly reach 2 pixels, which meets the requirements of relevant UAV missions.

Keywords: multiview; satellite and UAV image; joint description; image matching; neural network

Citation: Xu, C.; Liu, C.; Li, H.; Ye, Z.; Sui, H.; Yang, W. Multiview Image Matching of Optical Satellite and UAV Based on a Joint Description Neural Network. *Remote Sens.* **2022**, *14*, 838. https://doi.org/10.3390/rs14040838

Academic Editor: Fahimeh Farahnakian

Received: 19 December 2021
Accepted: 7 February 2022
Published: 10 February 2022

1. Introduction

Aviation and space-based remote sensing technology has been applied in many fields due to its advantages of macroscopic, rapid, and accurate object recognition [1]. Therefore, it has important theoretical significance and practical value for mining and is associated with different sensors (not simultaneously), different angles, and different resolutions of image information (space and sky images) to achieve high precision and high efficiency in regional dynamic monitoring, change detection, target recognition, positioning, and other visual tasks [2–5]. Space images mean the images captured by the airborne platform and sky images mean the images captured by the spaceborne platform. Among them, image matching is the key core technology, and the resulting matching effect directly affects and restricts the success or failure of the subsequent follow-up tasks.

Image matching technology refers to mapping an image to other images obtained under different conditions, such as different time phases, angles, and levels of illumination, through spatial transformation and the establishment of spatial correspondence relations among these images. It is the key technology of image processing and analysis and provides technical support for medical image analysis, industrial image detection, remote sensing image processing, and other fields [6]. Remote sensing image matching connects the subregions of different images that correspond to the same landform scene, which lays a foundation for follow-up operations such as remote sensing image registration, mosaic procedures, and fusion and can also provide supervisory information for scene analyses of remote sensing images [7]. Due to the observation of large aviation platform dip angles, there are significant differences between the viewing angles and scales of satellite and unmanned aerial vehicle (UAV) images, which brings great difficulties to the feature matching process for the satellite and UAV images. This is shown in Figure 1.

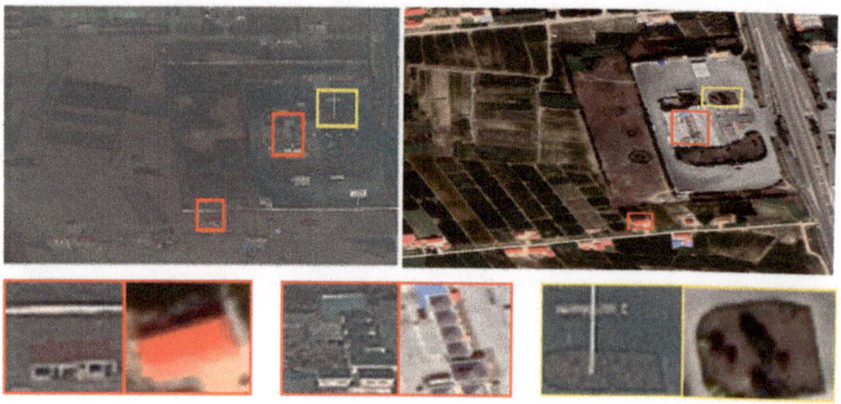

(**a**) View change 1 (**b**) View change 2 (**c**) Scale change

Figure 1. The UAV image is on the left and the satellite image is on the right. (**a**,**b**) show the difference in view between UAV images and satellite images. (**c**) shows the scale difference between UAV images and satellite images.

Due to differences in imaging mechanisms, illumination levels, time phases, and viewing angles, there are obvious nonlinear radiation distortions between UAV images and satellite images. Therefore, it is difficult to achieve reliable matching with multiview heterogeneous images by using only traditional artificial image-gradient-based operators (such as the scale-invariant feature transform (SIFT)) [8]. With the development of deep learning, convolutional neural networks (CNNs) have achieved great success in the field of image processing [9–11]. The convolutional layer in a CNN has strong feature extraction ability. Compared with artificially designed feature descriptors, CNN features can be trained by a network model to enable a deep network to find the most appropriate feature extraction process and representation form. Therefore, CNNs can be used for image matching to better solve the influence of nonlinear radiation distortion between images, which cannot be solved by the underlying gradient feature. During the process of network training, the parameters of the network layer are updated by monitoring information and a back-propagation function so that the CNN also has good robustness to deformation and noise. This paper proposes a joint description neural network specifically designed to match multiview satellite and UAV images. Compared with some traditional methods, the proposed method can achieve better results in the multiview satellite image and remote sensing image matching. First, the proposed method extracts features and filters them through a CNN. Second, the extracted features are expressed by hard and soft descriptions. Then, the loss function of the neural network is designed with a soft descriptor for neural network training. Finally, the hard description and soft description are combined as the final feature description, and the final matching result is obtained. The main contributions of this paper can be summarized as follows:

(1) A soft description method is designed for network training and auxiliary description.
(2) A high-dimensional hard description method is designed to ensure the matching accuracy of the model.

(3) The joint descriptor supplements the hard descriptor to highlight the differences between different features.

The rest of this article is organized as follows. In Section 2, the related works of image matching are briefly discussed. In Section 3, a neural network matching method is presented that includes feature detection, hard and soft descriptors, joint descriptors, multiscale models, and a training loss. In Section 4, the experimental results for this model are discussed. Finally, the conclusion is presented in Section 5.

2. Related Works

The existing image matching methods can be divided into gray-based matching methods and feature-based matching methods. These two kinds of methods as well as the image matching method based on deep learning and the improved method of multiperspective image matching will be reviewed and analyzed in the following sections. The practical image matching method is based on grayscale at the beginning. Due to the limitations of the method based on grayscale, the feature-based image matching method was proposed later, which greatly improves the applicability of image matching technology. In recent years, with the rapid development of deep learning technology, image matching methods based on deep learning are becoming more and more popular, which has brought the image matching technology to a new level. Its development is shown in Figure 2.

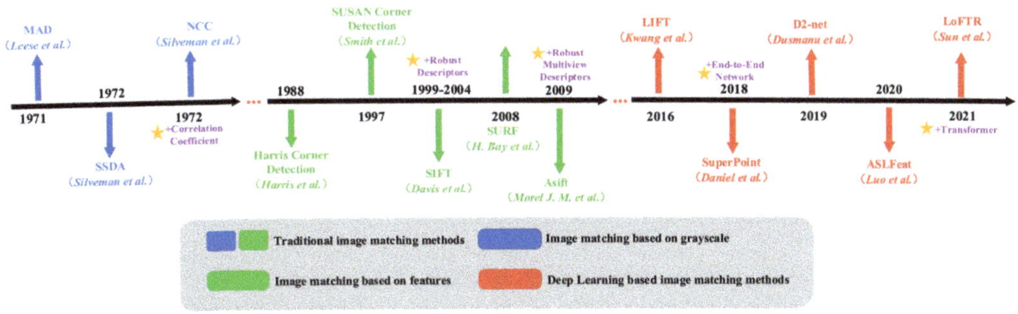

Figure 2. Image matching development history map. In the 1970s and 1980s, the main method of image matching was based on grayscale. By the end of the last century, feature-based image matching methods became popular. In recent years, with the development of deep learning technology, more and more image matching methods based on deep learning have been emerging.

2.1. Matching Method Based on Grayscale

The basic idea of image matching algorithms based on grayscale is to directly compare the grayscale values of image pixels one by one; this is the most basic type of matching method. Such an approach compares the similarity of all gray values of all pixels in the image and then uses a certain algorithm to search for the transformation model parameter value that maximizes or minimizes the similarity to judge the whole image. The similarity measurement functions commonly used in this kind of matching algorithm include the sum of squares, correlation, covariance, cross-correlation, and phase correlation functions of the gray difference between two images.

Image matching methods based on grayscale are most suitable for image pairs with only the rotation and scaling geometric relations. Leese [12] first proposed the multivariate alteration detection (MAD) algorithm in 1971, which is the basic image matching algorithm based on image gray levels. Subsequently, Silverman and Barnea [13] proposed the sequential similarity detection algorithm (SSDA) based on the MAD algorithm and then proposed the normalized cross-correlation (NCC) algorithm. Compared with other image matching algorithms based on the grayscale, the NCC algorithm has been proven to be the best approach for similarity evaluation, so the NCC algorithm has also been widely used. However, because the NCC algorithm uses the gray information of the whole input image for image matching, it consumes considerable time, thus reflecting its limitations in some applications requiring high real-time performance. Gray-based matching methods are sensitive to the grayscale differences between images, and they can only match images with linear positive grayscale characteristic correlations. In cases with large geometric disparities between images, this method often fails and it is difficult to use it to match multiview images [14].

Matching methods based on grayscale contain the information of all pixel points in the input image, so their matching accuracy rates are very high, but they also have many shortcomings and problems. (1) Because this type of method uses all image pixel points, the algorithmic complexity is high, and the matching speed is very slow. However, most matching algorithms require high real-time performance, which limits the application scope of this approach. (2) Because this class of algorithms is sensitive to brightness changes, its matching performance is greatly reduced for two images that are in the same scene but under different lighting conditions. (3) For two images with only rigid body transformations and affine transformations, the matching effects of these algorithms are good, but for images with serious deformation and occlusion issues, the matching performance is poor. (4) The algorithms exhibit poor antinoise performance.

2.2. Matching Method Based on Features

Feature-based image matching algorithms make up for the deficiencies of grayscale matching algorithms and have good effects on the matching results of image pairs with affine transformations and projection transformations. At the same time, because feature-based matching algorithms do not match the whole input image but rather extract a series of representative features from the image and then match the features between two images, the algorithmic complexity is greatly reduced, and the matching rate is faster. Feature-based image matching algorithms are typically used in applications requiring high real-time performance. Therefore, this type of algorithm has become a research hotspot in recent years. In 1988, Harris [15] proposed the Harris corner detection algorithm, and it was proven that the Harris corner is rotation invariant and robust to noise and brightness changes to a certain extent. In 1997, Smith and Brady [16] proposed Susan's corner detection method. In 1999, Davis Lowe et al. [17] proposed a SIFT descriptor-based detection method and improved the algorithm in 2004. The SIFT algorithm has been a hot research topic because of its high robustness and invariance to scaling, rotation, and other transformations. Bosch et al. [18] proposed the hue/saturation/value-SIFT (HSV-SIFT) algorithm due to the lack of color information in existing algorithms. The algorithm extracts feature points in each channel of the HSV color space and then connects the feature points in an end-to-end manner in three channels to form a 3×128-dimensional descriptor. Yan et al. [19] proposed reducing the dimensionality of the SIFT algorithm by using principal component analysis (PCA) to solve the problems regarding high SIFT dimensions and long matching times and formed the PCA-SIFT algorithm with low dimensions. Aiming at the sensitivity of the SIFT algorithm to affine transformation, Morel J M et al. [20] proposed the affine SIFT (ASIFT) algorithm with full affine invariance, which improved the matching accuracy of the algorithm for images with multiple perspectives. To improve remote sensing image registration technology, Pouriya and Hassan [21] proposed a sample consistency-based feature matching method built on sparse coding. This method can greatly improve the matching results of two images via SCSC through a joint checkpoint. In addition, the method exhibits excellent performance when many feature points are present or noise is observed. San J et al. [22] proposed a feature-based image matching method by taking advantage of the Delaunay triangulation. First, the Delaunay triangulation result and its corresponding map were used to form adjacent structures containing the randomly distributed feature points of an input image, and the image plane was divided into nearly equilateral triangle patches. Second, photometric and geometric constraints were implemented based on the constructed adjacent structures, and the influence of outliers on the algorithm's decision-making regarding the embedded lines was transmitted by combining hierarchical culling and left-right checking strategies to ensure the accuracy of the final matching results. Li et al. [23] proposed a method based on the concepts of local barycentric coordinates (LBCs) and matching coordinate matrices (MCMs) called locality affine-invariant feature matching (LAM). The LAM method first establishes a mathematical model based on LBCs to extract a good match-preserving local neighborhood structure. LAM then uses the extracted reliable communication to construct local MCMs and identifies the correctness of the residual match by minimizing the ranks of the MCMs. This method achieves excellent performance when matching real images with rigid and nonrigid images. Yu et al. [24] proposed an improved nonlinear SIFT framework algorithm, which combines spatial feature detection with local frequency domain description for synthetic aperture radar (SAR) image registration and optical image registration.

2.3. Multiview Space-Sky Image Matching Method

Under the condition of a large dip angle, the resulting image deformation is serious and traditional feature detection and description methods are often not applicable; especially in scenarios with extreme viewing angles, it is difficult to achieve reliable matching results. Gao et al. [25] proposed that there are two main methods for space-sky image matching at present. The first is the direct matching method, which directly calculates a feature descriptor for the input ground image and then realizes feature matching according to the similarity measure of the feature descriptor. The other approach is the matching method based on the geometric correction. This method first uses prior information to perform geometric correction on a vacant image, then generates composite images, eliminates or reduces the geometric deformation of the input space-sky image, and finally carries out feature matching between the composite images. In the field of photogrammetry, to overcome the matching problems of perspective and dimension changes, Hu et al. [26] proposed the use of a priori information, such as high-precision POS data, as auxiliary information and then performed geometric corrections on the obtained global image, which eliminated or reduced the effects of geometric deformation. Finally, traditional feature description and matching methods are used to match feature points. This kind of method can improve the image matching effect to some extent, but it relies on prior information, and the improvement yielded is limited because the global correction step has difficulty accurately describing the local geometric deformation between the compared images. Jiang et al. [27] proposed the idea that a certain number of matching points could be obtained through initial matching to calculate a geometric transformation model between pairs of stereo images in the absence of high-precision POS data, and then geometric correction could be carried out for these images. However, such methods rely on the initial matching results. In cases with more significant viewing angle and image scale differences with large dip angles from the sky to space, it is difficult for the existing methods to obtain reliable initial matching results for the subsequent geometric correction of the images and thus to ensure the reliability of the final matching results for points with the same labels.

2.4. Matching Method Based on Deep Learning

With the rise of artificial intelligence, methods based on deep learning have been introduced into the field of image feature matching. Kwang et al. [28] proposed a method called learned invariant feature transform (LIFT). This method is a pioneering approach in this field that combines three CNNs (corresponding to key point detection, direction estimation, and feature description) to perform image matching. Balntas et al. [29] proposed PN-Net, which adopts triplet network training. An image block triad T = {P1, P2, n} includes a positive sample pair (P1, P2) and negative sample pairs (P1, n) and (P2, n). A soft PN loss function is used to calculate the similarity between output network descriptors to ensure that the minimum negative sample pair distance is greater than the positive sample pair distance. Compared with other feature descriptors, PN-Net exhibits more efficient descriptor extraction and matching performance and can significantly reduce the time costs of training and execution. Daniel et al. [30] proposed a method called Super Point to train a full CNN consisting of an encoder and two decoders. The two decoders correspond to key point detection and key point feature description. Bhowmik et al. [31] proposed a new training method in which feature detectors were embedded in a complete visual pipeline, and learnable parameters were trained in an end-to-end manner. They used the principle of reinforcement learning to overcome the discrepancies of key point selection and descriptor matching. This training method has very few restrictions on learning tasks and can be used to predict any key point heat map and key point position descriptor architecture. Yuki et al. [32] proposed a novel end-to-end network structure, loss function, and training method to learn image matching (LF-Net). LF-Net uses the ideas of twin networks and Q-learning for reference; one branch generates samples and then trains the parameters of another branch. The network inputs a quarter video graphics array (QVGA) image, outputs a multiscale response distribution, and then processes the response distribution to predict the locations, scales, and directions of key points. Finally, it intercepts the local image input network to extract features. Jiamin S. et al. [33] proposed a method of local image feature matching based on the Transformer model, which operates under the idea that intensive pixel-level matching should be established at the coarse level first, and then fine matching should be refined at the fine level, rather than executing image feature detection, description, and matching first. The global acceptance fields provided by Transformer enable our approach to produce dense matches in low-texture areas where feature detectors typically have difficulty producing repeatable points of interest. Deep learning is also used for specific image matching. Lloyd et al. [34] proposed a three-step framework for the sparse matching of SAR and optical images, where a deep neural network encoded each step. Dusmanu et al. [35] proposed a method called D2Net, which uses more than 300,000 prematched stereo images for training. This method has made important progress in solving the problem of image matching in changing scenes and has shown great potential. However, the main purpose of these algorithmic models is to match close-up visible light ground images with light and visual angle changes, and they are mostly used for the three-dimensional reconstruction of buildings and visual navigation for vehicles. This paper attempts to propose a dense multiview feature extraction neural network specifically for multiview remote sensing image matching based on the idea of D2Net feature extraction.

In summary, the advantages and disadvantages of various type matching methods are compared in Table 1.

Table 1. Comparison of matching methods.

Method	Advantages	Disadvantages
NCC, MAD, SSDA (based on grayscale)	High matching accuracy rates.	Low efficiency, poor adaptability to scale, light, noise, etc.
SIFT, ASIFT, HSV-SIFT (based on features)	High adaptability to scale, illumination, and rotation.	Low adaptability to radiation distortion.
Refs. [26,27] (multiview space-sky image matching method)	High adaptability to large dip angle.	Dependent upon prior knowledge.
LIFT, SuperPoint, D2net (based on deep learning)	High feature extraction capability, strong adaptability to different factors through training.	Depending on the equipment, complex model, tedious training process.
Ours	High feature extraction capability, high adaptability to scale, large dip angle, radiation distortion, etc.	Depending on the equipment, at present, it cannot meet the needs of real-time processing.

3. Proposed Method

In this section, a dense multiview feature extraction neural network is proposed to solve the matching problem between space and sky images. Firstly, CNN is used to extract high-dimensional feature maps for heterologous images with large space and sky dip angles. Secondly, the salient feature points and feature vectors are selected from the obtained feature map, and the feature vector is used as the hard descriptor of the feature points. Meanwhile, based on the gradient information around the feature points and their multiscale information, soft descriptors for the feature points are constructed, which are also used in the neural network training process. Then, by combining the hard and soft descriptors, a joint feature point descriptor is obtained. Finally, the fast nearest neighbor search method (FLANN) [36] is used to match the feature points, and random sample consensus (RANSAC) [37] is used to screen out false matches. Figure 3 shows the structure of the proposed method.

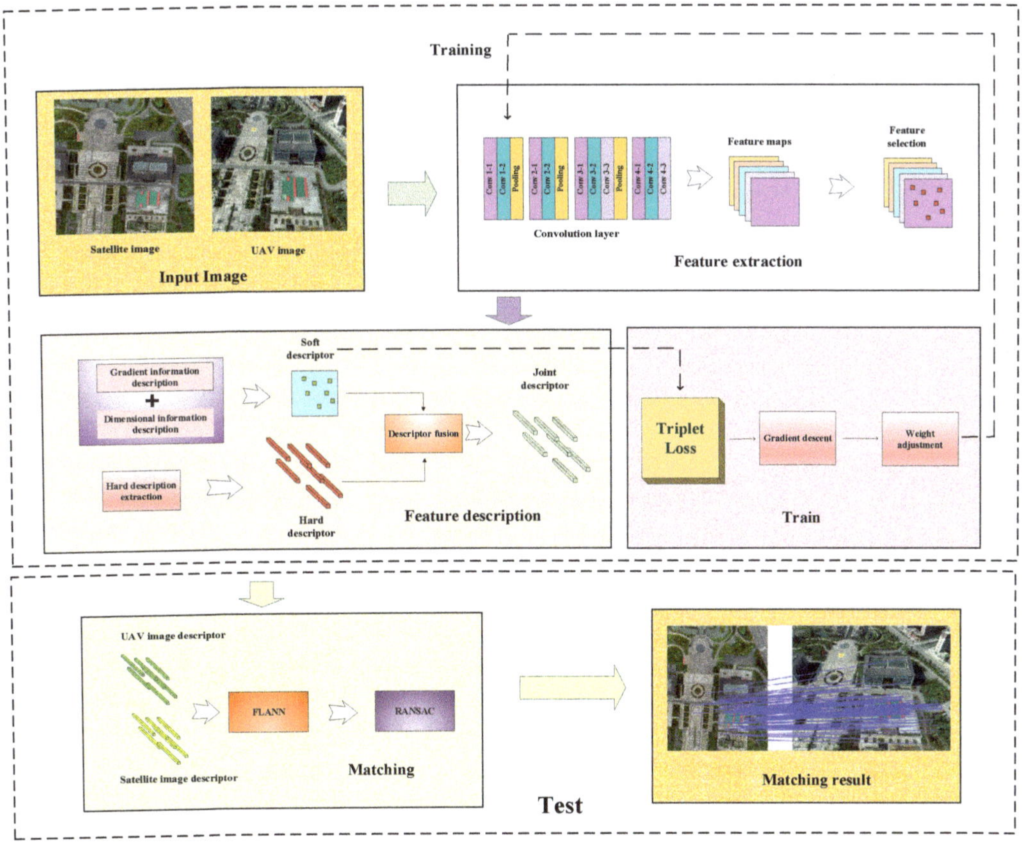

Figure 3. Flow chart of histogram of the proposed image matching method. After the input image is passed through the convolutional network, the feature map is obtained. Then, the salient feature points are screened from the feature map and the hard description is extracted. At the same time, a soft description is made for the salient feature points, which is also used in the loss function. Finally, the final descriptor is obtained by combining hard description and soft description.

3.1. Feature Detection and Hard Descriptor

In this section, the aim is to extract salient feature points and obtain their hard descriptor. In the first step, the proposed method uses the fast nearest neighbor search method (FLANN) [36] to match the feature points and uses random sample consensus (RANSAC) [37] to screen out false matches. We use a CNN to convolve the input image and obtain a 3D depth feature map D. The form of D is shown in Equation (1).

$$D \in R^{h \times w \times n}, D^k \in R^{h \times w} \tag{1}$$

where h is the height of the convoluted image, w is the width of the convoluted image, and n is the number of channels in the convolution output. The two-dimensional array of the output of the two-dimensional convolution layer can be regarded as a representation of the input at a certain level of spatial dimension (width and height). Therefore,$D^k(k = 1, \cdots, n)$ is equivalent to a 2D feature map that represents a feature in a certain direction.

To screen out more significant feature points in D, the feature point screening strategy adopted by the method in this paper is as follows: (1) The feature point is the most prominent in the channel direction of the high-dimensional feature map. (2) The feature point is also the most prominent feature point on the local plane of the feature map. So, ${D^k}_{ij}$ is required to be a local maximum in D^k and k is derived from Equation (2).

$$k = \underset{t}{\operatorname{argmax}} {D^t}_{ij} \tag{2}$$

${D^k}_{ij}$ is the feature value at point (i, j) of D^k. For a point $P(i, j)$ to be selected, the channel k with the maximum response value is firstly selected from n channel feature maps. Then, ${D^k}_{ij}$ is verified to be locally maximum. If the above two conditions are met, it means that $P(i, j)$ is obtained as the significant feature point through screening.

Then, the channel row vector at $P(i, j)$ is extracted from the feature map D as the hard descriptor $\hat{d}_{ij}$ of $P(i, j)$, and we apply L2 normalization on the hard descriptor, as shown in Equation (3).

$$\hat{d}_{ij} = \frac{d_{ij}}{\|d_{ij}\|_2} \tag{3}$$

However, the extrema of discrete space are not real extreme points. To obtain more accurate key point positions, the proposed method uses the SIFT algorithm for reference and adopts the method of local feature map interpolation and encryption to accurately perform subpixel-level positioning. Some points are removed by considering eliminating edge response and eliminating points with low contrast, and then the subpixel extreme points are accurately located by curve fitting. Finally, the precise coordinates of feature points are obtained. Additionally, the hard descriptor is also obtained by bilinear interpolation in the neighborhood.

3.2. Soft Descriptor

In this section, we attempt to introduce a soft descriptor for training and auxiliary description. During the training process, the descriptor is designed as a one-dimensional vector to be amenable for neural network backpropagation.

First, the proposed method extracts the gradient information of the salient feature points. A 3×3 matrix is constructed with the point D_{ij} as the center. The gradient information of the feature point is calculated according to the pixel values of the nine points in the matrix in the k dimension. Therefore, the gradient scores of these feature points are calculated. The gradient score α_{ij} containing the simple gradient information of point D_{ij} can be obtained by Equation (4).

$$\alpha_{ij} = \frac{e^{{D_{ij}}^k}}{\sum e^{{D_{i'j'}}^k}}, (i', j' = [i-1, i, i+1], [j-1, j, j+1]) \tag{4}$$

Then, the proposed method extracts the dimensional difference information of the salient feature points. Since the extracted salient feature points are relatively significant in some dimensions but not so significant in other dimensions, the differences among the salient feature points are highlighted according to these different pieces of information. Thus, the dimension scores of these feature points are calculated by Equations (5) and (6).

$$\overline{D_{ij}} = \frac{\sum_{m=1}^{n} {D_{ij}}^m}{n} \tag{5}$$

$$\beta_{ij} = 2 \times \frac{\sum_{m=1}^{n}\left(D_{ij}{}^{m} - \overline{D_{ij}}\right)^{2}}{n} \tag{6}$$

where $\overline{D_{ij}}$ is the average pixel value of the feature point D_{ij} in each dimension. The dimension score β_{ij} contains the dimension difference information of the feature point D_{ij}.

Finally, the proposed method constructs a soft descriptor from the gradient score and dimension score of point D_{ij}. This is because the product rule is well adaptable to input data of different scales. Since the above two feature scores are one-dimensional values, the final soft descriptor is obtained by multiplying the above two feature scores to highlight the differences among the significant feature points. Soft descriptor s_{ij} is derived from Equation (7).

$$s_{ij} = \alpha_{ij} \cdot \beta_{ij} \tag{7}$$

Soft descriptors have two functions. On the one hand, they are used as the evaluation basis for the training of neural networks; on the other hand, they are used as auxiliary parts of hard descriptors to make the subsequent descriptions more accurate.

3.3. Joint Descriptors

In this section, we attempt to introduce a way to combine hard and soft descriptions, as well as ways to adapt models to multiple scales.

Usually, the first few layers of the network have small receptive domains, and the features obtained are edges, corners, and other local features relative to the bottom layer, but the positioning accuracy is high. The deeper the network layers, the more abstract the extracted features are and the more global the information is. The more resistant the interference caused by geometric deformation and scale difference is, the worse the positioning accuracy is. Therefore, the use of a hard description as the main description results in deeper feature expression ability and ensures a certain positioning accuracy. At the same time, the use of soft descriptions as auxiliary descriptions strengthens the antijamming ability of joint descriptors.

Regarding the fusion of hard descriptors and soft descriptors, there are several strategies for combining them, such as the sum, product, and maximum rules. In this paper, we employ the product rule for similar reasons as those in Yang et al. [38]. First, utilizing the product rule to integrate hard descriptions and soft descriptions can better amplify the differences between the descriptions. Second, the product rule adapts well to input data with different scales and does not require heavy normalization of the data. The joint descriptor $J(d_{ij}, s_{ij})$ is calculated as Equation (8).

$$J(d_{ij}, s_{ij}) = d_{ij} \cdot s_{ij} \tag{8}$$

3.4. Multiscale Models

The CNN model uses training samples with different scales for training, and the feature descriptor can learn scale invariance to a certain extent, but it is also difficult to deal with situations involving large-scale changes. Therefore, this paper adopts the discrete image pyramid model to cope with large-scale changes.

Given an input image I, an image pyramid I^{ρ} containing four different resolutions $(\rho = 0.25, 0.5, 1, 2)$ is used to accommodate drastic changes in the resolutions of the two images. Each layer of the pyramid extracts the F^{ρ} of the feature map and then accumulates the fusion results according to Equation (9).

$$\widetilde{F}^{\rho} = F^{\rho} + \sum_{\gamma<\rho} F^{\gamma} \tag{9}$$

The feature descriptions of key points are extracted through the fusion feature graph $\widetilde{F}^{\rho}$ obtained by accumulation. Due to the different resolutions of pyramids, the low-resolution feature maps need to be linearly interpolated to the same size as that of the high-resolution feature maps before they can be accumulated. In addition, to prevent the detection of repetitive features at different levels, this paper starts from the coarsest scale and marks the detected positions. These positions are unsampled into a feature map with a higher scale as a template. To ensure the number of key points extracted from the feature map at low resolution, if the key points extracted from the feature map at a higher resolution fall into the template, they are discarded.

3.5. Training Loss

The purpose of the loss function is to judge the quality of the neural network through its output value so that the parameters of the neural network can be adjusted adaptively. Furthermore, the feature detector and feature descriptions can be optimized so that the next output result of the neural network is improved.

In this paper, the triple margin ranking loss (TMRL) is used as the loss function. During the process of feature detection, the feature points should have some uniqueness that allows them to adapt to the effects of environmental light and geometric differences. However, at the same time, during the process of feature description, we want the feature vector to be as unique as possible to find the homonymic image point. To address this problem, the triple distance sorting loss function enhances the uniqueness of the correlation descriptor by penalizing any uncorrelated descriptor that leads to a false match. Similar to D2Net, first, images I_1 and I_2 are given, and a pair of the corresponding feature points A and B are in I_1 and I_2, respectively, where A $\in I_1$ and B $\in I_2$. The distance between the soft descriptors of A and B is derived from Equation (10).

$$r = \sqrt{(s_A - s_B)^2} \tag{10}$$

s_A and s_B are soft descriptor values of A and B, respectively. At the same time, a pair of points N_1 and N_2 can be found, which are the point structures most similar to A and B, respectively. N_1 is derived from Equation (11).

$$N_1 = argmin\sqrt{(s_P - s_A)^2}, P \in I_1 \; and \; \sqrt{(\mathrm{P} - A)^2} > K \tag{11}$$

$\sqrt{(\mathrm{P} - A)^2}$ represents the pixel coordinate distance from the point to point. The distance should be greater than K to prevent N_1 from being adjacent to point A. N_2 is also obtained as in Equation (11). Then, the distances between points A and B and their unrelated approximate points are calculated by Equation (12).

$$p = min\left(\sqrt{(s_{N_1} - s_A)^2}, \sqrt{(s_{N_2} - s_B)^2}\right) \tag{12}$$

Finally, the triplet loss is derived from Equation (13).

$$Loss(I_1, I_2) = \sum_{c \in C} max\left(0, M + p(c)^2 - r(c)^2\right) \tag{13}$$

where M is the margin parameter, and the function of the margin parameter is to widen the gap between the matched point pair and the unmatched point pair. The smaller it is set, the more easily the loss value approaches zero, but it is difficult to distinguish between similar images. The larger it is set, the more difficult it is for the loss value to approach zero, which even leads to network nonconvergence.

In Equation (13), C is the set of corresponding points including A and B in image pair I_1 and I_2. The smaller the loss value is, the closer the value of the corresponding point descriptor is, and the greater the difference between it and the value of an irrelevant point descriptor. Therefore, the evolution of the neural network towards the direction of a smaller loss value means that it evolves towards the direction of more accurate matching.

For the CNN model to learn a pixel-level feature similarity expression under radiation and geometric differences, the training data must satisfy the following two conditions in addition to containing a sufficient quantity of points. First, the training images must have great radiometric and geometric differences. Then, the training images must have pixel-level correspondence. Similar to D2Net, we use the MegaDepth data set consisting of 196 different scenes reconstructed from more than a million internet photos using COLMAP.

3.6. Feature Matching Method

After the feature points and feature descriptors of the image are extracted in the third section, FLANN [36] method is used for feature matching. FLANN uses KDTree or Kmeans to conduct clustering modelling for features so that the nearest neighbor point can be found quickly. By comparing, screening the feature points and corresponding feature vectors of the input target images, FLANN finally establishes a mapping set to matching points. Since the proposed method extracts as many features as possible, a large number of mismatched point pairs are generated. Therefore, RANSAC [37] is used to screen out the mismatched point pairs. RANSAC randomly selects at least four samples from the matched data set and ensures that the four samples are not collinear to calculate the homography matrix. Then, RANSAC uses this model to test all the data and calculate the number of data points and the projection error (the cost function) that satisfy this model. If this model is optimal, the corresponding cost function is minimum.

3.7. Model Training Methods and Environment

The proposed method uses a VGG16 model pretrained on the ImageNet data set. The last dense feature extractor Conv4_3 in the network model is trained using the migration learning fine-tuning training method. The initial learning rate was set to 10^{-3}, and then reduced by half for every 10 epochs. For each pair of homonymous image points, a random image region of 256×256 pixels centered on the homonymous image point is selected and fed into the network for training.

In the experimental process, the proposed method is implemented in the PyTorch framework. The computer used for the experiments has a CPU of i9-10900X, a graphics card of NVIDIA TITAN RTX (24 GB video memory), and 32 GB of memory. The implementation language is Python, and the operating system is Windows 10.

4. Experiment and Results

4.1. Data

Here, four sets of experiments (see Figure 4 and Table 2) are designed to evaluate the proposed approach and compare it with previously developed methods. The areas where the images are taken are a square, school, gas station, and park, each with its own characteristics and representativeness. Finally, all sets are analyzed to provide comparative results.

4.2. Comparison of Image Matching Methods

To demonstrate the effectiveness of our approach, we use the proposed method, D2Net [35] (a mainstream deep learning image matching method) and ASIFT [20] (a classical multiview image matching method) to conduct control experiments on these four groups of data, as shown in Table 3.

The convolutional network layer of the proposed method mainly refers to the convolutional layer settings of Visual Geometry Group 16 (VGG16), as shown in Figure 5. In addition, the size of the feature extraction frame is 7×7. The D2Net parameter is set as its default parameter. The algorithm feature extraction and feature description components of ASIFT use the VLFeat library.

The number of correct matching points (NCM), matching accuracy (SR), root mean square error (RMSE), and matching consumption time (MT) were used to evaluate the performance of these algorithms, as shown in Figure 6.

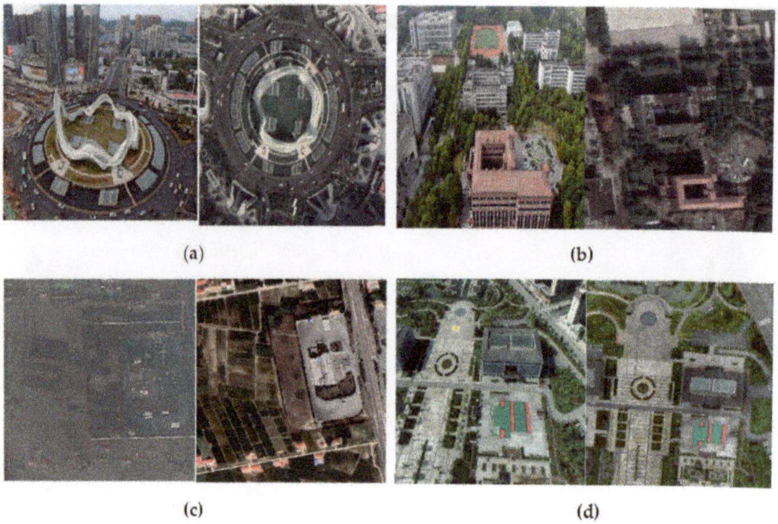

Figure 4. These are UAV images and the corresponding satellite remote sensing images. The left side of each group of images is UAV image, and the right side is the satellite remote sensing image. Each pair of images has obvious scale differences and perspective differences. In group (**a**) and group (**b**), the UAV images are low-altitude UAV images. In groups (**c**) and (**d**), the UAV images are high-altitude UAV images.

Table 2. Test data.

Test Data	Data Description		
	UAV Image	**Satellite Image**	**Study Area Description**
Group a	Sensor: UAV Resolution: 0.24 m Date: \ Size: 1080 × 811	Sensor: Satellite Resolution: 0.24 m Date: \ Size: 1080 × 811	The study area is located at Wuhan City, Hubei Province, China. The UAV image is taken by a small, low-altitude UAV in a square. The satellite image is downloaded from Google Satellite Images. There is a significant perspective difference between the two images, which increases the difficulty of image matching.
Group b	Sensor: UAV Resolution: 1 m Date: \ Size: 1000 × 562	Sensor: Satellite Resolution: 0.5 m Date: \ Size: 402 × 544	The study area is located at Hubei University of Technology, Wuhan, China. The UAV image is taken by a small, low-altitude UAV at the school. The satellite image is downloaded from Google Satellite Images. There is a large perspective difference between the two images, which increases the difficulty of image matching.
Group c	Sensor: UAV Resolution: 0.5 m Date: \ Size: 1920 × 1080	Sensor: Satellite Resolution: 0.24 m Date: \ Size: 2344 × 2124	The study area is located at Tongxin County, Gansu Province, China. The UAV image is taken by a large, high-altitude UAV at a gas station. The satellite image is downloaded from Google Satellite Images. Similarly, the two images have a significant perspective difference. Furthermore, these images are taken from different sensors, resulting in radiation differences that make matching more difficult.
Group d	Sensor: UAV Resolution: 0.3 m Date: \ Size: 800 × 600	Sensor: Satellite Resolution: 0.3 m Date: \ Size: 590 × 706	The study area is located at Anshun City, Guizhou Province, China. The UAV image is taken by a large, high-resolution UAV in a park. The satellite image is downloaded from Google Satellite Images. The linear features of the two images are distinct and rich. However, the shooting angles of the two images are quite different, which leads to difficulty during the image matching process.

Table 3. Test image matching.

Image\Method		NCM	SR	RMSE	MT
Group a	**Ours**	**141**	**18.6%**	**2.18**	**6.1 s**
	D2Net	118	15.6%	2.44	6.2 s
	ASIFT	0	-	-	-
Group b	**Ours**	**56**	**14.5%**	**2.13**	**5.1 s**
	D2Net	41	10.6%	2.57	5.4 s
	ASIFT	0	-	-	-
Group c	**Ours**	**259**	**18.1%**	**2.17**	**17.1 s**
	D2Net	124	8.7%	2.71	17.2 s
	ASIFT	0	-	-	-
Group d	**Ours**	**78**	**19.1%**	**2.23**	**5.8 s**
	D2Net	66	16.2%	2.49	6.1 s
	ASIFT	22	31.9%	3.21	8.2 s

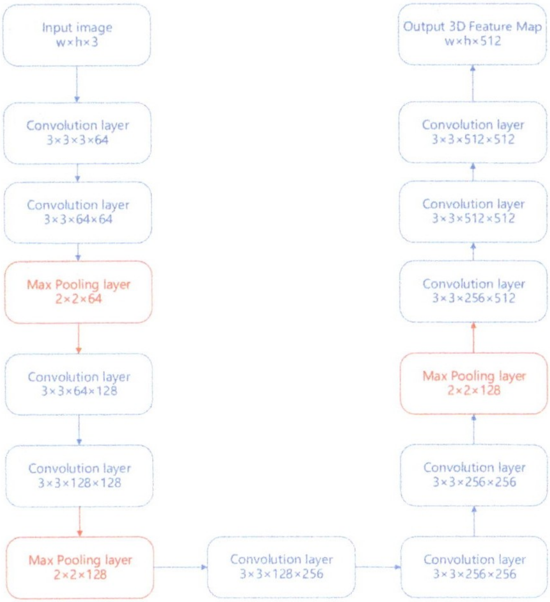

Figure 5. Configuration of the convolutional network layer in our joint description neural network for multiview satellite and UAV image matching.

NCM: NCM is the number of matched pairs on the whole image that satisfy Equation (14). This metric can reflect the performance of the matching algorithm.

$$\|H(x_i) - y_i\| \leq \varepsilon \tag{14}$$

where x_i, y_i denote the matching feature points to be judged, respectively. $\|H(x_i) - y_i\|$ is the reprojection error between the image corresponding matched point pairs and H is the true transformation parameter between the image pairs.

SR: SR is the percentage of NCM to all initial match points.

RMSE: RMSE can reflect the accuracy of the matching point, which is calculated by the following Equation (15).

$$RMSE = \frac{1}{NCM}\sum_i \|H(x_i) - y_i\| \tag{15}$$

This indicator reflects the position offset error of the matching point on the pixel.

MT: MT indicates the matching consumption time, reflecting the efficiency of the method.

Figures 6 and 7 intuitively show the matching effects of the proposed method and D2Net on the images in groups A, B, C, and D. Notably, compared with the D2Net results, the matching points obtained by the proposed method are more widely distributed. It can be intuitively seen from C that, in a case with large perspective, scale, and time phase differences, the proposed method yields a better matching effect than D2Net.

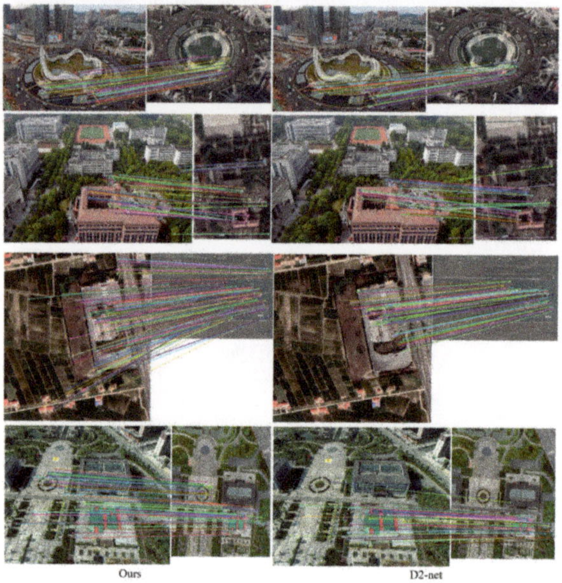

Figure 6. The matching effects of the proposed method and D2Net on groups (**a**–**d**).

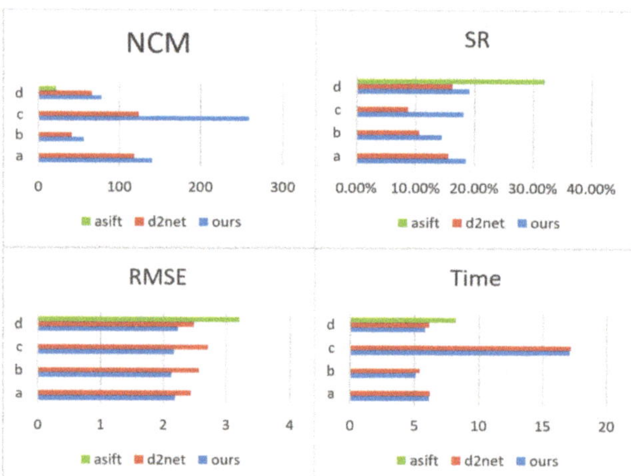

Figure 7. Quantitative comparisons of the proposed method, D2Net, and ASIFT on groups (**a**–**d**). The higher the NCM and SR, the better the matching performance. A smaller RMSE means higher matching accuracy. The smaller MT is, the higher the matching efficiency is. Based on the graph analysis, the proposed method has better matching performance and accuracy than the other two methods.

For ASIFT [20] method, due to the radiation differences and the fuzziness of the UAV images, it cannot work well in the image groups A, B, and C, and there are no matching points to be found. However, our method can effectively eliminate the influence of radiation difference; thus, good results can be achieved for these images, which highlights the effectiveness of matching UAV and remote sensing images from multiple perspectives. For the image group D, the radiation difference is not obvious, and ASIFT method is superior to our method on SR. However, based on the comparison of NCM, our method shows better and more stable matching performance.

Compared with D2Net, for image groups A, B, and D, a slight improvement can be achieved by the proposed method. However, for the image group C, there are more significant scale and perspective differences; thus, the advantages of our method are more obvious.

As can be seen from Figure 7, compared with the other two methods, our method has better matching accuracy and matching performance. This reflects the superiority of the joint description method. Hard description ensures a certain matching performance. Soft description and hard description complement each other, which makes the joint descriptor more specifically reflect the uniqueness of features.

In general, the proposed method can provide certain numbers of correctly matched points for all test image pairs, and the RMSEs of the matched points are approximately 2 to 3 pixels, which is a partial accuracy improvement over that of D2Net. Moreover, the ASIFT algorithm has difficulty matching the correct points for images with large perspective and scale differences. This shows that the proposed method has better adaptability for multiview satellite and UAV image matching.

4.3. Angle Adaptability Experiment

Notably, as the visual angle differences between the images increase, the matching difficulty becomes greater. To verify the feasibility of the proposed method for matching multiview UAV images and satellite images, we conducted experiments on UAV images and satellite images taken from different angles at the same location. The experimental results are shown in Figure 8.

(**a**)

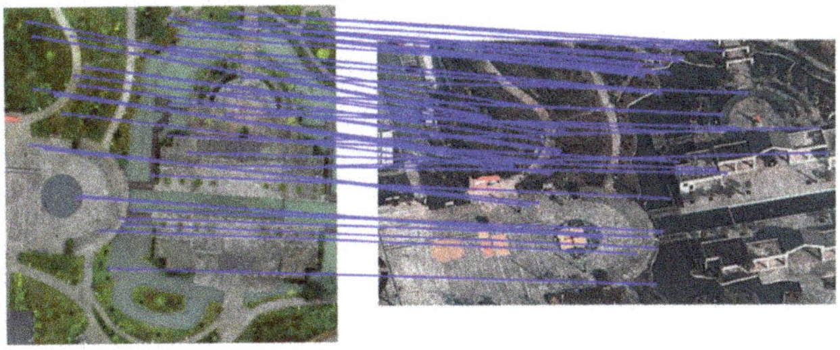

(**b**)

Figure 8. *Cont.*

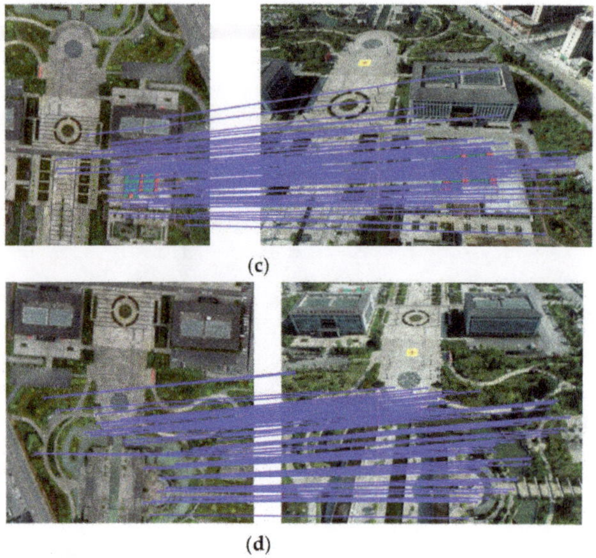

Figure 8. The results of experiments with UAV images and satellite images taken from different angles at the same locations. (**a**) The angle degree difference between this group of images is about 5°. (**b**) The angle degree difference between this group of images is about 10–15°. (**c**) The angle degree difference between this group of images is about 20–25°. (**d**) The angle degree difference between this group of images is about 30°.

Four sets of multi-angle experiments are shown in Figure 8. There are scale, phase, and viewing angle differences in each group of experimental images. These four sets of experimental images are well matched. It can be seen from these four experimental image pairs that although the viewing angle increases, the matching effect does not fail. In brief, the algorithm proposed in this paper is applicable to UAV image matching with satellite images when the tilt degree is less than or equal to 30 degrees.

4.4. Application in Image Geometric Correction

One of the main purposes of matching UAV images with remote sensing images is to correct UAV images and provide geographic information. Based on the correctly matched points determined in the previous section, a homographic transformation matrix is estimated, and then this matrix is used to correct the input UAV image. Figure 9 shows the results of correcting UAV images and assigning geographic information after matching them with the proposed method.

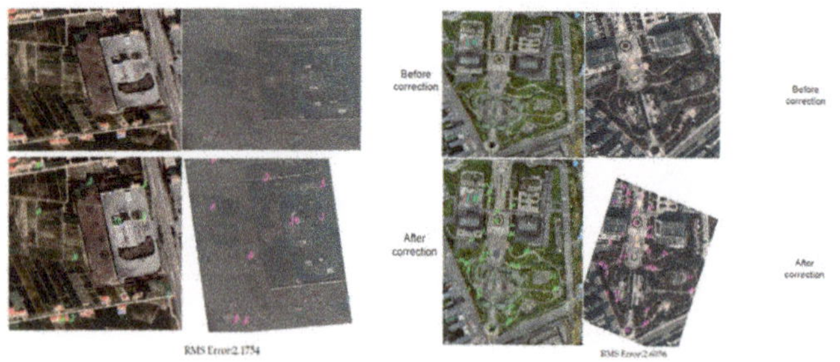

Figure 9. These are the results of selecting evenly distributed points from satellite images and corrected UAV images, and calculating the errors among them.

From the registration results, the registration effect for UAV and satellite images is improved due to the good matching correspondence. The registration accuracy nearly reaches 2 pixels, which can meet the needs of UAV reconnaissance target positioning.

5. Discussion

The method presented in this paper exhibits a good matching effect for multiview UAV and satellite images from the matching results. A certain number of relatively uniform distributions of correctly matched points were obtained by the proposed method, which can support the registration of UAV images. In addition, the proposed method exhibits good adaptability to viewing angle, scale, and time phase differences among multiview images. This shows that our designed joint descriptor makes our algorithm more robust for multiview, multiscale, and multitemporal images. However, due to the large number of convolutional computations required by deep feature learning, despite the use of GPU acceleration, the efficiency of feature extraction is not greatly improved relative to that of traditional feature extraction algorithms.

It is difficult to match multiview satellite images with UAV images due to the large time phase, perspective, and scale differences between these images. The method proposed in this paper uses joint description to make the resulting features more prominent, solving the situation in which the features are difficult to match due to the above problems. Experiments show that the proposed method is better than the traditional method in solving these matching difficulties. However, the proposed method also has the problem of a long matching time requirement, which makes it impossible to carry out real-time positioning and registration for UAV images. Thus, in the future, it will be important to accurately screen out the significant feature points to reduce the matching time. With the development of deep learning technology, image matching technology of multiview satellites and UAV should also make continuous progress from its development trend.

6. Conclusions

In this paper, an algorithm for multiview UAV and satellite image matching is proposed. This method is based on a joint description network. The developed joint descriptor includes a specifically designed hard descriptor and soft descriptor, among which the hard descriptor ensures the matching accuracy of the network, and the soft descriptor is used for network training and auxiliary description. According to experiments, the algorithm proposed in this paper can achieve good matching effects for multiview satellite images and UAV images in comparison with some popular methods. Moreover, the matching accuracy of the proposed method in optical satellite and UAV images nearly reaches 2 pixels, which meets the requirements of relevant UAV missions.

Author Contributions: Conceptualization, C.X.; methodology, C.L.; software, C.L.; validation, H.L., Z.Y. and H.S.; formal analysis, W.Y.; data curation, H.L.; writing—original draft preparation, C.L.; writing—review and editing, C.X.; visualization, C.X.; supervision, Z.Y.; project administration, C.X.; funding acquisition, C.X. All authors have read and agreed to the published version of the manuscript.

Funding: This research was funded by National Natural Science Foundation of China, grant number 41601443 and 41771457, Scientific Research Foundation for Doctoral Program of Hubei University of Technology (Grant No. BSQD2020056).

Institutional Review Board Statement: Not applicable.

Informed Consent Statement: Not applicable.

Data Availability Statement: The data presented in this study are available on request from the corresponding author. The raw/processed data required to reproduce these findings cannot be shared at this time as the data also forms part of an ongoing study.

Conflicts of Interest: The authors declare no conflict of interest. The funders had no role in the design of the study; in the collection, analyses, or interpretation of data; in the writing of the manuscript, or in the decision to publish the results.

References

1. Li, Y.; Chen, W.; Zhang, Y.; Tao, C.; Xiao, R.; Tan, Y. Accurate cloud detection in high-resolution remote sensing imagery by weakly supervised deep learning. *Remote Sens. Environ.* **2020**, *250*, 112045. [CrossRef]
2. Dou, P.; Chen, Y. Dynamic monitoring of land-use/land-cover change and urban expansion in Shenzhen using Landsat imagery from 1988 to 2015. *Int. J. Remote Sens.* **2017**, *38*, 5388–5407. [CrossRef]
3. Shi, W.; Zhang, M.; Zhang, R.; Chen, S.; Zhan, Z. Change detection based on artificial intelligence: State-of-the-art and challenges. *Remote Sens.* **2020**, *12*, 1688. [CrossRef]
4. Guo, Y.; Du, L.; Wei, D.; Li, C. Robust SAR Automatic Target Recognition via Adversarial Learning. *IEEE J. Sel. Top. Appl. Earth Obs. Remote Sens.* **2020**, *14*, 716–729. [CrossRef]
5. Guerra, E.; Munguía, R.; Grau, A. UAV visual and laser sensors fusion for detection and positioning in industrial applications. *Sensors* **2018**, *18*, 2071. [CrossRef]
6. Ma, J.; Jiang, X.; Fan, A.; Jiang, J.; Yan, J. Image matching from handcrafted to deep features: A survey. *Int. J. Comput. Vis.* **2021**, *129*, 23–79. [CrossRef]
7. Ye, Y.; Shan, J.; Hao, S.; Bruzzone, L.; Qin, Y. A local phase based invariant feature for remote sensing image matching. *ISPRS J. Photogramm. Remote Sens.* **2018**, *142*, 205–221. [CrossRef]
8. Manzo, M. Attributed relational sift-based regions graph: Concepts and applications. *Mach. Learn. Knowl. Extr.* **2020**, *2*, 13. [CrossRef]
9. Zhao, X.; Li, H.; Wang, P.; Jing, L. An Image Registration Method Using Deep Residual Network Features for Multisource High-Resolution Remote Sensing Images. *Remote Sens.* **2021**, *13*, 3425. [CrossRef]
10. Zeng, L.; Du, Y.; Lin, H.; Wang, J.; Yin, J.; Yang, J. A Novel Region-Based Image Registration Method for Multisource Remote Sensing Images via CNN. *IEEE J. Sel. Top. Appl. Earth Obs. Remote Sens.* **2020**, *14*, 1821–1831. [CrossRef]
11. Wang, S.; Quan, D.; Liang, X.; Ning, M.; Guo, Y.; Jiao, L. A deep learning framework for remote sensing image registration. *ISPRS J. Photogramm. Remote Sens.* **2018**, *145*, 148–164. [CrossRef]
12. Leese, J.A.; Novak, C.S.; Clark, B.B. An automated technique for obtaining cloud motion from geosynchronous satellite data using cross correlation. *J. Appl. Meteorol. Climatol.* **1971**, *10*, 118–132. [CrossRef]
13. Barnea, D.I.; Silverman, H.F. A class of algorithms for fast digital image registration. *IEEE Trans. Comput.* **1972**, *100*, 179–186. [CrossRef]
14. Zitova, B.; Flusser, J. Image registration methods: A survey. *Image Vis. Comput.* **2003**, *21*, 977–1000. [CrossRef]
15. Harris, C.G.; Stephens, M. A combined corner and edge detector. In Proceedings of the Alvey Vision Conference, Manchester, UK, 31 August–2 September 1988; Volume 15, p. 10-5244.
16. Smith, S.M.; Brady, J.M. SUSAN—A new approach to low level image processing. *Int. J. Comput. Vis.* **1997**, *23*, 45–78. [CrossRef]
17. Lowe, D.G. Object recognition from local scale-invariant features. In Proceedings of the Seventh IEEE International Conference on Computer Vision, Corfu, Greece, 20–25 September 1999; IEEE: Piscataway, NJ, USA, 1999; Volume 2, pp. 1150–1157.
18. Bosch, A.; Zisserman, A.; Munoz, X. Scene classification using a hybrid generative/discriminative approach. *IEEE Trans. Pattern Anal. Mach. Intell.* **2008**, *30*, 712–727. [CrossRef]
19. Ke, Y.; Sukthankar, R. PCA-SIFT: A more distinctive representation for local image descriptors. In Proceedings of the 2004 IEEE Computer Society Conference on Computer Vision and Pattern Recognition, Washington, DC, USA, 27 June–2 July 2004; IEEE: Piscataway, NJ, USA, 2004; Volume 2, p. 2.
20. Morel, J.M.; Yu, G. ASIFT: A new framework for fully affine invariant image comparison. *SIAM J. Imaging Sci.* **2009**, *2*, 438–469. [CrossRef]
21. Etezadifar, P.; Farsi, H. A New Sample Consensus Based on Sparse Coding for Improved Matching of SIFT Features on Remote Sensing Images. *IEEE Trans. Geosci. Remote Sens.* **2020**, *58*, 5254–5263. [CrossRef]
22. Jiang, S.; Jiang, W. Reliable image matching via photometric and geometric constraints structured by Delaunay triangulation. *ISPRS J. Photogramm. Remote Sens.* **2019**, *153*, 1–20. [CrossRef]
23. Li, J.; Hu, Q.; Ai, M. LAM: Locality affine-invariant feature matching. *ISPRS J. Photogramm. Remote Sens.* **2019**, *154*, 28–40. [CrossRef]
24. Yu, Q.; Ni, D.; Jiang, Y.; Yan, Y.; An, J.; Sun, T. Universal SAR and optical image registration via a novel SIFT framework based on nonlinear diffusion and a polar spatial-frequency descriptor. *ISPRS J. Photogramm. Remote Sens.* **2021**, *171*, 1–17. [CrossRef]
25. Gao, X.; Shen, S.; Zhou, Y.; Cui, H.; Zhu, L.; Hu, Z. Ancient Chinese Architecture 3D Preservation by Merging Ground and Aerial Point Clouds. *ISPRS J. Photogramm. Remote Sens.* **2018**, *143*, 72–84. [CrossRef]
26. Hu, H.; Zhu, Q.; Du, Z.; Zhang, Y.; Ding, Y. Reliable spatial relationship constrained feature point matching of oblique aerial images. *Photogramm. Eng. Remote Sens.* **2015**, *81*, 49–58. [CrossRef]
27. Jiang, S.; Jiang, W. On-Board GNSS/IMU Assisted Feature Extraction and Matching for Oblique UAV Images. *Remote Sens.* **2017**, *9*, 813. [CrossRef]

28. Yi, K.M.; Trulls, E.; Lepetit, V.; Fua, P. Lift: Learned invariant feature transform. In *European Conference on Computer Vision*; Springer: Cham, Switzerland, 2016; pp. 467–483.
29. Balntas, V.; Johns, E.; Tang, L.; Mikolajczyk, K. PN-Net: Conjoined triple deep network for learning local image descriptors. *arXiv* **2016**, arXiv:1601.05030.
30. DeTone, D.; Malisiewicz, T.; Rabinovich, A. Superpoint: Self-supervised interest point detection and description. In Proceedings of the IEEE Conference on Computer Vision and Pattern Recognition Workshops, Salt Lake City, UT, USA, 18–22 June 2018; pp. 224–236.
31. Bhowmik, A.; Gumhold, S.; Rother, C.; Brachmann, E. Reinforced Feature Points: Optimizing Feature Detection and Description for a High-Level Task. In Proceedings of the 2020 IEEE/CVF Conference on Computer Vision and Pattern Recognition (CVPR), Seattle, WA, USA, 14–19 June 2020; IEEE: Piscataway, NJ, USA, 2020.
32. Ono, Y.; Trulls, E.; Fua, P.; Yi, K.M. LF-Net: Learning local features from images. *arXiv* **2018**, arXiv:1805.09662.
33. Sun, J.; Shen, Z.; Wang, Y.; Bao, H.; Zhou, X. LoFTR: Detector-free local feature matching with transformers. In Proceedings of the IEEE/CVF Conference on Computer Vision and Pattern Recognition, Nashville, TN, USA, 19–25 June 2021; pp. 8922–8931.
34. Lhh, A.; Dm, B.; Slb, C.; Dtb, D.; Msa, E. A deep learning framework for matching of SAR and optical imagery. *ISPRS J. Photogramm. Remote Sens.* **2020**, *169*, 166–179.
35. Dusmanu, M.; Rocco, I.; Pajdla, T.; Pollefeys, M.; Sivic, J.; Torii, A.; Sattler, T. D2-net: A trainable cnn for joint description and detection of local features. In Proceedings of the IEEE Conference on Computer Vision and Pattern Recognition, Long Beach, CA, USA, 16–20 June 2019; pp. 8092–8101.
36. Megalingam, R.K.; Sriteja, G.; Kashyap, A.; Apuroop, K.G.S.; Gedala, V.V.; Badhyopadhyay, S. Performance Evaluation of SIFT & FLANN and HAAR Cascade Image Processing Algorithms for Object Identification in Robotic Applications. *Int. J. Pure Appl. Math.* **2018**, *118*, 2605–2612.
37. Li, H.; Qin, J.; Xiang, X.; Pan, L.; Ma, W.; Xiong, N.N. An efficient image matching algorithm based on adaptive threshold and RANSAC. *IEEE Access* **2018**, *6*, 66963–66971. [CrossRef]
38. Yang, T.Y.; Hsu, J.H.; Lin, Y.Y.; Chuang, Y.Y. Deepcd: Learning deep complementary descriptors for patch representations. In Proceedings of the IEEE International Conference on Computer Vision, Venice, Italy, 22–29 October 2017; pp. 3314–3332.

remote sensing

MDPI

Article

Logging Trail Segmentation via a Novel U-Net Convolutional Neural Network and High-Density Laser Scanning Data

Omid Abdi *, Jori Uusitalo and Veli-Pekka Kivinen

Department of Forest Sciences, University of Helsinki, Latokartanonkaari 7, 00014 Helsinki, Finland; jori.uusitalo@helsinki.fi (J.U.); veli.kivinen@helsinki.fi (V.-P.K.)
* Correspondence: omid.abdi@helsinki.fi; Tel.: +358-294158466

Abstract: Logging trails are one of the main components of modern forestry. However, spotting the accurate locations of old logging trails through common approaches is challenging and time consuming. This study was established to develop an approach, using cutting-edge deep-learning convolutional neural networks and high-density laser scanning data, to detect logging trails in different stages of commercial thinning, in Southern Finland. We constructed a U-Net architecture, consisting of encoder and decoder paths with several convolutional layers, pooling and non-linear operations. The canopy height model (CHM), digital surface model (DSM), and digital elevation models (DEMs) were derived from the laser scanning data and were used as image datasets for training the model. The labeled dataset for the logging trails was generated from different references as well. Three forest areas were selected to test the efficiency of the algorithm that was developed for detecting logging trails. We designed 21 routes, including 390 samples of the logging trails and non-logging trails, covering all logging trails inside the stands. The results indicated that the trained U-Net using DSM (k = 0.846 and IoU = 0.867) shows superior performance over the trained model using CHM (k = 0.734 and IoU = 0.782), DEM_{avg} (k = 0.542 and IoU = 0.667), and DEM_{min} (k = 0.136 and IoU = 0.155) in distinguishing logging trails from non-logging trails. Although the efficiency of the developed approach in young and mature stands that had undergone the commercial thinning is approximately perfect, it needs to be improved in old stands that have not received the second or third commercial thinning.

Keywords: U-Net; high-density laser scanning; logging trails; digital surface model; canopy height model; commercial thinning; semantic segmentation; convolutional neural networks

Citation: Abdi, O.; Uusitalo, J.; Kivinen, V.-P. Logging Trail Segmentation via a Novel U-Net Convolutional Neural Network and High-Density Laser Scanning Data. *Remote Sens.* **2022**, *14*, 349. https://doi.org/10.3390/rs14020349

Academic Editors: Fahimeh Farahnakian, Jukka Heikkonen and Pouya Jafarzadeh

Received: 3 December 2021
Accepted: 10 January 2022
Published: 13 January 2022

1. Introduction

In modern timber harvesting, logging trails are crucial entities for the accurate navigations of harvesters and forwarders to penetrate into forest stands for silvicultural operations [1] in the pathway of precision harvesting. However, spotting the accurate locations of old logging trails is one the major and most challenging tasks for forest owners or operators/drivers, particularly in the stands that have not undergone commercial thinning for a long period of time. Little is known about holistic solutions for the detection of logging trails using remote-sensing data. However, cutting-edge deep-learning based approaches using high-density laser scanning data may aid in solving this problem.

In Finland, rotation forest management (RFM) is the most common silvicultural method. It relies on three main phases: establishment, thinning, and final felling [2]. Normally, forest stands are thinned two to three times between the ages of 20 and 70 years [3,4]. Logging trails are determined with a width of 4–5 m and a spacing of 20–25 m in the first commercial thinning [1,3,5], which covers the entirety of a stand. However, some segments of a trail, an entire trail, or even a logging trail network may vanish on the ground over time, due to the regrowth of trees, the growth of seedlings, and the spreading of the crowns of trees on the trail surface. Therefore, spotting the initial locations of logging trails can be time-consuming and costly. Additionally, misinterpreting the original logging trail network in subsequent thinning operations may cause overcut of the growing stock.

In recent decades, airborne laser scanning (ALS) systems have become central to characterizing the 3D structure of forest canopies. These systems have provided cutting-edge applications and research in forestry, particularly in the areas of forest inventory and ecology [6–8]. Few studies have addressed the detection of logging trails using laser scanning data [9,10], while well-documented literature is available regarding the mapping of forest roads using either low-density laser scanning data or high-density laser scanning data [11–16]. The majority of these studies have used traditional methods based on edge detection, thresholding, or object-based segmentation to detect logging trails or forest roads under canopies via machine learning algorithms. Sherba et al. [10] presented a rule-based classification approach for detecting old logging roads using slope models derived from high-density LiDAR data in Marin County, California. They reported that some post-classification techniques such as LiDAR-derived flow direction raster and curvature increased the accuracy of detecting logging trails by dropping streams and gullies and adding ridge trails to the final classified layer. They emphasized that the high point density of LiDAR data has a significant influence on the accuracy of discriminating old logging trails from non-trail objects. Similarly, Buján et al. [16] proposed a pixel-based random forest approach to map paved and unpaved roads through numerous LiDAR-derived metrics in the forests of Spain. However, they concluded that the density of LiDAR points did not have a significant impact on the accuracy of the detection of roads using random forest. Lee et al. [9] extracted trails using the segmentation of canopy density derived from the airborne laser swath mapping (ALSM) data. They labeled the sharpened sightlines as trails that result from the visibility vectors between the canopies. The introduced approaches may show promising results but rely on heavy pre-processing and post-processing tasks. Typically, they are developed for a specific type of trail or road in a particular forest. Furthermore, the detection of a logging trail is more difficult than the detection of a forest road using these developed approaches, due to a lower geometric consistency, more complex background, and the occlusions of the canopy [17]. Therefore, the need to develop a versatile approach, such as deep learning methods with minimal processing and optimal efficiency for detecting logging trails from laser scanning data, is undeniable.

Recently, convolutional neural networks (CNNs), as one of the architectures of deep learning neural networks, have become the epicenter for image classification, semantic segmentation, pattern recondition, and object detection, in particular with the emerging high-resolution remote sensing data [18,19]. The standard architecture of a CNN encompasses a set of convolutional layers, pooling and non-linear operations [20]. The primary characteristics of a CNN are the spatial connectivity between the adjacent layers, sharing of the weights, acquiring features from low-spatial scale to high-spatial scale, and integrating the modules of feature extractions and classifiers [21]. Various successful CNN architectures have been developed for main road classification, such as U-Net [22] and GANs [23], and for main road area or centerline extractions, such as U-Net [24–29], ResNet [30], GANs [31], Y-Net [32], SegNet [33], and CasNet [34], which mostly were used very high-resolution satellite (VHR) images or UAV. Several studies have addressed the outperforming of deep learning-based approaches in forest applications, such as individual tree detection [35–38], species classification [35,39–42], tree characteristics extraction [43,44], and forest disturbances [45–48], mostly using VHR, UAV, or high-density laser scanning data. At present, little is known about the efficiency of the deep learning-based approaches on the extraction of logging trails or forest roads.

Tree occlusions and other noises hampered accurate road detection using the traditional road segmentation methods even using VHR images [17,49,50]. However, the CNN-based approaches could relatively alleviate the effects of complex background and the occlusion of trees [34,51]. Using high-density laser scanning data with the capability of penetrating into the canopy and reaching the ground surface may aid to solve these problems. Few studies explored the feasibility of CNN-based architectures in using laser scanning-derived metrics for detecting road networks [52,53]. Caltagirone et al. [52] developed a fast fully convolutional neural network (FCN) for road detection through the metrics of average elevation and density layers derived from laser scanning data. They reported excellent performance of this approach in detecting roads, particularly for real-time applications. Similarly, Verschoof-van der Vaart et al. [53] demonstrated the efficiency of CarcassonNet using a digital terrain model (DTM) derived from laser scanning data for detecting and tracing of archaeological objects such as historical roads in Netherlands.

Although the performance of CNNs methods for road extractions and its components have been well documented using VHR and UAV for public roads [51], this efficiency requires greater scrutiny in the more complex backgrounds, such as for detecting commercial forest roads or logging trails in forests, and with different data such as laser scanning data. Therefore, this study seeks to test the performance of U-Net, as one of the most popular architectures of CNNs, in integration with high-density laser scanning data for detecting logging trails, as one of the most complex networks regarding geometry and visibility in the mechanized forests of Finland.

The main purpose of this research is to develop an end-to-end deep learning-based approach that uses the metrics of high-density laser scanning data to automate the detection of logging trails in forest stands that have undergone commercial thinning. Specifically, we aim to comparatively evaluate the performance of a trained U-Net algorithm by using different derivatives of laser scanning datasets (i.e., canopy height and elevation-based models) for the detection of logging trails. We are also eager to investigate the performance of this approach to detect logging trails in young and mature stands with different development classes.

2. Materials and Methods

2.1. Description of the Study Area

We focused our research on the Kakkurinmaa, Länsi-Aure, and Karpanmaa forests in the municipalities of Parkano and Ikaalinen, Southern Finland. The Kakkurinmaa and Karpanmaa forests are owned by Finsilva Oyj, and the Länsi-Aure forest, as governmental public land, is managed by state-owned Metsähallitus. The forest areas are structured in spatially uniform forest stands, typically 3–10 hectares in size. The tree species are pine, spruce, and birch with a predominance of pine in the three regions. The stands are managed even-aged, and the age range of the stands is between 34 and 72 years. The height of trees ranges between 5 and 30 m. Forest stands are typically thinned 2–3 times during a rotation period in which around 25–30% of the trees are removed [4,54]. We classified forest stands concerning age, height, and thinning operations into four development categories to facilitate the detection of logging trails (Figure 1): (1) young stands before the first commercial thinning, (2) young stands that had experienced the first commercial thinning, (3) mature stands before the second commercial thinning, and (4) mature stands that had undergone the second or third thinning operation. Logging trails may be visible within Categories 2 and 4 stands (Figure 1b,d); however, in some development classes, for example, within Category 3 stands, old logging trails are very challenging to find (Figure 1c).

2.2. Data

We ordered a license to access the high-quality laser scanning data for the study area in 2020, under the framework of the National Land Survey of Finland (NLS). These data are the latest and most accurate laser scanning data that have been collected by the NLS in Finland. The density of data is at least 5 points per square meter, as the average distance between points is circa 40 cm. The mean altimetric error of the data is less than 10 cm and the mean error of horizontal accuracy is less than 45 cm [55]. To detect logging trails, we extracted the canopy height and the elevation metrics after processing the high-quality laser scanning data. The characteristics of the forest stands (e.g., species composition, age, height, and thinning history) and their boundaries were collected from the databases of Finsilva Oy and Metsähallitus. These data were used for the classification of the stands as described in Section 2.1. A further set of required data such as topographic maps and the time-series of orthophotos were also obtained from the open databases of the NLS [56].

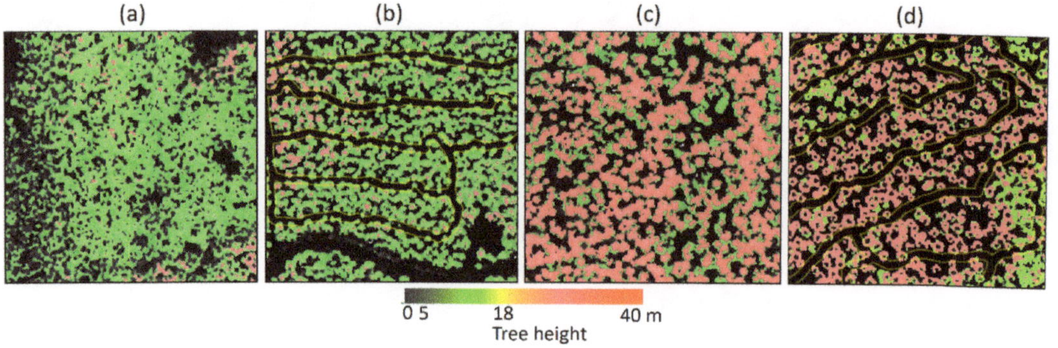

Figure 1. Forest stands regarding commercial thinning: (**a**) young stands before the first commercial thinning; (**b**) young stands after the first commercial thinning; (**c**) mature stands before the second commercial thinning; and (**d**) mature stands after the second/third commercial thinning. The logging trails are visible in Categories (**b**) and (**d**), but they are difficult to spot in Category (**c**).

We used these data to create the labeled dataset of logging trails for training the U-Net algorithm. In addition to the extensive ground-truth samplings of the logging trails to test the algorithm efficiency (Section 2.5), we visited the logging trails and recorded some tracks in three regions before creating the dataset of labels.

2.3. Training Datasets

We selected 44 laser scanning tiles of 1 × 1 km to create image and labeled datasets for training the deep-learning algorithm. After decompressing the laser scanning datasets, we merged the tiles and produced required data from the cloud points such as the height metrics and elevation models. The canopy height model (CHM) was utilized [57] with a spatial resolution of 0.5 m to estimate the total height of trees. The binning interpolation methods were adopted to derive a digital elevation models (DEMs) based on the minimum cell assignment types (DEM_{min}) (i.e., close to the terrain using the point clouds with minimum elevation) and the average digital elevation model (DEM_{avg}) as well as a digital surface model (DSM) based on the maximum cell assignment type [58]. For example, the assignment of each output cell was determined from the maximum value of point clouds that fall within its extent to form the DSM. The values of all the raster models were normalized between 0 and 255 using a min–max scaling method. Finally, we smoothed the raster layers by calculating their median value in a 3 × 3 neighborhood around each cell.

The labels of logging trails were generated from a variety of resources such as orthophotos (Figure 2a), trees height (Figure 2b) and profiles extracted from the laser scanning points (Figures 2c and 3). The ground elevation model was used to discriminate ditches and forest roads from the logging trails (Figure 2d). We created a total of 336 km of logging trails and then defined a 2 m buffer, as the width of a segment, from the centerline. The logging trails were converted into a binary image containing the cells with the labels 0 (non-trail) and 1 (trail) (Figure 2e).

The images and their corresponding labels were converted into the patches with a size of 256 × 256 cells (Figure 4a) before entering these into U-Net. In total, we selected 1888 image patches and their corresponding labels for training (75%) and validation (25%) of the U-Net. We excluded some image patches from training datasets that were in the areas selected for collecting test data, as described in Section 2.5.

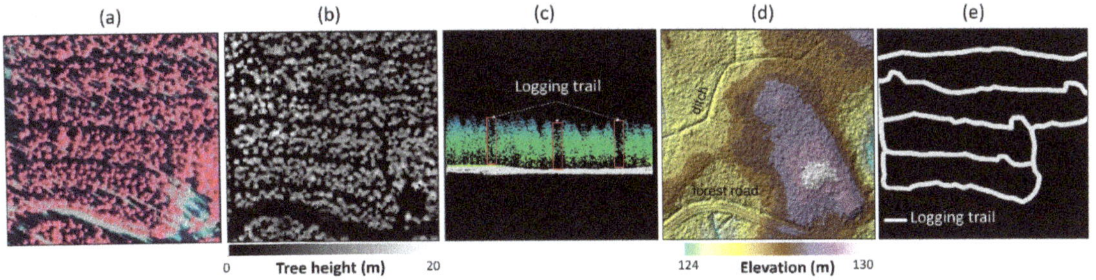

Figure 2. References comprising (**a**) near-infrared orthophotos and the derivatives of high-density laser scanning data such as (**b**) canopy height model, (**c**) tree profiles, and (**d**) the ground elevation model, used to produce the labeled datasets (**e**) from logging trails for training the U-Net convolutional neural network architecture. While the orthophoto, tree height, and tree profiles enhanced the visibility of logging trails, the digital terrain model heightened the ditches and roads that might inadvertently be digitized as logging trails during creation of the labeled dataset.

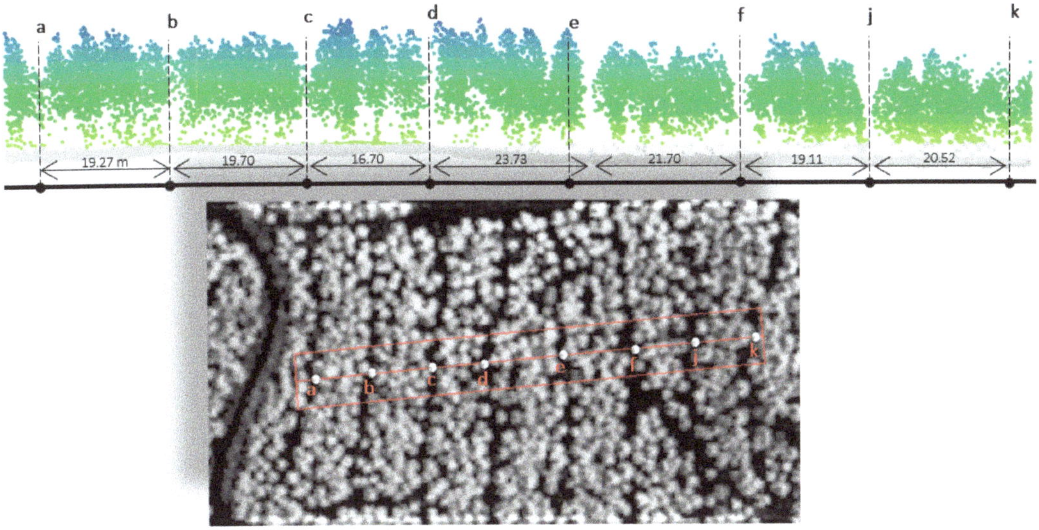

Figure 3. The profile of the cloud points of a laser scanning dataset within a young stand that has undergone its first commercial thinning (**a–f**,**j**,**k**). The intervals between two logging trails and their footprint are shown on the layers of canopy height and trees' profile.

2.4. U-Net Architecture

The U-Net is one of the cutting-edge architectures of the convolutional neural network for image segmentation due to its simple structure, ability to work with little training data, and high performance [59,60]. The U-Net concatenates low-level information and high-level semantic information that is derived from the convolutional layers. This strategy enables it to produce accurate prediction maps, even with limited training data [59]. The U-shaped structure of U-Net consists of a contraction path (encoder) and an expansion path (decoder). The extraction of low-level features and the reduction of spatial dimensions are implemented in the contraction path, while the spatial dimensions of the features are enhanced through a series of upward convolutions and concatenations in the expansion path. In the architecture of our U-Net (Figure 4b), the contraction path consists of four steps, each step comprising two 3 × 3 convolution layers. Each convolution layer is followed by an ReLU activation function and a batch normalization layer with a same-padded. The spatial dimensions of the features were reduced using a 2 × 2 max-pooling layer. The number of filters/features was doubled, while the spatial dimensions were halved at each contraction step. In our U-Net, the first and last convolution layers of the contraction path entail 16 and 128 filters, respectively. The expansion path consists of a sequence of upsampling of the features, followed by the transposed convolution layers with a stride 2. The upsampling layers combine the high-level features with the corresponding features in the contraction path using the intermediate concatenations. A bottleneck layer with 256 filters is located between the contraction and expansion blocks as well (Figure 4b). The output is a 1 × 1 convolutional layer with one dimension that is followed by a sigmoid activation function (Figure 4c).

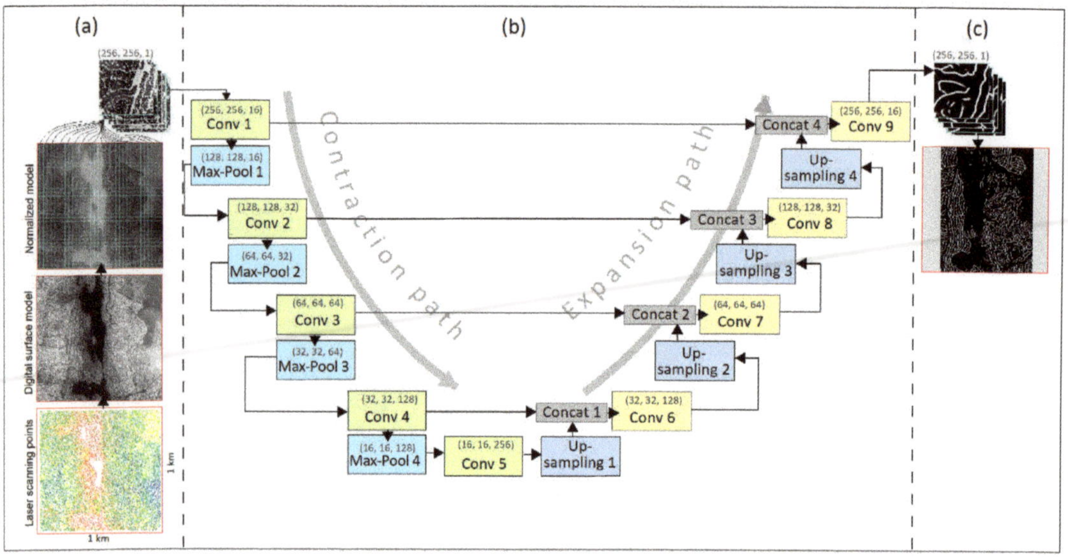

Figure 4. Architecture of the constructed U-Net for detecting logging trails using high-density laser scanning data: (**a**) preparation of a laser scanning tile for use in the U-Net to detect logging trails; (**b**) architecture of the designed U-Net, which includes the contraction path and the expansion path; and (**c**) predicted logging trails.

The U-Net architecture was constructed and trained in Python using the powerful Keras and TensorFlow libraries [61]. The model was trained using the GPU of NVIDIA Quadro RTX 4000 with 8 GB. We implemented the Hyperband algorithm in Keras Tuner to search the optimal set of hyperparameters for our algorithm [62], such as the optimization algorithm, learning rate, dropout rate, batch size, and loss function [20]. The model builder was used to define the search algorithm and hypertuned model. The model was trained using the training data and evaluated using the test data. Table A1 shows a number of tuned optimal values for the hyperparameters in training the U-Net. The minimum number of epochs was set at 100, and the early stop rule was implicated to stop the process of training, in case of overfitting. The cross-entropy loss function was set to monitor how poorly the U-Net was performing. The plots of accuracy and loss versus the epochs in the training of U-Net are provided in Figure A1.

Figure 5 shows an example of the predicted logging trails from DSM data, using the trained U-Net. The algorithm accepts an input layer (i.e., a DSM) with a fixed size (256, 256, 1). It produces different feature maps in the intermediate step, such as convolution, batch normal, dropout, and max-pooling layers. The convolutional layers generate several spatial features from small parts of the image, based on the defined number and size of the filters. The batch normalization layer normalizes the previous layers in the network. The dropout layer reduces the complexity of the network. The batch normalization and dropout layers act as regulators to avoid overfitting in the model. The max-pooling layer reduces the scale of the features in each step of the contraction path [63]. The output layer indicates the probability of existing logging trails by the fixed size, as the input layer. A few low-level feature maps generated from 32 filters (3 × 3) in the second block of the contraction path along with the obtained high-level feature maps during the expansion path with the same filters are shown in Figures 5b and 5c, respectively.

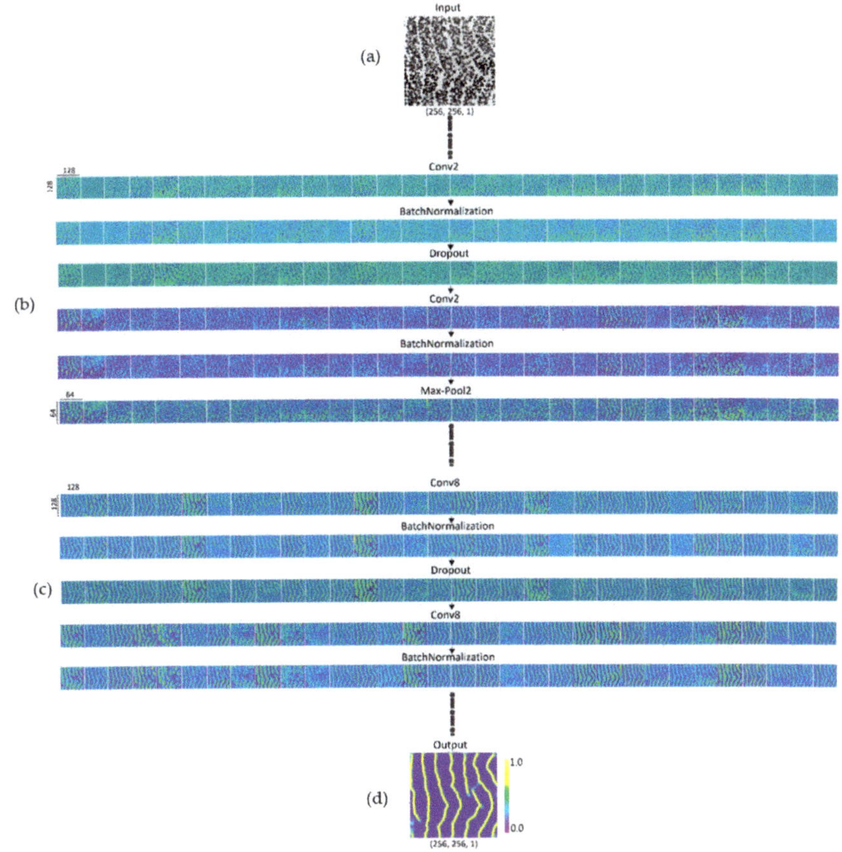

Figure 5. Visualization of different layers of the U-Net: (**a**) the input layer (e.g., a DSM derived from high-density laser scanning data) with a fixed size (256, 256, 1). A few intermediate feature maps such as convolutional layer, batch normalization, dropout, and max pooling generated from 32 filters (**b**) in the contraction path and (**c**) in the expansion path, and (**d**) the output layer of logging trails with the same size of the input layer.

2.5. Accuracy Assessment

2.5.1. Collecting Testing Data from Logging Trails

We selected some stands to collect testing data from logging trails in the Kakkurinmaa, Länsi-Aure, and Karpanmaa regions (Figure 6a). We designed 21 routes to collect the samples from segments to cover all of the logging trails within a stand (Figure 6b–d). Each route consisted of endpoints, trail segments, and edges (interval between two segment trails) (Figure 6f,g). The segments and edges indicated ground-truth trails and non-trails, respectively.

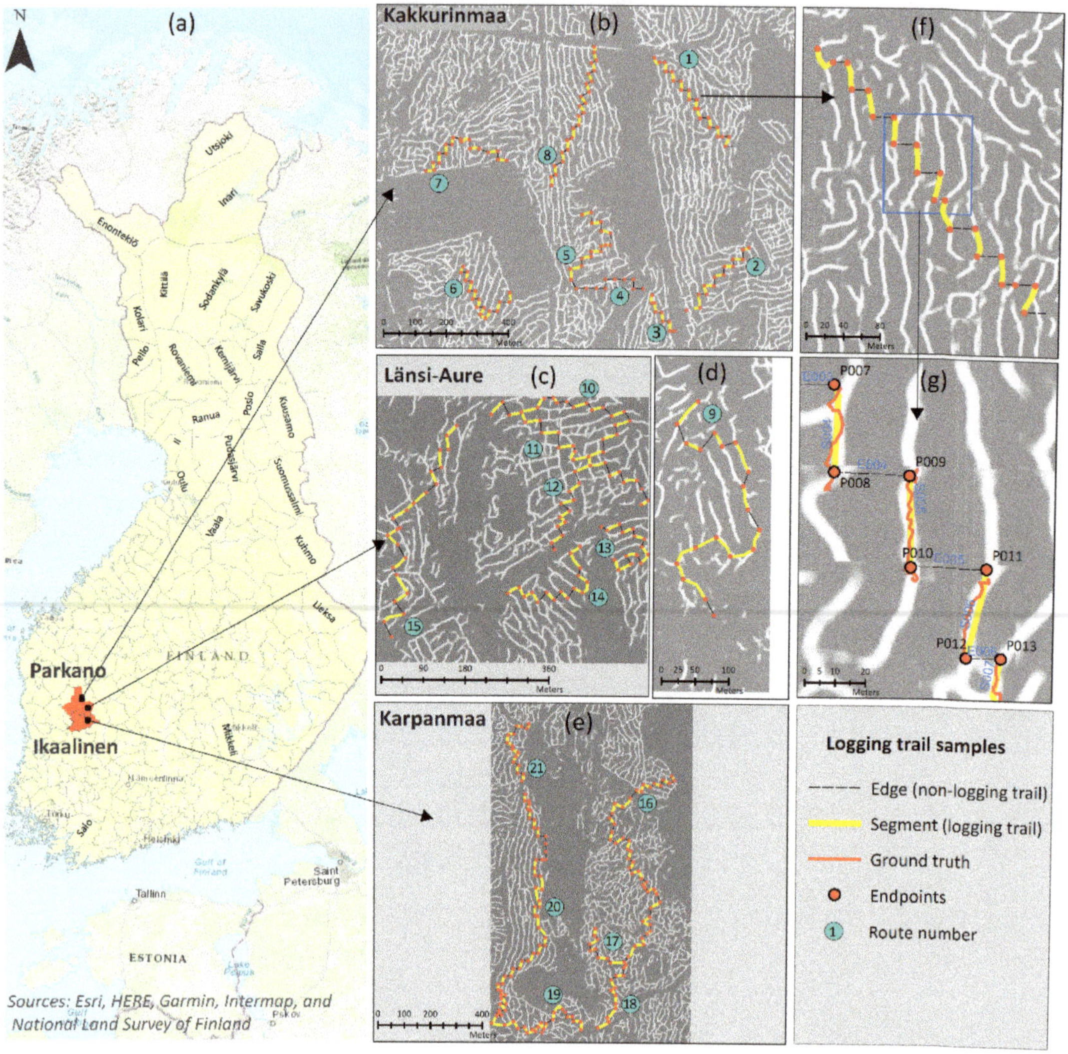

Figure 6. Collecting testing samples from logging and non-logging trails. (**a**) Selected forest stands for sampling from the logging trails in the Parkano and Ikaalinen areas in southern Finland; (**b**–**e**) designated routes for testing segments (logging trails) and edges (no logging trails) in the three selected sites; (**f**) an example of a designed route and (**g**) its components.

The length of an example sample segment trail was approximately 30 m; it may be longer in some cases, however, due to certain conditions such as existing connections or looped trails at the edges. Each segment has a start point and an endpoint that are both called endpoints. The positions of endpoints were converted into the GPS Exchange Format (GPX) and imported into a Garmin Oregon 750t GNSS receiver. The routes were reconstructed based on their corresponding endpoints and then navigated point by point with a PDOP (position dilution of precision) of less than 3 m. After finding the approximate location of an endpoint, the surveyor moved to the center of the trail and recorded the segment between the two endpoints using a Trimble GeoXT GNSS receiver. It also controlled the existence of any possible trails between two adjacent trails in the connector edges. The attributes of each endpoint, segment, and edge (e.g., PDOP, dominant tree species, existence trail, or other objects) were recorded. The data were transferred into GPS Pathfinder Office to correct errors based on the nearby GPS base stations to achieve an accuracy of less than 50 cm. The corrected data files were exported in shapefile format for use in assessing the accuracy of the predicted trails by the trained U-Net using the high-density laser scanning datasets.

2.5.2. Accuracy Metrics

A confusion matrix was constructed to assess the accuracy of the trained U-Net through the testing data in predicting logging trails using the laser scanning-derived datasets. The confusion matrix consisted of the number of the ground-truth samples that were labeled as logging trails on the ground and predicted as logging trails through the U-Net (*TP*), the number of samples that were labeled as non-logging trails and predicted as non-logging trails (*TN*), the number of samples that were labeled as logging trails but predicted as non-logging trails (*FN*), and the number of samples that were labeled as non-logging trails but predicted as logging trails (*FP*). Cohen's kappa, overall accuracy, intersection over union (*IoU*), and recall metrics were then derived from the confusion matrix to quantify the U-Net's performance in detecting logging trails from the canopy height and elevation models.

Cohen's kappa indicates the ratio of agreement after removing chance agreement [64,65]. It was calculated as Equation (1) [20] with respect to the observed accuracy (P_0) and the randomly expected accuracy (P_e).

$$Cohen's\ kappa = \frac{(P_0 - P_e)}{(1 - P_e)} \tag{1}$$

$$P_0 = \frac{TP + TN}{N} \tag{1a}$$

$$P_e = \frac{(TP + FN) \times (TP + FP)}{N^2} + \frac{(TN + FP) \times (TN + FN)}{N^2} \tag{1b}$$

where N is the total number of ground-truth samples.

The overall accuracy indicates the ratio of correct predictions for both logging trail and non-logging trail classes (Equation (2)).

$$Overall\ Accuracy = \frac{TP + TN}{TP + TN + FP + FN} \tag{2}$$

IoU expresses the similarity ratio between the predicted logging trails and the corresponding segments of ground truth samples (Equation (3)).

$$IoU = \frac{TP}{TP + FP + FN} \tag{3}$$

Recall expresses the perfection of the positive predictions. It is the proportion that a real instance of the target class (i.e., logging trails) can be correctly detected through the model (Equation (4)).

$$Recall = \frac{TP}{(TP + FN)} \tag{4}$$

3. Results

3.1. Performance of Trained Models

3.1.1. Detection Logging Trails in the Entire Forest

The results of the accuracy assessment of the trained U-Net using the CHM, DSM, and DEMs datasets in distinguishing logging trails from non-logging trails demonstrate the superior performance of the DSM (Table 1). The accuracy metrics show almost excellent performance of the U-Net using the DSM (k = 0.846 and *IoU* = 0.867), substantial performance using the CHM (k = 0.734 and *IoU* = 0.782), moderate performance using the DEM_{avg} (k = 0.528 and *IoU* = 0.587), and a slight performance using the DEM_{min} (k = 0.136 and *IoU* = 0.155). The values of *Recall* show the excellent performance of trained U-Net using the DSM (0.959) and the CHM (0.908) in detecting the logging trail class.

Table 1. The accuracy of the trained U-Net using the derivatives of high-density laser scanning data, including the canopy height model (CHM), the digital surface model (DSM), and the digital elevation models based on the average (DEM_{avg}) and minimum (DEM_{min}) values to distinguish the logging trails from the non-logging trails in three testing forests in southern Finland.

Metric	CHM	DSM	DEM_{avg}	DEM_{min}
Cohen's kappa	0.734	0.846	0.528	0.136
Overall accuracy	0.867	0.923	0.736	0.553
IoU	0.782	0.867	0.587	0.155
Recall	0.908	0.959	0.649	0.157

3.1.2. Detection Logging Trails in Different Stages of Commercial Thinning

The performance of the trained U-Net using the CHM, the DSM, and the DEMs varies in distinguishing logging trails from non-logging trails in the four classes of stand development (Figure 7) as well. Although the trained U-Net using CHM could distinguish significantly logging trails from non-logging trails in young stands after the first commercial thinning (k = 0.859 and *IoU* = 0.893) and in mature stands after the second/third commercial thinning (k = 0.834 and *IoU* = 0.876), it shows moderate performance in mature stands before the second commercial thinning (k = 0.438 and *IoU* = 0.505) (Figure 7a).

Similarly, the trained U-Net using DSM showed excellent performance to distinguish the logging trails from the non-logging trails in young stands (k = 0.953 and *IoU* = 0.963) and mature stands (k = 0.854 and *IoU* = 0.889) after receiving the commercial thinning operations. The efficiency of the trained model using DSM is higher than the trained model using CHM in mature stands before the second commercial thinning (k = 0.684 and *IoU* = 0.686) (Figure 7b).

The trained U-Net using DEM_{avg} showed moderate performance in detecting logging trails within thinned stands, with slightly better performance in the mature stands after receiving the commercial thinning (k = 0.542 and *IoU* = 0.667) (Figure 7c). The trained U-Net using DEM_{min} demonstrated a slight performance in all four stand classes. The accuracy values in the mature stands with commercial thinning is slightly better than other stands (k = 0.179 and *IoU* = 0.218) (Figure 7d).

3.2. Prediction of Logging Trails

Figure 8 shows some examples of predicted logging trails by trained U-Net using different datasets within different stages of commercial thinning. Logging trails were detected with high probabilities using both CHM (Figure 8b,c) and DSM (Figure 8f,g) datasets in young stands and mature stands that had undergone commercial thinning. The detected logging trail patterns were very similar by these two models. However, the trained model using DSM detected the trails under the canopy with a higher probability. In the old stands before the second commercial thinning, the trained U-Net, based on the both CHM (Figure 8d) and DSM (Figure 8h), predicted some segments of a trail with a high probability while other segments with a low probability. Typically, most of these segments are located in complex backgrounds that are clogged by regenerated trees or seedlings. However, this detection, even with a low probability, can be used to restore the original network of old logging trails in this type of stand.

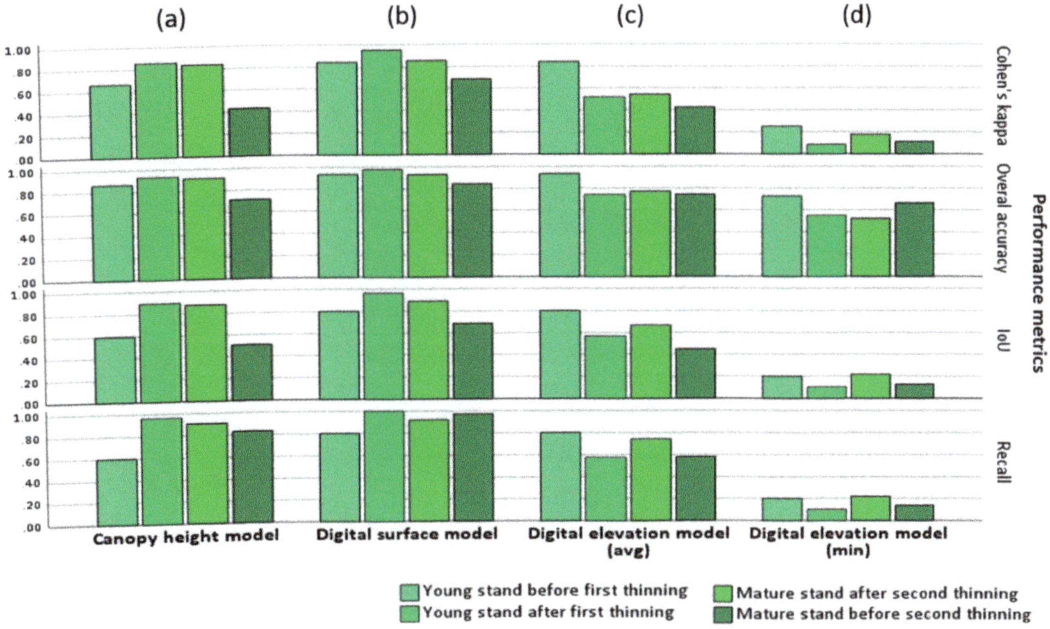

Figure 7. Comparison of the accuracy of the trained U-Net (**a**) using the canopy height model (CHM), (**b**) using the digital surface model (DSM), (**c**) using the average digital elevation model (DEM_{avg}), and (**d**) using the minimum digital elevation model (DEM_{min}) in detecting logging trails from non-logging trails in different stages of commercial thinning operations.

The trained U-Net using DEM_{avg} dataset for detecting logging trails, demonstrated a weak prediction in the young stands that had received the first thinning (Figure 8j), a relatively high prediction in the mature stands that had received the second thinning (Figure 8k), and a moderate prediction in the old stands (Figure 8l). The trained U-Net using DEM_{min} dataset only indicated a high prediction of logging trails in mature stands after a second or third commercial thinning (Figure 8o). As logging trails were not established in young stands before the first commercial thinning, the trained models did not predict any significant segments as part of a logging trail (Figure 8a,e,i,m).

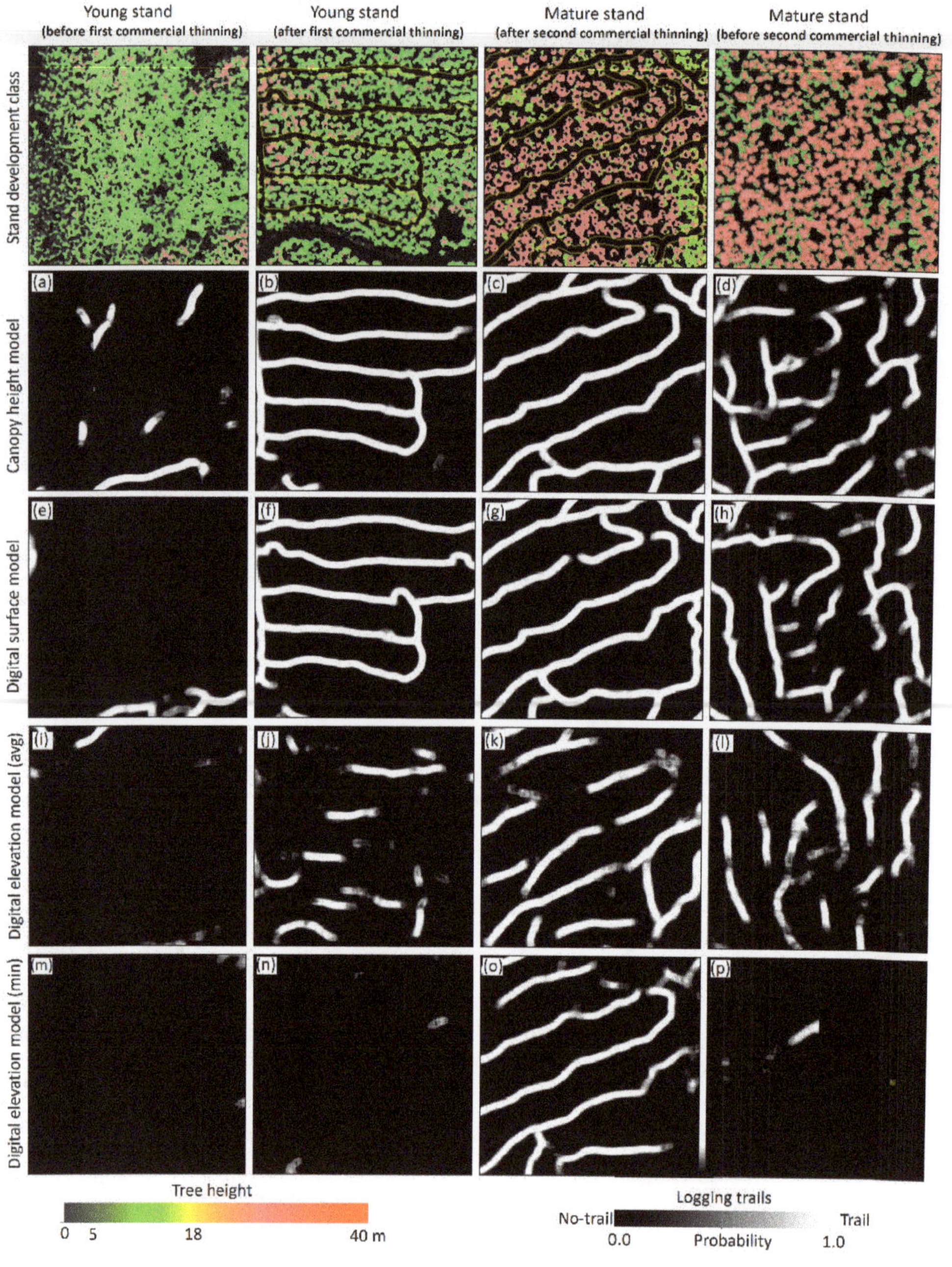

Figure 8. Comparison of the probability of prediction logging trails using U-Net in different forest development classes based on (**a–d**) the canopy height model (CHM), (**e–h**) digital surface model (DSM), and (**i–p**) digital elevation models (DEMs), in a patch with a size of 256 by 256. Although the U-Net using DSM and CHM showed high probability in detecting logging trails, using DEM_{min} and DEM_{avg}, it showed weak and moderate probabilities throughout forest stand classes except for mature stands that received the final commercial thinning operations.

4. Discussion

4.1. Distinguishing Logging Trails from Non-Logging Trails Using U-Net

The developed U-Net algorithm can distinguish logging trails from non-logging trails with almost perfect accuracy in the studied forest stands. The algorithm could precisely classify wide-open, polygonal spaces within the stands, such as forest storage areas and landing areas as a non-logging trail (Figure 9b). Nevertheless, few narrow corridors, mostly within the mature stands that were not thinned for a long time are predicted as logging trails (Figure 9f). Additionally, some linear features such as drainage ditches with geometric characteristics similar to logging trails (e.g., ditch width/cleaned area from tress) may be misidentified as logging trails in some stands (Figure 9g,h). We classified the testing samples of these objects as the *FP* samples in the confusion matrix during the performance assessment. However, the pattern of the corridors in the network and the geometric characteristics, such as their spacing and width, might cause the U-Net to recognize them as a logging trail. The forest roads are detected as non-logging trails in all stands; the specific geometry of a forest road and its texture on the DSM or CHM resulted in distinguishing it from a logging trail through the U-Net (Figure 9c). As previous studies reported the efficiency of U-Net in detection of road areas using VHR or UAV images [24–29], this study adds its efficiency in detection of logging trails using high-density laser scanning data as well. On the basis of traditional machine learning, some studies have extracted numerous metrics from laser scanning data to achieve accurate segments of roads under the canopy [10,16]. However, logging trail segmentation using our trained U-Net does not require laborious feature extractions or post-processing to detect the final trail using laser scanning-derived metrics. The developed end-to-end convolutional neural network approach obtains the image patches of the DSM or CHM, derived from laser scanning points, as inputs without extensive pre-processing and creates trail segments without requiring specific post-processing.

4.2. Detection of Logging Trails in Different Stages of Commercial Thinning

Using the CHM and DSM datasets, our algorithm perfectly detected logging trails in both young and mature stands that had undergone commercial thinning operations (Figure 7a,b). The misidentification of some drainage ditches as logging trails mainly occurred in these two types of stand; we recommend excluding these from the final network. Triangular irregular networks (TIN), which are derived from the laser scanning data, can significantly detect drainage ditches (Figure 9h) and solve this problem. Moreover, using the DSM, the U-Net was able to detect logging trails within mature stands that had not recently undergone a second commercial thinning. The logging trails in these stands do not form a continuous network, as opposed to stands that have undergone recent commercial thinning operations (Figure 8). Some segments of logging trails in the old stands are occluded by regenerated young trees (Figure 9i). The U-Net detected some of these clogged trails with a lower probability, however, which may aid in reconstruction of the original network of logging trails in these stands, for example, similar to the proposed approach [53] for restoring the network of historical roads through hollow roads detected by CarcassonNet and laser scanning-derived DTM.

4.3. Geometric Properties of the Predicted Logging Trails

The trained U-Net has sharpened the geometric properties of the logging-trail network as accurate as that of the labeled dataset used for its training. It recognized the pattern of a network within a stand (Figure 9) and attempted to keep the average spacing (i.e., 20–25 m) between the logging trails, while avoiding any overlap between them, particularly in the stands that were thinned (Figure 9a). The connection between the trails occurred at the endpoints or through intermediate trail connections that looped the trails (Figure 9d,e). The algorithm also detected those segments of a trail that were clogged by new trees, mostly in mature stands that were not thinned over a long period of time (Figure 9i). However, it did retain the overall pattern of a network, making it possible to restore the missing segments and the original network. Similarly, earlier studies reported the efficiency of some CNN-based algorithms, such as CasNet [34] and DH-GAN [66], for the extraction of some characterizations of main roads using VHR images.

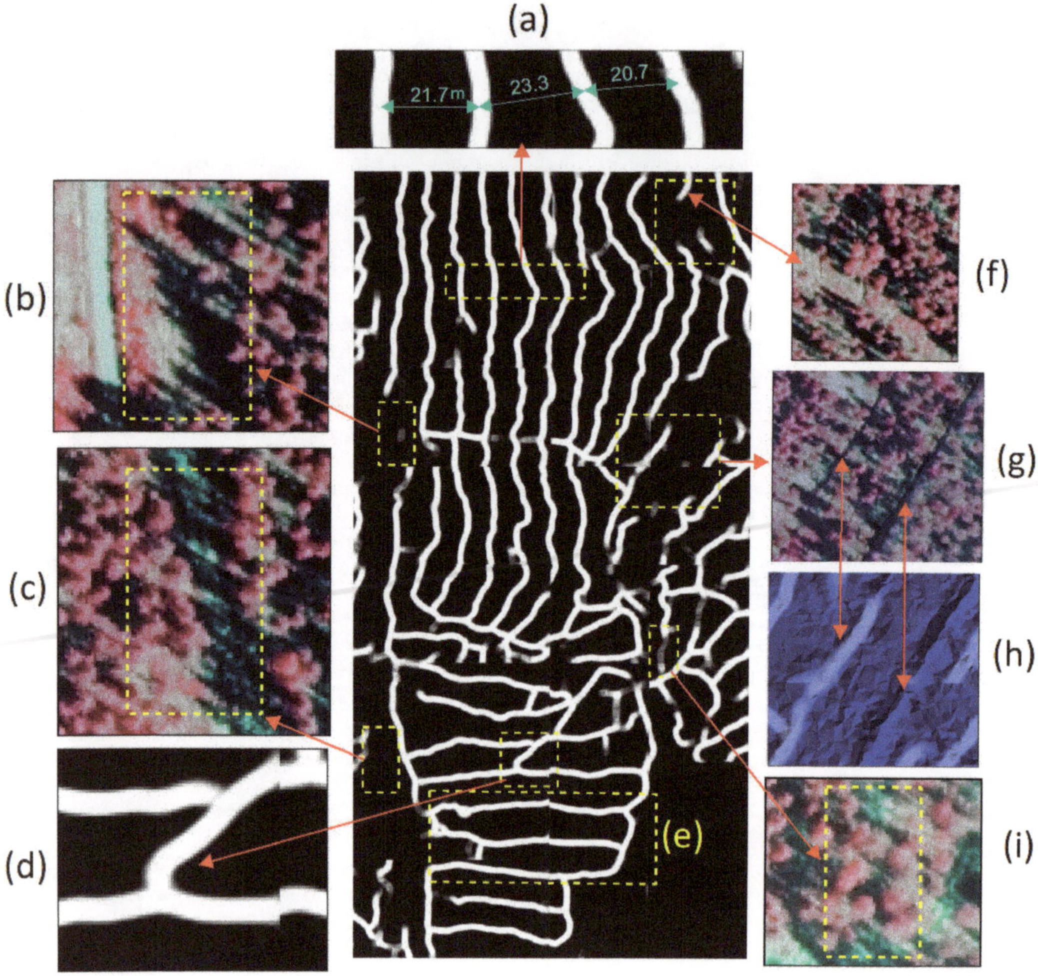

Figure 9. The ability of developed U-Net in detection the characteristics of logging trails: (**a**) patterns and geometric properties of the detected logging trails, such as trail spacing; (**d**) intermediate trail connections; and (**e**) looped trails through the U-Net and the DSM dataset. The algorithm correctly distinguished some complex features such as (**b**) landing areas and (**c**) forest roads as non-logging trails in the vicinity of the logging trails; (**f**) a corridor that was wrongly identified as a logging trail; (**g**,**h**) a deep ditch was detected as a non-logging trail and a shallow ditch that was detected as a logging trail; and (**i**) the occlusion of an old logging-trail by regenerated trees, although the algorithm was able to guess it as a logging trail with a lower probability.

No ground data was available to measure the accurate width of the logging trails. Therefore, we took the standard width of 4 m for a logging trail into account during the creation of the labeled dataset. We attempted to select trails that are visible in the set of our applied sources (e.g., orthophotos and tree profiles), particularly for the stands that had undergone commercial thinning. We randomly visited some of the logging trails within the selected sites to achieve the highest confidence in the created labeled dataset, before training the model.

The U-Net perfectly detected the features as logging trails when their width was close to the average value. For example, forest roads were classified as non-logging trails using this geometry by the U-Net. However, we could not find reliable labels in some complex stands, such as the mature stands that were not thinned for a long time. With the modern harvesting methods, the harvesters and forwarders are equipped with a computer system and a global navigation satellite system (GNSS) [67,68] that enables them to record the tracks of logging trails with an acceptable accuracy during thinning operations. We recommend employing this large dataset to train the deep learning-based algorithms to sharpen the detection of logging trails using high-quality laser scanning data, particularly in the complex stands.

To explore how well the developed U-Net algorithm performed with the datasets of high-quality laser scanning, we carried out a novel sampling method, with an extensive field survey from the predicted logging trails and non-logging trails in three selected forest sites. For this purpose, we collected adequate ground-truth samples (390) from the segments of predicted logging trails (with a size of circa 30 m) and the interval between two logging trails to check for possible missing trails that might not be detected by the algorithm (Figure 6). This surveying method enabled us to take samples from almost entire logging trails inside a stand; as a logging trail is designed as a continuous loop line starting from one side of the stand and continuing to the other side, so a segment of this line represents the existing or non-existence of the entire trail. It also enabled us to detect the non-logging trail objects either in the spot of predicted logging trails (i.e., segments) or the space between the trails (i.e., edges). Therefore, we maintained a balance between the samples of logging trails and non-logging trail objects, which is curtailed in the assessment of the efficiency of machine learning- or deep learning-based approaches to avoid unbalanced testing data and then miss-evaluation by the algorithm [69].

4.4. DEM Drawbacks in Detecting Logging Trails

Our research confirms the efficiency of the laser scanning derived metrics that sharpen the changes in the canopy structure of the trees, such as the DSM and DHM, in detecting logging trails (Figure 3). Therefore, the metrics, such as DEMs, which merely demonstrate the topographic characteristics of the ground surface, failed to recognize logging trails in the stands that had undergone commercial thinning. The earlier studies reported the efficiency of high-quality laser scanning data for detecting old logging trails or skid trails in harvested forests using DEM-derivatives, such as the morphological metrics [10,70]. Conversely, using the DEMs dataset (i.e., close to the ground surface), our research did not verify the efficiency of U-Net for detecting logging trails in forests that use harvesters or forwarders in commercial thinning (Figure 7). The soil damage created by harvesters and forwarders is reported to be less than that of skidders during the forest operations [71]. Forwarders carry a large volume of timber, but skidders drag the logs on the ground with several passes, which results in soil disturbance, compaction, and rutting [72]. Moreover, Finland's forest management regulations do not allow heavy soil disturbances, such as deep ruts (>10 cm), during commercial thinning. They recommend spreading logging residues on the logging trails, particularly on routes prone to rutting, to minimize soil damage [5]. The looped pattern of a logging trail network and retaining the optimal spacing between the trails within a stand, may aid to reduce the number of passes on some specific trails and then soil disturbances in forest operations as well. These practices lead to minimal alterations in the natural condition of the ground by logging trails. Therefore, logging trails are not expected to emerge in the DEM against the skid trails [70], abandoned logging trails [10], drainage ditches, or forest roads (Figure 9). Nevertheless, a few logging trails were detected using the DEMs dataset in the mature stands that had undergone commercial thinning (Figure 8). There is a good chance that increasing the number of passes by the machinery [73], using multifactional and heavier harvesters/forwarders [1], surging the weight of timber loads, and concentrating the forest operation during wet seasons [73], have all resulted in soil compression and then alteration in the natural ground (i.e., terrain).

4.5. Applications

Our findings and the procedure that we developed have several implications for precision harvesting and sustainable forest management during forest operations. A holistic network of old logging trails may lead to a better understanding of the patterns, geometric characteristics, efficiency, and drawbacks of the network. This understanding provides a new perspective on the designation of an optimal logging-trail network in the new stands, one that can minimize the costs of thinning operations and the damage to the soil and the trees left. This new perspective also provides a modification of the routes of a network that probably passes the soils with low bearing capacity due to a weakness in the design of the initial network or the deformation of the ground surface over time.

By having a network of old logging trails, the operators can import the routes into the computer system of the harvesters/forwarders for accurate navigation of the machines. Doing so decreases the costs of finding the old trails and prevents the overthinning of the stand, which may occur when removing trees for establishing new trails. This is a crucial step to approaching the aims of precision harvesting by minimizing the operation costs and preserving the forest landscape in modern forestry.

4.6. Outlook

Despite difficulties in finding reliable logging trails, we could collect acceptable patches of labeled datasets for training the U-Net algorithm. However, the datasets are limited to the Parkano and Ikaalinen areas, in Southern Finland. We strongly recommend employing a large dataset of logging trails that covers similar forest stands, with regard to commercial thinning, at least in the Nordic region for training the deep learning-based algorithm to achieve a versatile algorithm for the detection of logging trails.

The developed model performed with reasonable accuracy in detection of old logging trails in the mature stands that had not received the second thinning. However, detecting entire segments of a logging trail is still challenging in this type of stand. As mentioned earlier, providing an appropriate labeled dataset for improving the process of training the algorithm or testing the performances of other deep learning-based algorithms may aid in sharpening old logging trails in the mature stands.

In some stands, the drainage ditches hampered the efficiency of U-Net using the DSM or DHM to distinguish the logging trails through the semantic segmentation procedure that relies on the binary segmentation. We recommend testing high-level semantic segmentation or instance segmentation that discriminates different objects from each other [74]. However, this requires a larger labeled dataset based on the number of objects.

5. Conclusions

In this research, we presented an end-to-end U-Net convolutional neural network that uses high-density laser scanning-derived metrics for logging trail extraction. We carried out an extensive field survey to test the efficiency of the trained model based on three metrics (i.e., DSM, CHM, and DEMs) in forests with different commercial thinning. The trained U-Net using DSM was able to distinguish logging trails from the background with a high probability and very high performance, particularly in young and mature stands that had undergone commercial thinning. However, it needs to be improved for the very old stands that have not received second commercial thinning for a long time. The developed model can be used easily by the end-users, without heavy pre-processing of the laser scanning data or heavy post-processing of the outputs. We recommend creating a large labeled dataset from logging trials collected by harvesters during thinning operations and use them to train the deep-learning based algorithms. It would help to develop a versatile model that can extract logging trails in different forest management systems and different thinning stages, at least over the Nordic regions.

Author Contributions: Conceptualization, O.A., J.U. and V.-P.K.; methodology, O.A., J.U. and V.-P.K.; data provision, J.U.; data preparation, O.A.; software and programming, O.A.; field investigation and sampling, O.A., J.U. and V.-P.K.; visualization, O.A.; writing—original draft preparation, O.A.; writing—review and editing, J.U. and V.-P.K.; supervision, J.U. and V.-P.K.; project administration, J.U. All authors have read and agreed to the published version of the manuscript.

Funding: This work has been funded by the public-private partnership grant established for the professorship of forest operation and logistics at the University of Helsinki, grant number 7820148 and by the proof-of-concept-grant by the Faculty of Agriculture and Forestry, University of Helsinki, grant number 78004041. The APC was funded by University of Helsinki.

Institutional Review Board Statement: Not applicable.

Informed Consent Statement: Not applicable.

Data Availability Statement: Not applicable.

Acknowledgments: We would like thank Mikko Leinonen and Juho Luotola for assisting in the field operations. We would also like to express our gratitude to Finsilva Oyj and Metsähallitus for providing the access to their forest holdings and related forest inventory databases.

Conflicts of Interest: The authors declare no conflict of interest. The funders had no role in the design of the study; in the collection, analyses, or interpretation of data; in the writing of the manuscript, or in the decision to publish the results.

Appendix A

Table A1. U-Net tuned hyperparameters.

Parameter	Value
Kernels	16, 32, 64, 128, 256
Activation	RELU
Weight initializer	HeNormal
Max-pooling size	(2, 2)
Optimizer	Adam ($\beta 1 = 0.9$, $\beta 2 = 0.999$, $\varepsilon = 1 \times 10^{-7}$)
Learning rate	0.0008
Batch size	32
Dropout rate	[0.2, 0.4]

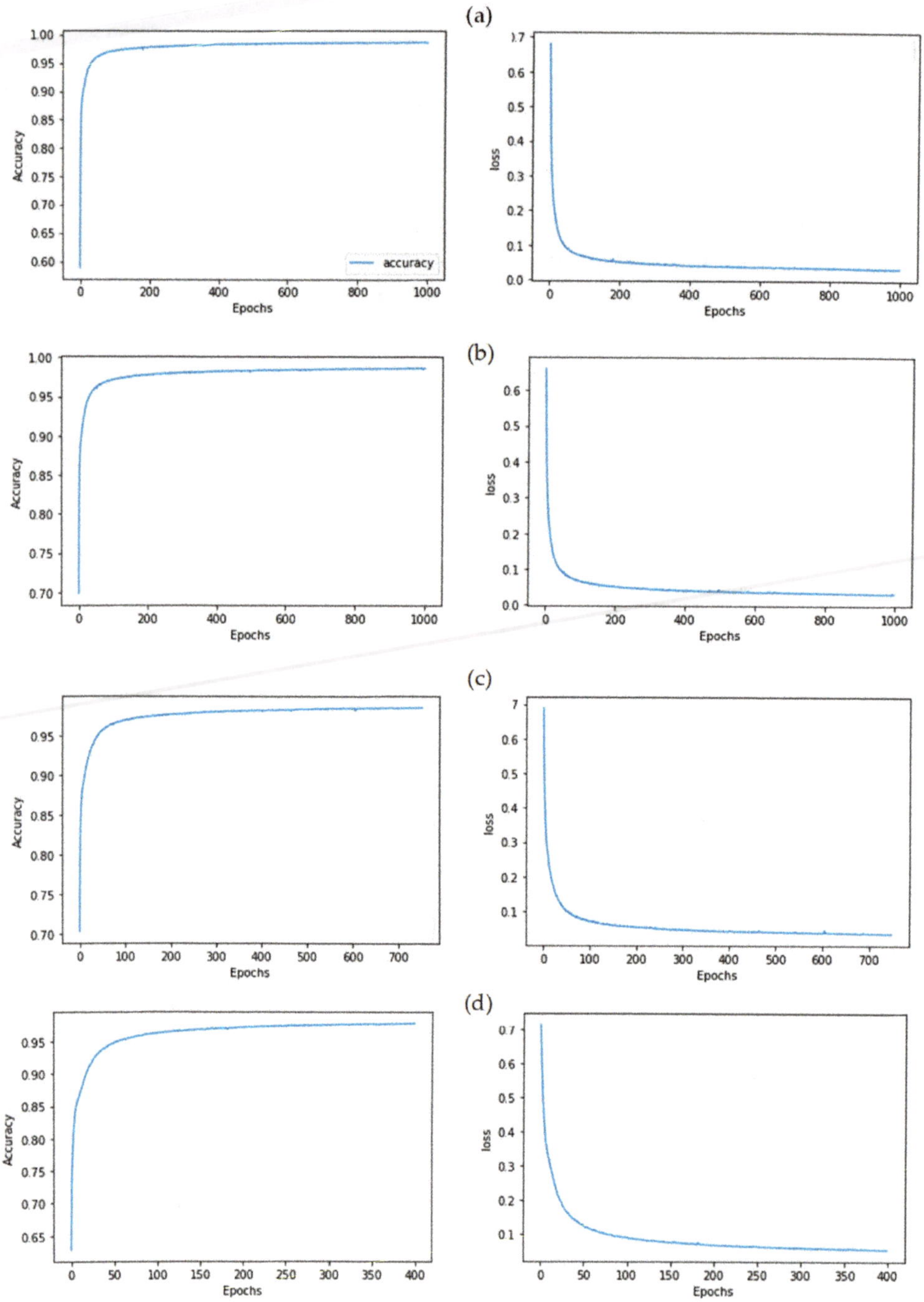

Figure A1. Accuracy and loss versus epochs during training of U-Net using (**a**) the DSM, (**b**) the CHM, (**c**) the DEM_{avg}, and (**d**) the DEM_{min} derived from high-density laser scanning data.

References

1. Uusitalo, J. *Introduction to Forest Operations and Technology*; JVP Forest Systems OY: Helsinki, Finland, 2010; ISBN 978-952-92-5269-5.
2. Pukkala, T.; Lähde, E.; Laiho, O. Continuous Cover Forestry in Finland—Recent Research Results. In *Continuous Cover Forestry*; Pukkala, T., von Gadow, K., Eds.; Springer: Dordrecht, The Netherlands, 2012; pp. 85–128, ISBN 978-94-007-2201-9.
3. Mielikaeinen, K.; Hakkila, P. Review of wood fuel from precommercial thinning and plantation cleaning in Finland. In *Wood Fuel from Early Thinning and Plantation Cleaning: An International Review*; Puttock, D., Richardson, J., Eds.; Vantaa Research Centre, Finnish Forest Research Institute: Vantaa, Finland, 1998; pp. 29–36, ISBN 9514016009.
4. Leinonen, A. *Harvesting Technology of Forest Residues for Fuel in the USA and Finland*; Valopaino Oy: Helsinki, Finland, 2004; ISBN 951-38-6212-7.
5. Äijälä, O.; Koistinen, A.; Sved, J.; Vanhatalo, K.; Väisänen, P. *Recommendations for Forest Management*; Tapio Oy: Helsinki, Finland, 2019. Available online: https://tapio.fi/wp-content/uploads/2020/09/Metsanhoidon_suositukset_Tapio_2019.pdf (accessed on 31 May 2021).
6. Maltamo, M.; Næsset, E.; Vauhkonen, J. *Forestry Applications of Airborne Laser Scanning: Concepts and Case Studies*; Maltamo, M., Næsset, E., Vauhkonen, J., Eds.; Springer: Dordrecht, The Netherlands, 2014; ISBN 978-94-017-8662-1.
7. Saukkola, A.; Melkas, T.; Riekki, K.; Sirparanta, S.; Peuhkurinen, J.; Holopainen, M.; Hyyppä, J.; Vastaranta, M. Predicting Forest Inventory Attributes Using Airborne Laser Scanning, Aerial Imagery, and Harvester Data. *Remote Sens.* **2019**, *11*, 797. [CrossRef]
8. Lin, C. Improved derivation of forest stand canopy height structure using harmonized metrics of full-waveform data. *Remote Sens. Environ.* **2019**, *235*, 111436. [CrossRef]
9. Lee, H.; Slatton, K.C.; Jhee, H. Detecting forest trails occluded by dense canopies using ALSM data. In Proceedings of the 2005 IEEE International Geoscience and Remote Sensing Symposium, Seoul, Korea, 25–29 July 2005; Institute of Electrical and Electronics Engineers (IEEE): Piscataway, NJ, USA, 2005; pp. 3587–3590, ISBN 0-7803-9050-4. Available online: https://ieeexplore.ieee.org/document/1526623 (accessed on 5 October 2021).
10. Sherba, J.; Blesius, L.; Davis, J. Object-Based Classification of Abandoned Logging Roads under Heavy Canopy Using LiDAR. *Remote Sens.* **2014**, *6*, 4043–4060. [CrossRef]
11. Ferraz, A.; Mallet, C.; Chehata, N. Large-scale road detection in forested mountainous areas using airborne topographic lidar data. *ISPRS J. Photogramm. Remote Sens.* **2016**, *112*, 23–36. [CrossRef]
12. Li, C.; Ma, L.; Zhou, M.; Zhu, X. Study on Road Detection Method from Full-Waveform LiDAR Data in Forested Area. In Proceedings of the Fourth International Conference on Ubiquitous Positioning, Indoor Navigation and Location Based Services (UPINLBS), Shanghai, China, 2–4 November 2016.
13. Hrůza, P.; Mikita, T.; Tyagur, N.; Krejza, Z.; Cibulka, M.; Procházková, A.; Patočka, Z. Detecting Forest Road Wearing Course Damage Using Different Methods of Remote Sensing. *Remote Sens.* **2018**, *10*, 492. [CrossRef]
14. Prendes, C.; Buján, S.; Ordoñez, C.; Canga, E. Large scale semi-automatic detection of forest roads from low density LiDAR data on steep terrain in Northern Spain. *iForest* **2019**, *12*, 366–374. [CrossRef]
15. Waga, K.; Tompalski, P.; Coops, N.C.; White, J.C.; Wulder, M.A.; Malinen, J.; Tokola, T. Forest Road Status Assessment Using Airborne Laser Scanning. *For. Sci.* **2020**, *66*, 501–508. [CrossRef]
16. Buján, S.; Guerra-Hernández, J.; González-Ferreiro, E.; Miranda, D. Forest Road Detection Using LiDAR Data and Hybrid Classification. *Remote Sens.* **2021**, *13*, 393. [CrossRef]
17. Kaiser, J.V.; Stow, D.A.; Cao, L. Evaluation of Remote Sensing Techniques for Mapping Transborder Trails. *Photogramm. Eng. Remote Sens.* **2004**, *70*, 1441–1447. [CrossRef]
18. Hoeser, T.; Kuenzer, C. Object Detection and Image Segmentation with Deep Learning on Earth Observation Data: A Review—Part I: Evolution and Recent Trends. *Remote Sens.* **2020**, *12*, 1667. [CrossRef]
19. Hoeser, T.; Bachofer, F.; Kuenzer, C. Object Detection and Image Segmentation with Deep Learning on Earth Observation Data: A Review—Part II: Applications. *Remote Sens.* **2020**, *12*, 3053. [CrossRef]
20. Kneusel, R.T. *Practical Deep Learning: A Python-Based Introduction*, 1st ed.; No Starch Press Inc.: San Francisco, CA, USA, 2021; ISBN 978-1-7185-0075-4.
21. Simonyan, K.; Zisserman, A. Very Deep Convolutional Networks for Large-Scale Image Recognition. 2014. Available online: https://arxiv.org/pdf/1409.1556 (accessed on 12 August 2021).
22. Constantin, A.; Ding, J.-J.; Lee, Y.-C. Accurate Road Detection from Satellite Images Using Modified U-net. In Proceedings of the IEEE Asia Pacific Conference on Circuits and Systems (APCCAS), Chengdu, China, 26–30 October 2018; Institute of Electrical and Electronics Engineers (IEEE): Piscataway, NJ, USA, 2018; pp. 423–426, ISBN 978-1-5386-8240-1.
23. Shi, Q.; Liu, X.; Li, X. Road Detection from Remote Sensing Images by Generative Adversarial Networks. *IEEE Access* **2018**, *6*, 25486–25494. [CrossRef]
24. Buslaev, A.; Seferbekov, S.; Iglovikov, V.; Shvets, A. Fully Convolutional Network for Automatic Road Extraction from Satellite Imagery. In Proceedings of the 2018 IEEE/CVF Conference on Computer Vision and Pattern Recognition Workshops (CVPRW), Salt Lake City, UT, USA, 18–22 June 2018; Institute of Electrical and Electronics Engineers (IEEE): Piscataway, NJ, USA, 2018; pp. 197–1973, ISBN 978-1-5386-6100-0.
25. Kestur, R.; Farooq, S.; Abdal, R.; Mehraj, E.; Narasipura, O.; Mudigere, M. UFCN: A fully convolutional neural network for road extraction in RGB imagery acquired by remote sensing from an unmanned aerial vehicle. *J. Appl. Remote Sens.* **2018**, *12*, 1. [CrossRef]

26. He, H.; Yang, D.; Wang, S.; Wang, S.; Liu, X. Road segmentation of cross-modal remote sensing images using deep segmentation network and transfer learning. *Ind. Robot.* **2019**, *46*, 384–390. [CrossRef]
27. Xin, J.; Zhang, X.; Zhang, Z.; Fang, W. Road Extraction of High-Resolution Remote Sensing Images Derived from DenseUNet. *Remote Sens.* **2019**, *11*, 2499. [CrossRef]
28. Xu, Y.; Xie, Z.; Feng, Y.; Chen, Z. Road Extraction from High-Resolution Remote Sensing Imagery Using Deep Learning. *Remote Sens.* **2018**, *10*, 1461. [CrossRef]
29. Zhang, Z.; Liu, Q.; Wang, Y. Road Extraction by Deep Residual U-Net. *IEEE Geosci. Remote Sens. Lett.* **2018**, *15*, 749–753. [CrossRef]
30. Doshi, J. Residual Inception Skip Network for Binary Segmentation. In Proceedings of the 2018 IEEE/CVF Conference on Computer Vision and Pattern Recognition Workshops (CVPRW), Salt Lake City, UT, USA, 18–22 June 2018; Institute of Electrical and Electronics Engineers (IEEE): Piscataway, NJ, USA, 2018; pp. 206–2063, ISBN 978-1-5386-6100-0.
31. Varia, N.; Dokania, A.; Senthilnath, J. DeepExt: A Convolution Neural Network for Road Extraction using RGB images captured by UAV. In Proceedings of the 2018 IEEE Symposium Series on Computational Intelligence (SSCI), Bangalore, India, 18–21 November 2018; Institute of Electrical and Electronics Engineers (IEEE): Piscataway, NJ, USA, 2018; pp. 1890–1895, ISBN 978-1-5386-9276-9.
32. Li, Y.; Xu, L.; Rao, J.; Guo, L.; Yan, Z.; Jin, S. A Y-Net deep learning method for road segmentation using high-resolution visible remote sensing images. *Remote Sens. Lett.* **2019**, *10*, 381–390. [CrossRef]
33. Panboonyuen, T.; Jitkajornwanich, K.; Lawawirojwong, S.; Srestasathiern, P.; Vateekul, P. Road Segmentation of Remotely-Sensed Images Using Deep Convolutional Neural Networks with Landscape Metrics and Conditional Random Fields. *Remote Sens.* **2017**, *9*, 680. [CrossRef]
34. Cheng, G.; Wang, Y.; Xu, S.; Wang, H.; Xiang, S.; Pan, C. Automatic Road Detection and Centerline Extraction via Cascaded End-to-End Convolutional Neural Network. *IEEE Trans. Geosci. Remote Sens.* **2017**, *55*, 3322–3337. [CrossRef]
35. Fujimoto, A.; Haga, C.; Matsui, T.; Machimura, T.; Hayashi, K.; Sugita, S.; Takagi, H. An End to End Process Development for UAV-SfM Based Forest Monitoring: Individual Tree Detection, Species Classification and Carbon Dynamics Simulation. *Forests* **2019**, *10*, 680. [CrossRef]
36. Miyoshi, G.T.; Arruda, M.d.S.; Osco, L.P.; Marcato Junior, J.; Gonçalves, D.N.; Imai, N.N.; Tommaselli, A.M.G.; Honkavaara, E.; Gonçalves, W.N. A Novel Deep Learning Method to Identify Single Tree Species in UAV-Based Hyperspectral Images. *Remote Sens.* **2020**, *12*, 1294. [CrossRef]
37. Ocer, N.E.; Kaplan, G.; Erdem, F.; Kucuk Matci, D.; Avdan, U. Tree extraction from multi-scale UAV images using Mask R-CNN with FPN. *Remote Sens. Lett.* **2020**, *11*, 847–856. [CrossRef]
38. Korznikov, K.A.; Kislov, D.E.; Altman, J.; Doležal, J.; Vozmishcheva, A.S.; Krestov, P.V. Using U-Net-Like Deep Convolutional Neural Networks for Precise Tree Recognition in Very High Resolution RGB (Red, Green, Blue) Satellite Images. *Forests* **2021**, *12*, 66. [CrossRef]
39. Schiefer, F.; Kattenborn, T.; Frick, A.; Frey, J.; Schall, P.; Koch, B.; Schmidtlein, S. Mapping forest tree species in high resolution UAV-based RGB-imagery by means of convolutional neural networks. *ISPRS J. Photogramm. Remote Sens.* **2020**, *170*, 205–215. [CrossRef]
40. Xi, Z.; Hopkinson, C.; Rood, S.B.; Peddle, D.R. See the forest and the trees: Effective machine and deep learning algorithms for wood filtering and tree species classification from terrestrial laser scanning. *ISPRS J. Photogramm. Remote Sens.* **2020**, *168*, 1–16. [CrossRef]
41. La Rosa, L.E.C.; Sothe, C.; Feitosa, R.Q.; de Almeida, C.M.; Schimalski, M.B.; Oliveira, D.A.B. Multi-task fully convolutional network for tree species mapping in dense forests using small training hyperspectral data. *ISPRS J. Photogramm. Remote Sens.* **2021**, *179*, 35–49. [CrossRef]
42. Seidel, D.; Annighöfer, P.; Thielman, A.; Seifert, Q.E.; Thauer, J.-H.; Glatthorn, J.; Ehbrecht, M.; Kneib, T.; Ammer, C. Predicting Tree Species From 3D Laser Scanning Point Clouds Using Deep Learning. *Front. Plant Sci.* **2021**, *12*, 635440. [CrossRef]
43. Ercanlı, İ. Innovative deep learning artificial intelligence applications for predicting relationships between individual tree height and diameter at breast height. *For. Ecosyst.* **2020**, *7*, 1–18. [CrossRef]
44. Qi, Y.; Dong, X.; Chen, P.; Lee, K.-H.; Lan, Y.; Lu, X.; Jia, R.; Deng, J.; Zhang, Y. Canopy Volume Extraction of Citrus reticulate Blanco cv. Shatangju Trees Using UAV Image-Based Point Cloud Deep Learning. *Remote Sens.* **2021**, *13*, 3437. [CrossRef]
45. Deng, X.; Tong, Z.; Lan, Y.; Huang, Z. Detection and Location of Dead Trees with Pine Wilt Disease Based on Deep Learning and UAV Remote Sensing. *AgriEngineering* **2020**, *2*, 19. [CrossRef]
46. Tran, D.Q.; Park, M.; Jung, D.; Park, S. Damage-Map Estimation Using UAV Images and Deep Learning Algorithms for Disaster Management System. *Remote Sens.* **2020**, *12*, 4169. [CrossRef]
47. Kislov, D.E.; Korznikov, K.A.; Altman, J.; Vozmishcheva, A.S.; Krestov, P.V. Extending deep learning approaches for forest disturbance segmentation on very high-resolution satellite images. *Remote Sens. Ecol. Conserv.* **2021**, *7*, 355–368. [CrossRef]
48. Qin, J.; Wang, B.; Wu, Y.; Lu, Q.; Zhu, H. Identifying Pine Wood Nematode Disease Using UAV Images and Deep Learning Algorithms. *Remote Sens.* **2021**, *13*, 162. [CrossRef]
49. Wang, M.; Li, R. Segmentation of High Spatial Resolution Remote Sensing Imagery Based on Hard-Boundary Constraint and Two-Stage Merging. *IEEE Trans. Geosci. Remote Sens.* **2014**, *52*, 5712–5725. [CrossRef]
50. Zhong, Y.; Zhu, Q.; Zhang, L. Scene Classification Based on the Multifeature Fusion Probabilistic Topic Model for High Spatial Resolution Remote Sensing Imagery. *IEEE Trans. Geosci. Remote Sens.* **2015**, *53*, 6207–6222. [CrossRef]

51. Abdollahi, A.; Pradhan, B.; Shukla, N.; Chakraborty, S.; Alamri, A. Deep Learning Approaches Applied to Remote Sensing Datasets for Road Extraction: A State-Of-The-Art Review. *Remote Sens.* **2020**, *12*, 1444. [CrossRef]
52. Caltagirone, L.; Scheidegger, S.; Svensson, L.; Wahde, M. Fast LIDAR-based road detection using fully convolutional neural networks. In Proceedings of the 2017 IEEE Intelligent Vehicles Symposium (IV), Los Angeles, CA, USA, 11–14 June 2017; pp. 1019–1024, ISBN 978-1-5090-4804-5.
53. Verschoof-van der Vaart, W.B.; Landauer, J. Using CarcassonNet to automatically detect and trace hollow roads in LiDAR data from the Netherlands. *J. Cult. Herit.* **2021**, *47*, 143–154. [CrossRef]
54. Staaf, K.A.G.; Wiksten, N.A. *Tree Harvesting Techniques*; Nijhoff: Dordrecht, The Netherlands, 1984; ISBN 978-90-247-2994-4.
55. National Land Survey of Finland (NLS). Laser Scanning Data 5 p. Available online: https://www.maanmittauslaitos.fi/en/maps-and-spatial-data/expert-users/product-descriptions/laser-scanning-data-5-p (accessed on 6 May 2021).
56. National Land Survey of Finland (NLS). NLS Orthophotos. Available online: https://tiedostopalvelu.maanmittauslaitos.fi/tp/kartta?lang=en (accessed on 1 May 2021).
57. Esri. Lidar Solutions in ArcGIS: Estimating Forest Canopy Density and Height. Available online: https://desktop.arcgis.com/en/arcmap/latest/manage-data/las-dataset/lidar-solutions-estimating-forest-density-and-height.htm (accessed on 1 June 2021).
58. Esri. Lidar Solutions in ArcGIS: Creating Raster DEMs and DSMs from Large Lidar Point Collections. Available online: https://desktop.arcgis.com/en/arcmap/latest/manage-data/las-dataset/lidar-solutions-creating-raster-dems-and-dsms-from-large-lidar-point-collections.htm (accessed on 1 June 2021).
59. Ronneberger, O.; Fischer, P.; Brox, T. U-Net: Convolutional Networks for Biomedical Image Segmentation. 18 May 2015. Available online: http://arxiv.org/pdf/1505.04597v1 (accessed on 15 July 2021).
60. Li, Y.; Li, W.; Xiong, J.; Xia, J.; Xie, Y. Comparison of Supervised and Unsupervised Deep Learning Methods for Medical Image Synthesis between Computed Tomography and Magnetic Resonance Images. *Biomed. Res. Int.* **2020**, *2020*, 5193707. [CrossRef]
61. Chollet, F. *Deep Learning with Python*; Manning Publications Co.: Shelter Island, NY, USA, 2018; ISBN 1617294438.
62. TensorFlow. Introduction to the Keras Tuner. Available online: https://www.tensorflow.org/tutorials/keras/keras_tuner (accessed on 21 August 2021).
63. Wlodarczak, P. *Machine Learning and Its Applications*, 1st ed.; CRC Press: Boca Raton, FL, USA, 2019; ISBN 978-1-138-32822-8.
64. Cohen, J. A Coefficient of Agreement for Nominal Scales. *Educ. Psychol. Meas.* **1960**, *20*, 37–46. [CrossRef]
65. Landis, J.R.; Koch, G.G. The Measurement of Observer Agreement for Categorical Data. *Biometrics* **1977**, *33*, 159. [CrossRef]
66. Costea, D.; Marcu, A.; Leordeanu, M.; Slusanschi, E. Creating Roadmaps in Aerial Images with Generative Adversarial Networks and Smoothing-Based Optimization. In Proceedings of the 2017 IEEE International Conference on Computer Vision Workshop (ICCVW), Venice, Italy, 22–29 October 2017; pp. 2100–2109, ISBN 978-1-5386-1034-3.
67. Kemmerer, J.; Labelle, E.R. Using harvester data from on-board computers: A review of key findings, opportunities and challenges. *Eur. J. For. Res.* **2021**, *140*, 1–17. [CrossRef]
68. Woo, H.; Acuna, M.; Choi, B.; Han, S. FIELD: A Software Tool That Integrates Harvester Data and Allometric Equations for a Dynamic Estimation of Forest Harvesting Residues. *Forests* **2021**, *12*, 834. [CrossRef]
69. Nguyen, M.H. Impacts of Unbalanced Test Data on the Evaluation of Classification Methods. *Int. J. Adv. Comput. Sci. Appl.* **2019**, *10*, 497–502. [CrossRef]
70. Affek, A.N.; Zachwatowicz, M.; Sosnowska, A.; Gerlée, A.; Kiszka, K. Impacts of modern mechanised skidding on the natural and cultural heritage of the Polish Carpathian Mountains. *For. Ecol. Manag.* **2017**, *405*, 391–403. [CrossRef]
71. Picchio, R.; Mederski, P.S.; Tavankar, F. How and How Much, Do Harvesting Activities Affect Forest Soil, Regeneration and Stands? *Curr. For. Rep.* **2020**, *6*, 115–128. [CrossRef]
72. Burley, J.; Evans, J.; Youngquist, J. *Encyclopedia of Forest Sciences*; Elsevier: Amsterdam, The Netherlands; Oxford, UK, 2004; ISBN 0-12-145160-7.
73. Sirén, M.; Ala-Ilomäki, J.; Mäkinen, H.; Lamminen, S.; Mikkola, T. Harvesting damage caused by thinning of Norway spruce in unfrozen soil. *Int. J. For. Eng.* **2013**, *24*, 60–75. [CrossRef]
74. Carvalho, O.L.F.d.; de Carvalho Júnior, O.A.; Albuquerque, A.O.d.; Bem, P.P.d.; Silva, C.R.; Ferreira, P.H.G.; Moura, R.d.S.d.; Gomes, R.A.T.; Guimarães, R.F.; Borges, D.L. Instance Segmentation for Large, Multi-Channel Remote Sensing Imagery Using Mask-RCNN and a Mosaicking Approach. *Remote Sens.* **2021**, *13*, 39. [CrossRef]

remote sensing

MDPI

Article

BiFDANet: Unsupervised Bidirectional Domain Adaptation for Semantic Segmentation of Remote Sensing Images

Yuxiang Cai [1], Yingchun Yang [1,*], Qiyi Zheng [1], Zhengwei Shen [2,3], Yongheng Shang [2,3], Jianwei Yin [1,4,5] and Zhongtian Shi [6]

1 College of Computer Science and Technology, Zhejiang University, Hangzhou 310027, China; caiyuxiang@zju.edu.cn (Y.C.); 22051229@zju.edu.cn (Q.Z.); zjuyjw@cs.zju.edu.cn (J.Y.)
2 Research Institute of Advanced Technology, Zhejiang University, Hangzhou 310027, China; zwshen@zju.edu.cn (Z.S.); yh_shang@zju.edu.cn (Y.S.)
3 Deqing Institute of Advanced Technology and Industrialization, Zhejiang University, Huzhou 313200, China
4 School of Software Technology, Zhejiang University, Ningbo 315048, China
5 China Institute for New Urbanization Studies, Huzhou 313000, China
6 Hangzhou Planning and Natural Resources Survey and Monitoring Center, Hangzhou 310012, China; zhongtian_shi@outlook.com
* Correspondence: yyc@zju.edu.cn

Abstract: When segmenting massive amounts of remote sensing images collected from different satellites or geographic locations (cities), the pre-trained deep learning models cannot always output satisfactory predictions. To deal with this issue, domain adaptation has been widely utilized to enhance the generalization abilities of the segmentation models. Most of the existing domain adaptation methods, which based on image-to-image translation, firstly transfer the source images to the pseudo-target images, adapt the classifier from the source domain to the target domain. However, these unidirectional methods suffer from the following two limitations: (1) they do not consider the inverse procedure and they cannot fully take advantage of the information from the other domain, which is also beneficial, as confirmed by our experiments; (2) these methods may fail in the cases where transferring the source images to the pseudo-target images is difficult. In this paper, in order to solve these problems, we propose a novel framework BiFDANet for unsupervised bidirectional domain adaptation in the semantic segmentation of remote sensing images. It optimizes the segmentation models in two opposite directions. In the source-to-target direction, BiFDANet learns to transfer the source images to the pseudo-target images and adapts the classifier to the target domain. In the opposite direction, BiFDANet transfers the target images to the pseudo-source images and optimizes the source classifier. At test stage, we make the best of the source classifier and the target classifier, which complement each other with a simple linear combination method, further improving the performance of our BiFDANet. Furthermore, we propose a new bidirectional semantic consistency loss for our BiFDANet to maintain the semantic consistency during the bidirectional image-to-image translation process. The experiments on two datasets including satellite images and aerial images demonstrate the superiority of our method against existing unidirectional methods.

Keywords: unsupervised domain adaptation; bidirectional domain adaptation; convolutional neural networks (CNNs); image-to-image translation; generative adversarial networks (GANs); remote sensing images; semantic segmentation

Citation: Cai, Y.; Yang, Y.; Zheng, Q.; Shen, Z.; Shang, Y.; Yin, J.; Shi, Z. BiFDANet: Unsupervised Bidirectional Domain Adaptation for Semantic Segmentation of Remote Sensing Images. *Remote Sens.* **2022**, *14*, 190. https://doi.org/10.3390/rs14010190

Academic Editors: Fahimeh Farahnakian, Jukka Heikkonen and Pouya Jafarzadeh

Received: 30 November 2021
Accepted: 28 December 2021
Published: 1 January 2022

1. Introduction

In the last few years, it has been possible to collect a mass of remote sensing images, thanks to the continuous advancement of remote sensing techniques. For example, Gaofen satellites can capture a large number of satellite images with high spatial resolution on a large scale. In remote sensing, such a large amount of data has offered many more capability for image analysis tasks; for example, semantic segmentation [1], change detection [2] and scene classification [3]. Among these tasks, the semantic segmentation of remote sensing images has become one of the most interesting and important research topics because it is widely used in many applications, such as dense labeling, city planning, urban management, environment monitoring, and so on.

For the semantic segmentation of remote sensing images, CNN [4] has become one of the most efficient methods in the past decades and several CNN models have shown their effectiveness, such as DeepLab [5] and its variants [6,7]. However, these methods have some limitations, because CNN-based architectures tend to be sensitive to the distributions and features of the training images and test images. Even though they give satisfactory predictions when the distributions of training and test images are similar [1], when we attempt to use this model to classify images obtained from other satellites or cities, the classification accuracy severely decreases due to different distributions of the source images and target images, as shown in Figure 1. In the literature, the aforementioned problem is known as domain adaptation [8]. In remote sensing, domain gap problems are often caused due to many reasons, such as illumination conditions, imaging times, imaging sensors, geographic locations and so on. These factors will change the spectral characteristics of objects and resulted in a large intra-class variability. For instance, the images acquired from different satellite sensors may have different colors, as shown in Figure 1a,b. Similarly, due to the differences of the imaging sensors, images may have different types of channels. For example, a few images may consist of near-infrared, green, and red channels while the others may have green, red, and blue bands.

In typical domain adaptation problems, the distributions of the source domain are different from those of the target domain. In remote sensing, we assume that the images collected from different satellites or locations (cities) are different domains. The unsupervised domain adaptation defines that only annotations of the source domain are available and aims at generating satisfactory predicted labels for the unlabeled target domain, even if the domain shift between the source domain and target domain is huge. To improve the performances of the segmentation models in aforementioned settings, one of the most common approaches in remote sensing is to diversify the training images of the source domain, by performing data augmentation techniques, such as random color change [9], histogram equalization [10], and gamma correction [11]. However, even if these methods slightly increase the generalization capabilities of the models, the improvement is unsatisfactory when there exists huge differences between the distributions of different domains. For example, it is difficult to adapt the classifier from one domain with near-infrared, red, and green bands to another one with red, green and blue channels by using simple data augmentation techniques. To overcome such limitation, a generative adversarial network [12] was applied to transfer images between the source and target domains and made significant progress in unsupervised domain adaptation for semantic segmentation [13,14]. These approaches based on image translation can be divided into two steps. At first, it learns to transfer the source images to the target domain. Secondly, the translated images and the labels for the corresponding source images are used to train the classifier which will be tested on the unlabeled source domain. When the first step reduce the domain shift, the second step can effectively adapt the segmentation model to the target domain. In addition, inverse translations which adapt the segmentation model from the target domain to the source domain have been implemented as well [15]. In our experiments, we find that these two translations in opposite directions should be complementary rather than alternative. Furthermore, such unidirectional (e.g., source-to-target) setting might ignore the information from the inverse direction. For example, Benjdira et al. [16] adapted the source classifier to the unlabeled target domain, they only simulated the distributions of the target images instead of making the target images fully participate in domain adaption. Therefore, these unidirectional methods cannot take full advantage of the information from the target domain. Meanwhile, the key to the domain adaptation methods based on image translation is the similarity between the distributions of the pseudo-target images and the target images. Given fixed image translation models, it will depend on the difficulty of converting between two domains: there might be some situations where transferring the target images to the source domain is more difficult, and situations where transferring the source images to the target domain is more difficult. By combining the two opposite directions, we will acquire an architecture more general than those unidirectional methods. Furthermore, the recent image translation network (e.g., CycleGAN [17]) is bidirectional so that we can usually obtain two image generators in the source-to-target and target-to-source directions when the training of the image translation model is done. We can use both of generators to make the best of the information from the two directions.

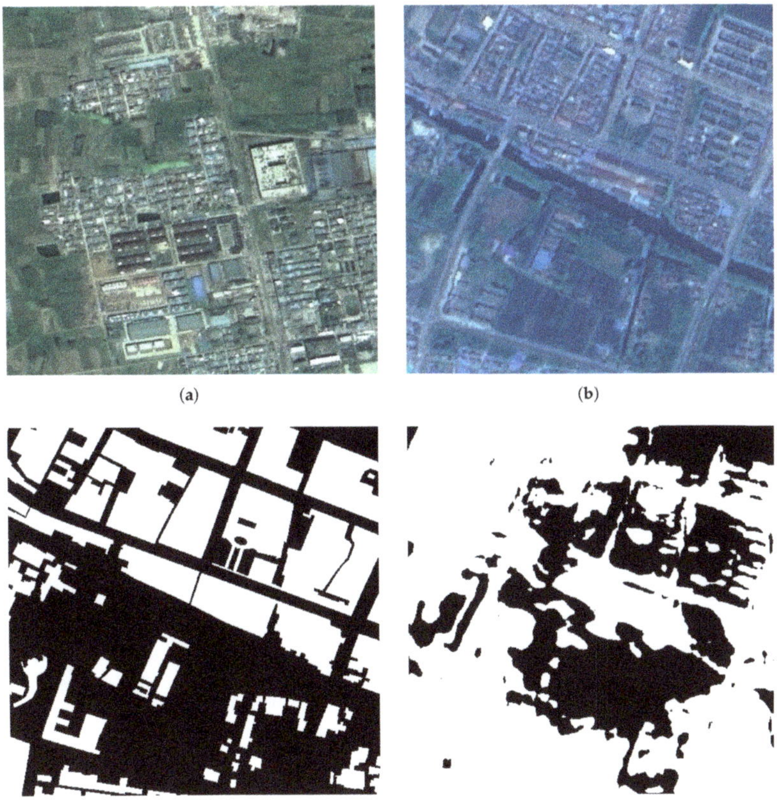

(a) (b) (c) (d)

Figure 1. An example of the domain adaptation. We show the source images and the target images which are obtained from different satellites, the label of the target image and the prediction of DeeplabV3+. In the label and the prediction, black and white pixels represent background and buildings respectively. (**a**) Source image. (**b**) Target image. (**c**) Label of the target image. (**d**) Prediction for the target image.

However, solving the aforementioned problems presents a few challenges. First, the transformed images and their corresponding original images must have the same semantic contents with the original images. For instance, if the image-to-image translation model replaces buildings with bare land during the translation, the labels of the original images cannot match the transformed images. As a result, semantic changes in any directions will affect our models. If the semantic changes occur in the source-to-target direction, the target domain classifier will have poor performance. If the approach replaces some objects with others in the target-to-source direction, the predicted labels of the source domain classifier would be unsatisfactory. Secondly, when we transfer the source images to the target domain, the data distributions of the pseudo-target images should be as similar as possible to the data distributions of the target images and the data distributions of the pseudo-source and source images should be similar as well. Otherwise, the transformed images of one domain cannot represent the other domain. Finally, the predicted labels of the two directions complement each other and the method of combining the labels is crucial because it will affect the final predicted labels. Simply combining the two predicted labels may leave out some correct objects or add some wrong objects.

In this article, we propose a new bidirectional model to address the above challenges. This framework involves two opposite directions. In the source-to-target direction, we generate pseudo-target transformed images which are semantically consistent with the original images. For this purpose, we propose a bidirectional semantic consistency loss to maintain the semantic consistency during the image translation. Then we employ the labels of the source images and their corresponding transformed images to adapt the segmentation model to the target domain. In the target-to-source direction, we optimize the source domain classifier to predict labels for the pseudo-source transformed images. These two classifiers may make different types of mistakes and assign different confidence ranks to the predicted labels. Overall the two classifiers are complementary instead of alternative. We make full use of them with a simple linear method which fuses their probability output.

Our contributions are as follows:

(1) We propose a new unsupervised bidirectional domain adaptation method, coined BiFDANet, for semantic segmentation of remote sensing images, which conducts bidirectional image translation to minimize the domain shift and optimizes the classifiers in two opposite directions to take full advantage of the information from both domains. At test stage, we employ a linear combination method to take full advantage of the two complementary predicted labels which further enhances the performance of our BiFDANet. As far as we know, BiFDANet is the first work on unsupervised bidirectional domain adaptation for semantic segmentation of remote sensing images.
(2) We propose a new bidirectional semantic consistency loss which effectively supervises the generators to maintain the semantic consistency in both source-to-target and target-to-source directions. We analyze the bidirectional semantic consistency loss by comparing it with two semantic consistency losses used in the existing approaches.
(3) We perform our proposed framework on two datasets, one consisting of satellite images from two different satellites and the other is composed of aerial images from different cities. The results indicate that our method can improve the performance of the cross-domain semantic segmentation and minimize the domain gap effectively. In addition, the effect of each component is discussed.

This article is organized as follows: Section 2 summarizes the related works. Section 3 presents the theory of our proposed framework. Section 4 describes the data set, the experimental design and discusses the obtained results, Section 5 provides the discussion and Section 6 draws our conclusions.

2. Related Work

2.1. Domain Adaptation

Tuia et al. [8] explained that in the research literature the adaptation methods could be grouped as: the selection of invariant features [18–21], the adaptation of classifiers [22–27], the adaptation of the data distributions [28–31] and active learning [32–34]. Here we focus on the methods of aligning the data distributions by performing image-to-image translation [35–39] between the different domains [40–43]. These methods usually match the data distributions of different domains by transferring the images from the source domain to the target domain. Next, the segmentation model is trained on the transferred images to classify the target images. In the fields of computer vision, Gatys et al. [40] raised a style transfer method to synthesizes fake images by combining the source contents with the target style. Similarly, Shrivastava et al. [41] generated realistic samples from synthetic images and the synthesized images could train a classification model on real images. Bousmalis et al. [42] learned the source-to-target transformation in the pixel space and transformed source images to target-like images. Taigman et al. [44] proposed a compound loss function to enforce the image generation network to transfer images from target to themselves. Hoffman et al. [14] used CycleGAN [17] to transfer the source images into the target style alternatively and transformed images were input into the classifier to improve its performance in the target domain. Zhao et al. [45] transformed fake images to the target domain which performed pixel-level and feature-level alignments with sub-domain aggregation. The segmentation model trained on such transformed images with the style of the target domain outperformed several unsupervised domain adaptation approaches. In remote sensing, Graph matching [46] and histogram matching [47] were employed to perform abovementioned image-to-image translation. Benjdira et al. [16] generated the fake target-like images by using CycleGAN [17], then the target-like images are used to adapt the source classifier to segment the target images. Similarly, Tasar et al. proposed ColorMapGAN [48], SemI2I [49] and DAugNet [50] to perform image-to-image translation between satellite image pairs to reduces the impact of domain gap. All the above mentioned methods focus on adapting the source segmentation model to the target domain without taking into account the opposite target-to-source direction that is beneficial.

2.2. Bidirectional Learning

Bidirectional learning was used to approach the neural machine translation problem [51,52], which train a language translation system in opposite directions of a language pair. Compared with unidirectional learning, it can improve the performance of the model effectively. Recently, bidirectional learning was applied to image-to-image translation problems as well. Li et al. [53] learned the image translation model and the segmentation adaptation model alternatively with a bidirectional learning method. Chen et al. [54] presented a bidirectional cross-modality adaptation method that aligned different domains from feature and image perspectives. Zhang et al. [55] adapted the model by minimizing the pixel-level and feature-level gaps. The theses method does not optimize the segmentation model in the target-to-source directions. Yang et al. [56] proposed a bi-directional generation network that trained a simple framework for image translation and classification from source to target and from target to source. Jiang et al. [57] proposed a bidirectional adversarial training method which performs adversarial trainings with generating adversarial examples from source to target and back. These methods only use bidirectional learning techniques in training process, but at test time, they do not make full use of two domains even if they have optimized the classifiers in both directions. Russo et al. [58] proposed a bidirectional image translation approach which trained two classifiers on different domains respectively and finally fuses the classification results. However, semantic segmentation task is more sensitive to pixel category while classification task focuses on image category. This proposed method can only be used to deal with the classification tasks, which can't apply to semantic segmentation tasks directly because it may not preserve the semantic contents.

3. Materials and Methods

The unsupervised domain adaptation assumes that the labeled source domain (X_S,Y_S) and unlabeled target domain X_T are available. The goal is to train a framework which correctly predicts the results for unlabeled target domain X_T.

The proposed BiFDANet consists of bidirectional image translation and bidirectional segmentation adaptation. It learns to transfer source images to the target domain and transfer target images to the source domain, and then optimizes the source classifier F_S and the target classifier F_T in two opposite directions. In this section, we detail how we transfer images between the source and target domain. And then we introduce how we adapt the classifier F_T to the target domain and optimize the classifier F_S in the target-to-source direction. Thereafter, we describe how we combine the two predicted results of the two classifiers F_S and F_T. Finally, we illustrate the implementations of the network architectures.

3.1. Bidirectional Image Translation

To perform bidirectional image translation between different domains, we use two generators and two discriminators based on GAN [12] architecture and we add two classifiers to extract the contents from the images. $G_{S \to T}$ denotes the target generator which generates pseudo-target images, $G_{T \to S}$ denotes the source generator which generates pseudo-source images. D_S, D_T denote the discriminators and F_S, F_T are the classifiers.

First of all, we want the source images x_s and the pseudo-source images $G_{T \to S}(x_t)$ to be drawn form similar data distributions, while the target images x_t and the pseudo-target images $G_{S \to T}(x_s)$ have similar data distributions. To deal with these issues, we enforce the data distributions of the pseudo-target images $G_{S \to T}(x_s)$ and the pseudo-source images $G_{T \to S}(x_t)$ to be similar to that of the target domain and the source domain respectively by applying adversarial learning (see Figure 2 blue portion). The discriminator D_S discriminates between the source images and the pseudo-source images while the discriminator D_T distinguishes the pseudo-target images from the target domain. We train the generators to fool the discriminators while the discriminators D_T and D_S attempt to classify the images from the target domain and the source domain. The adversarial loss for the target generator $G_{S \to T}$ and the discriminator D_T in the source-to-target direction is as follows:

$$\mathcal{L}_{adv}^{S \to T}(D_T, G_{S \to T}) = \mathbb{E}_{x_t \sim X_T}[\log D_T(x_t)] + \mathbb{E}_{x_s \sim X_S}[\log(1 - D_T(G_{S \to T}(x_s)))] \quad (1)$$

where $\mathbb{E}_{x_s \sim X_S}$, $\mathbb{E}_{x_t \sim X_T}$ are the expectation over x_s and x_t drawn by the distribution described by X_S and X_T respectively. $G_{S \to T}$ tries to generate the pseudo-target images $G_{S \to T}(x_s)$ which have data distributions similar to the that of the target images x_t, while D_T learns to discriminate the pseudo-target images from the target domain.

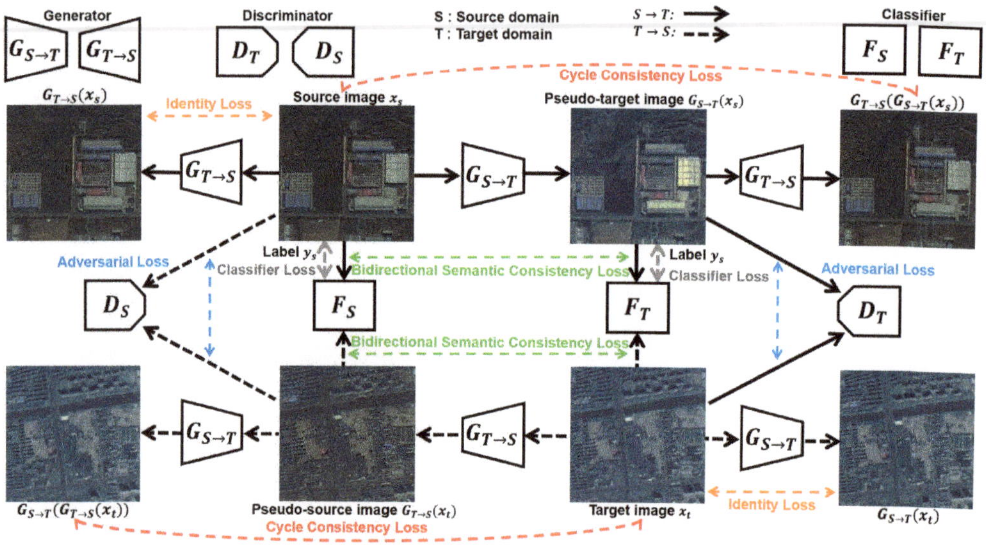

Figure 2. BiFDANet, training: The **top row** (black solid arrow) shows the source-to-target direction while the **bottom row** (black dashed arrow) shows the target-to-source direction. The colored dashed arrows correspond to different losses. The generator $G_{S \to T}$ transfers the images to the pseudo-target images while the generator $G_{T \to S}$ transfers the images to the source domain. D_S and D_T discriminate the images from the source domain and the target domain. F_S and F_T segment the images which are drawn from source domain and target domain, respectively.

This objective ensures that the pseudo-target images $G_{S\to T}(x_s)$ will resemble the images drawn from the target domain X_T. We use a similar adversarial loss in the target-to-source direction:

$$\mathcal{L}_{adv}^{T\to S}(D_S, G_{T\to S}) = \mathbb{E}_{x_s\sim X_S}[\log D_S(x_s)] + \mathbb{E}_{x_t\sim X_T}[\log(1 - D_S(G_{T\to S}(x_t)))] \quad (2)$$

This objective ensures that the pseudo-source images $G_{T\to S}(x_t)$ will resemble the images drawn from the source domain X_S. We compute the overall adversarial loss for the generators and the discriminators as:

$$\mathcal{L}_{adv}(D_S, D_T, G_{S\to T}, G_{T\to S}) = \mathcal{L}_{adv}^{S\to T}(D_T, G_{S\to T}) + \mathcal{L}_{adv}^{T\to S}(D_S, G_{T\to S}) \quad (3)$$

Another purpose is to maintain the original images and transformed images semantically consistent. Otherwise, the transformed images won't match the labels of the original images, and the performance of the classifiers would significantly decrease. To keep the semantic consistency between the transformed images and the original images, we define three constraints.

Firstly, we introduce a cycle-consistency constraint [17] to preserve the semantic contents during the translation process (see Figure 2 red portion). We encourage that transferring the source images from source to target and back reproduces the original contents. At the same time, transferring the target images from target to source and back to the target domain reproduces the original contents. These constraints are satisfied by imposing the cycle-consistency loss defined in the following equation:

$$\begin{aligned}\mathcal{L}_{cyc}(G_{S\to T}, G_{T\to S}) = \\ \mathbb{E}_{x_s\sim X_S}[\|G_{T\to S}(G_{S\to T}(x_s)) - x_s\|_1] + \mathbb{E}_{x_t\sim X_T}[\|G_{S\to T}(G_{T\to S}(x_t)) - x_t\|_1]\end{aligned} \quad (4)$$

Secondly, we require that $G_{T\to S}(x_s)$ for the source images x_s and $G_{S\to T}(x_t)$ for the target images x_t will reproduce the original images, thereby enforcing identity consistency (see Figure 2 orange portion). Such constraint is implemented by the identity loss defined as follows:

$$\begin{aligned}\mathcal{L}_{idt}(G_{S\to T}, G_{T\to S}) = \\ \mathbb{E}_{x_t\sim X_T}[\|G_{S\to T}(x_t) - x_t\|_1] + \mathbb{E}_{x_s\sim X_S}[\|G_{T\to S}(x_s) - x_s\|_1]\end{aligned} \quad (5)$$

The identity loss $\mathcal{L}_{idt}$ can be divided into two parts: the source-to- target identity loss Equation (6) and the target-to-source identity loss Equation (7). These two parts are as follows:

$$\mathcal{L}_{idt}^{S\to T}(G_{S\to T}) = \mathbb{E}_{x_t\sim X_T}[\|G_{S\to T}(x_t) - x_t\|_1] \quad (6)$$

$$\mathcal{L}_{idt}^{T\to S}(G_{T\to S}) = \mathbb{E}_{x_s\sim X_S}[\|G_{T\to S}(x_s) - x_s\|_1] \quad (7)$$

Thirdly, we enforce the transformed images to be semantically consistent with the original images. CyCADA [14] proposed the semantic consistency loss to maintain the semantic contents. The source images x_s and the transformed images $G_{S\to T}(x_s)$ are fed into the source classifier F_S pretrained on labeled source domain. However, since the transformed images $G_{S\to T}(x_s)$ are drawn from the target domain, the classifier trained on the source domain could not extract the semantic contents from the transformed images effectively. As a result, computing the semantic consistency loss in this way is not conducive to the image generation. In ideal conditions, the transformed images $G_{S\to T}(x_s)$ should be input to the target classifier F_T. However, it is impractical because the labels of the target domain aren't available. Instead of using the source classifier F_S to segment the transformed images $G_{S\to T}(x_s)$, MADAN [45] proposed to dynamically adapt the source classifier F_S to the target domain by taking the transformed images $G_{S\to T}(x_s)$ and the source labels as input. And then, they employed the classifier trained on the transformed domain as F_T, which performs better than the original classifier. The semantic consistency loss computed by

F_T would promote the generator $G_{S\to T}$ to generate images that preserve more semantic contents of the original images. However, MADAN only considers the generator $G_{S\to T}$ but ignores the generator $G_{T\to S}$ which is crucial to the bidirectional image translation. For bidirectional domain adaptation, we expect both source generator $G_{T\to S}$ and target generator $G_{S\to T}$ to maintain semantic consistency during image-to-image translation process. Therefore, we propose a new bidirectional semantic consistency loss (see Figure 2 green portion). The proposed bidirectional semantic consistency loss is:

$$\begin{aligned}\mathcal{L}_{sem}(G_{S\to T}, G_{T\to S}, F_S, F_T) = \\ \mathbb{E}_{x_s\sim X_S} KL(F_S(x_s)\|F_T(G_{S\to T}(x_s))) + \mathbb{E}_{x_t\sim X_T} KL(F_T(x_t)\|F_S(G_{T\to S}(x_t)))\end{aligned} \tag{8}$$

where $KL(\cdot\|\cdot)$ is the KL divergence.

Our proposed bidirectional semantic consistency loss can be divided into two parts: source-to-target semantic consistency loss Equation (9) and target-to-source semantic consistency loss Equation (10). These two parts are as follows:

$$\mathcal{L}_{sem}^{S\to T}(G_{S\to T}, F_T) = \mathbb{E}_{x_s\sim X_S} KL(F_S(x_s)\|F_T(G_{S\to T}(x_s))) \tag{9}$$

$$\mathcal{L}_{sem}^{T\to S}(G_{T\to S}, F_S) = \mathbb{E}_{x_t\sim X_T} KL(F_T(x_t)\|F_S(G_{T\to S}(x_t))) \tag{10}$$

3.2. *Bidirectional Segmentation Adaptation*

Our adaptation includes the source-to-target direction and the target-to-source direction as shown in Figure 2.

3.2.1. Source-to-Target Adaptation

To reduce the domain gap, we train the generator $G_{S\to T}$ with $\mathcal{L}_{adv}^{S\to T}$ Equation (1), $\mathcal{L}_{cyc}$ Equation (4), $\mathcal{L}_{idt}^{S\to T}$ Equation (6) and $\mathcal{L}_{sem}^{S\to T}$ Equation (9) to map the source images x_s to the pseudo-target images (see Figure 2, top row). Note that the labels of the transformed images $G_{S\to T}(x_s)$ won't be changed by the generator $G_{S\to T}$. Therefore, we can train the target classifier F_T with the transformed images $G_{S\to T}(x_s)$ and the ground truth segmentation labels of the original source images x_s (see Figure 2 gray portion). For C-way semantic segmentation, the classifier loss is defined as:

$$\begin{aligned}\mathcal{L}_{F_T}(G_{S\to T}(x_s), F_T) = \\ -\mathbb{E}_{G_{S\to T}(x_s)\sim G_{S\to T}(X_S)} \sum_{c=1}^{C} \mathbb{I}_{[c=y_s]} \log(softmax(F_T^{(c)}(G_{S\to T}(x_s))))\end{aligned} \tag{11}$$

where C denotes the category number of categories and $\mathbb{I}_{[c=y_s]}$ represents the corresponding loss only for class c.

Above all, the framework optimizes the objective function in the source-to-target direction as follows:

$$\begin{aligned}\min_{\substack{G_{S\to T}\\F_T}} \max_{D_T} \lambda_1 \mathcal{L}_{adv}(G_{S\to T}, D_T) + \lambda_2 \mathcal{L}_{cyc}(G_{S\to T}, G_{T\to S}) \\ + \lambda_3 \mathcal{L}_{idt}^{S\to T}(G_{S\to T}) + \lambda_4 \mathcal{L}_{sem}^{S\to T}(G_{S\to T}, F_T) + \lambda_5 \mathcal{L}_{F_T}(G_{S\to T}(x_s), F_T)\end{aligned} \tag{12}$$

3.2.2. Target-to-Source Adaptation

We take into account the opposite target-to-source direction and employ a symmetrical framework (Figure 2, black dashed arrow). In this direction, we optimize the generator $G_{T\to S}$ with $\mathcal{L}_{adv}^{T\to S}$ Equation (2), $\mathcal{L}_{cyc}$ Equation (4), $\mathcal{L}_{idt}^{T\to S}$ Equation (7) and $\mathcal{L}_{sem}^{T\to S}$ Equation (10) to map the target images x_t to the pseudo-source images $G_{T\to S}(x_t)$ (see Figure 2, bottom row). Then, we use the source classifier F_S to segment the pseudo-source images $G_{T\to S}(x_t)$ to compute the semantic consistency loss Equation (10) instead of the classifier loss because the ground truth segmentation labels for the target images are not available. The segmentation model F_S are trained using the labeled source images x_s with following classifier loss (see Figure 2 gray portion):

$$\mathcal{L}_{F_S}(X_S, F_S) = -\mathbb{E}_{x_s \sim X_S} \sum_{c=1}^{C} \mathbb{I}_{[c=y_s]} \log(softmax(F_S^{(c)}(x_s))) \tag{13}$$

Collecting the above components, the target-to-source part of the framework optimizes the objective function as follows:

$$\begin{aligned} \min_{\substack{G_{T\to S} \\ F_S}} \max_{D_S} & \lambda_1 \mathcal{L}_{adv}(G_{T\to S}, D_S) + \lambda_2 \mathcal{L}_{cyc}(G_{T\to S}, G_{S\to T}) \\ & + \lambda_3 \mathcal{L}_{idt}^{T\to S}(G_{T\to S}) + \lambda_4 \mathcal{L}_{sem}^{T\to S}(G_{T\to S}, F_S) + \lambda_6 \mathcal{L}_{F_S}(X_S, F_S) \end{aligned} \tag{14}$$

3.3. Bidirectional Domain Adaptation

Combining above two directions, we conclude with the complete loss function of BiFDANet:

$$\begin{aligned} \mathcal{L}_{BiFDANet}(G_{S\to T}, G_{T\to S}, D_S, D_T, F_S, F_T) = \\ \lambda_1 \mathcal{L}_{adv} + \lambda_2 \mathcal{L}_{cyc} + \lambda_3 \mathcal{L}_{idt} + \lambda_4 \mathcal{L}_{sem} + \lambda_5 \mathcal{L}_{F_T} + \lambda_6 \mathcal{L}_{F_S} \end{aligned} \tag{15}$$

where λ_1, λ_2, λ_3, λ_4, λ_5 and λ_6 control the interaction of the six objectives.

The training process corresponds to solving for the generators $G_{S\to T}$ and $G_{T\to S}$, the source classifier F_S and the target classifier F_T according to the optimization:

$$G_{S\to T}^*, G_{T\to S}^*, F_S^*, F_T^* = \arg \min_{F_S, F_T} \min_{\substack{G_{S\to T} \\ G_{T\to S}}} \max_{D_S, D_T} \mathcal{L}_{BiFDANet} \tag{16}$$

3.4. Linear Combination Method

The target classifier F_T is trained on the pseudo-target domain which have data distributions similar to the target domain and segment the target images. The source segmentation model F_S is optimized on the source domain and segment the pseudo-source images $G_{T\to S}(x_t)$. These two classifiers make different types of mistakes and assign different confidence ranks to the predicted labels. All in all, the predicted labels of the two classifiers are complementary instead of alternative. When addressing fusion, it is important to stress that we should remove the wrong objects from both predicted labels as much as possible and preserve the correct objects at the same time. For this purpose, we design a simple method which linearly combines their probability output as follows:

$$output = \lambda F_S(G_{T\to S}(x_t)) + (1-\lambda) F_T(x_t) \tag{17}$$

where λ is a hyperparameter in the range $(0, 1)$.

Then, we convert the probability output to the predicted labels. A schematic illustration of the linear combination method is shown in Figure 3.

3.5. Network Architecture

Our proposed BiFDANet consists of two generators, two discriminators and two classifiers.

We choose DeeplabV3+ [7] as the segmentation model and use ResNet34 [59] as the DeeplabV3+ backbone. The encoder applies atrous convolution at multiple scales to acquire multi-scale features. The decoder module which is simple yet effective provides the predicted results. We use dropout in the decoder module to avoid overfitting. Figure 4 shows the architecture of the classifier.

As shown in Figure 5, we use nine residual blocks for the generators which are used in [17]. Four convolutional layers are used to downsample the features, while four deconvolutional layers are applied to upsample the features. We use instance normalization rather than batch normalization and we apply ReLU to activate all layers.

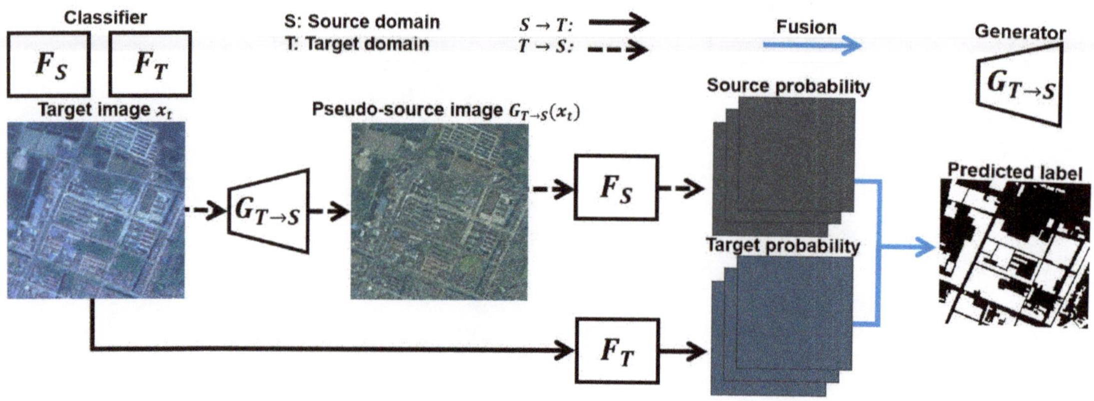

Figure 3. BiFDANet, test: the target classifier F_T and the source classifier F_S are used to segment the target images and the pseudo-source images respectively. And then the probability outputs are fused with a linear combination method and converted to the predicted labels.

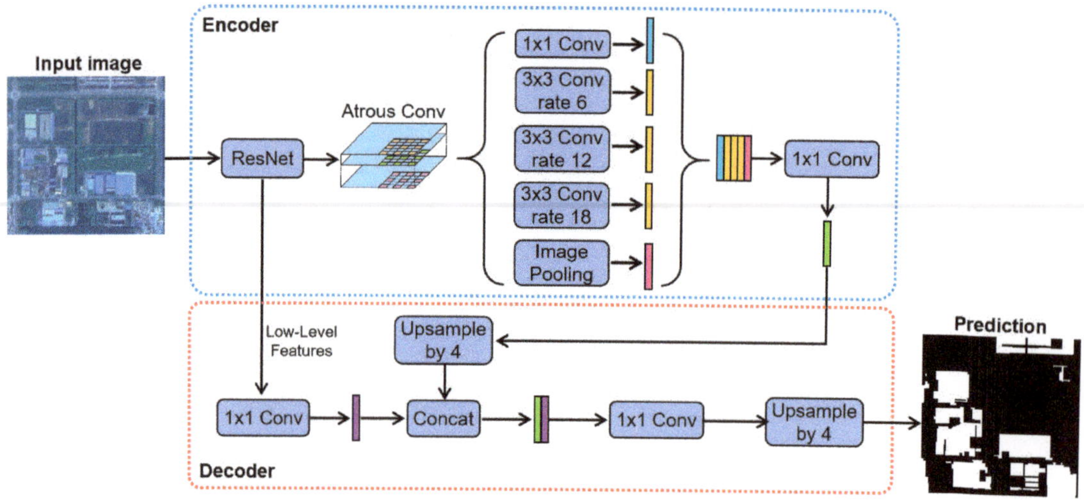

Figure 4. The architecture of the classifier (DeeplabV3+ [7]). The encoder acquires multi-scale features from the images while the decoder provides the predicted results from the multi-scale features and low-level features.

Similar to the discriminator in [17], we use five convolution layers for discriminators as shown in Figure 6. The discriminators encode the input images into a feature vector. Then, we compute the mean squared error loss instead of using Sigmoid to convert the feature vector into a binary output (real or fake). We use instance normalization rather than batch normalization. Unlike the generator, leaky ReLU is applied to activate the layers of the discriminator.

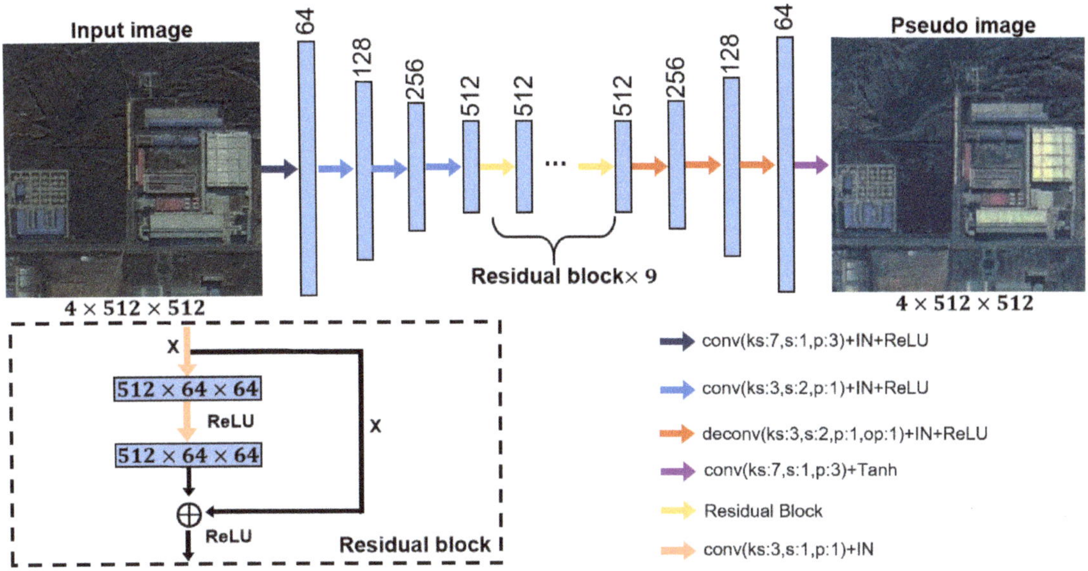

Figure 5. The architecture of the generator. ks, s, p and op correspond to kernel size, stride, padding and output padding parameters of the convolution and deconvolution respectively. ReLU and IN stand for rectified linear unit and instance normalization. The generator uses nine residual blocks.

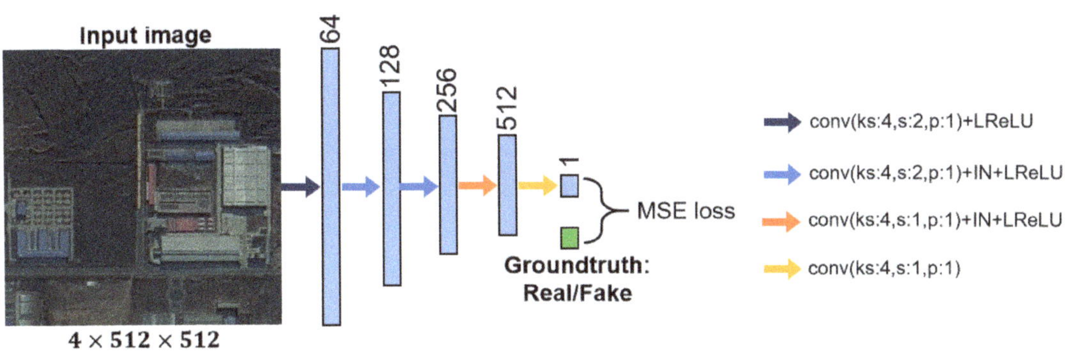

Figure 6. The architecture of the discriminator. LReLU and IN correspond to leaky rectified linear unit and instance normalization respectively. We use mean squared error loss instead of Sigmoid.

4. Results

In this section, we introduce the two datasets, illustrate the experimental settings, and analyse the obtained results both quantitatively and qualitatively.

4.1. Data Set

To conduct our experiments, we employ the Gaofen Satellite dataset and the ISPRS (WGII/4) 2D semantic segmentation benchmark dataset [60]. In the rest of this paper, we abbreviate the Gaofen Satellite data and the ISPRS (WGII/4) 2D semantic segmentation benchmark dataset to the Gaofen dataset and the ISPRS data set to simplify the description.

4.1.1. Gaofen Data Set

The Gaofen dataset consists of the Gaofen-1 (GF-1) satellite images and the Gaofen-1B (GF-1B) satellite images, which are civilian optical satellites of China and equipped with two sets of multi-spectral and panchromatic cameras. We reduce spatial resolution of the images to 2 m and convert the images to 10 bit. The images from both satellites contain 4 channels (i.e., red, green, blue and near-infrared). The labels of buildings are provided. We assume that only the labels of the source domain can be accessed. We cut the images and their labels into 512 × 512 patches. Table 1 reports the number of patches and the class percentages belonging to each satellite. Figure 7a,b show samples from the GF-1 satellite and the GF-1B satellite.

Table 1. Statistics For Data Set.

Image	# of Patches	Patch Size	Class Percentages
GF-1	2039	512 × 512	12.6%
GF-1B	4221	512 × 512	5.4%
Potsdam	4598	512 × 512	28.1%
Vaihingen	459	512 × 512	26.8%

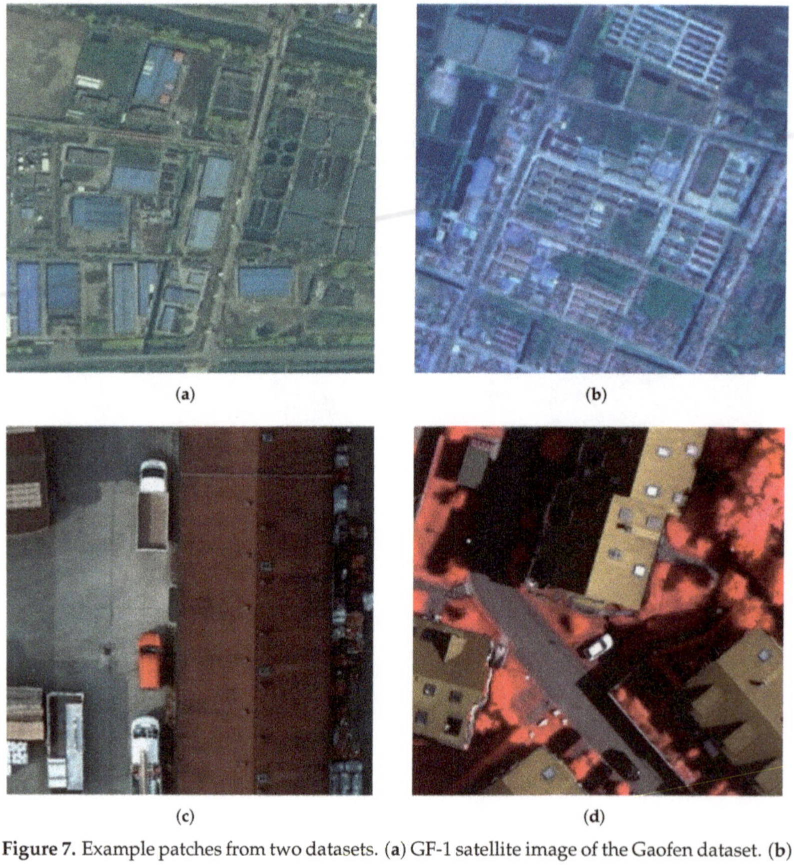

Figure 7. Example patches from two datasets. (**a**) GF-1 satellite image of the Gaofen dataset. (**b**) GF-1B satellite image of the Gaofen dataset. (**c**) Potsdam image of ISPRS dataset. (**d**) Vaihingen image of the ISPRS dataset.

4.1.2. ISPRS Data Set

This ISPRS dataset includes aerial images acquired from [61,62], which have been publicly available to the community. The Vaihingen dataset consists of images with a spatial resolution of 0.09 m and the spatial resolution of Potsdam dataset is 0.05 m. The Potsdam images contain red, green and blue channels while the Vaihingen images have 3 different channels (i.e., red, green and infrared). All images in both datasets are converted to 8 bit. Some images are manually labeled with land cover maps and the labels of impervious surfaces, buildings, trees, low vegetations and cars are provided. We cut the images and their labels into 512 × 512 patches. Table 1 reports the number of patches and the class percentages for the ISPRS dataset. Figure 7c,d show samples from each city.

4.1.3. Domain Gap Analysis

The domain shift between different domains is caused by many factors such as illumination conditions, camera angle, imaging sensors and so on.

In terms of the Gaofen data set, the same objects (e.g., buildings) have similar structures, but the colors of the GF-1 satellite images are different from the colors of the GF-1B satellite images as shown in Figure 7a,b. What's more, we depict the histograms to represent the data distributions of the two datasets. There are some differences between the histograms of the GF-1 satellite images and the GF-1B satellite images as shown in Figure 8a,b.

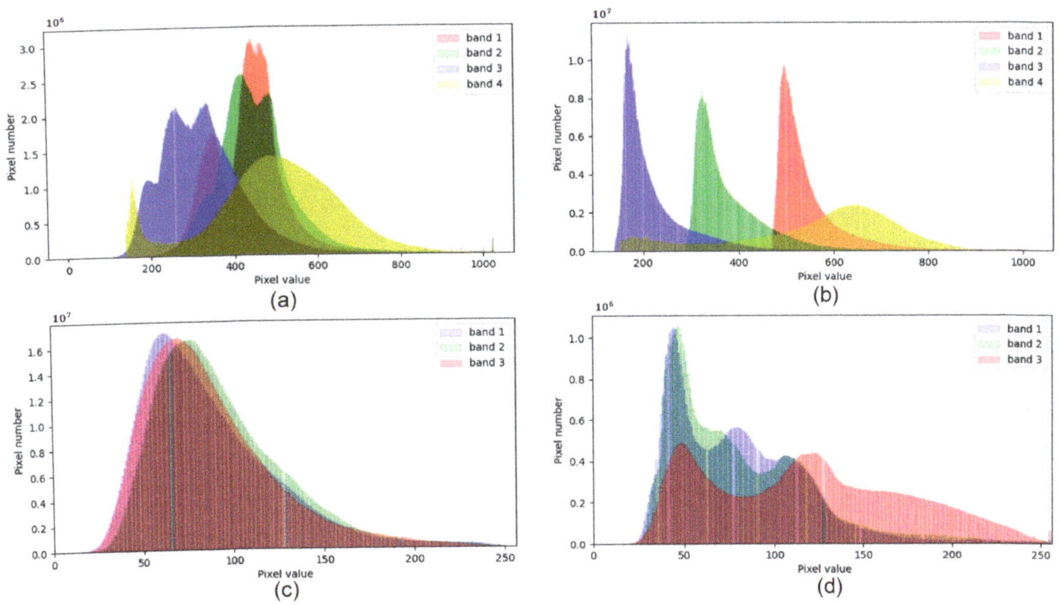

Figure 8. Color histograms of the Ganfen data set and the ISPRS data set. Different colors represent the histograms for different channels. (**a**) GF-1 images. (**b**) GF-1B images. (**c**) Potsdam images. (**d**) Vaihingen images.

In terms of the ISPRS dataset, the Potsdam images and the Vaihingen images have many differences, such as imaging sensors, spatial resolutions and structural representations of the classes. The Potsdam images and the Vaihingen images contain different kinds of channels due to the different imaging sensors, which results in the same objects in the two datasets being of different colors. For example, the vegetations and trees are green in the Potsdam dataset while the vegetations and trees are red color because of the infrared band. Besides, the Potsdam images and the Vaihingen images are captured using various spatial resolutions, which leads to the same objects being of different sizes. What's more, the structural representations of the same objects in the Potsdam dataset and Vaihingen dataset might be different. For example, there may be some differences between the buildings in different cities. At the same time, we depict the histograms to represent the data distributions of the Potsdam dataset and Vaihingen dataset as well. As shown in Figure 8c,d, the histograms of the Potsdam images are quite different from the histograms of the Vaihingen images.

4.2. Experimental Settings

We train BiFDANet in two stages. First, the training process minimizes the overall objective $\mathcal{L}_{BiFDANet}(G_{S\to T}, G_{T\to S}, D_S, D_T, F_S, F_T)$ without the bidirectional semantic consistency loss by setting λ_4 parameters in Equation (15) to 0. This is because, without a trained target segmentation model, the bidirectional semantic consistency loss would not be helpful in training process. The λ_1, λ_2, λ_3, λ_5 and λ_6 parameters in Equation (15) are set to 1, 10, 5, 10 and 10, respectively. We have found these values through repeated experiments. We train the framework for 100 epochs in this step. Second, after we obtain the well-trained target classifier, we add the bidirectional semantic consistency loss by setting λ_4 to 10 and the λ_1, λ_2, λ_3, λ_5 and λ_6 parameters in Equation (15) are the same as in the first step. We then optimize the network for 200 epochs. For all the methods, the networks are implemented in the PyTorch framework. We trained the models with Adam optimizer [63], using a batch size of 12. The learning rates for the generators, the discriminators and the classifiers are all set to 10^{-4}. At test time, the parameters to combine the segmentation models are $\lambda \in [0, 0.05, 0.1, 0.15, 0.2, ..., 0.95, 1]$ chosen on the validation set of 20% patches from the target domain.

4.3. Methods Used for Comparison

(1) DeeplabV3+ [7]: We do not apply any domain adaptation methods and directly segment the unlabeled target images with a DeeplabV3+ trained on the labeled source domain.

(2) Color Matching: For each channel of the images, we adjust the average brightness values of the source images to that of the target images. Then, we train the target segmentation model on the transformed domain.

(3) CycleGAN [17]: This method uses two generators G and F to perform image translation. The generator G learns to transfer the source images to the target domain while F learns to transfer the target images to the source domain. This method forces the transferring from source to target and back and transferring from target to source and back reproduce the original contents. Then the generated target-like images are used to train the target classifier.

(4) For BiFDANet, besides the full approach, we also give the results obtained by the segmentation models F_S and F_T before the linear combination method. At the same time, to show the effectiveness of the linear combination method, we also show the results obtained by simply taking the intersection or union of the two results.

For the above approaches, we use the same training parameters and architecture to make a fair comparison.

4.4. Evaluation Metrics

To evaluate all the methods quantitatively and comprehensively, we use scalar metrics included *Precision*, *Recall*, *F1-score (F1)* and *IoU* [64] defined as follows:

$$Precision = \frac{TP_b}{TP_b + FP_b} \tag{18}$$

$$Recall = \frac{TP_b}{TP_b + FN_b} \tag{19}$$

$$F1 = \frac{2 \times Precision \times Recall}{Precision + Recall} \tag{20}$$

$$IoU = \frac{TP_b}{TP_b + FN_b + FP_b} \tag{21}$$

where b denotes the category. FP (false positive) is the number of pixels which are classified as category b but do not belong to category b. FN (false negative) corresponds to the number of pixels which are category b but classified as other categories. TP (true positive) is the number of pixels which are correctly classified as category b and TN (true negative) corresponds to the number of pixels which are classified as other categories and belong to other categories. The aforementioned evaluation metrics are computed for each category (except the background). Especially, because we only segment buildings in our experiments, all the evaluation results we reported in tables are corresponding to the building (category).

4.5. Quantitative Results

To report fair and reliable results, we repeat training our framework and the comparison methods with the same parameters and architecture five times and depict the average precision, recall, F1-score and IoU values in Tables 2 and 3. Tables 2 and 3 show the comparison results on the Gaofen dataset and the ISPRS dataset, respectively. The DeeplabV3+ row includes results are corresponding to the no-adaptation case. For BiFDANet, we report the results obtained by the source classifier F_S and the target classifier F_T separately before the linear combination method and obtained by simply taking the intersection or union of the predicted results of the two classifiers F_S and F_T.

Table 2. Comparison results on Gaofen dataset. The best values are in bold.

Method	Source: GF-1, Target: GF-1B				Source: GF-1B, Target: GF-1			
	Recall (%)	Precision (%)	F1 (%)	IoU (%)	Recall (%)	Precision (%)	F1 (%)	IoU (%)
DeeplabV3+	**74.78**	16.60	27.16	15.72	2.14	70.07	4.17	2.13
Color matching	53.82	55.65	54.72	37.66	49.00	**83.64**	61.80	44.71
CycleGAN	54.72	67.31	60.37	43.24	60.74	75.12	67.17	50.57
BiFDANet F_S	58.56	69.34	63.50	46.52	71.65	72.21	71.93	56.17
BiFDANet F_T	61.82	67.00	64.31	47.39	71.81	73.69	72.74	57.16
$F_S \cap F_T$	57.12	**70.99**	63.31	46.31	67.90	75.77	71.62	55.79
$F_S \cup F_T$	60.92	68.11	64.32	47.40	71.94	73.88	72.90	57.36
BiFDANet	63.31	65.70	**64.48**	**47.58**	**75.57**	70.58	**72.99**	**57.47**

Table 3. Comparison results on ISPRS dataset. The best values are in bold.

Method	Source: Vaihingen, Target: Potsdam				Source: Potsdam, Target: Vaihingen			
	Recall (%)	Precision (%)	F1 (%)	IoU (%)	Recall (%)	Precision (%)	F1 (%)	IoU (%)
DeeplabV3+	30.10	17.81	22.37	12.59	29.64	33.16	31.30	18.55
Color matching	39.27	54.28	45.57	29.51	42.61	36.13	39.11	24.30
CycleGAN	61.13	55.86	58.38	41.22	49.75	66.44	56.90	39.76
BiFDANet F_S	68.82	61.62	65.02	48.17	59.00	75.39	66.20	49.47
BiFDANet F_T	56.90	62.39	59.52	42.37	60.44	76.70	67.60	51.06
$F_S \cap F_T$	52.35	**69.27**	59.63	42.48	53.60	**79.67**	64.09	47.15
$F_S \cup F_T$	**73.37**	57.63	64.55	47.66	59.95	77.12	67.46	50.90
BiFDANet	66.37	64.03	**65.18**	**48.35**	**65.83**	73.33	**69.38**	**53.12**

4.6. Visualization Results

Figures 9–12 depict the predicted results for DeeplabV3+, CycleGAN, color matching and BiFDANet. Our proposed BiFDANet which considers distribution alignment and bidirectional semantic consistency obtains the best predicted results, and the contours of the predicted buildings are more accurate than those acquired by color matching and CycleGAN.

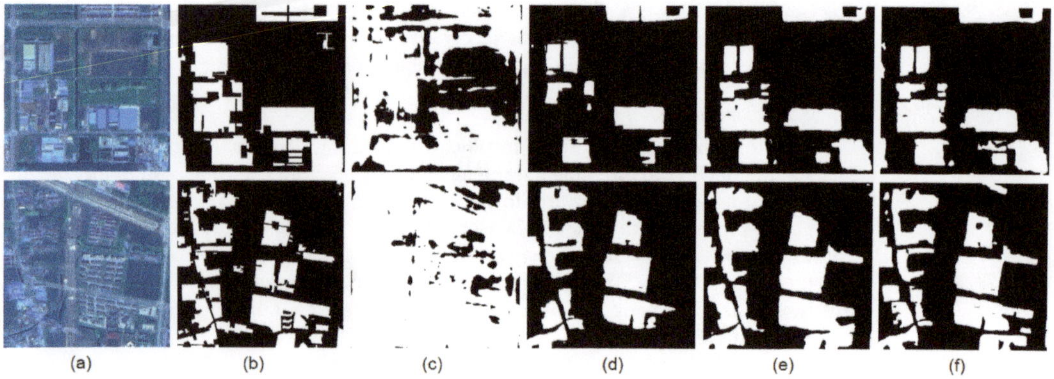

Figure 9. Segmentation results in GF-1 → GF-1B experiment. White and black pixels represent buildings and background. (**a**) GF-1B. (**b**) Label. (**c**) DeeplabV3+. (**d**) Color matching. (**e**) CycleGAN. (**f**) BiFDANet.

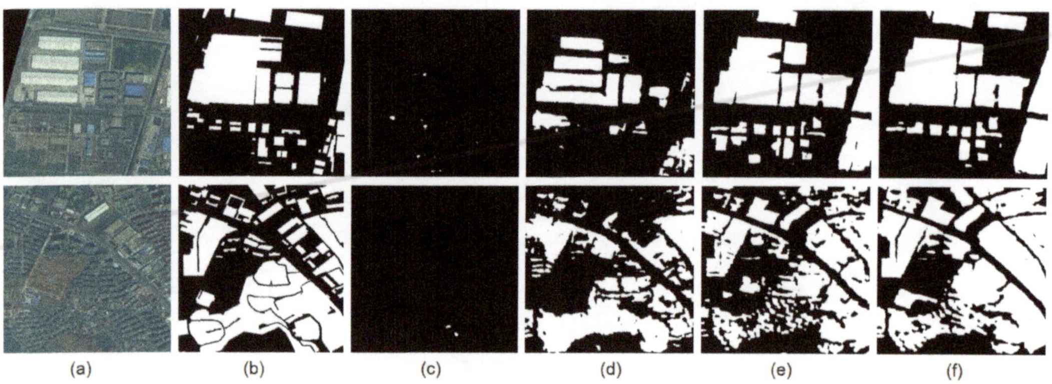

Figure 10. Segmentation results in GF-1B → GF-1 experiment. White and black pixels represent buildings and background. (**a**) GF-1. (**b**) Label. (**c**) DeeplabV3+. (**d**) Color matching. (**e**) CycleGAN. (**f**) BiFDANet.

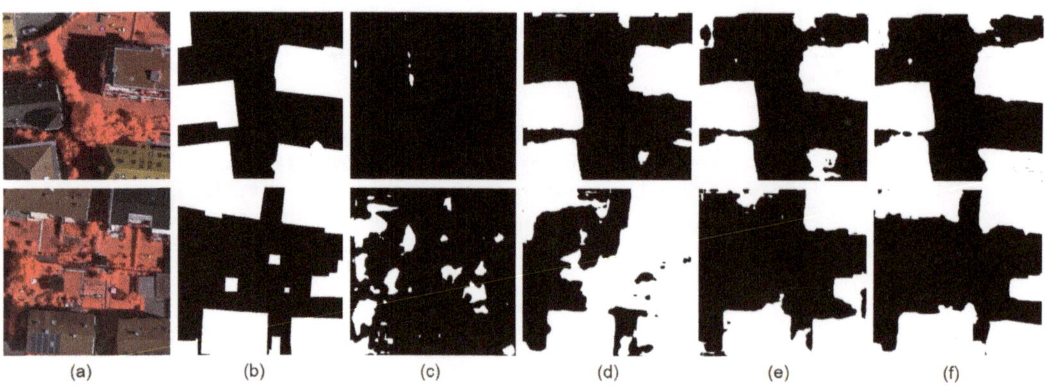

Figure 11. Segmentation results in Potsdam → Vaihingen experiment. White and black pixels represent buildings and background. (**a**) Vaihingen. (**b**) Label. (**c**) DeeplabV3+. (**d**) Color matching. (**e**) CycleGAN. (**f**) BiFDANet.

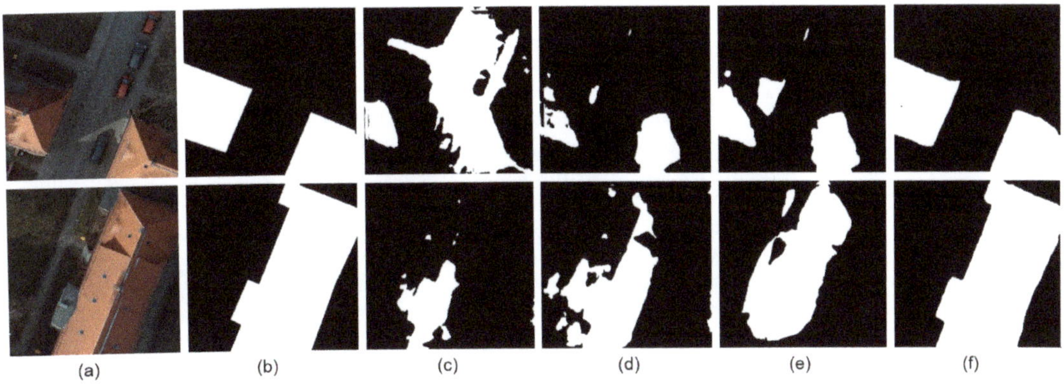

Figure 12. Segmentation results in Vaihingen → Potsdam experiment. White and black pixels represent buildings and background. (**a**) Potsdam. (**b**) Label. (**c**) DeeplabV3+. (**d**) Color matching. (**e**) CycleGAN. (**f**) BiFDANet.

5. Discussion

In this section, we compare our results with the compared methods in detail, and discuss the effect of our proposed bidirectional semantic consistency (BSC) loss and the roles of each component in our BiFDANet.

5.1. Comparisons with Other Methods

As shown in Tables 2 and 3, the DeeplabV3+ method which directly apply the source segmentation model to classify the target images performs worst in all settings. Color matching obtains a better performance than the DeeplabV3+ method, which indicates the effectiveness of domain adaptation for semantic segmentation of remote sensing images. CycleGAN perform better than both DeeplabV3+ and Color matching. Among all the compared methods, BiFDANet achieves the highest F1-score and IoU score in all settings. And the separate segmentation models F_S and F_T also significantly outperform the other adaptation methods. When combing the two segmentation models with the linear combination method, the performance of BiFDANet is further enhanced. Moreover, in the Vaihingen → Potsdam experiment, BiFDANet F_S performs much better than BiFDANet F_T. Because transferring from Vaihingen to Potsdam is more difficult than transferring from Potsdam to Vaihingen. There are far more Potsdam images than Vaihingen images, in some ways, the widely variable target domain (Potsdam) contains more variety of shapes and textures, and therefore it is more difficult to adapt the classifier from Vaihingen to Potsdam. Thanks to its bidirectionality which is disregarded in previous methods, BiFDANet achieves a performance gain of +7 percentage points while the gain in performance of BiFDANet F_T is only +1 percentage points. In this experiment, our proposed method makes full use of the information from the inverse target-to-source translation to produce much better results.

5.1.1. BiFDANet versus DeeplabV3+

There is no doubt that BiFDANet performs much better than DeeplabV3+ for all four cases. Because of the domain gap, there are some significant differences between the source domain and target domain. Without domain adaptation, the segmentation model cannot deal with the domain gap.

5.1.2. BiFDANet versus CycleGAN

In order to reduce the domain gap, CycleGAN and BiFDANet perform image-to-image translation to align data distribution of different domains. Figures 13–16 show some original images and the corresponding transformed images generated by color matching, CycleGAN and BiFDANet. As shown in Figures 13 and 14, it is obvious that the semantic contents of the images are changed by CycleGAN because there are no constraints for CycleGAN to enforce the semantic consistency during the image generation process. For instance, during the translation, CycleGAN replaces the buildings with bare land as shown in Figures 13 and 14 yellow rectangles. Besides, when generating transformed images, CycleGAN produces some buildings which do not exist before, as indicated in Figures 13 and 14 green rectangles. By contrast, the pseudo images transformed by BiFDANet and their corresponding original images have the same semantic contents and the data distributions of the pseudo images are similar to the data distributions of the target images. Similarly, as shown in Figure 15, we observe that there are some objects which look like red trees on the rooftops of the buildings as highlighted by green rectangles. At the same time, the pseudo images transformed by CycleGAN generates a few artificial objects in the outlined areas in Figure 15. What's more, in Figure 16, the pseudo images transformed by CycleGAN transfer the gray ground to the orange buildings, as highlighted by cyan rectangles. On the contrary, we do not observe aforementioned artificial objects and semantic inconsistency in the transformed images generated by BiFDANet in the vast majority of cases. Because the bidirectional semantic consistency loss enforces the classifiers to maintain semantic consistency during the image-to-image translation process. For CycleGAN, because the transformed images do not match the labels of the original images, the segmentation model F_T learns wrong information in training progress. Such wrong information may affect the performances of classifiers significantly. As a result, the domain adaptation methods with CycleGAN performs worse than our proposed method at test time, as confirmed by Figures 13–16.

5.1.3. BiFDANet versus Color Matching

Figures 13 and 14 illustrate that color matching can efficiently reduce the color difference between different domains. At the first sight, color matching works well. It preserves the semantic contents of the original source images in the transformed images, and the color of the target images is transferred to the transformed images. Besides, the transformed images generated by color matching look similar to the images generated by BiFDANet in Figure 14. However, in Tables 2 and 3, we can see that there are relatively big gaps between the performances of BiFDANet and color matching. The quantitative results for color matching are worse than the results for CycleGAN which can not keep semantic contents well. To better understand why there is such a difference in performance, we further analyse the differences between BiFDANet and color matching. The main problem of color matching is that it only tries to match the color of the images instead of considering the differences in features and data distributions. On the contrary, BiFDANet learns high-level features of the target images by using the discriminators to distinguish the features and data distributions of the pseudo-target transformed images from that of the original target images. In other word, the generators of BiFDANet generate pseudo-target transformed images whose high-level features and data distributions are similar to that of the target images. For this reason, our proposed BiFDANet outperforms color matching substantially.

Furthermore, to prove our point, we show color histograms of the GF-1 images, the pseudo GF-1 images generated by color matching and BiFDANet, and the GF-1B images, the pseudo GF-1B images generated by color matching and BiFDANet in Figure 17. And we depict color histograms of the Potsdam images, the pseudo Potsdam images generated by color matching and BiFDANet, and the Vaihingen images, the pseudo Vaihingen images generated by color matching and BiFDANet in Figure 18. Since the source domain and the target domain are drawn from different data distributions, the histograms of the pseudo-target images and the target images can't be exactly the same. However, we want them to be as similar as possible. Although color matching tries to match the color of the source images with the color of the target images, it doesn't learn the data distributions so that the histograms of the pseudo-target images are quite different from that of the target images.

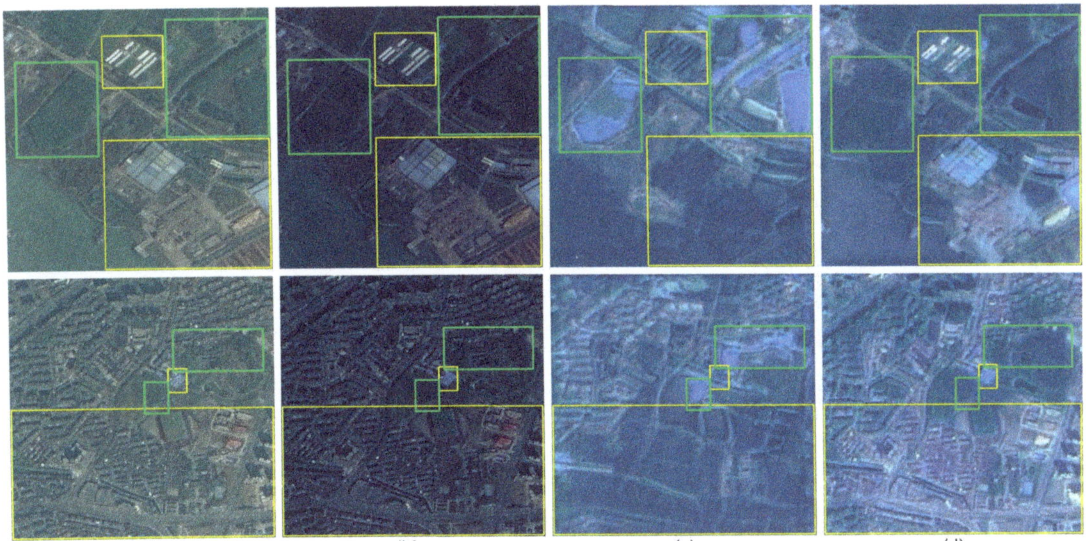

(a) (b) (c) (d)

Figure 13. GF-1 to GF-1B: Original GF-1 images and the transformed images which are used to train the classifier for GF-1B images. (**a**) GF-1 images. (**b**) Color matching. (**c**) CycleGAN. (**d**) BiFDANet (ours).

(a) (b) (c) (d)

Figure 14. GF-1B to GF-1: Original GF-1B images and the transformed images which are used to train the classifier for GF-1 images. (**a**) GF-1B images. (**b**) Color matching. (**c**) CycleGAN. (**d**) BiFDANet (ours).

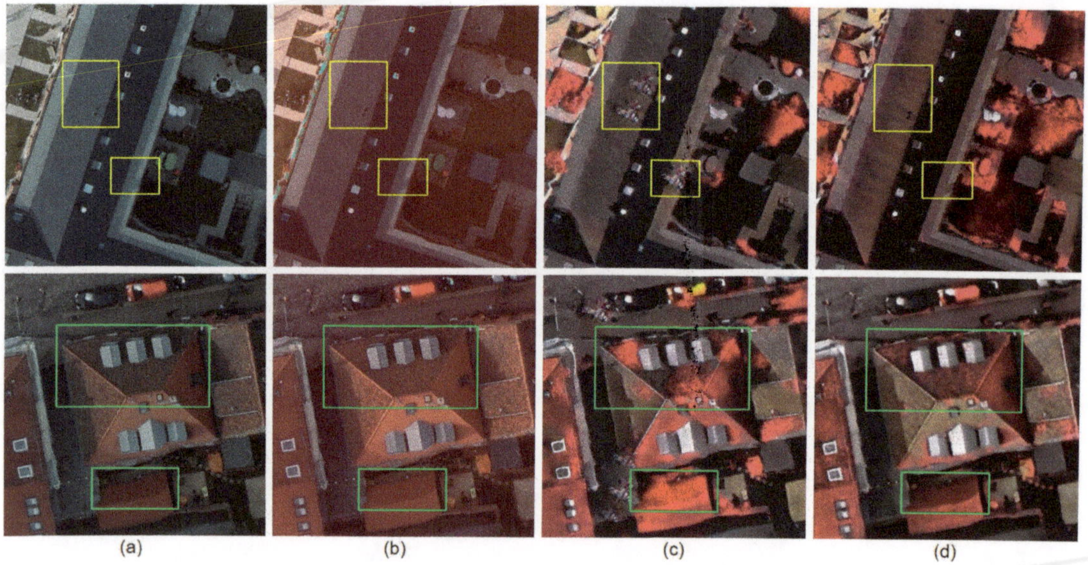

Figure 15. Potsdam to Vaihingen: Original Potsdam images and the transformed images which are used to train the classifier for Vaihingen images. (**a**) Potsdam images. (**b**) Color matching. (**c**) CycleGAN. (**d**) BiFDANet (ours).

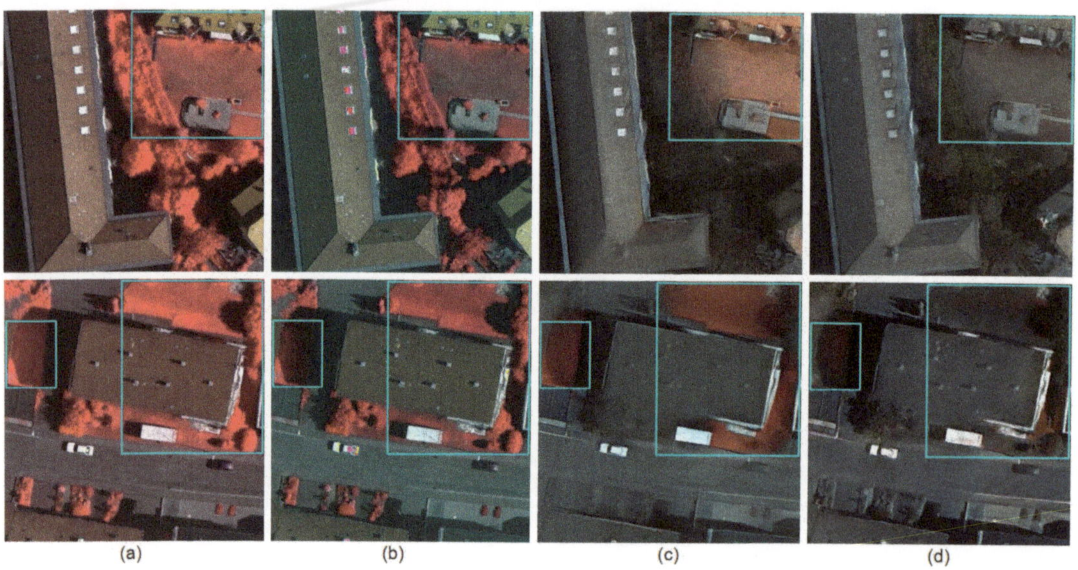

Figure 16. Vaihingen to Potsdam: Original Vaihingen images and the transformed images which are used to train the classifier for Potsdam images. (**a**) Vaihingen images. (**b**) Color matching. (**c**) CycleGAN. (**d**) BiFDANet (ours).

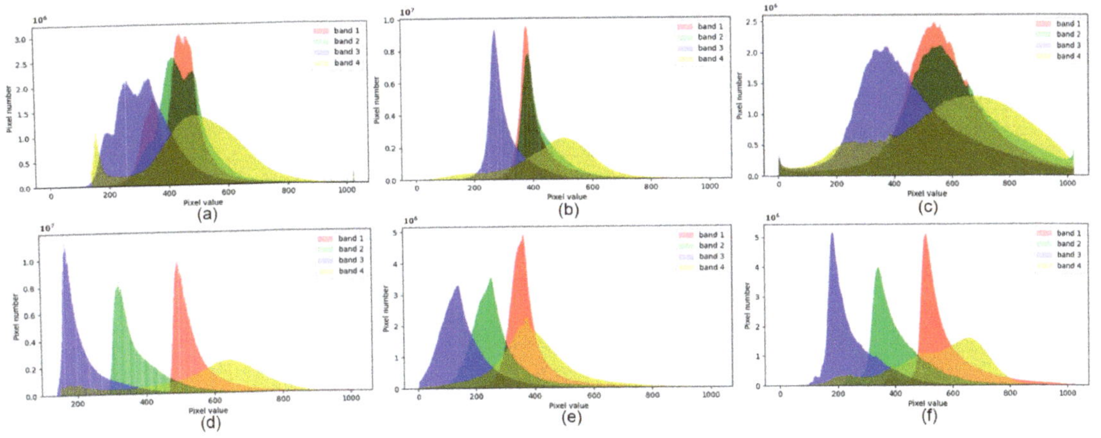

Figure 17. Color histograms of the Gaofen dataset. (**a**) GF-1. (**b**) Pseudo GF-1 transformed by color matching. (**c**) Pseudo GF-1 transformed by BiFDANet. (**d**) GF-1B. (**e**) Pseudo GF-1B transformed by color matching. (**f**) Pseudo GF-1B transformed by BiFDANet.

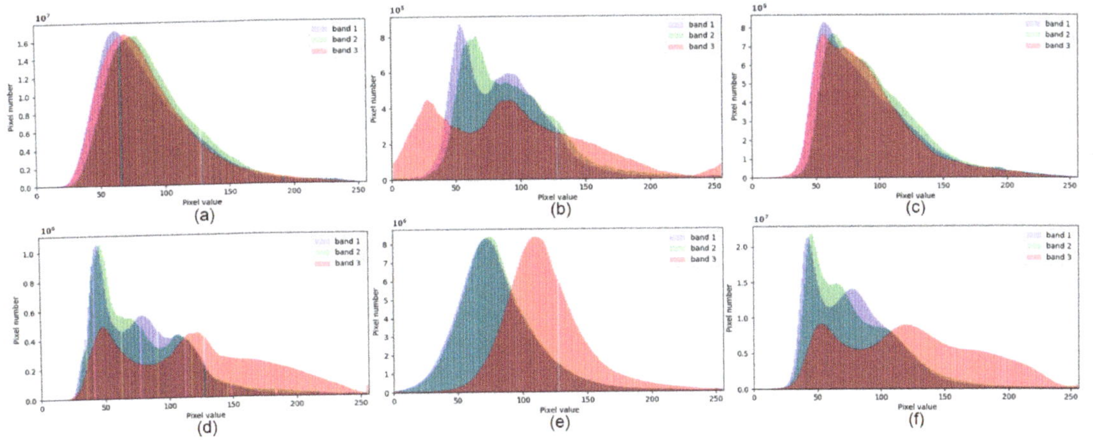

Figure 18. Color histograms of the ISPRS dataset. It is worth noting that Potsdam and Vaihingen have different kinds of bands. (**a**) Potsdam. (**b**) Pseudo Potsdam transformed by color matching. (**c**) Pseudo Potsdam transformed by BiFDANet. (**d**) Vaihingen. (**e**) Pseudo Vaihingen transformed by color matching. (**f**) Pseudo Vaihingen transformed by BiFDANet.

As shown in Figures 17 and 18, color matching does not match the data distributions of the pseudo-target images with the data distributions of the target images. For Gaofen dataset, there are still some differences between the histograms of the pseudo-target images generated by color matching and the real target images as shown in Figure 17. In contrast, the histograms of the pseudo-target images transformed by BiFDANet are similar to that of the real target images as shown in Figure 17. Thus the performances of BiFDANet are better than color matching. For ISPRS dataset, the histograms of the pseudo-target images generated by color matching are much different from the histograms of the target images as shown in Figure 18. In comparison, BiFDANet effectively matches the histograms of pseudo-target images with the histograms of the real target images, as shown in Figure 18. Therefore, the performance gap between BiFDANet and color matching becomes larger as confirmed by Figures 13–16.

5.1.4. Linear Combination Method versus Intersection and Union

In the GF-1 → GF-1B, Vaihingen → Potsdam and Potsdam → Vaihingen experiments, simply taking the intersection or union of the results of the two classifiers F_S and F_T obtains the highest precision values or recall values, these results prove that the two opposite directions are complementary instead of alternative. However, the F1-score values and IoU values can't achieve the highest by the intersection and union operation. In the Vaihingen → Potsdam and Potsdam → Vaihingen experiments, simply taking the intersection or union of the outputs of the two classifiers F_S and F_T results in performance degradation. It shows that the intersection operation and union operation of the two predicted results aren't always stable, because these methods may leave out some correct objects or introduce some wrong objects during the combination process. In comparison, the linear combination method leads to further improvements for all four experiments because the combination of probability output is more reliable.

5.2. Bidirectional Semantic Consistency Loss

We replace the bidirectional semantic consistency (BSC) loss in BiFDANet with semantic consistency (SC) loss [14] and dynamic semantic consistency (DSC) loss [45], and report the evaluation results in Tables 4 and 5.

As shown in Tables 4 and 5, we can see that for all adaptations in both directions on Gaofen data set and ISPRS data set, our proposed bidirectional semantic consistency loss achieves better results. It is worth noting that our framework with SC loss [14] and DSC loss [45] also performs well in the source-to-target direction, but the performance of $BiFDANet$ F_S degrades. This illustrates the necessity of the proposed bidirectional semantic consistency loss when optimizing the classifier F_S in the target-to-source direction. What's more, our framework with the proposed bidirectional semantic consistency (BSC) loss outperforms our framework with the dynamic semantic consistency (DSC) loss in the source-to-target direction even if the semantic constraints are the same in this direction. It shows that keeping semantic consistency in the target-to-source direction is helpful to maintain the semantic consistency in the source-to-target direction. At the same time, the source classifier F_S in our framework with semantic consistency loss [14] and dynamic semantic consistency loss [45] perform better than the source classifier F_S in our framework without semantic consistency loss even though there are no semantic constraints for these methods in the target-to-source direction. It means that the semantic consistency constraints in the source-to-target direction are also beneficial to preserve the semantic contents in the target-to-source direction. In conclusion, these two transferring directions promote each other to keep the semantic consistency.

5.3. Loss Functions

We study the roles of each part in BiFDANet in the Vaihingen → Potsdam experiment. We start from the base source-to-target GAN model with the adversarial loss $\mathcal{L}_{adv}$ and the classification loss $\mathcal{L}_{F_T}$. Then we test the symmetric target-to-source GAN model with the adversarial loss $\mathcal{L}_{adv}$ and the classification loss $\mathcal{L}_{F_S}$. We combine the two symmetric models that form a closed loop. In the next steps, we add the cycle consistency loss $\mathcal{L}_{cyc}$ and the identity loss $\mathcal{L}_{idt}$ in turn. Finally, the framework is completed by introducing the bidirectional semantic consistency loss $\mathcal{L}_{sem}$. The results are shown in Table 6. We can observe that all components help our framework to achieve better IoU and F1 scores, and the proposed bidirectional semantic consistency loss could further improve the performance of the models, which demonstrates the effectiveness of our bidirectional semantic consistency loss again.

Table 4. Evaluation results of different semantic consistency loss on Gaofen dataset. The best values are in bold.

Method		Source: GF-1, Target: GF-1B				Source: GF-1B, Target: GF-1			
		Recall (%)	Precision (%)	F1 (%)	IoU (%)	Recall (%)	Precision (%)	F1 (%)	IoU (%)
BiFDANet w/o	F_S	55.68	62.07	58.70	41.55	65.36	67.21	66.27	49.68
	F_T	52.97	70.69	60.56	43.43	65.80	70.63	68.13	51.53
	BiFDANet	54.83	68.83	61.04	43.92	67.10	69.86	68.45	51.87
BiFDANet w/SC	F_S	50.84	**73.68**	60.16	43.02	69.43	68.36	68.89	52.33
	F_T	57.76	68.39	62.63	45.59	65.28	**74.48**	69.58	53.35
	BiFDANet	56.10	71.21	62.76	45.73	66.67	73.20	69.78	53.59
BiFDANet w/DSC	F_S	53.66	69.36	60.51	43.38	68.14	70.36	69.23	52.84
	F_T	59.90	66.69	63.11	46.11	70.44	73.24	71.81	56.02
	BiFDANet	58.47	70.23	63.81	46.86	72.34	71.93	72.13	56.41
BiFDANet w/BSC	F_S	58.56	69.34	63.50	46.52	71.65	72.21	71.93	56.17
	F_T	61.82	67.00	64.31	47.39	71.81	73.69	72.74	57.16
	BiFDANet	**63.31**	65.70	**64.48**	**47.58**	**75.57**	70.58	**72.99**	**57.47**

Table 5. Evaluation results of different semantic consistency loss on ISPRS dataset. The best values are in bold.

Method		Source: Vaihingen, Target: Potsdam				Source: Potsdam, Target: Vaihingen			
		Recall (%)	Precision (%)	F1 (%)	IoU (%)	Recall (%)	Precision (%)	F1 (%)	IoU (%)
BiFDANet w/o	F_S	49.37	72.12	58.62	41.46	45.81	71.71	55.91	38.80
	F_T	44.60	73.89	55.63	38.53	47.30	72.32	57.19	40.05
	BiFDANet	51.39	68.75	58.81	41.66	48.41	72.67	58.11	40.96
BiFDANet w/SC	F_S	52.72	72.96	61.21	44.10	53.69	69.82	60.70	43.58
	F_T	49.71	72.20	58.88	41.73	56.05	72.89	63.37	46.38
	BiFDANet	53.83	71.97	61.59	44.50	60.35	67.40	63.68	46.71
BiFDANet w/DSC	F_S	58.93	67.35	62.86	45.84	58.01	68.30	62.74	45.70
	F_T	50.66	70.76	59.05	41.89	60.53	74.44	66.77	50.11
	BiFDANet	53.03	**77.84**	63.08	46.07	62.75	73.13	67.54	50.99
BiFDANet w/BSC	F_S	**68.82**	61.62	65.02	48.17	59.00	75.39	66.20	49.47
	F_T	56.90	62.39	59.52	42.37	60.44	**76.70**	67.60	51.06
	BiFDANet	66.37	64.03	**65.18**	**48.35**	**65.83**	73.33	**69.38**	**53.12**

Table 6. Evaluation results of each component on ISPRS dataset.

Source: Vaihingen, Target: Potsdam											
$S \rightarrow T$					$T \rightarrow S$					F1 (%)	IoU (%)
$\mathcal{L}_{F_T}$	$\mathcal{L}_{adv}$	$\mathcal{L}_{cyc}$	$\mathcal{L}_{idt}$	$\mathcal{L}_{sem}$	$\mathcal{L}_{F_S}$	$\mathcal{L}_{adv}$	$\mathcal{L}_{cyc}$	$\mathcal{L}_{idt}$	$\mathcal{L}_{sem}$		
✓	✓									35.67	18.65
					✓	✓				39.84	23.63
✓	✓				✓	✓				40.17	24.08
✓	✓	✓								55.24	38.16
					✓	✓	✓			56.73	39.64
✓	✓	✓			✓	✓	✓			57.04	40.06
✓	✓	✓	✓							54.36	37.83
					✓	✓	✓	✓		57.74	40.12
✓	✓	✓	✓		✓	✓	✓	✓		58.81	41.66
✓	✓	✓	✓	✓						58.44	41.54
					✓	✓	✓	✓	✓	63.96	47.08
✓	✓	✓	✓	✓	✓	✓	✓	✓	✓	65.18	48.35

6. Conclusions

In this article, we present a novel unsupervised bidirectional domain adaptation framework to overcome the limitations of the unidirectional methods for semantic segmentation in remote sensing. First, while the unidirectional domain adaptation methods do not consider the inverse adaptation, we take full advantage of the information from both domains by performing bidirectional image-to-image translation to minimize the domain shift and optimizing the source and target classifiers in two opposite directions. Second, the unidirectional domain adaptation methods may perform badly when transferring from one domain to the other domain is difficult. In order to make the framework more general and robust, we employ a linear combination method at test time, which linearly merge the softmax output of two segmentation models, providing a further gain in performance. Finally, to keep the semantic contents in the target-to-source direction which was neglected by the existing methods, we propose a novel bidirectional semantic consistency loss and supervise the translation in both directions. We validate our framework on two remote sensing datasets, consisting of the satellite images and the aerial images, where we perform a one-to-one domain adaptation in each dataset in two opposite directions. The experimental results confirm the effectiveness of our BiFDANet. Furthermore, the analysis reveals the proposed bidirectional semantic consistency loss performs better than other semantic consistency losses used in the previous approaches. In our future work, we will redesign the combination method to make our framework more robust and further improve the segmentation accuracy. What's more, in practical terms, the huge number of remote sensing images usually contain several domains, we will extend our approach to multi-source and multi-target domain adaptation.

Author Contributions: Conceptualization, Y.C.; methodology, Y.C. and Q.Z.; formal analysis, Y.Y. and Y.S.; resources, J.Y. and Z.S. (Zhongtian Shi); writing—original draft preparation, Y.C.; writing—review and editing, Y.Y., Y.S., Z.S. (Zhengwei Shen) and J.Y.; visualization, Y.C.; data curation, Z.S. (Zhengwei Shen); funding acquisition, J.Y. and Z.S. (Zhongtian Shi). All authors have read and agreed to the published version of the manuscript.

Funding: This work was funded by the National Natural Science Foundation of China under Grant 61825205 and Grant 61772459 and the Key Research and Development Program of Zhejiang Provence, China under grant 2021C01017.

Institutional Review Board Statement: Not applicable.

Informed Consent Statement: Not applicable.

Data Availability Statement: The satellite dataset presented in this study is available on request from China resources satellite application center and the aerial dataset used in our research are openly available; see reference [60–62] for details.

Acknowledgments: We acknowledge the National Natural Science Foundation of China (Grant 61825205 and Grant 61772459) and the Key Research and Development Program of Zhejiang Provence, China (grant 2021C01017).

Conflicts of Interest: The authors declare no conflict of interest.

References

1. Chen, Z.; Li, D.; Fan, W.; Guan, H.; Wang, C.; Li, J. Self-Attention in Reconstruction Bias U-Net for Semantic Segmentation of Building Rooftops in Optical Remote Sensing Images. *Remote Sens.* **2021**, *13*, 2524. [CrossRef]
2. Kou, R.; Fang, B.; Chen, G.; Wang, L. Progressive Domain Adaptation for Change Detection Using Season-Varying Remote Sensing Images. *Remote Sens.* **2020**, *12*, 3815. [CrossRef]
3. Ma, C.; Sha, D.; Mu, X. Unsupervised Adversarial Domain Adaptation with Error-Correcting Boundaries and Feature Adaption Metric for Remote-Sensing Scene Classification. *Remote Sens.* **2021**, *13*, 1270. [CrossRef]
4. Krizhevsky, A.; Sutskever, I.; Hinton, G.E. ImageNet Classification with Deep Convolutional Neural Networks. In Proceedings of the International Conference on Neural Information Processing Systems (NIPS), Lake Tahoe, NV, USA, 3–6 December 2012; pp. 1097–1105.

5. Chen, L.C.; Papandreou, G.; Kokkinos, I.; Murphy, K.; Yuille, A.L. DeepLab: Semantic Image Segmentation with Deep Convolutional Nets, Atrous Convolution, and Fully Connected CRFs. *IEEE Trans. Pattern Anal. Mach. Intell.* **2018**, *40*, 834–848. [CrossRef]
6. Chen, L.C.; Papandreou, G.; Schroff, F.; Adam, H. Rethinking Atrous Convolution for Semantic Image Segmentation. *arXiv* **2017**, arXiv:1706.05587.
7. Chen, L.C.; Zhu, Y.; Papandreou, G.; Schroff, F.; Adam, H. Encoder-Decoder with Atrous Separable Convolution for Semantic Image Segmentation. In Proceedings of the European Conference on Computer Vision (ECCV), Munich, Germany, 8–14 September 2018; pp. 801–818.
8. Tuia, D.; Persello, C.; Bruzzone, L. Domain Adaptation for the Classification of Remote Sensing Data: An Overview of Recent Advances. *IEEE Geosci. Remote Sens. Mag.* **2016**, *4*, 41–57. [CrossRef]
9. Buslaev, A.; Iglovikov, V.I.; Khvedchenya, E.; Parinov, A.; Druzhinin, M.; Kalinin, A.A. Albumentations: Fast and Flexible Image Augmentations. *Informatic* **2020**, *11*, 125. [CrossRef]
10. Stark, J.A. Adaptive Image Contrast Enhancement Using Generalizations of Histogram Equalization. *IEEE Trans. Image Process.* **2000**, *9*, 889–896. [CrossRef] [PubMed]
11. Huang, S.C.; Cheng, F.C.; Chiu, Y.S. Efficient Contrast Enhancement Using Adaptive Gamma Correction With Weighting Distribution. *IEEE Trans. Image Process.* **2013**, *22*, 1032–1041. [CrossRef] [PubMed]
12. Goodfellow, I.; Pouget-Abadie, J.; Mirza, M.; Xu, B.; Warde-Farley, D.; Ozair, S.; Courville, A.; Bengio, Y. Generative Adversarial Nets. In Proceedings of the International Conference on Neural Information Processing Systems (NIPS), Montreal, QC, Canada, 8–13 December 2014; pp. 2672–2680.
13. Sankaranarayanan, S.; Balaji, Y.; Jain, A.; Lim, S.N.; Chellappa, R. Unsupervised Domain Adaptation for Semantic Segmentation with GANs. *arXiv* **2017**, arXiv:1711.06969.
14. Hoffman, J.; Tzeng, E.; Park, T.; Zhu, J.Y.; Isola, P.; Saenko, K.; Efros, A.; Darrell, T. CyCADA: Cycle-Consistent Adversarial Domain Adaptation. In Proceedings of the 35th International Conference on Machine Learning (ICML), Stockholm, Sweden, 10–15 July 2018; pp. 1989–1998.
15. Tzeng, E.; Hoffman, J.; Saenko, K.; Darrell, T. Adversarial Discriminative Domain Adaptation. In Proceedings of the IEEE Conference on Computer Vision and Pattern Recognition (CVPR), Honolulu, HI, USA, 21–26 July 2017; pp. 7167–7176.
16. Benjdira, B.; Bazi, Y.; Koubaa, A.; Ouni, K. Unsupervised Domain Adaptation using Generative Adversarial Networks for Semantic Segmentation of Aerial Images. *Remote Sens.* **2019**, *11*, 1369. [CrossRef]
17. Zhu, J.Y.; Park, T.; Isola, P.; Efros, A.A. Unpaired Image-to-Image Translation Using Cycle-Consistent Adversarial Networks. In Proceedings of the IEEE International Conference on Computer Vision (ICCV), Venice, Italy, 22–29 October 2017; pp. 2223–2232.
18. Rida, I.; Al-Maadeed, N.; Al-Maadeed, S.; Bakshi, S. A comprehensive overview of feature representation for biometric recognition. *Multimed. Tools Appl.* **2020**, *79*, 4867–4890. [CrossRef]
19. Bruzzone, L.; Persello, C. A Novel Approach to the Selection of Spatially Invariant Features for the Classification of Hyperspectral Images With Improved Generalization Capability. *IEEE Trans. Geosci. Remote Sens.* **2009**, *47*, 3180–3191. [CrossRef]
20. Persello, C.; Bruzzone, L. Kernel-Based Domain-Invariant Feature Selection in Hyperspectral Images for Transfer Learning. *IEEE Trans. Geosci. Remote Sens.* **2016**, *54*, 2615–2626. [CrossRef]
21. Rida, I.; Al Maadeed, S.; Bouridane, A. Unsupervised feature selection method for improved human gait recognition. In Proceedings of the 2015 23rd European Signal Processing Conference (EUSIPCO), Nice, France, 31 August–4 September 2015; pp. 1128–1132.
22. Hoffman, J.; Wang, D.; Yu, F.; Darrell, T. FCNs in the Wild: Pixel-level Adversarial and Constraint-based Adaptation. *arXiv* **2016**, arXiv:1612.02649.
23. Tsai, Y.H.; Hung, W.C.; Schulter, S.; Sohn, K.; Yang, M.H.; Chandraker, M. Learning to Adapt Structured Output Space for Semantic Segmentation. In Proceedings of the IEEE Conference on Computer Vision and Pattern Recognition (CVPR), Salt Lake City, UT, USA, 18–22 June 2018; pp. 7472–7481.
24. Zhang, Y.; David, P.; Gong, B. Curriculum Domain Adaptation for Semantic Segmentation of Urban Scenes. In Proceedings of the IEEE International Conference on Computer Vision (ICCV), Venice, Italy, 22–29 October 2017; pp. 2020–2030.
25. Zhang, Y.; Qiu, Z.; Yao, T.; Liu, D.; Mei, T. Fully Convolutional Adaptation Networks for Semantic Segmentation. In Proceedings of the IEEE Conference on Computer Vision and Pattern Recognition (CVPR), Salt Lake City, UT, USA, 18–22 June 2018; pp. 6810–6818.
26. Bruzzone, L.; Prieto, D.F. Unsupervised Retraining of a Maximum Likelihood Classifier for the Analysis of Multitemporal Remote Sensing Images. *IEEE Trans. Geosci. Remote Sens.* **2001**, *39*, 456–460. [CrossRef]
27. Bruzzone, L.; Cossu, R. A Multiple-Cascade-Classifier System for a Robust and Partially Unsupervised Updating of Land-Cover Maps. *IEEE Trans. Geosci. Remote Sens.* **2002**, *40*, 1984–1996. [CrossRef]
28. Chen, Y.; Li, W.; Van Gool, L. ROAD: Reality Oriented Adaptation for Semantic Segmentation of Urban Scenes. In Proceedings of the IEEE Conference on Computer Vision and Pattern Recognition (CVPR), Salt Lake City, UT, USA, 18–22 June 2018; pp. 7892–7901.
29. Tasar, O.; Tarabalka, Y.; Giros, A.; Alliez, P.; Clerc, S. StandardGAN: Multi-source Domain Adaptation for Semantic Segmentation of Very High Resolution Satellite Images by Data Standardization. In Proceedings of the IEEE/CVF Conference on Computer Vision and Pattern Recognition Workshops, Seattle, WA, USA, 13–19 June 2020; pp. 192–193.

30. Zhang, L.; Zhang, L.; Tao, D.; Huang, X. Sparse Transfer Manifold Embedding for Hyperspectral Target Detection. *IEEE Trans. Geosci. Remote Sens.* **2014**, *52*, 1030–1043. [CrossRef]
31. Yang, H.L.; Crawford, M.M. Spectral and Spatial Proximity-Based Manifold Alignment for Multitemporal Hyperspectral Image Classification. *IEEE Trans. Geosci. Remote Sens.* **2016**, *54*, 51–64. [CrossRef]
32. Huang, H.; Huang, Q.; Krahenbuhl, P. Domain Transfer Through Deep Activation Matching. In Proceedings of the European Conference on Computer Vision (ECCV), Munich, Germany, 8–14 September 2018; pp. 590–605.
33. Demir, B.; Minello, L.; Bruzzone, L. Definition of Effective Training Sets for Supervised Classification of Remote Sensing Images by a Novel Cost-Sensitive Active Learning Method. *IEEE Trans. Geosci. Remote Sens.* **2014**, *52*, 1272–1284. [CrossRef]
34. Ghassemi, S.; Fiandrotti, A.; Francini, G.; Magli, E. Learning and Adapting Robust Features for Satellite Image Segmentation on Heterogeneous Data Sets. *IEEE Trans. Geosci. Remote Sens.* **2019**, *57*, 6517–6529. [CrossRef]
35. Liu, M.Y.; Breuel, T.; Kautz, J. Unsupervised Image-to-Image Translation Networks. In Proceedings of the International Conference on Neural Information Processing Systems (NIPS), Long Beach, CA, USA, 4–9 December 2017; pp. 700–708.
36. Huang, X.; Liu, M.Y.; Belongie, S.; Kautz, J. Multimodal Unsupervised Image-to-Image Translation. In Proceedings of the European Conference on Computer Vision (ECCV), Munich, Germany, 8–14 September 2018; pp. 172–189.
37. Lee, H.Y.; Tseng, H.Y.; Huang, J.B.; Singh, M.; Yang, M.H. Diverse Image-to-Image Translation via Disentangled Representations. In Proceedings of the European Conference on Computer Vision (ECCV), Munich, Germany, 8–14 September 2018; pp. 35–51.
38. Johnson, J.; Alahi, A.; Fei-Fei, L. Perceptual Losses for Real-Time Style Transfer and Super-Resolution. In Proceedings of the European Conference on Computer Vision (ECCV), Amsterdam, The Netherlands, 11–14 October 2016; pp. 694–711.
39. Ulyanov, D.; Lebedev, V.; Vedaldi, A.; Lempitsky, V.S. Texture Networks: Feed-forward Synthesis of Textures and Stylized Images. In Proceedings of the International Conference on Machine Learning (ICML), New York, NY, USA, 19–24 June 2016; p. 4.
40. Gatys, L.A.; Ecker, A.S.; Bethge, M. Image Style Transfer Using Convolutional Neural Networks. In Proceedings of the IEEE Conference on Computer Vision and Pattern Recognition (CVPR), Las Vegas, NV, USA, 27–30 June 2016; pp. 2414–2423.
41. Shrivastava, A.; Pfister, T.; Tuzel, O.; Susskind, J.; Wang, W.; Webb, R. Learning from Simulated and Unsupervised Images through Adversarial Training. In Proceedings of the IEEE Conference on Computer Vision and Pattern Recognition (CVPR), Honolulu, HI, USA, 21–26 July 2017; pp. 2107–2116.
42. Bousmalis, K.; Silberman, N.; Dohan, D.; Erhan, D.; Krishnan, D. Unsupervised Pixel-Level Domain Adaptation with Generative Adversarial Networks. In Proceedings of the IEEE Conference on Computer Vision and Pattern Recognition (CVPR), Honolulu, HI, USA, 21–26 July 2017; pp. 3722–3731.
43. Murez, Z.; Kolouri, S.; Kriegman, D.; Ramamoorthi, R.; Kim, K. Image to Image Translation for Domain Adaptation. In Proceedings of the IEEE Conference on Computer Vision and Pattern Recognition (CVPR), Salt Lake City, UT, USA, 18–22 June 2018; pp. 4500–4509.
44. Taigman, Y.; Polyak, A.; Wolf, L. Unsupervised Cross-Domain Image Generation. *arXiv* **2016**, arXiv:1611.02200.
45. Zhao, S.; Li, B.; Yue, X.; Gu, Y.; Xu, P.; Hu, R.; Chai, H.; Keutzer, K. Multi-source Domain Adaptation for Semantic Segmentation. In Proceedings of the International Conference on Neural Information Processing Systems (NIPS), Vancouver, BC, Canada, 8–14 December 2019.
46. Tuia, D.; Munoz-Mari, J.; Gomez-Chova, L.; Malo, J. Graph Matching for Adaptation in Remote Sensing. *IEEE Trans. Geosci. Remote Sens.* **2013**, *51*, 329–341. [CrossRef]
47. Rakwatin, P.; Takeuchi, W.; Yasuoka, Y. Restoration of Aqua MODIS Band 6 Using Histogram Matching and Local Least Squares Fitting. *IEEE Trans. Geosci. Remote Sens.* **2009**, *47*, 613–627. [CrossRef]
48. Tasar, O.; Happy, S.; Tarabalka, Y.; Alliez, P. ColorMapGAN: Unsupervised Domain Adaptation for Semantic Segmentation Using Color Mapping Generative Adversarial Networks. *IEEE Trans. Geosci. Remote Sens.* **2020**, *58*, 7178–7193. [CrossRef]
49. Tasar, O.; Happy, S.; Tarabalka, Y.; Alliez, P. SEMI2I: Semantically Consistent Image-to-Image Translation for Domain Adaptation of Remote Sensing Data. In Proceedings of the IEEE International Geoscience and Remote Sensing Symposium (IGARSS), Waikoloa, HI, USA, 26 September–2 October 2020; pp. 1837–1840.
50. Tasar, O.; Giros, A.; Tarabalka, Y.; Alliez, P.; Clerc, S. DAugNet: Unsupervised, Multisource, Multitarget, and Life-Long Domain Adaptation for Semantic Segmentation of Satellite Images. *IEEE Trans. Geosci. Remote Sens.* **2021**, *59*, 1067–1081. [CrossRef]
51. He, D.; Xia, Y.; Qin, T.; Wang, L.; Yu, N.; Liu, T.Y.; Ma, W.Y. Dual Learning for Machine Translation. In Proceedings of the International Conference on Neural Information Processing Systems (NIPS), Barcelona, Spain, 5–10 December 2016; pp. 820–828.
52. Niu, X.; Denkowski, M.; Carpuat, M. Bi-Directional Neural Machine Translation with Synthetic Parallel Data. In Proceedings of the 2nd Workshop on Neural Machine Translation and Generation, Melbourne, Australia, 20 July 2018; pp. 84–91.
53. Li, Y.; Yuan, L.; Vasconcelos, N. Bidirectional Learning for Domain Adaptation of Semantic Segmentation. In Proceedings of the IEEE Conference on Computer Vision and Pattern Recognition (CVPR), Long Beach, CA, USA, 16–20 June 2019; pp. 6936–6945.
54. Chen, C.; Dou, Q.; Chen, H.; Qin, J.; Heng, P.A. Unsupervised Bidirectional Cross-Modality Adaptation via Deeply Synergistic Image and Feature Alignment for Medical Image Segmentation. *IEEE Trans. Med. Imaging* **2020**, *39*, 2494–2505. [CrossRef] [PubMed]
55. Zhang, Y.; Nie, S.; Liang, S.; Liu, W. Bidirectional Adversarial Domain Adaptation with Semantic Consistency. In Proceedings of the Chinese Conference on Pattern Recognition and Computer Vision (PRCV), Xi'an, China, 8–11 November 2019; pp. 184–198.
56. Yang, G.; Xia, H.; Ding, M.; Ding, Z. Bi-Directional Generation for Unsupervised Domain Adaptation. In Proceedings of the AAAI Conference on Artificial Intelligence, New York, NY, USA, 7–12 February 2020; pp. 6615–6622.

57. Jiang, P.; Wu, A.; Han, Y.; Shao, Y.; Qi, M.; Li, B. Bidirectional Adversarial Training for Semi-Supervised Domain Adaptation. In Proceedings of the 29th International Joint Conference on Artificial Intelligence (IJCAI), Yokohama, Japan, 11–17 July 2020; pp. 934–940.
58. Russo, P.; Carlucci, F.M.; Tommasi, T.; Caputo, B. From Source to Target and Back: Symmetric Bi-Directional Adaptive GAN. In Proceedings of the IEEE Conference on Computer Vision and Pattern Recognition (CVPR), Salt Lake City, UT, USA, 18–22 June 2018; pp. 8099–8108.
59. He, K.; Zhang, X.; Ren, S.; Sun, J. Deep Residual Learning for Image Recognition. In Proceedings of the IEEE Conference on Computer Vision and Pattern Recognition (CVPR), Las Vegas, NV, USA, 27–30 June 2016; pp. 770–778.
60. Gerke, M. *Use of the Stair Vision Library within the ISPRS 2D Semantic Labeling Benchmark (Vaihingen)*; ResearcheGate: Berlin, Germany, 2014.
61. International Society for Photogrammetry and Remote Sensing. 2D Semantic Labeling Contest-Potsdam. Available online: http://www2.isprs.org/commissions/comm3/wg4/2d-sem-label-potsdam.html (accessed on 20 November 2021).
62. International Society for Photogrammetry and Remote Sensing. 2D Semantic Labeling-Vaihingen Data. Available online: http://www2.isprs.org/commissions/comm3/wg4/2d-sem-label-vaihingen.html (accessed on 20 November 2021).
63. Kingma, D.P.; Ba, J. A Method for Stochastic Optimization. In Proceedings of the International Conference on Learning Representations (ICLR), San Diego, CA, USA, 7–9 May 2015; pp. 1–13.
64. Csurka, G.; Larlus, D.; Perronnin, F.; Meylan, F. What is a good evaluation measure for semantic segmentation? In Proceedings of the British Machine Vision Conference (BMVC), Bristol, UK, 9–13 September 2013.

remote sensing

MDPI

Article

Semantic Segmentation and Analysis on Sensitive Parameters of Forest Fire Smoke Using Smoke-Unet and Landsat-8 Imagery

Zewei Wang [1,†], Pengfei Yang [1,†], Haotian Liang [1], Change Zheng [1,*], Jiyan Yin [2], Ye Tian [1] and Wenbin Cui [3]

1 School of Technology, Beijing Forestry University, Beijing 100083, China; wangzewei@bjfu.edu.cn (Z.W.); yangpf@sinovel.com (P.Y.); LHTht1998@bjfu.edu.cn (H.L.); tytoemail@bjfu.edu.cn (Y.T.)
2 China Fire and Rescue Institute, Beijing 102202, China; jkldora@126.com
3 Ontario Ministry of Northern Development, Mines, Natural Resources and Forestry, Sault Ste Marie, ON P6A 5X6, Canada; Wenbin.cui@ontario.ca
* Correspondence: zhengchange@bjfu.edu.cn
† These authors contributed equally to the work.

Abstract: Forest fire is a ubiquitous disaster which has a long-term impact on the local climate as well as the ecological balance and fire products based on remote sensing satellite data have developed rapidly. However, the early forest fire smoke in remote sensing images is small in area and easily confused by clouds and fog, which makes it difficult to be identified. Too many redundant frequency bands and remote sensing index for remote sensing satellite data will have an interference on wildfire smoke detection, resulting in a decline in detection accuracy and detection efficiency for wildfire smoke. To solve these problems, this study analyzed the sensitivity of remote sensing satellite data and remote sensing index used for wildfire detection. First, a high-resolution remote sensing multispectral image dataset of forest fire smoke, containing different years, seasons, regions and land cover, was established. Then Smoke-Unet, a smoke segmentation network model based on an improved Unet combined with the attention mechanism and residual block, was proposed. Furthermore, in order to reduce data redundancy and improve the recognition accuracy of the algorithm, the conclusion was made by experiments that the RGB, SWIR2 and AOD bands are sensitive to smoke recognition in Landsat-8 images. The experimental results show that the smoke pixel accuracy rate using the proposed Smoke-Unet is 3.1% higher than that of Unet, which could effectively segment the smoke pixels in remote sensing images. This proposed method under the RGB, SWIR2 and AOD bands can help to segment smoke by using high-sensitivity band and remote sensing index and makes an early alarm of forest fire smoke.

Keywords: forest fire; remote sensing; smoke segmentation; Smoke-Unet; attention mechanism; residual block; Landsat-8; band sensibility

Citation: Wang, Z.; Yang, P.; Liang, H.; Zheng, C.; Yin, J.; Tian, Y.; Cui, W. Semantic Segmentation and Analysis on Sensitive Parameters of Forest Fire Smoke Using Smoke-Unet and Landsat-8 Imagery. *Remote Sens.* **2022**, *14*, 45. https://doi.org/10.3390/rs14010045

Academic Editors: Fahimeh Farahnakian, Jukka Heikkonen and Pouya Jafarzadeh

Received: 4 November 2021
Accepted: 20 December 2021
Published: 23 December 2021

1. Introduction

The forest system, which occupied almost one third of the total land area, provides a variety of critical ecological services such as natural habitat, water conservation, timber products and maintaining biodiversity [1]. It also plays a central role in global carbon circle and energy balance [2,3]. However, the areas of global forests sharply declined at a rate of roughly 10 million hectares per year [4]. Wildfire is the principal threat in terrestrial ecosystems, and many evidences have proved that recent global warming and precipitation anomalies have made forests more susceptible to burning [5,6]. In the period of 2019–2020, the Amazon and South Australia faced the most severe wildfires, and these events have caused wide public concerns because of their considerable ecological and socioeconomic consequences such as consuming generous quantities of tropical rainforest, emitting great volumes of greenhouse gas and aerosols and altering the composition of the atmosphere.

Because smoke appeared at the earliest phase in wildfires, earlier detection and rapid identification of initial wildfire smoke are crucial for wildfire suppression and management to avoid the damages and negative impacts of wildfires [7]. Wildfire smoke is usually identified by means of manual observation, patrol of forest rangers, infrared and optical sensors of fire lookout towers and aviation monitoring. However, these techniques have shown ineffective, unsystematic, and geographical limit. Wildfires, caused by natural events (e.g., lightening and spontaneous combustion) or human-forcing activities, occurred in the remote regions, making it difficult and cost-consuming for accessibility and suppression. However, data from remote sensing satellites can provide continuous, frequent, and numerous systematic information with various spatial and temporal resolution at global scales, which may overcome several limitations of the conventional wildfire smoke observation methods [8].

Currently, the widely used remote sensing monitoring algorithms are mostly based on satellite remote sensing data of low and medium resolution (>250 m) [9,10], such as Advanced Very High Resolution Radiometer (AVHRR) [11–13], Moderate Resolution Imaging Spectroradiometer (MODIS) [14–16], etc., which has become an important business method to detect wildfire smoke for daily wildfire disaster monitoring in many countries around the world. However, the satellites with lower spatial resolution are unable to capture relevant information effectively at the early stage of forest fires due to too small initial burning area, and thus would cause the detection of early fire spots to be missed. Therefore, high-resolution satellite data are urgently needed to improve the accuracy of fire detection. Landsat-8 data can be publicly obtained and the resolution has increased by an order of magnitude, reaching 30 m, compared with Suomi National Polar-orbiting Partnership (S-NPP) and Visible Infrared Imaging Radiometer Suite (VIIRS) [17–20]. In addition, Operational Land Imager (OLI) and Thermal Infrared Sensor (TIRS) mounted on Landsat-8 can provide a new data source and capability allowing as small as 1 m^2 active fire to be observed [21]. Therefore, Landsat-8 data were used for wildfire smoke detection in this paper.

The satellite can carry many multispectral sensors and provide large amounts of multispectral data with more valuable information than RGB. Wildfire smoke presents different characteristics in different spectral ranges of remote sensing data and the choice of bands is crucial to smoke recognition. The wildfire smoke detection algorithms [22,23] of AVHRR mainly derived from band 3 (centered at 3.7 μm), band 4 (centered at 10.8μm) and band 5 (centered at 12 μm). The family of products [24,25] based on MODIS sensors primarily used two MIR bands (band 21 and band 22, centered at 3.96 μm) and TIR band 31 (centered at 11 μm). Data from band 4 (centered at 3.55~3.93 μm) and band 5 (centered at 10.5~12.4 μm) of VIIRS are used for tracking active fires [26–28]. Nevertheless the Landsat-8 wildfire smoke detection algorithm was based on the reflectance of band 7 (SWIR, centered at 2.2 μm), that is sensitive to thermal abnormality [29]. Therefore, the selection of the spectral range of remote sensing data is very important for smoke identification based on different spectral properties.

Due to the development of machine learning and data mining, several studies focused on the automatic retrieving smoke pixels. Li et al. [30] facilitated a neural network algorithm using AVHRR data to search smoke plumes but it failed when smoke pervades in the downwind area. As a powerful and popular machine learning approach, Support Vector Machine (SVM) is widely used in remote sensing task. The SVM classifiers can take advantage of combination of texture, color and other features of the remote sensing scene, and successfully distinguish the pixels contained smoke from non-smoke pixels [31–33]. Other machine learning techniques, such as K-means clustering, fisher linear classification [34] and BPNN algorithm [35], were used to discriminate smoke pixels. Nevertheless, it is still a challenge to extract smoke areas because of the wide range of shapes, color, texture, luminance and heterogeneous component of aerosol as well as diversity of cover types. In addition, with the development of remote sensing technology, a dramatically increasing satellites archive makes it no longer suitable for hand-crafted features of remote sensing data, and it is urgent to develop more automatic detection algorithms.

Deep learning, in the specific area of Convolutional Neural Networks (CNNs), is inspired by the working way of the human brain and recently has acquired many impressive achievements in many scientific fields such as image classification, object detection, and image segmentation. CNN can automatically extract features from data using a structure of multilayers. They are iteratively learning by forward propagation and backward derivation and updating parameters of kernels through complex nonlinear functions. The accuracies can be further improved by providing great amounts of input data, so it would be the best candidate for remote automated detection tasks. CNNs have successfully been employed in variety remote sensing fields such as road detection [36], cloud detection [37] and smoke classification [38]. Recent Unet-based methods [39] have also made good progress in the field of remote sensing [40,41]. However, remote sensing satellite data have many redundant bands so that too much information causes the wildfire smoke detection accuracy drop after the first rise and the detection efficiency decrease. How to reduce the interference of redundant information and make full use of the correlation of feature channels is a key problem on wildfire smoke detection based on remote sensing data.

The objective of this study was to propose a wildfire smoke detection algorithm of Landsat-8 satellite remote sensing imagery at the scene of a wildfire using multispectral data. First, a multispectral smoke dataset of Landsat-8 satellite at global scale, including the information from visible to TIRS1 infrared bands, was built in this paper. Second, a deep learning model, Smoke-Unet, based on Unet architecture incorporating with residual block [42] and attention mechanism [43], was proposed. Then, the performance of this algorithm on different region and various scale of wildfire smoke was evaluated by the experiments based on the abovementioned multispectral smoke dataset. Finally, to better extract the features of remote sensing smoke and reduce the redundancy of remote sensing data, the sensitivity of multiple bands was analyzed.

The main parts of this paper are structured as follows. Section 2 introduces the establishment of a multispectral smoke dataset of Landsat-8 satellite at a global scale, and a proposed deep learning model, Smoke-Unet, based on the Unet architecture incorporating with Attention mechanism and residual block, is presented in Section 3. To reduce the disturbance of the redundant information, the influence of different band combinations of multispectral data and remote sensing parameters on the accuracy of the algorithm are analyzed and the band sensitivity are evaluated in Section 4, and the conclusion is made in Section 5.

2. Data

2.1. *Landsat-8 Multispectral Data*

Landsat-8, carrying the OLI and the TIRS, was launched in 2013, and is operated by the US Geological Survey (USGS). As seen in Table 1, OLI is a nine-spectral-band push-broom sensor with spatial resolution of 30 m and 15 m for the panchromatic band, including near-infrared band (NIR) and Panchromatic (Pan). Standard terrain-corrected data (Level 1T) from OLI were used in this study.

2.2. *Study Area*

As shown in Figure 1, the various fire-prone ecosystems all over the world were selected as the study areas in this research, containing: (i) needleleaf trees of boreal forests in high latitude regions, such as Canada and Siberia; (ii) subtropical evergreen hard-leaved forest mixed conifer-broadleaf forests in Western America; (iii) dry sclerophyll woodland and open forest in Eastern Australia; (iv) tropical rainforest in the Amazon and Southeastern Asia; (v) tropical grasslands and savannas in Africa.

Table 1. Landsat-8 Satellite Parameters.

Payload Name	Band Number	Band Name	Spectral Range(nm)	Resolution(m)
OLI	1	Coastal	433~453	30
	2	Blue	450~515	30
	3	Green	525~600	30
	4	Red	630~680	30
	5	NIR	845~885	30
	6	SWIR1	560~660	30
	7	SWIR2	100~300	30
	8	Panchromatic	500~680	15
	9	Cirrus	1360~1390	30
TIRS	10	TIRS1	1060~1119	60
	11	TIRS2	1150~1251	60

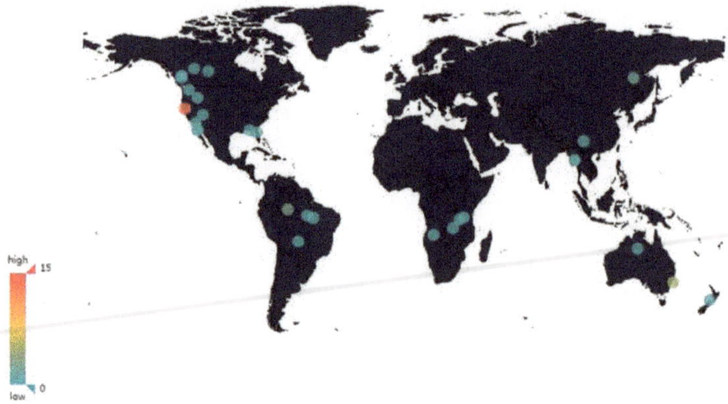

Figure 1. Spatial distribution of study regions in the datasets.

As seen in Figure 2, the study areas are located in Asia, North America, South America, Africa, etc. Considering that the frequent occurrence of wildfires in these areas is representative, the fire-prone regions in the USA, Canada, Brazil and Australia were selected as the primary research areas.

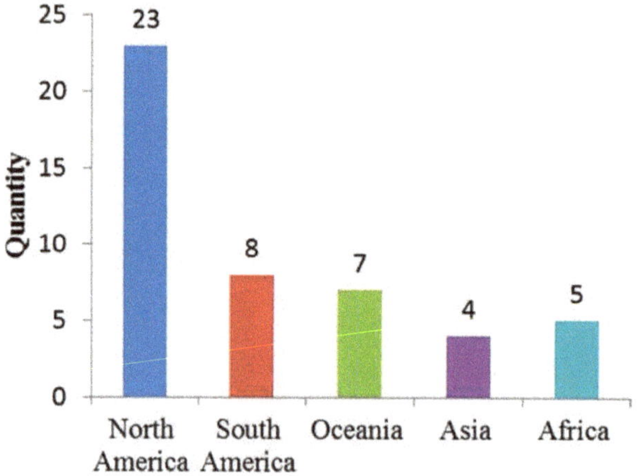

Figure 2. Different intercontinental data distribution.

As seen in Figure 3, the land cover data have 4 types, including ocean, city, bare soil and different kinds of vegetation (agricultural land, grassland, forest.)

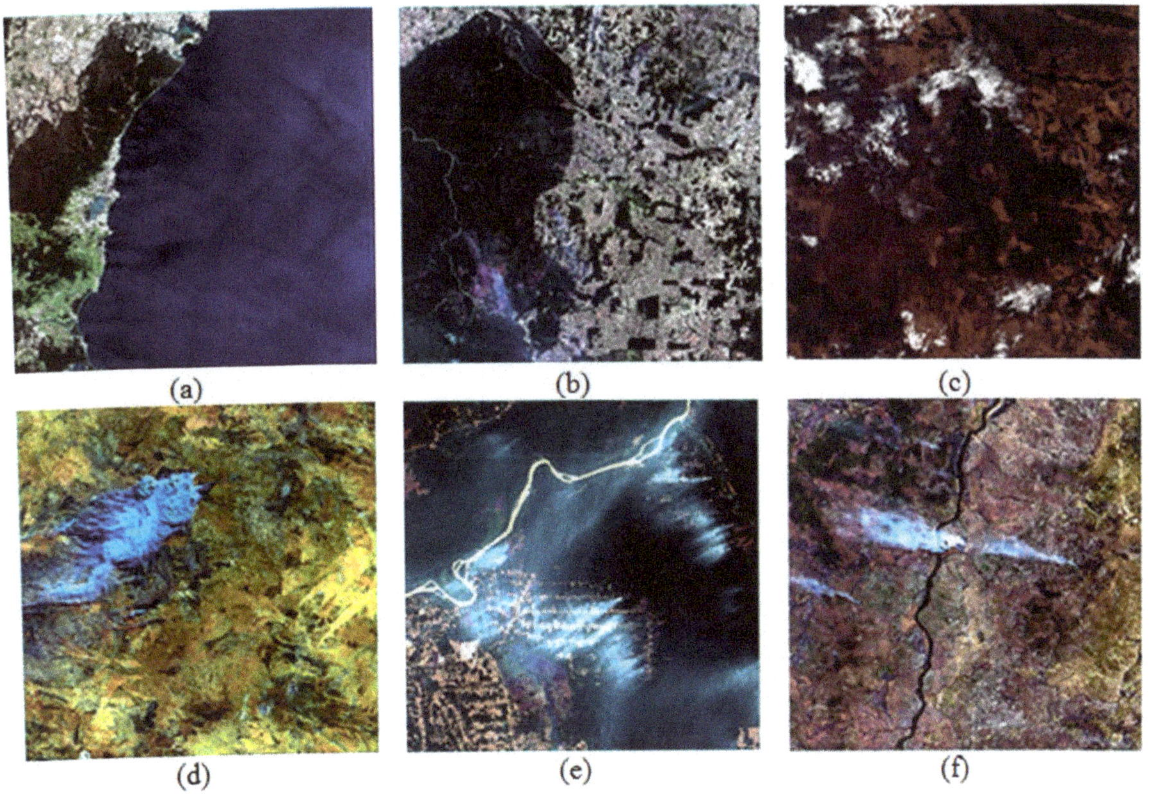

Figure 3. Different land cover types of datasets. (**a**) Ocean; (**b**) City; (**c**) Bare soil; (**d**) Agricultural land; (**e**) Grassland; (**f**) Forest. Different intercontinental data distribution.

2.3. Fire Seasons

Forest fires usually occur in the early stages of springs, autumns and winters due to the influence of climate. As a result of human activity, the wildfire occurrence in summers is dramatically increasing in North America and the Amazon [44,45]. In this study, the period of fire occurrence covered from 2013 to 2019, including different fire seasons, as shown in Figure 4.

2.4. Proportion of Smoke Pixel

Smoke concentration and the proportion of smoke pixels in one image are different with forest fire stage. At the beginning of fire, thin scattered smoke pixels account for a small amount in the image; however, in the middle stage of fire, the entire image is nearly occupied by densely spread smoke. The proportion distribution of smoke pixels is shown in Figure 5.

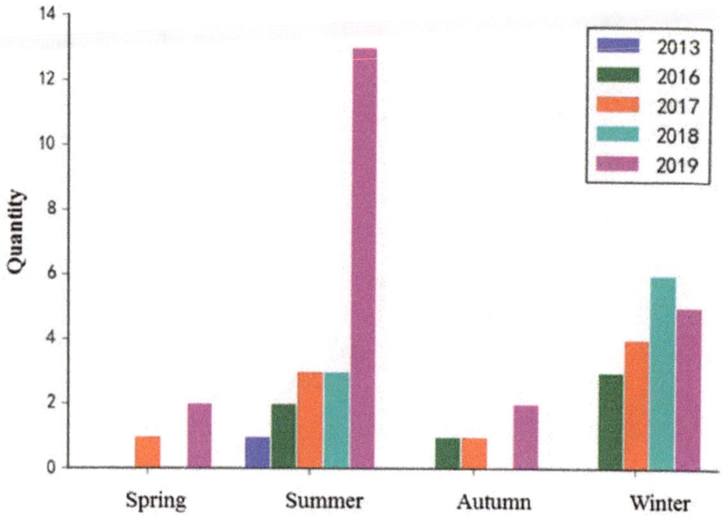

Figure 4. Period of fire occurrence.

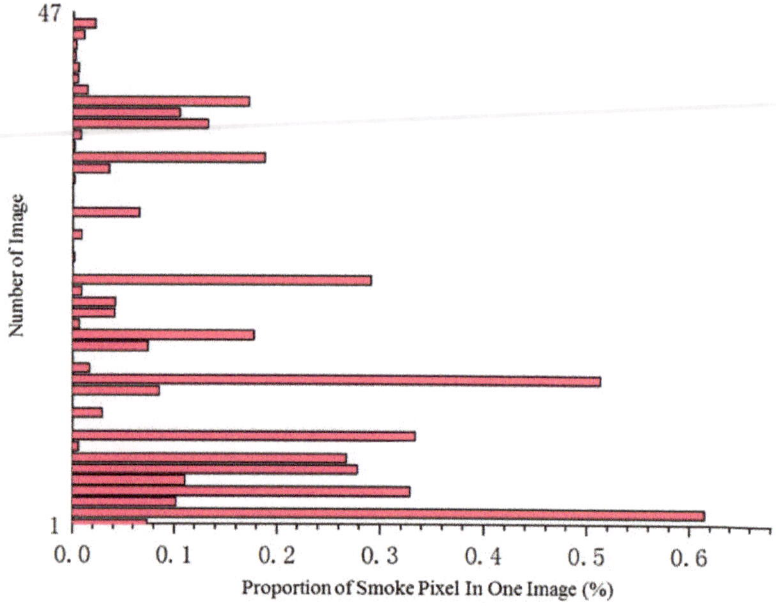

Figure 5. The proportion of smoke pixels of different images.

2.5. Training and Validation Dataset

To reduce overfitting, data augmentation was performed, including random cropping, vertical and horizontal mirroring operations on the images. As a result, the dataset in this study contains a total of 47 multispectral forest fire smoke images, composed of RGB, NIR, SWIR and mid-infrared bands. Thirty-four images are randomly selected as training data, 5 images are used as verification data, and 8 images are used as test data.

3. Methods

As a dense prediction problem, the task of smoke classification in satellite image is to make a prediction at each pixel. Based on the Unet network structure, Smoke-Unet, fused into residual blocks and attention model, was put forward to segment smoke in satellite images in this paper.

As seen in Figure 6, Smoke-Unet consists of a contraction path on the left side and an expansive path on the right side. The contracting path follows the typical architecture of a convolutional network. It consists of the repeated application of two 3 × 3 convolutions (padded convolutions), each followed by a linear unit (ELU) and a 2 × 2 max pooling operation with stride 1 for downsampling. At each downsampling step, we double the number of feature channels. Every step in the expansive path consists of an upsampling of the feature map followed by a 2 × 2 convolution ("up-convolution") that halves the number of feature channels, a concatenation with the correspondingly cropped feature map from the contracting path, and two 3 × 3 convolutions, each followed by a ELU. The cropping is necessary due to the loss of border pixels in every convolution. Because the resolution of the remote sensing image is smaller (one pixel for Landsat with a resolution of 30 m), downsampling will have a catastrophic effect on these local small target features, resulting in the problem of vanishing gradients for many network layers. Therefore, Smoke-Unet is designed to only downsample three times. The steps of convolution and downsampling are alternately performed three times to obtain a high-dimensional feature map and then the spatial resolution is restored through the three-time symmetrical convolution and upsampling operations. The feature map with the same resolution was fused through a skip connection to compensate for the loss of detail caused by downsampling.

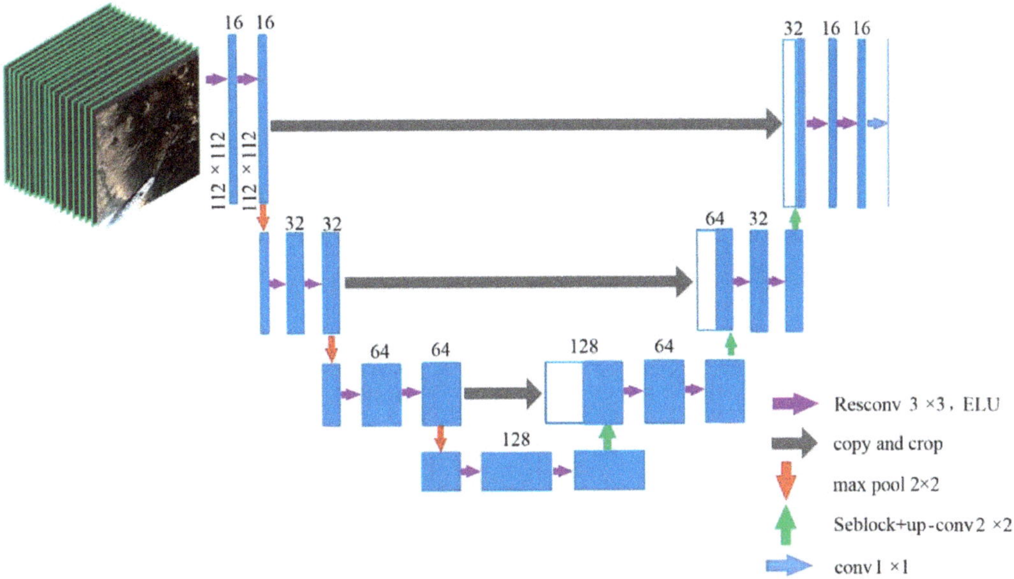

Figure 6. Smoke-Unet.

In order to improve the feature learning ability of the network, ResBlock, a residual block is added to the convolution block to enhance the feature extraction ability. The residual block with skip connection structure can enhance the robustness of the network and improve the performance of the network. The skips structure between layers can fuse coarse semantic and local appearance information. This skip feature is learned end-to-end to improve the semantics and spatial precision for the output. Remote sensors onboard satellite have so many spectral channels that too much irrelevant information leads to difficulty in extracting feature. In order to emphasize effective information and reduce the interference of invalid band information, the SEBlock module based on the attention mechanism is added to the Smoke-Unet network structure. In the attention model, the focus process can be imitated by setting the weight coefficient. The key attention areas can be set with larger weight coefficients, which represent the importance of the information in these areas, while other areas can be set with smaller coefficients to filter invalid information. Through considering different degree of importance for information, the efficiency and accuracy of information processing can be greatly improved. At the final layer, a 1×1 convolution is used to map each 16-component feature vector to final smoke class. In total, the network has 15 convolutional layers.

4. Results and Discussion

In this section, three kinds of semantic segmentation experiments were made on our dataset. By comparing the experimental results, the performance of Smoke-Unet was evaluated and the sensitivity of band and remote sensing parameters was analyzed.

4.1. Experimental Environment

The network structure uses the Keras architecture and several related image processing libraries, the programming language uses Python 3.5. The specific configuration is shown in Table 2.

Table 2. Deep learning environment configuration.

Programming Environment	Auxiliary Library	Hardware Configuration	Other Software
Python3.5 Tensorflow1.9 CUDA8.0 cuDNN10.0 Keras2.2.0	Shapely Opencv2.2 Tifffile0.12 Rasterio1.1.2 h5py2.6.0	CPU:InterE5-2620v3@2.4 GHz GPU:NVDIA TITAN X RAM:16 GB Numba0.26.0	ENVI5.3 ArcGIS10.3 Scikit_image0.12.3

4.2. Implementation Details

The input of the Smoke-Unet network is the multichannel remote sensing image and the index of the multi-remote sensing feature. The data have 13 channels, as shown in Table 3. The schematic diagram of the network is shown in Figure 6.

Table 3. Bands and remote index.

Number	Data Type	Item	Band
1	Band Data	Multispectral Band	1–7, 10
2	Band Data	Panchromatic Band	8
3	Remote Sensing Index	EVI	/
4	Remote Sensing Index	NBR	/
5	Remote Sensing Index	AOD	/
6	Remote Sensing Index	BT	/

During the model training, the back-propagation optimization algorithm uses the stochastic gradient descent (SGD) algorithm, the learning rate is 1×10^{-3}, the momentum is 0.9, the learning rate attenuation is 0.1, the loss function is the joint loss function, and the evaluation function is Jaccard similarity function. The batch size is 128. Considering the computing resources, there are 25 iterations in total, and shuffle is used to disrupt the order of training samples in each epoch. After each round of iteration is completed, the Jaccard coefficient, Accuracy, F1 and other indicators of the training set and the validation set are calculated.

4.3. Implementation Details

In the field of deep learning image segmentation, the similarity coefficient is an important indicator to measure the accuracy of image segmentation. Jaccard similarity coefficient is used in this paper to evaluate the similarity and difference between image targets. The larger the value of Jaccard, the more similar the two targets. For two sets A and B, the Jaccard coefficient is the ratio of the intersection and the union of the two, defined as:

$$J(A,B) = \frac{|A \cap B|}{|A \cup B|} = \frac{|A \cap B|}{|A| + |B| - |A \cap B|}, \tag{1}$$

$$0 \leq J(A,B) \leq 1, \tag{2}$$

4.4. Ablation and Comparative Analysis

In order to verify the role of residual block and attention mechanism of Smoke-Unet, the ablation experiments were made in wildfire smoke segmentation based on remote sensing satellite images. As shown in Table 4, Res-Unet means the network combined Unet with the residual module. Atten-Res-Unet means the network integrated the attention mechanism module with Res-Unet. The results of semantic segmentation were evaluated by metrics such as Jaccard, Accuracy, Recall and F1. In order to validate the effectivity more extensively, other common semantic segmentation networks such as FCN [46], Segnet [47] and PSPnet [48] have been compared. The results are compared in Table 4 and Figure 7.

Table 4. Ablation and comparative analysis of different models.

Network	Dataset	Loss	Jaccard	Accuracy	Recall	F1
Unet	Train	0.844	0.657	0.801	0.753	0.773
	Validation	1.889	0.699	0.694	0.781	0.735
Res-Unet	Train	0.851	0.690	0.805	0.829	0.813
	Validation	1.636	0.59	0.701	0.944	0.805
Atten-Res-Unet	Train	1.514	0.703	0.835	0.816	0.823
	Validation	1.926	0.654	0.696	0.894	0.782
FCN	Train	1.479	0.735	0.845	0.852	0.844
	Validation	1.974	0.58	0.711	0.811	0.758
Segnet	Train	1.532	0.712	0.831	0.835	0.828
	Validation	1.708	0.665	0.761	0.841	0.799
PSPnet	Train	1.406	0.748	0.845	0.871	0.851
	Validation	1.901	0.581	0.751	0.812	0.765
Smoke-Unet	Train	0.759	0.752	0.923	0.917	0.918
	Validation	1.134	0.644	0.725	0.838	0.775

It can be seen from Table 4 that Jaccard coefficient, accuracy, recall rate, F1 and other indicators of Smoke-Unet have been improved to varying degrees. Compared with the original Unet network architecture, the Jaccard coefficient on the training set is increased by 14.46% and the Jaccard coefficient on the verification set is reduced to a certain extent. The accuracy on the training set is increased by 15.23% and the accuracy on the validation set is increased by 4.47%. The recall rate on the training set was increased by 21.78% and the recall rate on the verification set was increased by 7.30%. F1 on the training set is increased by 18.76% and F1 on the validation set is increased by 5.44%. It can be concluded that the proposed network performs better than the original Unet network, and it can be seen from Table 4 that Smoke-Unet is better than other common semantic segmentation networks. The specific segmentation image is shown in Figure 7.

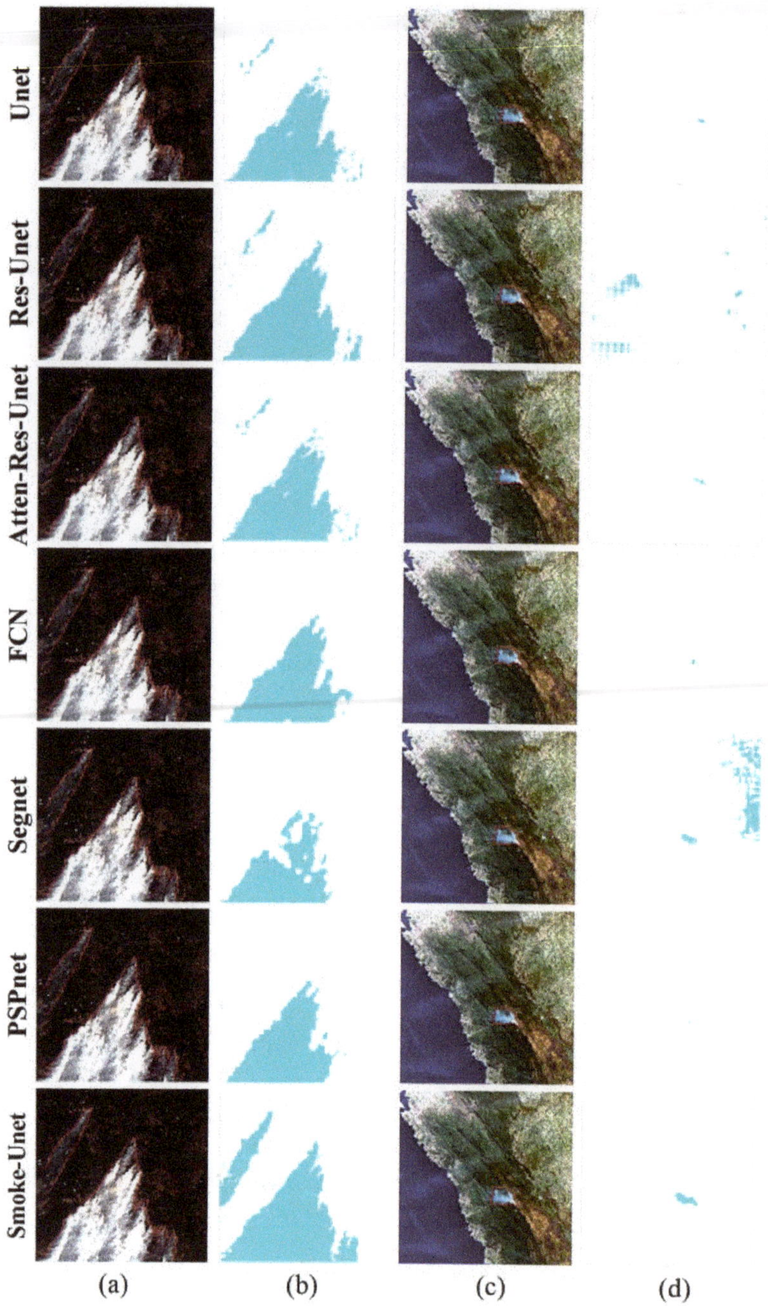

Figure 7. The results of segmentation of different networks. (**a**) Image acquired over British Columbia, Canada, on 4 August 2017, the smoke is depicted in red line area; (**b**) The segmentation results of smoke over British Columbia, the smoke pixels are depicted in aqua color; (**c**) Image acquired over New Zealand area, on 7 Feb 2019, the smoke is depicted in red line area; (**d**) The segmentation results of smoke over New Zealand area, the smoke pixels are depicted in aqua color.

In Figure 7a, the smoke contains a wide range of dense smoke and scattered diffuse thin smoke, and the land cover includes vegetation, bare soil, and some cirrus clouds. In Figure 7c, the smoke, located near the fire point, is thin and has a relatively small range, and the land cover includes sea water, seashore, bare land, vegetation and so on.

It can be seen from Figure 7b,d that the Unet network can roughly segment the smoke pixels in different images. In Figure 7b, Res-Unet can effectively segment the smoke pixels, because the number of smoke pixels in the diffusion area at the upper left of Figure 7b has increased, while in Figure 7d there is an over-segmentation by Res-Unet, and some pixels are incorrectly segmented as the smoke pixel. In Figure 7b, Atten-Res-Unet can effectively segment the smoke pixels, as the number of smoke pixels in the diffusion area at the upper left of Figure 7b has increased, while the under-segmentation exists in Figure 7d, resulting that some pixels are not identified. The segmentation effects using FCN, SegNet and PSPnet are worse than Unet-based methods. It can be seen from Figure 7b,d that the Smoke-Unet network has a better recognition performance than the other networks when segmenting a wide range of dense smoke and a small area of thin smoke.

4.5. Sensitivity Analysis

With the increasing number of high-resolution images and dimensional channels of data, the information redundancy generated by high-dimensionality makes it difficult to effectively utilize the rich information of remote sensing images. Based on the above-mentioned forest fire smoke detection algorithm, this section will analyze and discuss the influence of different band combinations of multispectral data and remote sensing parameters on the accuracy of the algorithm.

4.5.1. Sensitivity of Bands

In order to evaluate the band sensitivity, the segmentation experiments based on different band combination were made on our dataset. The data source distribution is shown in Table 5. The test images contain a large proportion of smoke, small proportion of smoke, the land cover includes bare land, vegetation, seashores and highly reflective ground.

Table 5. Details of different bands combination.

Number	Data Type	Data Dimension	Band
1	RGB	3	Band 2~4
2	RGB + NIR	4	Band 2~5
3	RGB + TIRS1	4	Band 2~4,10
4	RGB + SWIR2	4	Band 2~4,7
5	RGB + SWIR1 + SWIR2	5	Band 2~4,6,7
6	RGB + SWIR1 + NIR	5	Band 2~6
7	RGB + TIRS1 + SWIR2	5	Band 2~4,7,10
8	TIRS1	1	Band 10
9	NIR + SWIR1/2 + TIRS1	4	Band 5~7,10
10	SWIR1 + NIR + Blue	3	Band 2,5,6
11	Multiple	8	Band 1~7, Band 10
12	Multiple + Pan	9	Band 1~7, Band 10~11
13	All data	11	Band 1~11

From Table 6, Figures 8 and 9, it can be found that the segmentation result of smoke is the best when the input band is RGB and SWIR2. Compared to all the data bands as the input, Jaccard with the input of RGB and SWIR2 increases by 6.5%. When the input is all data source, it can effectively segment a wide range of smoke. However, compared with the segmentation result of the RGB data source, the smoke pixel with the input of all band data has the problem of under-segmentation for a small area of smoke, especially in the downwind diffusion area. It shows that too much data will interfere with the network parameter learning and degrade the performance of the network.

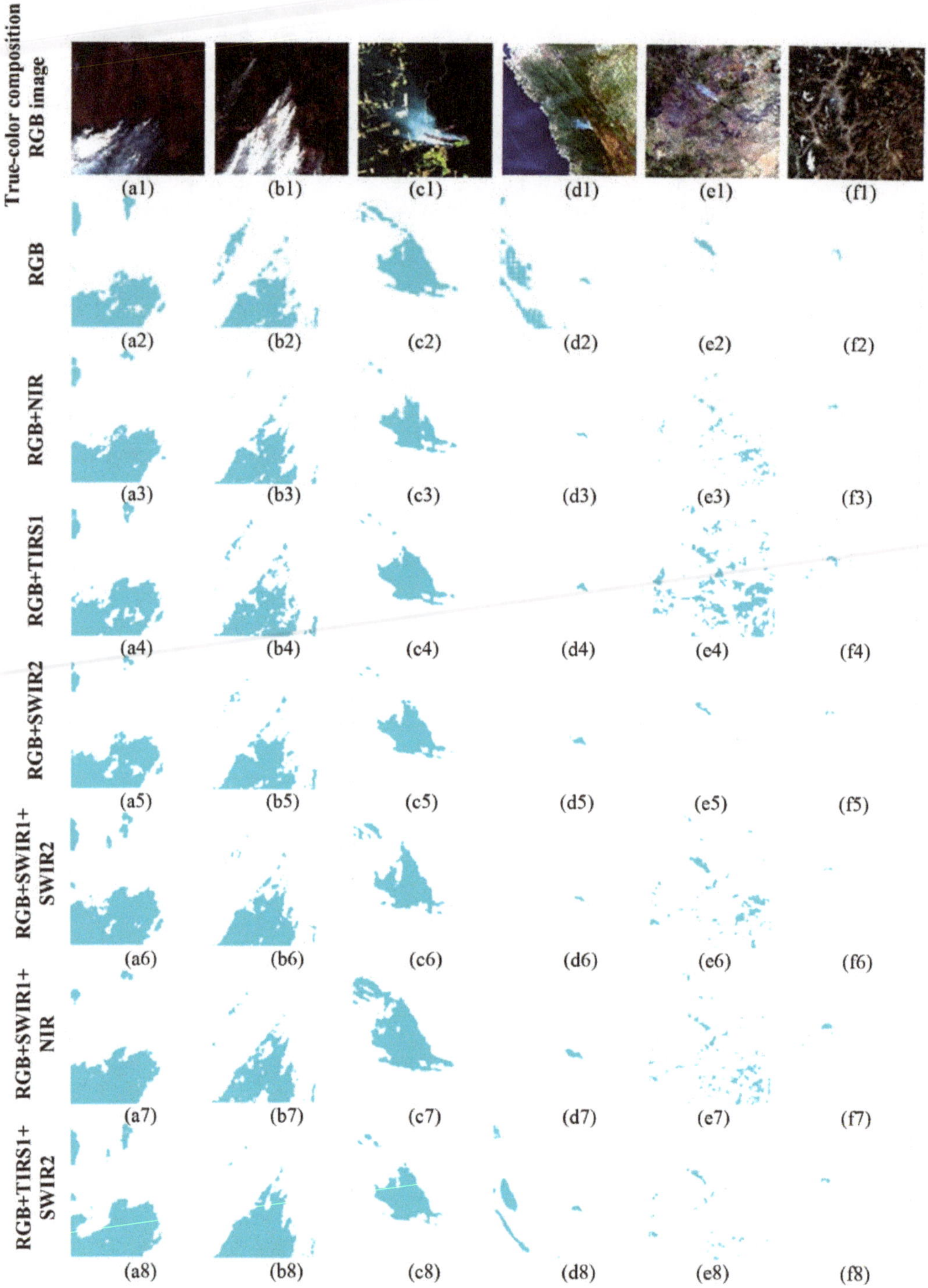

Figure 8. *Cont.*

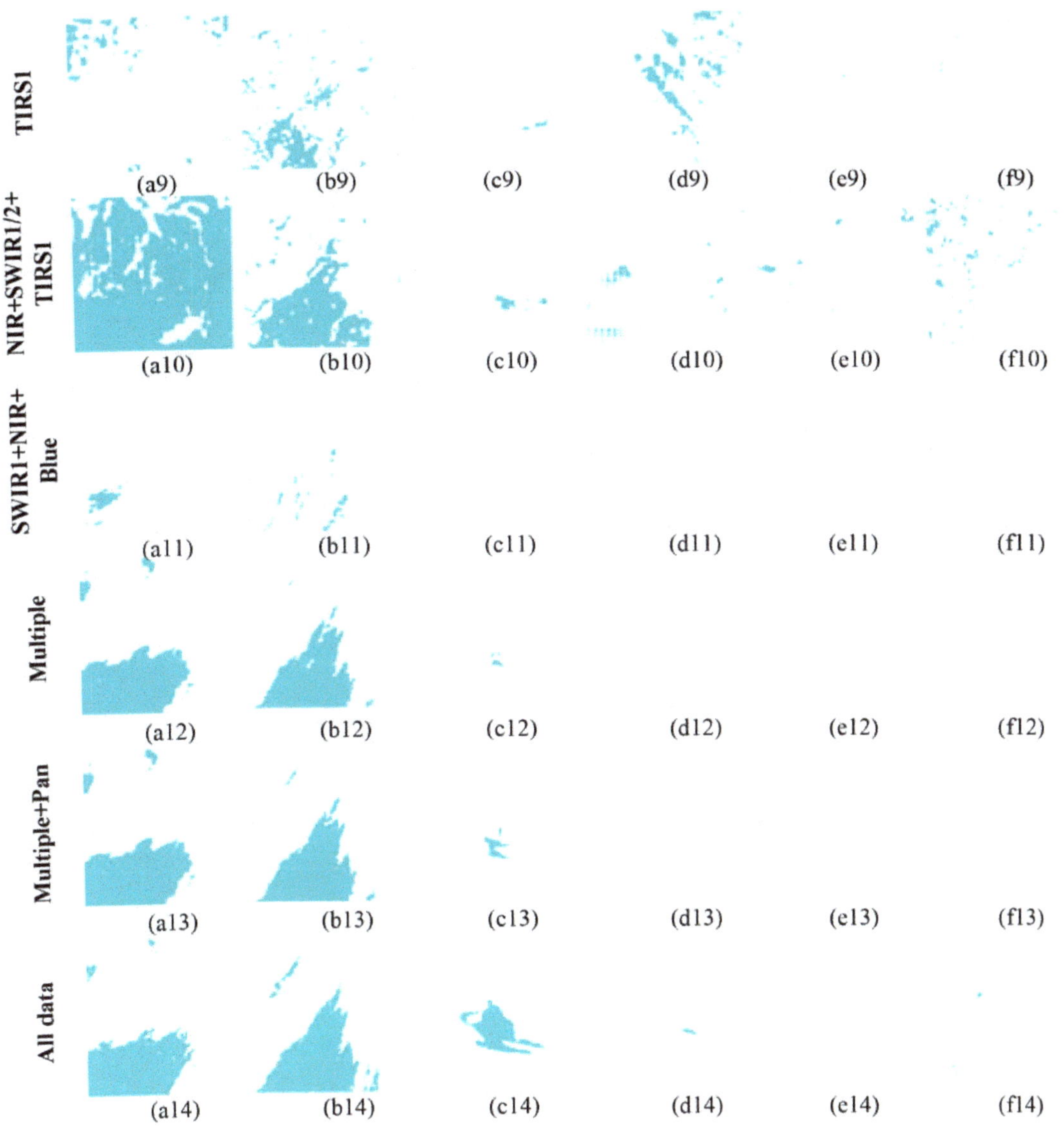

Figure 8. The first line shows true-color composition RGB images of smoke plumes. (**a1–a14**) Siberia area, Russia, on 17 March 2018; (**b1–b14**) British Columbia, Canada, on 4 August 2017; (**c1–c14**) Amazon region, Brazil, on 9 August 2019; (**d1–d14**) New Zealand area, on 7 Feb 2019; (**e1–e14**) Zambia, on 26 June 2017; (**f1–f14**) Liangshan region, China, on 21 May 2019. All rows except the first are segmentation results of smoke with different input data, the smoke pixels are depicted in aqua color.

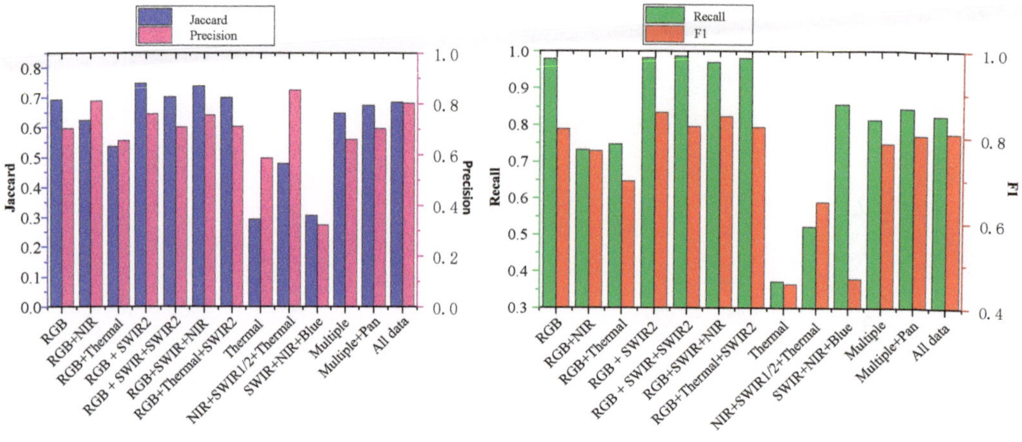

(a) (b)

Figure 9. The segmentation results of smoke with variety bands combination. (**a**) The result of Jaccard and Accuracy; (**b**) The result of recall and F1.

Table 6. The segmentation results of different bands combination.

Number	Data Type	Jaccard	Accuracy	Recall	F1
1	RGB	0.692	0.701	0.980	0.818
2	RGB + NIR	0.623	0.809	0.730	0.767
3	RGB + TIRS1	0.535	0.653	0.747	0.697
4	RGB + SWIR2	0.748	0.759	0.982	0.856
5	RGB + SWIR1 + SWIR2	0.701	0.707	0.988	0.824
6	RGB + SWIR1 + NIR	0.737	0.753	0.970	0.848
7	RGB + TIRS1 + SWIR2	0.700	0.709	0.981	0.823
8	TIRS1	0.294	0.585	0.371	0.455
9	NIR + SWIR1/2 + TIRS1	0.479	0.852	0.522	0.648
10	SWIR1 + NIR + Blue	0.305	0.322	0.855	0.468
11	Multiple	0.646	0.658	0.814	0.784
12	Multiple + Pan	0.673	0.701	0.844	0.804
13	All data	0.683	0.801	0.825	0.809

In order to better distinguish smoke from clouds, the spectral characteristics of smoke and cloud in different bands were compared. As shown in Figure 10, the image contains smoke (heavy smoke numbered 2; smoke near the fire point numbered 5; thin smoke in the diffusion area numbered 3 and 4) and clouds (numbered 1). To highlight the features, the logarithmic transformation was made to the image. The spectral characteristics of different objects in each band of the multispectrum are shown in Figure 11.

It can be seen from Figure 11a,b that clouds and dense smoke have very similar spectral characteristics in the RGB band (Band 3~5); therefore, it is difficult to distinguish dense smoke with clouds by the naked eye. However, the pixel values of the two are quite different in the SWIR2 band (Band 8), which may be the reason why the smoke pixels can be better distinguished by using RGB and SWIR2. From Figure 11b,c, it shows that the spectral characteristics of heavy smoke and thin smoke are greatly different, which makes the task of smoke recognition challenging.

4.5.2. Sensitivity of Remote Sensing Parameters

In order to evaluate the sensitivity of different remote sensing feature indexes to forest fire smoke, EVI, NBR, BT and AOD were respectively combined with RGB and SWIR2 as shown in Table 7 to evaluate the impact on the smoke segmentation.

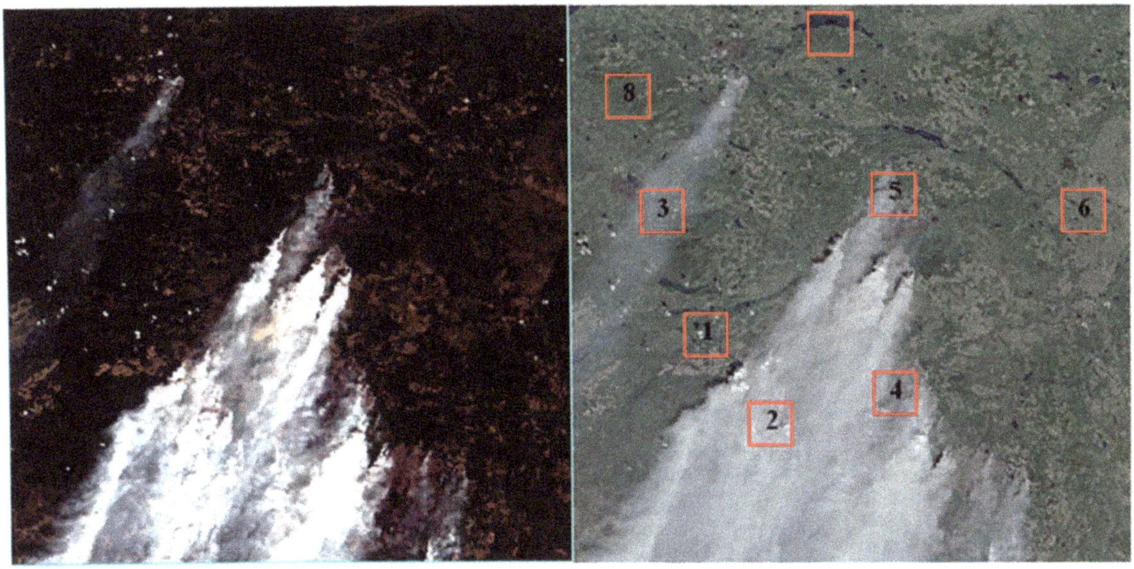

(**a**) (**b**)

Figure 10. The image of smoke acquired over British Columbia, Canada, on 4 August 2017. (**a**) The true-color composition image. (**b**) The image of smoke after logarithmic transformed. Different targets are marked with numbers 1 through 8. (1) The cloud; (2) The heavy smoke; (3) The thin smoke over area 3; (4) The thin smoke over area 4; (5) The smoke over the hot spot; (6) The soil; (7) The water; (8) The vegetation.

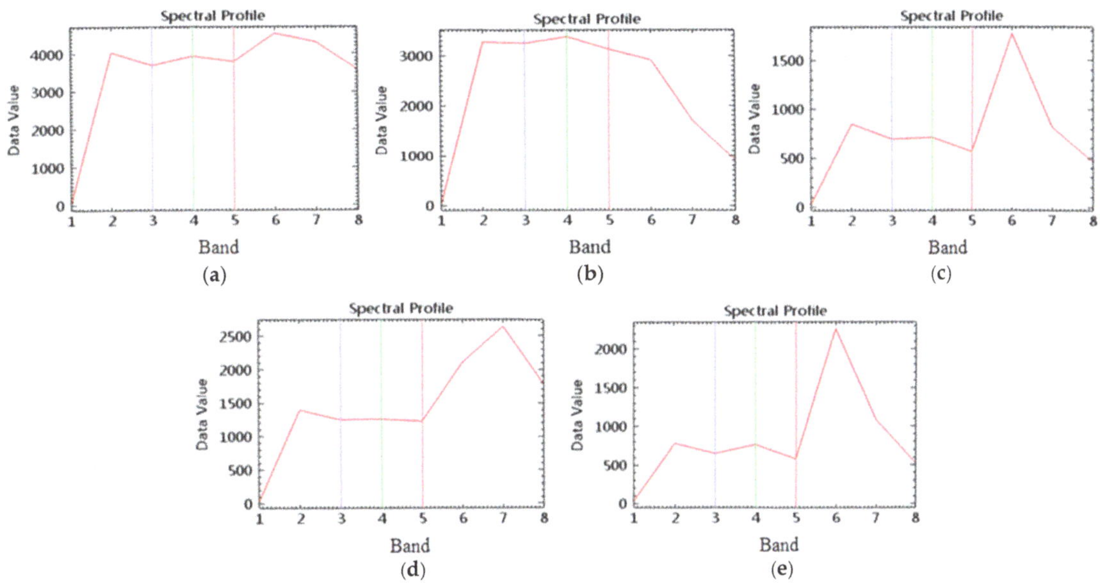

Figure 11. The spectral profile of different objects. (**a**) The profile of cloud on area 1; (**b**) The profile of heavy smoke on area 2; (**c**) The profile of thin smoke over the area 3; (**d**) The profile of thin smoke over the area 4; (**e**) The profile of smoke over the hot spot (the fire point) on area 5.

Table 7. Fusion of different remote sensing features.

Number	Data Type	Data Dimension
1	RGB + SWIR2 + EVI	5
2	RGB + SWIR2 + NBR	5
3	RGB + SWIR2 + BT	5
4	RGB + SWIR2 + AOD	5

As shown in Figure 12, both EVI and NBR do not contribute to forest fire smoke segmentation and BT help to identify high temperature abnormal points, resulting in under-segmentation of smoke pixels.

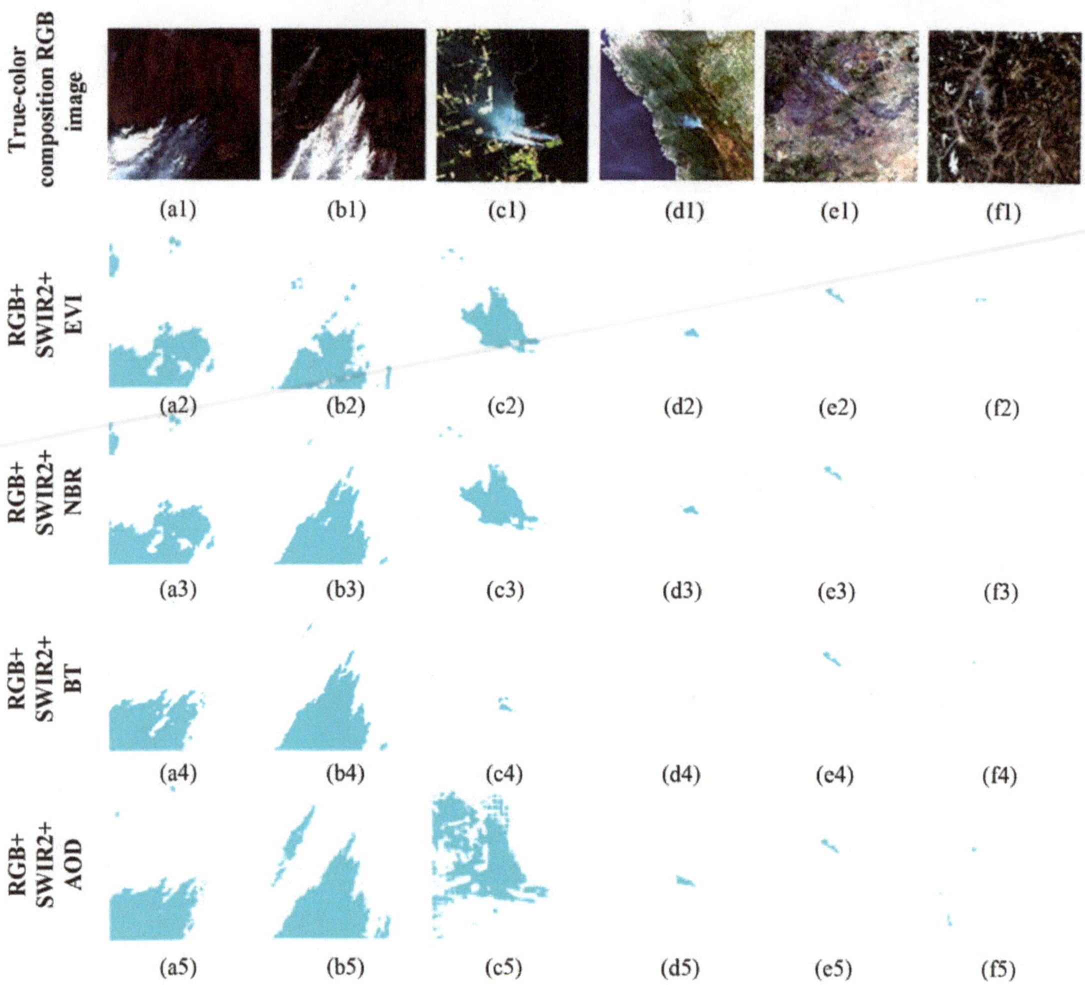

Figure 12. The first line is true-color composition RGB images of smoke plumes. (**a1–a5**) Siberia area, Russia on 17 Mar 2018; (**b1–b5**) British Columbia, Canada, on 4 August 2017; (**c1–c5**) Amazon region, Brazil, on 9 August 2019; (**d1–d5**) New Zealand area, on 7 February 2019; (**e1–e5**) Zambia, on 26 June 2017; (**f1–f5**) Liangshan region, China, on 21 May 2019. All rows except the first are segmentation results of smoke with multiple bands and remote sensing indexes, the smoke pixels are depicted in aqua color.

In Figure 12(c5), the upper left area is the smoke plume diffusion area, and a large number of smoke pixels that could not be identified by visual interpretation were segmented. This may be a result from the increasing aerosol concentration in this area due to the large amount of carbon oxides and nitrogen oxides contained in forest fire smoke. In Figure 12(f5), some mis-segmentation was made because much smaller smoke area and fewer smoke pixels are prone to be mis-recognized by image noise. Therefore, it can be concluded that the segmented smoke pixels significantly increase, especially for the thin smoke in the downwind diffusion zone, when AOD is added as the input of RGB and SWIR2.

5. Conclusions

In order to solve the difficulty of detecting forest fire smoke in remote sensing images, this study proposed the Smoke-Unet network to segment forest fire smoke and analyzed the sensitivity of remote sensing satellite data and remote sensing index used for wildfire detection. This paper first constructed a multispectral remote sensing smoke dataset containing different years, seasons, regions and land cover. Second, Smoke-Unet, which combined an improved Unet network with attention mechanism and residual block, was put forward in this paper and verified by comparing with other methods on the experiments. Third, the sensitivity of different spectral band combinations of multispectral data and the remote sensing index to the wildfire smoke segmentation were analyzed by the experiments. The results show that the smoke pixel accuracy rate using the proposed Smoke-Unet is 3.1% higher than that of Unet and RGB, SWIR2 and AOD bands are verified as the sensitive band combination and the remote sensing index for wildfire smoke segmentation, which could effectively segment the smoke pixels in remote sensing images. This proposed method under the RGB, SWIR2 and AOD bands can help to segment smoke by using high-sensitivity band and remote sensing index and makes an early alarm of forest fire smoke. However, some problems need to be further solved in subsequent studies. A large amount of mixed spectrum phenomenon in the diffusion area makes it much difficult to label thin smoke plume in the downwind direction by visual interpretation. How to exploit the feature-extraction advantages of deep learning methods to better interpret remote sensing images requires a lot of exploration.

Author Contributions: Conceptualization, Z.W. and P.Y.; data curation, P.Y.; formal analysis, P.Y.; funding acquisition, C.Z.; methodology, P.Y.; project administration, C.Z.; software, P.Y.; supervision, H.L., C.Z., J.Y., Y.T. and W.C.; validation, Z.W., P.Y., C.Z., J.Y., Y.T. and W.C.; visualization, Z.W. and P.Y.; writing—original draft, Z.W. and P.Y.; writing—review and editing, Z.W., H.L., C.Z., J.Y., Y.T. and W.C. All authors have read and agreed to the published version of the manuscript.

Funding: This research was funded by the National Natural Science Foundation of China, grant number 31971668.

Informed Consent Statement: Not applicable.

Data Availability Statement: Data available on request due to restrictions of privacy.

Conflicts of Interest: The authors declare no conflict of interest.

References

1. Sun, G.; Hallema, D.; Asbjornsen, H. Ecohydrological processes and ecosystem services in the Anthropocene: A review. *Ecol. Process.* **2017**, *6*, 35. [CrossRef]
2. Hansen, M.C.; Loveland, T.R. A review of large area monitoring of land cover change using Landsat data. *Remote Sens. Environ.* **2012**, *122*, 66–74. [CrossRef]
3. Houghton, R. Historic role of forests in the global carbon cycle. In *Carbon Dioxide Mitigation in Forestry and Wood Industry*; Springer: Berlin/Heidelberg, Germany, 1998; pp. 1–24.
4. Canadell, J.G.; Raupach, M.R. Managing forests for climate change mitigation. *Science* **2008**, *320*, 1456–1457. [CrossRef]
5. Bowman, D.M.J.S.; Balch, J.K.; Artaxo, P.; Bond, W.J.; Carlson, J.M.; Cochrane, M.A.; D'Antonio, C.M.; DeFries, R.S.; Doyle, J.C.; Harrison, S.P.; et al. Fire in the earth system. *Science* **2009**, *324*, 481–484. [CrossRef]

6. Allen, C.D.; Macalady, A.K.; Chenchouni, H.; Bachelet, D.; McDowell, N.; Vennetier, M.; Kitzberger, T.; Rigling, A.; Breshears, D.D.; Hogg, E.H.T. A global overview of drought and heat-induced tree mortality reveals emerging climate change risk for forests. *For. Ecol. Manag.* **2010**, *259*, 660–684. [CrossRef]
7. Hirsch, K.G.; Corey, P.N.; Martell, D.L. Using expert judgment to model initial attack fire crew effectiveness. *For. Sci.* **1998**, *44*, 539–549.
8. Korontzi, S.; McCarty, J.; Loboda, T.; Kumar, S.; Justice, C. Global distribution of agricultural fires in croplands from 3 years of Moderate Resolution Imaging Spectroradiometer (MODIS) data. *Glob. Biogeochem. Cycles* **2006**, *20*, GB2021. [CrossRef]
9. Dwyer, E.; Pinnock, S.; Gregoire, J.M.; Pereira, J.M.C. Global spatial and temporal distribution of vegetation fire as determined from satellite observations. *Int. J. Remote Sens.* **2000**, *21*, 1289–1302. [CrossRef]
10. Csiszar, I.; Denis, L.; Giglio, L.; Justice, C.O.; Hewson, J. Global fire activity from two years of MODIS data. *Int. J. Wildland Fire* **2005**, *14*, 117–130. [CrossRef]
11. Dozier, J. A method for satellite identification of surface temperature fields of subpixel resolution. *Remote Sens. Environ.* **1981**, *11*, 221–229. [CrossRef]
12. Matson, M.; Stephens, G.; Robinson, J. Fire detection using data from the NOAA-N satellites. *Int. J. Remote Sens.* **1987**, *8*, 961. [CrossRef]
13. Robinson, J.M. Fire from space: Global evaluation using infrared remote sensing. *Int. J. Remote Sens.* **1991**, *12*, 3–24. [CrossRef]
14. Kaufman, Y.J.; Tanré, D. Algorithm for Remote Sensing of Tropospheric Aerosol From MODIS. In *NASA MODIS Algorithm Theoretical Basis Document*; Goddard Space Flight Center: Greenbelt, MD, USA, 1998; Volume 85, pp. 3–68.
15. Giglio, L.; Descloitres, J.; Justice, C.O.; Kaufman, Y.J. An enhanced contextual fire detection algorithm for MODIS. *Remote Sens. Environ.* **2003**, *87*, 273–282. [CrossRef]
16. Giglio, L.; Schroeder, W.; Justice, C.O. The collection 6 MODIS active fire detection algorithm and fire products. *Remote Sens. Environ.* **2016**, *178*, 31–41. [CrossRef]
17. Justice, C.O.; Román, M.O.; Csiszar, I.; Vermote, E.F.; Wolfe, R.E.; Hook, S.J. Land and cryosphere products fromSuomi NPP VIIRS: Overview and status. *J. Geophys. Res.* **2013**, *118*, 9753–9765. [CrossRef] [PubMed]
18. Wolfe, R.E.; Lin, G.; Nishihama, M.; Tewari, K.P.; Tilton, J.C.; Isaacman, A.R. Suomi NPP VIIRS prelaunch and on-orbit geometric calibration and characterization. *J. Geophys. Res.* **2013**, *118*, 508–511, 521. [CrossRef]
19. Csiszar, I.; Schroeder, W.; Giglio, L.; Ellicott, E.; Vadrevu, K.P.; Justice, C.O. Active fires from the Suomi NPP Visible Infrared Imaging Radiometer Suite: Product status and first evaluation results. *J. Geophys. Res.* **2014**, *119*, 803–816. [CrossRef]
20. Schroeder, W.; Oliva, P.; Giglio, L.; Quaryle, B.; Lorenz, E.; Morelli, F. Active fire detection using Landsat-8/OLI data. *Remote Sens. Environ.* **2015**, *185*, 210–220. [CrossRef]
21. Li, Z.; Nadon, S.; Cihlar, J. Satellite-based detection of Canadian boreal forest fires:development and application of the algorithm. *Int. J. Remote Sens.* **2000**, *21*, 3057–3069. [CrossRef]
22. Li, Z.; Kaufman, Y.J.; Ichoku, C.; Fraser, R.; Trishchenko, A.; Giglio, L.; Yu, X. A Review of AVHRR-based Active Fire Detection Algorithms: Principles, Limitations, and Recommendations. 2000. Available online: http://www.fao.org/GTOS/gofc-gold/docs/fire_ov.pdf (accessed on 29 September 2021).
23. Csiszar, I.A.; Morisette, J.T.; Giglio, L. Validation of active fire detection from moderateresolution satellite sensors: The MODIS example in northern Eurasia. *Remote Sens.* **2006**, *44*, 1757–1764. [CrossRef]
24. Genet, H.; McGuire, A.D.; Barrett, K.; Breen, A.; Euskirchen, E.S.; Johnstone, J.F.; Yuan, F. Modeling the effects of fire severity and climate warming on active layer thickness and soil carbon storage of black spruce forests across the landscape in interior Alaska. *Environ. Res. Lett.* **2013**, *8*, 045016. [CrossRef]
25. Giglio, L.; Kendall, J.D.; Justice, C.O. Evaluation of global fire detection algorithms using simulated AVHRR infrared data. *Int. J. Remote Sens.* **1999**, *20*, 1947–1985. [CrossRef]
26. Schroeder, W.; Oliva, P.; Giglio, L.; Ivan, A.C. The New VIIRS375m active fire detection data product: Algorithm description and initial assessment. *Remote Sens. Environ.* **2014**, *143*, 85–96. [CrossRef]
27. Waigl, C.F.; Stuefer, M.; Prakash, A.; Ichoku, C. Detecting high and low-intensity fires in Alaska using VIIRS I-band data: An improved operational approach for high latitudes. *Remote Sens. Environ.* **2017**, *199*, 389–400. [CrossRef]
28. Elvidge, C.D.; Zhizhin, M.; Hsu, F.C.; Baugh, K.E. VIIRS Nightfire: Satellite Pyrometry at Night. *Remote Sens.* **2013**, *5*, 4423–4449. [CrossRef]
29. Cho, K.; Kim, Y.; Kim, Y. Disaggregation of Landsat-8 Thermal Data Using Guided SWIR Imagery on the Scene of a Wildfire. *Remote Sens.* **2018**, *10*, 105. [CrossRef]
30. Li, Z.; Khananian, A.; Fraser, R.H.; Cihlar, J. Automatic detection of fire smoke using artificial neural networks and threshold approaches applied to AVHRR imagery. *IEEE Trans. Geosci. Remote Sens.* **2001**, *39*, 1859–1870.
31. Garay, M.J.; Mazzoni, D.M.; Davies, R.; Diner, D. The application of support vector machines to the analysis of global datasets from MISR. In Proceedings of the Fourth Conference on Artificial Intelligence Applications to Environmental Science, San Diego, CA, USA, 9–13 January 2005.
32. Mazzoni, D.; Garay, M.J.; Davies, R.; Nelson, D. An operational MISR pixel classifier using support vector machines. *Remote Sens. Environ.* **2006**, *107*, 149–158. [CrossRef]
33. Mazzoni, D.; Logan, J.A.; Diner, D.; Kahn, R.; Tong, L.; Li, Q. A data-mining approach to associating MISR smoke plume heights with MODIS fire measurements. *Remote Sens. Environ.* **2007**, *107*, 138–148. [CrossRef]

34. Li, X.L.; Wang, J.; Song, W.G.; Ma, J.; Telesca, L.; Zhang, Y.M. Automatic Smoke Detection in MODIS Satellites Data based on K-means Clustering and Fisher Linear Discrimination. *Photogramm. Eng. Remote Sens.* **2014**, *80*, 971–982. [CrossRef]
35. Li, X.L.; Song, W.G.; Lian, L.; Wei, X. Forest Fire Smoke Detection Using Back-Propagation Neural Network Based on MODIS Data. *Remote Sens.* **2015**, *7*, 4473–4498. [CrossRef]
36. Zhang, Z.; Liu, Q.; Wang, Y. Road extraction by deep residual u-net. *IEEE Geosci. Remote Sens. Lett.* **2018**, *15*, 749–753. [CrossRef]
37. Jeppesena, J.H.; Jacobsena, R.H. A cloud detection method for landsat 8 images based on pcanet. *Remote Sens.* **2018**, *10*, 877.
38. Ba, R.; Chen, C.; Yuan, J.; Song, W.; Lo, S. SmokeNet:Satellites Smoke Scene Detection Using Convolutional Neural Network with Spatial and Channel-Wise Attetion. *Remote Sens.* **2019**, *11*, 1702. [CrossRef]
39. Ronneberger, O.; Fischer, P.; Brox, T. U-Net: Convolutional Networks for Biomedical Image Segmentation. 2015. Available online: https://arxiv.org/pdf/1505.04597.pdf (accessed on 29 September 2021).
40. Bao, Y.; Liu, W.; Gao, O.; Lin, Z.; Hu, Q. E-Unet++: A Semantic Segmentation Method for Remote Sensing Images. In Proceedings of the 2021 IEEE 4th Advanced Information Management, Communicates, Electronic and Automation Control Conference (IMCEC), Chongqing, China, 18–20 June 2021; pp. 1858–1862. [CrossRef]
41. Li, X.; Du, Z.; Huang, Y.; Tan, Z. A deep translation (GAN) based change detection network for optical and SAR remote sensing images. *ISPRS J. Photogramm. Remote Sens.* **2021**, *179*, 14–34. [CrossRef]
42. Maratkhan, A.; Ilyassov, I.; Aitzhanov, M.; Demirci, M.F.; Ozbayoglu, A.M. Deep learning-based investment strategy: Technical indicator clustering and residual blocks. *Soft Comput.* **2021**, *25*, 5151–5161. [CrossRef]
43. Kastner, S.; Ungerleider, L.G. Mechanisms of visual attention in the human cortex. *Annu. Rev. Neurosci.* **2000**, *23*, 315–341. [CrossRef] [PubMed]
44. Balshi, M.S.; McGuire, A.D.; Zhuang, Q.; Melillo, J.; Kicklighter, D.W.; Kasischke, E. The role of historical fire disturbance in the carbon dynamics of the pan-boreal region: A process-based analysis. *J. Geophys. Res.* **2007**, *112*, 1–18. [CrossRef]
45. Dennison, P.E.; Brewer, S.C.; Arnold, J.D.; Moritz, M.A. Large wildfire trends in the western United States, 1984–2011. *Geophys. Res. Lett.* **2014**, *41*, 2928–2933. [CrossRef]
46. Shelhamer, E.; Long, J.; Darrell, T. Fully Convolutional Networks for Semantic Segmentation. *IEEE Trans. Pattern Anal. Mach. Intell.* **2017**, *39*, 640–651. [CrossRef]
47. Badrinarayanan, V.; Kendall, A.; Cipolla, R. SegNet: A Deep Convolutional Encoder-Decoder Architecture for Image Segmentation. *IEEE Trans. Pattern Anal. Mach. Intell.* **2017**, *39*, 2481–2495. [CrossRef] [PubMed]
48. Zhao, H.; Shi, J.; Qi, X.; Wang, X.; Jia, J. Pyramid Scene Parsing Network. In Proceedings of the 2017 IEEE Conference on Computer Vision and Pattern Recognition (CVPR), Honolulu, HI, USA, 21–26 July 2017; pp. 6230–6239.

remote sensing

MDPI

Article

Pyramid Information Distillation Attention Network for Super-Resolution Reconstruction of Remote Sensing Images

Bo Huang, Zhiming Guo, Liaoni Wu *, Boyong He, Xianjiang Li and Yuxing Lin

School of Aerospace Engineering, Xiamen University, Xiamen 361102, China; huangbo@stu.xmu.edu.cn (B.H.); guozm@xmu.edu.cn (Z.G.); heboyong0220@stu.xmu.edu.cn (B.H.); lixianjiang@stu.xmu.edu.cn (X.L.); linyuxing@stu.xmu.edu.cn (Y.L.)

* Correspondence: wuliaoni@xmu.edu.cn

Abstract: Image super-resolution (SR) technology aims to recover high-resolution images from low-resolution originals, and it is of great significance for the high-quality interpretation of remote sensing images. However, most present SR-reconstruction approaches suffer from network training difficulties and the challenge of increasing computational complexity with increasing numbers of network layers. This indicates that these approaches are not suitable for application scenarios with limited computing resources. Furthermore, the complex spatial distributions and rich details of remote sensing images increase the difficulty of their reconstruction. In this paper, we propose the pyramid information distillation attention network (PIDAN) to solve these issues. Specifically, we propose the pyramid information distillation attention block (PIDAB), which has been developed as a building block in the PIDAN. The key components of the PIDAB are the pyramid information distillation (PID) module and the hybrid attention mechanism (HAM) module. Firstly, the PID module uses feature distillation with parallel multi-receptive field convolutions to extract short- and long-path feature information, which allows the network to obtain more non-redundant image features. Then, the HAM module enhances the sensitivity of the network to high-frequency image information. Extensive validation experiments show that when compared with other advanced CNN-based approaches, the PIDAN achieves a better balance between image SR performance and model size.

Keywords: attention mechanism; feature distillation; remote sensing; super-resolution

Citation: Huang, B.; Guo, Z.; Wu, L.; He, B.; Li, X.; Lin, Y. Pyramid Information Distillation Attention Network for Super-Resolution Reconstruction of Remote Sensing Images. *Remote Sens.* **2021**, *13*, 5143. https://doi.org/10.3390/rs13245143

Academic Editor: Lefei Zhang

Received: 10 November 2021
Accepted: 17 December 2021
Published: 17 December 2021

1. Introduction

High-resolution (HR) remote sensing imagery can provide rich and detailed information about ground features and this has led to it being widely used in various tasks, including urban surveillance, forestry inspection, disaster monitoring, and military object detection [1]. However, it is difficult to guarantee the clarity of remote sensing images because it can be restricted by the imaging hardware, transmission conditions, and other factors. Considering the high cost and time-consuming research cycle of hardware sensors, the development of a practical and inexpensive algorithm for HR imaging technology in the field of remote sensing is in great demand.

Single-image super-resolution (SISR) [2] aims to obtain an HR image from its corresponding low-resolution (LR) counterpart by using the intrinsic relationships between the pixels in an image. Traditional SISR methods can be roughly divided into three main categories: Interpolation- [3,4], reconstruction- [5,6], and example learning-based methods [7,8]. However, these approaches are not suitable for image SR tasks in the remote sensing field because of their limited ability to capture detailed features and the loss of a large amount of high-frequency information (edges and contours) in the reconstruction process.

With the flourishing development of deep convolutional neural networks (DCNNs) and big-data technology, promising results have been obtained in computer vision tasks.

Because of their end-to-end training strategy and powerful feature-reconstruction ability, DCNNs have been extensively applied in the domain of SR reconstruction in recent years [9–14]. Dong et al. [9] successfully introduced a CNN into the SR reconstruction task using a simple three-layer neural network, and they demonstrated that CNNs can directly learn end-to-end nonlinear mappings from LR images to their corresponding HR counterparts, achieving good results without the need for the manual features required by traditional methods. Kim et al. [10] proposed a 20-layer network for predicting residual images, and they verified that the SR model performance improves significantly when the number of structure layers is increased. Furthermore, Lim et al. [11] expanded the network to 69 layers by stacking more residual blocks, and this uses more features from each convolution layer to restore the image. Zhang et al. [12] designed a network using more than 400 layers, and this achieved obvious improvements for SISR by embedding a channel attention mechanism (CAM) [15] module into the residual block. Inspired by [9], Zeng et al. [14] employed two autoencoders to automatically extract hidden representations in LR and HR image patches. These methods have obtained promising results in SISR tasks however, there are still some limitations among CNN-based methods for the task of remote sensing SR reconstruction.

Firstly, the depth of the CNNs is important for image SR however, deeper networks are more difficult to train and require much greater computing resources. Moreover, this may result in the SR effect becoming saturated or even degraded, which illustrates that it is crucial to design a rational and efficient network that has a good balance between SR quality and model complexity.

Secondly, remote sensing images are more complex in terms of the spatial distribution of features and are richer in detailed information than natural images; moreover, the objects in remote sensing images have a relatively wide range of scales, which results in a requirement for the model to have a high restoration ability in high-frequency regions [16]. However, most existing CNN-based methods ignore the differing importance of different spatial areas, and this hinders the recovery of high-frequency information.

Thirdly, as the depth of a CNN increases, the feature information obtained in the different convolutional layers will be hierarchical in different receptive fields. Traditionally, a small-sized convolution kernel can extract low-frequency information, but this is not sufficient for the extraction of more detailed information. The work of [17] shows that applying convolutional layers with different receptive fields in the same layer can ensure the acquisition of low-frequency and high-frequency details of the source image. Therefore, the selection of suitable of receptive field and better utilization of hierarchical features should be considered when designing an SR network.

To address the urgent issues noted above, we propose a novel remote sensing SR image reconstruction network called a pyramid information distillation attention network (PIDAN), which includes a carefully designed pyramid information distillation attention block (PIDAB) that was inspired by information distillation networks (IDNs) [18]. An IDN reduces the network parameters by compressing the dimensions of its feature map, which increases the speed of processing while guaranteeing the restoration results. However, the ability of an IDN to differentially exploit different locations and channel features is still insufficient [19], which limits the further improvement of SR performance. Considering this, the PIDAB adopts a strategy of feature distillation, and its structure combines a pyramid convolution block and an attention mechanism.

A PIDAN consists of a shallow feature-extraction part, several PIDABs, and a reconstruction part. Each PIDAB is a single deep feature-extraction unit, and this contains a pyramid information distillation (PID) module, a hybrid attention mechanism (HAM) module, and a single channel compression (CC) unit. The PID can extract both deep and shallow features, and the HAM can restore high-frequency detailed information. The PID module utilizes an enhancement unit (EU) and a pyramid convolution channel split (PCCS) operation to gradually integrate the local short- and long-path features for reconstruction. The EU can be divided into two levels according to the inference order. In the first level, we use a shallow convolution network to obtain local short-path features. After the first level, the PCCS extracts the refined features by using convolution layers with different receptive fields in parallel. Then, a split operation is placed after each convolution layer, and this divides the feature channel into two parts: One for further enhancement in the second level to obtain long-path features, and another to represent reserved short-path features. In the second level of the EU, the HAM utilizes the short-path feature information by fusing a CAM and a spatial attention mechanism (SAM). Specifically, unlike the structure of a convolutional block attention module (CBAM) [20], in which the spatial feature descriptors are generated along the channel axis, our CAM and SAM are parallel branches that operate on the input features simultaneously. Finally, the CC unit is used for achieving a reduction of the channel dimensionality by taking advantage of a 1×1 convolution layer, as used in an IDN.

In summary, the main contributions of this work are as follows:

(1) Inspired by IDNs, we constructed an effective and convenient end-to-end trainable architecture, PIDAN, which is designed for SR reconstruction of remote sensing images. Our PIDAN structure consists of a shallow feature-extraction part, stacked PIDABs, and a reconstruction part. Compared with an IDN, a PIDAN recovers more high-frequency information.

(2) Specifically, we propose the PIDAB, which is composed of a PID module, a HAM module, and a single CC unit. Firstly, the PID module uses an EU and a PCCS operation to gradually integrate the local short- and long-path features for reconstruction. Secondly, the HAM utilizes the short-path feature information by fusing a CAM and SAM in parallel. Finally, the CC unit is used for achieving channel dimensionality reduction.

(3) We compared our PIDAN with other advanced SISR approaches using remote sensing datasets. The extensive experimental results demonstrate that the PIDAN achieves a better balance between SR performance and model complexity than the other approaches.

The remainder of this paper is organized as follows. Section 2 introduces previous works on CNN-based SR reconstruction algorithms and attention mechanism methods. Section 3 presents a detailed description of the PIDAN, Section 4 presents a verification of its effectiveness by experimental comparisons, and Section 5 concludes our work.

2. Related Works

2.1. CNN-Based SR Methods

The basic principle of SR methods based on deep learning technology is to establish a nonlinear end-to-end mapping relationship between an input and output through a multi-layer CNN. Dong et al. [9] were the first to apply a CNN to the image SR task, producing a system named SRCNN. This uses a bicubic interpolation operation to enlarge an LR image to the target size, then it fits the nonlinear mapping using three convolution layers before finally outputting an HR image. The SRCNN system provides great improvement in the SR quality when compared with traditional algorithms, but its training speed is very low. Soon after this, Dong et al. [21] reported the Faster-SRCNN, which increases the speed of SRCNN by adding a deconvolution layer. Inspired by [9], Zeng et al. [14] developed a data-driven model named, coupled deep autoencoder (CDA), which automatically learns the intrinsic representations of LR and HR image patches by employing two autoencoders. Shi et al. [22] investigated how to directly input an LR image into the network and developed the efficient sub-pixel convolutional neural network (ESPCN), which reduces the computational effort of the network by enlarging the image through the sub-pixel convolution layer, and this improves the training speed exponentially. The network structures of the above algorithms are simple and easy to implement. However, due to the use of a large convolution kernel, even a shallow network requires the calculation of a large number of parameters. Training is therefore difficult when the network is deepened and widened, and the SR reconstruction is thus not effective.

To reduce the difficulty of model training, Kim et al. [10] deepened the network to 20 layers using a residual-learning strategy [23]; their experimental results demonstrated that the deeper the network, the better the SR effect. Then, Kim et al. [24] proposed a deeply recursive convolutional network (DRCN), which applies recursive supervision to make the deep network easier to train. Based on DRCN, Tai et al. [25] developed a deep recursive residual network (DRRN), which introduces recursive learning into the residual branch, and this deepens the network without increasing computational effort and speeds up the convergence. Lai et al. proposed the deep Laplacian super-resolution network (LapSRN) [26], which predicts the sub-band residuals in a coarse-to-fine fashion. Tong et al. [27] employed the dense connected convolutional networks, which allows the reuse of feature maps from preceding layers, and alleviates the gradient vanishing problem by facilitating the information flow in the network. Zhang et al. [28] proposed a deep residual dense network (RDN), which combines the residual skip structure with the dense connections, and this fully utilizes the hierarchical features. Lim et al. [11] built an enhanced deep SR network (EDSR), which constructs a deeper CNN by stacking more residual blocks, and this takes more features from each convolution layer to restore the image. The EDSR expanded the network to 69 layers and won the NTIRE 2017 SR challenge. Yu et al. [29] proposed a wide activation SR (WDSR) network, which shows that simply expanding features before the rectified linear unit (ReLU) activation results in obvious improvements for SISR. Based on EDSR, Zhang et al. [12] built a deep residual channel attention network (RCAN) with more than 400 layers, and this achieves promising results by embedding the channel attention [15] module into the residual block. It is noteworthy that while increasing the network's depth may improve the SR effect, it also increases the computational complexity and memory consumption of the network, which makes it difficult to apply these methods to lightweight scenarios such as mobile terminals.

Considering this issue, many researchers have focused on finding a better balance between SR performance and model complexity when designing a CNN. Ahn et al. [30] proposed a cascading residual network (CARN), which was designed to be a high-performing SR model that implements a cascading mechanism to fuse multi-layer feature information. The IDN, which is a concise but effective SR network, was proposed by Hui et al. [18], and this uses a distillation module to gradually extract a large number of valid features. Profiting from this information distillation strategy, IDN achieves good performance at a moderate size. However, IDN treats different channel and spatial areas equally in LR feature space, and this restricts its feature representation ability.

2.2. Attention Mechanisms

For human perception, attention usually refers to the human visual system focusing on salient regions and adaptively processing visual information. Recently, many visual recognition tasks have tended to embed attention modules with networks to improve their performance. Hu et al. [15] proposed the squeeze-and-excitation network (SENet), which captures feature relationships by explicitly modeling interdependencies between channels. This ranked first in the ILSVRC 2017 classification competition. Motivated by SENet, Woo et al. [20] created the CBAM, which includes a SAM that can adaptively allocate weights in different spatial locations. Using the classical non-local means method [31], Wang et al. [32] developed a non-local (NL) block that can be plugged into a neural network. This uses a self-attention mechanism to directly model long-range dependencies instead of adopting multiple convolutions to obtain feature information with a larger receptive field. The NL block can thus provide rich semantic information for a network. Cao et al. [33] developed a global context block, which combines the simplified NL block and the squeeze-and-excitation (SE) block of SENet to reduce the computational effort while making full use of global contextual information.

Recently, several works have focused on introducing attention mechanisms to the SISR task. Inspired by SENet [15], Zhang et al. [12] produced the RCAN, which enhances the representation ability by using the channel attention mechanism to differentially treat the feature channels in each layer so that the reconstructed image contains more texture information. Zhang et al. [34] built a very deep residual non-local attention network, which includes residual local and non-local attention blocks as the basic building modules. This improves the local and non-local information learning ability using the hierarchical features. Anwar et al. [35] proposed a densely residual Laplacian network, which replaces the CAM with a proposed Laplacian module to learn features at multiple sub-band frequencies. Guo et al. [36] proposed a novel image SR approach named the multi-view aware attention network. This applies locally and globally aware attention to unequally deal with LR images. Dai et al. [37] proposed a deep second-order attention network, in which a second-order channel attention mechanism captures feature inter-dependencies by using second-order feature statistics. Hui et al. [38] proposed a contrast-aware channel attention mechanism, and this is particularly suited to low-level vision tasks such as image SR and image enhancement. Zhao et al. [39] proposed a pixel attention mechanism, which generates three-dimensional attention maps instead of a one-dimensional vector or a two-dimensional map, and this achieves better SR results with fewer additional parameters. Wang et al. [40] built a spatial pyramid pooling attention module via integrating the channel-wise and multi-scale spatial information, which is beneficial for capturing spatial context cues and then establishing the accurate mapping from low-dimension space to high-dimension space.

Considering that the previous promising results have benefited from the introduction of an attention mechanism, we propose PIDAN, which also includes an attention mechanism, to focus on extracting high-frequency details from images.

3. Methodology

In this section, we will describe PIDAN in detail. An overall graphical depiction of PIDAN is shown in Figure 1. Firstly, we will give an overview of the proposed network architecture. After this, we will present each module of the PIDAB in detail. Finally, we will give the loss function used in the training process. Here, we denote an initial LR input image and an SR output image as I_{LR} and I_{SR}, respectively.

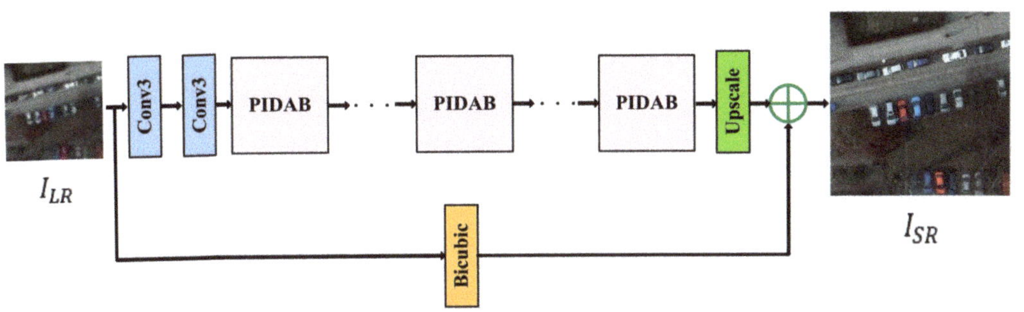

Figure 1. Overview of the PIDAN network structure.

3.1. Network Architecture

As shown in Figure 1, the PIDAN approach consists of a shallow feature-extraction part, a deep feature-extraction part (stacked PIDABs), and a reconstruction part. As with the operation of an IDN, the shallow features F_0 are extracted from the LR input via two convolutional layers:

$$F_0 = H_{\mathrm{SF}}(I_{\mathrm{LR}}), \tag{1}$$

where $H_{\mathrm{SF}}(\cdot)$ denotes two convolutional layers with a kernel size of 3×3 to extract C initial feature maps. The resulting F_0 contributes to the next deep feature-extraction part using the PIDABs. Moreover, the proposed PIDAB can be regarded as a basic component for residual feature extraction. The operation of the n-th PIDAB can be defined as:

$$F_{b,n} = H_{\text{PIDAB},n}(F_{b,n-1}), \tag{2}$$

where $H_{\text{PIDAB},n}(\cdot)$ denotes the function of the n-th PIDAB, and $F_{b,n-1}$ and $F_{b,n}$ are the inputs and outputs of the n-th PIDAB, respectively.

After obtaining the deep features of the LR images, an up-sampling operation aims to project these features into the HR space. Previous approaches, such as EDSR [11], RCAN [12], and the information multi-distillation network (IMDN) [38] have shown that a sub-pixel [22] convolution operation can reserve more parameters and achieve a better SR effect than other up-sampling approaches. Considering this, we used a transition layer with a 3 × 3 kernel and a sub-pixel convolution layer as our reconstruction part. This operator can be expressed as:

$$F_{\text{up}} = H_{\text{subpixel}}(H_{\text{A}}(F_{b,N})), \tag{3}$$

where $H_{\text{A}}(\cdot)$ denotes a convolutional layer with a convolution kernel size of 3 × 3, $H_{\text{subpixel}}(\cdot)$ denotes a sub-pixel convolution, $F_{b,N}$ is the output of the last PIDAB, and F_{up} is the upscaled feature maps.

Finally, using the idea of global residual learning [23], the output of the PIDAN I_{SR} is estimated by combining the up-sampled image F_{up} with the interpolated image using an element-wise summation. This can be formulated as:

$$I_{\text{SR}} = F_{\text{up}} + H_{\text{bicubic}}(I_{\text{LR}}), \tag{4}$$

where $H_{\text{bicubic}}(\cdot)$ denotes the bicubic interpolation operation.

3.2. PIDAB

In this section, we will present a description of the overall structure using a PIDAB. Figure 2 compares the PIDAB with the original IDB in an IDN. As noted, the PIDAB was developed using a PID module, a HAM module, and a CC unit. The PID module can extract both deep and shallow features, and the HAM module can restore high-frequency detailed information.

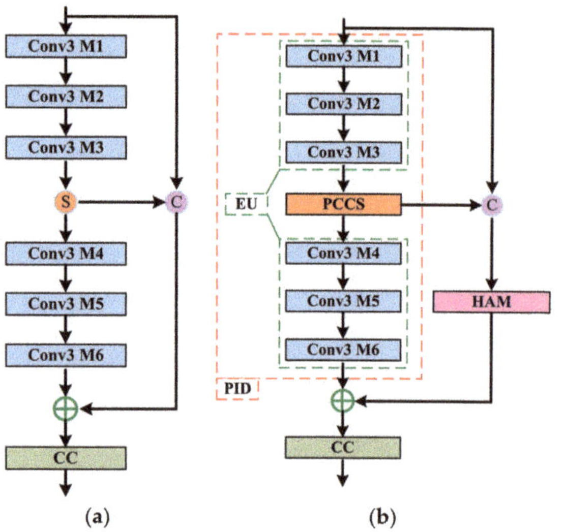

Figure 2. Illustrations of (**a**) original IDB structure of an IDN and (**b**) the PIDAB structure in a PIDAN.

3.2.1. PID Module

As shown in Figure 2b, the PID module consists of two parts: An EU and a PCCS component. The EU can be roughly divided into two modules, the upper shallow convolution network and the lower shallow convolution network. Each module has three cascaded convolutional layers with a convolution kernel size of 3 × 3; each of these is followed by a leaky rectified linear unit (LReLU) activation function, which is omitted here. We label the feature map dimensions of the i-th layer as M_i $(i = 1, \cdots, 6)$, and the relationship among the upper three convolutions can be formulated as:

$$M_3 - M_1 = M_1 - M_2 = m, \tag{5}$$

where m denotes the difference between the first layer and second layer or between the first layer and third layer. Simultaneously, the relationship among the lower three convolution layers can be described as:

$$M_4 - M_5 = M_6 - M_4 = m, \tag{6}$$

where $M_4 = M_3$. Supposing the input of this module is $F_{b,n-1}$, we have:

$$P_1^n = C_a(F_{b,n-1}), \tag{7}$$

where $F_{b,n-1}$ denotes the output of the $(n-1)$-th PIDAB (which is also the input of the n-th PIDAB), $C_a(\cdot)$ denotes the upper shallow convolution network in the enhancement unit, and P_1^n denotes the output of the upper shallow convolution network in the n-th PIDAB.

As shown in Figure 2a, in the original IDN, the output of the upper cascaded convolutional layers is split into two parts: One for further enhancement in the lower shallow convolution network to obtain the long-path features, and another to represent reserved short-path features via concatenation with the input of the current block. In PIDAN, to obtain more non-redundant and extensive feature information, a feature-purification component with parallel structures was designed.

The convolutional layers in the CNN can extract local features from a source image by automatically learning convolutional kernel weights during the training process. Therefore, choosing an appropriate size of convolution kernel is crucial for feature extraction. Traditionally, a small-sized convolution kernel can extract low-frequency information, but this is not sufficient for the extraction of more detailed information. Considering this, the PCCS component is proposed to extract the features of multiple receptive fields. In the pyramid structure, the size of the convolution kernel of each parallel branch is different, which allows the network to perceive a wider range of hierarchical features. As presented in Figure 3, the PCCS component is built from three parallel feature-purification branches and two feature-fusion operations.

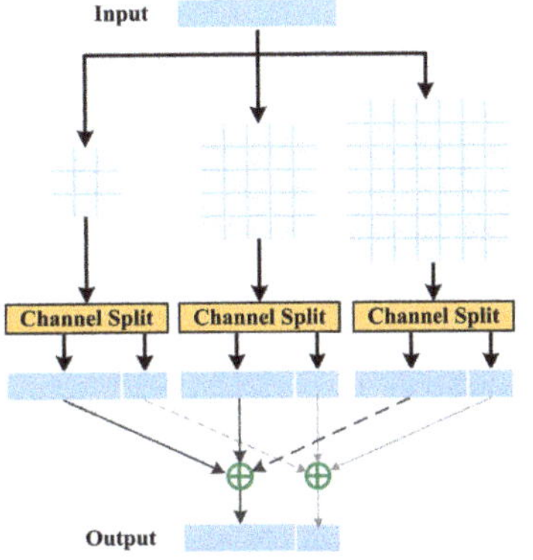

Figure 3. Structure of PCCS component.

For a PCCS component, assuming that the given input feature map is $P_1^n \in R^{C \times W \times H}$, the pyramid convolution layer operation is applied to the extraction of refined features with different kernel sizes. The split operation is performed after each feature-refinement branch, and this can split the channel into two parts. The process can be formulated as:

$$F^n_{\text{distilled\_1}}, F^n_{\text{remaining\_1}} = \text{Split}(\text{CL}^3_1(P^n_1)), \tag{8}$$

$$F^n_{\text{distilled\_2}}, F^n_{\text{remaining\_2}} = \text{Split}(\text{CL}^5_2(P^n_1)), \tag{9}$$

$$F^n_{\text{distilled\_3}}, F^n_{\text{remianing\_3}} = \text{Split}(\text{CL}^7_3(P^n_1)), \tag{10}$$

where: $\text{CL}^k_j(\cdot)$ denotes the j-th convolution layer (including an LReLU activation unit) with a convolution kernel size of $k \times k$; $\text{Split}(\cdot)$ denotes a channel-splitting operation similar to that used in an IDN; and $F^n_{\text{distilled\_}j}$ denotes the j-th distilled features; $F^n_{\text{remaining\_}j}$ denotes the j-th coarse features that will be further processed by the lower shallow convolution network in the n-th PIDAB, specifically, the number of channels of $F^n_{\text{distilled\_}j}$ is defined as $\frac{C}{s}$, therefore the number of channels of $F^n_{\text{remianing\_}j}$ is set to $\left(c - \frac{C}{s}\right)$.

All the distilled features and remaining features are then respectively added together:

$$F^n_{\text{distilled}} = F^n_{\text{distilled\_1}} + F^n_{\text{distilled\_2}} + F^n_{\text{distilled\_3}}, \tag{11}$$

$$F^n_{\text{remaining}} = F^n_{\text{remaining\_1}} + F^n_{\text{remaining\_2}} + F^n_{\text{remianing\_3}}. \tag{12}$$

Then, as shown in Figure 2b, $F^n_{\text{distilled}}$ will be concatenated with the input of the current PIDAB to obtain the retained short-path features:

$$R^n = f_{\text{concat}}(F^n_{\text{distilled}}, F_{b,n-1}), \tag{13}$$

where $f_{\text{concat}}(\cdot)$ denotes the concatenation operator, and R^n denotes partially retained local short-path information. We take $F^n_{\text{remaining}}$ as the input of the lower shallow convolution network, which obtains the long-path feature information:

$$P^n_2 = C_b(F^n_{\text{remaining}}), \tag{14}$$

where P^n_2 and $C_b(\cdot)$ denote the output and cascaded convolution layer operations of the lower shallow convolution network, respectively. As shown in Figure 2a, in the initial IDB structure of an IDN, the reserved local short-path information and the long-path information are summed before the CC unit. In PIDAN, to fully utilize the local short-path feature information, we embed an attention mechanism module to enable the network to focus on more useful high-frequency feature information and improve the SR effect. Therefore, before the CC unit, the fusion of short-path and long-path feature information can be formulated as:

$$P^n = P^n_2 + \text{HAM}(R^n), \tag{15}$$

where $\text{HAM}(\cdot)$ denotes the hybrid attention mechanism operation, which will be illustrated in detail in the next subsection.

3.2.2. HAM Module

In an IDN, the information distillation module is used to gradually extract a large number of valid features, and the intention of the channel-split operation is to combine short- and long-path hierarchical information. However, an IDN treats different channels and spatial areas equally in LR feature space, which restricts the feature representation ability of the network. Moreover, if sufficient features are not extracted in the short path, information learned later will also become inadequate. Considering that an attention mechanism can make a network pay more attention to high-frequency information, which is beneficial for the SR reconstruction task, we further utilize the extracted short-path features by fusing a CAM and SAM to construct a HAM, which makes the split operation yield better performance. Specifically, unlike the structure of a CBAM [20], in which the spatial feature descriptors are generated along the channel axis, our SAM and CAM are parallel branches that operate on the input features simultaneously. In this way, our HAM makes maximum use of the attention mechanism through self-optimization and mutual optimization of the channel and spatial attention during the gradient back-propagation process. The formula of the HAM is:

$$\mathrm{HAMF}(F) = \mathrm{CAM}(F) \otimes \mathrm{SAMF}(F) + F, \tag{16}$$

where: F denotes the input of the HAM; and CAM(·), SAM(·), and HAM(·) respectively denote the CAM, SAM, and HAM functions. Here $\otimes$ denotes element-wise multiplication between the CAM and SAM functions. Like an RCAN, short-skip connections are added to enable the network to directly learn more complex high-frequency information while improving the ease of model training. The structure of the HAM is presented in Figure 4.

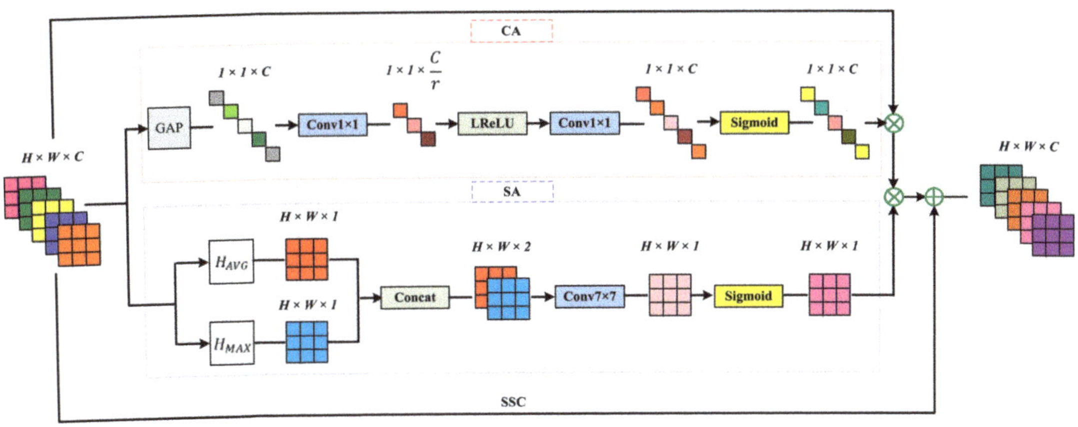

Figure 4. Overview of the HAM.

Channel Attention Mechanism

The high performance of CNNs for feature extraction has been demonstrated however, the standard convolution kernel treats different channels equally and is restricted by its convolutional calculation being translation invariant. This makes it difficult for the network to use contextual information to effectively learn features. A previous report has shown that the attention mechanism can help capture channel correlations between features [15]. In PIDAN, by following RCAN [12], we consider channel-wise information by using the global pooling average operation, which can transform the information in the global space into channel descriptors.

Suppose the input features F have C channels with size $H \times W$ (as shown in Figure 4). The global average pooling operation is adopted to obtain the channel descriptor (one-dimensional feature vector) of each feature map:

$$\mathrm{GAP}(C,1,1) = \frac{1}{H \times W} \sum_{i=1}^{H} \sum_{j=1}^{W} F(C,H,W). \tag{17}$$

After the pooling operation, we use a similar perceptron network as that used in a CBAM [20] to fully learn the nonlinear interactions between different channels. Specifically, we replace ReLU with LReLU activation. The calculation process of the CAM can be described:

$$\mathrm{CAM}(F) = \mathrm{Sigmoid}[W_U^{1\times1}(\mathrm{LReLU}(W_D^{1\times1}(\mathrm{GAP}(F))))] \otimes F, \tag{18}$$

where: $W_D^{1\times1}$ and $W_U^{1\times1}$ denote the weight matrices of two convolution layers with a kernel size of 1×1, in which the channel dimensions of the features are defined as C/r and C, respectively; SIGMOID[$\cdot$] and LReLU($\cdot$) denote the sigmoid and LReLU functions, respectively; and $\otimes$ denotes element-wise multiplication.

Spatial Attention Mechanism

Generally, the LR images have rich low-frequency information and valuable high-frequency information components. The difference between low-frequency information and high-frequency information is that the former is generally flat, while the latter is usually filled with edges, textures, and details in certain areas. Compared to low-frequency information, high-frequency information is usually more difficult to restore in the image SR task. Moreover, remote sensing images are more complex in their spatial distribution and richer in detailed information than natural images, which means that the designed SR network needs to show adequate perception of the high-frequency information regions. However, existing CNN-based algorithms usually ignore the variability of different spatial locations, and this tends to weaken the weight of high-frequency information. Considering this, in PIDAN, the SAM is designed to emphasize the attention to high-frequency areas, thus improving the accuracy of the SR algorithm.

As shown in Figure 4, we produce two efficient two-dimensional spatial feature descriptors by performing average-pooling and max-pooling operations:

$$\mathrm{AvgPool}(1, H, W) = \frac{1}{C}\sum_{k=1}^{C} F(C, H, W), \tag{19}$$

$$\mathrm{MaxPool}(1, H, W) = \max_{k=\{1,\cdots,k,\cdots,C\}} F(C, H, W). \tag{20}$$

These two spatial feature descriptors are then concatenated and convolved by a standard convolution layer, producing the spatial attention map. The calculation process of the SAM can be described as:

$$\mathrm{SAM}(F) = \mathrm{Sigmoid}[W_C^{7\times7}(\mathrm{Concat}(\mathrm{AvgPool}(F),\ \mathrm{MaxPool}(F)))], \tag{21}$$

where: Concat($\cdot$) denotes the feature-map concatenation operation; $W_C^{7\times7}(\cdot)$ denotes the weight matrix of a convolution layer with a kernel size of 7×7, which reduces the channel dimensions of the spatial feature maps to one; Sigmoid[$\cdot$] denotes the sigmoid function; and $\otimes$ denotes element-wise multiplication.

3.2.3. CC Unit

We realize the channel dimensionality reduction by taking advantage of a 1×1 convolution layer. Thus, the compression unit can be expressed as:

$$F_{b,n} = W_{\mathrm{CU}}^{1\times1}(P^n), \tag{22}$$

where: P^n denotes the result of the fusion of short- and long-path feature information in the n-th PIDAB; $F_{b,n}$ denotes the output of the n-th PIDAB; and $W_{\mathrm{CU}}^{1\times1}\otimes$ denotes the weight matrix of a convolution layer with a kernel size of 1×1, which compresses the number of channels of features to be consistent with the input of the n-th PIDAB.

Table 1 presents the network structure parameter settings of a PIDAB. It should be noted that: C is defined as 64 in line with an IDN; in the PID module, we set m as 16, and we define s as 4; and in the HAM module, the reduction ratio r is set as 16, consistent with an RCAN.

Table 1. PIDAB block parameter settings.

Structure Component	Layer	Input	Output
M1	Conv3 × 3	$H \times W \times 64$	$H \times W \times 48$
M2	Conv3 × 3	$H \times W \times 48$	$H \times W \times 32$
M3	Conv3 × 3	$H \times W \times 32$	$H \times W \times 64$
PCCS	Conv3 × 3	$H \times W \times 64$	$H \times W \times 64$
	Split	$H \times W \times 64$	$H \times W \times 48, H \times W \times 16$
	Conv5 × 5	$H \times W \times 64$	$H \times W \times 64$
	Split	$H \times W \times 64$	$H \times W \times 48, H \times W \times 16$
	Conv7 × 7	$H \times W \times 64$	$H \times W \times 64$
	Split	$H \times W \times 64$	$H \times W \times 48, H \times W \times 16$
	Sum	$H \times W \times 48, H \times W \times 48, H \times W \times 48$	$H \times W \times 48$
	Sum	$H \times W \times 16, H \times W \times 16, H \times W \times 16$	$H \times W \times 16$
	Concat	$H \times W \times 64, H \times W \times 16$	$H \times W \times 80$
HAM	GAP	$H \times W \times 80$	$1 \times 1 \times 80$
	Conv1 × 1	$1 \times 1 \times 80$	$1 \times 1 \times 5$
	Conv1 × 1	$1 \times 1 \times 5$	$1 \times 1 \times 80$
	Multiple	$H \times W \times 80, 1 \times 1 \times 80$	$H \times W \times 80$
	AvgPool	$H \times W \times 80$	$H \times W \times 1$
	MaxPool	$H \times W \times 80$	$H \times W \times 1$
	Concat	$H \times W \times 1, H \times W \times 1$	$H \times W \times 2$
	Conv7 × 7	$H \times W \times 2$	$H \times W \times 1$
	Multiple	$H \times W \times 80, H \times W \times 1$	$H \times W \times 80$
	Sum	$H \times W \times 80, H \times W \times 80, H \times W \times 80$	$H \times W \times 80$
M4	Conv3 × 3	$H \times W \times 48$	$H \times W \times 64$
M5	Conv3 × 3	$H \times W \times 64$	$H \times W \times 48$
M6	Conv3 × 3	$H \times W \times 48$	$H \times W \times 80$
	Sum	$H \times W \times 80, H \times W \times 80$	$H \times W \times 80$
CC unit	Conv1 × 1	$H \times W \times 80$	$H \times W \times 64$

3.3. Loss Function

In our approach, the gradient is updated by minimizing the difference between the reconstruction result and the real image. The loss function is one of the key factors affecting the performance of the network, and there are two commonly used loss functions in CNN-based SR algorithms, namely the $L1$ norm [11,18] and $L2$ norm [27]. Compared to the $L2$ norm, the $L1$ norm loss function tends to perceive more high-frequency detailed information and results in higher-quality test metrics. In line with the IDN approach [18], the minimum loss function was formulated as:

$$L(\Theta) = \frac{1}{N}\sum_{i=1}^{N} \|H_{\mathrm{PIDAN}}(Y_i;\Theta) - X_i\|_1, \quad (23)$$

where: N denotes the number of input images; $H_{\mathrm{PIDAN}}(\cdot)$ denotes the PIDAN network reconstruction process; Y_i denotes the reconstructed image; $\Theta = \{W_i, b_i\}$, which denote the weight and bias parameters that the network needs to learn; X_i denotes the corresponding HR image; and $\|\cdot\|_1$ denotes the $L1$ norm.

4. Experiments and Results

In this section, firstly, we demonstrate the experimental settings, including datasets, evaluation metrics, and training implementation details. Then, we report the experimental results and correlation analysis.

4.1. Settings

4.1.1. Dataset Settings

Following the previous work [41], we used the recently popular Aerial Image Dataset (AID) [42] for training. We augmented our training dataset using horizontal flipping, vertical flipping, and 90° rotation strategies. During the tests, to evaluate the trained SR model, we used two available remote sensing image datasets, namely, the NWPU VHR-10 [43] dataset and the Cars Overhead With Context (COWC) [44] dataset. In our experiments, the AID, NWPU VHR-10, and COWC datasets consisted of 10,000, 650, and 3000 images, respectively. Specifically, for the fast validation of the convergence speed of SR models, we constructed a new data set called FastTest10, which consists of 10 randomly selected samples from the NWPU VHR-10 dataset. The LR images were obtained by downsampling the corresponding HR label samples through bicubic interpolation with $\times 2$, $\times 3$, and $\times 4$ scale factors. Some examples from each of these remote sensing datasets are shown in Figure 5.

Figure 5. Examples of images in the three remote sensing datasets. In order, the top–bottom lines show samples from the AID, NWPU VHR-10, and COWC datasets.

4.1.2. Evaluation Metrics

We adopted the average peak signal-to-noise ratio (PSNR) [45] and structural similarity (SSIM) [46] as the SR reconstruction evaluation metrics. The PSNR measures the quality of an image by calculating the difference in pixel values between the reconstructed image and original HR image. The PSNR indicator mainly judges the similarity of the images from the perspective of the signal, and it is not completely consistent with human visual perception. Therefore, the SSIM was adopted because it models image distortion as a combination of three factors—luminance, contrast, and structure—so as to estimate the degree of similarity between two images from the perspective of overall image composition. Larger PSNR and SSIM values indicate a better SR image reconstruction result that is closer to the original image. Following the previous work in this field [9], SR is only performed on the luminance (Y) channel of the transformed YCbCr space.

4.1.3. Implementation Details

All experiments adopted the deep-learning framework PyTorch, and four Nvidia GTX-2080Ti GPUs were used to train all CNN models. The SR network was optimized with Adam [47] by setting $\beta_1 = 0.9$, $\beta_2 = 0.999$, and $\epsilon = 10^{-8}$. We set the initial learning rate to 10^{-4}, and this was decreased by a factor of 10 after every 500 epochs. The training for PIDAN was iterated for 1500 epochs in total. The batch size was set to 16. Patches with a size of 48 × 48 were randomly cropped from LR images as the input of the model, and the corresponding input HR label images were divided into 96 × 96, 144 × 144, and 192 × 192 sizes according to upscaling factors of ×2, ×3, and ×4, respectively.

4.2. Results and Analysis

4.2.1. Comparison with Other Approaches

We compared our PIDAN with the bicubic interpolation, SRCNN [9], very deep super resolution (VDSR) [10], LapSRN [26], DRCN [24], pixel attention network (PAN) [39], DRRN [25], WDSR [29], CARN [30], residual feature distillation network (RFDN) [48], IDN [18], and IMDN [38] approaches. Specifically, for a fair comparison, the number of PIDABs was set to four in line with the IDN approach. Table 2 shows quantitative comparisons using the NWPU VHR-10 and COWC datasets. The best performances are indicated in bold, and the second-best performances are indicated with an underline. Our PIDAN performed better than all other approaches in most datasets with upscaling factors of ×2, ×3, and ×4.

Table 2. Quantitative evaluation of PIDAN and other advanced SISR approaches. Bold indicates the optimal performance, and an underline indicates the second-best performance.

Method	NWPU VHR-10 PSNR/SSIM ×2	NWPU VHR-10 PSNR/SSIM ×3	NWPU VHR-10 PSNR/SSIM ×4	COWC PSNR/SSIM ×2	COWC PSNR/SSIM ×3	COWC PSNR/SSIM ×4
Bicubic	32.76031/0.8991	29.90444/0.8167	28.28280/0.7524	32.87844/0.9180	29.53540/0.8384	27.72172/0.7725
SRCNN	34.03260/0.9136	30.97869/0.8400	29.20195/0.7793	35.05635/0.9341	31.14172/0.8661	28.99814/0.8058
VDSR	34.46067/0.9196	31.46934/0.8517	29.62497/0.7931	35.81885/0.9401	31.89712/0.8788	29.62051/0.8220
LapSRN	34.24569/0.9169	31.26756/0.8468	29.67748/0.7942	35.48608/0.9375	31.62203/0.8741	29.70046/0.8236
DRCN	34.36621/0.9181	31.31746/0.8476	29.51012/0.7887	35.65558/0.9387	31.67424/0.8751	29.46399/0.8180
PAN	34.48577/0.9199	31.53275/0.8529	29.75737/0.7967	35.86121/0.9403	31.98120/0.8800	29.80853/0.8262
DRRN	34.57956/0.9213	31.59945/0.8548	29.85024/0.8002	36.01337/0.9417	32.08846/0.8820	29.85881/0.8272
WDSR	34.56984/0.9210	31.65636/0.8558	29.87613/0.8003	36.01360/0.9416	32.17758/0.8832	**30.00641/0.8305**
CARN	34.54988/0.9208	31.59971/0.8545	29.83102/0.7990	35.97727/0.9413	32.07578/0.8817	29.93067/0.8289
RFDN	34.55302/0.9207	31.61688/0.8548	29.81638/0.7984	35.99849/0.9413	32.14530/0.8826	29.91353/0.8285
IDN	34.56317/0.9210	31.61978/0.8550	29.83245/0.7989	35.99732/0.9415	32.12127/0.8823	29.92513/0.8286
IMDN	34.55570/0.9207	31.62651/0.8549	29.81952/0.7984	36.02204/0.9415	32.17454/0.8829	29.95087/0.8291
PIDAN	**34.59635/0.9215**	**31.66433/0.8559**	**29.87914/0.8005**	**36.09257/0.9423**	**32.23239/0.8840**	30.00399/0.8303

We take the NWPU VHR-10 dataset as an example. Compared with other SISR approaches, the PIDAN produces superior PSNR and SSIM values. Under the SR upscaling factor of ×2, the PSNR of the PIDAN is 0.01679 dB higher than that obtained with the second-best DRRN method and 0.03318 dB higher than that of the basic IDN; the SSIM of the PIDAN is 0.0002 higher than that obtained with the second-best DRRN method and 0.0005 higher than that of the IDN. Under the SR upscaling factor of ×3, the PSNR of the PIDAN is 0.00797 dB higher than that of the second-best WDSR method and 0.04455 dB than that of the IDN; the SSIM of the PIDAN is 0.0002 higher than that of the second-best WDSR method and 0.0009 higher than that of the IDN. Under the SR upscaling factor of ×4, the PSNR of the PIDAN is 0.00301 dB higher than that of the second-best WDSR method and 0.04669 dB than that of the IDN; the SSIM of the PIDAN is 0.0002 higher than that of the WDSR method and 0.0006 higher than that of the IDN.

We take the NWPU VHR-10 dataset as an example. Compared with other SISR approaches, the PIDAN produces superior PSNR and SSIM values. Under the SR upscaling factor of ×2, the PSNR of the PIDAN is 0.01679 dB higher than that obtained with the second-best DRRN method and 0.03318 dB higher than that of the basic IDN; the SSIM of the PIDAN is 0.0002 higher than that obtained with the second-best DRRN method and 0.0005 higher than that of the IDN. Under the SR upscaling factor of ×3, the PSNR of the PIDAN is 0.00797 dB higher than that of the second-best WDSR method and 0.04455 dB than that of the IDN; the SSIM of the PIDAN is 0.0002 higher than that of the second-best WDSR method and 0.0009 higher than that of the IDN. Under the SR upscaling factor of ×4, the PSNR of the PIDAN is 0.00301 dB higher than that of the second-best WDSR method and 0.04669 dB than that of the IDN; the SSIM of the PIDAN is 0.0002 higher than that of the WDSR method and 0.0006 higher than that of the IDN.

Next, we consider the COWC dataset as an example. Under the SR upscaling factor of ×2, the PSNR of the PIDAN is 0.07053 dB higher than that obtained with the second-best IMDN method and 0.09525 dB higher than that of the basic IDN; the SSIM of the PIDAN is 0.0006 higher than that obtained with the second-best DRRN method and 0.0008 higher than that of the IDN. Under the SR upscaling factor of ×3, the PSNR of the PIDAN is 0.05481 dB higher than that of the second-best WDSR method and 0.11112 dB higher than that of the IDN; the SSIM of the PIDAN is 0.0008 higher than that of the second-best WDSR method and 0.0017 higher than that of the IDN. Under the SR upscaling factor of ×4, the PSNR and SSIM of the PIDAN are both second-best, and the PSNR of the PIDAN is 0.00242 dB lower than that of the optimal WDSR method and 0.07886 dB higher than that of the IDN; the SSIM of the PIDAN is 0.0002 lower than that of the optimal WDSR method and 0.0017 higher than that of the IDN.

Figure 6 shows a comparison of the PSNR values between the PIDAN and DRRN, WDSR, CARN, RFDN, IDN, and IMDN networks using the FastTest10 dataset in the epoch range of 0 to 100. Compared to the other methods, the PIDAN converges faster and achieves better accuracy.

4.2.2. Model Size Analyses

We compared the model sizes of our PIDAN with other DCNN-based approaches. The results of an upscaling factor of ×2 SR on the COWC test set are shown in Figure 7. The x axis denotes the SR model size, with M indicating the number of parameters in millions, and the y axis denoting the average PSNR score. It can be concluded that our proposed PIDAN achieves an optimal PSNR score with a model parameter that is less than one-third of that of DRRN. This finding demonstrates that our PIDAN is relatively lightweight while ensuring a promising SR reconstruction performance.

4.2.3. Visual Effect Comparison

In addition to the comparison of the objective indicators, we also conducted evaluations in terms of the visual results. Figure 8 presents a visual comparison between the PIDAN and other advanced approaches using image samples from the COWC test sets with three upscaling factors, ×2, ×3, and ×4. Specifically, in each case, we enlarged a small rectangle area for a clearer presentation and comparison. As can be seen, the images reconstructed by the bicubic interpolation algorithm are the most blurred. Figure 8a shows that the PIDAN obtains more promising results with fewer jaggies and ringing artifacts, and meanwhile reconstructs clearer image contours than the compared advanced approaches. In Figure 8b, the reconstructed vehicle result obtained using PIDAN restores sharper edge details and maintains the maximum structural integrity with less distortion. Figure 8c shows that the PIDAN can reconstruct the parallel lines more completely and precisely than the other approaches. The PIDAN also obtains the highest quantitative analysis values when compared with the other advanced SISR approaches. These visual results indicate that our model recovers feature information with rich high-frequency details, producing better SR results.

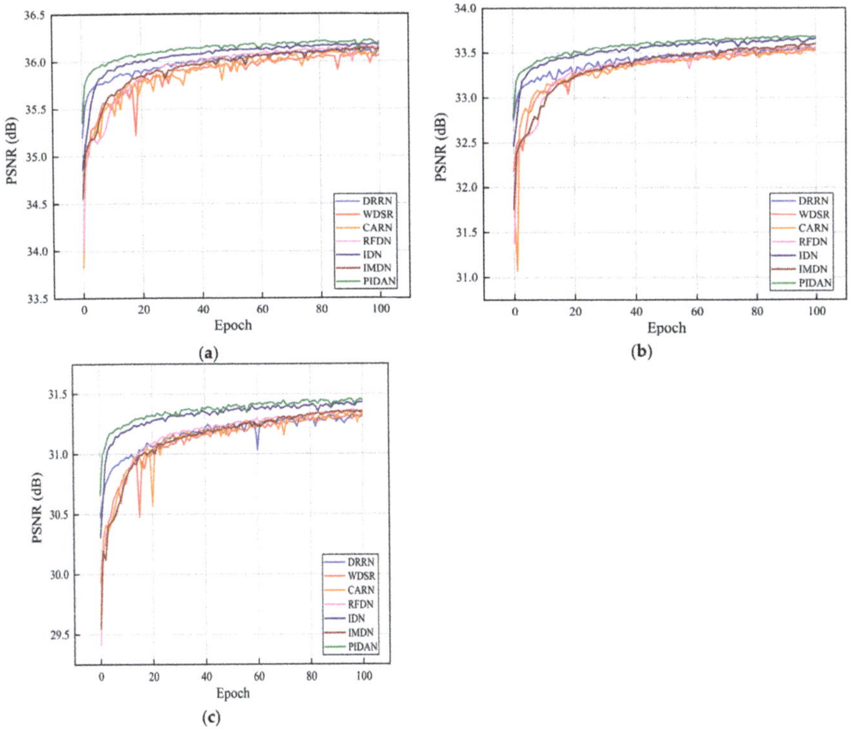

Figure 6. Performance curves for PIDAN and other methods using the FastTest10 dataset with scale factors of (**a**) ×2, (**b**) ×3, and (**c**) ×4.

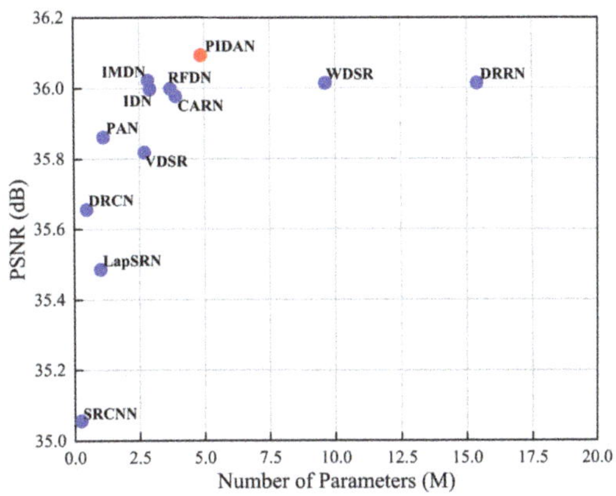

Figure 7. Comparison of model parameters and mean PSNR values of different DCNN-based methods.

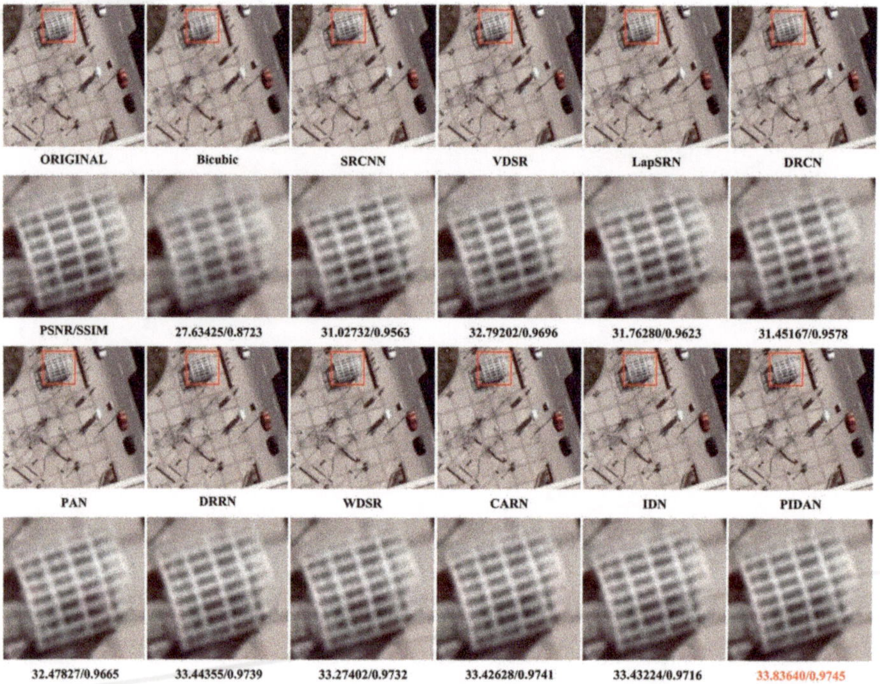

(**a**)

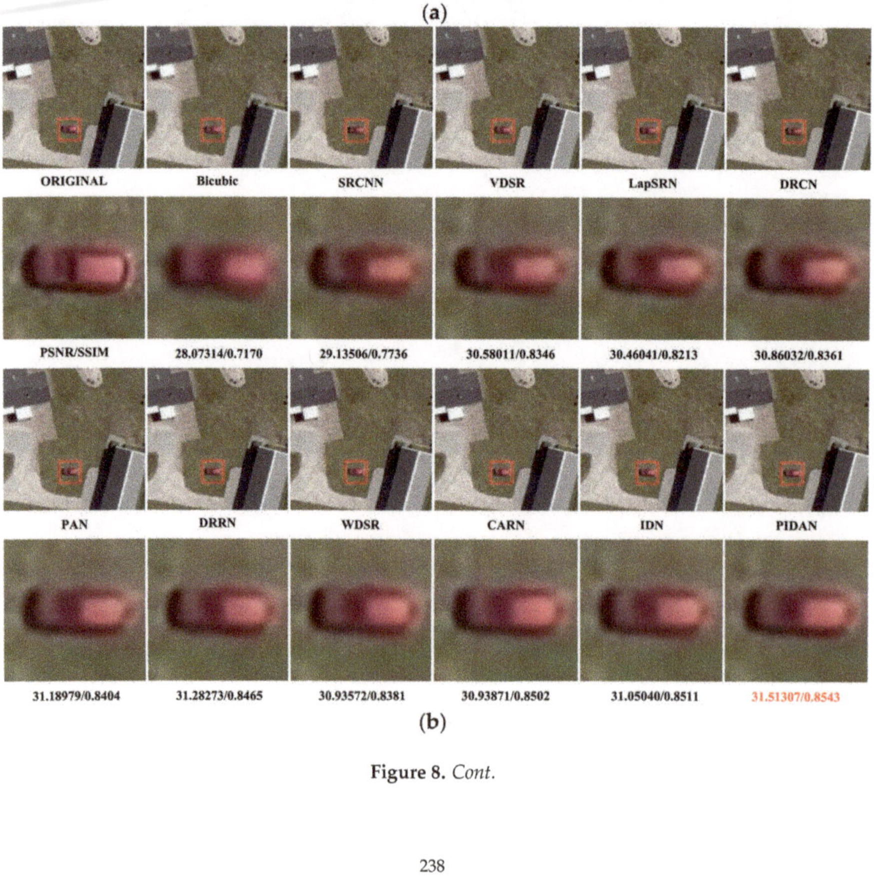

(**b**)

Figure 8. *Cont.*

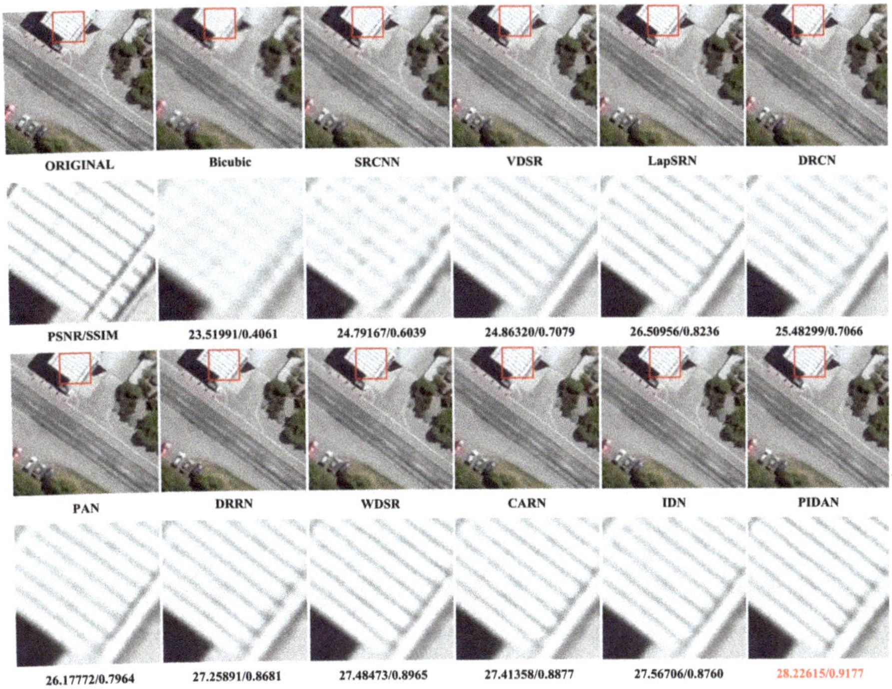

Figure 8. Visual comparison of SR results using samples from the COWC dataset with (**a**) upscaling factor ×2, (**b**) upscaling factor ×3, and (**c**) upscaling factor ×4.

4.2.4. Analysis of PIDAB

The PIDAB is the most critical aspect of the PIDAN. To demonstrate the necessity of the PCCS operation and the HAM in the PIDAB, we carried out a set of ablation experiments on the NWPU VHR-10 and COWC datasets. As shown in Table 3, when we removed PCCS and HAM, the PSNR scores on the two datasets were 34.55616 and 35.99601 dB, respectively. When we added the PCCS component, the PSNR scores were 34.58637 and 36.03984 dB; when we added the HAM module, the PSNR scores were 34.57436 and 36.03683 dB, respectively. With the addition of both PCCS and HAM, the PSNR scores for images from the NWPU VHR-10 and COWC datasets were 34.59635 and 36.09257 dB, respectively. We can conclude from Table 3 that the network structure with both PCCS and HAM yields optimal SR reconstruction results.

Table 3. Results of ablation study of PCCS and HAM. Bold indicates optimal performance.

Scale	PCCS	HAM	NWPU VHR-10 PSNR/SSIM	COWC PSNR/SSIM
×2	×	×	34.55616/0.9209	35.99601/0.9415
	√	×	34.58637/0.9214	36.03984/0.9419
	×	√	34.57436/0.9211	36.03683/0.9417
	√	√	**34.59635/0.9215**	**36.09257/0.9422**

The PCCS uses three convolution layers with different kernel sizes in parallel to obtain more non-redundant and extensive feature information from an image. Table 3 indicates that the PCCS component leads to performance gains (e.g., 0.03021 dB on NWPU VHR-10 and

0.04383 dB on COWC). This is mainly due to the PCCS, which makes the network flexible in processing feature information at different scales. Furthermore, we explored the influence of different convolution kernel settings in the PCCS components on the SR performance. Table 4 shows the experimental results of different convolution kernel settings with an upscaling factor of ×2. Broadly, the models with multiple convolutional kernels achieve better results than those with only a single convolutional kernel, and our PCCS obtains the best results owing to its three parallel progressive feature-purification branches.

Table 4. Results of comparison experiments using different convolution kernel settings in the PID component. Bold indicates optimal performance.

Scale	Kernel Size 3	Kernel Size 5	Kernel Size 7	NWPU VHR-10 PSNR/SSIM	COWC PSNR/SSIM
×2	×	×	×	34.55616/0.9209	35.99601/0.9415
	√	×	×	34.57641/0.9212	36.02632/0.9418
	×	√	×	34.57483/0.9212	36.01945/0.9418
	×	×	√	34.57012/0.9212	36.02009/0.9418
	√	√	×	34.57821/0.9212	36.02750/0.9418
	√	×	√	34.58540/0.9213	36.03357/0.9419
	×	√	√	34.58416/0.9214	36.02602/0.9419
	√	√	√	**34.58637/0.9214**	**36.03984/0.9419**

HAM generates more balanced attention information by adopting a structure that has both channel and spatial attention mechanisms in parallel. Table 3 indicates that the PCCS component leads to performance gains (e.g., 0.01820 dB on NWPU VHR-10 and 0.04082 dB on COWC). To further verify the effectiveness of the proposed HAM, we compared HAM with the SE block [15] and CBAM [20]. The SE block comprises a gating mechanism that obtains a completely new feature map by multiplying the obtained feature map with the response of each channel. Compared to the SE block, CBAM includes both channel and spatial attention mechanisms, which requires the network to be able to understand which parts of the feature map should have higher responses at the spatial level. Our HAM also includes channel and spatial attention mechanisms however, CBAM connects them serially while HAM accesses these two parts in parallel and combines them with the input feature map in a residual structure. As can be seen from Table 5, the addition of attention modules can improve the performance to different degrees. The effects of the dual attention modules are better than that of the SE block, which only adopts a CAM. Moreover, compared with CBAM, our HAM component leads to performance gains (e.g., 0.01000 dB on NWPU VHR-10 and 0.00662 dB on COWC). This finding illustrates that connecting a SAM and CAM in parallel is more effective for feature discrimination. These comparisons show that HAM in our PIDAB is advanced and effective.

Table 5. Results of comparison experiments using different attention modules. Bold indicates optimal performance.

Scale	Approach	NWPU VHR-10 PSNR/SSIM	COWC PSNR/SSIM
×2	/	34.55616/0.9209	35.99601/0.9415
	SE block	34.56088/0.9209	36.02749/0.9416
	CBAM	34.56436/0.9211	36.03021/0.9416
	HAM	**34.57436/0.9211**	**36.03683/0.9417**

4.2.5. Effect of Number of PIDABs

In this subsection, we report the results of adjusting the depth of the network by simply increasing the number of PIDAB. Specifically, numbers of PIDABs ranging from 4 to 20 were used. Figure 9 shows the performance with different numbers of PIDABs using the FastTest10 dataset in the epoch range 0 to 100. When simply increasing the value of N

to 20, the improvement increases, and a gain of approximately 0.08 dB is achieved when compared to the basic network ($N = 4$) with a scaling factor of $\times 2$, which demonstrates that the PIDAN can achieve a higher average PSNR with a larger number of PIDABs.

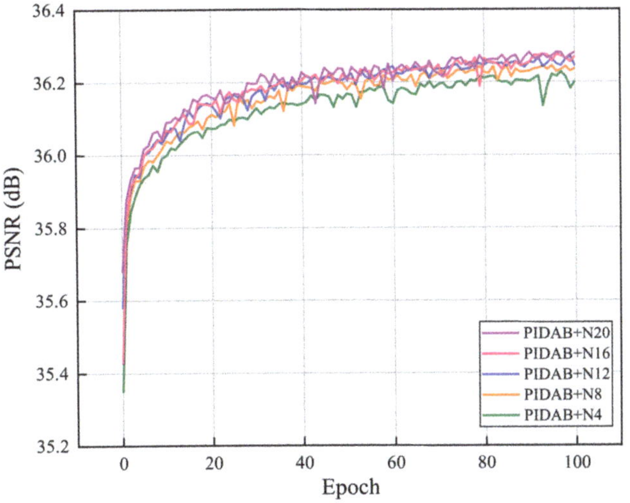

Figure 9. Performance curve for PIDAN with different numbers of PIDABs using the FastTest10 dataset with a scale factor of $\times 2$.

5. Conclusions

To achieve SR reconstruction of remote sensing images more efficiently, based on the IDN, we proposed a convenient but very effective approach named pyramid information distillation attention network (PIDAN). The main contribution of our work is the pyramid information distillation attention block (PIDAB), which is constructed as the building block of the deep feature-extraction part of the proposed PIDAN. To obtain more extensive and non-redundant image features, the PIDAB includes a pyramid information distillation module, which introduces a pyramid convolution channel split to allow the network to perceive a wider range of hierarchical features and reduce output feature maps, decreasing the model parameters. In addition, we proposed a hybrid attention mechanism module to further improve the restoration ability for high-frequency information. The results of extensive experiments demonstrated that the PIDAN outperforms other comparable deep CNN-based approaches and could maintain a good trade-off between the factors that affect practical application, including objective evaluation, visual quality, and model size. In future, we will further explore this approach in other computer vision tasks in remote sensing scenarios, such as object detection and recognition.

Author Contributions: Conceptualization, B.H. (Bo Huang); Investigation, B.H. (Bo Huang) and Y.L.; Formal analysis, B.H. (Bo Huang), Z.G. and B.H. (Boyong He); Validation, Z.G., B.H. (Boyong He) and X.L.; Writing—original draft, B.H. (Bo Huang); Supervision, L.W.; Writing—review & editing B.H. (Bo Huang) and L.W. All authors have read and agreed to the published version of the manuscript.

Funding: This work was supported in part by the National Natural Science Foundation of China (no. 51276151).

Institutional Review Board Statement: Not applicable.

Informed Consent Statement: Not applicable.

Data Availability Statement: The data presented in this study are available on request from the corresponding author.

Acknowledgments: The authors would like to thank the anonymous reviewers for their valuable comments and helpful suggestions.

Conflicts of Interest: The authors declare no conflict of interest.

References

1. Som-ard, J.; Atzberger, C.; Izquierdo-Verdiguier, E.; Vuolo, F.; Immitzer, M. Remote sensing applications in sugarcane cultivation: A review. *Remote Sens.* **2021**, *13*, 4040. [CrossRef]
2. Glasner, D.; Bagon, S.; Irani, M. Super-resolution from a single image. In Proceedings of the 2009 IEEE 12th International Conference on Computer Vision, Kyoto, Japan, 29 September–2 October 2009; pp. 349–356.
3. Chang, H.; Yeung, D.; Xiong, Y. Super-resolution through neighbor embedding. In Proceedings of the 2004 IEEE Computer Society Conference on Computer Vision and Pattern Recognition, Washington, DC, USA, 27 June–2 July 2004; pp. 275–282.
4. Zhang, L.; Wu, X. An edge-guided image interpolation algorithm via directional filtering and data fusion. *IEEE Trans. Image Process.* **2006**, *15*, 2226–2238. [CrossRef] [PubMed]
5. Zhang, K.; Gao, X.; Tao, D.; Li, X. Single image super-resolution with non-local means and steering kernel regression. *IEEE Trans. Image Process.* **2012**, *21*, 4544–4556. [CrossRef] [PubMed]
6. Protter, M.; Elad, M.; Takeda, H.; Milanfar, P. Generalizing the nonlocal-means to super-resolution reconstruction. *IEEE Trans. Image Process.* **2009**, *18*, 36–51. [CrossRef] [PubMed]
7. Freeman, W.; Jones, T.; Pasztor, E. Example-based super-resolution. *IEEE Comput. Graph. Appl.* **2002**, *22*, 56–65. [CrossRef]
8. Mu, G.; Gao, X.; Zhang, K.; Li, X.; Tao, D. Single image super resolution with high resolution dictionary. In Proceedings of the 2011 18th IEEE Conference on Image Processing, Brussels, Belgium, 11–14 September 2011; pp. 1141–1144.
9. Dong, C.; Loy, C.; He, K.; Tang, X. Image super-resolution using deep convolutional networks. *IEEE Trans. Pattern Anal. Mach. Intell.* **2015**, *38*, 295–307. [CrossRef] [PubMed]
10. Kim, J.; Lee, J.; Lee, K. Accurate image super-resolution using very deep convolutional networks. In Proceedings of the Conference on Computer Vision and Pattern Recognition, Las Vegas, NY, USA, 27–30 June 2016; pp. 1646–1654.
11. Lim, B.; Son, S.; Kim, H.; Nah, S.; Lee, K.M. Enhanced deep residual networks for single image super-resolution. In Proceedings of the IEEE Conference on Computer Vision and Pattern Recognition Workshops, Honolulu, HI, USA, 21–26 July 2017; pp. 1132–1140.
12. Zhang, Y.; Li, K.; Li, K.; Wang, L.; Zhong, B.; Fu, Y. Image super-resolution using very deep residual channel attention networks. In Proceedings of the European Conference on Computer Vision, Munich, Germany, 8–14 September 2018; pp. 286–301.
13. Ledig, C.; Theis, L.; Huszár, F.; Caballero, J.; Cunningham, A.; Acosta, A.; Aitken, A.; Tejani, A.; Totz, J.; Wang, Z.; et al. Photorealistic single image super-resolution using a generative adversarial network. In Proceedings of the Conference on Computer Vision and Pattern Recognition, Honolulu, HI, USA, 21–26 July 2017; pp. 4681–4690.
14. Zeng, K.; Yu, J.; Wang, R.; Li, C.; Tao, D. Coupled deep autoencoder for single image super-resolution. *IEEE Trans. Cybern.* **2017**, *47*, 27–37. [CrossRef] [PubMed]
15. Hu, J.; Shen, L.; Sun, G. Squeeze-and-excitation networks. In Proceedings of the IEEE Conference on Computer Vision and Pattern Recognition, Salt Lake City, UT, USA, 18–22 June 2018; pp. 7132–7141.
16. Zhang, D.; Shao, J.; Li, X.; Shen, H.T. Remote sensing image super-resolution via mixed high-order attention network. *IEEE Trans. Geosci. Remote Sens.* **2021**, *59*, 5183–5196. [CrossRef]
17. Szegedy, C.; Liu, W.; Jia, Y.; Sermanet, P.; Reed, S.; Anguelov, D.; Erhan, D.; Vanhoucke, V.; Rabinovich, A. Going deeper with convolutions. In Proceedings of the IEEE Conference on Computer Vision and Pattern Recognition, Boston, MA, USA, 7–12 June 2015; pp. 1–9.
18. Hui, Z.; Wang, X.; Gao, X. Fast and accurate single image super-resolution via information distillation network. In Proceedings of the IEEE Conference on Computer Vision and Pattern Recognition, Salt Lake City, UT, USA, 18–22 June 2018; pp. 723–731.
19. Dun, Y.; Da, Z.; Yang, S.; Qian, X. Image super-resolution based on residually dense distilled attention network. *Neurocomputing* **2021**, *443*, 47–57. [CrossRef]
20. Woo, S.; Park, J.; Lee, J.Y. *CBAM: Convolutional Block Attention Module*; Springer: Cham, Switzerland, 2018; p. 112211.
21. Dong, C.; Loy, C.; Tang, X. Accelerating the super-resolution convolutional neural network. In Proceedings of the European Conference on Computer Vision, Amsterdam, The Netherlands, 8–16 October 2016; pp. 391–407.
22. Shi, W.; Caballero, J.; Huszar, F.; Totz, J.; Aitken, A.; Bishop, R.; Rueckert, D.; Wang, Z. Real-time single image and video super-resolution using an efficient sub-pixel convolutional neural network. In Proceedings of the IEEE Conference on Computer Vision and Pattern Recognition, Las Vegas, NV, USA, 27–30 June 2016; pp. 1874–1883.
23. He, K.M.; Zhang, X.Y.; Ren, S.Q.; Sun, J. Deep residual learning for image recognition. In Proceedings of the IEEE Conference on Computer Vision and Pattern Recognition, Las Vegas, NV, USA, 27–30 June 2016; pp. 770–778.
24. Kim, J.; Lee, J.; Lee, K. Deeply-recursive convolutional network for image super-resolution. In Proceedings of the Conference on Computer Vision and Pattern Recognition, Las Vegas, NV, USA, 26 June–1 July 2016; pp. 1637–1645.
25. Tai, Y.; Yang, J.; Liu, X. Image super-resolution via deep recursive residual network. In Proceedings of the IEEE Conference on Computer Vision and Pattern Recognition, Honolulu, HI, USA, 21–26 July 2017; pp. 3147–3155.
26. Lai, W.; Huang, J.; Ahuja, J.; Yang, M. Deep Laplacian pyramid networks for fast and accurate super-resolution. In Proceedings of the Conference on Computer Vision and Pattern Recognition, Honolulu, HI, USA, 21–26 July 2017; pp. 624–632.

27. Tong, T.; Li, G.; Liu, X.; Gao, Q. Image super-resolution using dense skip connections. In Proceedings of the IEEE International Conference on Computer Vision, Venice, Italy, 22–29 October 2017; pp. 4809–4817.
28. Zhang, Y.; Tian, Y.; Kong, Y.; Zhong, B.; Fu, Y. Residual dense network for image super-resolution. In Proceedings of the IEEE Conference on Computer Vision and Pattern Recognition, Salt Lake City, UT, USA, 18–23 June 2018; pp. 2472–2481.
29. Yu, J.; Fan, Y.; Yang, J.; Xu, N.; Wang, Z.; Wang, X.; Huang, T. Wide activation for efficient and accurate image super-resolution. *arXiv* **2018**, arXiv:1808.08718.
30. Ahn, N.; Kang, B.; Sohn, K.A. Fast, accurate, and lightweight super-resolution with cascading residual network. In Proceedings of the European Conference on Computer Vision, Munich, Germany, 8–14 September 2018; pp. 252–268.
31. Buades, A.; Coll, B.; Morel, J. A non-local algorithm for image denoising. In Proceedings of the IEEE Computer Society Conference on Computer Vision and Pattern Recognition, San Diego, CA, USA, 20–25 June 2005; pp. 60–65.
32. Wang, X.; Girshick, R.; Gupta, A.; He, K. Non-local neural networks. In Proceedings of the IEEE/CVF Conference on Computer Vision and Pattern Recognition, Salt Lake City, UT, USA, 18–23 June 2018; pp. 7794–7803.
33. Cao, Y.; Xu, J.; Lin, S.; Wei, F.; Hu, H. GCNet: Non-local networks meet squeeze-excitation networks and beyond. In Proceedings of the IEEE/CVF International Conference on Computer Vision Workshop, Seoul, Korea, 27–28 October 2019; pp. 1971–1980.
34. Zhang, Y.; Li, K.; Li, K.; Zhong, B.; Fu, Y. Residual non-local attention networks for image restoration. *arXiv* **2019**, arXiv:1903.10082.
35. Anwar, S.; Barnes, N. Densely residual Laplacian super-resolution. *arXiv* **2019**, arXiv:1906.12021. [CrossRef] [PubMed]
36. Guo, J.; Ma, S.; Guo, S. MAANet: Multi-view aware attention networks for image super-resolution. *arXiv* **2019**, arXiv:1904.06252.
37. Dai, T.; Cai, J.; Zhang, Y.; Xia, S.-T.; Zhang, L. Second-order attention network for single image super-resolution. In Proceedings of the IEEE Conference on Computer Vision and Pattern Recognition, Long Beach, CA, USA, 16–20 June 2019; pp. 11065–11074.
38. Hui, Z.; Gao, X.; Yang, Y.; Wang, X. Lightweight image super-resolution with information multi-distillation network. In Proceedings of the 27th ACM International Conference on Multimedia, Nice, France, 21–25 October 2019; Volume 10, pp. 2024–2032.
39. Zhao, H.; Kong, X.; He, J.; Qiao, Y.; Dong, C. Efficient image super-resolution using pixel attention. *arXiv* **2020**, arXiv:2010.01073.
40. Wang, H.; Wu, C.; Chi, J.; Yu, X.; Hu, Q.; Wu, H. Image super-resolution using multi-granularity perception and pyramid attention networks. *Neurocomputing* **2021**, *443*, 247–261. [CrossRef]
41. Huang, B.; He, B.; Wu, L.; Guo, Z. Deep residual dual-attention network for super-resolution reconstruction of remote sensing images. *Remote Sens.* **2021**, *13*, 2784. [CrossRef]
42. Xia, G.-S.; Hu, J.; Hu, F.; Shi, B.; Bai, X.; Zhong, Y.; Zhang, L. AID: A benchmark data det for performance evaluation of aerial scene classification. *IEEE Trans. Geosci. Remote Sens.* **2017**, *55*, 3965–3981. [CrossRef]
43. Cheng, G.; Zhou, P.; Han, J. Learning rotation-invariant convolutional neural networks for object detection in VHR optical remote sensing images. *IEEE Trans. Geosci. Remote Sens.* **2016**, *54*, 7405–7415. [CrossRef]
44. Mundhenk, T.N.; Konjevod, G.; Sakla, W.A.; Boakye, K. A large contextual dataset for classification, detection and counting of cars with deep learning. In *European Conference on Computer Vision*; Springer: Cham, Switzerland, 2016; pp. 785–800.
45. Horé, A.; Ziou, D. Image quality metrics: PSNR vs. SSIM. In Proceedings of the International Conference on Computer Vision, Istanbul, Turkey, 23–26 August 2010; pp. 2366–2369.
46. Wang, Z.; Bovik, A.C.; Sheikh, H.R.; Simoncelli, E.P. Image quality assessment: From error visibility to structural similarity. *IEEE Trans. Image Process.* **2004**, *13*, 600–612. [CrossRef] [PubMed]
47. Kingma, D.P.; Ba, J.L. Adam: A method for stochastic optimization. In Proceedings of the 3rd International Conference on Learning Representations, San Diego, CA, USA, 7–9 May 2015.
48. Liu, J.; Tang, J.; Wu, G. Residual feature distillation network for lightweight image super-resolution. *arXiv* **2020**, arXiv:2009.11551.

remote sensing

MDPI

Article

Deep Learning Triplet Ordinal Relation Preserving Binary Code for Remote Sensing Image Retrieval Task

Zhen Wang [1,2,*], Nannan Wu [1], Xiaohan Yang [1], Bingqi Yan [1] and Pingping Liu [2,3]

1 School of Computer Science and Technology, Shandong University of Technology, Zibo 255000, China; 21505020635@stumail.sdut.edu.cn (N.W.); 21505020639@stumail.sdut.edu.cn (X.Y.); 19805010885@stumail.sdut.edu.cn (B.Y.)
2 Key Laboratory of Symbolic Computation and Knowledge Engineering of Ministry of Education, Jilin University, Changchun 130012, China; liupp@jlu.edu.cn
3 School of Computer Science and Technology, Jilin University, Changchun 130012, China
* Correspondence: wzh@sdut.edu.cn

Abstract: As satellite observation technology rapidly develops, the number of remote sensing (RS) images dramatically increases, and this leads RS image retrieval tasks to be more challenging in terms of speed and accuracy. Recently, an increasing number of researchers have turned their attention to this issue, as well as hashing algorithms, which map real-valued data onto a low-dimensional Hamming space and have been widely utilized to respond quickly to large-scale RS image search tasks. However, most existing hashing algorithms only emphasize preserving point-wise or pair-wise similarity, which may lead to an inferior approximate nearest neighbor (ANN) search result. To fix this problem, we propose a novel triplet ordinal cross entropy hashing (TOCEH). In TOCEH, to enhance the ability of preserving the ranking orders in different spaces, we establish a tensor graph representing the Euclidean triplet ordinal relationship among RS images and minimize the cross entropy between the probability distribution of the established Euclidean similarity graph and that of the Hamming triplet ordinal relation with the given binary code. During the training process, to avoid the non-deterministic polynomial (NP) hard problem, we utilize a continuous function instead of the discrete encoding process. Furthermore, we design a quantization objective function based on the principle of preserving triplet ordinal relation to minimize the loss caused by the continuous relaxation procedure. The comparative RS image retrieval experiments are conducted on three publicly available datasets, including UC Merced Land Use Dataset (UCMD), SAT-4 and SAT-6. The experimental results show that the proposed TOCEH algorithm outperforms many existing hashing algorithms in RS image retrieval tasks.

Keywords: remote sensing image retrieval; hashing algorithm; binary code; triplet ordinal relation preserving; cross entropy

Citation: Wang, Z.; Wu, N.; Yang, X.; Yan, B.; Liu, P. Deep Learning Triplet Ordinal Relation Preserving Binary Code for Remote Sensing Image Retrieval Task. *Remote Sens.* **2021**, *13*, 4786. https://doi.org/10.3390/rs13234786

Academic Editors: Jukka Heikkonen, Fahimeh Farahnakian and Pouya Jafarzadeh

Received: 26 September 2021
Accepted: 23 November 2021
Published: 26 November 2021

1. Introduction

With the rapid development of satellite observation technology, both the amount and the quality of remote sensing (RS) images have improved dramatically. An era of remote sensing image big data has arrived. An increasing number of researchers are focusing on the task of large-scale RS image retrieval, due to its broad applications, such as disaster prevention, soil erosion monitoring, disaster rescue scenario and short-term weather forecasting [1–5]. The content-based image retrieval (CBIR) [6,7] method extracts feature information representing RS image content and finds similar RS images by comparing the distance values among their feature information. However, the feature information in CBIR is always represented as high dimensional float point data and it is difficult to directly compute the similarity relationship based on the original high dimensional feature information. Fortunately, hashing methods [1–5,8,9] can map high dimensional float point data into compact binary codes and return the approximate nearest neighbors according to Hamming distance; this measure effectively improves the retrieval speed. In summary, the content-based image retrieval method assisted by hashing algorithms enables the efficient and effective retrieval of target remote sensing images from a large-scale dataset.

In recent years, many hashing algorithms [10–14] have been proposed to achieve the approximate nearest neighbor (ANN) search task, due to its advantage of computation and storage. According to the learning framework, the existing hashing algorithms can be roughly divided into two types: the shallow model [12–14] and the deep model [10,11,15,16]. Conventional shallow hashing algorithms, such as locality sensitive hashing (LSH) [14], spectral hashing (SH) [17], iterative quantization hashing (ITQ) [13] and k-means hashing (KMH) [12], have been applied to various approximate nearest neighbor search tasks, including image retrieval. Locality sensitive hashing [14] is a kind of data-independent method, which learns hashing functions without a training process. LSH [14] randomly generates linear hashing functions and encodes data into binary codes according to their projection signs. Spectral hashing (SH) [17] utilizes a spectral graph to represent the similarity relationship among data points. The binary codes in SH are generated by partitioning a spectral graph. Iterative quantization hashing [13] considers the vertexes of a hyper cubic as encoding centers. ITQ [13] rotates the principal component analysis (PCA) projected data and maps the rotated data to the nearest encoding center. The encoding centers in ITQ are fixed and they are not adaptive to the data distribution [12]. To fix this problem, k-means hashing [12] learns the encoding centers by simultaneously minimizing the quantization error and the similarity loss. KMH [12] encodes the data as the same binary code as the nearest center. For the image search task, the shallow model first learns the high dimensional features, such as scale-invariant feature transform (SIFT) [18] or a holistic representation of the spatial envelope (GIST) [19], then retrieves similar images by mapping these features into the compact Hamming space. In contrast, the deep learning model enables end-to-end representation learning and hash coding [10,11,20–22]. In particular, the deep learning to hash, such as deep Cauchy hashing (DCH) [11] and twin-bottleneck hashing (TBH) [10], proves crucial to jointly learn, thereby similarly preserving the representations and control quantization error of converting continuous representations to binary codes. Deep Cauchy hashing [11] defines a pair-wise similarity preserving restriction based on Cauchy distribution and it heavily penalizes the similar image pairs with large Hamming distance. Twin-bottleneck hashing [10] proposes a code-driven graph to represent the similarity relationship among data points and aims to minimize the loss between the original data and decoded data. These deep learning to hash methods have shown state-of-the-art results for many datasets.

Recently, many hashing algorithms have been applied to the large-scale RS image search task [1–5]. Partial randomness hashing [23] maps RS images into a low dimensional Hamming space by both the random and well-trained projection functions. Demir et al. [24] proposed two kernel-based methods to learn hashing functions in the kernel space. Liu et al. [25] fully utilized the supervised deep learning framework and hashing learning to generate the binary codes of RS images. Li et al. [25] carried out a comprehensive study of DHNN systems and aimed to introduce the deep neural network into the large-scale RS image search task. Fan et al. [26] proposed a distribution consistency loss (DCL) to capture the intra-class distribution and inter-class ranking. Both deep Cauchy hashing [11] and the distribution consistency loss functions [26] employ pairwise similarity [15] to describe the relationship among data. However, the similarity relationship among RS images is more complex. In this paper, we propose the triplet ordinal cross entropy hashing (TOCEH) to deal with the large-scale RS image search task. The flowchart of the proposed TOCEH is shown in Figure 1.

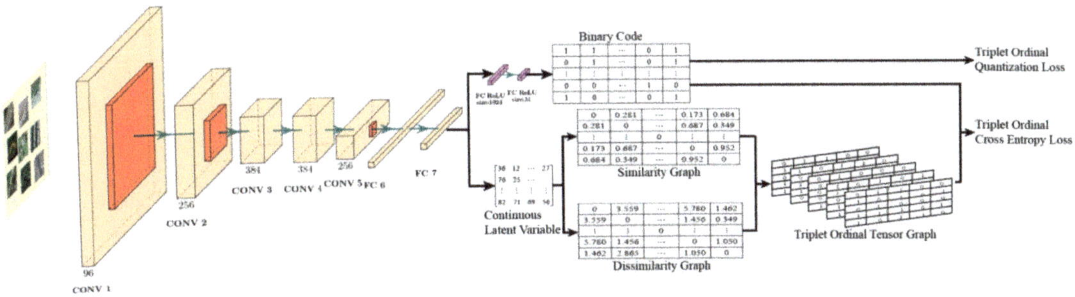

Figure 1. Flowchart of the proposed TOCEH algorithm. Firstly, to represent the image content, we use the Alexnet, including five convolutional (CONV) networks and two fully connected (FC) networks, to learn the continuous latent variable. Secondly, the triplet ordinal relation is computed by the tensor product of the similarity and dissimilarity graphs. Thirdly, two fully connected layers with the activation function of ReLU are utilized to generate the binary code. To guarantee the performance, we define the triplet ordinal cross entropy loss to minimize the inconsistency between the triplet ordinal relations in different spaces. Furthermore, we design the triplet ordinal quantization loss to reduce the loss caused by the relaxation mechanism.

As shown in Figure 1, the TOCEH algorithm consists of two parts: the triplet ordinal tensor graph generation part and the hash code learning part. In part 1, we first utilize the AlexNet [27] pre-trained on the ImageNet dataset [28] to extract the 4096-dimension image feature information of the target domain RS images. Then, we separately compute the similarity and dissimilarity graph among the high dimensional features. Finally, we establish the triplet ordinal tensor graph representing the ordinal relation among any triplet RS images. Part 2 utilizes two fully connected layers to generate binary codes. During the training process, we define two excellent objection functions, including the triplet ordinal cross entropy loss and the triplet ordinal quantization loss to guarantee the performance of the obtained binary codes and utilize the back-propagation mechanism to optimize the variables of the deep neural network. The main contributions of the proposed TOCEH are summarized as follows:

1. The learning procedure of TOCEH takes into account the triplet ordinal relations, rather than the pairwise or point-wise similarity relations, which can enhance the performance of preserving the ranking orders of approximate nearest neighbor retrieval results from the high dimensional feature space to the Hamming space.
2. TOCEH establishes a triplet ordinal graph to explicitly indicate the ordinal relationship among any triplet RS images and preserves the ranking orders by minimizing the inconsistency between the probability distribution of the given triplet ordinal relation and that of the ones derived from binary codes.
3. We conduct comparative experiments on three RS image datasets: UCMD, SAT-4 and SAT-6. Extensive experimental results demonstrate that TOCEH generates highly concentrated and compact hash codes, and it outperforms some existing state-of-the-art hashing methods in large-scale RS image retrieval tasks.

The rest of this paper is organized as follows. Section 2 introduces the proposed TOCEH algorithm. Section 2.1 shows the important notation. The hash learning problem is stated in Section 2.2. The tensor graph representing the triplet ordinal relation among RS images is introduced in Section 2.3. We provide the formulation of triplet ordinal cross entropy loss and triplet ordinal quantization loss in Sections 2.4 and 2.5, respectively. The extensive experimental evaluations are presented in Section 3. Finally, we set out a conclusion in Section 4.

2. Triplet Ordinal Cross Entropy Hashing

2.1. Notation

In this paper, we use the letters B and X to separately represent the data matrix in the Hamming and Euclidean spaces. The columns in the data matrix are denoted as the letters with subscript. The important notations are summarized in Table 1.

Table 1. The important notations used in this paper.

Notation	Description
B	Compact binary code matrix
B_i, B_j, B_k	The i-th, j-th, k-th column in B
$H(\cdot)$	Hashing function
X	Data matrix in the Euclidean space
x_i, x_j, x_k	The i-th, j-th, k-th column in X
G	Triplet ordinal graph in the Euclidean space
$\hat{G}$	Triplet ordinal relation in the Hamming space
g_{ijk}	The entry (i, j, k) in G
S	Similarity graph
DS	Dissimilarity graph
N	The number of training samples
L	The number of k-means centers
$P(\cdot)$	Probability distribution function
$d_h(\cdot,\cdot)$	Hamming distance function
M	Binary code length
$\mathbf{1}$	The binary matrix with all values of 1

2.2. Hashing Learning Problem

The purpose of the hashing algorithm [3,10,11] is to learn the hashing function $H(\cdot)$, mapping the high dimensional float point data x into the compact Hamming space as defined in Equation (1). $B(x)$ represents the compact binary code of x.

$$B(x) = (sign(H(x) - 0.5) + 1)/2 \tag{1}$$

With the assistance of the obtained hashing function $H(\cdot)$, we can encode RS image content as compact binary code and efficiently achieve RS image search task according to their Hamming distances [1–5,23–25]. Furthermore, to guarantee the quality of the RS image search result, we expect the triplet ordinal relation among RS images in the Hamming space to be consistent with that in the original space [29,30]. To illustrate this requirement, a simple example is provided below. Here, x_i, x_j and x_k separately represent RS image content information. In the original space, the image pair (x_i, x_j) is more similar than the image pair (x_j, x_k). After mapping them into the Hamming space, the Hamming distance of the data pair (x_i, x_j) should be smaller than that of the data pair (x_j, x_k). This constraint is defined as in Equation (2).

$$\begin{aligned} & ||H(x_i) - H(x_j)||_1 \leq ||H(x_k) - H(x_j)||_1 \\ s.t. \quad & ||x_i - x_j||_2^2 \leq ||x_k - x_j||_2^2 \end{aligned} \tag{2}$$

The constraint in Equation (2) guarantees that the ranking order of the retrieval result in the Hamming space is consistent with that in the Euclidean space. Thus, the hashing algorithm, satisfying the triplet ordinal relation preserving constraint, can achieve RS image ANN search tasks [31–35].

2.3. Triplet Ordinal Tensor Graph

To learn the triplet ordinal relation preserving hashing functions, the first problem is how to efficiently compute the probability distribution of the triplet ordinal relation among the training set in the original space.

Generally, we select the triplet data (x_i, x_j, x_k) from the training set to compute their ordinal relation, where the data pair (x_i, x_j) has a small Euclidean distance value and (x_j, x_k) is considered as the dissimilar data pair. However, this mechanism needs to randomly select triplet samples and compare the distance values among all data points. It has a high time complexity and costly memory. Furthermore, it is difficult to define the similar and dissimilar data pairs for the problem without supervised information.

In this paper, to solve the above problem, we employ a tensor ordinal graph G to represent the ordinal relation among the triplet images (x_i, x_j, x_k). We establish the tensor ordinal graph G by tensor production and each entry in G is calculated as $G(ij, jk) = S(i, j) \cdot DS(j, k)$. $S(i, j)$ is the similarity graph as defined in Equation (3). A larger value of $S(i, j)$ means the data pair (x_i, x_j) is more similar. $DS(i, j)$ is the dissimilarity graph and its value is calculated as $DS(i, j) = 1/S(i, j)$.

$$S(i,j) = \begin{cases} 0, & i = j \\ e^{-||x_i - x_j||_2^2 / 2\sigma^2}, & otherwise \end{cases} \tag{3}$$

We further process G to obey the binary distribution as in Equation (4). g_{ijk} is the entry of $G(i, j, k)$.

$$\begin{cases} g_{ijk} = 1, & G(i,j,k) > 1 \\ g_{ijk} = 0, & G(i,j,k) \leq 1 \end{cases} \tag{4}$$

Given N training samples, the size of the similarity graph and dissimilarity graph is $N \times N$. The tensor product of the two graphs is shown in Figure 2, and its size is $N^2 \times N^2$. However, the proposed TOCEH only concerns the relative similarity relationship among the data pairs (x_i, x_j) and (x_j, x_k). The corresponding elements are marked blue. There are N rectangles and each rectangle contains $N \times N$ elements. We pick up these elements and restore them into a matrix with the size of $N \times N \times N$.

	x_1x_1	x_1x_2	$\cdots$	x_1x_N	x_2x_1	x_2x_2	$\cdots$	x_2x_N	$\cdots$	x_Nx_1	x_Nx_2	$\cdots$	x_Nx_N
x_1x_1	1	1	$\cdots$	0	0	0	$\cdots$	0	$\cdots$	1	1	$\cdots$	0
x_1x_2	0	1	$\cdots$	1	0	1	$\cdots$	1	$\cdots$	1	0	$\cdots$	0
$\vdots$	$\vdots$	$\vdots$	$\ddots$	$\vdots$	$\vdots$	$\vdots$	$\cdots$	$\vdots$	$\cdots$	$\vdots$	$\vdots$	$\cdots$	$\vdots$
x_1x_N	1	0	$\cdots$	1	1	1	$\cdots$	0	$\cdots$	1	1	$\cdots$	0
x_2x_1	0	0	$\cdots$	1	0	0	$\cdots$	1	$\cdots$	0	0	$\cdots$	1
x_2x_2	0	1	$\cdots$	0	0	0	$\cdots$	0	$\cdots$	1	1	$\cdots$	1
$\vdots$	$\vdots$	$\vdots$	$\ddots$	$\vdots$	$\vdots$	$\vdots$	$\ddots$	$\vdots$	$\ddots$	$\vdots$	$\vdots$	$\ddots$	$\vdots$
x_2x_N	0	0	$\cdots$	0	1	1	$\cdots$	0	$\cdots$	1	0	$\cdots$	0
$\vdots$	$\vdots$	$\vdots$	$\ddots$	$\vdots$	$\vdots$	$\vdots$	$\ddots$	$\vdots$	$\ddots$	$\vdots$	$\vdots$	$\ddots$	$\vdots$
x_Nx_1	1	0	$\cdots$	1	1	1	$\cdots$	0	$\cdots$	0	0	$\cdots$	1
x_Nx_2	0	1	$\cdots$	0	0	0	$\cdots$	1	$\cdots$	1	1	$\cdots$	1
$\vdots$	$\vdots$	$\vdots$	$\ddots$	$\vdots$	$\vdots$	$\vdots$	$\ddots$	$\vdots$	$\ddots$	$\vdots$	$\vdots$	$\ddots$	$\vdots$
x_Nx_N	1	0	$\cdots$	1	1	1	$\cdots$	0	$\cdots$	0	1	$\cdots$	1

Figure 2. The marked elements are picked up to restore in a matrix with the size of $N \times N \times N$.

Finally, the ordinal relation among any triplet items can be represented by the triplet ordinal graph G, as defined in Equation (5).

$$\begin{cases} S(i,j) > S(k,j), & g_{ijk} = 1 \\ S(i,j) \leq S(k,j), & g_{ijk} = 0 \end{cases} \tag{5}$$

To illustrate the cases defined in Equation (5), a simple explanation is provided below. For the triplet item (x_i, x_j, x_k), the value of the (ij, kj)-th entry is $G(ij, kj) = S(i, j) \cdot DS(k, j) = S(i, j)/S(k, j)$. If the triplet ordinal relation is $S(i, j) > S(k, j)$, we have $G(ij, kj) > 1$ and $g_{ijk} = 1$; otherwise, we have $G(ij, kj) \leq 1$ and $g_{ijk} = 0$. Thus, the value in G can correctly indicate the true ordinal relation among any triplet items.

As described above, we can establish a tensor ordinal graph G with size N^3 to represent the triplet ordinal relation among N images. In practice, during the training procedure, we use L ($L \ll N$) k-means centers to establish the tensor ordinal graph, which can reduce the training time complexity.

2.4. Triplet Ordinal Cross Entropy Loss

In this section, we define $\hat{G}$ as RS images' triplet ordinal relation in the Hamming space. As discussed in Section 2.2, an ideal hashing algorithm should minimize the inconsistency between $\hat{G}$ and G. In this paper, the above requirement is achieved by minimizing the cross entropy value, as defined in Equation (6).

$$\min CEH(G, \widehat{G}) = \min - P(G) \log P(\widehat{G}) \tag{6}$$

$P(G)$ defined in Equation (7) computes the probability distribution of RS images' triplet ordinal relation in the Euclidean space.

$$\begin{cases} w_{ijk} = \frac{T_1}{T} & g_{ijk} = 1 \\ w_{ijk} = \frac{T_0}{T} & g_{ijk} = 0 \end{cases} \tag{7}$$

The definitions of T_1, T_0 and T are shown in Equation (8). T_1 is the number of samples with a value of 1 in the matrix G and T_0 is the number of samples with a value of 0 in the matrix G. T is the total number of the elements in the matrix G.

$$\begin{aligned} T_1 &= \sum_{i,j,k=1}^{N} g_{i,j,k} \\ T_0 &= \sum_{i,j,k=1}^{N} (1 - g_{i,j,k}) \\ T &= \sum_{i,j,k=1}^{N} \left|2 \cdot g_{i,j,k} - 1\right| \end{aligned} \tag{8}$$

$P(\hat{G})$ is a conditional probability of the triplet ordinal relation with given binary codes. As the samples are independent from each other, we calculate $P(\hat{G})$ by Equation (9).

$$P(\widehat{G}) = \Pi_{i,j,k=1}^{N} P(g_{ijk} \middle| B_i, B_j, B_k) \tag{9}$$

$P(g_{ijk} \mid B_i, B_j, B_k)$ is the probability of the triplet images satisfying the ordinal relation g_{ijk}, and the samples' are assigned the binary codes (B_i, B_j, B_k). The definition is shown in Equation (10).

$$P(g_{ijk}|B_i, B_j, B_k) = \begin{cases} \phi(d_h(B_k, B_j) - d_h(B_i, B_j)), & g_{ijk} = 1 \\ 1 - \phi(d_h(B_k, B_j) - d_h(B_i, B_j)), & g_{ijk} = 0 \end{cases} \tag{10}$$

We further rewrite the definition of $P(g_{ijk} \mid B_i, B_j, B_k)$ as in Equation (11).

$$P(g_{ijk}|B_i, B_j, B_k) = \phi(d_h(B_k, B_j) - d_h(B_i, B_j))^{g_{ijk}} (1 - \phi(d_h(B_k, B_j) - d_h(B_i, B_j)))^{1-g_{ijk}} \tag{11}$$

$d_h(\cdot,\cdot)$ returns the Hamming distance and $\phi(\cdot)$ computes the probability value. If $g_{ijk} = 1$, the probability value should be close to 1 as $d_h(B_k, B_j)$-$d_h(B_i, B_j)$ gets larger and the probability value should be close to 0 as $d_h(B_k, B_j)$-$d_h(B_i, B_j)$ gets smaller. The characteristic of the function $(\cdot)$ is shown in Figure 3.

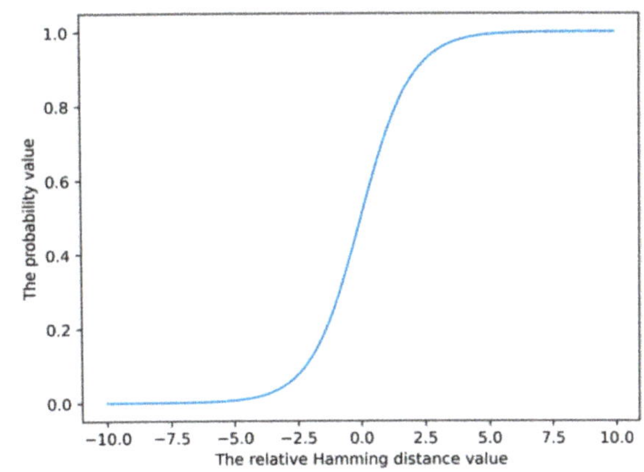

Figure 3. The characteristic of the function (·).

In this paper, the sigmoid function is considered as the function (·) as in Equation (12).

$$\phi(d_h(B_k, B_j) - d_h(B_i, B_j)) = \frac{1}{1 + e^{-\alpha(d_h(B_k,B_j) - d_h(B_i,B_j))}} \tag{12}$$

By merging Equations (7), (9), (11) and (12) into Equation (6), we reach the final triplet ordinal relation preserving objective function, as shown in Equation (13).

$$\begin{aligned} L &= -w_{ijk} \log \Pi_{i,j,k=1}^{N} P(g_{ijk} \big| B_i, B_j, B_k) \\ &= \sum_{i,j,k=1}^{N} -w_{ijk} \log P(s_{ijk} \big| B_i, B_j, B_k) \\ &= \sum_{i,j,k=1}^{N} -w_{ijk} \log \left(\frac{1}{1+e^{-\alpha(d_h(B_k,B_j)-d_h(B_i,B_j))}}\right)^{g_{ijk}} \left(1 - \frac{1}{1+e^{-\alpha(d_h(B_k,B_j)-d_h(B_i,B_j))}}\right)^{1-g_{ijk}} \\ &= \sum_{i,j,k=1}^{N} w_{ijk}(g_{ijk} \log(1 + e^{-\alpha(d_h(B_k,B_j)-d_h(B_i,B_j))}) + (1-g_{ijk}) \log(1 + \frac{1}{e^{-\alpha(d_h(B_k,B_j)-d_h(B_i,B_j))}})) \\ &= \sum_{i,j,k=1}^{N} w_{ijk}(g_{ijk} \log(e^{-\alpha(d_h(B_k,B_j)-d_h(B_i,B_j))}) + \log(1 + \frac{1}{e^{-\alpha(d_h(B_k,B_j)-d_h(B_i,B_j))}})) \end{aligned} \tag{13}$$

2.5. Triplet Ordinal Quantization Loss

Generally, the sign function is adopted to map the real-valued data output by the last layer of deep neural network into binary codes. However, it generates discrete values and makes the objective function non-deterministic polynomial (NP) hard for optimization [20,36]. To fix this problem, the continuous tanh(·) function is utilized instead of the sign(·) function in this paper. Furthermore, to minimize the quantization loss caused by the continuous relaxation procedure, we expect the output of the tanh(·) function to be close to ±1. Here, we utilize the triplet ordinal cross entropy to formulate the quantization loss. We define the binary code obtained by the tanh(·) function as ${B^i}_{tah}$. B_{ref} is the reference binary code. The ideal encoding result is **1**. Thus, we formulate the quantization loss Q as in Equation (14).

$$\begin{aligned} Q &= \sum_{i=1}^{N} -\log P(1 | (||B^i_{tah}||, \mathbf{1}, ||B_{ref}||)) \\ &= \sum_{i=1}^{N} -\log \phi(-d_h(||B^i_{tah}||, \mathbf{1}) + \delta) \\ &= \sum_{i=1}^{N} \log(1 + e^{-\alpha(-d_h(||B^i_{tah}||,\mathbf{1})+\delta)}) \end{aligned} \tag{14}$$

In Equation (14), the triplet ordinal relation among ($||{B^i}_{tah}||$, **1** and $||B_{ref}||$) is defined as 1 and it indicates that the data pair ($||{B^i}_{tah}||$, **1**) is more similar than the data pair ($\mathbf{1}$, $||B_{ref}||$). Therefore, to minimize the quantization loss, the Hamming distance of the data pair ($||B_{tah}||$, $\mathbf{1}$) should be smaller than the Hamming distance $\delta = d_h(||B_{ref}||, \mathbf{1})$. During the training procedure, we tune the value of δ to balance the optimization complexity and the approximation performance. A small δ value let the encoding results be close to the output of sign function and the training process will become hard. In contrast, a large δ value creates low optimization complexity, but it leads to poor approximation results.

After applying the continuous relaxation mechanism, we compute the Hamming distance of one data pair by Equation (15). $\otimes$ computes the sum of bitwise production value. $f_8(\cdot)$ represents the output of the deep neural network's last layer.

$$d_h(B_i, B_j) = \frac{1}{2}(M - \tanh(f_8(x_i)) \otimes \tanh(f_8(x_j))) \tag{15}$$

Finally, we utilize the back propagation mechanism to optimize the variables of the deep neural network by simultaneously minimizing the triplet ordinal relation cross entropy loss in Equation (13) and the quantization loss in Equation (14).

3. Experimental Setting and Results

In this section, we introduce the comparative experimental setting and evaluate the approximate nearest neighbor search performance of the proposed TOCEH and some state-of-the-art hashing methods.

3.1. Datasets

The comparative experiments are conducted on three large-scale RS image datasets, including UC Merced land use dataset (UCMD) [37], SAT-4 dataset [38] and SAT-6 dataset [38]. The details of these three RS image datasets are introduced below.

1. UCMD [37] stores aerial image scenes with a human label. There are 21 land cover categories, and each category includes 100 images with the normalized size of 256 × 256 pixels. The spatial resolution of each pixel is 0.3 m. We randomly choose 420 images as query samples and the remaining 1680 images are utilized as training samples.
2. The total number of images in SAT-4 [38] is 500k and it includes four broad land cover classes: barren land, grass land, trees and other. The size of images is normalized to 28 × 28 pixels and the spatial resolution of each pixel is 1 m. We randomly select 400k images to train the network and the other 100k images to test the ANN search performance.
3. The SAT-6 [38] dataset contains 405k images covering barren land, buildings, grassland, roads, trees and water bodies. These images are normalized to 28 × 28 pixels size and the spatial resolution of each pixel is 1 m. We randomly select 81k images as query set and the other 324k images as training set.

Some sample images of the above three datasets are shown in Figures 4–6, and the statistics are summarized in Table 2.

3.2. Experimental Settings and Evaluation Matrix

To verify the ANN search performance of the proposed TOCEH method, many state-of-the-art hashing methods, including locality sensitive hashing (LSH) [14], spectral hashing (SH) [17], iterative quantization hashing method (ITQ) [13], k-means hashing (KMH) [12], partial randomness hashing (PRH) [23], deep variational binaries (DVB) [39], deep hashing (DH) [40], DeepBit [41], deep Cauchy hashing (DCH) [11] and twin-bottle neck hashing (TBH) [10] are utilized as the baseline methods. LSH [14], SH [17], ITQ [13] and KMH [12] belong to the shallow methods. During the ANN search experiments, we extract the content information from RS images by AlexNet and the features are represented as 4096-dimension float point data. Then, these shallow hashing methods map the 4096-dimension features into the compact Hamming space and achieve the ANN search task according to the Hamming distance. DCH [11], TBH [10], DVB [39], DH [40], DeepBit [41] and the proposed TOCEH are deep learning hashing methods. They directly generate the RS image's binary feature using an end-to-end mechanism.

Figure 4. Sample images of the UCMD dataset.

Figure 5. Sample images of the SAT-4 dataset.

Figure 6. Sample images of the SAT-6 dataset.

Table 2. Statistics and several parameter settings of three datasets.

	UCMD	SAT4	SAT6
Class Number	21	4	6
Image Size	256 × 256	28 × 28	28 × 28
Dataset Size	2100	500,000	405,000
Training Set	1470	400,000	360,000
Query Set	630	100,000	45,000
Ground Truth	100	1000	1000

The training process and comparative experiments are conducted on a high-performance computer with GPU Tesla T4 16 GB, CPU Intel Xeon 6242R 3.10 GHz and 64 GB RAM.

To evaluate the ANN search performance, two widely used standards, mean average precision (mAP) and recall curves, are employed in this paper.

The recall curve represents the fraction of the positive samples that are successfully retrieved. The definition of recall is shown in Equation (16). $\#(\cdot)$ returns the number of samples.

$$recall = \frac{\#(retrieved\ positive\ samples)}{\#(all\ positive\ samples)} \tag{16}$$

Mean average precision value expresses the return rate of positive samples as defined in Equation (17). $|total|$ is the total number of retrieved samples. K_i returns the number of positive samples of the *i*-th query sample. *rank*(*j*) is the ranking number of the *j*-th positive sample in the retrieved results.

$$mAP = \frac{1}{|total|} \sum_{i=1}^{|total|} \frac{1}{K_i} \sum_{j=1}^{K_i} \frac{j}{rank(j)} \tag{17}$$

3.3. Experimental Results

3.3.1. Qualitative Analysis

In this section, we show the qualitative image search results on the UCMD dataset [37]. The proposed TOCEH and the other seven state-of-the-art methods separately map the image content information into 64-, 128- and 256-bit binary code. The images with minimal Hamming distance to the query sample are returned as retrieval results and the false images are marked with red rectangles, as shown in Figures 7–9.

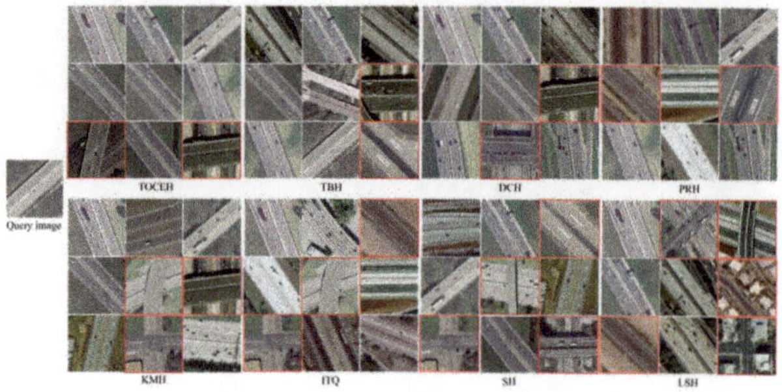

Figure 7. The RS image retrieval results on the UCMD dataset, and the length of the binary code is 64. The false images are marked with red rectangles.

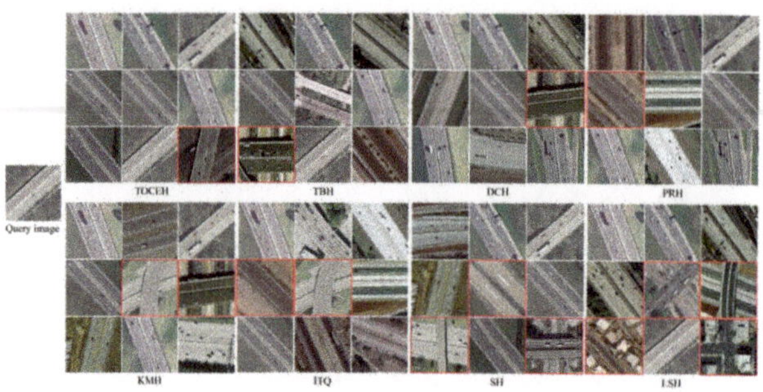

Figure 8. The RS image retrieval results on the UCMD dataset, and the length of the binary code is 128. The false images are marked with red rectangles.

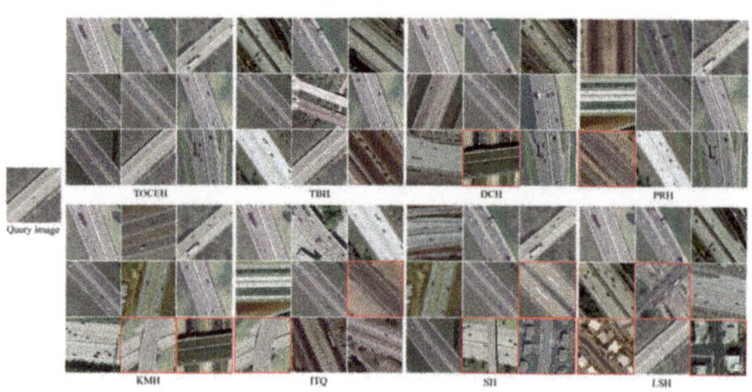

Figure 9. The RS image retrieval results on the UCMD dataset, and the length of the binary code is 256. The false images are marked with red rectangles.

From the RS image retrieval results, we intuitively know that TOCEH owns the best retrieval results. When encoding RS image content as a 64-bit binary code in Figure 6, TOCEH and TBH [10] return two false positive images. Correspondingly, the number of false images retrieved by the other six methods is larger than two. Furthermore, the false RS images' ranking position in TOCEH is higher than that in TBH [10], which gives TOCEH a larger mAP value. In Figure 7, the length of the binary code is 128. One RS image is incorrectly returned by TOCEH, TBH [10], DCH [11] and PRH [23], and the false image has a relatively higher ranking position in TOCEH. As the number of binary bits increases to 256, only TOCEH and TBH [10] retrieve no false image, as shown in Figure 8.

3.3.2. Quantitative Analysis

In this section, we adopt *recall* curves and *mAP* to quantitatively analyze the ANN search performance of the proposed TOCEH and the other seven state-of-the-art hashing methods. These hashing methods separately generate 64-, 128-, and 256-bit binary code to represent the image content. The *mAP* values are in Tables 3–5. The recall curves are shown in Figures 10–12.

Table 3. Comparison of *mAP* with different binary code lengths on UCMD.

	TOCEH	TBH	DVB	DCH	DeepBit	PRH	DH	KMH	ITQ	SH	LSH
64-bit	0.3914	0.3415	0.3261	0.2917	0.2657	0.2462	0.2296	0.2135	0.1986	0.1724	0.1637
128-bit	0.5479	0.4638	0.4259	0.3963	0.3781	0.3527	0.3467	0.2816	0.2462	0.2015	0.1842
256-bit	0.5837	0.4975	0.4757	0.4319	0.4197	0.3746	0.3528	0.3168	0.2673	0.2351	0.2148

Table 4. Comparison of *mAP* with different binary code lengths on SAT-4.

	TOCEH	TBH	DVB	DCH	DeepBit	PRH	PRH	KMH	ITQ	SH	LSH
64-bit	0.7011	0.5768	0.5271	0.4862	0.4522	0.4361	0.4139	0.3946	0.3657	0.3482	0.3407
128-bit	0.7236	0.6124	0.5537	0.4986	0.4794	0.4528	0.4385	0.4173	0.3856	0.3724	0.3615
256-bit	0.7528	0.6345	0.6149	0.5128	0.5068	0.4857	0.4653	0.4361	0.4285	0.4152	0.3986

Table 5. Comparison of *mAP* with different binary code lengths on SAT-6.

	TOCEH	TBH	DVB	DCH	DeepBit	PRH	DH	KMH	ITQ	SH	LSH
64-bit	0.7124	0.5826	0.5446	0.4936	0.4725	0.4586	0.4352	0.4125	0.3764	0.3695	0.3628
128-bit	0.7351	0.6268	0.5841	0.5174	0.4921	0.4795	0.4596	0.4281	0.3927	0.3864	0.3752
256-bit	0.7842	0.6527	0.6261	0.5394	0.5175	0.4972	0.4628	0.4516	0.4359	0.4238	0.4175

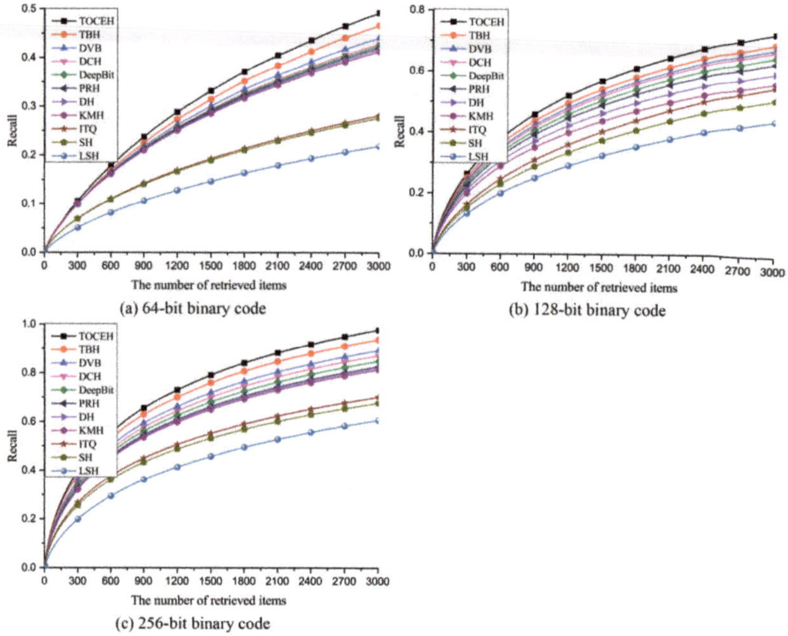

Figure 10. The recall curves of all comparative methods on UCMD; the data are separately encoded as (**a**) 64-, (**b**) 128- and (**c**) 256-bit binary code.

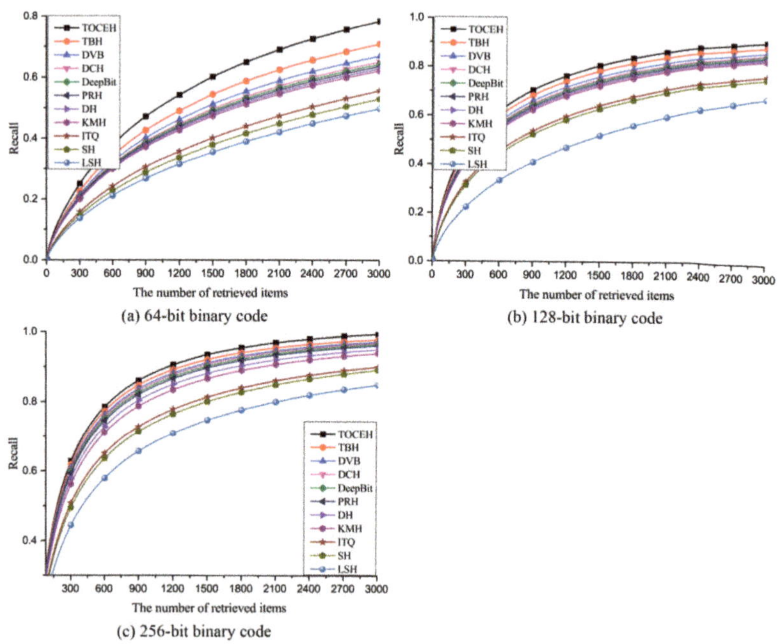

Figure 11. The recall curves of all comparative methods on SAT-4 and the data are separately encoded as (**a**) 64-, (**b**) 128- and (**c**) 256-bit binary code.

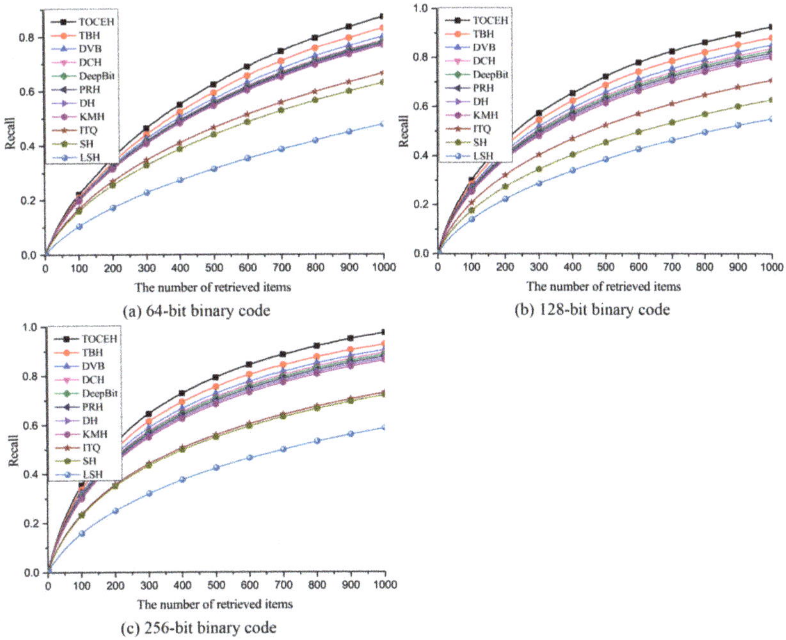

Figure 12. The recall curves of all comparative methods on SAT-6 and the data are separately encoded as (**a**) 64-, (**b**) 128- and (**c**) 256-bit binary code.

From the quantitative results, we know TOCEH achieves the best ANN search performance. LSH [14], the data-independent hashing algorithm, randomly generates hashing projection functions without a training process. As a result, the ANN search performance of LSH cannot drastically improve as the number of binary bits increases [9]. In contrast, the proposed TOCEH and the other nine comparative hashing methods utilize a machine learning mechanism to obtain the hashing functions, which are adaptive to the training data distribution. Thus, these machine-learning-based hashing algorithms achieve a better ANN search performance than LSH. SH [17] establishes a spectral graph to measure the similarity relation among samples, and divides the samples into different cluster groups by spectral graph partition. Then, SH [17] assigns the same code to the samples in the same group. For a large-scale RS image dataset, the time complexity of establishing a spectral graph would be high. Both ITQ [13] and KMH [12] first learn encoding centers, then assign the samples as the same binary code as their nearest center. ITQ [13] considers the fixed vertexes of a hyper cubic as centers, but they are not well adapted to the training data distribution. KMH [12] learns the encoding centers with minimal quantization loss and similarity loss by a k-means iterative mechanism. This measure effectively helps KMH improve the ANN search performance. To balance the training complexity and ANN search performance, PRH [23] employs the partial randomness and partial learning strategy to generate hashing functions. LSH [14], SH [17], ITQ [13], KMH [12] and PRH [23] belong to the shallow hashing algorithms, and their performances relate to the quality of the intermediate high dimensional features. To eliminate this effect, TOCEH, TBH [10], DVB [39], DH [40], DeepBit [41] and DCH [11] adopt a deep learning framework to learn the end-to-end binary feature, which can further boost the ANN search performance. The classical DH [40] proposes three constraints at the top layer of the deep network: the quantization loss, balance bits and independent bits. However, the pair-wise similarity preserving or the triplet ordinal relation preserving is not considered in DH. This may lead a poor performance of DH. The same problem also exists in DeepBit [41]. However, DeepBit augments the training data with different rotations and further updates the parameters of the network. This measure helps DeepBit to obtain a better ANN search performance than DH. For most deep hashing, it is hard to unveil the intrinsic structure of the whole sample space by simply regularizing the output codes within each single training batch. In contrast, the conditional auto-encoding variational Bayesian networks are introduced in DVB to exploit the feature space structure of the training data using the latent variables. DCH [11] pre-trains a similarity graph and expects that the probability distribution in the Hamming space should be consistent with that in the Euclidean space. TBH [10] abandons the process of the pre-computing similarity graph and embeds it in the deep neural network. TBH aims to preserve the similarity between the original data and the data decoded from the binary feature. Both TBH [10] and DCH [11] aim to preserve the pair-wise similarity, and it is difficult to capture the hyper structure among RS images. TOCEH establishes a tensor graph representing the triplet ordinal relation among RS images in both Hamming space and Euclidean space. During the training process, TOCEH expects that the triplet ordinal relation graphs have the same distribution in different spaces. Thus, it can enhance the ability of preserving the Euclidean ranking orders in the Hamming space. As discussed above, TOCEH can achieve the best RS image retrieval results.

3.3.3. Ablation Experiments

To guarantee the ANN search performance of the obtained binary codes, the TOCEH algorithm proposes two key components: the triplet ordinal cross entropy loss and the triplet ordinal quantization loss. Here, we conduct the comparative experiments to analyze these two components. TOCEL only utilizes the triplet ordinal cross entropy loss as the objective function for deep learning binary code. The deep hashing TOQL only employs the triplet ordinal quantization loss as the objective function. TOCEH, TOCEL and TOQL separately map the data into 64- and 128-bit binary code. The ANN search results are shown in Figures 13–15.

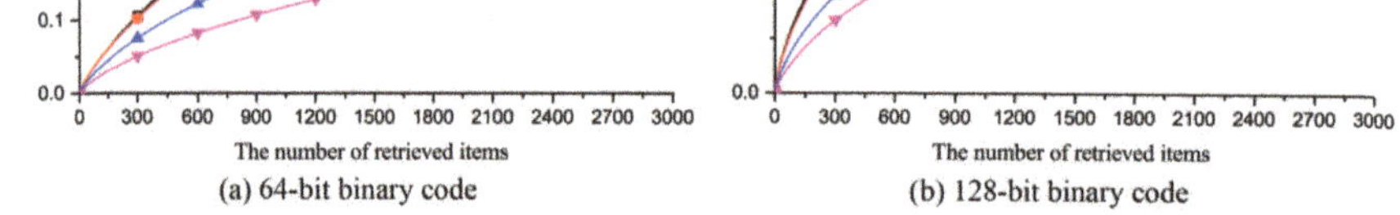

Figure 13. The ablation experiments on UCMD. The data are separately encoded as (**a**) 64- and (**b**) 128-bit binary code.

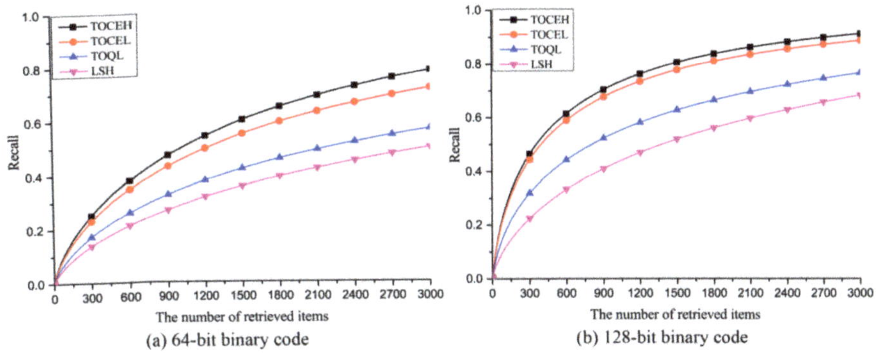

Figure 14. The ablation experiments on SAT-4. The data are separately encoded as (**a**) 64- and (**b**) 128-bit binary code.

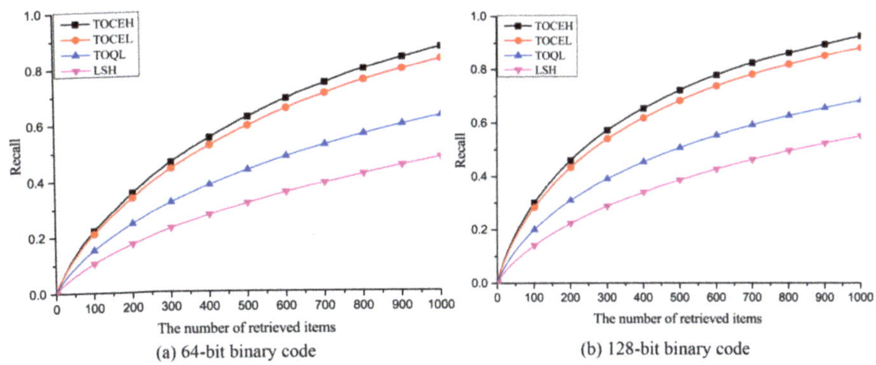

Figure 15. The ablation experiments on SAT-6. The data are separately encoded as (**a**) 64- and (**b**) 128-bit binary code.

From the comparative results, we know that both the triplet ordinal cross entropy loss and the triplet ordinal quantization loss play important roles in improving the performance of TOCEH. The triplet ordinal cross entropy loss minimizes the inconsistency between the probability distributions of the triplet ordinal relations in different spaces. For example, the data pair (x_i, x_j) is more similar than data pair (x_j, x_k) in the Euclidean space. Then, to minimize the triplet ordinal cross entropy loss, it should be a larger probability to assign x_i and x_j as similar binary codes. Without the triplet ordinal cross entropy loss, TOQL randomly generates the samples' binary codes. LSH algorithm also randomly generates the hashing functions. Thus, the ANN search performance of TOQL is almost the same as that of LSH. To fix the NP hard problem of the objective function, we apply the continuous relaxation mechanism to the binary encoding procedure. Furthermore, we define the triplet ordinal quantization loss to minimize the loss between the binary codes and the corresponding continuous variable. Without the triplet ordinal quantization loss, the difference between the optimized variables and the binary encoding results would become larger in TOCEL. Thus, TOCEL has a relatively inferior ANN search performance. As discussed above, both the triplet ordinal cross entropy loss and the triplet ordinal quantization loss are necessary for the TOCEH algorithm.

4. Conclusions

In this paper, to boost the RS image search performance in the Hamming space, we propose a novel deep hashing method called triplet ordinal cross entropy hashing (TOCEH) to learn an end-to-end binary feature of an RS image. Generally, most of the existing hashing methods place emphasis on preserving point-wise or pair-wise similarity.

In contrast, TOCEH establishes a tensor graph to capture the triplet ordinal relation among RS images and defines the triplet ordinal relation preserving problem as the formulation of minimizing the cross entropy value. Then, TOCEH achieves the aim of preserving triplet ordinal relation by minimizing the inconsistency between the probability distributions of the triplet ordinal relations in different spaces. During the training process, to avoid the NP hard problem, we apply continuous relaxation to the binary encoding process. Furthermore, we define a quantization function based on the triplet ordinal relation preserving restriction, which can reduce the loss caused by the continuous procedure. Finally, the extensive comparative experiments conducted on three large-scale RS image datasets, including UCMD, SAT-4 and SAT-6, show that the proposed TOCEH outperforms many state-of-the-art hashing methods in RS image search tasks.

Author Contributions: Conceptualization, Z.W. and P.L.; methodology, Z.W. and N.W.; software, P.L. and X.Y.; validation, N.W., X.Y. and B.Y.; formal analysis, Z.W. and N.W.; investigation, P.L. and X.Y.; resources, B.Y.; data curation, B.Y.; writing—original draft preparation, Z.W.; writing—review and editing, P.L.; visualization, N.W. and X.Y.; supervision, Z.W. and P.L.; project administration, Z.W. and P.L.; funding acquisition, Z.W. All authors have read and agreed to the published version of the manuscript.

Funding: This research was funded by the National Natural Science Foundation of China, grant number 61841602, the Natural Science Foundation of Shandong Province of China, grant number ZR2018PF005, and the Fundamental Research Funds for the Central Universities, JLU, grant number 93K172021K12.

Acknowledgments: The authors express their gratitude to the institutions that supported this research: Shandong University of Technology (SDUT) and Jilin University (JLU).

Conflicts of Interest: The authors declare no conflict of interest.

References

1. Cheng, Q.; Gan, D.; Fu, P.; Huang, H.; Zhou, Y. A Novel Ensemble Architecture of Residual Attention-Based Deep Metric Learning for Remote Sensing Image Retrieval. *Remote Sens.* **2021**, *13*, 3445. [CrossRef]
2. Shan, X.; Liu, P.; Wang, Y.; Zhou, Q.; Wang, Z. Deep Hashing Using Proxy Loss on Remote Sensing Image Retrieval. *Remote Sens.* **2021**, *13*, 2924. [CrossRef]
3. Shan, X.; Liu, P.; Gou, G.; Zhou, Q.; Wang, Z. Deep Hash Remote Sensing Image Retrieval with Hard Probability Sampling. *Remote Sens.* **2020**, *12*, 2789. [CrossRef]
4. Kong, J.; Sun, Q.; Mukherjee, M.; Lloret, J. Low-Rank Hypergraph Hashing for Large-Scale Remote Sensing Image Retrieval. *Remote Sens.* **2020**, *12*, 1164. [CrossRef]
5. Han, L.; Li, P.; Bai, X.; Grecos, C.; Zhang, X.; Ren, P. Cohesion Intensive Deep Hashing for Remote Sensing Image Retrieval. *Remote Sens.* **2020**, *12*, 101. [CrossRef]
6. Hou, Y.; Wang, Q. Research and Improvement of Content Based Image Retrieval Framework. *Int. J. Pattern. Recogn.* **2018**, *32*, 1850043.1–1850043.14. [CrossRef]
7. Liu, Y.; Zhang, D.; Lu, G.; Ma, W.Y. A survey of content-based image retrieval with high-level semantics. *Pattern. Recogn.* **2007**, *40*, 262–282. [CrossRef]
8. Wang, J.; Zhang, T.; Song, J.; Sebe, N.; Shen, H.T. A Survey on Learning to Hash. *IEEE Trans. Pattern. Anal.* **2018**, *40*, 769–790. [CrossRef]
9. Wang, J.; Liu, W.; Kumar, S.; Chang, S.F. Learning to Hash for Indexing Big Data—A Survey. *Proc. IEEE* **2016**, *104*, 34–57. [CrossRef]
10. Shen, Y.; Qin, J.; Chen, J.; Yu, M.; Liu, L.; Zhu, F.; Shen, F.; Shao, L. Auto-encoding twin-bottleneck hashing. In Proceedings of the Conference on Computer Vision and Pattern Recognition, Seattle, WA, USA, 13–19 June 2020; pp. 2815–2824.
11. Cao, Y.; Long, M.; Liu, B.; Wang, J. Deep cauchy hashing for hamming space retrieval. In Proceedings of the Conference on Computer Vision and Pattern Recognition, Salt Lake City, UT, USA, 18–23 June 2018; pp. 1229–1237.
12. He, K.; Wen, F.; Sun, J. K-means hashing: An affinity-preserving quantization method for learning binary compact codes. In Proceedings of the Conference on Computer Vision and Pattern Recognition, Portland, OR, USA, 23–28 June 2013; pp. 2938–2945.
13. Gong, Y.; Lazebnik, S.; Gordo, A.; Perronnin, F. Iterative Quantization: A Procrustean Approach to Learning Binary Codes for Large-Scale Image Retrieval. *IEEE Trans. Pattern. Anal.* **2013**, *35*, 2916–2929. [CrossRef]
14. Datar, M.; Immorlica, N.; Indyk, P.; Mirrokni, V.S. Locality-sensitive hashing scheme based on p-stable distributions. In Proceedings of the 20th ACM Symposium on Computational Geometry, Brooklyn, NY, USA, 8–11 June 2004; pp. 253–262.

15. Cao, Y.; Liu, B.; Long, M.; Wang, J. HashGAN: Deep learning to hash with pair conditional Wasserstein GAN. In Proceedings of the Conference on Computer Vision and Pattern Recognition, Salt Lake City, UT, USA, 18–23 June 2018; pp. 1287–1296.
16. Liu, H.; Wang, R.; Shan, S.; Chen, X. Deep supervised hashing for fast image retrieval. In Proceedings of the Conference on Computer Vision and Pattern Recognition, Las Vegas, NV, USA, 27–30 June 2016; pp. 2064–2072.
17. Weiss, Y.; Torralba, A.; Fergus, R. Spectral hashing. In Proceedings of the Advances in Neural Information Processing Systems, Vancouver, BC, Canada, 8–11 December 2008; pp. 1753–1760.
18. Lowe, D.G. Distinctive image features from scale-invariant keypoints. *Int. J. Comput. Vis.* **2004**, *60*, 91–110. [CrossRef]
19. Oliva, A.; Torralba, A. Modeling the shape of the scene: A holistic representation of the spatial envelope. *Int. J. Comput. Vis.* **2001**, *42*, 145–175. [CrossRef]
20. Shen, F.; Xu, Y.; Liu, L.; Yang, Y.; Huang, Z.; Shen, H.T. Unsupervised Deep Hashing with Similarity-Adaptive and Discrete Optimization. *IEEE Trans. Pattern. Anal.* **2018**, *40*, 3034–3044. [CrossRef] [PubMed]
21. Wang, Y.; Song, J.; Zhou, K.; Liu, Y. Unsupervised deep hashing with node representation for image retrieval. *Pattern. Recogn.* **2021**, *112*, 107785. [CrossRef]
22. Zhang, M.; Zhe, X.; Chen, S.; Yan, H. Deep Center-Based Dual-Constrained Hashing for Discriminative Face Image Retrieval. *Pattern. Recogn.* **2021**, *117*, 107976. [CrossRef]
23. Li, P.; Ren, P. Partial Randomness Hashing for Large-Scale Remote Sensing Image Retrieval. *IEEE Geosci. Remote Sens.* **2017**, *14*, 1–5. [CrossRef]
24. Demir, B.; Bruzzone, L. Hashing-Based Scalable Remote Sensing Image Search and Retrieval in Large Archives. *IEEE Trans. Geosci. Remote Sens.* **2016**, *54*, 892–904. [CrossRef]
25. Li, Y.; Zhang, Y.; Huang, X.; Zhu, H.; Ma, J. Large-Scale Remote Sensing Image Retrieval by Deep Hashing Neural Networks. *IEEE Trans. Geosci. Remote Sens.* **2017**, *56*, 950–965. [CrossRef]
26. Fan, L.; Zhao, H.; Zhao, H. Distribution Consistency Loss for Large-Scale Remote Sensing Image Retrieval. *Remote Sens.* **2020**, *12*, 175. [CrossRef]
27. Krizhevsky, A.; Sutskever, I.; Hinton, G.E. ImageNet classification with deep convolutional neural networks. In Proceedings of the NIPS, Lake Tahoe, NV, USA, 3–6 December 2012; pp. 1106–1114.
28. Russakovsky, O.; Deng, J.; Su, H.; Krause, J.; Satheesh, S.; Ma, S.; Huang, Z.; Karpathy, A.; Khosla, A.; Bernstein, M.S.; et al. ImageNet Large Scale Visual Recognition Challenge. *Int. J. Comput. Vis.* **2015**, *115*, 211–252. [CrossRef]
29. Wang, Z.; Sun, F.Z.; Zhang, L.B.; Wang, L.; Liu, P. Top Position Sensitive Ordinal Relation Preserving Bitwise Weight for Image Retrieval. *Algorithms* **2020**, *13*, 18. [CrossRef]
30. Liu, H.; Ji, R.; Wang, J.; Shen, C. Ordinal Constraint Binary Coding for Approximate Nearest Neighbor Search. *IEEE Trans. Pattern Anal.* **2019**, *41*, 941–955. [CrossRef] [PubMed]
31. Liu, H.; Ji, R.; Wu, Y.; Liu, W. Towards optimal binary code learning via ordinal embedding. In Proceedings of the 30th AAAI Conference on Artificial Intelligence, Phoenix, AZ, USA, 12–17 February 2016; pp. 1258–1265.
32. Wang, J.; Liu, W.; Sun, A.X.; Jiang, Y.G. Learning hash codes with listwise supervision. In Proceedings of the IEEE International Conference on Computer Vision, Sydney, NSW, Australia, 1–8 December 2013; pp. 3032–3039.
33. Norouzi, M.; Fleet, D.J.; Salakhutdinov, R. Hamming distance metric learning. In Proceedings of the Advances in Neural Information Processing Systems, Lake Tahoe, NV, USA, 3–6 December 2012; pp. 1061–1069.
34. Wang, Q.; Zhang, Z.; Luo, S. Ranking preserving hashing for fast similarity search. In Proceedings of the International Conference on Artificial Intelligence, Buenos Aires, Argentina, 25–31 July 2015; pp. 3911–3917.
35. Liu, L.; Shao, L.; Shen, F.; Yu, M. Discretely coding semantic rank orders for supervised image hashing. In Proceedings of the Computer Vision and Pattern Recognition, Honolulu, HI, USA, 21–26 July 2017; pp. 5140–5149.
36. Chen, S.; Shen, F.; Yang, Y.; Xu, X.; Song, J. Supervised hashing with adaptive discrete optimization for multimedia retrieval. *Neurocomputing* **2017**, *253*, 97–103. [CrossRef]
37. Yang, Y.; Newsam, S.D. Bag-of-visual-words and spatial extensions for land-use classification. In Proceedings of the 18th SIGSPATIAL International Conference on Advances in Geographic Information Systems, San Jose, CA, USA, 3–5 November 2010; pp. 270–279.
38. Basu, S.; Ganguly, S.; Mukhopadhyay, S.; DiBiano, R.; Karki, M.; Nemani, R.R. DeepSat: A learning framework for satellite imagery. In Proceedings of the 23rd SIGSPATIAL International Conference on Advances in Geographic Information Systems, Bellevue, WA, USA, 3–6 November 2015; pp. 1–10.
39. Shen, Y.; Liu, L.; Shao, L. Unsupervised Binary Representation Learning with Deep Variational Networks. *Int. J. Comput. Vis.* **2019**, *127*, 1614–1628. [CrossRef]
40. Liong, V.E.; Lu, J.; Wang, G.; Moulin, P.; Zhou, J. Deep hashing for compact binary codes learning. In Proceedings of the Conference on Computer Vision and Pattern Recognition, Boston, MA, USA, 7–12 June 2015; pp. 1063–6919.
41. Lin, K.; Lu, J.; Chen, C.S.; Zhou, J. Learning compact binary descriptors with unsupervised deep neural networks. In Proceedings of the Conference on Computer Vision and Pattern Recognition (CVPR), Las Vegas, NV, USA, 27–30 June 2016; pp. 1063–6919.

remote sensing

MDPI

Article

An Improved Swin Transformer-Based Model for Remote Sensing Object Detection and Instance Segmentation

Xiangkai Xu, Zhejun Feng, Changqing Cao *, Mengyuan Li, Jin Wu, Zengyan Wu, Yajie Shang and Shubing Ye

School of Physics and Optoelectronic Engineering, Xidian University, 2 South TaiBai Road, Xi'an 710071, China; xkxu@stu.xidian.edu.cn (X.X.); zhjfeng@mail.xidian.edu.cn (Z.F.); myli151024@stu.xidian.edu.cn (M.L.); jinw9824@stu.xidian.edu.cn (J.W.); zywu_21@stu.xidian.edu.cn (Z.W.); 20051212174@stu.xidian.edu.cn (Y.S.); sbye@stu.xidian.edu.cn (S.Y.)

* Correspondence: chqcao@mail.xidian.edu.cn

Abstract: Remote sensing image object detection and instance segmentation are widely valued research fields. A convolutional neural network (CNN) has shown defects in the object detection of remote sensing images. In recent years, the number of studies on transformer-based models increased, and these studies achieved good results. However, transformers still suffer from poor small object detection and unsatisfactory edge detail segmentation. In order to solve these problems, we improved the Swin transformer based on the advantages of transformers and CNNs, and designed a local perception Swin transformer (LPSW) backbone to enhance the local perception of the network and to improve the detection accuracy of small-scale objects. We also designed a spatial attention interleaved execution cascade (SAIEC) network framework, which helped to strengthen the segmentation accuracy of the network. Due to the lack of remote sensing mask datasets, the MRS-1800 remote sensing mask dataset was created. Finally, we combined the proposed backbone with the new network framework and conducted experiments on this MRS-1800 dataset. Compared with the Swin transformer, the proposed model improved the mask AP by 1.7%, mask AP_S by 3.6%, AP by 1.1% and AP_S by 4.6%, demonstrating its effectiveness and feasibility.

Keywords: instance segmentation; object detection; Swin transformer; remote sensing image; cascade mask R-CNN

Citation: Xu, X.; Feng, Z.; Cao, C.; Li, M.; Wu, J.; Wu, Z.; Shang, Y.; Ye, S. An Improved Swin Transformer-Based Model for Remote Sensing Object Detection and Instance Segmentation. *Remote Sens.* **2021**, *13*, 4779. https://doi.org/10.3390/rs13234779

Academic Editors: Fahimeh Farahnakian, Jukka Heikkonen and Pouya Jafarzadeh

Received: 19 October 2021
Accepted: 22 November 2021
Published: 25 November 2021

Publisher's Note: MDPI stays neutral with regard to jurisdictional claims in published maps and institutional affiliations.

1. Introduction

With the continuous advancement of science and technology, remote sensing technology is eagerly developing. The feature information contained in remote sensing images has become more abundant, and a large amount of valuable information can be extracted from it and used for scientific and technological research. Machine learning based on probability and statistics usually requires complex feature description and suffers from obvious deficiencies when dealing with complex object detection and segmentation problems [1,2]. The deep structure and feature learning capabilities of deep learning achieved great success in the field of image processing, and a large number of scholars also applied it to the field of remote sensing object detection and instance segmentation [3,4]. Remote sensing image object detection and segmentation tasks have an important research significance and value for the development of aviation and remote sensing fields, and have broad application prospects in many practical scenarios, such as marine monitoring, ship management and control, and ground urban planning. In urban planning, the extraction of relevant urban metrics is important for characterizing urban typologies, and image segmentation based on deep learning is optimal for the extraction of road features in marginal areas located in urban environments [5].

Instance segmentation has become an important, complex and challenging field in machine vision research. Instance segmentation can be defined as a technology that simultaneously solves the problem of object detection and semantic segmentation. As with semantic segmentation, it not only has the characteristics of pixel level classification, but also has the characteristics of object detection, where different instances must be located, even if they are of the same type. Figure 1 shows the differences and relationships among object detection, semantic segmentation and instance segmentation.

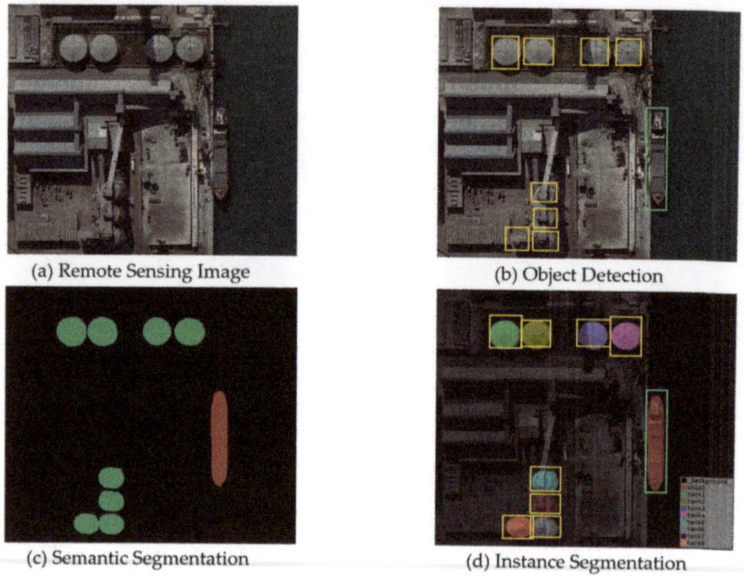

Figure 1. Examples of remote sensing image (**a**), object detection (**b**), semantic segmentation (**c**), and instance segmentation (**d**).

Since the emergence of the two-stage object detection algorithm, various object detection and segmentation algorithms based on convolutional neural networks (CNNs) have emerged, such as the region-based CNN (R-CNN), Faster R-CNN [6], and Mask R-CNN [7]. In recent years, although there are many excellent algorithms, such as the path aggregation network (PANet) [8], Mask Score R-CNN [9], Cascade Mask R-CNN [10] and segmenting objects by locations (SOLO) [11], typical problems remain, such as inaccurate segmentation edges and the establishment of global relations. If the long-range dependencies are captured by dilated convolution or by increasing the number of channels, dimensional disasters will occur due to the expansion of the model.

CNNs are useful for extracting local effective information, but they lack the ability to extract long-range features from global information. Inspired by the use of self-attention in the transformer [12] and in order to mine long-range correlation dependencies in text, many computer vision tasks propose the use of self-attention mechanisms to effectively overcome the limitations of CNNs. Self-attention mechanisms can obtain relationships between long-range elements faster and attend over different regions of the image and integrate information across the entire image. Vision transformer (ViT) [13] is a representative state-of-the-art (SOTA) work in the field of image recognition. It only uses a self-attention mechanism, which makes the image recognition rate far higher than models based on CNNs. End-to-end object detection with transformers (DETR) [14] first involved the use of transformers in high-level vision. This adds positional information to supplement image features and inputs them in the transformer structure to obtain the predicted class label and bounding box. Although transformer-based algorithms have greatly improved the object detection effect, there are still serious problems in the CV field:

1. Low detection performance for small-scale objects, and weak local information acquisition capabilities.

2. The current transformer-based framework is mostly used for image classification, but it is difficult for a single-level transformer to produce good results for the instance segmentation of densely predicted scenes. This has a great impact on object detection and instance segmentation in remote sensing images with a high resolution, a complex background, and small objects.

In order to solve these problems, there are a few works applying ViT models to the dense vision tasks of object detection and semantic segmentation via direct upsampling or deconvolution but with a relatively lower performance [15,16]. Wang et al. [17] proposed a backbone transformer for dense prediction, named "Pyramid Vision Transformer (PVT)", which designed a shrinking pyramid scheme to reduce the traditional transformer's sequence length. However, its calculation complexity is too large, which is quadratic to image size. Therefore, we chose the Swin transformer [18] as the prototype for our design of the backbone network. The Swin transformer builds a hierarchical transformer and performs self-attention calculations in the window area without overlap. The computational complexity is greatly reduced, and it is linearly related to the size of the input image. As a general-purpose visual backbone network, the Swin transformer achieves SOTA performance in tasks such as image classification, object detection, and semantic segmentation. However, the impact of the Swin transformer on context information encoding is limited; it needs to be improved for remote sensing image tasks.

In this paper, we first designed a local perception block and inserted it into each stage. Through the characteristics of dilated convolution, the block extracts a large range of local information from the image, and strengthens the network's learning of local correlation and structural information. We call the improved backbone network the "Local Perception Swin Transformer" (LPSW for short). Secondly, in order to enhance the object detection and instance segmentation of remote sensing images, inspired by the hybrid task cascade (HTC) [19], we designed the spatial attention interleaved execution cascade (SAIEC) network framework. We applied the ideas of the interleaved execution and mask information flow into Cascade Mask R-CNN. Both bounding box regression and mask prediction were combined in a multi-tasking manner. We also added an improved spatial attention module to the mask head, which helps the mask branch to focus on meaningful pixels and suppress meaningless pixels. Finally, we combined the designed LPSW backbone network with the SAIEC framework to form a new network model that achieves a higher accuracy in remote sensing object detection and instance segmentation tasks.

The main contributions of this paper can be summarized as follows:

1. In order to overcome the shortcomings of CNNs' poor ability to extract global information, we chose the Swin transformer as a basic backbone network to build a network model for remote sensing image object detection and instance segmentation.
2. According to the characteristics of remote sensing images, we propose a local perception Swin transformer (LPSW) backbone network. The LPSW combines the advantages of CNNs and transformers to enhance local perception capabilities and improve the detection accuracy of small-scale objects.
3. The spatial attention interleaved execution cascade (SAIEC) network framework is proposed. The mask prediction of the network is enhanced through the multi-tasking manner and the improved spatial attention module. Finally, the LPSW is inserted into the designed network framework as the backbone to establish a new network model that further improves the accuracy of model detection and segmentation.
4. Based on the shortage of existing remote sensing instance segmentation datasets, we selected a total of 1800 multi-object types of images from existing public datasets for annotation and created the MRS-1800 remote sensing mask dataset as the experimental resource for this paper.

2. Related Works

In this section, we introduce some previous works related to object detection and instance segmentation. For comparative analysis, we divide the content into CNN-based and transformer-based object detection and segmentation-related network models.

2.1. CNN-Based Object Detection and Instance Segmentation

In recent years, CNN-based object detection models have developed rapidly. The current object detection algorithms based on deep learning can be divided into two-stage object detection algorithms and single-stage object detection algorithms. Two-stage object detection is mainly represented by a series of regional convolutional neural network (Region-CNN, R-CNN) algorithms: the spatial pyramid pooling network (SPP-Net) [20] solves the problem of redundant operations; Fast R-CNN [21] based on R-CNN and SPP-Net proposes the concept of an region of interest (ROI) pooling layer, which can map the feature maps of different sizes of candidate regions to fixed-size feature maps; Faster R-CNN [6] uses the CNN-based region proposal network (RPN) to replace the selective search algorithm. The RPN can take an image feature map as an input, and then output a series of candidate regions. The single-stage object detection algorithm directly uses a single network to predict the category and location of the object of interest, mainly represented by the you only look once (YOLO) [22] series of algorithms. The single-shot multibox detector (SSD) [23] uses multiple-scale feature maps to perform detection tasks. On the basis of a feature pyramid network (FPN) [24], Tsung-Yi Lin et al. proposed Retinanet [25], which further improved the performance of the single-stage object detection algorithm.

At present, CNN-based instance segmentation algorithms can be divided into two main types: The top-down method and the bottom-up method. Compared with the top-down instance segmentation algorithm, the bottom-up algorithm usually has lower accuracy and more computation, such as the Proposal-Free [26] network.

The top-down method is based on the object detection algorithm. First, the object detection algorithm is used to find the bounding box of the object, semantic segmentation is then performed within the bounding box of each object, and, finally, each segmentation result is output as an instance. In the single-stage instance segmentation algorithm, inspired by YOLO, SOLO [11] directly decouples the instance segmentation problem into category prediction and instance mask generation problems. There is no need to generate bounding boxes during the prediction process. SOLO V2 [27] makes a further adjustment; Center-Mask [28] adds a head network to predict the mask to the single-order end object detection algorithm, FCOS [29], to complete instance segmentation. Although these methods have a certain speed advantage over the two-step method, they are usually unable to achieve the accuracy of the two-step method. In terms of the two-stage algorithm, He Kaiming et al. proposed Mask R-CNN [7], a simple and effective instance segmentation framework. Mask R-CNN adds a mask branch to the head network of Faster R-CNN. Additionally, the original classification branch and regression branch are juxtaposed with the mask branch. Inspired by Mask R-CNN, Shu Liu [8] et al. proposed PANet, which makes full use of shallow network features for instance segmentation; Mask Scoring R-CNN [9], on the basis of the Mask R-CNN, expands with an additional mask branch in order to obtain a more accurate mask. Cascade Mask R-CNN [10] combines Mask R-CNN with Cascade R-CNN, which slightly improves detection accuracy, but it is still unsatisfactory in mask prediction. The key reason for this is that the ability of the CNN to capture long-range features is relatively weak, and the problem of establishing the global relations in the image has not been solved.

2.2. Transformer-Based Object Detection and Instance Segmentation

Transformers are deep neural networks mainly based on the self-attention mechanism [12], and were originally applied in the field of natural language processing and later extended to computer vision tasks. Compared with the CNN network, the advantage of the transformer lies in the use of self-attention to capture global contextual information to establish a long-range dependence on the object, thereby extracting more powerful features. The structure of the self-attention mechanism is shown in Figure 2. For each element in the input sequence, it will generate Q (query), K (key), and V (value) through three learning matrices. In order to determine the relevance between an element and other elements in the sequence, the dot product is calculated between the Q vector of this element with the K vectors of other elements. The results determine the relative importance of patches in the sequence. Then, the results of the dot product are then scaled and fed into a softmax. Finally, the value of the vector for each patch embedding is multiplied by the output of the softmax to find the patch with the high attention scores.

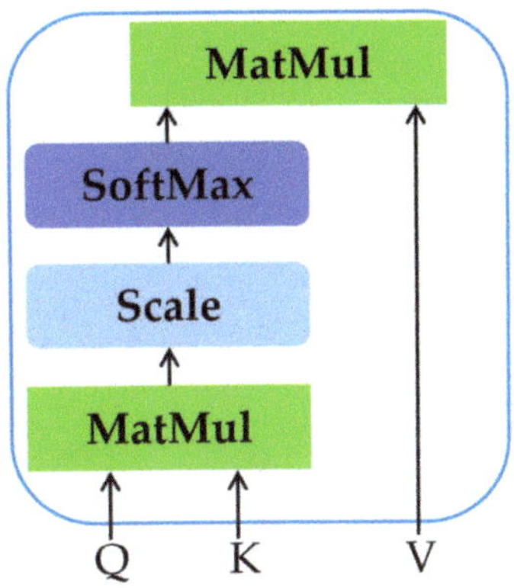

Figure 2. Structure of self-attention mechanism.

In 2020, Carion et al. [14] combined the CNN and the transformer to propose a complete end-to-end DETR object detection framework, applying transformer architecture to object detection for the first time. Zhu [30] et al. proposed the Deformable DETR model that draws on the variable convolutional neural network. Zheng et al. [31] proposed the end-to-end object detection with adaptive clustering transformer (ACT) to reduce the computational complexity of the self-attention module. DETR can naturally extend the panoramic segmentation task by attaching a mask head to the decoder and obtaining competitive results. Wang et al. [32] proposed a transformer-based video instance segmentation (VisTR) model, which takes a series of images as inputs and generates corresponding instance prediction results. Although these models perform well in object detection tasks, they still have many shortcomings. For example, the detection speed of the DETR series models is slow, and the detection performance of small objects is not effective.

For remote sensing images, the image resolution is high, which increases the calculation size of the transformer models. Remote sensing images usually have complex background information and variable object scales, and the training effect of a single-level transformer network is not effective. Based on the above problems, the Swin transformer [18] was proposed to solve the problems of a high amount of computation and the poor detection effect of dense objects, but it still has weak local information acquisition capabilities.

Therefore, for the object detection and instance segmentation of remote sensing images, we need to exploit both the advantages of CNNs to address the underlying vision and those of transformers to address the relationship between visual elements and objects. We need to then design a novel backbone network and detection framework and focus on enhancing the mask prediction ability to improve the detection and segmentation accuracy of remote sensing images.

3. Materials and Methods

This section focuses on the designed network structure. As shown in Figure 3, the model feeds the input image to the local perception Swin transformer (LPSW) backbone network. After the feature map is generated, it is sent to the spatial attention interleaved execution cascade (SAIEC) network model after the FPN structure. The back-end of the model performs feature map classifications, bounding box regression, and instance segmentation tasks. In our model, each bounding box is divided into object and non-object regions. The detailed information of each module is introduced below:

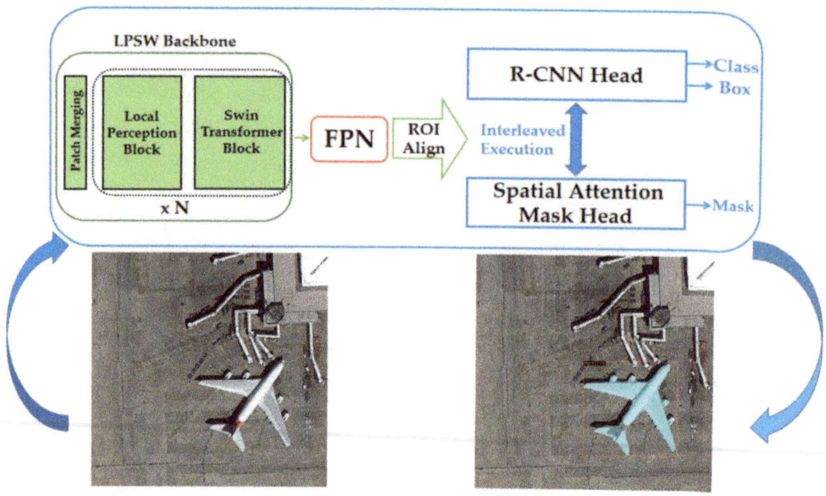

Figure 3. Flow chart of the designed model, which combines the proposed local perception Swin transformer (LPSW) backbone network with the spatial attention interleaved execution cascade (SAIEC) network framework and includes feature pyramid network (FPN) and region of interest (ROI) structures. The new network model can accurately complete remote sensing image object detection and instance segmentation tasks.

3.1. Local Perception Swin Transformer (LPSW) Backbone

The flow chart of the proposed local perception Swin transformer (LPSW) backbone network is shown in Figure 4. The Swin transformer provides four versions of the model, which, from large to small [18], are Swin-T, Swin-S, Swin-B and Swin-L. Taking into account the particularity and computational complexity of remote sensing images, this paper introduces Swin-T. Each stage has 2, 2, 6, and 2 blocks, respectively.

Similar to ViT, it first splits an input RGB image into non-overlapping patches by patch partition layer. Each patch is treated as a "token" and its feature is set as a concatenation of the raw pixel RGB values. The Swin transformer contains four stages to produce a different number of tokens. Given an image with a size of $H \times W$, a token is a raw pixel concatenation vector of an RGB image patch with the size of 4×4. A linear embedding is employed on this token to map it in a vector with the dimension C. Stages 1, 2, 3, and 4 produce $\frac{H}{4} \times \frac{W}{4}$, $\frac{H}{8} \times \frac{W}{8}$, $\frac{H}{16} \times \frac{W}{16}$, and $\frac{H}{32} \times \frac{W}{32}$ tokens, respectively. Each stage consists of a patch merging block (a combination of a patch partition layer and a linear embedding layer), local perception block, and some Swin transformer blocks.

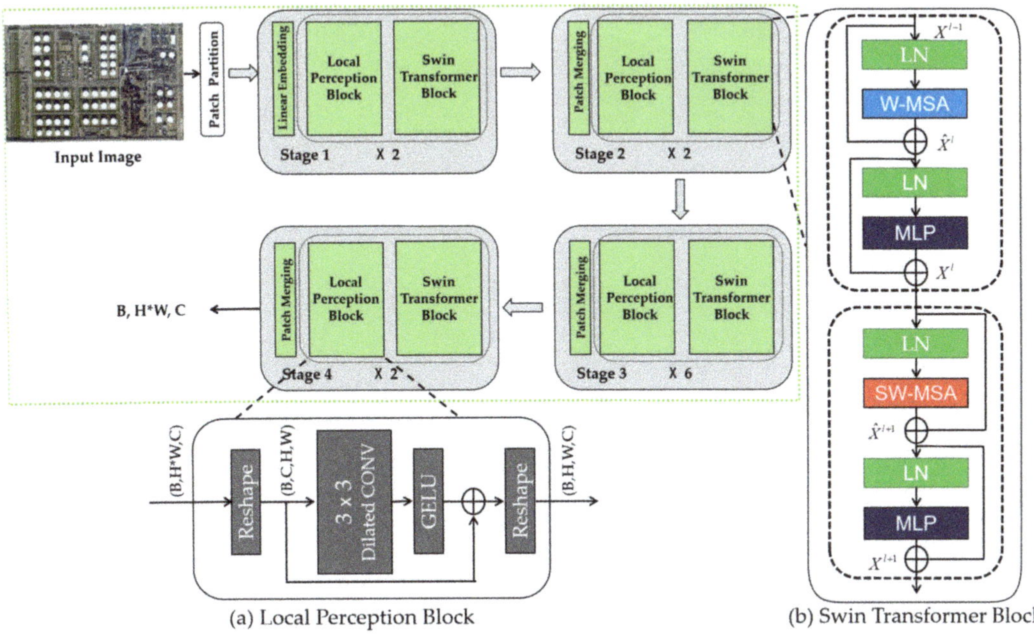

Figure 4. The architecture of the local perception Swin transformer (LPSW). (**a**) The detailed structure of the local perception block; (**b**) the detailed structure of the Swin transformer block.

3.1.1. Swin Transformer Block

The Swin transformer block is the core part of the Swin transformer algorithm. The detailed structure is shown in Figure 4b. The block is composed of window multi-head self-attention (W-MSA), shifted windows multi-head self-attention (SW-MSA) and multilayer perceptron (MLP). Inserting a layernorm (LN) layer in the middle makes the training more stable and uses a residual connection after each module. This part can be expressed as Equation (1):

$$\begin{aligned}
\hat{X}^l &= W - MSA\left(LN\left(X^{l-1}\right)\right) + X^{l-1} \\
X^l &= MLP\left(LN\left(\hat{X}^l\right)\right) + \hat{X}^l \\
\hat{X}^{l+1} &= SW - MSA\left(LN\left(X^l\right)\right) + X^l \\
X^{l+1} &= MLP\left(LN\left(\hat{X}^{l+1}\right)\right) + \hat{X}^{l+1}
\end{aligned} \tag{1}$$

3.1.2. W-MSA and SW-MSA

Compared with the Multi-Head Self Attention (MSA) [12] in the traditional ViT, the W-MSA in the Swin transformer block controls the calculation area in a window as a unit (window size is set to 7 by default). This reduces the amount of network calculations and reduces the complexity to a linear ratio of the image size, as shown in Figure 5. MSA lacks connections across windows. The position of SW-MSA is connected to the W-MSA layer. Therefore, SW-MSA is required to provide a different window segmentation method after W-MSA to realize cross-window communication.

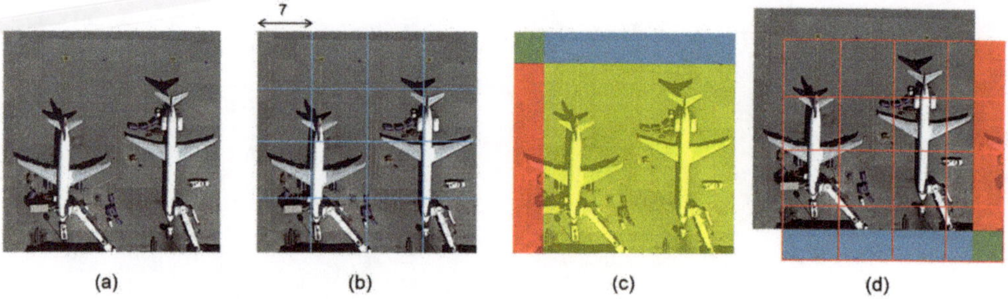

Figure 5. The mechanism of action of the shifted windows. (**a**) The input image; (**b**) Window segmentation (window size is set to 7) of the input image through the window multi-head self-attention (W-MSA); (**c**) Action of the shifted windows; (**d**) A different window segmentation method through the shifted windows multi-head self-attention (SW-MSA).

The result of window segmentation of the input image through W-MSA is shown in Figure 5b. Each cycle of the image is moved up and left by half the size of the window, and the blue and red areas in Figure 5c are then moved to the lower and right sides of the image, respectively, as shown in Figure 5d. On the basis of these shifts, the window is divided according to W-MSA, and SW-MSA has a window segmentation method different from W-MSA.

3.1.3. Local Perception Block (LPB)

Position encoding in a transformer can easily fail to detect the local correlation and structural information of the image. Although the Swin transformer has a shift window scheme of sequential layers in a hierarchical structure, a large range of spatial context information is still not well encoded. In order to alleviate this problem, we proposed the local perception block (LPB), which is inserted in front of the Swin transformer block. The composition of the local perception block is shown in Figure 4a.

Considering that the data flow in the Swin transformer consists of vectors instead of feature maps in traditional CNNs, in the LPB, it firstly reshapes a group of vector features into a spatial feature map. For example, a token (B, H * W, C) is reshaped as a feature map (B, C, H, W). A layer of 3×3 dilated convolution (dilation = 2) and a GELU activation function is then added, and a residual connection is used to increase the extraction of spatial local features while keeping the receptive field sufficiently large. Finally, the feature map is reshaped to (B, H, W, C) and sent to the Swin transformer block.

Through the characteristics of dilated convolution, the receptive field of the spatial image is increased, such that a large range of contextual information can be coded well at different scales. Dilated convolution was proposed by Yu and Koltun [33] in 2015. Compared with the traditional convolution operation, dilated convolution supports the expansion of the receptive field. It is worth noting that the traditional 3×3 convolutions each have a 3×3 field. If it is a dilated convolution (dilation = 2) with the same kernel size, the receptive field is 7×7. Therefore, dilated convolution can extend the corresponding field without a loss of feature resolution.

3.2. *Spatial Attention Interleaved Execution Cascade (SAIEC)*

The proposal of Cascade R-CNN mainly defines the input intersection over union (IoU) threshold of positive and negative samples at different stages. The detector pays more attention to the positive samples within the threshold because of the difference in IoU input at each stage. The output IoU threshold is better than the input IoU threshold, which provides better positive samples for the next stage. Each stage is in a progressive relationship, such that the detector effect can gradually improve. Cascade Mask R-CNN is a product that directly combines Mask R-CNN and Cascade R-CNN. Although it improves in box AP, it does not improve significantly in mask AP. Therefore, inspired by the HTC algorithm, we improve Cascade Mask R-CNN and propose the spatial attention interleaved execution cascade (SAIEC), a new framework of instance segmentation. The specific improvement methods are as follows.

3.2.1. Interleaved Execution and Mask Information Flow

We improved the network head of Cascade Mask R-CNN, as shown in Figure 6. Although Cascade R-CNN forces two branches into each stage, there is no interaction between the two branches during the training process, and they are executed in parallel. Therefore, we propose the interleaved execution; that is, in each stage, the box branch is executed first, and the updated bounding box predictions are then passed to the mask branch to predict the mask, as shown in Figure 6b. In the figure, F represents the features of the backbone network, P is the ROI Align or ROI pooling, and Bi and Mi denote the box and mask head at the i-th stage. This not only increases the interaction between different branches in each stage, but also eliminates the gap between training and testing processes.

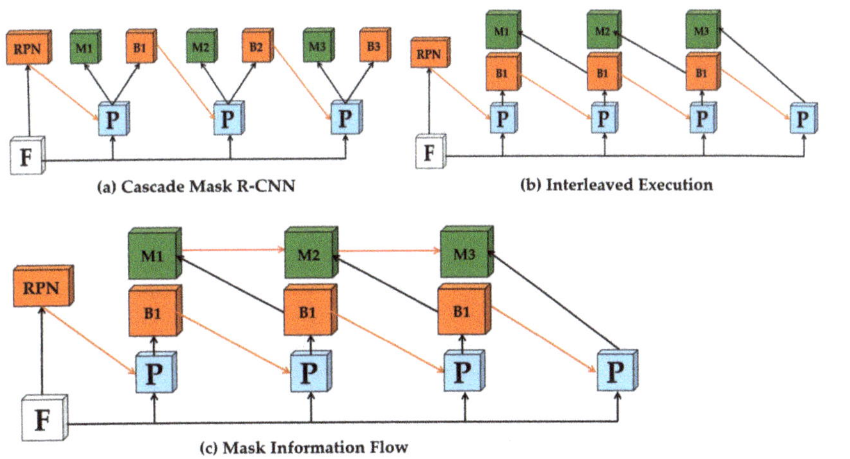

Figure 6. The Cascade Mask R-CNN network head improvement process. (**a**) The Cascade Mask R-CNN network head; (**b**) The addition of the interleaved execution in the network head; (**c**) The final network head structure after adding Mask Information Flow.

At the same time, in the Cascade Mask R-CNN, only the current stage in the box branch has an impact on the next stage, and the mask branch between different stages does not have any direct information flow. In order to solve this problem, we added a connection between adjacent mask branches, as shown in Figure 6c. We provided mask information flow for the mask branch so that M_{i+1} could obtain the features of M_i. The specific implementation is shown above in the red part of Figure 7. We used the feature of Mi to perform feature embedding through a 1×1 convolution, and then entered it into M_{i+1}. In this way, M_{i+1} could obtain the characteristics of not only the backbone, but also the previous stage.

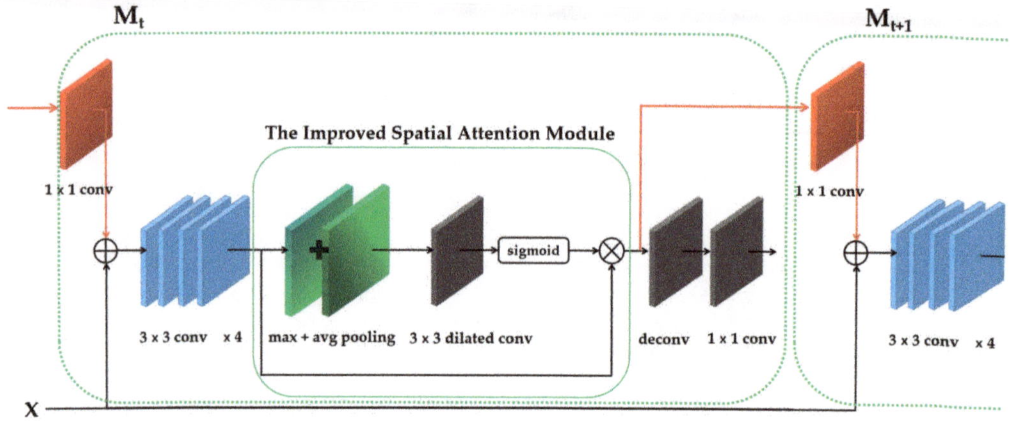

Figure 7. Structure of the spatial attention mask head. It includes the improved spatial attention module, which helping to focus on objects and suppressing noise.

3.2.2. Spatial Attention Mask Head

The attention method [34] helps one to focus on important features and suppress unnecessary noise. Inspired by the spatial attention mechanism [35], we designed the spatial attention mask head, using the spatial attention module to guide the mask head, in order to highlight meaningful pixels and suppress useless pixels. As shown in Figure 7, we improved on the original mask head. We designed an improved spatial attention module and inserted it before transposed convolution. In the spatial attention mask head, the resized local features need to pass through four 3×3 convolution layers with 256 channels, and then pass through the improved spatial attention module. The improved spatial attention module first generates pooled features P_{max} and P_{avg} by both average and max pooling operations, respectively, along the channel axis, and then aggregates them via concatenation. This is followed by a 3×3 dilated convolution layer and is normalized by the sigmoid function. The computation process is summarized as follows:

$$X_{sa} = X_i \otimes sigmoid(D_{3\times3}(P_{\max} \circ P_{avg})) \quad (2)$$

where $\otimes$ denotes element-wise multiplication, X_{sa} is the attention-guided feature map, $D_{3\times3}$ is the 3×3 conv layer, and $\circ$ represents the concatenate operation. Afterwards, 2×2 deconv is used for upsampling and 1×1 conv is used to predict the category of the specific mask. By combining the above structures, we completed the design of the mask branch in the SAIEC framework. The spatial attention mask head not only effectively improves the cross-stage information communication in the network, but also adds a spatial attention mechanism to help with focusing on objects and suppressing noise.

4. Results

4.1. Dataset

There are many conventional object detection datasets. Models that are trained based on conventional datasets do not perform well on remote sensing images. The main reason is the particularity of remote sensing images, and few datasets are related to remote sensing image object detection and instance segmentation. Therefore, we selected images from three public datasets (Object Detection in Optical Remote Sensing Images (DIOR) [36], High Resolution Remote Sensing Detection (HRRSD) [37], and convolutional neural networks for object detection in VHR optical remote sensing images (NWPU VHR-10) [38]) to produce new remote sensing image object detection and instance segmentation datasets. The research group of the Western University of Technology proposed a large-scale benchmark dataset "DIOR" for object detection in optical remote sensing images, which consists of 23,463 images and 190,288 object examples and is based on deep learning. The image size is 800 × 800 pixel, and the resolution ranges from 0.5 m to 30 m. The aerospace remote sensing object detection dataset "NWPU VHR-10," annotated by Northwestern Polytechnical University, has a total of 800 images, including 650 of the objects and 150 background images. Objects include: airplanes, ships, oil tanks, baseball fields, and nets. There are 10 categories of courts, basketball courts, track and field arenas, ports, bridges, and vehicles. HRRSD is a dataset produced by the Optical Image Analysis and Learning Center of the Xi'an Institute of Optics and Fine Mechanics, Chinese Academy of Sciences for research on object detection in high-resolution remote sensing images. The image resolution ranges from 500 × 500 pixels to 1400 × 1000 pixels.

We selected high-resolution images from these three public datasets for manual annotation, and performed data enhancement on the labeled dataset by vertically flipping, horizontally flipping, rotating, and cutting to create the MRS-1800 remote sensing mask dataset. We merged these three classic remote sensing datasets together, which can be regarded as a means of data enhancement and expansion. This approach allowed our dataset to contain more styles and sizes of remote sensing images, making the dataset more challenging. Training our model in this way can help overcome the overfitting problem, thereby improving the robustness and generalization ability of the model.

The MRS-1800 dataset has a total of 1800 remote sensing images. The size of the images varies and the dataset contains a variety of detection objects. The detection objects are divided into three categories: planes, ships, and storage tanks. The specific information of the dataset is shown in Table 1.

Table 1. Number distribution of datasets and class.

Dataset	Dior	Hrrsd	Nwpu Vhr-10	Statistics
Number	403	1093	304	1800
Class	Plane	Ship	Storage tank	
Number	674	687	557	

Figure 8 shows part of the images and mask information of the MRS-1800 dataset. Different sizes of high-resolution images contain different types of objects. We used LabelMe 4.5.9 (Boston, MA, USA) to mark the image with mask information and generate the corresponding "json" files. The dataset contains planes, ships, and storage tanks of different sizes. A total of 16,318 objects were collected, and the object sizes include three types: large, medium and small (ranging from 32 × 32 pixels to 500 × 500 pixels), and the numbers of these types are evenly distributed. We used 1440 images as the training set, 180 images as the validation set, and 180 images as the test set, according to the 8:1:1 allocation ratio.

4.2. Experiments and Analysis

Throughout the experiment, we used a computer equipped with a Geforce RTX 3060 GPU (12 G) as the hardware platform for the experiment. We used pytorch as the DL framework, and the compilation environment was python 3.8 and pytorch 1.8.1. We used multiple classic frameworks such as Mask R-CNN [7], Sparse R-CNN [39], Cascade Mask R-CNN [10], DETR [14], and so on. Additionally, we used Resnet-50 (R-50), the Swin transformer and LPST backbone networks. Suitable pre-training models were chosen to train the self-made dataset, MRS-1800.

We used the same settings in training for the proposed models: multi-scale training (the input size was adjusted so that the short side was between 480 and 800, and the long side was, at most, 1333), the AdamW [40] optimizer (the initial learning rate was 0.00001, the weight decay was 0.05, and the batch size was 1), and 3× scheduling (50 epochs with the learning rate decayed by 10× at 27 epochs). We chose some deep learning indicators as our experimental evaluation criteria, such as frames per second (FPS), AR_S (the average recall measurement value of object frames smaller than 32 × 32 pixels), average precision (AP), AP_{50} (AP measurement value when the IoU threshold is 0.5), AP_{75} (AP measurement value when the IoU threshold is 0.75), AP_S (the AP measurement value of object frames smaller than 32 × 32 pixel), and their mask counterparts: mask AP, mask AP_{50}, mask AP_{75}, and mask AP_S. AP and AR are averaged over multiple intersection over union (IoU) values, where the IoU threshold value ranges from 0.5 to 0.95, with a stride of 0.05. Mask AP is used to comprehensively evaluate the effectiveness of the instance segmentation model. The difference from box AP is only that the objects of the IoU threshold are different. The box AP functions in the standard ordinary ground truth and the IoU value of the prediction box, while the mask AP functions in the ground truth mask and the mask IoU of the prediction mask.

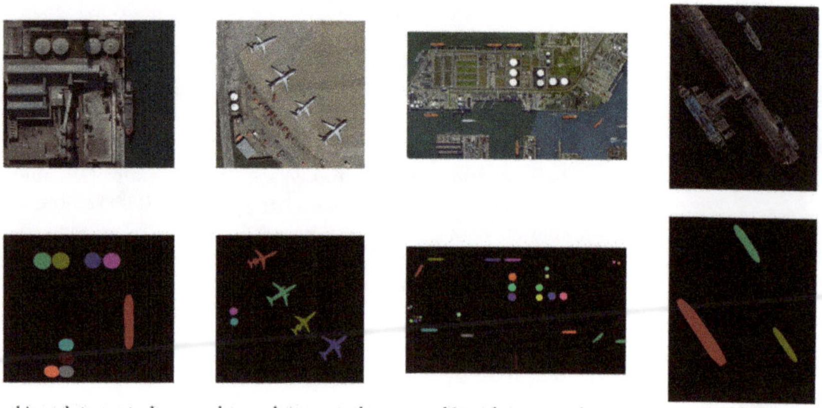

Figure 8. MRS-1800 dataset display. The top row is the remote sensing images of different sizes randomly selected in the dataset, and the next row contains corresponding mask images produced with LabelMe.

Figure 9 shows the mask loss function graph during the training of the network model we designed. It can be seen that the network model is still under-fitting during the first 38 k steps (27 epochs), and the loss function fluctuates greatly. We adjusted the learning rate in time after 38 k steps to avoid overfitting. The training loss value after the final step was 0.03479.

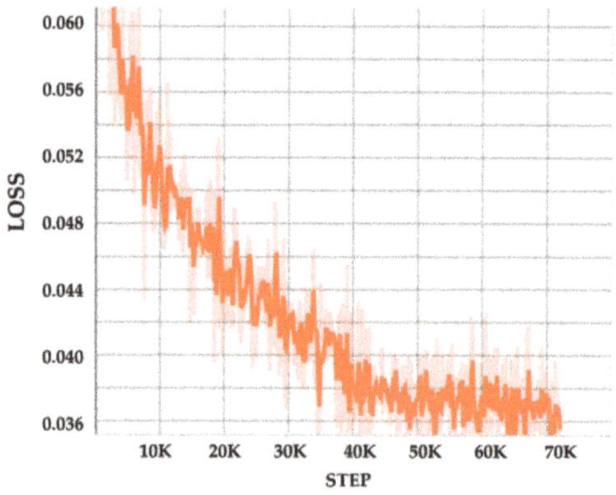

Figure 9. The training mask loss function diagram of the LPSW backbone using the SAIEC framework on the dataset.

4.3. Ablation Experiment

We performed a number of ablation experiments to gradually verify each component in the proposed method in this section. We analyzed and compared the data trained on the MRS-1800 dataset. The specific experiments are as follows:

4.3.1. Study for Optimizer and Initial Learning Rate

The optimizer plays an important role in deep learning. We first conducted ablation experiments on the selection of the optimizer and the corresponding parameter values. Commonly used optimizers for object detection are the SGD [41] and the AdamW [40]. We chose Cascade Mask R-CNN as the network framework and the Swin transformer as the backbone network, using the SGD and the AdamW optimizers for experiments. At the same time, in order to explore the influence of the optimizer's initial learning rate parameters on the experiment, we set the initial learning rate to 1×10^{-4}, 1×10^{-5}, and 1×10^{-6} for comparison experiments.

It can be seen from Table 2 that the overall performance of the AdamW is better than that of the SGD, and AP can increase by more than 8% by replacing the optimizer. In addition, it can be drawn from the table that when the initial learning rate is 1×10^{-5}, the model can achieve the highest detection accuracy. Therefore, we can conclude that the Swin transformer can achieve a better performance when the AdamW optimizer is used for model training and the initial learning rate is 1×10^{-5}.

Table 2. The results of optimizers and learning rate ablation study.

Method	Optimizer	Learning Rate	AP^{box}	AP^{mask}
Swin-T	SGD	1×10^{-4}	60.1	33.9
		1×10^{-5}	69.2	52.1
		1×10^{-6}	53.6	41.5
	AdamW	1×10^{-4}	73.9	58.0
		1×10^{-5}	77.2	60.7
		1×10^{-6}	75.0	58.4

4.3.2. Experiment for the Swin Transformer and LPST Backbone

We inserted the Swin transformer (Swin-T) and LPST as a new backbone network into typical object detection frameworks: Mask R-CNN and Cascade Mask R-CNN, for object detection and instance segmentation experiments. We compared them with traditional convolutional networks (Sparse R-CNN, PANet, and Mask Scoring R-CNN) and previous transformer networks (DETR). The experimental results are shown in Table 3.

Table 3. Detection and segmentation performance of different methods.

		Various Frameworks									
Method	Backbone	AP^{box}	AP^{box}_{50}	AP^{box}_{75}	AP^{box}_{s}	AP^{mask}	AP^{mask}_{50}	AP^{mask}_{75}	AP^{mask}_{s}	AR_S	FPS
Mask R-CNN	R-50	69.0	91.5	83.3	31.6	57.2	90.5	58.9	25.0	44.1	11.5
	Swin-T	75.5	92.8	88.1	44.6	60.9	91.7	66.6	34.1	47.2	8.6
	LPST	75.8	93.1	88.0	46.6	60.4	92.1	65.8	36.2	49.2	8.1
Cascade Mask R-CNN	R-50	72.1	91.0	83.3	31.3	56.6	90.3	57.7	32.9	38.5	8.4
	Swin-T	77.2	92.7	87.6	41.5	60.7	91.4	66.3	31.7	45.5	5.4
	LPST	77.4	93.0	88.0	46.7	61.3	91.7	68.3	36.8	50.0	5.1
Mask Scoring	R-50	71.9	91.5	84.5	40.3	60.7	90.4	67.4	32.4	43.5	11.4
Sparse R-CNN	R-50	73.9	91.0	83.8	35.4					39.4	13.4
PANet	R-50	71.6	91.8	84.5	35.3					38.3	12.1
DETR	R-50	65.3	86.7	74.3	21.4					29.7	15.1

Table 3 shows that, compared with the traditional CNN models, in each framework, the use of the Swin transformer and the LPSW as the backbone network has a greater improvement in the various indicators of the experimental results. Compared with the previous transformer network, the experimental result of Swin-T based on Cascade Mask R-CNN is 11.9% AP and 20.1% APs higher than DETR, which is sufficient to prove the superiority of the Swin transformer. It overcomes the shortcoming of the transformer's poor small-scale objects detection and slow convergence.

At the same time, we compared the LPSW with Swin-T using the same basic framework. The experimental results show that, after using the LPSW, the experimental indicators are improved: when using the Cascade Mask R-CNN framework, APs increased by 5.2%, mask AP_S increased by 5.1%, ARs increased by 4.5%, and mask AP and AP increased by 0.6% and 0.2%, respectively. The data show that, for the Swin transformer, the LPSW significantly improved the detection and segmentation of small-scale objects without a significant reduction in the inference speed. Due to the large number of small objects in remote sensing images, this improvement was exactly what was necessary.

The result generated by the traditional Cascade Mask R-CNN, the Swin-T, and LPSW are shown in Figures 10–12. Compared with the traditional CNN network, the Swin transformer pays more attention to the learning of global features; particularly, the detection ability of image edge objects was greatly improved. As shown in the enlarged images on the right side of Figures 10 and 11, Cascade Mask R-CNN has a low confidence in terms of the detection of ships in the upper right of the image, and false detection objects appeared. The Swin transformer does not detect false objects for the same edge detection area, and the confidence of object detection increases.

Compared with the Swin transformer, the LPSW pays more attention to local features. As shown in Figures 11 and 12, the most obvious difference between the two images is that the LPSW eliminates the false detection of white buildings in the lower part of the image. In addition, the number of real objects detected by the LPSW increases, and the confidence of object detection also improves.

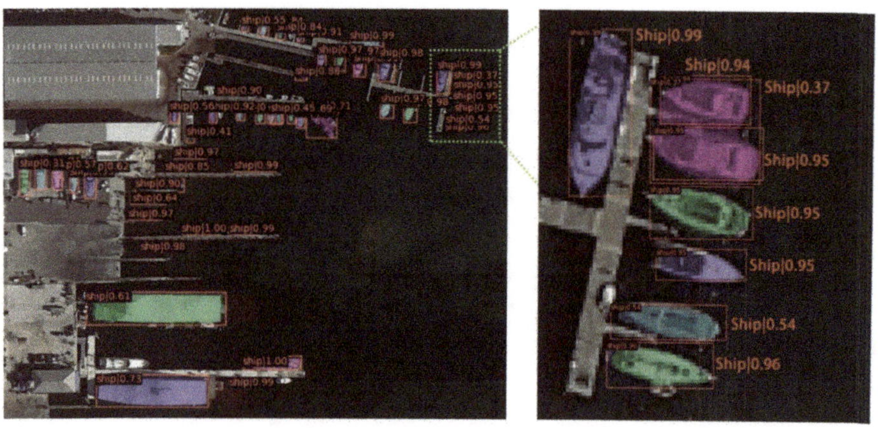

Figure 10. The results of Cascade Mask R-CNN using the Resnet-50 backbone.

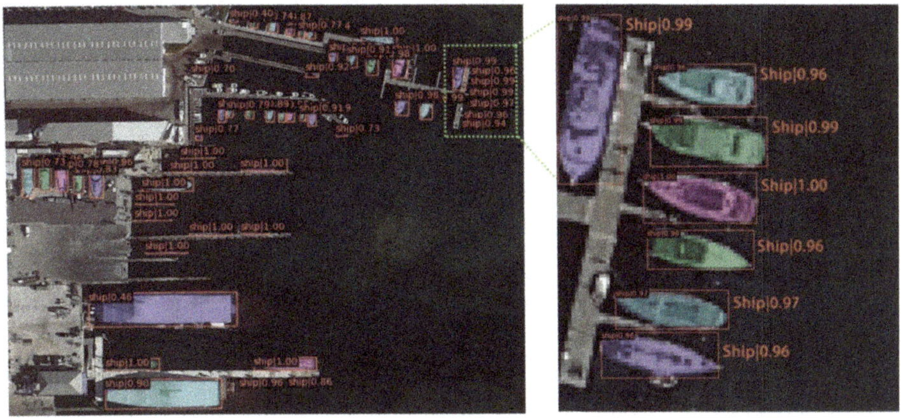

Figure 11. The results of Cascade Mask R-CNN using the Swin transformer backbone.

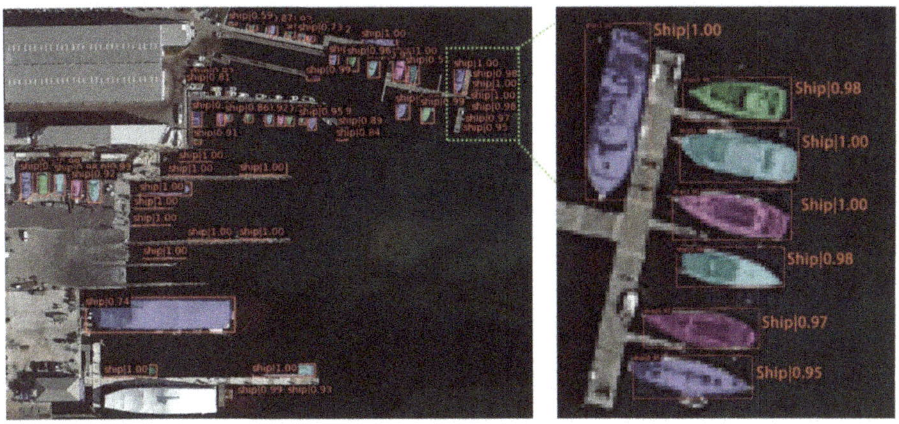

Figure 12. The results of Cascade Mask R-CNN using the LPSW backbone.

4.3.3. Experience for SAIEC and the New Network Model

We used the newly designed SAIEC network framework to perform object detection and instance segmentation on remote sensing images. The MRS-1800 dataset was used, and the backbone network used the LPSW and Swin-T. In order to verify the effectiveness of the improved model designed, we compared the experimental results with data in Section 4.3.1. At the same time, we compared and analyzed the designed model with the current SOTA object detection model on the COCO dataset (the Swin transformer using an HTC framework) [18].

Since this paper improves Cascade Mask R-CNN and the Swin transformer, respectively, we considered Swin-T using the Cascade Mask R-CNN framework as the baseline. It can be concluded from Table 4 that, compared with the baseline, the object detection and instance segmentation model we designed (the SAIEC network framework using the LPSW backbone) saw an improvement in all indicators. Among them, mask AP increased by 1.7%, mask AP_{75} increased by 4.0%, mask AP_S increased by 3.6%, AP increased by 1.1%, AP_S increased by 4.6%, and AR_S increased by 7.7%.

Table 4. Performance comparison of each part of the improved model.

Method	AP^{box}	AP^{box}_{50}	AP^{box}_{75}	AP^{box}_{s}	AP^{mask}	AP^{mask}_{50}	AP^{mask}_{75}	AP^{mask}_{s}	AR_S	FPS
Cascade Mask R-CNN (Swin-T) *baseline*	77.2	92.7	87.6	41.5	60.7	91.4	66.3	31.7	45.5	5.4
Cascade Mask R-CNN (LPSW)	77.4	93.0	88.0	46.7 (+5.2)	61.3	91.7	68.3	36.8 (+5.1)	50.0	5.1
HTC (Swin-S [18])	77.8	93.3	88.1	46.6	61.9	92.4	68.8	35.9	51.8	4.6
HTC (Swin-T)	77.4	92.7	88.2	41.7	61.6	91.9	69.7	31.4	49.6	5.4
SAIEC (Swin-T)	77.8	93.2	88.7	43.4	62.3	92.0	69.4	33.7	50.0	5.5
SAIEC (LPSW) (ours)	78.3 (+1.1)	93.0	88.7	46.1 (+4.6)	62.4 (+1.7)	92.3	70.3 (+4.0)	35.3 (+3.6)	53.2 (+7.7)	5.1

The data show that the network model we designed greatly improved the detection and segmentation of small-scale objects in remote sensing images. The increase in the detection rate of small-scale objects affects the improvement of AP_{75} and mask AP_{75}. Compared with the current SOTA network (the Swin transformer using an HTC framework), the indicators of the model designed in this article are similar or even surpassed, and the inference speed is higher (5.1 FPS vs. 4.6 FPS). The above experimental data demonstrate the advantages of the model proposed in this paper in remote sensing image object detection and instance segmentation.

Figure 13 shows the remote sensing image segmentation results of traditional Cascade Mask R-CNN, the Swin transformer using Cascade Mask R-CNN and the network proposed in this paper. It can be seen from the figure that Cascade R-CNN is not ideal in terms of overall segmentation effect or edge detail processing. Although the Swin transformer is optimized for the overall segmentation effect, it does not accurately present the details of the edge. In contrast, it can be seen from the figure that the network model proposed in this paper shows good results in remote sensing images, and the details at the edges are well segmented.

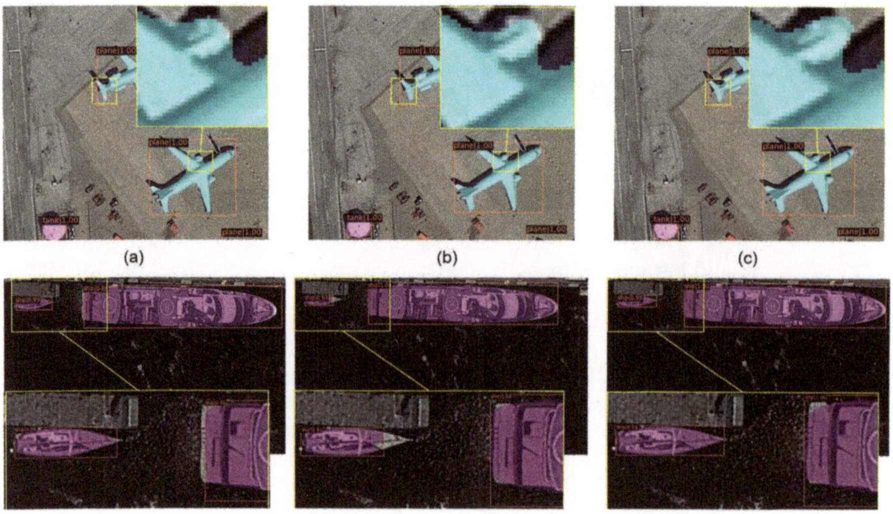

Figure 13. Segmentation results of remote sensing images by various networks. (**a**–**c**) Detection results of the traditional Cascade Mask R-CNN, the Swin transformer using Cascade Mask R-CNN and the LPSW using SAIEC.

5. Discussion

Because convolutional neural networks (CNN) have shown defects in the object detection of remote sensing images. We innovatively introduced the Swin transformer as the basic detection network, and designed the LPSW backbone network and SAIEC network framework for improvement. Experimental results show that the new network model we designed can greatly improve the detection effect of small-scale objects in remote sensing images and can strengthen the segmentation accuracy of multi-scale objects. However, it is worth noting that our experiment was only conducted on the MRS-1800 dataset due to the lack of mature and open remote sensing mask datasets, which may be limited in number and type. Moreover, our research on the improvement and promotion of the model inference speed is not sufficient. Generally, the processed images will be affected by uncertain factors [42]; however, it is also necessary to use fuzzy preprocessing techniques on images. In future research, we will focus on solving the above problems. First, we will search for and create more remote sensing mask datasets containing more object types, and use more realistic and representative datasets to validate our new models. Secondly, designing a lightweight network model to improve the inference speed without the loss of detection accuracy will be our next research direction.

6. Conclusions

Remote sensing image object detection and instance segmentation tasks have important research significance for the development of aviation and remote sensing fields, and have broad application prospects in many practical scenarios. First, we created the MRS-1800 remote sensing mask dataset, which contains multiple types of objects. Second, we introduced the Swin transformer into remote sensing image object detection and instance segmentation. This paper improved the Swin transformer based on the advantages and disadvantages of transformers and CNNs, and we designed the local perception Swin transformer (LPSW) backbone network. Finally, in order to increase the mask prediction accuracy of remote sensing image instance segmentation tasks, we designed the spatial attention interleaved execution cascade (SAIEC) network framework. Experimental conclusions can be drawn for the MRS-1800 remote sensing mask dataset: (1) According to experiments, the SAIEC model using the LPSW as the backbone can improve mask AP by 1.7%, mask AP_S by 3.6%, AP by 1.1%, and AP_S by 4.6%. (2) The innovative combination of CNNs and transformers' advantages in capturing local information and global information can significantly improve the detection and segmentation accuracy of small-scale objects. Inserting the interleaved execution structure and the improved spatial attention module into the mask head can help to suppress noise and enhance the mask prediction of the network. (3) Compared with the current SOTA model in the COCO dataset, the model proposed in this paper also demonstrates important advantages.

Author Contributions: Conceptualization, Z.F., C.C. and X.X.; methodology, X.X.; software, C.C.; validation, M.L., J.W. and Z.W.; formal analysis, S.Y. and Y.S.; investigation, X.X.; resources, Z.F.; data curation, X.X. and Z.F.; writing—original draft preparation, X.X.; writing—review and editing, Z.F. and C.C.; visualization, C.C.; supervision, M.L.; project administration, X.X.; funding acquisition, X.X. All authors have read and agreed to the published version of the manuscript.

Funding: This research received no external funding.

Institutional Review Board Statement: Not applicable.

Informed Consent Statement: Not applicable.

Data Availability Statement: The data used to support the findings of this study are available from the corresponding author upon request.

Acknowledgments: The authors thank the team of optical sensing and measurement of Xidian University for their help. This research was supported by the National Natural Science Foundation of Shaanxi Province (Grant No.2020 JM-206), the National Defense Basic Research Foundation (Grant No.61428060201) and the 111 project (B17035).

Conflicts of Interest: The authors declare no conflict of interest.

References

1. Cao, C.; Wang, B.; Zhang, W.; Zeng, X.; Yan, X.; Feng, Z.; Liu, Y.; Wu, Z. An Improved Faster R-CNN for Small Object Detection. *IEEE Access* **2019**, *7*, 1. [CrossRef]
2. Zhu, W.T.; Xie, B.R.; Wang, Y.; Shen, J.; Zhu, H.W. Survey on Aircraft Detection in Optical Remote Sensing Images. *Comput. Sci.* **2020**, *47*, 1–8.
3. Wu, J.; Cao, C.; Zhou, Y.; Zeng, X.; Feng, Z.; Wu, Q.; Huang, Z. Multiple Ship Tracking in Remote Sensing Images Using Deep Learning. *Remote Sens.* **2021**, *13*, 3601. [CrossRef]
4. Li, X.Y. Object Detection in Remote Sensing Images Based on Deep Learning. Master's Thesis, Department Computer Application Technology, University of Science and Technology of China, Hefei, China, 2019.
5. Hermosilla, T.; Palomar, J.; Balaguer, Á.; Balsa, J.; Ruiz, L.A. Using street based metrics to characterize urban typologies. *Comput. Environ. Urban Syst.* **2014**, *44*, 68–79. [CrossRef]
6. Ren, S.; He, K.; Girshick, R.; Sun, J. Faster R-CNN: Towards Real-Time Object Detection with Region Proposal Networks. *IEEE Trans. Pattern Anal. Mach. Intell.* **2017**, *39*, 1137–1149. [CrossRef] [PubMed]
7. He, K.; Gkioxari, G.; Dollár, P.; Girshick, R. Mask R-CNN. In Proceedings of the IEEE ICCV, Venice, Italy, 22–29 October 2017; pp. 2980–2988.
8. Liu, S.; Qi, L.; Qin, H.; Shi, J.; Jia, J. Path Aggregation Network for Instance Segmentation. In Proceedings of the 2018, IEEE/CVF Conference on Computer Vision and Pattern Recognition (CVPR), Salt Lake City, UT, USA, 18–23 June 2018; Institute of Electrical and Electronics Engineers (IEEE): New York, NY, USA, 2018; pp. 8759–8768. [CrossRef]
9. Huang, Z.; Huang, L.; Gong, Y.; Huang, C.; Wang, X. Mask Scoring R-CNN. In Proceedings of the IEEE/CVF CVPR, Long Beach, CA, USA, 16–20 June 2019; pp. 6409–6418.
10. Dai, J.F.; He, K.M.; Sun, J. Instance-aware semantic segmentation via multi-task network cascades. In Proceedings of the IEEE Conference on Computer Vision and Pattern Recognition, Las Vegas, NV, USA, 27–30 June 2016.
11. Wang, X.; Kong, T.; Shen, C.; Jiang, Y.; Li, L. Solo: Segmenting objects by locations. In Proceedings of the European Conference on Computer Vision, Glasgow, UK, 23–28 August 2020; pp. 649–665.
12. Vaswani, A.; Shazeer, N.; Parmar, N.; Uszkoreit, J.; Jones, L.; Gomez, A.N.; Kaiser, Ł.; Polosukhin, I. Attention is all you need. In Proceedings of the Advances in Neural Information Processing Systems, Long Beach, CA, USA; 2017; pp. 5998–6008.
13. Dosovitskiy, A.; Beyer, L.; Kolesnikov, A.; Weissenborn, D.; Zhai, X.; Unterthiner, T.; Dehghani, M.; Minderer, M.; Heigold, G.; Gelly, S.; et al. An image is worth 16 × 16 words: Transformers for image recognition at scale. In Proceedings of the 9th International Conference on Learning Representations (ICLR 2021), Virtual Event, Austria, 3–7 May 2021.
14. Nicolas, C.; Francisco, M.; Gabriel, S.; Nicolas, U.; Alexander, K.; Sergey, Z. End-to-End Object Detection with Transformers. In Proceedings of the 16th ECCV, Glasgow, UK, 23–28 August 2020; pp. 213–229.
15. Touvron, H.; Cord, M.; Douze, M.; Massa, F.; Sablayrolles, A.; Jegou, H. Training data-efficient image transformers & distillation through attention. In Proceedings of the 38th ICML, Virtual Event, 18–24 July 2021; pp. 10347–10357.
16. Zheng, S.; Lu, J.; Zhao, H.; Zhu, X.; Luo, Z.; Wang, Y.; Fu, Y.; Feng, J.; Xiang, T.; Torr, P.H.S.; et al. Rethinking semantic segmentation from a sequence-to-sequence perspective with transformers. In Proceedings of the IEEE Conference on Computer Vision and Pattern Recognition (CVPR), Virtual, 19–25 June 2021; pp. 6881–6890.
17. Raffel, C.; Shazeer, N.; Roberts, A.; Lee, K.; Narang, S.; Matena, M.; Zhou, Y.Q.; Li, W.; Liu, P.J. Exploring the limits of transfer learning with a unified text-to-text transformer. *J. Mach. Learn. Res.* **2020**, *21*, 1–67.
18. Liu, Z.; Lin, Y.; Cao, Y.; Hu, H.; Wei, Y.; Zhang, Z.; Lin, S.; Guo, B. Swin transformer: Hierarchical vision transformer using shifted windows. *arXiv* **2021**, arXiv:2103.14030. Available online: https://arxiv.org/abs/2103.14030 (accessed on 19 October 2021).
19. Chen, K.; Pang, J.M.; Wang, J.Q.; Xiong, Y.; Li, X.X.; Sun, S.Y.; Feng, W.F.; Liu, Z.W.; Shi, J.P.; Wangli, O.Y.; et al. Hybrid Task Cascade for Instance Segmentation. In Proceedings of the IEEE CVPR, Long Beach, CA, USA, 15–21 June 2019; pp. 4974–4983.
20. He, K.; Zhang, X.; Ren, S.; Sun, J. Spatial Pyramid Pooling in Deep Convolutional Networks for Visual Recognition. *IEEE Trans. Pattern Anal. Mach. Intell.* **2015**, *37*, 1904–1916. [CrossRef] [PubMed]
21. Girshick, R. Fast R-CNN. In Proceedings of the 2015 IEEE International Conference on Computer Vision (ICCV), Santiago, Chile, 7–13 December 2015; pp. 1440–1448.
22. Redmon, J.; Divvala, S.; Girshick, R.; Farhadi, A. You Only Look Once: Unified, Real-Time Object Detection. In Proceedings of the IEEE CVPR, Las Vegas, NV, USA, 27–30 June 2016; pp. 779–788.
23. Liu, W.; Anguelov, D.; Erhan, D.; Szegedy, C.; Reed, S.E.; Fu, C.Y.; Berg, A.C. SSD: Single Shot MultiBox Detector. In Proceedings of the IEEE ECCV, Amsterdam, Netherlands, 11–14 October 2016; pp. 21–37.
24. Lin, T.; Dollar, P.; Girshick, R.; He, K.; Hariharan, B.; Belongie, S. Feature Pyramid Networks for Object Detection. In Proceedings of the IEEE Conference on Computer Vision and Pattern Recognition, Honolulu, HA, USA, 21–26 July 2017; pp. 2117–2125.
25. Lin, T.Y.; Goyal, P.; Girshick, R.; He, K.; Dollár, P. Focal Loss for Dense Object Detection. In Proceedings of the IEEE International Conference on Computer Vision, Venice, Italy, 22–29 October 2017; pp. 2980–2988.
26. Liang, X.; Lin, L.; Wei, Y.C.; Shen, X.H.; Yang, J.C.; Yan, S.C. Proposal-Free Network for Instance-Level Object Segmentation. *IEEE Trans. Pattern Anal. Mach. Intell.* **2017**, *40*, 2978–2991. [CrossRef] [PubMed]

27. Wang, X.L.; Zhang, R.F.; Kong, T.; Li, L.; Shen, C.H. SOLOv2: Dynamic and Fast Instance Segmentation. *arXiv* **2020**, arXiv:2003.10152. Available online: https://arxiv.org/abs/2003.10152v3 (accessed on 19 October 2021).
28. Lee, Y.; Park, J. Centermaslc: Real-Time Anchor-Free Instance Segmentation. In Proceedings of the the IEEE Conference on Computer Vision and Pattern Recognition, Seattle, WA, USA, 13–19 June 2020; pp. 13906–13915.
29. Tian, Z.; Shen, C.; Chen, H.; He, T. Fcos: Fully Convolutional One-Stage Object Detection. In Proceedings of the the IEEE International Conference on Computer Vision, Seoul, Korea, 27 October–2 November 2019; pp. 9627–9636.
30. Zhou, X.Z.; Su, W.J.; Lu, L.W.; Li, B.; Wang, X.G.; Dai, J.F. Deformable DETR: Deformable Transformers for End-to-End Object Detection. In Proceedings of the 9th International Conference on Learning Representations (ICLR), Virtual Event, Austria, 3–7 May 2020.
31. Zheng, M.H.; Gao, P.; Wang, X.G.; Li, H.S.; Dong, H. End-to-End Object Detection with Adaptive Clustering Transformer. *arXiv* **2020**, arXiv:2011.09315. Available online: https://arxiv.org/abs/2011.09315 (accessed on 19 October 2021).
32. Wang, Y.Q.; Xu, Z.L.; Wang, X.L.; Shen, C.H.; Cheng, B.S.; Shen, H.; Xia, H.X. End-to-End Video Instance Segmentation with Transformers. In Proceedings of the IEEE Conference on Computer Vision and Pattern Recognition (CVPR), Virtual, 19–25 June 2021; pp. 8741–8750.
33. Yu, F.; Koltun, V. Multi-scale context aggregation by dilated convolutions. In Proceedings of the 4th International Conference on Learning Representations (ICLR), San Juan, Puerto Rico, 2–4 May 2016.
34. Woo, S.; Park, J.; Lee, J.Y.; Kweon, I. Cbam: Convolutional block attention module. In Proceedings of the ECCV, Munich, Germany, 8–14 September 2018; pp. 3–19.
35. Zhu, X.Z.; Cheng, D.Z.; Zhang, Z.; Lin, S.; Dai, J.F. An empirical study of spatial attention mechanisms in deep networks. In Proceedings of the ICCV, Seoul, Korea, 27 October–2 November 2019; pp. 6687–6696.
36. Li, K.; Wang, G.; Cheng, G.; Meng, L.Q.; Han, J.W. Object Detection in Optical Remote Sensing Images: A Survey and A New Benchmark. *arXiv* **2019**, arXiv:1909.00133. Available online: https://arxiv.org/abs/1909.00133v2 (accessed on 19 October 2021). [CrossRef]
37. Zhang, Y.L.; Yuan, Y.; Feng, Y.C.; Lu, X.Q. Hierarchical and Robust Convolutional Neural Network for Very High-Resolution Remote Sensing Object Detection. *IEEE Trans. Geosci. Remote Sens.* **2019**, *57*, 5535–5548. [CrossRef]
38. Gong, C.; Zhou, P.; Han, J. Learning Rotation-Invariant Convolutional Neural Networks for Object Detection in VHR Optical Remote Sensing Images. *IEEE Trans. Geosci. Remote Sens.* **2016**, *12*, 7405–7415.
39. Sun, P.Z.; Zhang, R.F.; Jiang, Y.; Kong, T.; Xu, C.F.; Zhan, W.; Tomizuka, M.; Li, L.; Yuan, Z.H.; Wang, C.H.; et al. Sparse R-CNN: End-to-End Object Detection with Learnable Proposals. In Proceedings of the IEEE Conference on Computer Vision and Pattern Recognition (CVPR), Virtual, 19–25 June 2021; pp. 14454–14463.
40. Loshchilov, I.; Hutter, F. Decoupled weight decay regularization. In Proceedings of the 7th International Conference on Learning Representations (ICLR), New Orleans, LA, USA, 6–9 May 2019.
41. Robbins, H.; Monro, S. A stochastic approximation method. *Ann. Math. Stat.* **1951**, *22*, 400–407. [CrossRef]
42. Versaci, M.; Calcagno, S.; Morabito, F.C. Fuzzy Geometrical Approach Based on Unit Hyper-Cubes for Image Contrast Enhancement. In Proceedings of 2015 IEEE International Conference on Signal and Image Processing Applications (ICSIPA), Kuala Lumpur, Malaysia, 19–21 October 2015; pp. 488–493.

remote sensing

MDPI

Article

A Dense Encoder–Decoder Network with Feedback Connections for Pan-Sharpening

Weisheng Li *, Minghao Xiang and Xuesong Liang

College of Computer Science and Technology, Chongqing University of Posts and Telecommunications, Chongqing 400065, China; 2016211957@stu.cqupt.edu.cn (M.X.); s190231008@stu.cqupt.edu.cn (X.L.)
* Correspondence: liws@cqupt.edu.cn

Abstract: To meet the need for multispectral images having high spatial resolution in practical applications, we propose a dense encoder–decoder network with feedback connections for pan-sharpening. Our network consists of four parts. The first part consists of two identical subnetworks, one each to extract features from PAN and MS images, respectively. The second part is an efficient feature-extraction block. We hope that the network can focus on features at different scales, so we propose innovative multiscale feature-extraction blocks that fully extract effective features from networks of various depths and widths by using three multiscale feature-extraction blocks and two long-jump connections. The third part is the feature fusion and recovery network. We are inspired by the work on U-Net network improvements to propose a brand new encoder network structure with dense connections that improves network performance through effective connections to encoders and decoders at different scales. The fourth part is a continuous feedback connection operation with overfeedback to refine shallow features, which enables the network to obtain better reconstruction capabilities earlier. To demonstrate the effectiveness of our method, we performed several experiments. Experiments on various satellite datasets show that the proposed method outperforms existing methods. Our results show significant improvements over those from other models in terms of the multiple-target index values used to measure the spectral quality and spatial details of the generated images.

Keywords: convolutional neural network; double-stream structure; feedback; encoder–decoder network; dense connections

Citation: Li, W.; Xiang, M.; Liang, X. A Dense Encoder–Decoder Network with Feedback Connections for Pan-Sharpening. *Remote Sens.* **2021**, *13*, 4505. https://doi.org/10.3390/rs13224505

Academic Editors: Fahimeh Farahnakian, Jukka Heikkonen and Pouya Jafarzadeh

Received: 12 October 2021
Accepted: 6 November 2021
Published: 9 November 2021

1. Introduction

Satellite technology has developed rapidly since the last century, and remote sensing satellite images have gained widespread attention and applications in many fields. They provide an important reference for applications in digital maps, urban planning, disaster prevention and control, emergency rescue, and geological observations [1–4].

In most practical applications, remote sensing images with high spatial resolution and high spectral resolution are required. Given the physical structure of satellite sensors, a single sensor is unable to achieve this. Earth-observation satellites, such as Quick-Bird, IKONOS, and World-View, are equipped with sensors for obtaining high-spatial-resolution images for single bands and multispectral sensors for obtaining low-spatial-resolution images for multiple bands, which are acquired as panchromatic (PAN) and multispectral (MS) images, respectively.

In order to fully utilise all of the information available in the two types of images, PAN and MS images are usually fused using a pan-sharpening algorithm to simultaneously generate images having PAN image spatial resolution as well as the corresponding MS image spectral resolution. This results in images with high spatial resolution and high spectral resolution, which practical applications need.

Owing to the need for high-quality remote sensing images in practical applications, many researchers have studied varied directions related to pan-sharpening algorithms:

(1) component substitution (CS) [5–8], (2) multiresolution analysis (MRA) [9–13] (3) model-based algorithms [14–20], and (4) algorithms for deep learning. The representative CS algorithms are principal component analysis (PCA) [5], intensity-hue-saturation (IHS) transform [6], Gram–Schmidt (GS) sharpening [7], and partial substitution (PRACS) [8]. These methods all adopt the core idea of the CS method, namely to first rely on the MS image in another space to separate the spatial-structure component and the spectral-information component, then match the PAN image and spatial-structure component using histograms and complete the replacement or partial replacement. This makes the PAN image have the same mean and variance as the spatial component. Finally, the pan-sharpening task is completed through an inverse transformation operation. These methods can achieve good results when PAN images are highly correlated with MS images, but owing to spectral differences between MS and PAN images, CS methods often encounter spectral-preservation problems and suffer from spectral distortion. Methods based on MRA are more straightforward than CS-based methods; these extract details from the PAN images and then inject them into the upsampled MS images. This approach makes the quality of the output image sensitive to the details of the injection, which makes the image blurred, while excessive detail injection leads to artifacts and spectral distortion. Decimated wavelet transform [9], atrous wavelet transform [10], Laplacian Pyramid [11], curvelet [12], and non-subsampled contourlets transform [13] are examples of this approach. The hybrid method combines the advantages of the CS and MRA methods to improve the spectral distortion and fuzzy spatial-detail deficiencies, resulting in better fusion results.

Model-based methods are mainly based on the mapping relationship between MS images, PAN images, and the desired high-resolution multispectral (HRMS) images. If pan-sharpening can be viewed as an inverse problem, the PAN and MS images can be understood as degraded versions of the HRMS images and can be recovered through optimization procedures. As considerable information is lost during the degradation process, this is an unsettled problem. The general practice is to introduce prior constraints and regularization methods into formulas to fuse the images and thus to solve this ill-posed inverse problem. Representative algorithms include sparsity regularization [14], Bayesian posterior probability [15], and variational models [16]. A hierarchical Bayesian model to fuse many multiband images with various spectral and spatial resolutions is proposed [17]. An online coupled dictionary learning (OCDL) [18], and two fusion algorithms [19] that incorporate the contextual constraints into the fusion model via MRF models have been proposed. As these methods are highly dependent on regularization terms, the resulting solutions are sometimes unstable [20]. These methods have much more temporal complexity than many other algorithms, but they can make immense progress in gradient information extraction.

In recent years, with the rapid development of artificial intelligence, algorithms based on deep learning methods have achieved impressive results in various image-processing domains. In the field of computer vision, CNNs have been successfully applied to a large number of domains, including target detection [21], medical segmentation [22], image fusion [23], and image reconstruction [24]. Due to the superior feature-representation capabilities of deep convolutional neural networks, many researchers have used the technique for pan-sharpening [25,26].

To some extent, image super-resolution reconstruction is a task associated with whole-chromatic sharpening, as super-resolution and euchromatic sharpening are both designed to improve image resolution. However, there are some differences between them, as the former is usually a single-input, single-output process, while the latter is a multiple-input, single-output case. Therefore, in earlier work, the PAN image and the MS image are usually cascaded together in the input grid for training, treating the pan-sharpening task as an image-regression task. Inspired by the super-resolution work based on CNN [27], Masi et al. [28] followed the three-layer CNN architecture in SRCNN to implement pan-sharpening and increase input by introducing nonlinear radiation exponents. This is the first application of pan-sharpening in the generalised sharpening field. In light of the significant improvement of the network training effect due to the residual structure, Rao et al. [29]

proposed RCNNP, a residual convolutional neural network for pan-sharpening, which continued to use a three-layer network structure when the idea of jump connections was introduced to help the network with training. Wei et al. [30] designed a deep residual network (DRPNN) to complete the pan-sharpening task, and they extended the depth of the network to eleven layers, which improved the network performance. Based on these three papers, He et al. [31] proposed two networks employing detail-injection ideas while clarifying the role of CNN in the pan-sharpening task from a theoretical perspective and clearly explaining the effectiveness of adding residual structure for pan-sharpening network improvement.

Although earlier CNN-based methods achieved better results than previous methods, they did not take into account the importance of spatial and spectral retention in the fusion process, treating it as a black-box learning process. To enhance the network's ability to retain both spatial and spectral information, Yang et al. [32] proposed a deep network architecture for pan-sharpening (PanNet), which differs from the other methods. To preserve the spectral information, they propose a method, called spectral mapping, that directly mapps the upsampled multispectral images to the network output for lossless propagation. To enhance the network's focus on the spatial structure in PAN images, PanNet, unlike the previous work, chose to train the network in high-frequency domains. This idea from an earlier work helped them achieve remarkable results, but it had some limitations. It is generally believed in the pan-sharpening field that PAN and MS images contain different information. PAN images are the carriers of geometric-detail (spatial) information, while MS images provide the spectral information required to fuse the images. Although PanNet trains the network in the high-frequency domain, it still inputs PAN images and MS images after cascading into the network. This operation prevents the network from completely extracting different features contained in PAN and MS images and allows the network to effectively utilise varied spatial information and spectral information. Concurrently, it only uses a simple residual structure that complements the extraction of image features at various scales and lacks the ability to more efficiently recover details from the features. As the network outputs the fusion results directly through a convolutional layer, the network cannot make full use of all the features extracted by various residual blocks, affecting the final fusion effect.

In this study, we are inspired by the ideas of the detail-injection network and image super-resolution reconstruction network. We propose a dense encoder–decoder network with feedback connections for pan-sharpening. As the CNN methods in earlier works either viewed euchromatic sharpening as a super-resolution problem [29,30] or used a CNN as a tool to extract spatial details [31,32], they generate results with good visual quality, but spectral distortion or artifacts still exist. This is mainly because it is almost impossible to individually extract features representing spatial or spectral information from the input network by stacking the PAN image and the MS information together. To address this issue, we choose to perform image fusion at the feature level rather than at the pixel level, as in earlier works. We use a dual-stream network structure to extract features from the PAN and MS images separately, which allows the network to efficiently extract the desired spatial information and spectral information without interference. To extract richer and efficient multiscale features from images, we input efficient multiscale feature-extraction modules from the two-stream network. Given the powerful multilevel feature-extraction, fusion, and reconstruction capabilities of the encoder–decoder, the extracted multiscale features are encoded and decoded based on the idea of dense connections. The shallow networks are limited by the receptive field size and can only extract coarse features, which we have repeated in subsequent networks, owing to the idea of dense connections, which partly limits the learning power of the network. We, therefore, introduce a feedback-connectivity mechanism that transfers deep features back to the shallow network through long-jump connections to optimise coarse low-level features and improve early reconstruction capability by completing preliminary reconstructed-image correction for some incorrect features in the early network. Concurrently, we follow the idea of detail injection, using the fusion results of the network as the detail branch and low resolution multispectral (LRMS) images as the approximate branch. Both can help the network obtain excellent HRMS images.

In conclusion, the main contributions of this study are as follows:

1. We propose a multiscale feature-extraction block with an attention mechanism to address the issue of insufficient network extraction ability to extract diverse scales, which can not only effectively extract multiscale features but also utilise feature information between multiple channels. In addition, the spatial and channel-attention mechanisms can effectively enhance the acquisition of important features to the network so as to help the fusion and reconstruction of the later network.
2. We propose an efficient feature-extraction block with two-way residuals, which stacks three multiscale feature-extraction blocks, enables the network to extract multiscale features at different depths, and maps low-level features to high-level space with two jump connections for the purpose of collecting more information.
3. We use a network structure with a multilayer encoder and decoder combined with dense connections to complete the task of integrating and reconstructing the extracted multiscale spatial and spectral information. As the task of the deep network is to encode the semantic information and abstract information of images, it is difficult for the network to recover texture, boundary, and colour information directly from advanced features, but shallow networks are excellent at identifying such detailed information. We inject low-level features into high-level features via long-jump connections, making it easier for the network to recover fine real images, while numerous dense connection operations bring the feature graph at the semantic level in the encoder closer to the feature graph in the decoder.
4. We inject HRMS images from the previous subnetwork into the shallow structure of the latter subnetwork, complete the feedback connectivity operation, and attach the loss function to each subnetwork to ensure that correct deep information can be transmitted backwards in each iteration and the network can obtain better reconstruction capabilities earlier.

The rest of this article is arranged as follows. We present the relevant CNN-based work that inspired us in Section 2 and analyse networks that have achieved significant results in the current pan-sharpening work based on CNN. Section 3 introduces the motivation of our proposed dense encoder–decoder network with feedback connections and explains in detail the structure of each part of the network. In Section 4, we show the experimental results and compare them with other methods. We discuss the validity of the various structures in the network in Section 5 and summarise the paper in Section 6.

2. Background and Related Work

2.1. Convolutional Neural Networks

Based on work in other fields, it is shown that better results can be obtained by increasing the depth and width of the network [33,34]. However, blindly increasing the depth of the network does not improve the network effectively. Worse, the problem of gradient explosion and gradient extinction occurs during training with increasing network depth, hampering networks with deeper and more complex structures. To overcome this difficulty, He et al. [35] proposed a residual learning framework to reduce the difficulty of network optimization and reduce degradation problems so that a deeper network structure could be used in the task. The advent of ResNet made network optimization simpler and allowed researchers to design deeper and more complex network structures to improve results. Based on this work, Huang et al. [36] proposed the intensive connection network (DenseNet) by fully injecting simple features of shallow networks into deep networks, achieving better performance than ResNet but requiring fewer parameters and lower computational costs.

Olaf et al. [23] proposed a U-Net network with a fully symmetrical encoder–decoder structure. The encoder structure in the first half of the network obtains multiscale features by reducing the spatial dimension, and the decoder structure in the second half progressively recovers the details and spatial dimensions of the image. The loss of information during downsampling is compensated for by adding a shortcut connection between the encoder and the decoder, which helps the decoder to better fix the details of the target. This network structure has provided immense inspiration to other researchers. Zhou et al. [37] proposed the U-Net++ network based on the U-Net network, introducing the idea of dense connectivity into the network. They took advantage of long and short connections to allow the network to grasp various levels of features and integrate them through a feature superposition manner while adding a shallower U-Net structure to ensure smaller differences in feature-graph scaling at fusion. Huang et al. [38] improved the U-Net structure from another angle, and U-Net 3+ redesigned the jump connection compared to U-Net and U-Net++. To enhance the network's ability to explore full-scale information, they proposed full-scale jump connections, where each decoder layer in U-Net 3+ incorporates feature maps from small-scale and same-scale features in the encoder and large-scale features from the decoder, where fine-grained and coarse-grained semantics enable the network to produce more accurate location perception and boundary-enhanced images.

These network structures, which have achieved remarkable results in other fields, have considerably inspired researchers performing pan-sharpening work and have been applied to the core ideas of these networks in recent CNN-based pan-sharpening work, achieving good results.

2.2. CNN-Based Pan-Sharpening

Inspired by the idea of traditional pan-sharpening methods to improve the structural consistency of fusion images by using the Qualcomm information of PAN images, Yang et al. [32] proposed a network structure called PanNet. Inspired by enhanced network performance in U-Net [37], RBDN [39] and GoogLeNet [34] that enhanced the multiscale feature grasping of networks, Fu et al. [40] presented an improved approach based on the original structure of PanNet. As the introduction of extensive pooling operations to obtain abstract features results in irreparable loss of spatial information, the network used to perform pan-sharpening does not expand the receptive field after downsampling images by pooling operations to obtain multiscale features. However, removing pooling operations slows down the increase in receptive fields. Simultaneously, because PanNet uses high-frequency information as input, it is equivalent to only fine details and edges being input into the network, and extracting multiscale features in a hierarchical way leads to limited multiscale representation ability of the network. To overcome this difficulty, they proposed a grouped multiscale expansion block based on expansion convolution [41] to extract the multiscale representation at the fine-granular level.

As PAN images are the carriers of spatial information in pan-sharpening work while MS images provide spectral information, recent work abandoned the practice of stacking PAN images and MS input networks as in earlier works [28–32], instead extracting features separately and choosing to fuse images in the feature domain rather than the pixel domain. Liu et al. [42] proposed a dual-stream fusion network for pan-sharpening where, to make full use of the spatial and spectral information in the image, they used two identical subnetworks to extract complementary information and features of PAN and MS images. To recover fine and realistic details from the extracted features, they introduced the encoder–decoder structure from U-Net [37] into pan-sharpening. Furthermore, to enhance the network to utilise all levels of features, the encoder was added to the decoder and connected to the corresponding feature maps to inject more details lost during downsampling. In a subsequent work, Liu et al. [43] proposed an improvement on TFNet, called ResTFNet, that further improves the performance of the proposed network by using basic residual blocks instead of the continuous convolutional layer in TFNet. Inspired by the dual-stream network structure, Fu et al. [44] proposed a network structure called TPNwFB that, after extracting spatial and spectral information, introduces a feedback connectivity mechanism to implement a subnetwork iterative process using recurrent structures, which allows strong-deep feature backflow to modify poor low-level features.

In TPNwFB, input features are iteratively upsampled and downsampled in TPNwFB to achieve a reverse projection mechanism, enabling feature-extraction blocks to generate more powerful features. As early networks using MSE loss-constraint networks made images too smooth and lost edge information, TFNet, ResTFNet, and TPNwFB were trained using MAE loss-constraint networks.

Liu et al. [45] used a dual-stream network to extract PAN and MS image features and an encoder–decoder structure for fusion and reconstruction of images. They also introduces the idea of generating an adversarial network for the first time in pan-sharpening work, proposing a network called PSGAN. In this GAN-based model, the generator attempts to generate images similar to the ground truth values, while the discriminator attempts to distinguish between the generated images and the HRMS images. PSGAN builds a generator through a dual-stream network that generates high-quality HRMS images using encoders and decoders, and then introduces a five-layer structured network as a discriminator. Shao et al. [46] reference a PSGAN network by proposing a network structure called RED-cGAN. Unlike the former, RED-cGAN discards the operation of up and downsampling in the network and replaces additional constraints as an input discriminator from an LRMS image for a PAN image. The two models differ from other methods by using multiple loss functions to constrain network learning rather than network training using MSE or MAE loss functions alone.

Zhang et al. [47] proposed a multilevel dense neural network for pan-sharpening. They made some modifications to the original DenseNet to enable it to complete the pan-sharpening task. They combined dual-stream and densely connected networks. To make full use of spatial and spectral information, the network in the hierarchical feature extraction and image reconstruction fraction consists of up to 83 convolutional layers, deep networks that have never been used in other pan-sharpening work. Li et al. [48] proposed to obtain higher performance HRMS images by using a network structure called MDECNN. They adopted a similar idea to PanNet to train the network in the high-frequency domain and enhance the spectral information of the image by spectral mapping but used a two-stream network to extract features for the PAN and MS images separately. Moreover, in their network, the feature information of the PAN image is extracted by using a multiscale feature-extraction module, and a parallel expansion of convolutional blocks is used to obtain the features of the various receptive fields of the image. MDECNN encodes and decodes U-Net-like structures and designs dense encoding blocks to comprehensively image deep images with a symmetric structure with the same number of encoders and decoders but discards upsampling and downsampling operations in the U-Net network and replaces the jump connections in the encoder and decoder for dense connections between all convolutional layers. The network is constrained by a mixed loss function, which is a combination of MSE loss and MAE loss. The loss of spectral information is constrained by MSE loss, and MAE is used as a constraint on spatial loss.

3. Proposed Network

In this section, we detail the specific structure of the DEDwFB model presented in this study. As we use a detail-injection network, our proposed network has clear interpretability. The use of dense and feedback connections in the network gives the network excellent early ability to reconstruct images, while effective feature reuse helps the network alleviate the challenge of gradient disappearance and gradient explosion during gradient transmission, giving the network very good performance against overfitting. We give a detailed description of each part of the proposed network framework. As shown in Figures 1 and 2, our model consists of two branches. One includes the LRMS image-approximation branch, which provides most of the spectral information and a small amount of spatial information needed to fuse the images, while the other is the detailed branch used to extract spatial details. This structure has clear physical interpretability, and the presence of approximate branching forces CNN to focus on learning the section information needed to complement LRMS images, which would reduce uncertainty in network training.

The detail branch has a structure similar to the encoder–decoder system, consisting of a two-path network, multiscale feature-extraction networks, feature-fusion and recovery networks, feedback connectivity structures, and image-reconstruction networks.

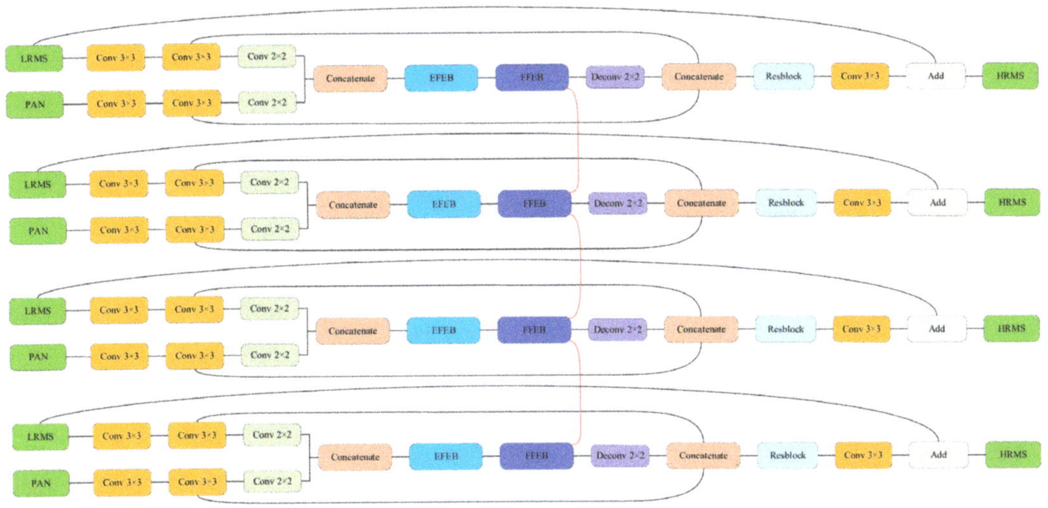

Figure 1. Detailed structure of the proposed multistage dense encoder–decoder network with feedback connections. Red lines denote the feedback connections.

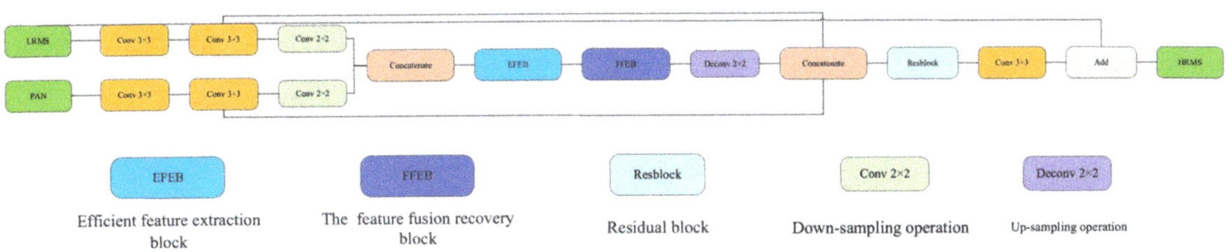

Figure 2. Specific structure of each subnet.

3.1. Two-Path Network

In pan-sharpening, it is widely accepted that the PAN and MS images contain different information. PAN images are the carriers of geometrical detail information, while MS images provide spectral information for the fusion images. The goal of pan-sharpening is to combine spatial details and spectral information to generate new HRMS images.

Although PAN images are considered carriers of spatial information, they may also contain spectral information. Similarly, the spatial information required for the HRMS image is also present in the MS image. To make full use of the information of PAN and MS images, we rely on CNN to fully extract the varied spatial and spectral information in the images and to perform feature-fusion reconstruction and image-recovery work in the feature domain.

We used two identical network results to extract features from the PAN and MS images separately. One network took single-band PAN images (size $H \times W \times 1$) as input, while the other network used multiband MS images (size $H \times W \times N$) as input. Before entering the network, we upsampled the MS images by transposition convolution to make them the same size as the PAN image. Each subnetwork consists of two separate convolutional layers and a subsampling layer, each followed by a parametric rectified linear unit (PReLU). The downsampling operation improves the robustness of the input image to certain perturbations while obtaining features of translation invariance, rotation invariance, and scale invariance and reduces the risk of overfitting. Most CNN architectures utilise maximum or average pooling for downsampling, but pooling results in an irreparable loss of spatial information, which is unacceptable for pan-sharpening. Therefore, throughout the network, we use a convolutional kernel of step 2 for downsampling rather than simple pooling. The two-path network consists of two branches, each including two $Conv_{3,64}(\cdot)$ layers and one $Conv_{2,32}(\cdot)$ layer. We use $Conv_{f,n}(\cdot)$ to represent convolution layers with size $f \times f$ convolution kernels and n channels and use $\delta(\cdot)$ to represent the PReLU activation function, f_{MS}, while f_{PAN} represents the extracted MS and PAN image features, respectively, and $\otimes$ represents the concatenation operation:

$$f_{MS} = \delta(Conv_{2,32}(\delta(Conv_{3,64}(\delta(Conv_{3,64}(I_{LRMS}))))), \quad (1)$$

$$f_{PAN} = \delta(Conv_{2,32}(\delta(Conv_{3,64}(\delta(Conv_{3,64}(I_{PAN}))))), \quad (2)$$

$$f_{P+M} = f_{MS} \otimes f_{PAN}, \quad (3)$$

3.2. Multiscale Feature-Extraction Network

Remote sensing images contain a large number of large-scale objects, such as buildings, roads, vegetation, mountains, and water bodies, as well as vehicles, ships, pedestrians, and municipal facilities. In order to obtain more accurate HRMS images, our network needs to have the ability to fully capture features having different scales from the PAN and MS images. The depth and width of the network have a clear effect on the network's ability to acquire multiscale features. With a deeper network structure, the network can learn richer feature information and context-related mapping. Owing to the emergence of the ResNet [35] network structure, optimizing the network training process by adding skip connections effectively solves the issues of gradient explosion, gradient disappearance, and training difficulties as the network structure deepens, ensuring that we can use deeper networks to obtain features at various scales. The inception structure proposed by an earlier study [34] fully extends the width of the network so that the network can acquire features of various scales at the same depth.

Inspired by the idea of enhancing network feature extraction by extending network depth and width, we propose an efficient feature-extraction block (EFEB) to help the network efficiently acquire features at various scales. As shown in Figure 3, EFEB consists of three identical multiscale feature-extraction blocks (MFEB) with attention mechanisms and two jump connections. MFEB can help the network acquire local multiscale features by extending network width at a single depth, while EFEB uses multiple MFEB features at various depths. As each MFEB output contains different features and makes full use of these different hierarchical features, we use a simple hierarchical feature-fusion structure that maps low-level features to advanced space through two jump connections, giving EFEB more efficient multiscale feature grasping.

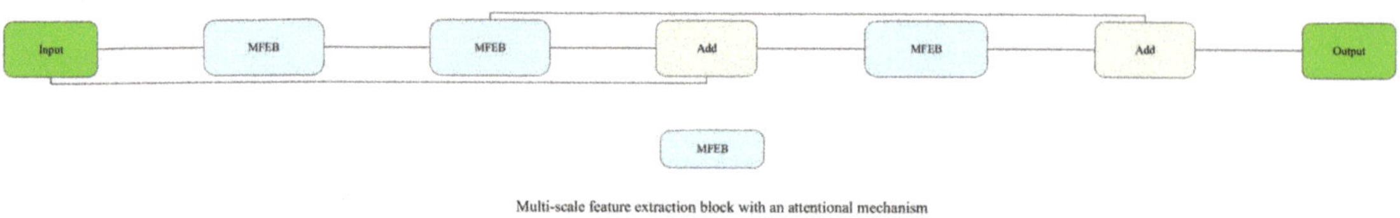

Figure 3. Specific structure of the efficient feature-extraction block.

Inspired by GoogLeNet, MFEB was designed to expand the ability of the network to obtain multiscale features using a structure shown in Figure 4. To obtain features at different scales in the same level of the network, we used four parallel branches for separate feature extraction. On each clade, we used convolutional nuclei of sizes $3 \times 3, 5 \times 5, 7 \times 7,$ and 9×9, respectively, to obtain receptive fields at different scales. However, this results in high computational costs, which increases the training difficulty of the network. Inspired by the structural improvement work of PanNet in a study [40], we chose to similarly use the dilated convolution [41] operation to expand the receptive field of small-scale convolutional kernels without additional parameters. As void convolution is a sparse sampling method, with a mesh effect when multiple void convolutions are superimposed, some pixels are not utilised at all while losing the continuity and correlation of information. This results in a lack of correlation between features obtained from distant convolution, which severely affects the quality of the last-obtained HRMS images. To mitigate this concern, we introduce Res2Net [49]'s idea to improve the dilated convolution.

We used a dilated convolution block on each branch to gain more contextual information using a 3×3 layer and set the expansion rate to 1, 2, 3, and 4, equivalent to our use of convolutional kernels of sizes $3 \times 3, 5 \times 5, 7 \times 7,$ and 9×9 but using a minimal number of parameters. To further expand the receptive field and obtain more sufficient multiscale features, we processed the features using a convolutional layer of 3×3 on each clade.

To mitigate the issue of grid effects caused by dilated convolution and the lack of correlation between the extracted features, we connected the output of the former branch to the next branch by jumping, which is repeated several times until the outputs of all branches are processed. This allows for different scale features to be effectively complementary and the loss of detailed features and semantic information to be avoided as large-scale convolutional kernels can be dominated by multiple small-scale convolutional cores. Jump connections between branches allow each branch to have continuous receptive fields of 3, 5, 7, and 9, respectively, while avoiding information loss from continuous use of dilated convolution. Finally, we fused the results from the four pathway cascades through a 1×1 convolutional layer. We then used spatial and channel-attention mechanisms through compressed spatial information to measure channel importance and compressed channel information to obtain measures of spatial location importance. Indicators indicate the importance of different feature channels and spatial locations that can help the network enhance features more important to the current task. To better preserve intrinsic information, the output features are fused to the original input in a similar manner, and the jump connections across the module effectively reduce training difficulty and possible degradation. This procedure can be defined as:

$$x = \delta(Conv_{1,64}(f_{3\times3} \otimes f_{5\times5} \otimes f_{7\times7} \otimes f_{9\times9})), \tag{4}$$

$$F_{CSE}(x) = \sigma(Conv_{1,64}(\delta(Conv_{1,32}(\mu(x))))), \tag{5}$$

$$F_{SSE}(x) = \sigma(Conv_{1,1}(x)), \tag{6}$$

$$F_{MFEB} = F_{CSE}(x) * x + F_{SSE}(x) * x + x, \tag{7}$$

We use $Conv_{f,n}(\cdot)$ to represent convolution layers with size $f \times f$ convolution kernels and n channels.$\delta(\cdot)$,$\delta(\cdot)$, and $\mu(\cdot)$ represent the sigmoid activation functions, PReLU activation function, and global average pooling layer, respectively. $F_{CSE}(x)$ and $F_{SSE}(x)$ represent the measures of channel importance and the measures of spatial location importance, respectively. Furthermore, x represents multiscale features extracted from four branches with different-scale receptive fields, and $\otimes$ represents the concatenation operation.

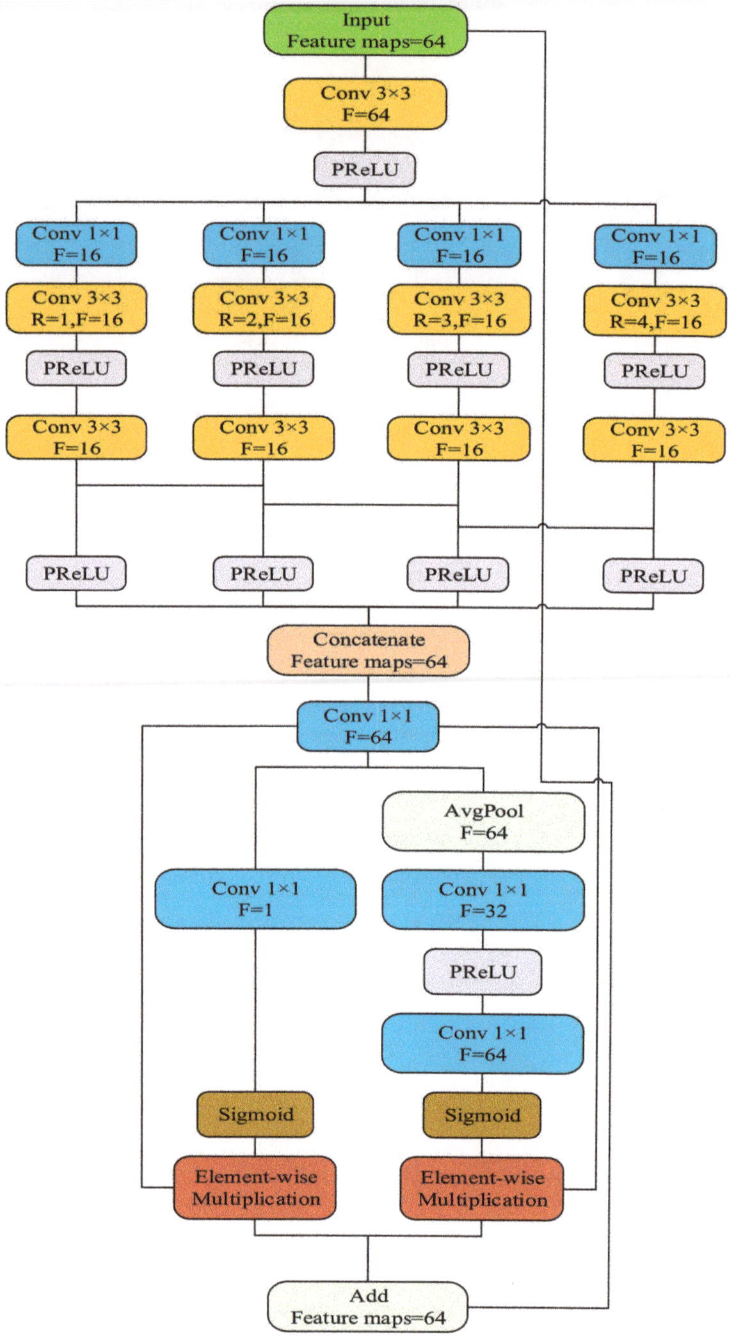

Figure 4. Detailed structure of the multiscale feature-extraction block.

3.3. Feature Fusion and Recovery Networks

To effectively fuse the various levels of extracted multiple-scale features and recover high-quality HRMS images, we propose a feature-fusion and recovery block (FFRB) composed of densely connected encoders and decoders. The concrete structures of the FFRB and residual block are shown in Figure 5. CNN-based pan-sharpening approaches, such as TFNet [42], ResTFNet [43], PSGAN [45], and RED-cGAN [46] adopt a fully symmetric encoder–decoder framework structure and achieve remarkable results. Unlike these works on network design based on the U-Net [23] infrastructure, we are inspired by U-Net++ [37] and U-Net3+ [38] to propose more complex but more efficient encoder–decoder structures.

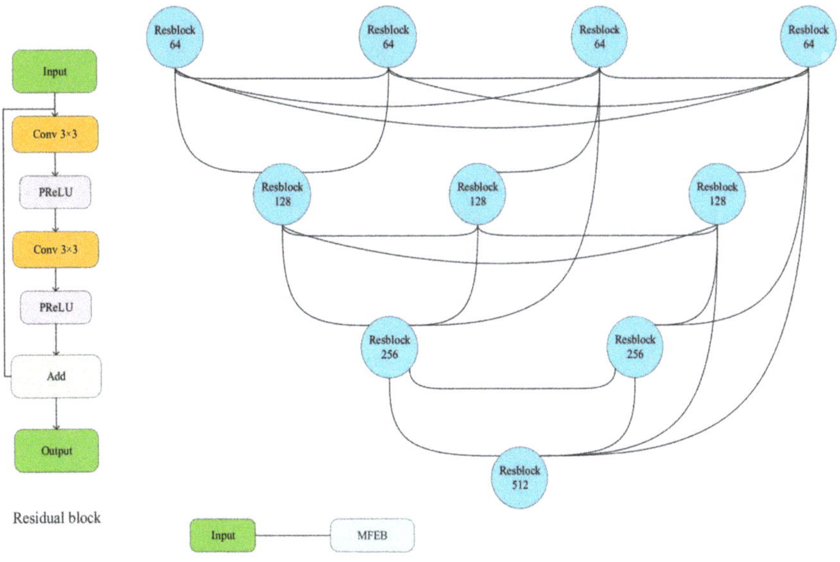

Figure 5. Structure of the proposed residual block and the feature-fusion recovery block.

Owing to the different size of the receptive field, the shallow structure of the network focuses on capturing some simple features, such as boundary, colour, and texture information, whereas deep structures are good at capturing semantic information and abstract features. The downsampling operation improves the robustness of the input image to certain perturbations while obtaining features of translation invariance, rotation invariance, and scale invariance and reducing the risk of overfitting. Continuous downsampling can increase the receptive-field size and help the network fully capture multiscale features. The downsampling operation helps the encoder fuse and encode features at different levels, the edge and detail information of the image are recovered through the upsampling operation and decoder, and the reconstruction of the fusion image was initially completed. However, multiple downsampling and upsampling operations can cause edge information and small-scale object loss. The complex-encoded semantic and abstract information also poses substantial difficulties for the decoder.

As shown in Figure 5, we used four residual blocks and three downsampling operations to compose the encoder network. Unlike other fully symmetrical encoder–decoder structures in the work, we used six residual blocks to constitute the decoder network and add an upsampling layer before each decoder. In the network, we doubled the number of channels of the feature graph by each subsampled layer and halve the number of feature-graph channels at each upsampled layer. As we changed the number of channels after each downsampling and upsampling, given that the jump connection of the residual block requires input and output with the same number of channels, we changed the number of channels via a 1×1 convolutional layer.

To effectively compensate for the information lost in multiple downsampling and upsampling operations and to reduce the difficulty for the decoder to recover features from highly complex and abstract information, we introduced the idea of dense connectivity in the encoder-decoder structure, adding dense connectivity between encoders and decoders with the same size of the feature graph, which not only places the encoder and decoder at a similar semantic level but also improves the ability of the network to resist overfitting. Different levels of features focus on different information but are consistent with the importance of completing pan-sharpening, and in order to obtain higher precision images while enhancing the ability of the network to explore full-scale information and make full use of all levels of features, we also added dense connections between decoders acting on the same encoder. The input to each decoder is composed of feature maps in encoders and decoders with the same scale and large scale that capture fine-grained and coarse-grained semantics at the full scale.

3.4. Feedback Connection Structure

Li et al. [50] carefully designed a feedback block to extract powerful high-level representations for low-level computer-vision tasks and transmit high-level representations to perfect low-level functions. Fu et al. [44] added this feedback connection mechanism for super-resolution tasks to the network for pan-sharpening. They enable the feature-extraction block to generate more powerful features by iterating the information in each subnetwork to the same module of the next subnetwork, iteratively up and downsampling the input features to achieve the feedback connectivity mechanism.

Our proposed network has a similar structure to that of TPNwFB, which consists of four identical subnetworks, each with a specific structure, as shown in Figure 2. Compared to feedforward connections, each network layer can only accept information from the previous layer, and the shallow network cannot access useful information from the deep network, so it can only extract the underlying features, lacking sufficient context information and abstract fields. Feedback connections can input features that have already completed the initial reconstruction as depth information into the next subnetwork. The high-level information transmitted can complement the semantic and abstract information lacking in low-level features, correct the misinformation carried in low-level features, correct some previous states, and provide the network with significant early reconstruction capability.

3.5. Image Reconstruction Network

We reconstructed the images from the recovered features using a residual block and a convolution layer of 3 × 3. We upsampled the recovered features to the same scale as the PAN image and injected them into the residual block after they were stacked with the features extracted by the two-path network, which helps compensate for the information lost by the network during convolution while effectively reducing the training difficulty of the network. Finally, the detailed features needed to complement the LRMS images were recovered by a convolutional layer and interacted with the LRMS in the approximate branch to generate high-quality HRMS images. This procedure can be defined as:

$$I_{out} = I_{LRMS} + \delta(Conv_{3,4}(F_{RB}(Deconv_{2,64}(F_{FEEB}(\cdot)) \otimes f_{PAN} \otimes f_{MS}))), \quad (8)$$

We use $\otimes$ to represent cascading operations. $Conv_{f,n}(\cdot)$ and $Deconv_{f,n}(\cdot)$ represent convolutional and deconvolutional layers, respectively, and f and n represent the size and number of channels of convolutional kernels. $F_{RB}(\cdot)$ and $F_{FEEB}(\cdot)$ represent the residual blocks and the feature-fusion reconstruction blocks, respectively.

3.6. Loss Function

The L2 loss function may cause local minimization problems and result in artifacts in the image-smoothing region. Simultaneously, the L1 loss function yields a good minimum, and the L1 loss function retains the spectral information, such as colour and brightness, better than the L2 loss function. Therefore, we chose the L1 loss function to optimise the parameters of the proposed network. We attached the loss function to each subnet, ensuring that the information passed to the latter subnetwork in the feedback connection is valid:

$$loss = \frac{1}{N}\sum_{i=1}^{N} |\Phi(X_p^{(i)}, X_m^{(i)}; \theta) - Y^{(i)}|_1, \quad (9)$$

where $X_p^{(i)}$, $X_m^{(i)}$ and $Y^{(i)}$ represent a set of training samples; $X_p^{(i)}$ and $X_m^{(i)}$ refer to the PAN image and low-resolution MS image, respectively; $Y^{(i)}$ represents high-resolution MS images; Φ represents the entire network; and θ is the parameter in the network.

4. Experiments and Analysis

In this section, we demonstrate the effectiveness and superiority of the proposed method through experiments on the QuickBird, WorldView-2, WorldView-3, and IKONOS datasets. In early experiments, the best model is selected for experiments by comparing and evaluating the training and test results of various network parameter models. Finally, the visual and objective metrics of our best model are compared with several existing traditional algorithms and CNN methods to demonstrate the superior performance of the proposed method.

4.1. Datasets

For QuickBird data, the spatial resolution of the MS image is 2.44 m, the spatial resolution of the PAN image is 0.61 m, and the MS image has four bands, i.e., blue, green, red, and near-infrared (NIR) bands, with a spectral resolution of 450–900 nm. For WorldView-2 and WorldView-3 data, the spatial resolutions of the MS images are 1.84 m and 1.24 m, respectively, the spatial resolutions of the PAN images are 0.46 m and 0.31 m, respectively, the MS image has eight bands, i.e., coastal, blue, green, yellow, red, edge, NIR and NIR 2 bands, and the spectral resolutions of the images are 400–1040 nm. For IKONOS data, the spatial resolution of the MS image is 4 m, the spatial resolution of the PAN image is 1 m, and the MS image has four bands, i.e., blue, green, red, and near-NIR bands, with a spectral resolution of 450–900 nm.

The network architecture in this study was implemented using the PyTorch deep learning framework and trained on an NVIDIA RTX 2080Ti GPU. The training time for the entire program was approximately eight hours. We used the Adam optimisation algorithm to minimise the loss function and optimise the model. We set the learning rate to 0.001 and the exponential decay factor to 0.8. The LRMS and PAN images were both downsampled by Wald's protocol in order to use the original LRMS images as the ground truth images. The image patch size was set to 64 $\times$ 64 and the batch size to 64. To facilitate visual observation, the red, green, and blue bands of the multispectral images were used as imaging bands of RGB images to form colour images. The results are presented using ENVI. In the calculation of image-evaluation indexes, all the bands of the images were used simultaneously.

Considering that different satellites have different properties, the models were trained and tested on all four datasets. Each dataset is divided into two subsets, namely the training and test sets, between which the samples do not overlap. The training set was used to train the network, and the test set was used to evaluate the performance. The sizes of the training and test sets for the four datasets are listed in Table 1. We used a separate set of images as a validation set to assess differences in objective metrics and to judge the quality of methods from a subjective visual perspective, each consisting of original 256 $\times$ 256 MS images and original 1024 $\times$ 1024 PAN images.

Table 1. Size of training and test sets for different satellite datasets.

Dataset	Total Numbers	Train Set	Validation Set
QuickBird	950	750	200
WorldView-2	750	600	150
WorldView-3	1300	1000	300
IKONOS	160	144	16

4.2. Evaluation Indexes

We contrast the performance of different algorithms through two different types of experiments, i.e., simulation experiments with HRMS images as a reference and real experiments without HRMS images as a reference, because in the actual application scenarios of remote sensing images, there is often a lack of HRMS images. In order to more objectively evaluate and analyse the performance of different algorithms in different aspects of different datasets, we selected ten objective evaluation indicators according to the characteristics of simulation experiments and real experiments. Depending on whether or not reference images are used, they can be divided into reference indicators and non-reference indicators.

The universal image quality index [51], averaged over the bands (Q_avg) and its four-band extension, Q4 [52] represents the quality of each band and the quality of all the bands, respectively. The relative global dimensional synthesis error (ERGAS) [32], also known as the relative overall two-dimensional comprehensive error, is generally used as the overall quality index. The relative average spectral error (RASE) [42] estimates the overall spectral quality of the pan-sharpened image. Structural similarity (SSIM) [53] is a measure of similarity between two images. The correlation coefficient (CC) [43] is a widely used index for measuring the spectral quality of pan-sharpened images. It calculates the correlation coefficient between the generated image and the corresponding reference image. The spectral angle mapper (SAM) [54] measures the spectral distortion of the pan-sharpened image compared with the reference image. It is defined as the angle between the spectral vectors of the pan-sharpened image and the reference image in the same pixel. The closer Q_avg, Q4, SSIM, and SCC are to 1, the better the fusion results, while the lower SAM, RASE, and ERGAS are, the better the fusion quality.

To evaluate these methods in the full-resolution case, we used the reference-free mass index (QNR) [55] and its spatial index (DS), as well as the spectral index ($D\lambda$) for quantitative evaluation. QNR primarily reflects the fusion performance with no real reference values and is composed of Ds and $D\lambda$. The Ds index being close to 0 indicates good structural performance; the $D\lambda$ index being close to 0 shows good fusion in the spectrum; and a QNR value close to 1 indicates the original full-colour pan-sharpening performance. As these metrics rely heavily on raw MS and PAN images, often, quantifying the similarity of certain components in the fusion images to low-resolution observations would bias these indicator estimates, and for this reason, some methods can generate images with high QNR values but poor quality.

4.3. Simulated Experiments and Real Experiments

To verify the effectiveness and reliability of the proposed network, we performed simulated and real experiments on different datasets. Some representative traditional and deep learning-based algorithms were selected from four datasets, and performance was compared between different methods by subjective visual and objective metrics. The selected traditional algorithms include the CS-based methods, such as IHS [5], PRACG [8], HPF [56], and GS [7]. Among the MRA-based methods, DWT [9] and GLP [57] were considered. One model-based method, PPXS [58], was considered. We selected five deep learning-based methods as contrast objects, including PNN [28], DRPNN [30], PanNet [40], ResTFNet [43], and TPNwFB [44].

4.3.1. Experiment with QuickBird Dataset

The fusion results using the QuickBird dataset with four bands are shown in Figure 6. Figure 6a–c shows the HRMS, LRMS, and PAN (with a resolution of 256 × 256 pixels), Figure 6d–j shows the fusion results of the traditional algorithms, and Figure 6k–p shows the fusion results of the deep learning methods.

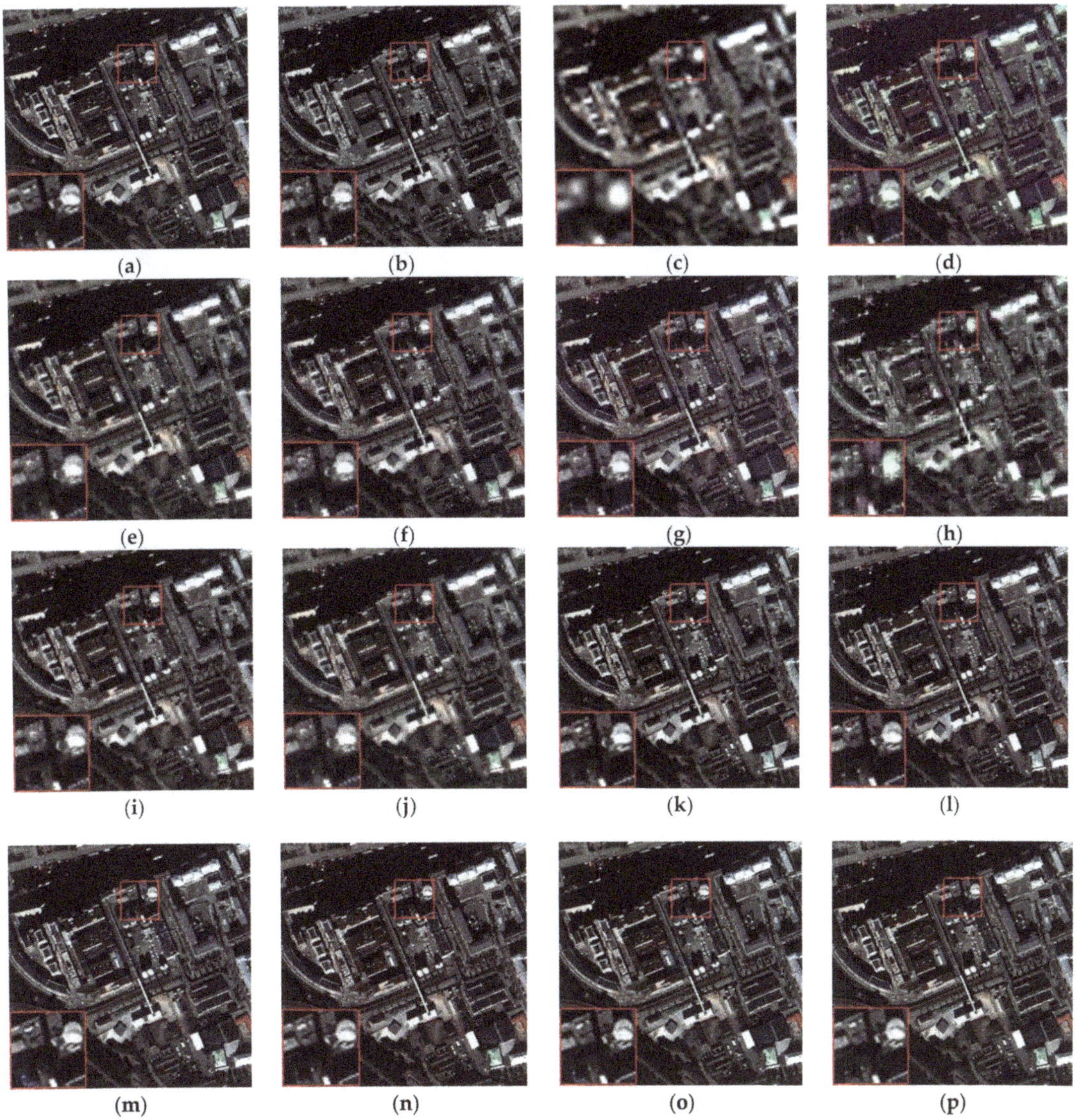

Figure 6. Results using the QuickBird dataset with four bands (resolutions of 256 × 256 pixels): (**a**) reference image; (**b**) PAN; (**c**) LRMS; (**d**) IHS; (**e**) PRACS; (**f**) HPF; (**g**) GS; (**h**) DWT; (**i**) GLP; (**j**) PPXS; (**k**) PNN; (**l**) DRPNN; (**m**) PanNet; (**n**) ResTFNet; (**o**) TPNwFB; (**p**) ours.

Based on the analysis of all the fused and contrast images, it can be intuitively observed that the fused images of the seven non-deep learning methods have obvious colour differences. These images have distinct spectral distortions, with some ambiguity in the edges of the image. Significant artifacts appear around moving objects. Among these methods, the spectral distortion of the DWT image is the most severe. The IHS fusion image has an obvious detail loss in the obvious part of the changing spectral information. The spatial distortion of the PPXS is the most severe, and the fusion image presents a very vague effect. GLP and GS present significant edge blur in the spectral distortion region, and the PRACS method presents artifacts in the image edges, while HPF images show slight blur and edge-texture blur on the image. The deep learning methods show good fidelity to spectral and spatial information on the QuickBird dataset, and it is difficult to determine the texture details of image generation through subjective vision. Therefore, we further compared the following metrics and objectively analysed the advantages and disadvantages of each fusion method. Table 2 lists the results of objective analysis of each method according to the index values.

Table 2. Evaluations using the QuickBird dataset (best result is in bold).

Method	SAM↓	RASE↓	Q_AVE↑	ERGAS↓	CC↑	Q4↑	SSIM↑
IHS	7.3370	29.2116	0.6930	7.7931	0.9245	0.8383	0.6968
PRACS	6.6502	27.0441	0.6985	7.2882	0.9287	0.8693	0.7003
HPF	6.1590	26.5007	0.7199	7.1123	0.9308	0.8795	0.7177
GS	6.7736	28.6871	0.6995	7.6727	0.9282	0.8421	0.7047
DWT	12.6372	39.1140	0.5688	9.9968	0.8361	0.7731	0.5492
GLP	6.2712	26.1510	0.7300	7.0190	0.9329	0.8872	0.7305
PPXS	6.3972	37.0457	0.4738	9.8349	0.8606	0.7126	0.4433
PNN	4.8988	20.4170	0.7949	5.4583	0.9612	0.9259	0.8060
DRPNN	4.0506	16.5490	0.8340	4.4543	0.9738	0.9527	0.8519
PanNet	3.8544	14.0295	0.8497	3.7743	0.9808	0.9627	0.8664
ResTFNet	2.9400	12.1735	0.8834	3.2852	0.9858	0.9739	0.9031
TPNwFB	2.5072	10.0468	0.9072	2.7214	0.9909	0.9822	0.9263
ours	**1.7930**	**6.6668**	**0.9495**	**1.7914**	**0.9958**	**0.9913**	**0.9577**

Objective evaluation metrics show that deep learning-based methods show significantly better performance than conventional methods in terms of evaluating spectral information as well as the metrics for measuring spatial quality. Among traditional methods, the HPF method achieves the best results on the overall metrics, but there is still a huge gap compared to those using deep learning. The HPF and GLP methods differ only slightly in other metrics, but the HPF method outperforms the GLP method in maintaining spectral information, while GLP's spatial details are better. With extremely severe spectral distortion and ambiguous spatial detail, the DWT band exhibits extremely poor performance across all metrics. The PPXS RASE index evaluation outperforms only the serious DWT, shows spatial distortion, and the fusion image is fuzzy. However, it has a good retention of spectral information. In CNN-based methods, affected by the network structure, the more complex networks can achieve better results in general. As only the three-layer network structure was used, even when the nonlinear radiation metrics were introduced with added input, PNN showed the worst performance in the deep learning-based approach. Networks using dual-stream structures achieve significantly superior performance over PNN, DRPNN, and PanNet, bringing the texture details and spectral information of the fused images closer to the original image. Although our proposed network and TPNwFB use feedback connectivity, we use a more efficient feature-extraction structure. Therefore, whether one indicator evaluates spatial or spectral information, the proposed neural network outperforms all compared fusion methods, without obvious artifacts or spectral distortion in the fusion results. These results demonstrate the effectiveness of our proposed method.

4.3.2. Experiment with WorldView-2 Dataset

The fusion results using the WorldView-2 dataset with four bands are shown in Figure 7. Figure 7a–c shows the HRMS, LRMS, and PAN (with a resolution of 256 × 256 pixels), Figure 7d–j shows the fusion results of the traditional algorithms, and Figure 7k–p shows the fusion results of the deep learning methods.

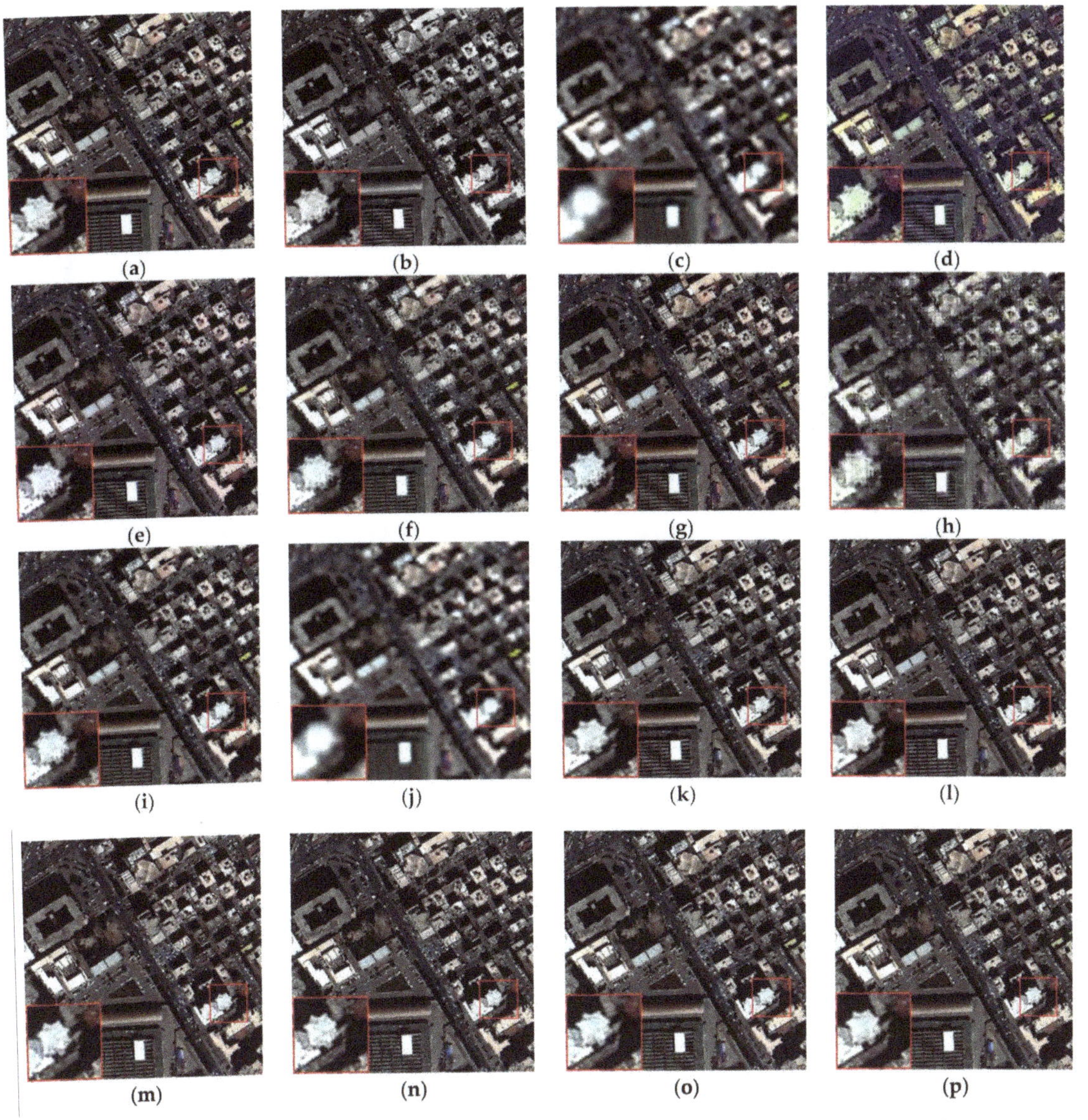

Figure 7. Results using the WorldView-2 dataset with four bands (resolutions of 256 × 256 pixels): (**a**) reference image; (**b**) PAN; (**c**) LRMS; (**d**) IHS; (**e**) PRACS; (**f**) HPF; (**g**) GS; (**h**) DWT; (**i**) GLP; (**j**) PPXS; (**k**) PNN; (**l**) DRPNN; (**m**) PanNet; (**n**) ResTFNet; (**o**) TPNwFB; (**p**) ours.

It is intuitively seen from the graph that the fusion images of non-deep learning methods have distinct colour differences compared to the reference images, and the results of traditional methods are affected by more serious spatial blurring than deep learning-based methods. PRACS and GLP partially recover better spatial details and spectral information, obtaining better subjective visual effects than other conventional methods. However, it is still affected by spectral distortion and artifacts. Through visual observation, it is intuitive that deep learning-based methods do better in the preservation of spectral information than conventional methods.

Table 3 presents the results of objective analysis of each method according to the index values. On the WorldView-2 dataset, images produced using conventional algorithms and fusion images produced based on deep learning algorithms do not show significant gaps in various metrics, but the latter still performs better from all perspectives.

Table 3. Evaluations using the WorldView-2 dataset (best result is in bold).

Method	SAM↓	RASE↓	Q_AVE↑	ERGAS↓	CC↑	Q4↑	SSIM↑
IHS	5.6371	25.9189	0.7103	6.4711	0.9003	0.8176	0.6712
PRACS	4.9892	24.8504	0.7471	6.0704	0.9056	0.8615	0.7070
HPF	4.7316	23.7913	0.7413	5.8646	0.9115	0.8643	0.6950
GS	5.1635	25.6432	0.7210	6.3201	0.9030	0.8286	0.6827
DWT	8.0542	31.1777	0.6142	7.8327	0.8368	0.7677	0.5529
GLP	4.8826	23.4767	0.7563	5.7863	0.9163	0.8732	0.7124
PPXS	5.0452	29.1005	0.5384	7.2093	0.8638	0.7565	0.4580
PNN	4.4631	20.0271	0.8148	4.9370	0.9390	0.9055	0.7846
DRPNN	4.3753	19.8093	0.8161	4.8780	0.9404	0.9075	0.7851
PanNet	4.4901	20.6826	0.8078	5.1074	0.9343	0.9003	0.7765
ResTFNet	4.2802	18.9940	0.8213	4.6836	0.9447	0.9107	0.7912
TPNwFB	4.0258	17.9753	0.8413	4.4353	0.9505	0.9216	0.8124
ours	**3.7506**	**16.5804**	**0.8643**	**4.0970**	**0.9584**	**0.9346**	**0.8407**

Unlike other methods, PanNet chose to train networks in the high-frequency domain, still inevitably causing a loss of information, even with spectral mapping. Owing to the differences between datasets, it is harder to train deep learning-based methods on WorldView-2 datasets than on other datasets. This results in PanNet failing to achieve satisfactory results on the objective evaluation indicators. Notably, the networks using the feedback connectivity mechanism yielded significantly better results than other methods, with better objective evaluation of metrics, indicating that the fusion images are more similar to ground truth. On each objective evaluation metric, our proposed method exhibits good quality in terms of spatial detail and spectral fidelity.

4.3.3. Experiment with WorldView-3 Dataset

The fusion results using the WorldView-3 dataset with four bands are shown in Figure 8. Figure 8a–c shows the HRMS, LRMS, and PAN (with a resolution of 256 × 256 pixels), Figure 8d–j shows the fusion results of the traditional algorithms, and Figure 8k–p shows the fusion results of the deep learning methods. Table 4 presents the results of objective analysis of each method according to the index values.

On the WorldView-3 dataset, non-deep learning methods are still affected by spectral distortion, which is particularly evident with buildings. The DWT fusion images exhibit the most severe spectral distortion and a loss of spatial detail. The IHS fusion images show partial details of some spectral distortion regions and fuzzy artifacts of the road-vehicle regions. The HPF, GS, GLP, and PRACS methods show good performance in the overall spatial structure, but they show distortion and ambiguity in spectrum and detail. The HPF and GS methods can show colours closer to the reference image, but the edges and details of the house are accompanied by artifacts visible to the naked eye. Spectral distortion in non-deep learning methods leads to local detail loss, with distortion and blurring of vehicle and building edges. Deep learning-based methods all reflect a better retention of spectral and spatial information as a whole.

Figure 8. Results using the WorldView-3 dataset with four bands (resolutions of 256 × 256 pixels): (**a**) reference image; (**b**) PAN; (**c**) LRMS; (**d**) IHS; (**e**) PRACS; (**f**) HPF; (**g**) GS; (**h**) DWT; (**i**) GLP; (**j**) PPXS; (**k**) PNN; (**l**) DRPNN; (**m**) PanNet; (**n**) ResTFNet; (**o**) TPNwFB; (**p**) ours.

Table 4. Evaluations using the WorldView-3 dataset (best result is in bold).

Method	SAM↓	RASE↓	Q_AVE↑	ERGAS↓	CC↑	Q4↑	SSIM↑
IHS	3.9227	20.0131	0.8249	5.0851	0.9532	0.9167	0.7991
PRACS	3.9758	17.9972	0.8500	4.4154	0.9577	0.9437	0.8194
HPF	3.3183	17.7482	0.8369	4.4816	0.9580	0.9407	0.8002
GS	3.5870	19.7825	0.8341	5.0001	0.9546	0.9229	0.8091
DWT	7.4893	29.8107	0.6770	7.5423	0.8853	0.8337	0.6257
GLP	3.3455	16.9436	0.8564	4.2733	0.9652	0.9489	0.8255
PPXS	3.5409	24.1764	0.7045	6.1892	0.9202	0.8763	0.6456
PNN	3.0606	11.3623	0.9219	2.8347	0.9828	0.9752	0.9095
DRPNN	2.9469	11.0848	0.9276	2.7820	0.9836	0.9774	0.9157
PanNet	2.6216	10.9912	0.9288	2.7574	0.9840	0.9773	0.9170
ResTFNet	2.6916	11.3202	0.9317	2.8295	0.9831	0.9764	0.9207
TPNwFB	2.6904	11.1373	0.9257	2.7867	0.9835	0.9769	0.9125
ours	**2.4029**	**9.9737**	**0.9421**	**2.4939**	**0.9868**	**0.9813**	**0.9326**

To further compare the performance of the various methods, we analysed them using objective evaluation measures for different networks. Although PPXS achieved good evaluation on SAM, it has an obvious gap in terms of other metrics and other methods. The HPF and GLP methods show performance similar to that of deep learning methods on SAM metrics, achieving good results in preserving spatial information and yielding better spectral information in the fused results over other non-deep learning methods. However, they still have a large gap on RASE and ERGAS and the methods using CNN, indicating that there are more detailed blurs and artifacts in the fused images.

Among the CNN methods, PanNet showed the best performance, with superior results using high-frequency domains on the WorldView-3 dataset. ResTFnet and TPNwFB achieved similar performance, in addition to TPNwFB, still showing better performance in SSIM indicators, which shows that feedback connection operations in the network still play an important role. Compared with all the contrast methods, our proposed network more effectively retains the spectral and spatial information in the image, yielding good fusion results. Based on all the evaluation measures, the proposed method significantly outperforms the existing fusion methods, demonstrating the effectiveness of the proposed method.

4.3.4. Experiment with the IKONOS Dataset

The fusion results using the IKONOS dataset with four bands are shown in Figure 9. Figure 9a–c shows the HRMS, LRMS, and PAN (with a resolution of 256 × 256 pixels), Figure 9d–j shows the fusion results of the traditional algorithms, and Figure 9k–p shows the fusion results of the deep learning methods. Table 5 presents the results of objective analysis of each method according to the index values.

Table 5. Evaluations using the IKONOS dataset (best result is in bold).

Method	SAM↓	RASE↓	Q_AVE↑	ERGAS↓	CC↑	Q4↑	SSIM↑
IHS	3.1691	13.8400	0.3860	3.1599	0.9427	0.4741	0.4089
PRACS	2.8249	12.7932	0.4800	2.6011	0.9513	0.6675	0.5197
HPF	2.7730	13.5253	0.4683	2.7728	0.9458	0.6389	0.4950
GS	2.8089	14.0234	0.4487	2.8821	0.9411	0.6032	0.4896
DWT	9.4846	22.7378	0.3183	5.4503	0.8553	0.2945	0.3417
GLP	2.7788	13.5999	0.4852	2.8028	0.9455	0.6458	0.5083
PPXS	2.7693	12.7035	0.4065	2.5701	0.9535	0.6351	0.4725
PNN	2.4621	8.2089	0.7088	1.8787	0.9801	0.8057	0.7508
DRPNN	2.3908	8.6174	0.7147	1.9280	0.9786	0.8121	0.7521
PanNet	1.8269	5.6283	0.7899	1.3172	0.9909	0.8862	0.8210
ResTFNet	0.6309	1.4935	0.9512	0.4399	0.9994	0.9747	0.9659
TPNwFB	1.2008	3.3423	0.8842	0.8731	0.9968	0.9375	0.9069
ours	**0.4096**	**1.0310**	**0.9680**	**0.2973**	**0.9997**	**0.9824**	**0.9802**

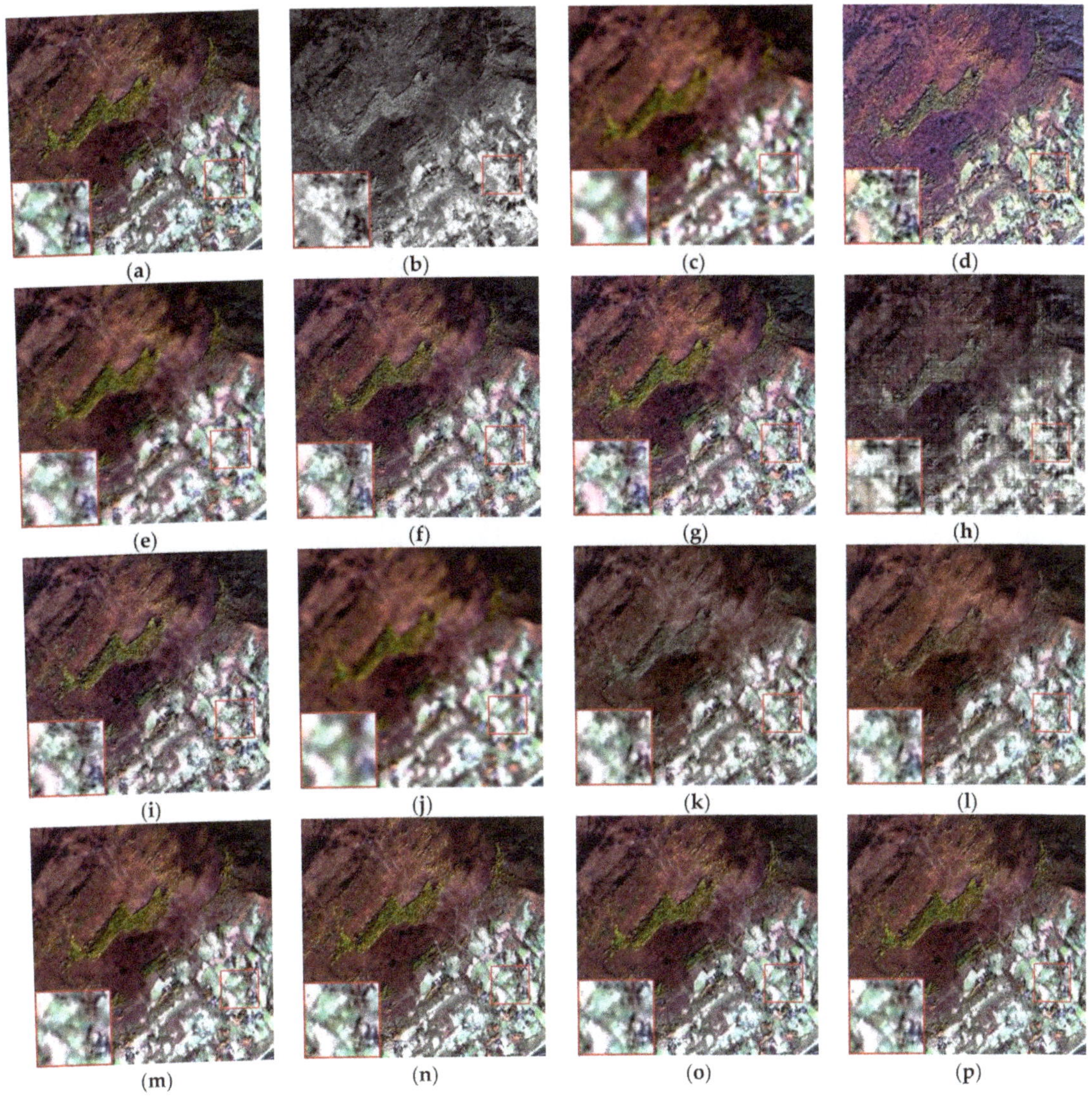

Figure 9. Results using the IKONOS dataset with four bands (resolutions of 256 × 256 pixels): (**a**) reference image; (**b**) PAN; (**c**) LRMS; (**d**) IHS; (**e**) PRACS; (**f**) HPF; (**g**) GS; (**h**) DWT; (**i**) GLP; (**j**) PPXS; (**k**) PNN; (**l**) DRPNN; (**m**) PanNet; (**n**) ResTFNet; (**o**) TPNwFB; (**p**) ours.

All conventional methods produce images with apparent spectral distortion and blur or loss of edge detail. It is clear from the figure that the images obtained using the PNN and DRPNN methods have significant spectral distortion. At the same time, given that the spatial structure is too smooth and a lot of edge information is lost, the index value objectively shows the advantages and disadvantages of various methods, and the overall effect of deep learning is significantly better than that of traditional methods. These data suggest that networks with an encoder–decoder structure have better performance than other structures. ResTFNet obtained significantly superior results using this dataset. Through our proposal that the network-generated images closest approach the original image, the evaluation metrics clearly show the effectiveness of the method.

4.3.5. Experiment with WorldView-3 Real Dataset

For the full-resolution experiment, we used the model trained by the reduced-resolution experiment and the real data as the input to generate fused images. In this experiment, we directly input MS and PAN images into models without any resolution reduction, which guarantees the ideal full-resolution experimental results and follows a similar approach to those used by the other models.

The fusion results using the WorldView-3 Real dataset with four bands are shown in Figure 10. Figure 10a,b shows the LRMS and PAN (with a resolution of 256 × 256 pixels), Figure 10c–i shows the fusion results of the traditional algorithms, and Figure 10j–o shows the fusion results of the deep learning methods. Table 6 presents the results of objective analysis of each method according to the index values.

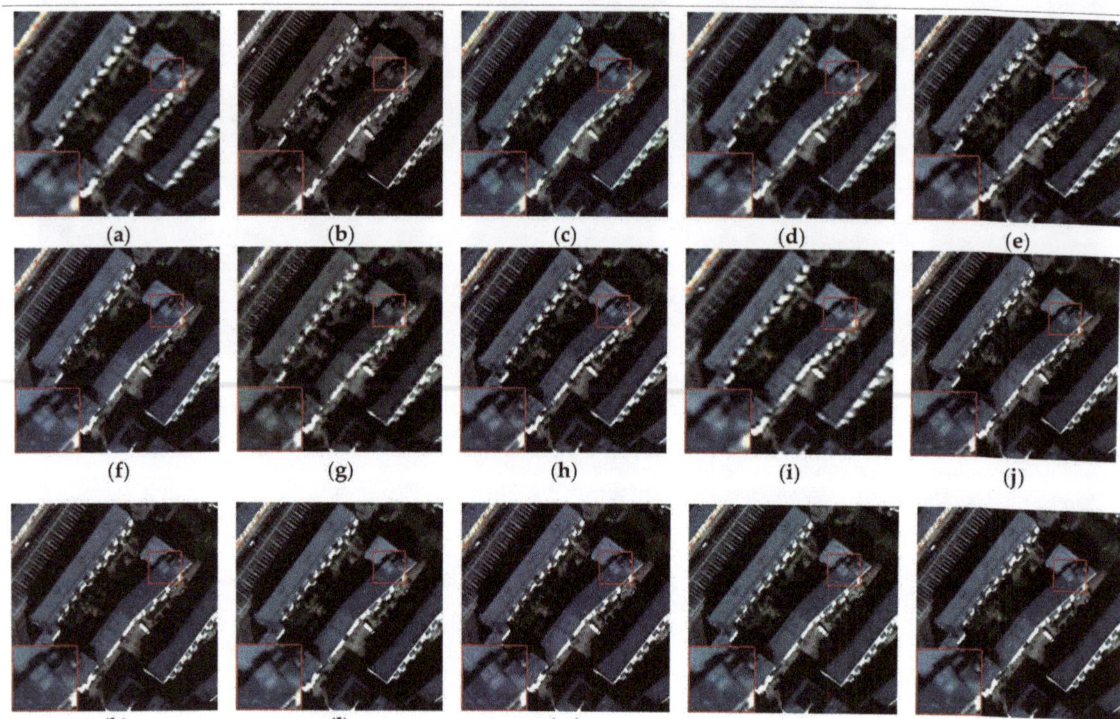

Figure 10. Results using the WorldView-3 Real dataset with four bands (resolutions of 256 × 256 pixels): (**a**) LRMS; (**b**) PAN; (**c**) IHS; (**d**) PRACS; (**e**) HPF; (**f**) GS; (**g**) DWT; (**h**) GLP; (**i**) PPXS; (**j**) PNN; (**k**) DRPNN; (**l**) PanNet; (**m**) ResTFNet; (**n**) TPNwFB; (**o**) ours.

By observing the fusion images, it is found that DWT and IHS show obvious spectral distortion. Although in the GS and GLP methods, the overall spatial structure information is well preserved, local information is lost. The merged images in the PRACS method were too smooth, resulting in severe loss of edge detail.

TPNwFB and our proposed method have the best overall performance and can demonstrate practical utility in using feedback connection operations in the network. An analysis of objective data shows that the index values of PPXS are significantly better than other methods in Dλ but decreased slightly in QNP and Ds. Deep learning-based methods show a certain performance gap in non-deep learning methods. However, given the extremely simple network structure of PNN and DRPNN, satisfactory results are not achieved. Considering three indicators, our proposed network achieves better results in full-resolution experiments, conclusively demonstrating that the proposed innovation plays a positive role in generalised sharpening.

Table 6. Evaluations using the WorldView-3 Real Dataset (best result is in bold).

Method	QNP↑	Dλ↓	Ds↓
IHS	0.6315	0.0794	0.3140
PRACS	0.8041	0.0287	0.1721
HPF	0.6710	0.1067	0.2488
GS	0.6426	0.0708	0.3084
DWT	0.6119	0.2875	0.1412
GLP	0.6755	0.1082	0.2425
PPXS	0.8936	**0.0063**	0.1008
PNN	0.7134	0.1080	0.2003
DRPNN	0.7515	0.0715	0.1907
PanNet	0.8052	0.0790	0.1257
ResTFNet	0.8805	0.0509	0.0723
TPNwFB	0.9116	0.0511	**0.0393**
ours	**0.9213**	0.0201	0.0598

4.3.6. Experiment with QuickBird Real Dataset

The fusion results using the QuickBird Real dataset with four bands are shown in Figure 11. Figure 11a,b shows the LRMS and PAN (with a resolution of 256 × 256 pixels), Figure 11c–i shows the fusion results of the traditional algorithms, and Figure 11j–o shows the fusion results of the deep learning methods. Table 7 presents the results of objective analysis of each method according to the index values.

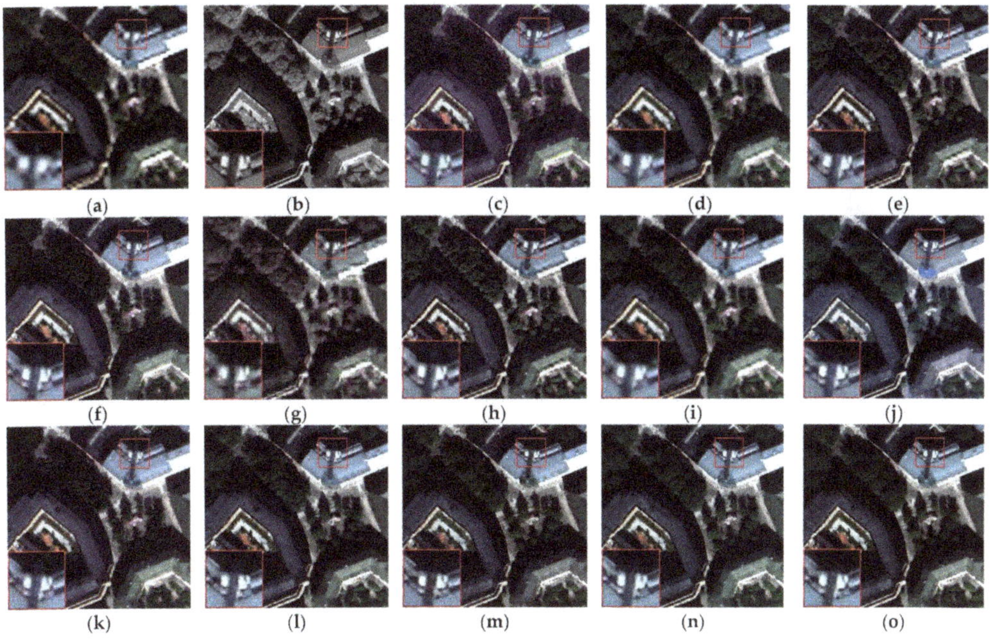

Figure 11. Results using the QuickBird Real dataset with four bands (resolutions of 256 × 256 pixels): (**a**) LRMS; (**b**) PAN; (**c**) IHS; (**d**) PRACS; (**e**) HPF; (**f**) GS; (**g**) DWT; (**h**) GLP; (**i**) PPXS; (**j**) PNN; (**k**) DRPNN; (**l**) PanNet; (**m**) ResTFNet; (**n**) TPNwFB; (**o**) ours.

Table 7. Evaluations using the QuickBird Real Dataset (best result is in bold).

Method	QNP↑	Dλ↓	Ds↓
IHS	0.6096	0.1173	0.3093
PRACS	0.8293	0.0374	0.1384
HPF	0.6468	0.1216	0.2636
GS	0.6418	0.0920	0.2932
DWT	0.5647	0.3273	0.1605
GLP	0.6512	0.1198	0.2601
PPXS	0.8743	**0.0061**	0.1203
PNN	0.7766	0.1871	0.0446
DRPNN	0.8178	0.0979	0.0935
PanNet	0.8236	0.0817	0.1031
ResTFNet	0.9211	0.0426	**0.0379**
TPNwFB	0.9090	0.0495	0.0437
Proposed	**0.9311**	0.0205	0.0494

PRACS and PPXS obtain better visual effects in non-deep learning methods with sufficient retention of spectral information but still lack effective retention of detail compared to deep learning methods. Among the deep learning methods, ResTFNet and our proposed method achieved the best results on the whole, with full and effective retention of spatial details and spectral colour and comprehensive analysis of three objective evaluation indicators. The use of encoder–decoder structure in the network structure can effectively improve the performance of the network in real experiments.

4.3.7. Processing Time and Model Size

As shown in Table 8, for different deep learning methods, our proposed method had the longest processing time in the test mode. Our method also has a far greater number of parameters than the other methods. The data clearly show that the more complex the model, the more time it takes to generate a single fusion image; however, a more complex structure can achieve better performance results. Our method is mainly designed to optimize the structure from the perspective of improving the effect of the fusion result. The issue of optimizing the network runtime was not considered.

Table 8. Different deep learning methods for processing time and model size.

Method	Time (S)	Model Size (MB)
PNN	1.92	0.31
DRPNN	2.08	3.19
PanNet	2.22	2.06
ResTFNet	2.49	8.55
TPNwFB	2.82	52.3
Proposed	3.13	210

5. Discussion

5.1. Discussion of EFEB

In this subsection, we examine the influence of each part of the model through ablation learning in order to obtain the best performance of the model. To obtain high-quality HRMS images, we propose a dense encoder–decoder network with feedback connections for pan-sharpening. In the network, we use an efficient feature-extraction module to fully capture features at different scales in networks of different depths and widths. To increase the depth of the network, we used three MFEBs. In each MFEB, we increased the width of the network by using four branches with different receptive fields.

To validate the effectiveness of our proposed EFEB and to explore the impact of combinations using different receptive field branches on the fusion results, we performed comparative experiments on them using four datasets. We performed experiments using convolutional kernel combinations with different receptive field sizes while retaining three MEFB and four branches in each block, from which the best receptive field scale was selected for combination. Experiments demonstrate that the highest-performing multiscale modules can be obtained by using structures with an expansion rate of {1,2,3,4}. We used four branches with receptive field sizes of 3, 5, 7, and 9, separately, although if we increased the parameters and the number of calculations, we would obtain noticeably better results. The experimental results are presented in Table 9.

Table 9. Quantitative evaluation results of multiscale feature-extraction modules with different combinations are shown in bold.

Scale	SAM↓	RASE↓	Q_AVE↑	ERGAS↓	CC↑	Q4↑	SSIM↑
1123	2.0460	7.4101	0.9369	2.0083	0.9949	0.9897	0.9498
1124	2.0284	7.4482	0.9425	1.9891	0.9948	0.9899	0.9539
1125	2.1016	7.3502	0.9356	1.9812	0.9949	0.9897	0.9485
1223	2.1681	7.6609	0.9295	2.0630	0.9944	0.9890	0.9453
1224	2.2350	7.8802	0.9207	2.1199	0.9941	0.9879	0.9402
1225	2.0571	7.2789	0.9379	1.9671	0.9949	0.9903	0.9509
1233	1.9660	6.6951	0.9392	1.8075	0.9958	0.9913	0.9532
1234	**1.7930**	**6.6668**	**0.9495**	**1.7914**	**0.9958**	**0.9913**	**0.9577**
1235	1.8182	6.6792	0.9487	1.7930	0.9958	0.9914	0.9579
1333	2.1834	7.6122	0.9229	2.0516	0.9945	0.9889	0.9424
1334	1.9818	7.1717	0.9431	1.9291	0.9952	0.9906	0.9543
1335	2.2714	8.0409	0.9193	2.1526	0.9940	0.9879	0.9391

To validate the effectiveness of EFEB across the model, we compared the networks using EFEB to those not using this module on four datasets. The objective evaluation indicators are listed in Table 10. Using EFEB increases the width and depth of the network to extract richer feature information and to identify additional mapping relationships that meet expectations. Elimination of multiscale modules results in a lack of multiscale feature learning and detail learning, which hampers the extraction of more efficient features in the current task, thus reducing image-reconstruction capabilities. EFEB demonstrates the effectiveness of multiple-enhancing network performance in experiments on all four datasets.

Table 10. Quantitative evaluation results of different structures using different datasets. In A, a contrasting network without EFEB. In B, our network is used.

Scale	SAM	RASE	Q_AVE	ERGAS	CC	Q4	SSIM
QuickBird (A)	2.4643	8.5049	0.9135	2.2910	0.9932	0.9863	0.9335
QuickBird (B)	1.7930	6.6668	0.9495	1.7914	0.9958	0.9913	0.9577
WorldView-2 (A)	3.8236	16.6670	0.8622	4.1180	0.9578	0.9332	0.8386
WorldView-2 (B)	3.7506	16.5804	0.8643	4.0970	0.9584	0.9346	0.8407
WorldView-3 (A)	2.4399	10.2544	0.9402	2.5637	0.9861	0.9804	0.9302
WorldView-3 (B)	2.4029	9.9737	0.9421	2.4939	0.9868	0.9813	0.9326
IKONOS (A)	0.4096	1.0310	0.9680	0.2973	0.9997	0.9824	0.9802
IKONOS (B)	0.6157	1.4487	0.9558	0.4389	0.9996	0.9831	0.9748

5.2. Discussion of FFRB

In the network, we used a network structure with a multilayer encoder and decoder combined with dense connections to complete the task of integrating and reconstructing the extracted multiscale spatial and spectral information. In contrast with other two-stream networks for pan-sharpening, which used encoder–decoder structures to decode only the results after the last level encoding, we decoded the results after each level encoding. We also added sufficient dense connections between the encoder and the decoder, which is a further improvement of the conventional symmetric encoder–decoder structure.

To demonstrate that the dense connection between the encoder and the decoder is valid, we retrained a network for comparison on four datasets that retained the same number of encoders and decoders as our proposed network but did not use the dense connection operation. The experimental results are presented in Table 11.

Table 11. Quantitative evaluation results of different structures using different datasets. In A, a contrasting network is used. In B, our network is used.

Scale	SAM	RASE	Q_AVE	ERGAS	CC	Q4	SSIM
QuickBird (A)	2.8675	10.8443	0.8930	2.9372	0.9888	0.9788	0.9152
QuickBird (B)	1.7930	6.6668	0.9495	1.7914	0.9958	0.9913	0.9577
WorldView-2 (A)	3.8805	17.6535	0.8488	4.3531	0.9529	0.9260	0.8222
WorldView-2 (B)	3.7506	16.5804	0.8643	4.0970	0.9584	0.9346	0.8407
WorldView-3 (A)	2.4125	10.2680	0.9396	2.5737	0.9860	0.9803	0.9294
WorldView-3 (B)	2.4029	9.9737	0.9421	2.4939	0.9868	0.9813	0.9326
IKONOS (A)	0.7847	1.9036	0.9430	0.5412	0.9990	0.9689	0.9582
IKONOS (B)	0.6157	1.4487	0.9558	0.4389	0.9996	0.9831	0.9748

Through objective indicators on four datasets, it is clear that we injected low-level features into advanced features through long-jump connections, improved the ability of the network to make full use of all features, reduced information loss during upsampling and downsampling, reduced differences in semantic feature level in the encoder and decoder, reduced the difficulty of network training, and improved the network's ability to recover fine real images.

5.3. Discussion of Feedback Connections

In the network, to obtain better reconstruction power earlier, we introduced feedback connectivity operations to refine deep features in the previous subnetwork by iterating exactly the same network four times into the shallow network structure. As the number of iterations of the subnet had significant effects on the final result, we evaluated the network with different numbers of iterations using the QuickBird dataset. The experimental results are presented in Table 12.

Table 12. Results of the network quantitative evaluation with different iterations. The best performance is shown in bold.

Scale	SAM	RASE	Q_AVE	ERGAS	CC	Q4	SSIM
1	2.7088	9.1094	0.9039	2.4473	0.9923	0.9841	0.9276
2	2.4039	8.4655	0.9214	2.2745	0.9931	0.9861	0.9361
3	2.0831	7.3411	0.9402	1.9763	0.9948	0.9898	0.9509
4	**1.7930**	**6.6668**	**0.9495**	**1.7914**	**0.9958**	**0.9913**	**0.9577**
5	2.0550	7.1303	0.9379	1.9180	0.9952	0.9903	0.9504

We trained a network with the same four subnet structures and attached the loss function to each subnet, but we disconnected the feedback connection between each subnetwork. A comparison of the resulting indexes is presented in Table 13. Although the two networks trained under exactly the same conditions, there is a clear gap in their relative performance, and the feedback connection significantly improves performance and gives the network good early reconstruction capability.

Table 13. Quantitative evaluation results of different structures using different datasets. In A, a contrasting network. In B, our network is used.

Scale	SAM	RASE	Q_AVE	ERGAS	CC	Q4	SSIM
QuickBird (A)	2.6883	8.7127	0.9040	2.3564	0.9927	0.9854	0.9274
QuickBird (B)	1.7930	6.6668	0.9495	1.7914	0.9958	0.9913	0.9577
WorldView-2 (A)	4.2092	18.5268	0.8379	4.5671	0.9489	0.9198	0.8102
WorldView-2 (B)	3.7506	16.5804	0.8643	4.0970	0.9584	0.9346	0.8407
WorldView-3 (A)	2.5027	9.9731	0.9384	2.4939	0.9869	0.9813	0.9284
WorldView-3 (B)	2.4029	9.9737	0.9421	2.4939	0.9868	0.9813	0.9326
IKONOS (A)	0.6362	1.6218	0.9557	0.4448	0.9993	0.9750	0.9691
IKONOS (B)	0.6157	1.4487	0.9558	0.4389	0.9996	0.9831	0.9748

6. Conclusions

In this paper, we proposed a dense encoder–decoder network with feedback connections for pan-sharpening based on the practical demand for high-quality HRMS images. We adopted a network structure that has achieved remarkable results in other image-processing fields for pan-sharpening and combined it with knowledge in the remote sensing image field to effectively improve the network structure. Our proposed DEDwFB structure, which significantly improves the depth and width of the network, improves its ability to grasp large-scale features and reconstruct images, effectively improving the quality of fusion images.

We aimed to achieve two goals: spectral information preservation and spatial information preservation in pan-sharpening. PAN and LRMS were therefore chosen to process separate images using dual-stream structures, without interference, taking advantage of diverse information in the two images. Efficient feature-extraction blocks sufficiently increase the network's ability to grab features from different scales of receptive fields and fully recover higher-quality images from scratch-to features through an encoder–decoder network with dense connectivity mechanisms. Feedback mechanisms help networks refine low-level information through powerful deep features and help shallow networks obtain useful information from coarse reconstructed HRMS.

Experiments on four datasets demonstrate that the structure we used in the network is very efficient for obtaining higher-quality fusion images than other methods. As our proposed network has replicated feature extraction and image fusion reconstruction structures, the network can obtain better results when processing images with more complex information. The method is better at processing spectroscopic and spatially informative images, and complex network structures and dense jump connections can efficiently capture rich features from dense buildings, dense vegetation, and large amounts of transportation, which helps to produce satisfactory high-quality fusion images.

Author Contributions: Data curation, W.L.; formal analysis, W.L.; methodology, W.L. and M.X.; validation, M.X.; visualization, M.X. and X.L.; writing—original draft, M.X.; writing—review and editing, M.X. and X.L. All authors have read and agreed to the published version of the manuscript.

Funding: This research was funded by the National Natural Science Foundation of China (no. 61972060, U171321, and 62027827), the National Key Research and Development Program of China (no. 2019YFE0110800), and the Natural Science Foundation of Chongqing (cstc2020jcyj-zdxmX0025 and cstc2019cxcyljrc-td0270).

Data Availability Statement: Data sharing is not applicable to this article.

Acknowledgments: The authors would like to thank all of the reviewers for their valuable contributions to our work.

Conflicts of Interest: The authors declare no conflict of interest.

References

1. Wang, R.S.; Xiong, S.Q.; Ni, H.F.; Liang, S.N. Remote sensing geological survey technology and application research. *Acta Geol. Sinica* **2011**, *85*, 1699–1743.
2. Li, C.Z.; Ni, H.F.; Wang, J.; Wang, X.H. Remote Sensing Research on Characteristics of Mine Geological Hazards. *Adv. Earth Sci.* **2005**, *1*, 45–48.
3. Yin, X.K.; Xu, H.L.; Fu, H.Y. Application of remote sensing technology in wetland resource survey. *Heilongjiang Water Sci. Technol.* **2010**, *38*, 222.
4. Wang, Y.; Wang, L.; Wang, Z.Y.; Yu, Y. Research on application of multi-source remote sensing data technology in urban engineering geological exploration. In *Land and Resources Informatization*; Oriprobe: Taipei City, Taiwan, 2021; pp. 7–14.
5. Tu, T.-M.; Su, S.-C.; Shyu, H.-C.; Huang, P.S. A new look at IHS-like image fusion methods. *Inf. Fusion* **2001**, *2*, 177–186. [CrossRef]
6. Kwarteng, P.; Chavez, A. Extracting spectral contrast in Landsat Thematic Mapper image data using selective principal component analysis. *Photogramm. Eng. Remote Sens.* **1989**, *55*, 339–348.
7. Laben, C.A.; Brower, B.V. Process for Enhancing the Spatial Resolution of Multispectral Imagery Using Pan-Sharpening. U.S. Patent 6,011,875, 4 January 2000.

8. Choi, J.; Yu, K.; Kim, Y. A new adaptive component-substitution-based satellite image fusion by using partial replacement. *IEEE Trans. Geosci. Remote Sens.* **2011**, *49*, 295–309. [CrossRef]
9. Zhou, J.; Civco, D.L.; Silander, J.A. A wavelet transform method to merge Landsat TM and SPOT panchromatic data. *Int. J. Remote Sens.* **1998**, *19*, 743–757. [CrossRef]
10. Nunez, J.; Otazu, X.; Fors, O.; Prades, A.; Pala, V.; Arbiol, R. Multiresolution-based image fusion with additive wavelet decomposition. *IEEE Trans. Geosci. Remote Sens.* **1999**, *37*, 1204–1211. [CrossRef]
11. Burt, P.J.; Adelson, E.H. The Laplacian pyramid as a compact image code. *IEEE Trans. Commun.* **1983**, *3*, 532–540. [CrossRef]
12. Ghahremani, M.; Ghassemian, H. Remote-sensing image fusion based on Curvelets and ICA. *Int. J. Remote Sens.* **2015**, *36*, 4131–4143. [CrossRef]
13. Shah, V.P.; Younan, N.H.; King, R.L. An Efficient Pan-Sharpening Method via a Combined Adaptive PCA Approach and Contourlets. *IEEE Trans. Geosci. Remote Sens.* **2008**, *46*, 1323–1335. [CrossRef]
14. Fei, R.; Zhang, J.; Liu, J.; Du, F.; Chang, P.; Hu, J. Convolutional sparse representation of injected details for pansharpening. *IEEE Geosci. Remote Sens.* **2019**, *16*, 1595–1599. [CrossRef]
15. Yin, H. PAN-guided cross-resolution projection for local adaptive sparse representation-based pansharpening. *IEEE Trans. Geosci. Remote Sens.* **2019**, *57*, 4938–4950. [CrossRef]
16. Palsson, F.; Sveinsson, J.R.; Ulfarsson, M.O. A new pansharpening algorithm based on total variation. *IEEE Geosci. Remote Sens.* **2014**, *11*, 318–322. [CrossRef]
17. Wei, Q.; Dobigeon, J.N.; Tourneret, Y. Bayesian fusion of multiband images. *IEEE J. Sel. Top. Signal Process.* **2015**, *9*, 1117–1127. [CrossRef]
18. Guo, M.; Zhang, H.; Li, J.; Zhang, L.; Shen, H. An Online Coupled Dictionary Learning Approach for Remote Sensing Image Fusion. *IEEE J. Sel. Top. Appl. Earth Obs. Remote Sens.* **2014**, *7*, 1284–1294. [CrossRef]
19. Xu, M.; Chen, H.; Varshney, P.K. An Image Fusion Approach Based on Markov Random Fields. *IEEE Trans. Geosci. Remote Sens.* **2011**, *49*, 5116–5127.
20. Hallabia, H.; Hamam, H. An Enhanced Pansharpening Approach Based on Second-Order Polynomial Regression. In Proceedings of the 2021 International Wireless Communications and Mobile Computing (IWCMC), Harbin City, China, 28 June–2 July 2021; IEEE: Piscataway, NJ, USA, 2021; pp. 1489–1493.
21. Girshick, R.; Donahue, J.; Darrell, T.; Malik, J. Rich feature hierarchies for accurate object detection and semantic segmentation. In Proceedings of the IEEE Conference on Computer Vision and Pattern Recognition, Columbus, OH, USA, 23–28 June 2014; pp. 580–587.
22. Ronneberger, O.; Fischer, P.; Brox, T. U-Net: Convolutional Networks for Biomedical Image Segmentation. In Proceedings of the Lecture Notes in Computer Science, Munich, Germany, 5–9 October 2015; Volume 9351, pp. 234–241.
23. Li, H.; Wu, X.J. DenseFuse: A Fusion Approach to Infrared and Visible Images. *IEEE Trans. Image Process.* **2019**, *28*, 2614–2623. [CrossRef]
24. Dong, C.; Loy, C.C.; He, K.; Tang, X. Learning a deep convolutional network for image super-resolution. In Proceedings of the European Conference on Computer Vision, Zurich, Switzerland, 6–12 September 2014; pp. 184–199.
25. Vitale, S.; Scarpa, G. A detail-preserving cross-scale learning strategy for CNN-based pansharpening. *Remote Sens.* **2020**, *12*, 348. [CrossRef]
26. Azarang, A.; Kehtarnavaz, N. Image fusion in remote sensing by multi-objective deep learning. *Int. J. Remote Sens.* **2020**, *41*, 9507–9524. [CrossRef]
27. Dong, C.; Loy, C.; He, K.; Tang, X. Image super-resolution using deep convolutional networks. *IEEE Trans. Pattern Anal. Mach. Intell.* **2016**, *38*, 295–307. [CrossRef]
28. Masi, G.; Cozzolino, D.; Verdoliva, L.; Scarpa, G. Pansharpening by convolutional neural networks. *Remote Sens.* **2016**, *8*, 594. [CrossRef]
29. Rao, Y.Z.; He, L.; Zhu, J.W. A Residual Convolutional Neural Network for Pan-Sharpening. In Proceedings of the International Workshop on Remote Sensing with Intelligent Processing, Shanghai, China, 18–21 May 2017.
30. Wei, Y.; Yuan, Q.; Shen, H.; Zhang, L. Boosting the accuracy of multispectral image pansharpening by learning a deep residual network. *IEEE Geosci. Remote Sens. Lett.* **2017**, *14*, 1795–1799. [CrossRef]
31. He, L.; Rao, Y.; Li, J.; Chanussot, J.; Plaza, J.; Zhu, J.; Li, B. Pansharpening via Detail Injection Based Convolutional Neural Networks. *IEEE J. Sel. Top. Appl. Earth Observ. Remote Sens.* **2019**, *12*, 1188–1204. [CrossRef]
32. Yang, J.; Fu, X.; Hu, Y.; Huang, Y.; Ding, X.; Paisley, J. PanNet: A deep network architecture for pan-sharpening. In Proceedings of the IEEE International Conference on Computer Vision, Venice, Italy, 22–29 October 2017; pp. 5449–5457.
33. Simonyan, K.; Zisserman, A. Very Deep Convolutional Networks for Large-Scale Image Recognition. *arXiv* **2015**, arXiv:1409.1556.
34. Szegedy, C.; Vanhoucke, V.; Ioffe, S.; Shlens, J.; Wojna, Z. Rethinking the inception architecture for computer vision. In Proceedings of the IEEE Conference on Computer Vision and Pattern Recognition, Las Vegas, NV, USA, 27–30 June 2016; pp. 2818–2826.
35. He, K.; Zhang, X.; Ren, S.; Sun, J. Deep residual learning for image recognition. In Proceedings of the IEEE Conference on Computer Vision and Pattern Recognition, Las Vegas, NV, USA, 27–30 June 2016; pp. 770–778.
36. Huang, G.; Liu, Z.; Van, L.; Weinberger, K.Q. Densely Connected Convolutional Networks. In Proceedings of the IEEE Conference on Computer Vision and Pattern Recognition, Honolulu, HI, USA, 21–26 July 2017; pp. 4700–4708.

37. Zhou, Z.; Siddiquee, M.; Tajbakhsh, N. UNet++: A Nested U-Net Architecture for Medical Image Segmentation. In *Deep Learning in Medical Image Analysis and Multimodal Learning for Clinical Decision Support*; Springer: Cham, Switzerland, 2018; pp. 3–11.
38. Huang, H.M.; Lin, L.F.; Tong, R.F.; Hu, H.J.; Zhang, Q.W.; Iwamoto, Y.; Han, X.H.; Chen, Y.W. U-Net3+: A Full-Scale Connected Unet for Medical Image Segmentation. In Proceedings of the IEEE International Conference on Acoustics, Speech, and Signal Processing, Barcelona, Spain, 4–8 May 2020; pp. 1055–1059.
39. Santhanam, V.; Morariu, V.I.; Davis, L.S. Generalized deep image to image regression. In Proceedings of the IEEE Conference on Computer Vision and Pattern Recognition, Honolulu, HI, USA, 21–26 July 2017; pp. 5609–5619.
40. Fu, X.; Wang, W.; Huang, Y.; Ding, X.; Paisley, J. Deep Multiscale Detail Networks for Multiband Spectral Image Sharpening. *IEEE Trans. Neural Netw. Learn. Syst.* **2020**, *32*, 2090–2104. [CrossRef]
41. Yu, F.; Koltun, V. Multiscale context aggregation by dilated convolutions. *arXiv* **2015**, arXiv:1511.07122.
42. Liu, X.; Liu, Q.; Wang, Y. Remote sensing image fusion based on two-stream fusion network. In Proceedings of the 24th International Conference on Multimedia Modeling, Bangkok, Thailand, 5–7 February 2018; Springer: Berlin/Heidelberg, Germany, 2018; pp. 1–12.
43. Liu, X.; Liu, Q.; Wang, Y. Remote sensing image fusion based on two-stream fusion network. *Inf. Fusion* **2020**, *55*, 1–15. [CrossRef]
44. Fu, S.; Meng, W.; Jeon, G. Two-Path Network with Feedback Connections for Pan-Sharpening in Remote Sensing. *Remote Sens.* **2020**, *12*, 1674. [CrossRef]
45. Liu, X.; Wang, Y.; Liu, Q. PSGAN: A generative adversarial network for remote sensing image pan-sharpening. In Proceedings of the IEEE International Conference on Image Processing, Athens, Greece, 7–10 October 2018; pp. 873–877.
46. Shao, Z.; Lu, Z.; Ran, M.; Fang, L.; Zhou, J.; Zhang, Y. Residual encoder-decoder conditional generative adversarial network for pansharpening. *IEEE Geosci. Remote Sens. Lett.* **2019**, *17*, 1573–1577. [CrossRef]
47. Zhang, L.P.; Li, W.S.; Shen, L.; Lei, D.J. Multilevel dense neural network for pan-sharpening. *Int. J. Remote Sens.* **2020**, *41*, 7201–7216. [CrossRef]
48. Li, W.S.; Liang, X.S.; Dong, M.L. MDECNN: A Multiscale Perception Dense Encoding Convolutional Neural Network for Multispectral Pan-Sharpening. *Remote Sens.* **2021**, *13*, 3.
49. Gao, S.; Cheng, M.; Zhao, K.; Zhang, X.; Yang, M.; Torr, P.H.S. Res2Net: A new multiscale backbone architecture. *IEEE Trans. Pattern Anal. Mach. Intell.* **2019**, *43*, 652–662. [CrossRef] [PubMed]
50. Li, Z.; Yang, J.; Liu, Z.; Yang, X.; Jeon, G.; Wu, W. Feedback Network for Image Super-Resolution. In Proceedings of the IEEE Conference on Computer Vision and Pattern Recognition, Long Beach, CA, USA, 15–20 June 2019; pp. 3867–3876.
51. Wang, Z.; Bovik, A.C. A universal image quality index. *IEEE Signal Process. Lett.* **2002**, *9*, 81–84. [CrossRef]
52. Alparone, L.; Baronti, S.; Garzelli, A.; Nencini, F. A global quality measurement of pan-sharpened multispectral imagery. *IEEE Geosci. Remote Sens. Lett.* **2004**, *1*, 313–317. [CrossRef]
53. Wang, Z. Image Quality Assessment: From Error Visibility to Structural Similarity. *IEEE Trans. Image Process.* **2004**, *13*, 600–612. [CrossRef]
54. Yuhas, R.H.; Goetz, A.F.; Boardman, J.W. Discrimination among semi-arid landscape endmembers using the spectral angle mapper (SAM) algorithm. In Proceedings of the Summaries 3rd Annual JPL Airborne Geoscience Workshop, Pasadena, CA, USA, 1–5 June 1992; pp. 147–149.
55. Alparone, L.; Aiazzi, B.; Baronti, S.; Garzelli, A.; Nencini, F.; Selva, M. Multispectral and panchromatic data fusion assessment without reference. *Photogramm. Eng. Remote Sens.* **2008**, *74*, 193–200. [CrossRef]
56. Witharana, C.; Civco, D.L.; Meyer, T.H. Evaluation of pansharpening algorithms in support of earth observation based rapidmapping workflows. *Appl. Geogr.* **2013**, *37*, 63–87. [CrossRef]
57. Otazu, X.; Gonzalez-Audicana, M.; Fors, O.; Nunez, J. Introduction of sensor spectral response into image fusion methods. Application to wavelet-based methods. *IEEE Trans. Geosci. Remote Sens.* **2005**, *43*, 2376–2385. [CrossRef]
58. Shi, Y.; Wanyu, Z.; Wei, L. Pansharpening of Multispectral Images based on Cycle-spinning Quincunx Lifting Transform. In Proceedings of the IEEE International Conference on Signal, Information and Data Processing, Chongqing, China, 11–13 December 2019; pp. 1–5.

remote sensing

MDPI

Article

DisasterGAN: Generative Adversarial Networks for Remote Sensing Disaster Image Generation

Xue Rui [1], Yang Cao [2], Xin Yuan [1], Yu Kang [1,2,3] and Weiguo Song [1,*]

1 State Key Laboratory of Fire Science, University of Science and Technology of China, Hefei 230026, China; ruixue27@mail.ustc.edu.cn (X.R.); yx98314@mail.ustc.edu.cn (X.Y.); kangduyu@ustc.edu.cn (Y.K.)
2 Department of Automation, University of Science and Technology of China, Hefei 230026, China; forrest@ustc.edu.cn
3 Institute of Advanced Technology, University of Science and Technology of China, Hefei 230088, China
* Correspondence: wgsong@ustc.edu.cn

Abstract: Rapid progress on disaster detection and assessment has been achieved with the development of deep-learning techniques and the wide applications of remote sensing images. However, it is still a great challenge to train an accurate and robust disaster detection network due to the class imbalance of existing data sets and the lack of training data. This paper aims at synthesizing disaster remote sensing images with multiple disaster types and different building damage with generative adversarial networks (GANs), making up for the shortcomings of the existing data sets. However, existing models are inefficient in multi-disaster image translation due to the diversity of disaster and inevitably change building-irrelevant regions caused by directly operating on the whole image. Thus, we propose two models: disaster translation GAN can generate disaster images for multiple disaster types using only a single model, which uses an attribute to represent disaster types and a reconstruction process to further ensure the effect of the generator; damaged building generation GAN is a mask-guided image generation model, which can only alter the attribute-specific region while keeping the attribute-irrelevant region unchanged. Qualitative and quantitative experiments demonstrate the validity of the proposed methods. Further experimental results on the damaged building assessment model show the effectiveness of the proposed models and the superiority compared with other data augmentation methods.

Keywords: GAN; image generation; data augmentation; remote sensing disaster image

Citation: Rui, X.; Cao, Y.; Yuan, X.; Kang, Y.; Song, W. DisasterGAN: Generative Adversarial Networks for Remote Sensing Disaster Image Generation. *Remote Sens.* **2021**, *13*, 4284. https://doi.org/10.3390/rs13214284

Academic Editors: Fahimeh Farahnakian, Jukka Heikkonen and Pouya Jafarzadeh

Received: 13 September 2021
Accepted: 20 October 2021
Published: 25 October 2021

1. Introduction

Rapid detection and assessment after the occurrence of disaster play a very important role in humanitarian assistance and disaster recovery. The applications of deep-learning models in remote sensing have attracted much attention recently. Among them, as the building damage assessment data set represented by the xBD data set [1] has been open source, researchers have proposed several building detection and damage assessment models based on deep neural networks (DNNs) [2–4]. DNNs such as convolutional neural networks (CNNs) need a substantial amount of training data. Compared with the large data sets of natural images, the limited labeled remote sensing data becomes an obstacle to train a DNN well, especially in building damage data sets. Moreover, there is an obvious class imbalance in the xBD data set; specifically, the sample size of the damaged buildings in the three categories (minor damage, major damage, and destroyed) is far less than that of the no-damage buildings [1]. This problem makes it difficult for the model to extract the features of buildings damaged by different types of disasters, thus affecting the accuracy of the assessment model.

The fact proves that, among the existing models of damage building assessment based on the xBD data set, the accuracy of minor damage and major-damage categories is obviously lower than that of the no-damage category, which means that minor damage and major damage classes belong to the hard classes [1–4]. To address this problem, scholars also put forward several data augmentation strategies to improve the class imbalance. To be more specific, Shen et al. [2] apply the CutMix as a data augmentation method that combines the hard-classes images with random images to reconstruct new samples, Hao et al. [3] adopt the common data augmentation method such as horizontal flipping and random cropping during training, and Boin et al. [4] mitigate class imbalance with oversampling. Although the aforementioned methods have a certain effect on improving the accuracy of hard classes, in fact, these are deformation and reorganization of the original samples; more seriously, these may degrade the quality of images, thus affecting the rationality of the features extracted by the feature extractor. Essentially, the above methods do not add new samples and rely on human decisions and manual selection of data transformations, whereas it takes much manpower and material resources to collect and process remote sensing images of damaged buildings to make new samples.

Recently, generative adversarial networks (GANs) [5] and their variants have been widely used in the field of computer vision, such as image-to-image translation [6–8] and image attribute editing [9–12]. GANs aim to fit the real distribution of data by a Min-Max game theory. The standard GAN contains two parts: the generator G and discriminant D, by adversarial training, making the generator generate images gradually close to the real images. In this way, GAN has become an effective framework to generate random data distribution models so that scholars naturally associate that GAN can learn the data distribution of data samples and generate samples as close as possible to the training data distribution. In fact, this trait can be used as the data augmentation method. It is not uncommon to generate images using GAN as a data augmentation strategy currently [13–16], which also has been proven effective in different computer vision tasks.

Moreover, scholars also use GAN-based models to translate or edit satellite images in remote sensing fields [17–19]. Specifically, Li et al. [17] designed a translation model based on GAN to translate optical images to SAR images, which reduces the gap between two types of images. Benjdira et al. [18] design an algorithm that reduces the domain shift influence using GAN, considering that the images in the target domain and source domain are usually different. Moreover, Iqbal et al. [19] propose domain adaptation models to better train built-up segmentation models, which is also motivated by GAN methods.

The remote sensing images in xBD [1] data set have unique characteristics, which are quite different from natural images or other satellite images data sets. First, the remote sensing images include seven different types of disasters, and each class of disaster has its own traits, such as the way to destroy buildings. Second, the remote sensing images are collected from different countries and different events so that the density and damage level of buildings may be various. In order to design effective image generation models, we need to consider the disaster types and the traits of damaged buildings. However, the existing GAN-based models are inefficient in the multi-attribute image translation task; specifically, it is generally necessary to build several different models for every pair of image attributes. This problem is not conducive to the rapid image generation of multiple disaster types. In addition, most existing models directly operate on the whole image, which inevitably changes the attribute-irrelevant region. Nevertheless, the data augmentation for specific damaged buildings typically needs to consider the building region. Thus, to solve both problems in existing GAN-based image generation and more adapt to remote sensing disaster image generation tasks, we try to propose two image generation models that aim at generating disaster images with multiple disaster types and concentrating on different damaged buildings, respectively.

In recent image generation studies, StarGAN [6] has proven to be effective and efficient in multi-attribute image translation tasks; moreover, SaGAN [10] can only alter the attribute-specific region with the guidance of the mask in face. Inspired by these, we propose the algorithm called DisasterGAN, including two models: disaster translation GAN and damaged building generation GAN. The main contributions of this paper are as follows:

(1) Disaster translation GAN is proposed to realize multiple disaster attributes image translation flexibly using only a single model. The core idea is to adopt an attribute label representing disaster types and then take in as inputs both images and disaster attributes, instead of only translating images between two fixed domains such as the previous models.
(2) Damaged building generation GAN implements specified damaged building attribute editing, which only changes the specific damaged building region and keeps the rest region unchanged. Exactly, mask-guided architecture is introduced to keep the model only focused on the attribute-specific region, and the reconstruction loss further ensures the attribute-irrelevant region is unchanged.
(3) To the best of our knowledge, DisasterGAN is the first GAN-based remote sensing disaster images generation network. It is demonstrated that the DisasterGAN method can synthesize realistic images by qualitative and quantitative evaluation. Moreover, it can be used as a data augmentation method to improve the accuracy of the building damage assessment model.

The rest of this paper is organized as follows. Section 2 shows the related research about the proposed method. Section 3 introduces the detailed architecture of the two models, respectively. Then, Section 4 describes the experiment setting and shows the results quantitatively and qualitatively, while Section 5 discusses the effectiveness of the proposed method and verifies the superiority compared with other data augmentation methods. Finally, Section 6 makes a conclusion.

2. Related Work

In this section, we will introduce the related work from four aspects, which are close to the proposed method.

2.1. Generative Adversarial Networks

Since GANs [5] has been proposed, GANs and their variants [20,21] have shown remarkable success in a variety of computer vision tasks, specifically, image-to-image translation [6], image completion [7,8,12], face attribute editing [9,10], image super-resolution [22], etc. GANs aim to fit the real distribution of data by a Min-Max game theory. The standard GAN consists of a generator and a discriminator, and the idea of GANs training is based on adversarial learning to train generator and discriminator simultaneously. The goal of the generator is to generate realistic images, whereas the discriminator is trained to distinguish the generated images and true images. For the original GAN, it has problems that the training process is unstable, and the generated data is not controllable. Therefore, scholars put forward conditional generative adversarial network (CGAN) [23] as the extension of GAN. Additional conditional information (attribute labels or other modalities) was introduced in the generator and the discriminator as the condition for better controlling the generation of GAN.

2.2. Image-to-Image Translation

GAN-based image-to-image translation task has received much attention in the research community, including paired image translation and unpaired image translation. Nowadays, image translation has been widely used in different computer vision fields (i.e., medical image analysis, style transfer) or the preprocessing of downstream tasks (i.e., change detection, face recognition, domain adaptation). There have been some typical models in recent years, such as Pix2Pix [24], CycleGAN [7], and StarGAN [6]. Pix2Pix [24] is the early image-to-image translation model, which learns the mapping from the input and the output through the paired images. It can translate the images from one domain to another domain, and it is demonstrated in synthesizing photos from label maps, reconstructing objects from edge maps tasks. However, in some practical tasks, it is difficult to obtain paired training data, so that CycleGAN [7] is proposed to solve this problem. CycleGAN can translate images without paired training samples due to the cycle consistency loss.

Specifically, CycleGAN learns two mappings: $G : X \rightarrow Y$ (from source domain to target domain) and the inverse mapping $F : Y \rightarrow X$ (from target domain to source domain), while cycle consistency loss tries to enforce $F(G(X)) \approx X$. Moreover, scholars find that the aforementioned models can only translate images between two domains. So StarGAN [5] is proposed to address the limitation, which can translate images between multiple domains using only a single model. StarGAN adopts attribute labels of the target domain and extra domain classifier in the architecture. In this way, the multiple domain image translation can be effective and efficient.

2.3. *Image Attribute Editing*

Compared with the image-to-image translation, we also need to focus on more detailed part translation in the image instead of the style transfer or global attribute in the whole image. For example, the above image translation models may not apply in the eyeglasses and mustache editing in the face [25]. We pay attention to face attribute editing tasks such as removing eyeglasses [9,10] and image completion tasks such as filling the missing regions of the images [12]. Zhang et al. [10] propose a spatial attention face attribute editing model that only alters the attribute-specific region and keeps the rest unchanged. The model includes an attribute manipulation network for editing face images and a spatial attention network for locating specific attribute regions. In addition, as for the image completion task, Iizuka et al. [12] propose a global and locally consistent image completion model. With the introduction of the global discriminator and local discriminator, the model can generate images indistinguishable from the real images in both overall consistency and details.

2.4. *Data Augmentation*

Training a suitable deep-learning model is inseparable from a large amount of labeled data, especially in supervised learning. However, it is difficult to collect large data in some tasks. Standard data augmentation is usually based on geometric transformations, such as color transformations, cropping, flipping [13]. Moreover, using GANs to generate images as a data augmentation has attracted much attention recently, which is common in person re-identification [14,15], license plate recognition [16], few-shot classifier [13]. The GAN-based data augmentation model can directly learn the data distribution, which generates samples that are enforced to be close to the training data distribution [13]. To be more exact, Zhong et al. [10] use CycleGAN [7] to transfer labeled training images to each camera. In this way, the original training data set has been augmented. The model is demonstrated effective, which can be used as a data augmentation method to eliminate camera style differences in person re-identification. Wu et al. [16] propose PixTextGAN, which can generate synthetic license plate images with reasonable text details to enrich the existing license plate data set, thus improving the license plate recognition accuracy. Similar to the above tasks, adequate remote sensing images that used for training building damage assessment model is difficult to collect. In order to model the complex traits of damage, a large amount of damaged building data is indispensable. That is the motivation of our research, proposing a reasonable GAN model as a data augmentation strategy.

In conclusion, we introduce these four aspects of related work in order to make readers better understand the motivation and background of our proposed method. Specifically, the proposed method DisterGAN includes disaster translation GAN and damaged building generation GAN, which may be regarded as image-to-image translation and image attribute editing tasks, respectively. Moreover, we also try to generate damaged building images to make up for the limitation of the existing data as a data generation method.

3. Methods

In this section, we will introduce the proposed remote sensing image generation models, including disaster translation GAN and damaged building generation GAN. The aim of disaster translation GAN is to generate the post-disaster images with disaster attributes, while the damaged building generation GAN is to generate post-disaster images with building attributes.

3.1. Disaster Translation GAN

We first describe the framework of disaster translation GAN. The architecture is shown in Figure 1. Our model is inspired by StarGAN [6], which is introduced simply in Section 2.2. Then, we discuss the objective function and architecture in detail.

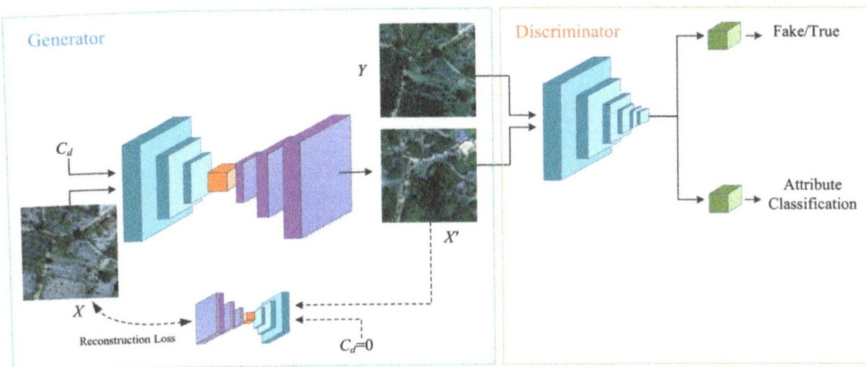

Figure 1. The architecture of disaster translation GAN, including generator G and discriminator D. D has two objectives, distinguishing the generated images from the real images and classifying the disaster attributes. G takes in as input both the images and target disaster attributes and generates fake images, with the inverse process that reconstructing original images with fake images given the original disaster attributes.

3.1.1. Proposed Framework

The goal of disaster translation GAN is to learn mapping functions between disaster images among different disaster attributes. As shown in Figure 1, pre-disaster images X and post-disaster images Y are the paired images. Each image has the corresponding disaster attribute C_d. C_d means the disaster type of the image; thus, the C_d of the X can be defined as 0 uniformly, and the C_d of Y can be defined as $C_d = \{1, 2, 3, 4, 5, 6, 7\}$ according to 7 types of disasters, respectively. The detailed information of C_d can be seen in Section 4.1. As for the generator, the mapping $G(X, C_d) \rightarrow Y$ translates X into Y conditioned on the target disaster attribute C_d. In addition, we introduce the discriminator D_{src} with an auxiliary classifier D_{cls}, where D_{src} aims to distinguish between Y and generated images and X' and D_{cls} aims to classify the images.

To achieve this, we train the D and the G with the following training process. (a) Train D to distinguish between true images and fake images and classify the images. (b) G takes as input both the X and the target attributes C_d, then outputs fake images. (c) G tries to generate images indistinguishable from the real images and classifiable as the target attributes by D. (d) G tries to reconstruct the original images from the fake images and the original attributes.

3.1.2. Objective Function

Disaster translation GAN is trained with the objective function including three types of loss function, i.e., the adversarial loss, the attribute classification loss, and the reconstruct loss, which are introduced as follows, respectively.

Adversarial Loss. To make the generated images indistinguishable from the real images, we adopt the strategy of adversarial learning to train the generator and the discriminator simultaneously. The adversarial loss is defined as

$$L_{adv} = E_X[\log D_{src}(X)] + E_{X,C_d}[\log(1 - D_{src}(X'))], \tag{1}$$

where the $D_{src}(X)$ is the probability distribution over sources given by D. The generator G and the discriminator D are adversarial to each other. The training of the G makes the adversarial loss as small as possible, while the D tries to maximize it.

Attribute Classification Loss. As mentioned above, our goal is to translate the pre-disaster images into the generated images of attributes C_d. Therefore, the attributes not only need to be correctly generated but also need to be correctly classified. To achieve this, we adopt attribute classification loss when we optimize both the generator and the discriminator. Specifically, we adopt the real images and their true corresponding attributes to optimize the discriminator and use the target attributes and the generated images to optimize the generator. The specific formula is shown below.

$$L^D_{cls} = E_{X,C_d}[-\log D_{cls}(C_d|Y)], \tag{2}$$

where $D_{cls}(c_d|Y)$ represents a probability distribution over attribute labels computed by D. In the experiment, the X and Y are both real images, in order to simplify the experiment, only the Y are inputted as the real images, and the corresponding attributes are target attributes. By optimizing this objective function, the classifier of discriminator can learn to identify the attribute.

Similarly, we use the generated images X' to optimize the generator so that it can generate images that can be identified as the corresponding attribute, as defined below

$$L^G_{cls} = E_{X,C_d}[-\log D_{cls}(C_d|X')]. \tag{3}$$

Reconstruction Loss. With the use of adversarial loss and attribute classification loss, the generated images can be as realistic as true images and be classified to their target attribute. However, these losses cannot guarantee that the translation only takes place in the attribute-specific part of the input. Based on this, construction loss is proposed to solve this problem, which is also used in CycleGAN [15].

$$L_{rec} = E_{X,C^g_d,C_d}[\|X - G(G(X,C_d),C^g_d)\|_1] \tag{4}$$

Here, C^g_d represents the original attribute of inputs. G is adopted twice, first to translate an original image into the one with the target attribute, then to reconstruct the original image from the translated image, for the generator to learn to change only what is relevant to the attribute.

Overall, the objective function of the generator and discriminator are shown as below:

$$\min L_D = -L_{adv} + \lambda_{cls} L^D_{cls} \tag{5}$$

$$\min L_G = L_{adv} + \lambda_{cls} L^G_{cls} + \lambda_{rec} L_{rec}, \tag{6}$$

where the $\lambda_{cls}, \lambda_{rec}$ is the hyper-parameters to balance the attribute classification loss and reconstruction loss, respectively. In this experiment, we adopt $\lambda_{cls} = 1, \lambda_{rec} = 10$.

3.1.3. Network Architecture

The specific network architecture of G and D are shown in Tables 1 and 2. I, O, K, P, and S, respectively, represent the number of input channels, the number of output channels, kernel size, padding size, and stride size. IN represents instance normalization, and ReLU and Leaky ReLU are the activation functions. The generator takes as input an 11-channel tensor, consisting of an input RGB image and a given attribute value (8-channel), then outputs RGB generated images. Moreover, in the output layer of the generator, Tanh is adopted as an activation function, as the input image has been normalized to [−1, 1]. The classifier and the discriminator share the same network except for the last layer. For the discriminator, we use the output structure such as PatchGAN [24], and we output a probability distribution over attribute labels by the classifier.

Table 1. Architecture of the generator.

Layer	Generator, G
L1	Conv(I11, O64, K7, P3, S1), I N, ReLU
L2	Conv(I64, O128, K4, P1, S2), IN, ReLU
L3	Conv(I128, O256, K4, P1, S2), IN, ReLU
L4	Residual Block(I256, O256, K3, P1, S1)
L5	Residual Block(I256, O256, K3, P1, S1)
L6	Residual Block(I256, O256, K3, P1, S1)
L7	Residual Block(I256, O256, K3, P1, S1)
L8	Residual Block(I256, O256, K3, P1, S1)
L9	Residual Block(I256, O256, K3, P1, S1)
L10	Deconv(I256, O128, K4, P1, S2), IN, ReLU
L11	Deconv(I128, O64, K4, P1, S2), IN, ReLU
L12	Conv(I64, O3, K7, P3, S1), Tanh

Table 2. Architecture of the discriminator.

Layer	Discriminator, D
L1	Conv(I3, O64, K4, P1, S2), Leaky ReLU
L2	Conv(I64, O128, K4, P1, S2), Leaky ReLU
L3	Conv(I128, O256, K4, P1, S2), Leaky ReLU
L4	Conv(I256, O512, K4, P1, S2), Leaky ReLU
L5	Conv(I512, O1024, K4, P1, S2), Leaky ReLU
L6	Conv(I1024, O2048, K4, P1, S2), Leaky ReLU
L7	src: Conv(I2048, O1, K3, P1, S1); cls: Conv(I2048, O8, K4, P0, S1) [1];

[1] src and cls represent the discriminator and classifier, respectively. These are different in L7 while sharing the same first six layers.

3.2. Damaged Building Generation GAN

In the following part, we will introduce the damaged building generation GAN in detail. The whole structure is shown in Figure 2. The proposed model is motivated by SaGAN [10].

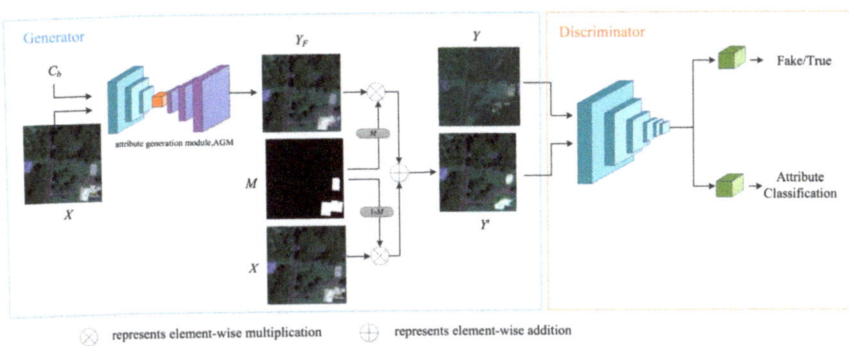

Figure 2. The architecture of damaged building generation GAN, consisting of a generator G and a discriminator D. D has two objectives, distinguishing the generated images from the real images and classifying the building attributes. G consists of an attribute generation module (AGM) to edit the images with the given building attribute, and the mask-guided structure aims to localize the attribute-specific region, which restricts the alternation of AGM within this region.

3.2.1. Proposed Framework

The training data of the model includes pre-disaster images X, post-disaster images Y, and the corresponding building attributes C_b. Among them, C_b means whether the image contains damaged buildings; specifically, the C_b of the X can be defined as 0 uniformly while the C_b of Y is expressed as $C_b = \{0, 1\}$ according to whether there are damaged buildings in the image. The specific information of data can refer to Section 4.1.

We train generator G to translate the X into the generated images Y' with target attributes C_b, formula as below:

$$Y' = G(X, C_b) \tag{7}$$

As Figure 2 shows, we can see the attribute generation module (AGM) in G, which we define as F. F takes as input both the pre-disaster images X and the target building attributes C_b, outputting the images Y_F, defined as:

$$Y_F = F(X, C_b) \tag{8}$$

As for the damaged building generation GAN, we only need to focus on the change of damaged buildings. The changes in the background and undamaged buildings are beyond our consideration. Thus, to better pay attention to this region, we adopt the damaged building mask M to guide the damaged building generation. The value of the mask M should be 0 or 1; specially, the attribute-specific regions should be 1, and the rest regions should be 0.

As the guidance of M, we only reserve the change of attribute-specific regions, while the attribute-irrelevant regions remain unchanged as the original image, formulated as follows:

$$Y' = G(X, C_b) = X \cdot (1 - M) + Y_F \cdot M \tag{9}$$

The generated images Y' should be as realistic as true images. At the same time, Y' should also correspond to the target attribute C_b as much as possible. In order to improve the generated images Y', we train discriminator D with two aims, one is to discriminate the images, and the other is to classify the attributes C_b of images, which are defined as D_{src} and D_{cls} respectively. Moreover, the detailed structure of G and D can be seen in Section 3.2.3.

3.2.2. Objective Function

The objective function of damaged building generation GAN includes adversarial loss, attribute classification loss, and reconstruction loss. We will cover that in this section. It should be emphasized that the definitions of these losses are basically the same as these in Section 3.1.2, so we provide a simple introduction in this section.

Adversarial Loss. To generate synthetic images indistinguishable from real images, we adopt the adversarial loss for the discriminator D

$$L_{src}^{D} = E_Y[\log D_{src}(Y)] + E_{Y'}\left[\log(1 - D_{src}(Y'))\right], \tag{10}$$

where Y is the real images, to simplify the experiment, we only input the Y as the real images, Y' is the generated images, $D_{src}(Y)$ is the probability that the image discriminates to the true images.

As for the generator G, the adversarial loss is defined as

$$L_{src}^{G} = E_{Y'}\left[-\log D_{src}(Y')\right], \tag{11}$$

Attribute Classification Loss. The purpose of attribute classification loss is to make the generated images closer to being classified as the defined attributes. The formula of D_{cls} can be expressed as follows for the discriminator

$$L_{cls}^{D} = E_{Y, c_b^g}\left[-\log D_{cls}(c_b^g | Y)\right] \tag{12}$$

where C_b^g is the attributes of true images, and $D_{cls}(c_b^g|Y)$ represents the probability of an image being classified as the attribute C_b^g. The attribute classification loss of G can be defined as

$$L_{cls}^G = E_{Y'}[-\log D_{cls}(c_b|Y')] \tag{13}$$

Reconstruction Loss. The goal of reconstruction loss is to keep the image of the attribute-irrelevant region mentioned above unchanged. The definition of reconstruction loss is as follows

$$L_{rec}^G = \lambda_1 E_{X,c_b^g,c_b}[(\left\|X - G(G(X,c_b),c_b^g)\right\|_1] + \lambda_2 E_{X,c_b^g}[(\left\|X - G(X,c_b^g)\right\|_1] \tag{14}$$

where c_b^g is the attribute of the original images, while c_b is the target attribute and λ_1, λ_2 are the hyper-parameters. We adopt $\lambda_1 = 1, \lambda_2 = 10$ in this experiment. To be more specific, the first part can be understood that the input image returns to the original input after being transformed twice by the generator; that is, the first generated images $Y' = G(X, c_b)$ input the generator again to make $G(Y', c_b^g)$ as close as possible to X. The second part is to guarantee that input image X is not modified when edited by its own attribute c_b^g.

Overall, the objective function of the generator and discriminator are shown below

$$\min L_G = L_{src}^G + L_{cls}^G + L_{rec}^G \tag{15}$$

$$\min L_D = L_{src}^D + L_{cls}^D \tag{16}$$

3.2.3. Network Architecture

The specific network architecture of the attribute generation module (AGM) and D are shown in Tables 3 and 4. The definition of I, O, K, P, S, IN, ReLU, and Leaky ReLU can be seen in Section 3.1.3. The AGM takes as input a 4-channel tensor, including an input RGB image and a given attribute value, then outputs RGB generated image.

Table 3. Architecture of attribute generation module (AGM).

Layer	Attribute Generation Module, AGM
L1	Conv(I4, O32, K7, P3, S1), I N, ReLU
L2	Conv(I32, O64, K7, P3, S1), I N, ReLU
L3	Conv(I64, O128, K4, P1, S2), IN, ReLU
L4	Conv(I128, O256, K4, P1, S2), IN, ReLU
L5	Residual Block(I256, O256, K3, P1, S1)
L6	Residual Block(I256, O256, K3, P1, S1)
L7	Residual Block(I256, O256, K3, P1, S1)
L8	Residual Block(I256, O256, K3, P1, S1)
L9	Deconv(I256, O128, K4, P1, S2), IN, ReLU
L10	Deconv(I128, O64, K4, P1, S2), IN, ReLU
L11	Deconv(I64, O32, K4, P1, S2), IN, ReLU
L12	Conv(I32, O3, K7, P3, S1), Tanh

Table 4. Architecture of the discriminator.

Layer	Discriminator, D
L1	Conv(I3, O16, K4, P1, S2), Leaky ReLU
L2	Conv(I16, O32, K4, P1, S2), Leaky ReLU
L3	Conv(I32, O64, K4, P1, S2), Leaky ReLU
L4	Conv(I64, O128, K4, P1, S2), Leaky ReLU
L5	Conv(I128, O256, K4, P1, S2), Leaky ReLU
L6	Conv(I256, O512, K4, P1, S2), Leaky ReLU
L7	Conv(I512, O1024, K4, P1, S2), Leaky ReLU
L8	src: Conv(I1024, O1, K3, P1, S1); cls: Conv(I1024, O1, K2, P0, S1) [1];

[1] src and cls represent the discriminator and classifier, respectively. These are different in L8 while sharing the same first seven layers.

4. Experiments and Results

In this section, we first introduce the data set, then illustrate implementation details and show the visualization results of the models, respectively. Next, we perform a quantitative evaluation index (FID) to evaluate the generated images.

4.1. Data Set

Our research is based on the open-source xBD data set [1], which is the largest damaged building remote sensing data set for building damage assessment so far. The assessment of building damage is a joint evaluation standard based on the existing disaster assessment standard [26,27], which classifies the damaged buildings into four categories (no damage, minor damage, major damage, destroyed). The data source of the xBD data set comes from Maxar/DigitalGlobe open data program, consisting of remote sensing images with RGB bands, a resolution equal to or less than 0.8 m GSD. For better generalization of the model, developers choose seven different types of disaster events in various parts of the world. The complete xBD data set contains 22,068 remote sensing images with the size of 1024 × 1024, covering 19 different disaster events and 850,736 buildings, seeing more information in the work of [1].

To adapt to the model training in this study, we have performed a series of processing on the xBD data set and obtained two new data sets (disaster data set and building data set). First, we crop each original remote sensing image (size of 1024 × 1024) to 16 remote sensing images (size of 256 × 256), getting 146,688 pairs of pre-disaster and post-disaster images. Then, labeling each image with the disaster attribute according to the types of disasters, specifically, the disaster attribute of the pre-disaster image is 0 ($C_d = 0$), and the attribute of the post-disaster image can be seen in Table 5 in detail. In the disaster translation GAN, we do not need to consider the damaged building, so the location and damage level of buildings will not be given in the disaster data set. The specific information of the disaster data set is shown in Table 5, and the samples of the disaster data set are shown in Figure 3.

Table 5. The statistics of disaster data set.

Disaster Types	Volcano	Fire	Tornado	Tsunami	Flooding	Earthquake	Hurricane
C_d	1	2	3	4	5	6	7
Number/Pair	4944	90,256	11,504	4176	14,368	1936	19,504

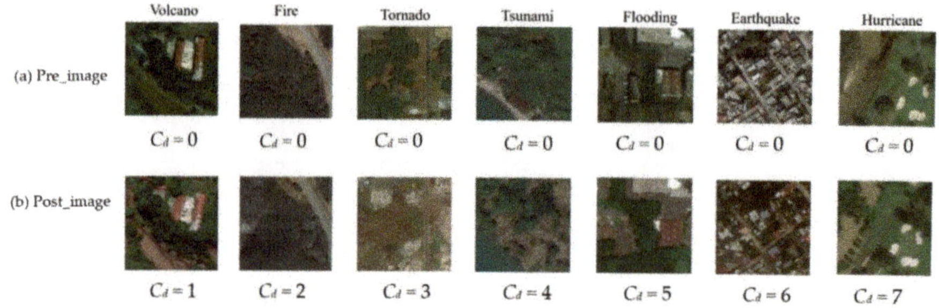

Figure 3. The samples of disaster data set, (**a**,**b**) represent the pre-disaster and post-disaster images according to the seven types of disaster, respectively, each column is a pair of images.

Based on the disaster data set, in order to train damaged building generation GAN, we further screen out the images containing buildings, then obtain 41,782 pairs of images. In fact, the damaged buildings in the same damage level may look different based on the disaster type and the location; moreover, the data of different damage levels in the xBD data set are insufficient, so we only classify the building into two categories for our tentative research. We simply label buildings as damaged or undamaged; that is, we label the building attributes of post-disaster images (C_b) as 1 only when there are damaged buildings in the post-disaster image. Moreover, we label the other post-disaster images and the pre-disaster image as 0. Then, comparing the buildings of pre-disaster and post-disaster images in the position and damage level of buildings to obtain the pixel-level mask, the position of damaged buildings is marked as 1 while the undamaged buildings and the background are marked as 0. Through the above processing, we obtain the building data set. The statistical information is shown in Table 6, and the samples are shown in Figure 4.

Table 6. The statistics of building data set.

Damage Level	Including Damaged Buildings	Undamaged Buildings
C_b	1	0
Number/Pair	24,843	16,948

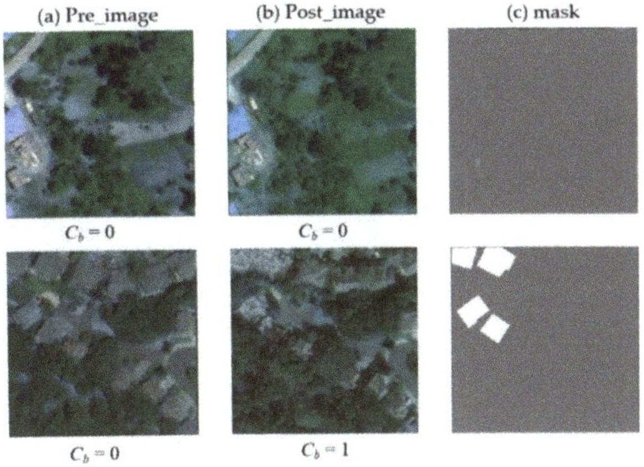

Figure 4. The samples of building data set. (**a–c**) represent the pre-disaster, post-disaster images, and mask, respectively, each row is a pair of images, while two rows in the figure represent two different cases.

4.2. *Disaster Translation GAN*

4.2.1. Implementation Details

To stabilize the training process and generate higher quality images, gradient penalty is proposed and has proven to be effective in the training of GAN [28,29]. Thus, we introduce this item in the adversarial loss, replacing the original adversarial loss. The formula is as follows. For more details, please refer to the work of [22,23].

$$L_{adv} = E_X[D_{src}(X)] - E_{X,C_d}[D_{src}(G(X,C_d))] - \lambda_{gp} E_{\hat{x}}[(\|\nabla_{\hat{x}} D_{src}(\hat{x})\|_2 - 1)^2] \quad (17)$$

Here, $\hat{x}$ is sampled uniformly along a straight line between a pair of real and generated images. Moreover, we set $\lambda_{gp} = 10$ in this experiment.

We train disaster translation GAN on the disaster data set, which includes 146,688 pairs of pre-disaster and post-disaster images. We randomly divide the data set into training set (80%, 117,350) and test set (20%, 29,338). Moreover, we use Adam [30] as an optimization algorithm, setting $\beta_1 = 0.5$, $\beta_2 = 0.999$. The batch size is set to 16 for all experiments, and the maximum epoch is 200. Moreover, we train models with a learning rate of 0.0001 for the first 100 epochs and linearly decay the learning rate to 0 over the next 100 epochs. Training takes about one day on a Quadro GV100 GPU.

4.2.2. Visualization Results

Single Attributes-Generated Image. To evaluate the effectiveness of the disaster translation GAN, we compare the generated images with real images. The synthetic images generated by disaster translation GAN and real images are shown in Figure 5. As shown in this, the first and second rows display the pre-disaster image (Pre_image) and post-disaster image (Post_image) in the disaster data set, while the third row is the generated images (Gen_image). We can see that the generated images are very similar to real post-disaster images. At the same time, the generated images can not only retain the background of pre-disaster images in different remote sensing scenarios but also introduce disaster-relevant features.

Figure 5. Single attributes-generated images results. (**a**–**c**) represent the pre-disaster, post-disaster images, and generated images, respectively, each column is a pair of images, and here are four pairs of samples.

Multiple Attributes-Generated Images Simultaneously. In addition, we visualize the multiple attribute synthetic images simultaneously. The disaster attributes in the disaster data set correspond to seven disaster types, respectively (volcano, fire, tornado, tsunami, flooding, earthquake, and hurricane). As shown in Figure 6, we get a series of generated images under seven disaster attributes, which are represented by disaster names, respectively. Moreover, the first two rows are the corresponding pre-disaster images and the post-disaster images from the data set. As can be seen from the figure, there are a variety of disaster characteristics in the synthetic images, which means that model can flexibly translate images on the basis of different disaster attributes simultaneously. More importantly, the generated images only change the features related to the attributes without changing the basic objects in the images. That means our model can learn reliable features universally applicable to images with different disaster attributes. Moreover, the synthetic images are indistinguishable from the real images. Therefore, we guess that the synthetic disaster images can also be regarded as the style transfer under different disaster backgrounds, which can simulate the scenes after the occurrence of disasters.

Figure 6. Multiple attributes-generated images results. (**a**,**b**) represent the real pre-disaster images and post-disaster images. The images (**c**–**i**) belong to generated images according to disaster types volcano, fire, tornado, tsunami, flooding, earthquake, and hurricane, respectively.

4.3. *Damaged Building Generation GAN*

4.3.1. Implementation Details

Same to the gradient penalty introduced in Section 4.2.1, we have made corresponding modifications in the adversarial loss of damaged building generation GAN, which will not be specifically introduced.

We train damaged building generation GAN on building data set, which includes 41,782 pairs of pre-disaster and post-disaster images. We randomly divided building data set into a training set (90%, 37,604) and test set (20%, 4178). We use Adam [24] to train our model, setting $\beta_1 = 0.5$, $\beta_2 = 0.999$. The batch size is set to 32, and the maximum epoch is 200. Moreover, to train the model stably, we train the generator with a learning rate of 0.0002 while training the discriminator with 0.0001. Training takes about one day on a Quadro GV100 GPU.

4.3.2. Visualization Results

In order to verify the effectiveness of damaged building generation GAN, we visualize the generated results. As shown in Figure 7, the first three rows are the pre-disaster images (Pre_image), the post-disaster images (Post_image), and the damaged building labels (Mask), respectively. The fourth row is the generated images (Gen_image). It can be seen that the changed regions of the generated images are obvious, meanwhile preserving attribute-irrelevant regions unchanged such as the undamaged buildings and the background. Furthermore, the damaged buildings generate by combining the original features of the building and the surrounding, which are also as realistic as true images. However, we also need to point out clearly that the synthetic damaged buildings are lacking in textural detail, which is the key point of model optimization in the future.

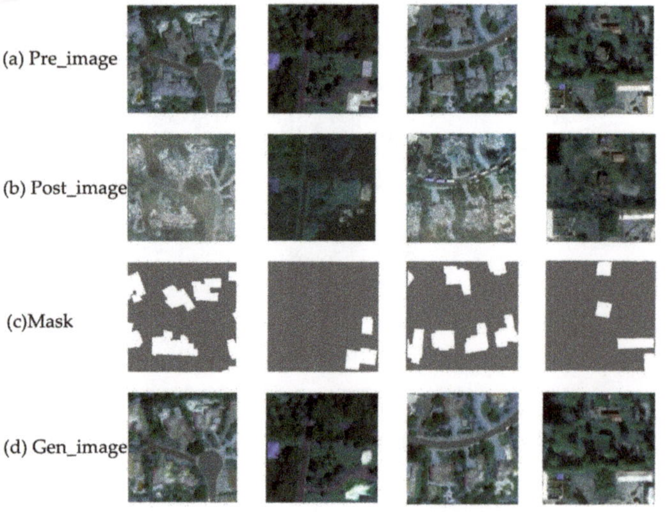

Figure 7. Damaged building generation results. (**a**–**d**) represent the pre-disaster, post-disaster images, mask, and generated images, respectively. Each column is a pair of images, and here are four pairs of samples.

4.4. *Quantitative Results*

To better evaluate the images generated by the proposed models, we choose the common evaluation metric Fréchet inception distance (FID) [31]. FID measures the discrepancy between two sets of images. Exactly, the calculation of FID is based on the features from the last average pooling layer of the ImageNet-pretrained Inception-V3 [32]. For each test image from the original attribute, we first translate it into a target attribute using 10 latent vectors, which are randomly sampled from the standard Gaussian distribution. Then, calculate FID between the generated images and real images in the target attribute. The specific formula is as follows

$$d^2 = \|\mu_1 - \mu_2\|^2 + Tr(C_1 + C_2 - 2(C_1C_2)^{1/2}), \tag{18}$$

where (μ_1, C_1) and (μ_2, C_2) represent the mean and covariance matrix of the two distributions, respectively.

As mentioned above, it should be emphasized that the model calculating FID bases on the pretrained ImageNet, while there are certain differences between the remote sensing images and the natural images in ImageNet. Therefore, the FID is only for reference, which can be used as a comparison value for other subsequent models of the same task.

For the models proposed in this paper, we calculate the FID value between the generated images and the real images based on the disaster data set and building data set, respectively. We carried out five tests and averaged the results to obtain the FID value of disaster translation GAN and damaged building generation GAN, as shown in Table 7.

Table 7. FID distances of the models.

Evaluation Metric	Disaster Translation GAN	Damaged Building Generation GAN
FID	31.1684	21.7873

5. Discussion

In this part, we investigate the contribution of data augmentation methods, considering whether the proposed data augmentation method is beneficial for improving the accuracy of building damage assessment. To this end, we adopt the classical building damage assessment Siamese-UNet [33] as the evaluation model, which is widely used in building damage assessment based on the xBD data set [3,34,35]. The code of the assessment model (Siamese-UNet) has been released at https://github.com/TungBui-wolf/xView2-Building-Damage-Assessment-using-satellite-imagery-of-natural-disasters, last accessed date: 21 October 2021).

In the experiments, we use DisasterGAN, including disaster translation GAN and damaged building generation GAN, to generate images, respectively. We compare the accuracy of Siamese-UNet, which trains on the augmented data set and the original data set, to explore the performance of the synthetic images. First, we select the images with damaged buildings as augmented samples. Then, we augment these samples into two samples, that is, expanding the data set with the corresponding generated images that take in as input both the pre-disaster images and the target attributes. The damaged building label of the generated images is consistent with the corresponding post-disaster images. The building damage assessment model is trained by the augmented data set, and the original data set is then tested on the same original test set.

In addition, we try to compare the proposed method with other data augmentation methods to verify the superiority. Different data augmentation methods have been proposed to solve the limited data problem [36]. Among them, geometric transformation (i.e., flipping, cropping, rotation) is the most common method in computer vision tasks. Cutout [37], Mixup [38], CutMix [39] and GridMask [40] are also widely adopted. In our experiment, considering the trait of the building damage assessment task, we choose geometric transformation and CutMix as the comparative methods. Specifically, we follow the strategy of CutMix in the work of [2], which verifies that CutMix on hard classes (minor damage and major damage) gets the best result. As for geometric transformation, we use horizontal/vertical flipping, random cropping, and rotation in the experiment.

The results are shown in Table 8, where the evaluation metric F1 is an index to evaluate the accuracy of the model. F1 takes into account both precision and recall. It is used in the xBD data set [1], which is suitable for the evaluation of samples with class imbalance. As shown in Table 8, we can observe that further improvement for all damage levels in the data augmentation data set. To be more specific, the data augmentation strategy on hard classes (minor damage, major damage, and destroyed) boosts the performance (F1) better. In particular, major damage is the most difficult class based on the result in Table 8, while the F1 of major damage level is improved by 46.90% (0.5582 vs. 0.8200) with the data augmentation. Moreover, the geometric transformation only improves slightly, while the results of CutMix are also worse than the proposed method. The results show that the data augmentation strategy is clearly improving the accuracy of the building damage assessment model, especially in the hard classes, which demonstrates that the augmented strategy promotes the model to learn better representations for those classes.

Table 8. Effect of data augmentation by disaster translation GAN.

Evaluation Metric	Original Data Set (Baseline)	Geometric Transformation	CutMix	Disaster Translation GAN	Improvement
F1_no-damage	0.9480	0.9480	0.9490	0.9493	0.0013 (0.14%)
F1_minor-damage	0.7273	0.7274	0.7502	0.7620	0.0347 (4.77%)
F1_major-damage	0.5582	0.5590	0.6236	0.8200	0.2618 (46.90%)
F1_destoryed	0.6732	0.6834	0.7289	0.7363	0.0631 (9.37%)

As for the building data set, the data is enhanced in the same way as above by the damaged building generation GAN. Then, we obtain the augmented data set and the original data set. It needs to be noted that we only classify the damage level of the building into damaged and undamaged. The minor damage, major damage, and destroyed class in the original data are classified as damaged uniformly. The building damage assessment model is trained in the original data set, and the augmented data set is then tested on the same original test set. The results are shown in Table 9. We can clearly observe that there is an obvious improvement in damaged classes compared with the undamaged class. Compared with the geometric transformation and CutMix, the proposed method has proven effectiveness and superiority.

Table 9. Effect of data augmentation by damaged building generation GAN.

Evaluation Metric	Original Data Set (Baseline)	Geometric Transformation	CutMix	Damaged Building Generation GAN	Improvment
F1_undamaged	0.9433	0.9444	0.9511	0.9519	0.0086 (0.91%)
F1_damaged	0.7032	0.7432	0.7553	0.7813	0.0781 (11.11%)

6. Conclusions

In this paper, we propose a GAN-based remote sensing disaster images generation method DisasterGAN, including the disaster translation GAN and damaged building generation GAN. These two models can translate disaster images with different disaster attributes and building attributes, which have proven to be effective by quantitative and qualitative evaluations. Moreover, to further validate the effectiveness of the proposed models, we employ these models to synthesize images as a data augmentation strategy. Specifically, the accuracy of hard classes (minor damage, major damage, and destroyed) are improved by 4.77%, 46.90%, and 9.37%, respectively, by disaster translation GAN. damaged building generation GAN further improves the accuracy of damaged class (11.11%). Moreover, this GAN-based data augmentation method is better than the comparative method.

Future research can be devoted to combined disaster types and subdivided damage levels, trying to optimize the existing disaster image generation model.

Author Contributions: X.R., W.S., Y.K. and Y.C. conceived and designed the experiments; X.R. performed the experiments; X.R., X.Y. and Y.C. analyzed the data; X.R. proposed the method and wrote the paper. All authors have read and agreed to the published version of the manuscript.

Funding: This research was funded by The National Key Research and Development Program of China," Study on all-weather multi-mode forest fire danger monitoring, prediction and early-stage accurate fire detection ".

Acknowledgments: The authors are grateful for the producers of the xBD data set and the Maxar/ DigitalGlobe open data program (https://www.digitalglobe.com/ecosystem/open-data, last accessed date: 21 October 2021).

Conflicts of Interest: The authors declare no conflict of interest.

Abbreviations

The following abbreviations are used in this manuscript:

GAN	generative adversarial network
DNN	deep neural network
CNN	convolutional neural network
G	generator
D	discriminator
SAR	synthetic aperture radar
FID	Fréchet inception distance
F1	F1 measure

References

1. Gupta, R.; Hosfelt, R.; Sajeev, S.; Patel, N.; Goodman, B.; Doshi, J.; Heim, E.; ChoseT, H.; Gaston, M. Creating xBD: A dataset for assessing building damage from satellite imagery. In Proceedings of the Computer Vision and Pattern Recognition Conference Workshops, Long Beach, CA, USA, 16–20 June 2019; pp. 10–17.
2. Shen, Y.; Zhu, S.; Yang, T.; Chen, C. Cross-Directional Feature Fusion Network for Building Damage Assessment from Satellite Imagery. In Proceedings of the Neural Information Processing Systems Workshops, Vancouver, BC, Canada, 6–12 December 2020.
3. Hao, H.; Baireddy, S.; Bartusiak, E.R.; Konz, L.; Delp, E.J. An Attention-Based System for Damage Assessment Using Satellite Imagery. *arXiv* **2020**, arXiv:2004.06643.
4. Boin, J.B.; Roth, N.; Doshi, J.; Llueca, P.; Borensztein, N. Multi-class segmentation under severe class imbalance: A case study in roof damage assessment. In Proceedings of the Neural Information Processing Systems Workshops, Vancouver, BC, Canada, 6–12 December 2020.
5. Goodfellow, I.; Pouget-Abadie, J.; Mirza, M.; Xu, B.; Warde-Farley, D.; Ozair, S.; Courville, A.; Bengio, Y. Generative adversarial nets. In Proceedings of the Advances in Neural Information Processing Systems (NIPS), Montreal, QC, Canada, 13 December 2014; pp. 2672–2680.
6. Choi, Y.; Choi, M.; Kim, M.; Ha, J.-W.; Kim, S.; Choo, J. StarGAN: Unified Generative Adversarial Networks for Multi-domain Image-to-Image Translation. In Proceedings of the Computer Vision and Pattern Recognition Conference (CVPR), Salt Lake City, UT, USA, 18–22 June 2018; pp. 8789–8797. [CrossRef]
7. Zhu, J.-Y.; Park, T.; Isola, P.; Efros, A.A. Unpaired Image-to-Image Translation Using Cycle-Consistent Adversarial Networks. In Proceedings of the IEEE International Conference on Computer Vision (ICCV), Venice, Italy, 22–29 October 2017; pp. 2242–2251. [CrossRef]
8. Jiang, Y.; Gong, X.; Liu, D.; Cheng, Y.; Fang, C.; Shen, X.; Yang, J.; Zhou, P.; Wang, Z. EnlightenGAN: Deep Light Enhancement Without Paired Supervision. *IEEE Trans. Image Process.* **2021**, *30*, 2340–2349. [CrossRef] [PubMed]
9. Lee, Y.-H.; Lai, S.-H. ByeGlassesGAN: Identity Preserving Eyeglasses Removal for Face Images. In Proceedings of the European Conference on Computer Vision (ECCV), Glasgow, UK, 23–28 August 2020; pp. 243–258. [CrossRef]
10. Zhang, G.; Kan, M.; Shan, S.; Chen, X. Generative Adversarial Network with Spatial Attention for Face Attribute Editing. In Proceedings of the European Conference on Computer Vision (ECCV), Munich, Germany, 8–14 September 2018; pp. 422–437. [CrossRef]
11. Choi, Y.; Uh, Y.; Yoo, J.; Jung, W.H. StarGAN v2: Diverse Image Synthesis for Multiple Domains. In Proceedings of the Computer Vision and Pattern Recognition Conference (CVPR), Seattle, WA, USA, 16–20 June 2020; pp. 8185–8194.
12. Iizuka, S.; Simo-Serra, E.; Ishikawa, H. Globally and locally consistent image completion. *ACM Trans. Graph. (TOG)* **2017**, *36*, 1–14. [CrossRef]

13. Mounsaveng, S.; Vazquez, D.; Ayed, I.B.; Pedersoli, M. Adversarial Learning of General Transformations for Data Augmentation. *arXiv* **2019**, arXiv:1909.09801.
14. Zhong, Z.; Liang, Z.; Zheng, Z.; Li, S.; Yang, Y. Camera Style Adaptation for Person Re-identification. In Proceedings of the Computer Vision and Pattern Recognition Conference (CVPR), Salt Lake City, UT, USA, 18–22 June 2018; pp. 5157–5166.
15. Huang, S.W.; Lin, C.T.; Chen, S.P. AugGAN: Cross Domain Adaptation with GAN-based Data Augmentation. In Proceedings of the European Conference on Computer Vision (ECCV), Munich, Germany, 8–14 September 2018; pp. 731–744.
16. Wu, S.; Zhai, W.; Cao, Y. PixTextGAN: Structure aware text image synthesis for license plate recognition. *IET Image Process.* **2019**, *13*, 2744–2752. [CrossRef]
17. Li, X.; Du, Z.; Huang, Y.; Tan, Z. A deep translation (GAN) based change detection network for optical and SAR remote sensing images. *ISPRS J. Photogramm. Remote Sens.* **2021**, *179*, 14–34. [CrossRef]
18. Benjdira, B.; Bazi, Y.; Koubaa, A.; Ouni, K. Unsupervised Domain Adaptation using Generative Adversarial Networks for Semantic Segmentation of Aerial Images. *Remote Sens.* **2019**, *11*, 1369. [CrossRef]
19. Iqbal, J.; Ali, M. Weakly-supervised domain adaptation for built-up region segmentation in aerial and satellite imagery. *ISPRS J. Photogramm. Remote Sens.* **2020**, *167*, 263–275. [CrossRef]
20. Li, Z.; Wu, X.; Usman, M.; Tao, R.; Xia, P.; Chen, H.; Li, B. A Systematic Survey of Regularization and Normalization in GANs. *arXiv* **2020**, arXiv:2008.08930.
21. Li, Z.; Xia, P.; Tao, R.; Niu, H.; Li, B. Direct Adversarial Training: An Adaptive Method to Penalize Lipschitz Continuity of the Discriminator. *arXiv* **2020**, arXiv:2008.09041.
22. Ledig, C.; Theis, L.; Huszar, F.; Caballero, J.; Cunningham, A.; Acosta, A.; Aitken, A.; Tejani, A.; Totz, J.; Wang, Z.; et al. Photo-realistic single image super-resolution using a generative adversarial network. In Proceedings of the IEEE Conference on Computer Vision and Pattern Recognition (CVPR), Honolulu, HI, USA, 21–26 July 2017; pp. 105–114.
23. Mirza, M.; Osindero, S. Conditional Generative Adversarial Nets. *arXiv* **2014**, arXiv:1411.1784.
24. Isola, P.; Zhu, J.Y.; Zhou, T. Image-to-Image Translation with Conditional Adversarial Networks. In Proceedings of the IEEE Conference on Computer Vision and Pattern Recognition (CVPR), Honolulu, HI, USA, 21–26 July 2017; pp. 5967–5976.
25. Tao, R.; Li, Z.; Tao, R.; Li, B. ResAttr-GAN: Unpaired deep residual attributes learning for multi-domain face image translation. *IEEE Access* **2019**, *7*, 132594–132608. [CrossRef]
26. Federal Emergency Management Agency. Damage assessment operations manual: A guide to assessing damage and impact. Technical report, Federal Emergency Management Agency, Apr. 2016. Available online: https://www.fema.gov/sites/default/files/2020-07/Damage_Assessment_Manual_April62016.pdf (accessed on 21 October 2021).
27. Federal Emergency Management Agency. Hazus Hurricane Model Uer Guidance. Technical Report, Federal Emergency Management Agency, Apr. 2018. Available online: https://www.fema.gov/sites/default/files/2020-09/fema_hazus_hurricane_user-guidance_4.2.pdf (accessed on 21 October 2021).
28. Arjovsky, M.; Chintala, S.; Bottou, L. Wasserstein generative adversarial networks. In Proceedings of the 34th International Conference on Machine Learning (ICML), Sydney, Australia, 6–11 August 2017; pp. 214–223.
29. Gulrajani, I.; Ahmed, F.; Arjovsky, M.; Dumoulin, V.; Courville, A. Improved training of wasserstein gans. In Proceedings of the Neural Information Processing Systems, Long Beach, CA, USA, 4–10 December 2017; pp. 5767–5777.
30. Kingma, D.; Ba, J. Adam: A method for stochastic optimization. *arXiv* **2014**, arXiv:1412.6980.
31. Heusel, M.; Ramsauer, H.; Unterthiner, T.; Nessler, B.; Hochreiter, S. Gans trained by a two time-scale update rule converge to a local nash equilibrium. In Proceedings of the Neural Information Processing Systems, Long Beach, CA, USA, 4–10 December 2017; pp. 6629–6640.
32. Szegedy, C.; Vanhoucke, V.; Ioffe, S.; Shlens, J.; Wojna, Z. Rethinking the inception architecture for computer vision. In Proceedings of the IEEE Conference on Computer Vision and Pattern Recognition (CVPR), Las Vegas, NV, USA, 26 June–1 July 2016; pp. 2818–2826.
33. Daudt, R.C.; Le, S.B.; Boulch, A. Fully convolutional siamese networks for change detection. In Proceedings of the IEEE International Conference on Image Processing (ICIP), Athens, Greece, 7–10 October 2018; pp. 4063–4067.
34. Bosch, M.; Conroy, C.; Ortiz, B.; Bogden, P. Improving emergency response during hurricane season using computer vision. In Proceedings of the SPIE Remote Sensing, Online, 21–25 September 2020; Volume 11534, p. 115340H. [CrossRef]
35. Benson, V.; Ecker, A. Assessing out-of-domain generalization for robust building damage detection. *arXiv* **2020**, arXiv:2011.10328.
36. Shorten, C.; Khoshgoftaar, T.M. A survey on Image Data Augmentation for Deep Learning. *J. Big Data* **2019**, *6*, 1–48. [CrossRef]
37. Devries, T.; Taylor, G.W. Improved Regularization of Convolutional Neural Networks with Cutout. *arXiv* **2017**, arXiv:1708.04552.
38. Zhang, H.; Cisse, M.; Dauphin, Y.N.; Lopez-Paz, D. Mixup: Beyond Empirical Risk Minimization. *arXiv* **2018**, arXiv:1710.09412.
39. Yun, S.; Han, D.; Oh, S.J.; Chun, S.; Choe, J.; Yoo, Y. CutMix: Regularization Strategy to Train Strong Classifiers with Localizable Features. *arXiv* **2019**, arXiv:1905.04899.
40. Chen, P.; Liu, S.; Zhao, H.; Jia, J. GridMask Data Augmentation. *arXiv* **2020**, arXiv:2001.040862020.

remote sensing

MDPI

Article

SGA-Net: Self-Constructing Graph Attention Neural Network for Semantic Segmentation of Remote Sensing Images

Wenjie Zi †, Wei Xiong †, Hao Chen *,†, Jun Li and Ning Jing

Department of Cognitive Communication, College of Electronic Science and Technology, National University of Defense Technology, Changsha 410000, China; ziwenjie@nudt.edu.cn (W.Z.); xiongwei@nudt.edu.cn (W.X.); junli@nudt.edu.cn (J.L.); ningjing@nudt.edu.cn (N.J.)
* Correspondence: hchen@nudt.edu.cn
† These authors contributed equally to this work.

Abstract: Semantic segmentation of remote sensing images is always a critical and challenging task. Graph neural networks, which can capture global contextual representations, can exploit long-range pixel dependency, thereby improving semantic segmentation performance. In this paper, a novel self-constructing graph attention neural network is proposed for such a purpose. Firstly, ResNet50 was employed as backbone of a feature extraction network to acquire feature maps of remote sensing images. Secondly, pixel-wise dependency graphs were constructed from the feature maps of images, and a graph attention network is designed to extract the correlations of pixels of the remote sensing images. Thirdly, the channel linear attention mechanism obtained the channel dependency of images, further improving the prediction of semantic segmentation. Lastly, we conducted comprehensive experiments and found that the proposed model consistently outperformed state-of-the-art methods on two widely used remote sensing image datasets.

Keywords: self-constructing graph; semantic segmentation; remote sensing

Citation: Zi, W.; Xiong, W.; Chen, H.; Li, J.; Jing, N. SGA-Net: Self-Constructing Graph Attention Neural Network for Semantic Segmentation of Remote Sensing Images. *Remote Sens.* **2021**, *13*, 4201. https://doi.org/10.3390/rs13214201

Academic Editor: Filiberto Pla

Received: 5 September 2021
Accepted: 15 October 2021
Published: 20 October 2021

1. Introduction

Semantic segmentation of remote sensing images aims to assign each pixel in an image with a definite object category [1], which is an urgent issue in ground object interpretation [2]. It has become one of the most crucial methods for traffic monitoring [3], environmental protection [4], vehicle detection [5], and land use assessment [6]. Remote sensing images are usually composed of various objects, highly imbalanced ground, and intricate variations in color texture, which bring challenges to the semantic segmentation of remote sensing images. Before the time of deep learning to display the distribution of vegetation and land cover, the superpixel was often used as measure for drawing features from multi-spectral images. However, hand-crafted descriptors are challenging tthe flexibility of these indices.

The convolutional neural network (CNN) [7] is widely used for the semantic segmentation of images. To achieve a better performance, CNN-based models regularly use multi-scale and deep CNN architectures to acquire information from multi-scale receptive fields and derive local patterns as much as possible. Owing to the restriction of the convolutional kernel, CNN-based models can only capture the dependency of pixels from the limited receptive field rather than the entire image.

CNN-based models have no ability to model the global dependency of each two pixels. However, a graph includes the connection of two nodes, so a graph neural network-based (GNN-based) model can capture the long-range global spatial correlation of pixels. There is no doubt that the traditional form of an image can be converted to a graph structure [8]. In this way, the graph can model the spatial relationship of each two pixels. In contrast, CNN can only obtain information from the limited receptive field. The adjacency matrix of

GNNs can represent the global relationship of images, which can contain more information than CNN-based models. Hence, we adopted a GNN to carry out semantic segmentation.

Nevertheless, a GNN does not ultimately demonstrate a strong point and is seldom used for dense prediction tasks because of the lack of prior knowledge of the adjacency matrix. Previous attempts [9–11] used prior knowledge-based manually generated static graphs, which did not fit each image well. A graph obtained by a neural network, is called "A self-constructing graph". Compared with these methods, a self-constructing graph can adjust itself and reflect the features of each remote sensing image.

Attention mechanisms [12] are added within the convolutional frameworks to improve the semantic segmentation performance in remote sensing images. Every true color image has RGB channels, and the RGB channels of objects have a potential correlation, which can be used to get a better semantic segmentation. The convolutional block attention module (CBAM) [13] adopts two kinds of non-local attention modules to the top of the atrous convolutional neural network: channel attention and spatial attention, respectively. CBAM achieves a competitive segmentation performance in the corresponding dataset. The channel attention mechanism can acquire the correlation among channels, improving the performance of semantic segmentation in remote sensing images. Every pixel has several channels, and each has a different importance for different kinds of pixels. Our channel attention mechanism could model the channels correlation to a large extent, inhibiting or enhancing the corresponding channel in different tasks, respectively.

In this paper, we propose a self-constructing graph attention neural network (SGA-Net) to implement the semantic segmentation of remote sensing images to model global dependency and meticulous spatial relationships between long-range pixels. The main contributions of this paper are as follows:

- Incorporating GATs into self-constructing graphs enhances long-range dependencies between pixels.
- A channel linear attention mechanism to catch th correlation among channel outputs of the graph neural network and further improve performance of the proposed GNN-based model.
- Comprehensive experiments on two widely used datasets in which our framework outperformed the state-of-the-art approaches on the F1 score and mean IoU .

The rest of this paper is organized as follows, the related work is showed in Section 2. Section 3 presents that the details of our architecture SGA-Net. The experiments and corresponding analyses are showed in Section 4, and Section 5 presents the conclusion.

2. Related Work

2.1. Semantic Segmentation

The rise of convolutional neural networks (CNNs) marks a significant improvement in semantic segmentation. The fully convolutional network (FCN), which widely consists of the encoder–decoder module has dominated pixel-to-pixel semantic segmentation [14]. The FCN dominates semantic segmentation, and one with an encoder-decoder module can segment images at the pixel level by deconvolutional and upsampling layers, promoting the development of semantic segmentation. Compared with the FCN, the U-Net [15] applies multi-scale strategies to withdraw contextual patterns and perform semantic segmentation better. Owing to the use of multi-scale context patterns, U-Net can derive a better prediction result than the FCN. Segnet [16] proposes max-pooling indices to enhance location information, which can improve segmentation performance. Deeplab V1 [17] proposes atrous convolutions, which can enlarge the receptive field without increasing the number of parameters. Compared with Deeplab V1, Deeplab V2 [18] presents atrous spatial pyramid pooling (ASPP) modules that consist of atrous convolutions with different sampling rates. Because it uses information from a multi-scale rates receptive field, Deeplab V2 has better prediction than Deeplab V1. The above methods are all supervised models. FESTA [19] is a semi-supervised learning CNN-based model that encodes and regularizes image features and spatial relations. Compared to FESTA, our proposed method extracts long-range spatial dependency and channels correlation to perform segmentation, and our proposed method is a GNN-based model. There are also models of non-grid convolutions for semantic segmentation. Deformable convolution [20] adds 2D offsets to the regular grid sampling locations in the standard convolution, which enhances the geometric transformation modeling capability of CNN. Deformable convolution is still limited in capturing long-range structured relationships. DGMN [21] obtains long-range structured relationships by constructing a dynamic graph. Our proposed model also adopts the idea of a dynamic graph to obtain global long-range correction of remote sensing images. HG-CNNs [22] is a heterogeneous grid convolutional neural network that constructs a data-adaptive graph structure from the convolutional layer by microclustering and assembling features into the graph. Our proposed model also constructs a data-adaptive graph, but the graph structure is extracted by convolutional operation from the high-level feature map.

2.2. Graph Neural Network

Recently, the GNN has become popular due to its success in many fields, such as natural language processing [23], social networks [24], reinforcement learning [25], computer vision [26]. There are lots of natural datasets of graph structures, recommender systems [27], protein networks [28] and knowledge graphs [29]. More and more GNN variants are produced and applied to various fields. In the beginning, only datasets in the form of graphs [10,30] were entered into graph neural networks. However, in a GNN neatly arranged matrix forms like remote sensing images can be extracted and transformed into diffferent kinds of graph structures [8]: convolutional networks, auto-encoders, attention networks (GATs) and isomorphism networks [31]. A GAT [32] and GCN are crucial branches of a GNN. Gao et al. [33] performed action recognition by using structured prior knowledge in the form of knowledge graphs. Yan et al. [34] completed skeleton-based action recognition with spatial-temporal graph convolutional networks (STGCNs) that auto-learn spatial and temporal patterns. Wang et al. [35] proposed a graph-based, language-guided attention mechanism that can clearly reveal inter-object properties and relationships with flexibility. GNN-based models (ASTGCN) [36] are used to predict traffic flow. Liu et al. [8] adopted a GCN to conduct experiences of semantic segmentation in remote sensing images, and the GCN adjacency matrix is built by neural networks. A GCN can simultaneously perform end-to-end learning of node feature information and structure information. In comparison, a GAT proposes a weighted summation of neighboring node features using an attention mechanism. The weights of neighboring node features entirely depend on the node features and are independent of the graph structure. GraphSAGE [37] solves the GCN and GAT memory explosion problem by neighbori sampling for the large-scale graph. GNN-based models are used in a variety of applications.

2.3. Attention Mechanisms

With the publication of the paper in [12], attention mechanisms became more and more popular and attractive. Fu et al. [38] propose a dual attention network (DANet) that can adaptively learn local and global dependency to conduct semantic segmentation. Huang et al. [39] propose channelized axial attention (CAA) to integrate channel and axial attention seamlessly. CAA is similar to DANet in double-attention mechanisms, and these models have a competitive result in the corresponding dataset. CAA pays attention to channel and axial attention, DANet focuses on local and global attention. Compared with multi-attention mechanism, Tao et al. [40] propose a multi-scale attention mechanism that improves the accuracy of semantic segmentation. Transformer [12] is used to solve natural language processing, which is entirely based on the multi-head self-attention mechanism. Dosovitskiy et al. [41] adopt a transformer into the task of image classification, achieving excellent prediction results in many small- and medium-image recognition benchmarks.

3. Methods

In this section, we introduce the details of the model SGA-Net. An overview of the framework is presented in Figure 1 and consists of a feature maps extraction network, self-constructing graph attention network and a channel linear attention mechanism. The four SGA-Nets are shared weights. First, ResNet50 was employed as the backbone of the feature extraction network to acquire feature maps of remote sensing images, and X was denoted as the feature maps. Second, to ensure geometric consistency, feature maps were rotated by several degrees—90, 180 and 270. In addition, X_{90}, X_{180} and X_{270} indicated the feature maps multi-views, where the index was the degree rotation. Third, multi-view feature maps were used to obtain self-constructing graphs A_0, A_1, A_2 and A_3 by a convolution neural network, separately. Fourth, these self-constructing graphs were fed into a neural network based on a GAT to extract the long-range dependency of pixels. Fifth, This network is called the self-constructing graph attention network and the outputs were used for inputs into channel linear attention, the ouputs of which were added to predict the final results. The adjacency matrix A is a high-level feature map of the corresponding remote sensing image feature map, and the projected remote sensing features maps in a specific dimension are defined as nodes. Therefore, the features maps X are defined as the features of nodes. A_{ij} indicating the weight of the edge between node i and node j. We focused on the SGA-Net below.

3.1. Self-Constructing Graph Attention Network

The self-constructing graph is an undirected graph that shows the spatial similarity relationship of feature maps in remote images. The self-constructing graph is extracted by a neural network, instead of prior knowledge. Every image is unique; thus, models based on a self-constructing graph can be fitted for each remote sensing image very well.

The input image is denoted as I, where $I \in \mathbb{R}^{C \times H \times W}$, H and W present the hight and width of corresponding image respectively, and C denotes the number of channels. The high-level feature maps is used as X, where $X \in \mathbb{R}^{H' \times W' \times C'}$, H', W' and C' indicate that the number of height, width and channels, respectively. Next, we applied a convolutional neural network and dropout layer to extract the latent embedding space S of every remote sensing image, where $S \in \mathbb{R}^{N \times E}$, $N = H' \times W'$, where E is the number of the classification.

As we can see from Figure 2, which shows the latent embedding space S of buildings, cars, roads, trees and grass, respectively. S of buildings indicated that they are brighter than other objects: the higher the gray value, the greater the spatial similarity. In general, the same kind of features have the greatest spatial similarity relationship. The adjacency matrix was defined as $A = \text{ReLU}(\text{matmul}(S, S^T))$, which highlighted and enhanced the differences between the target class and other categories. Since it does not arise from prior knowledge, but directly from the output of neural network the adjacency matrix is called the "self-constructing adjacency matrix ", which captures the distributions of the features in remote sensing images. Our model followed the convention of the variational auto-encoder [42] to learn the mean matrix M and the standard deviation matrix D, where $M \in \mathbb{R}^{N \times E}$ and $D \in \mathbb{R}^{N \times E}$, and E denotes the number of the classification. The details of the mean matrix M and logarithm of the standard deviation matrix D are as follows:

$$\begin{aligned} M' &= \text{Flatten}\left(\text{Conv}_{3\times 3,\,\text{padding}=1}(X)\right) \\ M &= \text{Dropout}(\text{p} = 0.2)(M') \end{aligned} \tag{1}$$

$$\begin{aligned} D' &= \text{Flatten}\ (\text{Conv}_{1\times 1})(X) \\ \log(D) &= \text{Dropout}(\text{p} = 0.2)(D') \end{aligned} \tag{2}$$

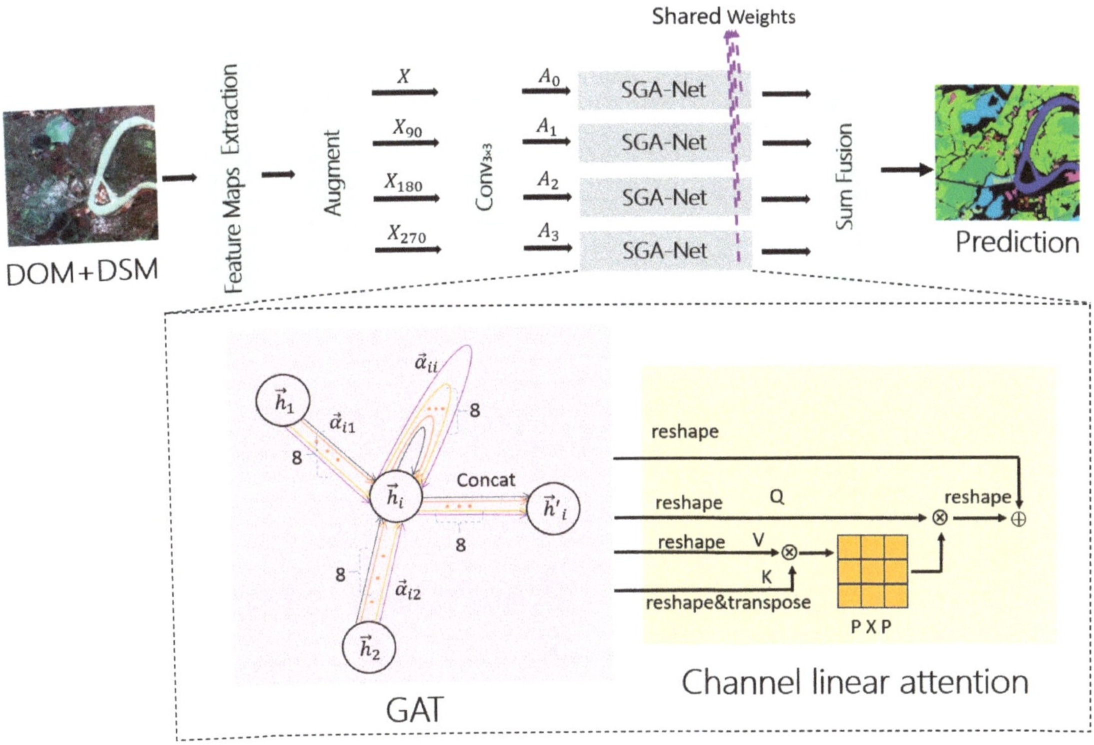

Figure 1. In the flow chart of our model for semantic segmentation, ResNet50 was selected as the feature maps extraction network of our model; $Conv_{3\times3}$ means the convolution operation with kernel size 3; SGA-Net denotes the self-constructing graph attention network and channel linear attention mechanism; GAT is graph attention network, and Q, K, V of channel linear attention mechanism indicate query, key and value, respectively. X denotes the feature input, X_{90}, X_{180} and X_{270} indicate the feature maps multi-views, where the index is the rotation degree, and A_0, A_1, A_2 and A_3 present the adjacency matrix of the self-constructing graph of corresponding feature maps. $\vec{h}_i$ means initial feature vector of each node, where $i \in [1,3]$; $\vec{\alpha}$ represents the correlation coefficient; Concat denotes a concatenating operation; P indicates the number of channels, and $\vec{h}'_i$ indicates the output of self-constructing graph attention neural network.

The latent embedding space $S = M + \log(D) \cdot \alpha$, where $\alpha \in \mathbb{R}^{N\times E}$ is an auxiliary noise variable that obeys standard normal distribution ($\alpha \sim \mathcal{N}_{N\times E}(\mathbf{0}, \mathbf{I})$). The adjacency matrix A was generated by an inner product operation between the transpose of the latent space embedding S^T and itself S, where $A \in \mathbb{R}^{N\times N}$ and A_{ij} denotes the spatial similarity relationship between node i and j.

$$A = \mathrm{ReLU}(\mathrm{matmul}(\mathrm{S}, \mathrm{S}^T)) \tag{3}$$

A therefore can indicate the spatial similarity relation of each two nodes of the latent embedding space S. However, the CNN receptive field was restricted by the kernel size, and the CNN did not have the ability to present a spatial similarity relation between each two nodes. A in our model is not traditional binary but weighted and undirected.

The calculation of the SGA-Net was the same as for all kinds of attention mechanisms. The first step was computing the attention coefficient, and the last was aggregating the sum of weighted features [12]. For node i, the similarity coefficient between its neighbour nodes j and itself was calculated, where $i \in \mathbb{N}$ and $j \in \mathbb{N}$. The details of the similarity coefficient are as follows:

$$e_{ij} = \mathbf{a}([U \cdot \vec{h}_i, U \cdot \vec{h}_j]) \tag{4}$$

where U is the learnable weight matrix, $\vec{h}_i$ indicates the node feature of node i, $h = (\vec{h}_1, \vec{h}_2, \cdots, \vec{h}_N)$, $\vec{h}_i \in \mathbb{R}^{N \times F}$, where F denotes the number of features in each node and $\vec{h} = X$, and $\mathbf{a}$ indicates the operation of self-attention, which is inner product, and the self-constructing adjacency matrix A is set as a mask. Thus, $e_{ij} \in \mathbb{R}^{N \times N}$. Next, we computed the attention coefficient $\vec{\alpha}_{ij}$ as follows:

$$\vec{\alpha}_{ij} = \frac{\exp(\text{LeakyReLU}(e_{ij}))}{\sum_{k \in N} \exp(\text{LeakyReLU}(e_{ik}))} \tag{5}$$

We applied an 8-head graph attention network to enhance the predictive capability of the model and make it more stable iduring training to improve the framework performance.

$$\vec{h}'_i = \|_{l=1}^{L} \sigma\left(\sum_{j \in \mathcal{N}_i} \vec{\alpha}_{ij}^k U^k \vec{h}_j\right) \tag{6}$$

where $\|$ indicates the operation of concatenating, and L is the number of attention, *sigma* is the activate function sigmoid, and $\mathcal{N}_i$ indicates some neighborhood nodes of the node i in the graph, and $\vec{\alpha}_{ij}^k$ is the normalized attention coefficients computed by the kth attention mechanism $\mathbf{a}^{(k)}$, and the $U^{(k)}$ indicates the kth corresponding input weight matrix. Specifically, $L = 8$ and we use an 8-head graph attention network in the work.

Figure 2. Latent embedding space of buildings, cars, roads, trees and low-vegetation present the latent embedding space of these categories separately.

3.2. Channel Linear Attention

Each channel of the high level features could be regarded as the special response of a category, and different responses have intrinsic independencies. The channels of each category had their own distinctive feature and correlations. Exploiting the inter-correlations among channels of images can improve the performance of specific semantic features. Therefore, we adopted a channel attention module to explore correlations among channels.

Suppose the query matrix is Q, the key matrix is K and the value matrix is V. In addition, all of Q, K and $V \in \mathbb{R}^{K \times P}$, where $P = H \times W$, and these are learnable parameters.

In addition, suppose the output of SGA-Net is $\vec{H}$, where $\vec{H} \in \mathbb{R}^{K\times P}$. The detail of the channel linear attention is as follows:

$$D(Q,K,V) = \vec{H} + \frac{V + \left(\frac{Q}{\|Q\|_2}\right)\left(\left(\frac{K}{\|K\|_2}\right)^T V\right)}{N + \left(\frac{Q}{\|Q\|_2}\right)\left(\frac{K}{\|K\|_2}\right)^T} \tag{7}$$

where N denotes the number of nodes. $D(Q,K,V) \in \mathbb{R}^{K\times P}$. The equation highlights the input of a GAT, and emphasizes the importance of the K, Q and V at the same time. The channel linear attention can model the importance of different channels in a different task.

3.3. Loss Function

There is no doubt that A_{ii} ought to be greater than 0 and close to 1; hence, we introduced a diagonal log regularization term to improve the prediction which was defined as:

$$\gamma = \sqrt{1 + \frac{n}{\sum_{i=1}^{n} A_{ii} + \epsilon}} \tag{8}$$

$$\mathcal{L}_{dl} = -\frac{\gamma}{n^2}\sum_{i=1}^{n} \log\left(|A_{ii}|_{[0,1]} + \epsilon\right) \tag{9}$$

where the subscript [0, 1] indicates that A_{ii} is clamped to [0, 1], and ϵ is a fixed and small positive tiny parameter and ($\epsilon = 10^{-5}$). We adopted the Kullback–Leibler divergence, which measures the difference between the distribution of latent variables and the unit Gaussian distribution [42] to be the part of loss function, and the details of Kullback–Leibler divergence were as follows:

$$\mathcal{L}_{kl} = -\frac{1}{2NK}\sum_{i=1}^{N}\sum_{j=1}^{K}\left(1 + \log(D_{ij})^2 - M_{ij}^2 - (D_{ij})^2\right) \tag{10}$$

where D is the standard deviation matrix. In addition, we adopted an adaptive multi-class weighting (ACW) loss function [26] to address the highly imbalanced distribution of the classes. The detail of $\mathcal{L}_{acw}$ is as follows:

$$\mathcal{L}_{acw} = \frac{1}{|Y|}\sum_{i\in Y}\sum_{j\in C} \tilde{w}_{ij} \cdot p_{ij} - \log(\text{MEAN}\{d_j \mid j \in C\}) \tag{11}$$

where Y includes all the labeled pixels and d_j denotes the dice coefficient:

$$d_j = \frac{2\sum_{i\in Y} y_{ij}\tilde{y}_{ij}}{\sum_{i\in Y} y_{ij} + \sum_{i\in Y} \tilde{y}_{ij}} \tag{12}$$

where $y_{i,j}$ and $\tilde{y}_{i,y}$ denote the ijth ground truth and prediction of class j respectively. p_{ij} is positive and negative balanced factor of node i and node j and its detail as follows:

$$p = (y - \tilde{y})^2 - \log\left(\frac{1 - ((y - \tilde{y})^2)}{1 + (y - \tilde{y})^2}\right) \tag{13}$$

$\tilde{w}_{ij}$ is a weight about the frequency of all categories, and the detail of it as follows:

$$\tilde{w}_{ij} = \frac{w_j^t}{\sum_{j\in C}\left(w_j^t\right)} \cdot (1 + y_{ij} + \tilde{y}_{ij}) \tag{14}$$

$$w_j^t = \frac{\text{MEDIAN}\left(\left\{f_j^t \mid j \in C\right\}\right)}{f_j^t + \epsilon} \quad (15)$$

$$f_j^t = \frac{\hat{f}_j^t + (t-1)\cdot f_j^{t-1}}{t} \quad (16)$$

where ϵ is a fixed parameter and $\epsilon = 10^{-5}$; C indicates the number of class; t is the iteration number; f_j^t represents the pixel sum of class j at the tth training step, which can be computed as $\frac{\text{SUM}(y_j)}{\sum_{j\in C}\text{SUM}(y_j)}$, and when $t = 0$, $f_j^t = 0$.

For refining the final prediction result, we adopted the sum of three kinds of loss function as the final loss function in our framework, which are $\mathcal{L}_{kl}$, $\mathcal{L}_{dl}$, and $\mathcal{L}_{acw}$ respectively. The loss function can be formulated as below:

$$Loss = \mathcal{L}_{kl} + \mathcal{L}_{dl} + \mathcal{L}_{acw} \quad (17)$$

4. Experiments

4.1. Datasets

We used two public benchmark the ISPRS 2D semantic labeling contest datasets as our datasets. The ISPRS datasets consisted of aerial images in two German cities: Potsdam and Vaihingen. They are labeled with six common land cover classes:impervious surfaces, buildings, low vegetation, trees, cars and clutter.

- Potsdam: The Potsdam datasets (https://www2.isprs.org/commissions/comm2/wg4/benchmark/2d-sem-label-potsdam/, accessed on 3 September 2021) comprised 38 tiles of a ground resolution of 5 cm with size 6000 × 6000 pixels. Moreover, these tiles consisted of four channel images—Red-Green-Blue-Infrared (RGB-IR)—and the dataset contained both digital surface model (DSM) and normalized digital surface model (nDSM) data. Of these tiles, 14 were used as hold-out test images: 2 were used as validation images, and 12 were used as training data. Furthermore, to compare with other models fairly, we only used RGB images as experience data in this paper.
- Vaihingen: The Vaihingen dataset (https://www2.isprs.org/commissions/comm2/wg4/benchmark/2d-sem-label-vaihingen/, accessed on 3 September 2021) consists of 33 tiles of varying size with a ground resolution of 9cm, of which 17 tiles are used as hold-out test images, 2 tiles are used as validation set, and the rest tiles are taken as training set. In addition, these tiles contain Infrared-Red-Green (IRRG) 3-channel images. In addition, the dataset includes DSM and nDSM. To compare other works fairly, we only apply 3-channel IRRG data in these frameworks in this paper.

4.2. Evaluation Metrics

To acquire reasonable and impartial results, we adopted the mean Intersection over Union (mIoU), the F1 score (F1) and accuracy (Acc) to evaluate performance, all of which are widely applied in semantic segmentation. In addition, based on the accumulated confusion matrix, these evaluation indicators were computed as:

$$\text{mIoU} = \frac{1}{N}\sum_{k=1}^{N}\frac{TP_k}{TP_k + FP_k + FN_k}, \quad (18)$$

$$\text{F1} = 2 \times \frac{\text{precision} \times \text{recall}}{\text{precision} + \text{recall}}, \quad (19)$$

$$\text{Acc} = \frac{\sum_{k=1}^{N} TP_k + TN_k}{\sum_{k=1}^{N} TP_k + FP_k + TN_k + FN_k} \quad (20)$$

where TP_k, FP_k, TN_k, and FN_k are the true positive, false positive, true negative, and false negatives, respectively, and k indicates the number of object index. Acc was computed for all categories except for clutter.

4.3. Experimental Setting

We achieved the proposed SGA-Net as well as all baselines working with PyTorch on a Linux cluster. Models were trained in a single Nvidia GeForce RTX 3090 with a batch size of 5. We applied AMSGrad [43] with adam as the optimizer with weight decay 2×10^{-5}. The weight decay was used in all learnable parameters except batch-norm and bias parameters. Polynomial learning rate (LR) decay was $\left(1 - \frac{cur-iter}{max-iter}\right)^{0.9}$ with the maximum iterations of 10^8, and learning rate decay set to 0.9. The learning rate of the bias parameters is $2 \times$ LR. The initial learning rate was set to $\frac{1.5 \times 10^{-4}}{\sqrt{3}}$. We sampled the patches of size 512×512 as input, and set the node size of graph to 1024×1024.

4.4. Baselines and Comparison

Our model was compared with several works as follows:

- **DDCM** [44]: This is a CNN-based model that consists of dense dilated convolutions merged with varying dilation rates. It can enlarge the receptive fields effectively. Moreover, this model can obtain fused global and local context information to raise the discriminative capability for the surroundings.
- **MSCG-Net** [26]: This method is a self-constructing graph convolutional network that applies neural networks to build graphs from the input of high-level features instead of prior knowledge. In addition, it is a GNN-based model. The feature maps extraction network of our entire framework was similar to a MSCG-Net, but our model used a self-constructing graph to input a GAT, and its outputs were input channel linear attention.
- **DANet** [45]: This framework includes the position and the channel attention mechanisms. The position attention mechanism can learn the spatial relationship of features, and the channel attention mechanism can obtain the channel dependency of images. It is an attention-based method.
- **DUNet** [46]: The model uses redundancy in the label space of semantic segmentation and can recover the pixel-level prediction from low-resolution results of CNNs. It is a CNN-based model.
- **DeeplabV3** [47]: This method captures multi-scale backgrounds by multi-scale cascading or parallel dilated convolution, which can improve the prediction of semantic segmentation. In addition, it is a CNN-based framework.

4.4.1. Prediction on Potsdam Dataset

We compared our model with five baselines on the Potsdam dataset. Table 1 presents the evaluation metrics of prediction in semantic segmentation. Obviously, Table 1 shows that the proposed SGA-Net outperformed the other models.

The SGA-Net was 3.4% higher than the MSCG-Net in mean F1 score, because a self-constructing graph attention network can acquire long-range global spatial dependency of images and channel linear attention to obtain a correlation among all channels. In addition, the proposed framework outperformed other model, which showed that the self-constructing graph had the ability to extract the spatial dependency of images well. In fact, we applied a self-constructing graph, obtained by neural network rather than prior knowledge, to a GAT. Our model performed better than DANet for prediction in all categories, indicating that a self-constructing graph attention neural network can dig the global long-range spatial correlation of nodes for the channel linear attention. Moreover, the multiviews of feature maps in remote sensing images can ensure the geometric consistency of spatial patterns. The reasons for the 3% improvement in average F1 score and 2.6% improvement in mIoU of SGA-Net over Deeplab V3 were that the self-constructing graph neural network obtaied the spatial similarity of each two nodes, and the channel linear attention mechanism captured the correlation among the channel outputs of the graph neural network. The GAT modeled the dependencies between each two nodes, thereby increasing information entropy about spatial correlation. The channel linear attention mechanism enhanced or inhibited the corresponding channel in different tasks. Furthermore, multi-views also can get more information about initial images, which has the ability to support predicting remote sensing images.

Table 1. The experimental results on the Potsdam dataset (bold: best; underline: runner-up).

Method	Road Surf	Buildings	Low Veg.	Trees	Cars	Mean F1	Acc	mIoU
MSCG-Net (GNN-based)	<u>0.907</u>	<u>0.926</u>	0.851	0.872	0.911	0.893	0.959	0.807
DANet (Attention-based)	<u>0.907</u>	0.922	0.853	0.868	0.919	0.894	0.959	0.807
Deeplab V3 (CNN-based)	0.905	0.924	0.850	0.870	<u>0.939</u>	0.897	0.958	0.806
DUNet (CNN-based)	<u>0.907</u>	0.925	0.853	0.869	0.935	0.898	0.959	<u>0.808</u>
DDCM (CNN-based)	0.901	0.924	<u>0.871</u>	<u>0.890</u>	0.932	<u>0.904</u>	<u>0.961</u>	<u>0.808</u>
SGA-Net (GNN-based)	**0.927**	**0.958**	**0.886**	**0.896**	**0.968**	**0.927**	**0.964**	**0.832**

Figure 3 shows the ground truth and predictions of all methods in tile5_15, and trhat the SGA-Net overmatched all baselines in the Potsdam dataset. The figure shows the overall predicting capability of our method in remote sensing images. For example, our model predicted surfaces better than that of MSCG-Net, while the proposed model outperformed all baselines in predicting buildings. The above phenomena illustrated that our framework modeled regularly shaped grounds well. Figure 4 is the result of predicting details from all baselines and the SGA-Net. The black boxes highlight the difference of results among ground truth, baselines and the SGA-Net. The first row shows that the proposed framework did much better predicting buildings compared to the other models, demonstrating that the SGA-Net can model global spatial dependency and channel correlation of remote sensing images.

The second row shows that the SGA-Net outperformed all baselines in predicting trees and buildings, which indicates that the SGA-Net can extract channel correlation in images well. The third row shows that the SGA-Net surpassed the other frameworks in predicting surfaces and low-vegetation. In addition, the last row shows that our model was superior to the other models for predicting trees and low-vegetation. The above phenomena illustrate that self-constructing graph attention network can capture long-range global spatial dependency of images, and the channel linear attention mechanism can acquire a correlation of images among channels. In addition, multiviews feature maps can ensure geometric consistency, improving the performance of predicting semantic segmentation in remote sensing images.

In conclusion, Figure 4 shows that the SGA-Net had a better performance predicting buildings, trees, low-vegetation, cars and surfaces in detail, demonstrating SGA-Net has powerful prediction in the semantic segmentation of remote sensing images.

4.4.2. Prediction on Vaihingen Dataset

We compared our framework with these five baselines on Vaihingen dataset, Table 2 presents the evaluation metrics of prediction in all models. The result showed that the mean F1 score of the SGA-Net was higher than that of the other methods, indicating the powerful ability of prediction in remote sensing images.

To be specific, the F1 score of our model for road surfaces, buildings and cars exceeded all baselines, and accuracy was higher than in other models. Because the SGA-Net contains a self-constructing graph attention neural network and a channel linear attention mechanism, the framework can model the spatial dependency and channel correlation of remote sensing images. Furthermore, because the self-constructing graph attention neural network has the ability to obtain a long-range global spatial correlation of the regular grounds, the predicting result of buildings and cars from the SGA-Net surpassed all baselines. The reason for bad performance on low-vegetation and trees is that the two kinds of grounds are surrounded by many others, leading to poor extraction of spatial dependency by the self-constructing graph. The similarity of tree colors to low-vegetation and the fact that the SGA-Net captures long-range dependencies results in a segmentation performance for trees that is slightly worse than some other methods. The distribution of low-vegetation is more scattered than other objects, and the proposed model cannot extract a very complex spatial relationship of low-vegetation, leading to a poorer performance than DDCM in semantic segmentation.

Table 2. The experimental results on the Vaihingen dataset (bold: best; underlined: runner-up).

Method	Road Surf	Buildings	Low Veg.	Trees	Cars	Mean F1	Acc	mIoU
MSCG-Net (GNN-based)	0.906	0.924	0.816	0.887	0.820	0.870	0.955	0.796
DANet (Attention-based)	0.905	0.934	**0.833**	0.887	0.761	0.859	0.955	0.797
Deeplab V3 (CNN-based)	0.911	0.927	0.819	0.886	0.818	0.872	0.956	0.800
DUNet (CNN-based)	0.910	0.927	0.817	0.887	0.843	0.877	0.955	0.801
DDCM (CNN-based)	0.927	0.953	**0.833**	**0.890**	0.883	0.898	0.963	**0.828**
SGA-Net (GNN-based)	**0.932**	**0.955**	0.826	0.884	**0.928**	**0.905**	**0.965**	0.826

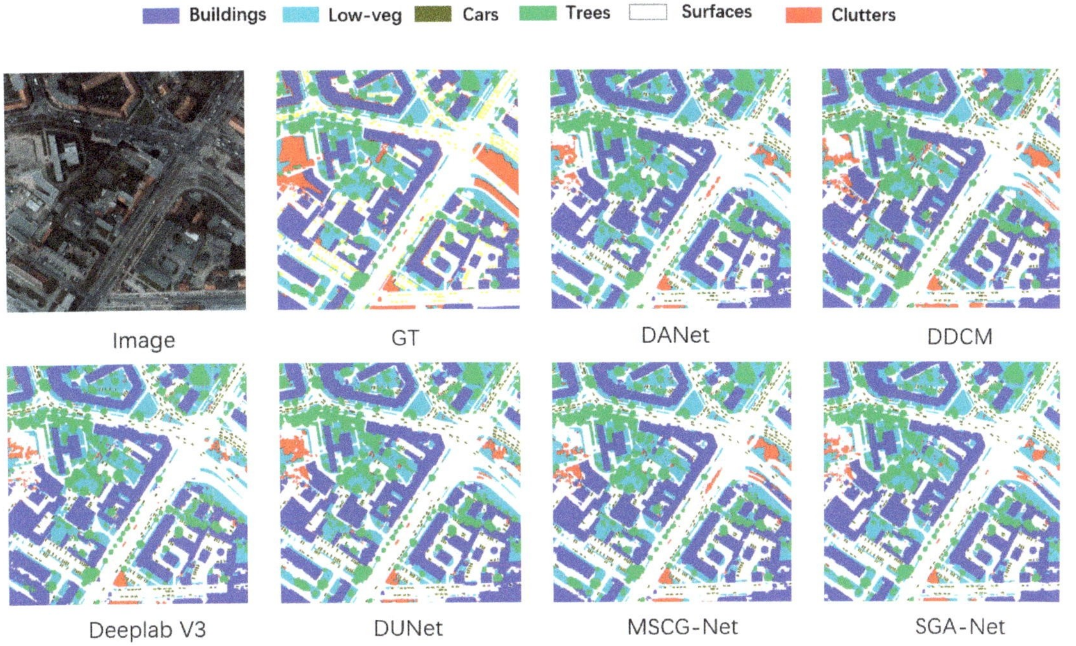

Figure 3. Visualization of tile5_15 in the Potsdam dataset.

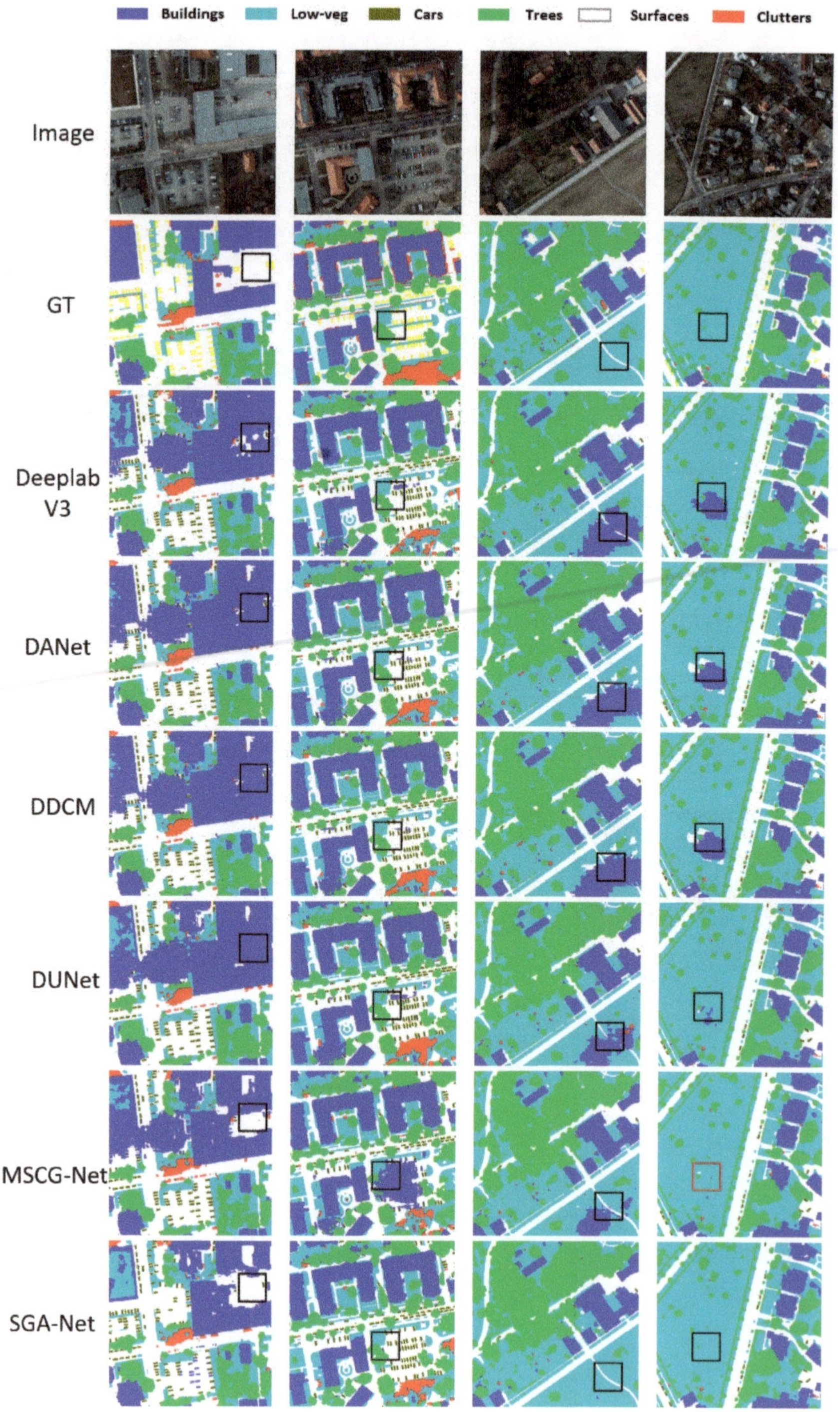

Figure 4. Visualization of prediction detail in the Potsdam dataset.

In addition, Figure 5 shows that the proposed model had a good overall prediction performance. In particular, this figure distinctly indicates that the predicting results of buildings and cars from the SGA-Net surpassed all models, showing that multi-views feature maps can enhance prediction capability, and a self-constructing graph can mine long-range spatial dependency for each image. Additionally, Figure 6 shows the details of the prediction results of the Vaihingen dataset. Because the self-constructing graph attention network can acquire the spatial dependency of each two nodes, the top three rows of Figure 6 indicate that the predictive buildings of the SGA-Net performed better than all baselines, and the last row shows that the predicting trees of our model were much better than other frameworks.

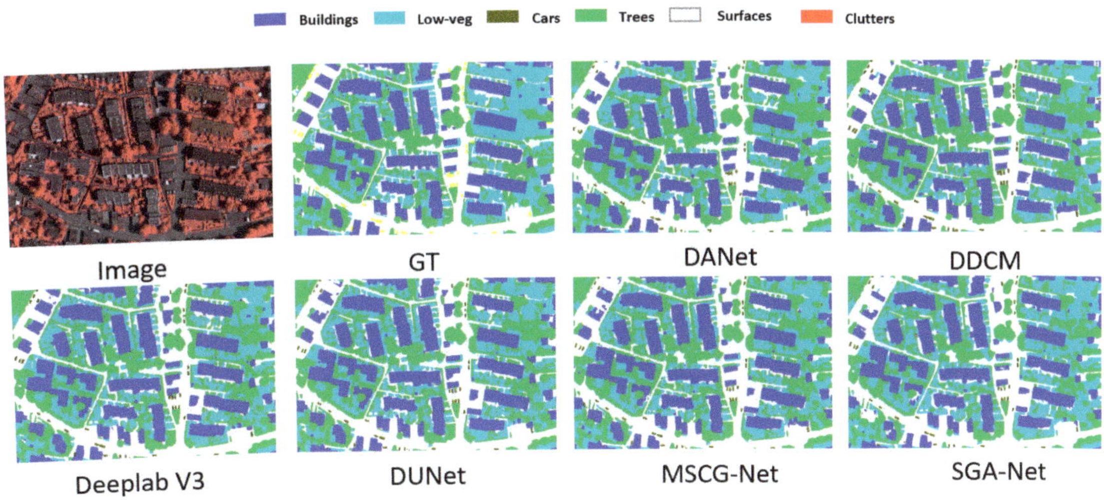

Figure 5. Visualization of tile35 in the Vaihingen dataset.

4.5. Ablation Studies

We conducted ample ablation experimentation to prove the effectiveness of the self-constructing graph neural network and channel linear attention mechanism (SGA-Net) in the proposed framework. Following the main experience as closely as possible, ResNet50 was selected as the baseline and feature extraction layers in our framework. To research the effectiveness of each model component further, we compared the SGA-Net with its variants as follows:

- ResNet50 [48]: a CNN-based neural network adopted as the feature extraction component of the proposed model.
- SGA-Net-ncl: To validate the effectiveness of the self-constructing graph neural network, we directly removed the channel linear attention mechanism from the framework.
- SGA-Net-one: To validate the effect of geometric consistency, we removed the branch roads of X_{90}, X_{180} and X_{270}.
- SGA-Net: our whole SGA-Net framework .

As can be seen from Table 3, the performance of the SGA-Net-ncl significantly overmatched the baseline of ResNet50, thereby showing how effectively a self-constructing graph can model the long-range global spatial correlation of images and get a competitive result. The SGA-Net outperformed ResNet50 and SGA-Net-ncl in two datasets, which shows that channel linear attention has ability to derive a correlation among channel outputs of a graph neural network, and further improve performance of the proposed model. The SGA-Net surpassed SGA-Net-one in predicting remote sensing images, showing that the rotation of images can keep geometric consistency, which improves image prediction performance.

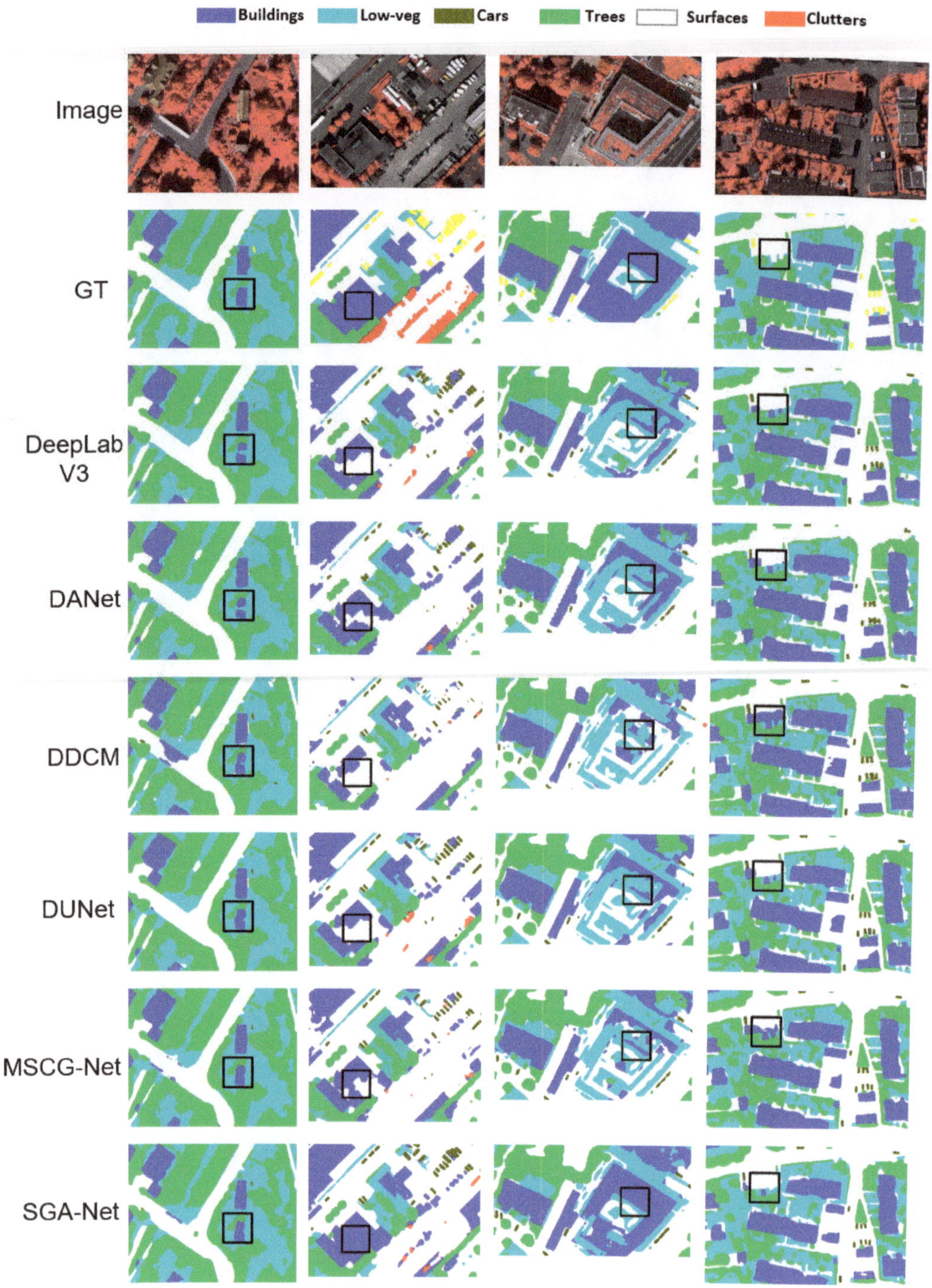

Figure 6. Visualization of prediction detail in the Vaihingen dataset.

Table 3. The ablation study about SGA-Net.

Dataset	Method	Mean F1	Acc	mIoU
Vaihingen	ResNet50	0.826	0.944	0.753
	SGA-Net-ncl	0.849	0.946	0.761
	SGA-Net-one	0.876	0.948	0.798
	SGA-Net	**0.905**	**0.965**	**0.826**
Potsdam	ResNet50	0.873	0.934	0.783
	SGA-Net-ncl	0.906	0.960	0.821
	SGA-Net-one	0.912	0.957	0.825
	SGA-Net	**0.927**	**0.964**	**0.832**

From Figures 7 and 8, we know that the performance of the SGA-Net-ncl surpassed ResNet50 and that the SGA-Net outperformed the baselines of the ablation study in two real-world datasets. Owing to long-range global spatial dependency extraction by a self-constructing graph attention network, the SGA-Net-ncl had a better prediction result than ResNet50. Moreover, channel linear attention acquired a correlation among the channel outputs of the graph neural network, which is why the SGA-Net was superior to the SGA-Net-ncl in semantic segmentation.

From Figure 9, we know the target object had a strong similarity with the same object. On the right of Figure 9, the target object is a building, and the color of the building region is red, meaning that the target pixel had a strong similarity with these pixeles of the building region. On the left of Figure 9, the target objects are low-vegetation and road, and the color of all cars is blue, indicating a low similarity. This picture shows that our attention mechanism works.

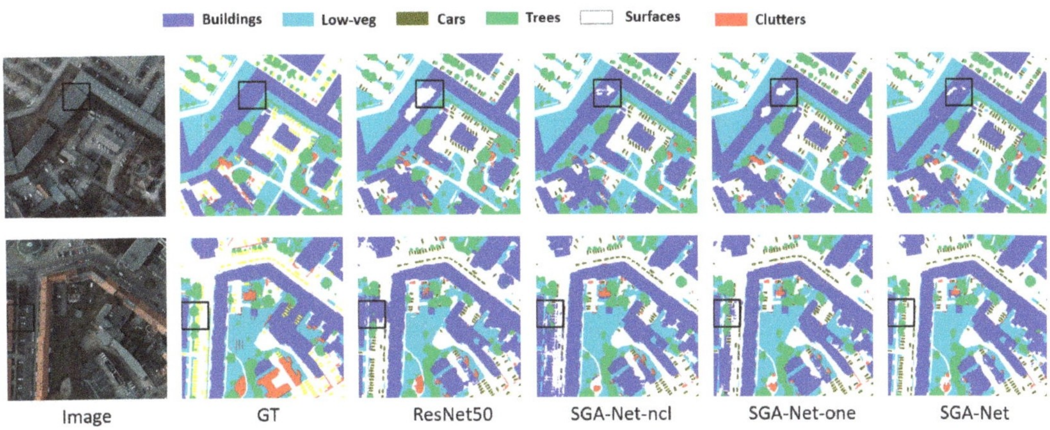

Figure 7. Visualization in the ablation study of Potsdam dataset.

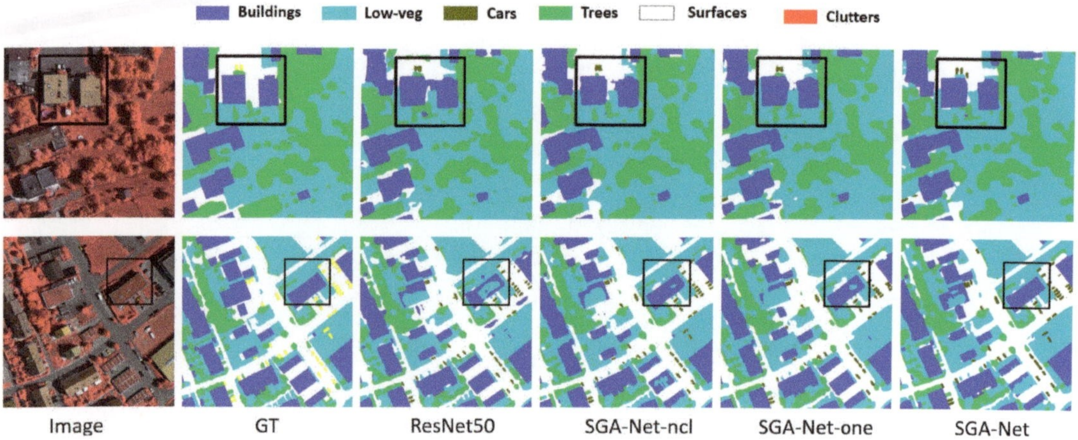

Figure 8. Visualization in the ablation study of Vaihingen dataset.

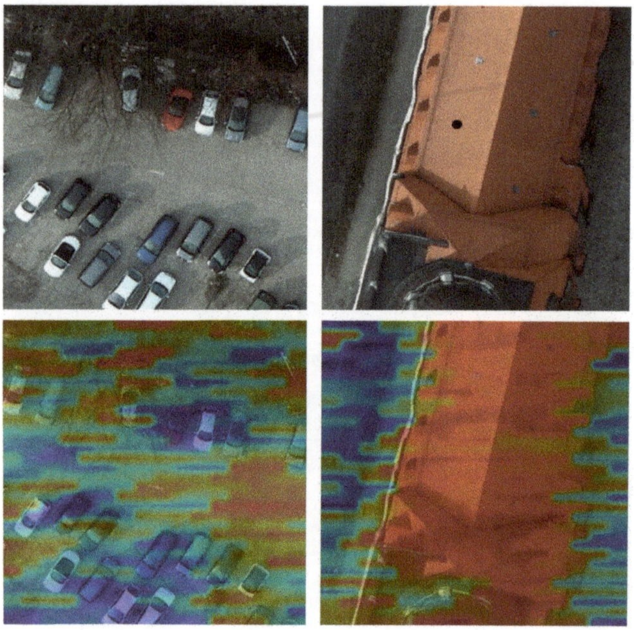

Figure 9. Visualization of the attention mechanism. The black dot is the target pixel or object. The red pixel color indicates that the target pixel is very similar to this pixel, and the blue color indicates that the target pixel is strongly different to this pixel.

5. Conclusions

In this paper, we proposed a novel model, SGA-Net, which includes a self-constructing graph attention network and a channel linear attention. The Self-constructing graph was obtained from feature maps of images rather than prior knowledge or elaborately designed manual static graphs. In this way, the global dependency of pixels can be extracted efficiently from high-level feature maps and present pixel-wise relationships of the remote sensing images. Then, a self-constructing graph attention network was proposed that aligned with the actual situation by using current and neighboring nodes. After that, a channel linear attention mechanism was designed to obtain the channel dependency of images and further improve the prediction performance of semantic segmentation. Comprehensive experiments were conducted on the ISPRS Potsdam and Vaihingen datasets to prove the effectiveness of our whole framework. Ablation studies demonstrated the validity of the self-constructing graph attention network to extract the spatial dependency of remote sensing images and the usefulness of channel linear attention mechanisms for mining correlation among channels. The SGA-Net achieved competitive performance for semantic segmentation in the ISPRS Potsdam and Vaihingen datasets.

In future research, we will re-evaluate the high-level feature map and the attention mechanism to improve the segmentation accuracy. Furthermore, we would like to employ our model to train other remote sensing images.

Author Contributions: Conceptualization, W.Z. and W.X.; Methodology, W.Z. and H.C.; Software, W.Z.; Validation, H.C., W.X. and N.J.; Data Curation, N.J.; Writing—Original Draft Preparation, W.Z.; Writing—Review and Editing, W.Z. and J.L.; Supervision, W.X.; Project Administration, H.C. All authors have read and agreed to the published version of the manuscript.

Funding: The work in this paper is supported by the National Natural Science Foundation of China (41871248, 41971362, U19A2058) and the Natural Science Foundation of Hunan Province No. 2020JJ3042.

Institutional Review Board Statement: Not applicable.

Informed Consent Statement: Not applicable.

Conflicts of Interest: The authors declare no conflict of interest.

References

1. Ignatiev, V.; Trekin, A.; Lobachev, V.; Potapov, G.; Burnaev, E. Targeted change detection in remote sensing images. In Proceedings of the Eleventh International Conference on Machine Vision (ICMV 2018), Munich, Germany, 1–3 November 2018; Volume 11041, p. 110412H.
2. Liu, Y.; Chen, H.; Shen, C.; He, T.; Jin, L.; Wang, L. ABCNet: Real-Time Scene Text Spotting with Adaptive Bezier-Curve Network. In Proceedings of the IEEE/CVF Conference on Computer Vision and Pattern Recognition (CVPR), Seattle, WA, USA, 16–18 June 2020.
3. Panero Martinez, R.; Schiopu, I.; Cornelis, B.; Munteanu, A. Real-time instance segmentation of traffic videos for embedded devices. *Sensors* **2021**, *21*, 275. [CrossRef] [PubMed]
4. Balado, J.; Martínez-Sánchez, J.; Arias, P.; Novo, A. Road environment semantic segmentation with deep learning from MLS point cloud data. *Sensors* **2019**, *19*, 3466. [CrossRef] [PubMed]
5. Behrendt, K. Boxy vehicle detection in large images. In Proceedings of the IEEE/CVF International Conference on Computer Vision Workshops, Seoul, Korea, 27–28 October 2019.
6. Helber, P.; Bischke, B.; Dengel, A.; Borth, D. Eurosat: A novel dataset and deep learning benchmark for land use and land cover classification. *IEEE J. Sel. Top. Appl. Earth Obs. Remote Sens.* **2019**, *12*, 2217–2226. [CrossRef]
7. Krizhevsky, A.; Sutskever, I.; Hinton, G.E. Imagenet classification with deep convolutional neural networks. *Adv. Neural Inf. Process. Syst.* **2012**, *25*, 1097–1105. [CrossRef]
8. Liu, Q.; Kampffmeyer, M.; Jenssen, R.; Salberg, A.B. Self-constructing graph neural networks to model long-range pixel dependencies for semantic segmentation of remote sensing images. *Int. J. Remote Sens.* **2021**, *42*, 6187–6211. [CrossRef]
9. Wang, Y.; Sun, Y.; Liu, Z.; Sarma, S.E.; Bronstein, M.M.; Solomon, J.M. Dynamic graph cnn for learning on point clouds. *ACM Trans. Graph.* **2019**, *38*, 1–12. [CrossRef]
10. Qi, X.; Liao, R.; Jia, J.; Fidler, S.; Urtasun, R. 3d graph neural networks for rgbd semantic segmentation. In Proceedings of the IEEE International Conference on Computer Vision, Venice, Italy, 22–29 October 2017; pp. 5199–5208.
11. Liang, X.; Hu, Z.; Zhang, H.; Lin, L.; Xing, E.P. Symbolic graph reasoning meets convolutions. In Proceedings of the 32nd International Conference on Neural Information Processing Systems, Montréal, QC, Canada, 3–8 December 2018; pp. 1858–1868.
12. Vaswani, A.; Shazeer, N.; Parmar, N.; Uszkoreit, J.; Jones, L.; Gomez, A.N.; Kaiser, Ł.; Polosukhin, I. Attention is all you need. *Adv. Neural Inf. Process. Syst.* **2017**, 5998–6008.
13. Woo, S.; Park, J.; Lee, J.Y.; Kweon, I.S. Cbam: Convolutional block attention module. In Proceedings of the European conference on computer vision (ECCV), Munich, Germany, 8–14 September 2018; pp. 3–19.
14. Ben-Cohen, A.; Diamant, I.; Klang, E.; Amitai, M.; Greenspan, H. Fully convolutional network for liver segmentation and lesions detection. In *Deep Learning and Data Labeling for Medical Applications*; Springer: Berlin/Heidelberg, Germany, 2016; pp. 77–85.

15. Ronneberger, O.; Fischer, P.; Brox, T. U-net: Convolutional networks for biomedical image segmentation. In Proceedings of the International Conference on Medical Image Computing and Computer-Assisted Intervention, Munich, Germany, 5–9 October 2015; pp. 234–241.
16. Badrinarayanan, V.; Kendall, A.; Cipolla, R. Segnet: A deep convolutional encoder-decoder architecture for image segmentation. *IEEE Trans. Pattern Anal. Mach. Intell.* **2017**, *39*, 2481–2495. [CrossRef]
17. Liang-Chieh, C.; Papandreou, G.; Kokkinos, I.; Murphy, K.; Yuille, A. Semantic Image Segmentation with Deep Convolutional Nets and Fully Connected CRFs. In Proceedings of the International Conference on Learning Representations, San Diego, CA, USA, 7–9 May 2015.
18. Chen, L.C.; Papandreou, G.; Kokkinos, I.; Murphy, K.; Yuille, A.L. Deeplab: Semantic image segmentation with deep convolutional nets, atrous convolution, and fully connected crfs. *IEEE Trans. Pattern Anal. Mach. Intell.* **2017**, *40*, 834–848. [CrossRef]
19. Hua, Y.; Marcos, D.; Mou, L.; Zhu, X.X.; Tuia, D. Semantic segmentation of remote sensing images with sparse annotations. *IEEE Geosci. Remote Sens. Lett.* **2021**. [CrossRef]
20. Dai, J.; Qi, H.; Xiong, Y.; Li, Y.; Zhang, G.; Hu, H.; Wei, Y. Deformable convolutional networks. In Proceedings of the IEEE International Conference on Computer Vision, Venice, Italy, 22–29 October 2017; pp. 764–773.
21. Zhang, L.; Xu, D.; Arnab, A.; Torr, P.H. Dynamic graph message passing networks. In Proceedings of the IEEE/CVF Conference on Computer Vision and Pattern Recognition, Seattle, WA, USA, 13–19 June 2020; pp. 3726–3735.
22. Hamaguchi, R.; Furukawa, Y.; Onishi, M.; Sakurada, K. Heterogeneous Grid Convolution for Adaptive, Efficient, and Controllable Computation. In Proceedings of the IEEE/CVF Conference on Computer Vision and Pattern Recognition, Nashville, TN, USA, 19–25 June 2021; pp. 13946–13955.
23. Yao, L.; Mao, C.; Luo, Y. Graph convolutional networks for text classification. In Proceedings of the AAAI Conference on Artificial Intelligence, Honolulu, HI, USA, 27 January–1 February 2019; Volume 33, pp. 7370–7377.
24. Wang, H.; Xu, T.; Liu, Q.; Lian, D.; Chen, E.; Du, D.; Wu, H.; Su, W. MCNE: An end-to-end framework for learning multiple conditional network representations of social network. In Proceedings of the 25th ACM SIGKDD International Conference on Knowledge Discovery & Data Mining, Anchorage, AK, USA, 4–8 August 2019; pp. 1064–1072.
25. Liu, Y.; Wang, W.; Hu, Y.; Hao, J.; Chen, X.; Gao, Y. Multi-agent game abstraction via graph attention neural network. In Proceedings of the AAAI Conference on Artificial Intelligence, New York, NY, USA, 7–12 February 2020; Volume 34, pp. 7211–7218.
26. Liu, Q.; Kampffmeyer, M.C.; Jenssen, R.; Salberg, A.B. Multi-view Self-Constructing Graph Convolutional Networks with Adaptive Class Weighting Loss for Semantic Segmentation. In Proceedings of the IEEE/CVF Conference on Computer Vision and Pattern Recognition Workshops, Seattle, WA, USA, 14–19 June 2020; pp. 44–45.
27. Su, Y.; Zhang, R.; Erfani, S.; Xu, Z. Detecting Beneficial Feature Interactions for Recommender Systems. In Proceedings of the 34th AAAI Conference on Artificial Intelligence (AAAI), Virtually, 2–9 February 2021.
28. Liu, B.; Li, C.C.; Yan, K. DeepSVM-fold: Protein fold recognition by combining support vector machines and pairwise sequence similarity scores generated by deep learning networks. *Brief. Bioinform.* **2020**, *21*, 1733–1741. [CrossRef] [PubMed]
29. Lampropoulos, G.; Keramopoulos, E.; Diamantaras, K. Enhancing the functionality of augmented reality using deep learning, semantic web and knowledge graphs: A review. *Vis. Inf.* **2020**, *4*, 32–42. [CrossRef]
30. Zi, W.; Xiong, W.; Chen, H.; Chen, L. TAGCN: Station-level demand prediction for bike-sharing system via a temporal attention graph convolution network. *Inf. Sci.* **2021**, *561*, 274–285. [CrossRef]
31. Zhou, J.; Cui, G.; Hu, S.; Zhang, Z.; Yang, C.; Liu, Z.; Wang, L.; Li, C.; Sun, M. Graph neural networks: A review of methods and applications. *AI Open* **2020**, *1*, 57–81. [CrossRef]
32. Xie, Y.; Zhang, Y.; Gong, M.; Tang, Z.; Han, C. Mgat: Multi-view graph attention networks. *Neural Netw.* **2020**, *132*, 180–189. [CrossRef]
33. Gao, J.; Zhang, T.; Xu, C. I know the relationships: Zero-shot action recognition via two-stream graph convolutional networks and knowledge graphs. In Proceedings of the AAAI Conference on Artificial Intelligence, Honolulu, HI, USA, 27 January–1 February 2019; Volume 33, pp. 8303–8311.
34. Yan, S.; Xiong, Y.; Lin, D. Spatial temporal graph convolutional networks for skeleton-based action recognition. In Proceedings of the AAAI Conference on Artificial Intelligence, New Orleans, LA, USA, 2–7 February 2018; Volume 32.
35. Wang, P.; Wu, Q.; Cao, J.; Shen, C.; Gao, L.; Hengel, A.v.d. Neighbourhood watch: Referring expression comprehension via language-guided graph attention networks. In Proceedings of the IEEE/CVF Conference on Computer Vision and Pattern Recognition, Long Beach, CA, USA, 15–20 June 2019; pp. 1960–1968.
36. Guo, S.; Lin, Y.; Feng, N.; Song, C.; Wan, H. Attention based spatial-temporal graph convolutional networks for traffic flow forecasting. In Proceedings of the AAAI Conference on Artificial Intelligence, Honolulu, HI, USA, 27 January–1 February 2019; Volume 33, pp. 922–929.
37. Hamilton, W.L.; Ying, R.; Leskovec, J. Inductive representation learning on large graphs. In Proceedings of the 31st International Conference on Neural Information Processing Systems, Long Beach, CA, USA, 4–9 December 2017; pp. 1025–1035.
38. Fu, J.; Liu, J.; Tian, H.; Li, Y.; Bao, Y.; Fang, Z.; Lu, H. Dual attention network for scene segmentation. In Proceedings of the IEEE/CVF Conference on Computer Vision and Pattern Recognition, Long Beach, CA, USA, 16–20 June 2019; pp. 3146–3154.
39. Huang, Y.; Jia, W.; He, X.; Liu, L.; Li, Y.; Tao, D. CAA: Channelized Axial Attention for Semantic Segmentation. *arXiv* **2021**, arXiv:2101.07434.
40. Tao, A.; Sapra, K.; Catanzaro, B. Hierarchical multi-scale attention for semantic segmentation. *arXiv* **2020**, arXiv:2005.10821.

41. Dosovitskiy, A.; Beyer, L.; Kolesnikov, A.; Weissenborn, D.; Zhai, X.; Unterthiner, T.; Dehghani, M.; Minderer, M.; Heigold, G.; Gelly, S. An Image is Worth 16 × 16 Words: Transformers for Image Recognition at Scale. *arXiv* **2020**, arXiv:2010.11929.
42. Kingma, D.P.; Welling, M. Auto-Encoding Variational Bayes. *arXiv* **2013**, arXiv:1312.6114.
43. Tran, P.T.; Phong, L.T. On the convergence proof of amsgrad and a new version. *IEEE Access* **2019**, *7*, 61706–61716. [CrossRef]
44. Kampffmeyer, M.; Jenssen, R.; Salberg, A.B. Dense dilated convolutions merging network for semantic mapping of remote sensing images. In Proceedings of the 2019 Joint Urban Remote Sensing Event (JURSE), Vannes, France, 22–24 May 2019; pp. 1–4.
45. Xue, H.; Liu, C.; Wan, F.; Jiao, J.; Ji, X.; Ye, Q. Danet: Divergent activation for weakly supervised object localization. In Proceedings of the IEEE/CVF International Conference on Computer Vision, Seoul, Korea, 27–28 October 2019; pp. 6589–6598.
46. Tian, Z.; He, T.; Shen, C.; Yan, Y. Decoders matter for semantic segmentation: Data-dependent decoding enables flexible feature aggregation. In Proceedings of the IEEE/CVF Conference on Computer Vision and Pattern Recognition, Long Beach, CA, USA, 16–20 June 2019; pp. 3126–3135.
47. Florian, L.C.C.G.P.; Adam, H. Rethinking Atrous Convolution for Semantic Image Segmentation. *arXiv* **2017**, arXiv:1706.05587.
48. He, K.; Zhang, X.; Ren, S.; Sun, J. Deep residual learning for image recognition. In Proceedings of the IEEE Conference on Computer Vision and Pattern Recognition, Las Vegas, NV, USA, 27–30 June 2016; pp. 770–778.

remote sensing

Article

SSSGAN: Satellite Style and Structure Generative Adversarial Networks

Javier Marín [1,*] and Sergio Escalera [2,3]

1 Satellogic, Carrer de Bailèn, 3, 1st Floor, 08010 Barcelona, Spain
2 Department of Mathematics and Informatics, Universitat de Barcelona, Gran via de les Corts Catalanes 585, 08007 Barcelona, Spain; sergio@maia.ub.es
3 Computer Vision Center, Building O, Campus UAB, Bellaterra (Cerdanyola), 08193 Barcelona, Spain
* Correspondence: javier.marin@satellogic.com

Abstract: This work presents Satellite Style and Structure Generative Adversarial Network (SSGAN), a generative model of high resolution satellite imagery to support image segmentation. Based on spatially adaptive denormalization modules (SPADE) that modulate the activations with respect to segmentation map structure, in addition to global descriptor vectors that capture the semantic information in a vector with respect to Open Street Maps (OSM) classes, this model is able to produce consistent aerial imagery. By decoupling the generation of aerial images into a structure map and a carefully defined style vector, we were able to improve the realism and geodiversity of the synthesis with respect to the state-of-the-art baseline. Therefore, the proposed model allows us to control the generation not only with respect to the desired structure, but also with respect to a geographic area.

Keywords: aerial image generation; satellite image generation; generative adversarial network; deep learning; structure map; style vector; high resolution image

check for updates

Citation: Marín, J.; Escalera, S. SSSGAN: Satellite Style and Structure Generative Adversarial Networks. *Remote Sens.* **2021**, *13*, 3984. https://doi.org/10.3390/rs13193984

Academic Editor: Fahimeh Farahnakian

Received: 4 August 2021
Accepted: 29 September 2021
Published: 5 October 2021

1. Introduction

The commercialization and the advancement of the geospatial industry has led to an explosive amount of remote sensing data being collected to characterize our changing planet Earth. Public and private industries are taking advantage of this increasing availability of information in order to perform analytics and obtain more precise information about geographic areas in order to support decisions and automatize technology. Due to the increasing revisiting frequency of recently launched satellites and a fine pixel resolution (up to 30 cm per pixel, commercial ones), satellite imagery has become of interest because computer vision algorithms can capture the presence of objects in an automatic and efficient manner at a large scale. Commonly studied computer vision tasks, such as semantic and instance segmentation, object detection or height estimation, aim to address problems such as land cover classification, precision agriculture, flood detection, building, road or car detection, and at the same time help to provide information about geographic zones that can improve agriculture, navigation, retail, smart city technologies, 3D precise world reconstruction or even assistance after natural disasters.

State-of-the-art methods comprise mostly of deep learning algorithms. With the presentation of AlexNet [1] in 2012 as the winner of the ImageNet LSVRC-2012 competition by a large margin, deep networks have dominated the scene of computer vision. Due to their large number of parameters, they present a high complexity that means they require a high volume of data to correctly extract latent features from imagery, key to achieving outstanding results.

Particularly, in the field of geo-informatics and remote sensing, datasets are usually sparse, expensive and difficult to collect when it comes to tasks that require high to very high resolution images (from 1 to 0.05 m). To overcome this situation of the scarcity of images, a commonly used technique is transfer learning. This approach to training consists of using pre-trained weights as a starting point in order to improve performance and decrease training time. Pre-training is done with a highly varied high-volume dataset, so the network can extract low-level features of the world. Then, this pre-trained model is trained again with a smaller task-specific available dataset that is known as fine-tuning. This tuning can be performed by a variety of strategies that range from the most basic ones, such as freezing most of the low level layers (layers that have learnt primitive low level features) and only tuning the shallow layers, to more complex schemes that apply different learning rates to different layers.

The idea is the model to take advantage of low-level extracted features to learn more easily task-specific features in the fine-tuning. Generally, public pre-trained models are trained in datasets such as ImageNet [2] or similar ones that consist of labeled images used in visual recognition tasks (ground level visualization). Those pre-trained models are applied in totally different domains, obtaining an increment in performance with respect to training the network from scratch. For example ImageNet presents completely different visual features with respect to satellite images. Aerial-imagery contains the presence of high-frequency details and a background clutter that heavily depends on the environment, geographic zone, weather conditions, illumination, sensors and pixel resolutions. Those factors constitute a challenge itself for computer vision models to work well in a variety of cities, countries, regions, continents or even pixel resolutions.

The performance of algorithms varies markedly across geographies, image qualities and resolutions. The performance of a model applied in new areas depends, on one hand, on the target texture and topology related to cultural regions and countries [3]. Other crucial characteristics present in the image are the geographic location, weather and type of terrain. An image taken from a rural area totally differs from an urban area or from the coast. Even a specific rural area contains a different biome from a rural area of a different country/region. These points explain why it is really difficult to train a general deep network that works well with images of different locations. Additionally to the image content characteristics, there are image technical characteristics related to the methodology of extraction, such as the type of sensor, radiometry, off-nadir angle, or the atmospheric conditions at the top layers of the atmosphere.

Supervised learning techniques that use deep networks are usually trained with a large number of classes that can go from tens to thousands of labels. Thus, labeling satellite imagery is a fundamental step in the training of deep networks. Depending on the quality of labels and the resolution of the images, the cost of annotating scenes varies. Generally, the most quality satellite imagery labeling is performed by trained professionals with knowledge of GIS and geographic imagery, making this demanding annotation process slow and costly. Even the cost is tightly related to the resolution of the images; as the spatial resolution of the image increases, the cost of annotation grows accordingly. This produces a scarcity of public datasets and a bias towards most developed urban regions that have enough resources to afford this data acquisition. Scientists should make a careful selection and analysis of the datasets before starting the data annotation phase and they should also pay special attention to the quality of the labels.

When a study or research presents a model claiming to efficiently extract and detect a specific target, it usually implies that they are presenting a model trained with a dataset with specific geographic, cultural and quality conditions that perform well. In order to overcome such necessity, one possibility can be to generate a large collection of diverse synthetic images with their corresponding labels. In this case, it would be necessary to contemplate the different characteristics mentioned before, so the resulting satellite images can augment efficiently in those desired directions.

In this work, we present Satellite Style and Structured Generative Adversarial Network (SSSGAN) to generate realistic synthetic imagery (see Figure 1) based on publicly available ground truth (to get access to the models and code, please contact the authors). Particularly, we propose the use of a conditional generative adversarial network (GAN) model capable of generating synthetic satellite images constrained by two components: (1) a semantic style description of the scene; and (2) a segmentation map that defines the structure of the desired output in terms of object classes. By this way the structure and the style constraint are decoupled so the user can easily generate novel synthetic images by defining a segmentation mask of the desired foot print labels and then selecting the proportion of semantic classes expressed as number of a vector in addition to the selection of the region or city. With this generation rule the model can capture and express variability present in the satellite imagery while at the same time provides an easy-to-use generation mechanism with high expressiveness. In this work, our key contributions are as follows:

- Development of a GAN model capable of producing highly diverse satellite imagery;
- Presentation of a semantic global vector descriptor dataset based on Open Street Maps (OSM). We analyse and categorize a set of 11 classes that semantically describes the visual features that are present in satellite imagery, leveraging the public description of this crowdsourced database;
- Evaluation and study that describe the different effects of the proposed mechanisms.

Figure 1. Synthetic images generated by SSSGAN.

1.1. Related Work

Synthetic image generation is an active research topic in the field of computer vision. A vast variety of models have been developed in the past years since the presentation in 2014 of generative adversarial networks (GAN) [4]. Even though, before and after GANS, there were numerous classical and deep learning methods, the increasing support and improvement of GAN models made this state-of-the-art technique achieve outstanding results where the synthetic generated images are hardly distinguishable from the real ones.

As mentioned before, Generative Adversarial Networks (GANS) have stated the baseline for deep generative learning. The model consists of two parts: a generator and a discriminator. The generator learns to generate synthetic, realistic images while it is trying to fool the discriminator that is responsible for distinguishing between real or fake generated images. This learning process consists of finding equilibrium in a two-player minimax game where each iteration of the generator G gets better at capturing the real data distribution thanks to the feedback of the discriminator D, which at the same time is also learning important features that help to distinguish whether the input image came from the training distribution or not. Mathematically, the generator G learns to map a latent random vector z to a generated sample tensor and tries to maximize the probability D of making a mistake, that is to say, minimizes $log(1 - D(G(z)))$. On the other hand , the opposite happens to D; it tries to maximize the probability of assigning the correct label $log(D(x))$ and $log(1 - D(G(z)))$, where x is a real image and z is the latent vector

$$min_G max_D L_{\text{GAN}}(G, D) = \mathop{\mathbb{E}}_{x \sim p_{data}(x)}[\log D(x)] + \mathop{\mathbb{E}}_{z \sim p_z(x)}[\log(1 - D(G(z)))]. \quad (1)$$

From a slightly different point of view, this process can be seen as minimizing the distance between distributions. In other words, the generator tries to approximate to the real latent distribution of images by mapping from a completely random distribution. During the training process, the Jhensen–Shannon distance is applied, measuring how far the approximated distribution is from the real one. As it is optimizing the models using gradient descent, this gradient information is back-propagated to the generator. Despite the fact that they mathematically demonstrate that there is a unique solution where D outputs 0.5 for every output and G recovers the latent training data distribution, these models are unstable during training, making it laborious to train. The problem arises due to the unfair competition between generator and discriminator generating mode collapse problems, discriminators shielding infinity predictions and generators producing blank images or always producing the same sample [5]. Moreover, the basic algorithm is capable of generating up to 64×64 images but runs into instabilities if the size is increased. Resolution of the generated image is an important topic to address since most of the geographic and visual properties are better expressed in high-resolution so it can be used in remote-sensing applications.

Having presented the cornerstone and basics of GANs, multiple models and different variations and flavours came up, providing novel techniques, loss functions, layers or applications. Particularly, some studies such as DCGAN [6], which immediately came after the original GANs paper, added a convolutional neural network layer (CNN) in order to increase the stability of synthetic image generation. Despite it proving to generate larger images of 128×128 pixels, studies such as [7] report that it is not sufficient due to insufficient detail in satellite images. They also include a similar analysis to that of [4] about the input latent space, demonstrating that generators are capable of disentangling latent space dimensions by mapping particular dimensions to particular features of the generated images. Advanced techniques, such as in [5], provide new methods for training such as feature matching included in the loss, changing the objective of the loss function from maximizing the discriminator output to reducing the distance between intermediate feature maps of the discriminator extracted from real images and generator images. By doing this, the generator is forced to generate samples that produce the same feature maps in the discriminator as the real images, similar to perceptual losses [8]. They also further analyse the problem of mode collapse by proposing many strategies, such as the mini batch discriminator, where the discriminator has information from other images included in the batch, and they also propose historical averaging that adds weight to the costs and they even suggest a semi-supervised technique that trains the discriminator with labeled and unlabeled data.

Progressive Growing GAN (PGGAN) [9] proposes a method that gradually trains the generator and the discriminator until they are capable of producing large resolution images of 512×512 and 1024×1024. Their method starts by training the generator on images of 4×4 pixels, and by gradually adding new layers to double the generated resolution until it is capable of generating high-res images. In addition, they propose a couple of techniques that further stabilize the training and provide variation such as a minibatch standard deviation layer at the end of the discriminator, helping it to compute statistics of the batch, they propose a weight initialization and a scaling factor during runtime, and, inspired by [10], they implement a Wasserstein gradient penalty as a loss function. They propose a novel metric called Sliced Wasserstein Distance (SWD) that allows the performance of a multi scale statistical similarity between distributions of local real and fake image patches drawn from a Laplacian pyramid, providing granular quantitative results at different scales of the generated image.

In addition to the generation of large images, researchers propose novel architectures for more complex applications such as image-to-image translation, mapping from an image to an output image (conditioned generation). Pix2Pix [11] and Pix2PixHD [12] are among the first to address both problems—the image-to-image translation and high-resolution generation. Ref. [11] proposes a PatchGAN discriminator that is highly involved in posterior GAN research. The PatchGAN discriminator is applied in patches at different scales and then its outputs are averaged to produce one scalar. In combination with L1 loss that captures low-frequency information, this model, which uses fewer parameters, focuses on the high frequencies contained in each patch. Its successor, Pix2PixHD [12], is able to produce images up to 2048 $\times$ 1024 pixels with a novel multi-scale generator and discriminator, and by retaking the ideas of [5] by adding perceptual pre-trained loss. Similar to [9], they divide the training in what they refer to as a coarse-to-fine generator. This generator G is divided into two U-Net models—global generator G_1 and local enhancer G_2. First, G_1 is trained in order to learn global characteristics at the 1024 $\times$ 512 scale. In the second phase, G_2 is added with the particularity that the encoder part is added at the beginning of G_1 and the decoder part is added at the end, leaving the G_1 in the middle. In this case, D is divided into three PatchGANs that operate at different scales. The image is downsampled in order to generate a pyramid of three scales. Then, each D_i operates at different scales with different receptive fields, the coarse scale with a large receptive field leads to global images while the finer scale leads to finer details. The final contribution is the instance level feature embedding, a mechanism to control the generation. First, they train an encoder to find a low-dimension feature vector that corresponds to a real image. Then, they train G and D with this vector and the instance map as the conditional input. After a K-means analysis to find the cluster descriptor of each feature, the user is able to control the generation in coordination of the interpretation that the G is assigned to each dimension.

CycleGAN [13] proposes a model that learns to translate an image from one source domain to a target domain, distressing the necessity of having two paired source and target datasets. This is done by adding an inverse mapping model in the loss that reverts the first transformation applied to the input, called cycle consistency. Additionally, they reuse PatchGAN [11] as a discriminator. They conclude that, by applying the cycle loop in addition to PatchGAN, they are able to reach higher image sizes. PSGAN Progressive Structured GAN [14] is a work that adds conditionally to PGGAN. Their network is able to generate high-resolution anime characters by providing the skeleton structure of the character as an input. They take up the progressing growth by imposing the skeleton map at different scale levels while the generator and the discriminator are growing. StyleGAN [15] is GAN designed for style transfer purposes that can deal with higher resolutions and control the generation by learning high-level attributes and stochastic variations, allowing the control of the style of synthesis. They use a progressive training in conjunction with Adaptive Instance normalization layers and Wassertein gradient penalty in addition to the original GAN loss. This adapted generator learns a latent space domain and how to control features at different scales. The Perceptual Adversarial network, PAN, [16] is a general framework that is also capable of performing high-resolution image-to-image translation. Their proposal also relies on feature matching of the D, encouraging the generated images to have similar high-level features to the real ones while at the same time they use the output of D as the classical GAN loss.

Finally, we describe SPADE [17], a model that generates photorealistic imagery given a semantic map. They propose a spatially adaptive denormalization module (SPADE module), a conditional normalization layer that uses the input segmentation map to modulate the activation of the normalization layer at different scales of the generation. They demonstrate that batch normalization layers drown the signal, so they de-normalize the signal at each scale level by using SPADE layers. These layers consist simply of a convolutional layer that extracts the features of the input map and then learn by two other convolutional layers the scaling parameter at each spatial position and the scale and bias according to the input map structure. By the addition of this simple modulation and residual blocks, they obtain consistent local semantic image translations that outperform previous models such as pix2pixHD and at the same time they remove the necessity of using an encoder–decoder network. They also comment that taking a progressive growing approach makes no difference in their technique. As a discriminator, they reuse the multi-scale PatchGAN [12] with the last term replaced by Hinge loss.

In the field of remote sensing, there are not many studies focused generally on image-to-image translation using GAN. In [7], the authors described the process of applying PGAN to synthetically generate satellite images of rivers and the necessity of high-resolution image generation for remote sensing applications that can capture particular high-frequency details of this kind of image that we mentioned at the beginning of this work. Most of the work that uses GAN for remote sensing applications is conducted for cloud removal [18] or super resolution applications with GAN [19] and without GAN [20] that put special emphasis in the usage of dense skip or residual connections to propagate high-frequency signals that is particularly present in this kind of image. Works such as [21] evaluated models trained with synthetic images and demonstrated the improvement of adding them, but they do not delve into synthetic image generation techniques.

At the moment of this work, there are no vast formal studies specifically applied to the image-to-image translation of generating satellite images conditioned into the segmentation map. Despite there being works that conduct similar tasks [11,13], they rely on generally translating satellite footprints to real images as a usage example rather than conducting a complete study of these challenging tasks. It is important to remark that there are a couple of companies, such as OneView.ai (https://one-view.ai/, accessed on 3 October 2021), that base their entire business model on providing synthetic image generation services for enriching training datasets by including in their pipeline their own developed GAN model to generate synthetic images from small datasets.

1.2. Problem Formulation

Before going deeper with more complex concepts and ideas, we first provide a high-level introduction about the principal ideas around this work. Let us start by considering we have $C = [0, \ldots, K]$, which represents K possible classes and 0 for the background. Let $m \in L^{HxW}$ be a segmentation map, a matrix where each position (x, y) contains a $k \in L$ the index of a class and H and W are the height and width of the image, respectively. Let $s = (v : r)$ a the $(V + R)$-dimensional semantic global vector, where each dimension of the first V-dimensional represents a proportion of one of V semantic global classes. The remaining R-dimensional vector is a categorical (one-hot encoding) vector that represents the categorical class of the region. In this way, each scene is represented by a matrix M and a vector s. We present a deep neural network G that is capable of generating a satellite image I by receiving as an input M and s. Each pixel position (i, j) of the resulting I corresponds to the label of position (x, y) of m. Particularly, in this work, we simplify the problem by choosing one class segmentation map despite the fact that it could be easily adapted to more classes. We decided that M would be a building footprint map due to dataset availability and it was more than enough to validate the model and demonstrate the simplicity of generation. For the first V-dimensional part of the semantic global vector, we carefully defined 17 classes that express the number of visual cues, land use and styles relative to classes such as forest, industrial, road, and so forth (in the following sections we will explain this more in detail). We selected four cities, with remarked style, cultural and geographic properties for the second categorical R-dimensional part of the vector. We ended up with a model that—given a binary M mask with the shape and position of the buildings and a global semantic vector s that defines content related to style such as number of roads, forests, industrial land use zones and so forth, and the city/region—is capable of generating a satellite image that contains all the stylish visual cues in addition to buildings with the exact same position and shape as defined by the mask. With this control mechanism, a user can define their own segmentation mask, or can even modify the region or the amount of semantic classes for the same mask, helping it to efficiently augment a dataset with varied region/culture synthetic satellite imagery.

Finally, the model consists of a generator G of a GAN that is modified from a SPADE model [17] and a discriminator D Figure 2. Mask m and vector s are passed to generator G for generating a synthetic scene to fool the discriminator that is responsible for discerning between synthetic images and real ones.

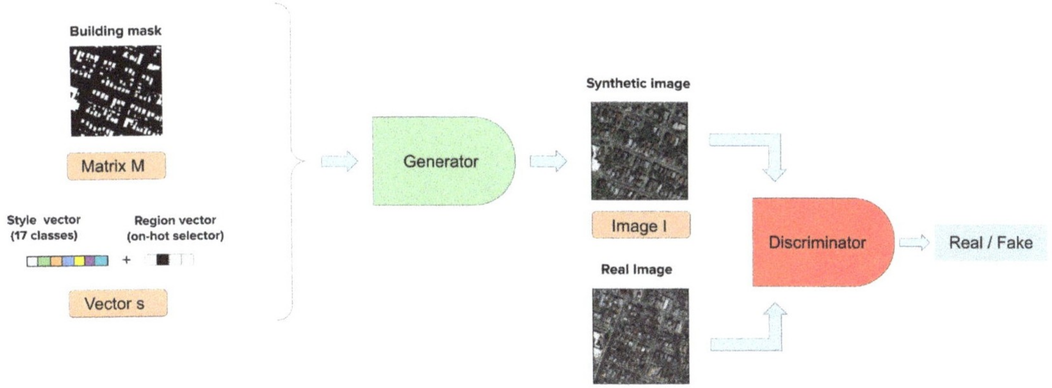

Figure 2. SPADE high level diagram. The generator takes the building footprint mask (m matrix) and the semantic global vector. It generates the synthetic image I and it is passed to the discriminator to determine if it is fake or real.

1.3. Research Questions

The objective of this work is to propose a simple mechanism that leverages the information of public geographic databases to enhance the geographic properties of a synthetic image generated via a GAN model. While defining this mechanism, we wanted to evaluate if that enhancement would help to enrich satellite synthetic generation with finer details and properties, using a simple representation such as a 17-dimensional vector. Therefore, the main research question we address in this work is:

How can a GAN model be modified to accept rich style satellite specific properties while at the same time this information comes in a small-dimensional representation?

Additionally, this work responds to the following subsequent questions:

- *How to leverage public annotation resources such as Open Street Maps to provide style information?*
- *How to define visual distinct land cover properties?*
- *Is the prior knowledge of region and style improving expressiveness of the GAN model?*

2. Datasets

In this section, we describe in detail the datasets we used for training the GAN model and for the development of the semantic global vector descriptor.

2.1. Inria Aerial Image Labeling Dataset

Inria Aerial Image Labeling Dataset (Inria) [22] is a high-resolution dataset designed for pixel-wise building segmentation (Figure 3). It consists of high-resolution objectified color imagery with spatial resolution of 0.3 m/pixel that covers 810 km^2 of 5 cities (inn the training dataset):

- Vienna, Austria
- Lienz, Austria
- Chicago, USA
- Kitsap county of Washington, USA
- Austin, Texas, USA

Segmentation maps are binary images where a 1 in position (i, j) means that the pixel belongs to a building and 0 that it belongs to the background class. This dataset became of interest because besides containing the structure segmentation map of buildings, its images cover a large variety of dissimilar urban and not-urban areas with different types of population, culture and urbanisation, ranging from highly urbanized Austin, Texas to the rural Tyrol region in Austria. The dataset was designed with the objective of evaluating the generalization capabilities of training in a region and extending it to images with varying illuminations, urban landscapes and times of the year. As we were interested only in the labeled images, we discarded the test set and focused on the above-mentioned cities. In consequence, our dataset consisted of 45 images of 3000×3000 pixels.

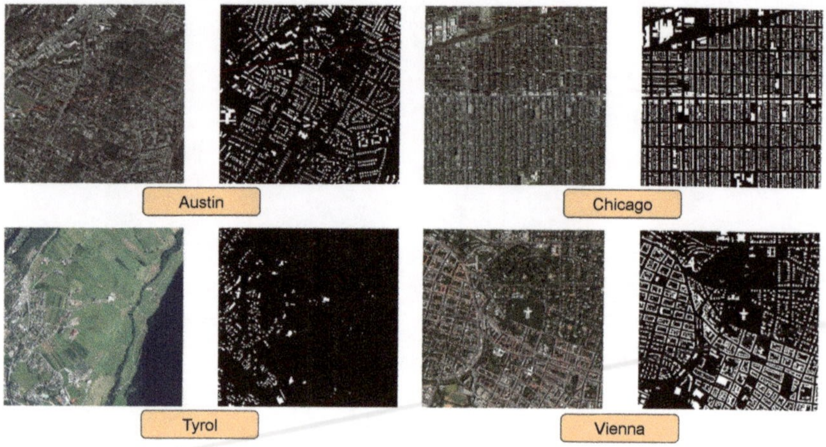

Figure 3. Inria building datasets sample [23].

2.2. *Open Street Map (OSM)*

In 2005, OpenStreetMap (OSM) [24] was created as an open and collaborative database that provides geodata and geo-annotations. In the past few years, OSM has been widely used in several applications in geosciences, Earth observation and environmental sciences [25–27]. Basically, it consists of a free editable map of the world that allows its more than two million users to annotate it or to provide collected data to enrich the OSM geo-information database. Its data primarily consist of annotations at multiple semantic levels that are expressed in keys (categories) and values. Under each key they provide finer grained information in different formats depending on the object of the annotation. For example, they provide annotations of land use that describe the human usage of an area as a polygon in a geojson. Another example is the annotation of roads; they structure the road network as a graph. There are many ways to access its data such as an API or dedicated public or private geo-servers that digest and renderize the data. In our case, we decided to use a public open source server that renders and compiles all the interested information for a specific area.

Therefore, we decided to download the render for each of the images using a rasterized tile server *rasterized tile server* (https://github.com/gravitystorm/openstreetmap-carto/) that provides cartographic style guidelines, see Section 3.3 for more details). As we have the source code of the server, we have the mapping between pixel colour and category. We ended up listing more than 200 categories present in the render and we were capable of reducing it to only 11 classes for the global semantic vector. We will explain this procedure more in detail in the following section.

3. Methods

This section explains the methods used in this study. We will start from a more detailed analysis of the baseline model SPADE [17]. Then, we will delineate the proposed

architecture modifications in order to develop SSSGAN. Next, we describe the creation of the global semantic vector. Finally, we present the metrics we used for evaluation.

3.1. SPADE

As previously explained, SPADE [17] proposed a conditional GAN architecture capable of generating high-resolution photorealistic images from a semantic segmentation map. They stated that, generally, image-to-image GANS receive the input at the beginning of the network, and consecutive convolutions and normalizations tend to wash away semantic and structural information, producing blurry and unaligned images. They propose to modulate the signal of the segmentation map at different scales of the network, producing better fidelity and alignment with the input layouts. In the following subsections, we will explain different key contributions of the proposed model.

Spatially-Adaptive Denormalization

The Spatially-Adaptive Layer is the novel contribution of this work. They demonstrated that spatial semantic information is washed away due to sequences of convolutions and batch normalization layers [28]. In order to avoid this, they propose to add these SPADE blocks that denormalize the signal in the function of the semantic map input, helping to preserve semantic spatial awareness such as semantic style and shape. Let $m \in \mathbb{L}^{HxW}$ be the segmentation mask whereas H and W are the height and width, respectively, and $\mathbb{L}$ is a set of labels that refers to each class. Let h^i be the activation of the i-th layer of a CNN. Let C^i, H^i and W^i be the channels, height and width of the i-th layer, respectively. Assuming that the batch normalization layer is applied channel wise, and obtainμ^i_c and σ^i_c for each channel $c \in C^i$ and i-th layer. The SPADE layer denormalization operation could be expressed as follows, if we consider $y \in H^i$, $x \in W^i$ and $n \in N$ be the batch size:

$$\gamma^i_{c,y,x}(m)\frac{h^i_{n,c,y,x} - \mu^i_c}{\sigma^i_c} + \beta^i_{c,y,x}(m), \tag{2}$$

where μ^i_c and σ^i_c are the batch normalization parameters computed channel-wise for the batch N:

$$\mu^i_c = \frac{1}{NH^iW^i}\sum_{n,y,x} h^i_{n,c,y,x} \tag{3}$$

$$\sigma^i_c = \sqrt{\frac{1}{NH^iW^i}\sum_{n,y,x}(h^i_{n,c,y,x} - (\mu^i_c)^2)}. \tag{4}$$

The role of the SPADE layer is to learn the scale $\gamma^i_{c,y,x}(m)$ and bias term $\beta^i_{c,y,x}(m)$ with respect to the mask m, which they call modulation parameters in Figure 4. It is interesting to put special emphasis on the fact that modulation parameters depend on the location (x, y), thus providing spatial awareness. This spatial awareness is what differentiates this modulation with respect to batch normalization that does not consider spatial information. Those modulation parameters are expressed as a functional because the SPADE layer passes m through a series of two convolutional layers in order to learn these spatially aware parameters. The structure of layers can be seen in Figure 4.

Having defined the SPADE block, the authors reformulate the common generator architecture that uses encoder–decoder architectures [11,12]. They remove the encoder layer since the mask is not fed in the beginning of the architecture. They decided to downsample the segmentation at different scales, and fed them via SPADE blocks after each batch normalization. Then they divided the network into four upscaling segments, where the last one generates an image with the size of the mask. Each segment that defines a scale level is composed of convolutional and upscaling layers followed by SPADE residual blocks. Each SPADE residual block consists of two consecutive blocks of SPADE layers (that ingest segmentation masks that have the same dimensions as the assigned to the SPADE residual block), followed by the RELU activation layer and a 3 × 3 convolution (Figure 5). In this

way, they removed the encoder and ingested information about the shape and structure of the map at each scale, obtaining a lightweight generator with fewer parameters.

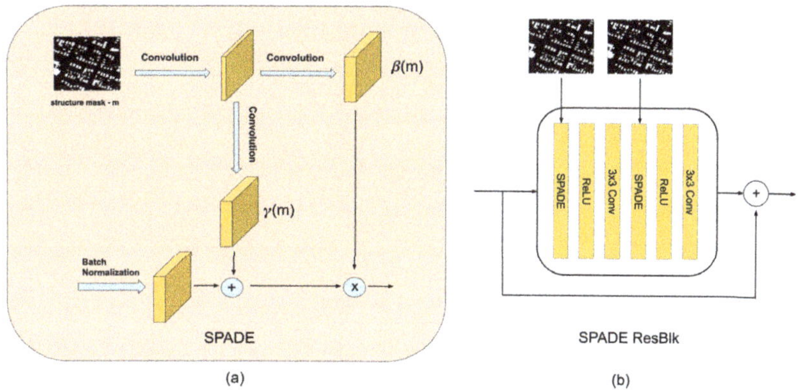

Figure 4. (**a**) SPADE block internal architecture. (**b**) SPADE Residual block (SPADE ResBlk).

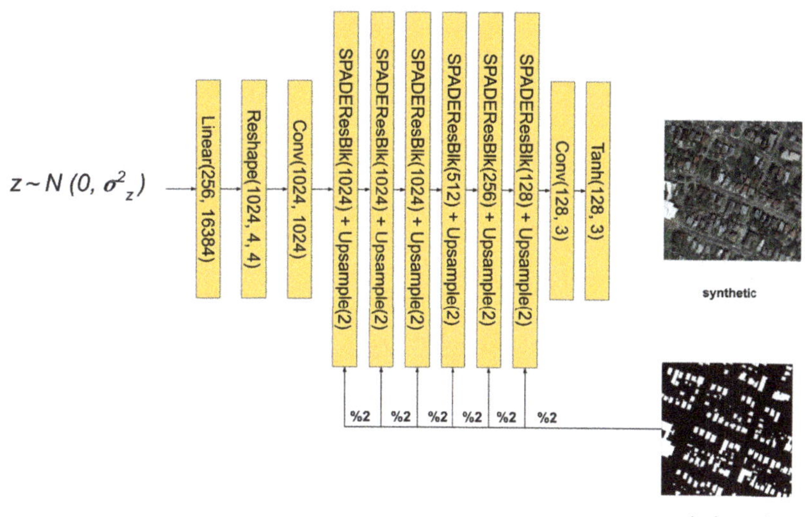

Figure 5. SPADE main architecture.

As a discriminator, they decided to use a Pix2PixHD multiscale PatchGAN discriminator [12]. The task of differentiating high-resolution real images from fake ones represents a special challenge for D, since it needs to have a large receptive field that would increase network complexity. To address this problem, they used three identical PatchGAN discriminators at three different scales (factor of 2) D_1, D_2 and D_3. The one that operates at the coarsest scale has larger global knowledge of the image while the one that operates at the finest scale forces the generator to produce finer details, hence the loss function is the following, where k refers to the index of the three different scales:

$$\min_{G} \max_{D} L(G, D) = \sum_{k=1,2,3} L_{\text{GAN}}(G, D_k). \quad (5)$$

Another particularity is that they did not use the classical GAN loss function. Instead, they used the least squared loss [29] term modification in addition to Hinge loss [30], and were demonstrated to provide more stable training and to avoid the vanishing gradient problems provided by the usage of the logistic function. Therefore, their adapted loss function is shown as follows:

$$L_D(G,D) = \mathbb{E}_{x\sim p_{\text{data}}(x)}[\min(0,-1+D(x))] + \mathbb{E}_{z\sim p(z)}[\min(0,-1-D(G(z)))] \quad (6)$$

$$L_G(G,D) = \mathbb{E}_{z\sim p(z)}[D(G(z)).] \quad (7)$$

Additionally, they used feature matching loss functions that we will not use in our experiments.

Finally, PatchGAN [11] is the lightweight discriminator network that is used at each scale Figure 6. It was developed with the idea that the discriminator focuses on high-frequency details while L_1 focuses on low frequencies. In consequence, they restricted the discriminator to look at particular $N \times N$ patches to decide if it is real or fake. Consequently, the discriminator is convolved through the image by averaging its prediction of each $N \times N$ patch into a single scalar. This allows the discriminator to have fewer parameters and focus on granular details and composition of the generated image. In fewer words, this discriminator is a simple ensemble of lightweight discriminators that reduce the input to a unique output that defines the probability of being real or fake. The authors interpreted this loss as a texture/style loss.

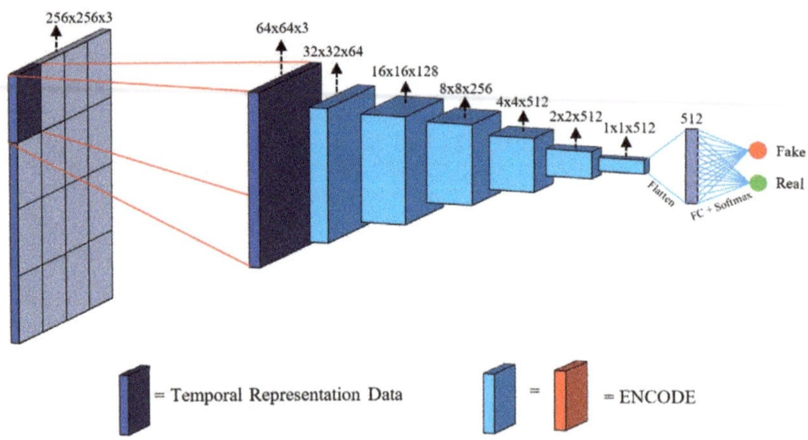

Figure 6. PatchGAN diagram [31]. The entire discriminator is applied to $N \times N$ patches. Then, the model is convolved over the image and their results are averaged in order to obtain a single scalar.

3.2. SSSGAN

Having studied the principal component of SPADE in detail, we were able to spot its weak points for being used in our study. The key idea of SPADE is to provide spatial semantic modulation through the SPADE layers. That property is useful for guaranteeing spatial consistency in the synthesis related to the structural segmentation map, which in our case is the building footprint, it does not apply to the global semantic vector. Our objective is to ingest global style parameters through the easy-to-generate global semantic vector, which allows the user to define the presence of semantic classes while avoiding the necessity of generating a mask with the particular location of these classes. As the semantic vector does not have spatial applicability, it cannot be concatenated, neither fed through the SPADE layer. On the other hand, we can think of this vector as a human-interpretable

and already disentangled latent space. Hence, we force the network to adapt this vector as a latent space.

We replace the latent random space generator of the SPADE model for a sequence of layers that receives the global semantic vector as an input (Figure 7). In order to ingest this information, we first generate the global vector by concatenating the first V classes, the 17 visual classes and the one hot encoding vector that defines the region or area (R-dimensional). The vector goes through three consecutive multi layer perceptron (MLP) blocks of 256, 1024 and 16,384 neurons followed by an activation function. The resulting activations are reshaped to an $1024 \times 4 \times 4$ activation volume tensor. That volume is passed to a convolutional layer and a batch normalization layer. The output is then passed to a SPADE layer that modulates this global style information with respect to the structure map. Ref. [17] suggests that the style information tends to be washed away as the network goes deeper. As a consequence, we decided to add skip connections between each of the scale blocks in channel-wise concatenation, similar to DenseNet [32]. In this way each scale block can receive the collective knowledge of previous stages, allowing the flow of the original style information . At the same time, it allows us to divide information in the way the SPADE block can focus on high-frequency spatial details, extremely important in aerial images, while the skip branch allows the flow of style and low-frequency information [20,20]. In addition, reduction blocks are added (colored in green in Figure 7) that reduce the channel dimension, which is increased by the concatenation. This helps to stack more layers for the dense connections without a significant increment of memory. Thus, it is extremely important to add those layers. Besides all of that, this structure helps to establish the training process because the dense connections also allow the gradient to be easily propagated to the lower layers, even allowing deeper network structures. This dense connection is applied by passing the volume input of each scale block with the output volume of the SPADE layer block. As the concatenation increases the channel (hence the complexity of the model), a 1×1 convolution layer is applied to reduce the volume.

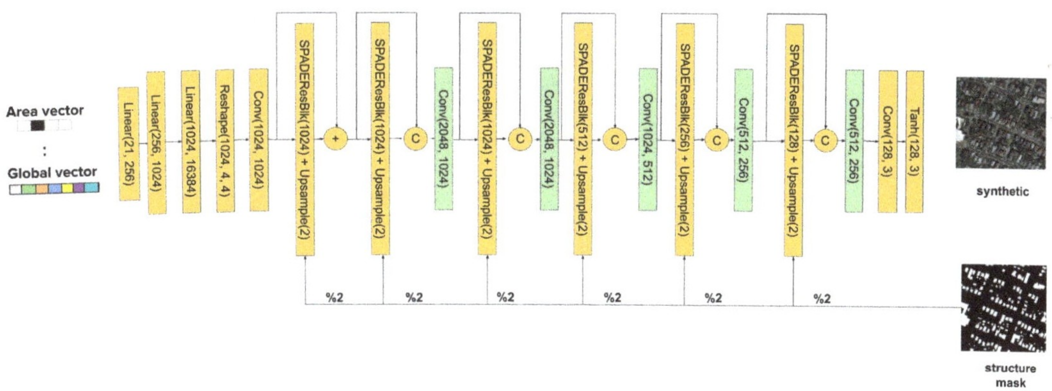

Figure 7. SSSGAN architecture. Every linear layer is followed implicitly by Relu activation. The structure mask is downsampled to its half resolution each time it is fed to a SPADEResBlock (referenced by '%2' label).

3.3. Global Semantic Vector

In this section, we describe the creation of the global semantic vector. The key idea is to obtain a semantic description of the image that may help the generator to distinguish and pay attention to key properties present on the satellite image and it also helps to modify the image generation. In order to obtain a description, the principal idea of this work is that it can be easily generated from the OSM tags. This crowdsource tagged map is available publicly and it offers tags named by category for land use, roads, places, services. First, we download tags related to the areas of interest: Chicago, Vienna, Austin, Kitsap and

Tyrol. These tags come in multiple formats, for example, land use is defined by polygons while roads are defined as a graph. We obtained more than 150 values so we decided to rasterize these tags and then to define the value that corresponds to each pixel. After that process, we analysed the results and we found different problems regarding the labels. The first problem is that urban zones were more densely and finely detailed tagged than urban zones. For example, Vienna had much more detail in tags that even individual trees were tagged (Figure 8a), while in the Tyrol region there were zones that were not even tagged. The second and more important problem was that there was no homogeneous definition of one tag in the same region or image. For example, in Chicago there were zones tagged as residential, while at the other side of the road—which has the same visual appearance—it was tagged as land (Figure 8b). Moreover, we noticed that all images of Kitsap were not annotated at all, there were roads and residential zones that were missed (Figure 8c). Finally, we come up with similar conclusions to [3], a work that only used land use information. Labels refer to human activities that are performed in specific zones. Those activities sometimes may be expressed with different visual characteristics at ground level, but from the aerial point of view those zones do not contain visual representative features. The clear example is the distinction between commercial and retail. The official definition in OSM is ambiguous, commercial refers to areas for commercial purposes, while retail is for zones where there are shops. Besides this ambiguity in definition, both areas express buildings with flat grey roofs in the aerial perspective.

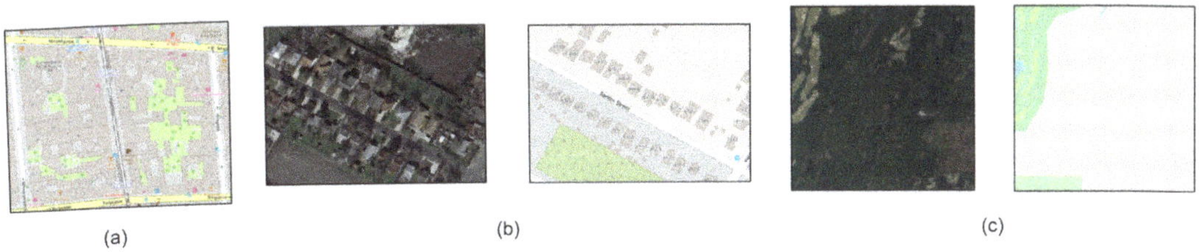

Figure 8. (**a**) Detailed annotation of the urban area of Vienna (**b**) Residential area from Chicago, from one side of the annotated road is defined as residential and the other side is not annotated, despite both belongs to the same visual residential cues (**c**) Area of Kitsap without annotation.

Having studied all of these problems in detail, we decided to perform a manual inspection of the data and we defined a series of conventions that help to aid the previously mentioned problems. The principal idea is to create a vector in an automatic way that digests all semantic visual information so it facilitates the model to put attention on particular visual characteristics. For that reason, we decided to group all of these categories in 17 classes that have a clear visual representation despite the use. In that case, classes such as commercial and retail will constitute the same class, since while we were doing the manual visual inspection we decided that those classes are visually indistinguishable. We manually corrected zones that were not labeled and we defined a unique label for ambiguous zones, fixing the problem with residential and land labeled zones. Finally, we decided to remove images from the Kitsap region from the detest due to the scarcity in label information.

At the end of this process, we ended with the 17 classes expressed in Table 1. In order to compute the vector, we defined an index or position to each class in the vector. Having this grouping rule of classes, we processed each image by counting the amount of pixels that belong to each class (taking into account the priority of the class) and then we normalized the vector to sum 1, obtaining a distribution of classes.

More conclusions were obtained from this analysis that also coincide with the ones expressed in [3]. A specific land use, such as residential or commercial, varies in visual characteristics from region to region due to architectural and cultural factors. In order to help the network to distinguish these cultural properties, and at the same time control the generation, we added to this vector a one hot encoding selector that defines the region: Chicago, Austin, Vienna or Tyrol.

Table 1. CNN Performance Long Format.

Semantic Category	OSM Tag	Index
grass	grass, heath, golf_course, farmland, ...	1
forest	forest, forest-text, orchard, scrub, ...	2
residential	residential, residential-line, land-color, ...	3
commercial	commercial, commercial-line, retail, ...	4
industrial	industrial, industrial-line, wastewater_plant, ...	5
parking	garages, parking, ...	6
construction	construction, construction_2, built-up-z12, quarry, ...	7
sports	pitch	8
highway	motorway, primary, secondary, ...	9
rail	motorway, primary, secondary, ...	10
road	living_street, residential, ...	11
footway	footway, pedestrian, ...	12
religious	cemetery, religius, ...	13
motorway	motorway, trunk, ...	14
water	water, ...	15
allotments	allotments, ...	16
block	block, ...	17

3.4. Metrics

We decided to employ two state-of-the-art perceptual metrics used in [9,17]. Since there is no ground truth, the quality of generated images is difficult to evaluate. Perceptual metrics try to provide a quantitative answer of how close the generator managed to understand and reproduce the target distribution of real images. The following metrics provide a scalar that represents the distance between distributions, and indirectly they are accessing how perceptually close the generated images are to the real ones.

3.4.1. Frechet Inception

Frechet Inception Distance (FID) [17,33] is commonly used in GAN works for measuring their image quality generation. It is a distance that measures the distance between synthetically generated images and the real distribution. Its value refers to how similar two sets of images are in terms of vector features extracted by Inception V3 model [34] trained for classification. Each image is passed through the Inception V3, and the last pooling layer prior to the output classification is extracted obtaining a feature vector of 2048 activations. These vectors are summarized as a multivariate Gaussian, computing the mean and covariance of each dimension for each image in each group. Hence, a multivariate Gaussian is obtained for each group, real and synthetic images. The resulting Frechet distance between these Gaussian distributions is the resulting score for FID. A lower score means that the two distributions are close, the generator has managed to efficiently emulate the latent real distribution.

3.4.2. Sliced Wasserstein

Sliced Wasserstein Distance (SWD) is an efficient approximation to the earth mover distance between distributions. Briefly speaking, despite being computationally inefficient, earth mover distance provides the vertical distance difference between distribution, giving an idea of the differences between densities. Ref. [9] comments that metrics such as MS-SSIM are useful for detecting coarse errors such as mode collapse, but fails to detect fine-detailed variations in color and textures. Consequently, they propose to build a Laplacian pyramid for each of the real and generated images, from 16×16 pixel and doubling resolution until the pyramid reaches the original dimensions. Basically, each level of the pyramid is a downsampled version of the upper level. This pyramid was constructed, having in mind that a perfect generator will synthesize similar image structures at different scales. Then, they select 16,384 images for each distribution and extract 128 patches of 7 × 7 with three RGB channels (descriptors) for each Laplacian level. This process ends up with 2.1 M of descriptors for each distribution. Each patch is normalized with respect to each color channel's mean and standard deviation. After that, the Sliced Wasserstein Distance is applied to both sets, real and generated. Lowering the distance means that patches between both distributions are statistically similar.

Therefore, this metric provides a granular quality description at each scale. Patches at 16 × 16 similarity indicate if the sets are similar in large-scale structures, while larger scale provides more information of finer details, color or textures similarities and pixel-level properties.

4. Results

In this section, we show quantitative and qualitative results using the INRIA dataset along with our global semantic vector descriptor. We start in Section 4.1 by describing the setup of the experiment. In Section 4.2, we show the quantitative results by performing a simple ablation study. Finally, in Section 4.3, we present some qualitative results, by showing how a change in the global vector changes the style of synthesised images.

4.1. Implementation Details

The original SPADE was trained on an NVIDIA DGX1 with eight 32 GB V100 GPUs [17]. In our case, we train our network with our network with eight NVIDIA 1080Ti of 11 GB each one. This difference in terms of computational resources made us reduce the batch down to 24 images, instead of 96. Usually, training with larger batch sizes should help stabilize the training and produce better results. Regardless of this aspect, we show our approach is able to improve the generation´s expressiveness and variety with respect to the baseline, while changing the style and domain of the generated images.

We applied a learning rate of 0.0002 to both, the generator and the discriminator. We used ADAM optimizer with $\beta_1 = 0$ and $\beta_2 = 0.9$. Additionally we applied data a few data augmentation that consisted of simply random 180° and 90° rotations. Original images were cropped to 256 × 256 patches with an overlap of 128 pixels that provide more variability. We trained each network for the same amount of 50 epochs.

4.2. Quantitative Analysis

We trained the original SPADE implementation as a baseline, which we used for referencing any quantitative improvement provided by our proposal. Then, we decided to evaluate our main architecture by using only the global semantic vector. Finally, we conducted the full approach that uses the global semantic vector and the dense connections scheme. We applied our two aforementioned metrics, Frechet (FID) Inception Metric and Sliced Wasserstein distance (SWD), for obtaining quantitative results. Table 2 shows the comparison results between different versions of the model. By a great margin, we can appreciate that the full implementation of SSSGAN, which uses the complete global semantic vector, outperforms the original baseline. The model could reduce, by more than a half, almost all the metrics. The reduction of the FID from 53.19 to 22.35 suggests that the generation was closer to estimating the latent distribution of the real images than the original baseline in general and global features. Moreover, it provides a more granular and detailed perspective about the generator performance at different scales. SSSGAN was able to reduce by a 56% at the original scale, an impressive 76.5% at 128 × 128 scale, a 67.6% at a 64 × 64 scale, a 64.3% at a 32 × 32 scale and a 45.8% at 16 × 16 scale the SWD score. Our hypothesis is that, by forcing the generator to understand the already disentangled space for humans, we are providing more prior knowledge about the real distribution of the real images. During the training process, the generator can assign a correlation between the presence of particular features of the image and an increment of the global vector value.

In this way the generator could produce more variable synthetic images generations and it could capture finer details structures at different scales. The generator not only reduces each metric, it could reach a constant performance in almost every scale, by learning how to generate closer to reality scale specific features.

Intermediate results that use only the semantic vector suggest that this approach provides variability to the image generation. Even though the absence of dense connections considerably reduced every score, the signal of the style that is fed into the beginning of the networks gets washed out by consistently activations modulation performed by SPADE blocks, that modulates activation only with respect to the structure of the buildings. The addition of dense connections before the modulation helps to propagate the style signal efficiently to each of the scales.

Table 2. Performance different SSSGAN versions with respect to SPADE baseline.

Model	FID	SWD-256	SWD-128	SWD-64	SWD-32	SWD-16
baseline	53.19	338.53	474.08	486.17	474.08	665.29
semantic(ours)	38.17	207.64	206.29	241.29	232.02	404.96
semantic+dense(ours)	**22.36**	**148.93**	**111.89**	**153.75**	**173.86**	**355.16**

Model comparison. Baseline refers to the original SPADE implementation [17]. While "semantic" refers to the SSSGAN with only style vector and "semantic+dense" is the full model SSSGAN with global semantic vector and dense connection structure.

4.3. Qualitative Analysis

In this section, we show a comparison between SPADE baseline, SSSGAN with only semantic information, and the full version of SSSGAN. From Figures 9–16, we see those networks compared in addition to the segmentation building mask, the semantic map to provide an idea of the proportion of the semantic classes and the three most influential classes of the semantic global vector. Qualitatively speaking, the full version of SPADE was able to perform a simple relation between shape of buildings and region in order to generate more consistent scenes. For example, when a structure mask is presented with the characteristic shape of Vienna's building (Figure 16), SSSGAN can infer from the shape of building besides the information of the global vector and the region of the intended image, and generate region-specific features of the region like tree shapes, illumination, and the characteristic orange roof. Most of the time, SPADE generated flat surfaces with an absence of fine details, textures and illumination (Figure 9, Figure 10, Figure 14 or Figure 15). Another aspect to remark on relates to the style of the region is that SSSGAN was able to remarkably capture Tyrol style images (Figures 14 and 15) with large light green meadows, trees, illumination and roads.

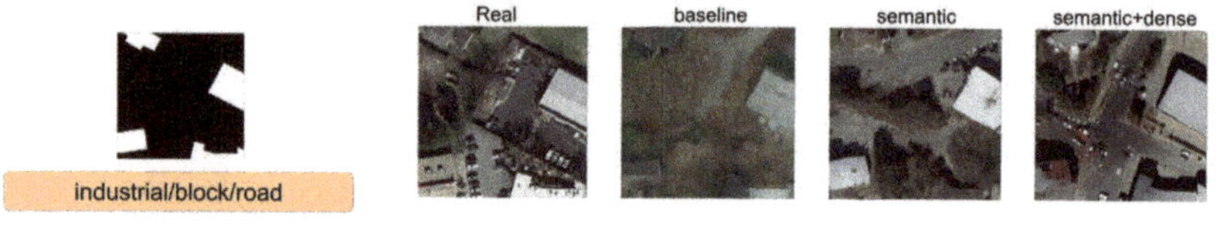

Figure 9. Visual comparison of Austin area. Mask of building footprint and main semantic classes of the vector are shown as reference.

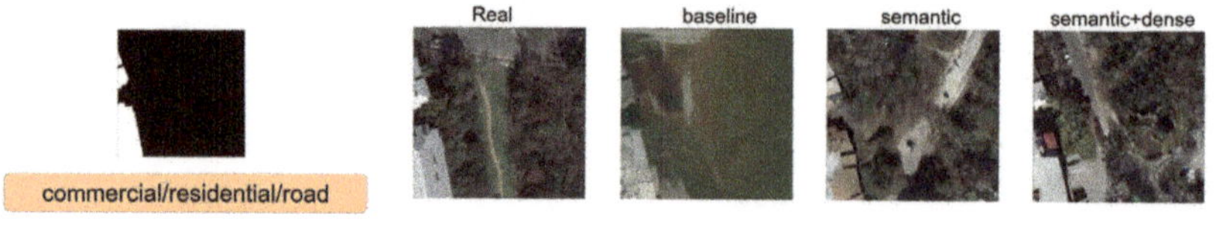

Figure 10. Visual comparison of Austin area. Mask of building footprint and main semantic classes of the vector are shown as reference.

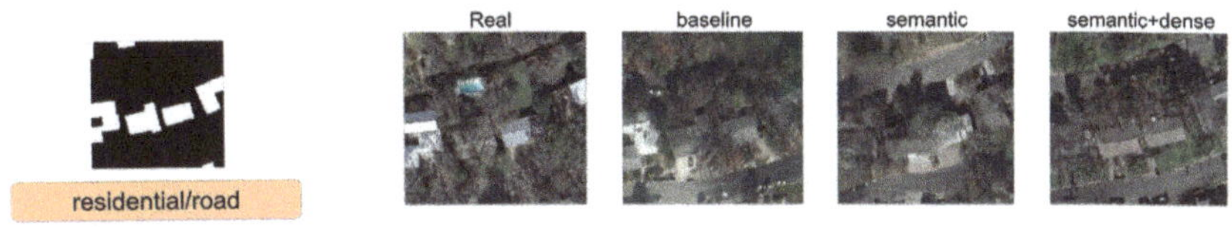

Figure 11. Visual comparison of Austin area. Mask of building footprint and main semantic classes of the vector are shown as a reference.

Generally speaking, SSSGAN demonstrated its vast ability to capture style and context related to each of the four regions. For example, in contrast to the baseline, SSSGAN was able to produce a detailed grass style of Tyrol and differentiate subtle tree properties of Austin and Chicago. In general, visual inspection of the generated images suggest that SSSGAN was able to capture railway track, roads and even the consistent generation of cars as in Figure 9.

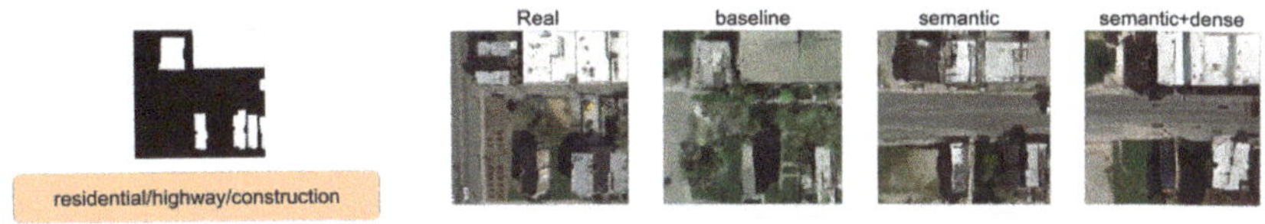

Figure 12. Visual comparison of Chicago area. Mask of building footprint and main semantic classes of the vector are shown as reference.

Another remarkable point is the consistent shadowing of the scenes; it can be appreciated in every scene that the network is able to generate consistent shadows among every salient feature such as trees or buildings. Finally, we can see that networks have difficulties in generating long rectified lines. The reason is that the building mask contains imperfect annotated boundaries that the networks reproduce and the adversarial learning procedure does not detect and therefore do not know how to overcome.

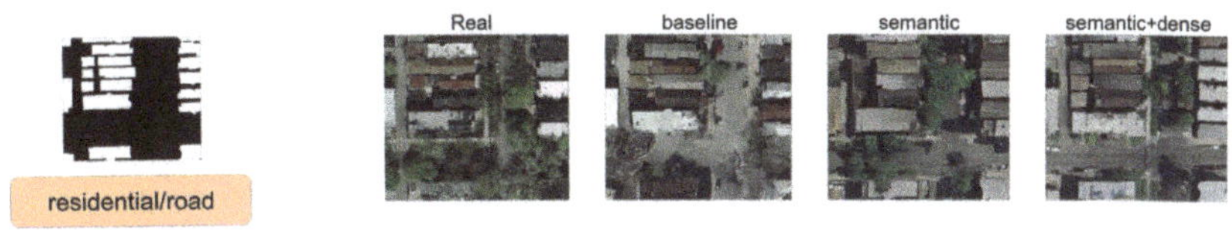

Figure 13. Visual comparison of Chicago area. Mask of building footprint and main semantic classes of the vector are shown as a reference.

Figure 14. Visual comparison of Tyrol area. Mask of building footprint and main semantic classes of the vector are shown as a reference.

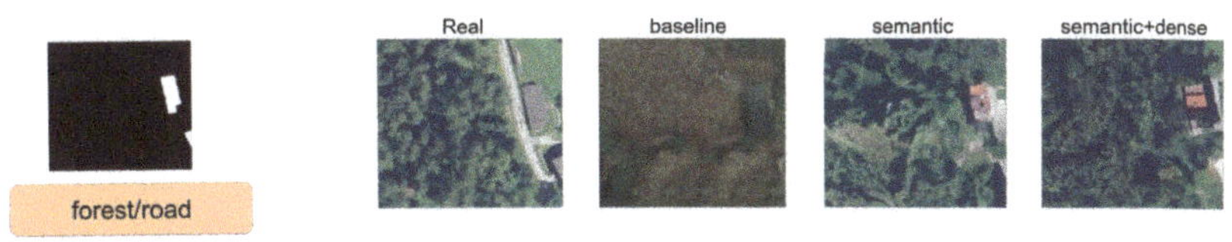

Figure 15. Visual comparison of Tyrol area. Mask of building footprint and main semantic classes of the vector are shown as a reference.

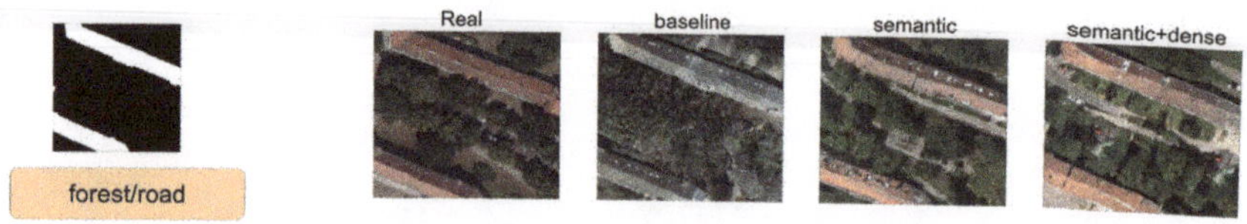

Figure 16. Visual comparison of Vienna area. Mask of building footprint and main semantic classes of the vector are shown as a reference.

In Figure 17, we show the generation capabilities. We increment the presence of four categories while diminishing the others in a Chicago building footprint. At the same time, we show how this generation mechanism is expressed in each region by changing the one hot encoded area vector. Efficiently, we see how each row contains a global style color palette related to the region. For example, the row of the Tyrol region in Figure 17 presents a global greenish style that is common in that region, while the trow of Chicago presents brownish and diminished colors. The increment of forest efficiently Figure 18 increases the presence of trees while the increment of industrial category tends to generate grey flat roofs over the buildings. It is important to remark that the style of the semantic category is captured, despite it does not show enough realism due to incompatibilities of building shapes with this specific style. For instance, when increasing industrial over a mask of residential houses of Chicago, the network is able to detect buildings and provide them a grey tonality, but is not providing finer details to these roofs because it is not relating the shape and dimensions of that building with respect to the increased style. Nevertheless, we can efficiently corroborate changes in style and textures by manipulating the semantic global vector.

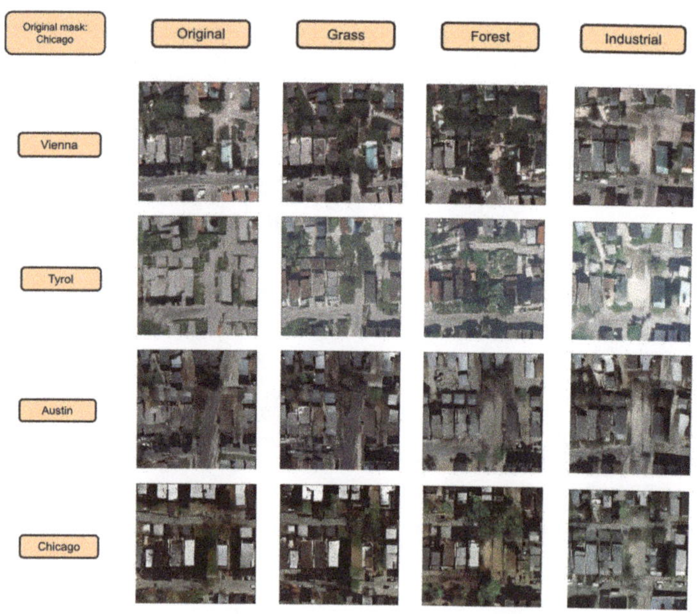

Figure 17. Original building footprint is from Chicago. Each row shows the generation for that Chicago footprint mask in different regions. First column uses the original global semantic vector. Second column the grass category is increased. Third column forest class is increased. Finally, the fourth column industrial class is increased.

Figure 18. Finer observation of the increment of grass class and forest class.

Finally, we show in Figure 19 some negative results. In Figure 19a, we show two different cases using our semantic+dense model. On the left, the model fails at generating cars. The top example marked in red seems to be a conglomerate of pixels rather than a row of cars. On the other side, the bottom example marked in red seems like an uncompleted car. On the right image, the transition between buildings and the ground is not properly defined. Figure 19b shows a clear example where the semantic model is actually performing better than the semantic+dense version. In the latter case, the division that usually splits the roof in half (in a Vienna-scenario) tends to disappear along the roof. Moreover, one can hardly see the highway. Figure 19c shows an example where both semantic and semantic+dense, fail at properly generating straight and consistent roads. Thus, although in general, the results look promising, small objects and buildings geometry could be further improved. Hence, to mitigate some of the failures we were describing above, a geometrical constraint for small objects and buildings could be incorporated into the model, either during the training phase or as a post-processing stage.

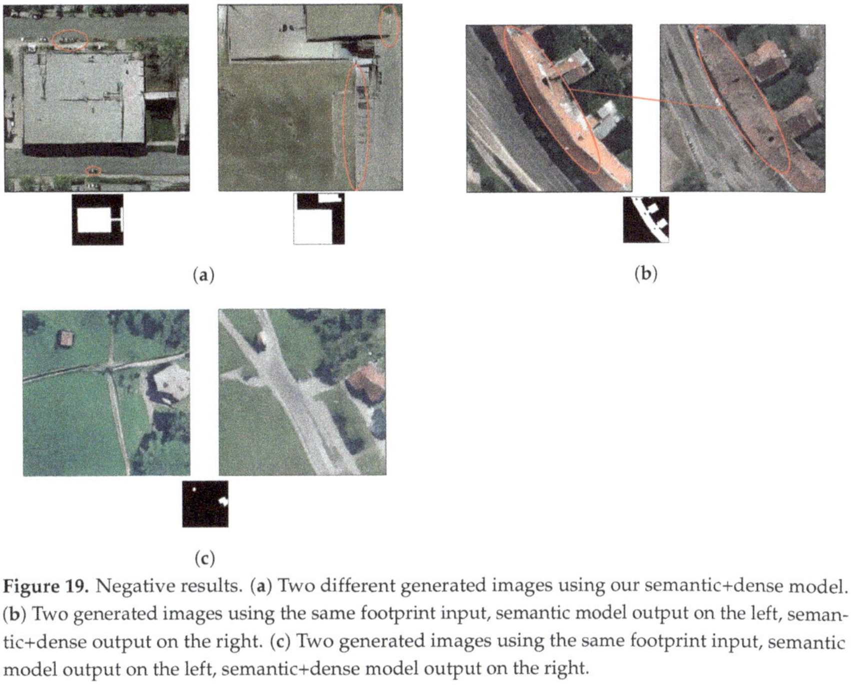

Figure 19. Negative results. (**a**) Two different generated images using our semantic+dense model. (**b**) Two generated images using the same footprint input, semantic model output on the left, semantic+dense output on the right. (**c**) Two generated images using the same footprint input, semantic model output on the left, semantic+dense model output on the right.

4.4. Conclusions

Global high resolution images with corresponding ground truth are difficult to obtain due to the infrastructure and cost required to acquire and label them, respectively. In order to overcome this issue, we present a novel method, SSSGAN, which integrates a mechanism capable of generating realistic satellite images, improving the semantic features generation by leveraging publicly available crowd sourced data from OSM. These static annotations, which purely describe a scene, can be used to enhance satellite image generation by encoding it in the global semantic vector. We also demonstrate that the use of this vector, in addition to the architecture proposed in this work, permits SSSGAN to effectively increase the expressiveness capabilities of the GAN model. In the first place, we manage to outperform the SPADE model in terms of FID and SWD metrics, meaning the generator was able to better approximate the latent real distribution of real images. By evaluating the SWD metric at multiple scales, we further show the consistent increment in terms of diversity at different scale levels of the generation, from fine to coarse details. In the qualitative analysis, we perform a visual comparison between the baseline and our model, comparing the increment in diversity and region-culture styles. We finish our analysis by showing the effectiveness of manipulating the global semantic vector. This brings to light the vast potential of the proposed approach. We hope this work will encourage future synthetic satellite image generation studies that will help with a better understanding of our planet.

Author Contributions: Conceptualization, J.M. and S.E.; methodology, J.M. and S.E; validation, J.M. and S.E.; investigation, J.M. and S.E.; writing—review and editing, J.M. and S.E.; supervision, J.M. and S.E. All authors have read and agreed to the published version of the manuscript.

Funding: This work was supported by the European Regional Development Fund (ERDF) and the Spanish Government, Ministerio de Ciencia, Innovación y Universidades—Agencia Estatal de Investigación—RTC2019-007434-7; and partially supported by the Spanish project PID2019-105093GB-I00 (MINECO/FEDER, UE) and CERCA Programme/Generalitat de Catalunya), and by ICREA under the ICREA Academia programme.

Data Availability Statement: The data used in this work was publicly available.

Acknowledgments: Due to professional conflicts, one of the contributors of this work, Emilio Tylson, requested to not appear in the list of authors. He was involved in the following parts: validation, software, investigation, data curation, and writing—original draft preparation. We would also like to thank Guillermo Becker, Pau Gallés, Luciano Pega and David Vilaseca for their valuable input.

Conflicts of Interest: The authors declare no conflict of interest. The funders had no role in the design of the study; in the collection, analyses, or interpretation of data; in the writing of the manuscript, or in the decision to publish the results.

References

1. Krizhevsky, A.; Sutskever, I.; Hinton, G.E. Imagenet classification with deep convolutional neural networks. *Adv. Neural Inf. Process. Syst.* **2012**, *25*, 1097–1105. [CrossRef]
2. Deng, J.; Dong, W.; Socher, R.; Li, L.J.; Li, K.; Li, F.-F. Imagenet: A large-scale hierarchical image database. In Proceedings of the 2009 IEEE Conference on Computer Vision and Pattern Recognition, Miami, FL, USA, 20–25 June 2009; pp. 248–255.
3. Albert, A.; Kaur, J.; Gonzalez, M.C. Using convolutional networks and satellite imagery to identify patterns in urban environments at a large scale. In Proceedings of the 23rd ACM SIGKDD International Conference on Knowledge Discovery and Data Mining, Halifax, NS, USA, 13–17 August 2017; pp. 1357–1366.
4. Goodfellow, I.J.; Pouget-Abadie, J.; Mirza, M.; Xu, B.; Warde-Farley, D.; Ozair, S.; Courville, A.; Bengio, Y. Generative adversarial networks. *arXiv* **2014**, arXiv:1406.2661.
5. Salimans, T.; Goodfellow, I.; Zaremba, W.; Cheung, V.; Radford, A.; Chen, X. Improved techniques for training gans. *arXiv* **2016**, arXiv:1606.03498.
6. Radford, A.; Metz, L.; Chintala, S. Unsupervised representation learning with deep convolutional generative adversarial networks. *arXiv* **2015**, arXiv:1511.06434.
7. Gautam, A.; Sit, M.; Demir, I. Realistic River Image Synthesis using Deep Generative Adversarial Networks. *arXiv* **2020**, arXiv:2003.00826.

8. Zhang, R.; Isola, P.; Efros, A.A.; Shechtman, E.; Wang, O. The unreasonable effectiveness of deep features as a perceptual metric. In Proceedings of the IEEE Conference on Computer Vision and Pattern Recognition, Salt Lake City, UT, USA, 18–23 June 2018; pp. 586–595.
9. Karras, T.; Aila, T.; Laine, S.; Lehtinen, J. Progressive growing of gans for improved quality, stability, and variation. *arXiv* **2017**, arXiv:1710.10196.
10. Arjovsky, M.; Chintala, S.; Bottou, L. Wasserstein generative adversarial networks. In Proceedings of the 34th International Conference on Machine Learning, Sydney, NSW, Australia, 6–11 August 2017; pp. 214–223.
11. Isola, P.; Zhu, J.Y.; Zhou, T.; Efros, A.A. Image-to-image translation with conditional adversarial networks. In Proceedings of the IEEE Conference on Computer Vision and Pattern Recognition, Honolulu, HI, USA, 21–26 July 2017; pp. 1125–1134.
12. Wang, T.C.; Liu, M.Y.; Zhu, J.Y.; Tao, A.; Kautz, J.; Catanzaro, B. High-resolution image synthesis and semantic manipulation with conditional gans. In Proceedings of the IEEE Conference on Computer Vision and Pattern Recognition, Salt Lake City, UT, USA, 18–23 June 2018; pp. 8798–8807.
13. Zhu, J.Y.; Park, T.; Isola, P.; Efros, A.A. Unpaired image-to-image translation using cycle-consistent adversarial networks. In Proceedings of the IEEE International Conference on Computer Vision, Venice, Italy, 22–28 October 2017; pp. 2223–2232.
14. Hamada, K.; Tachibana, K.; Li, T.; Honda, H.; Uchida, Y. Full-body high-resolution anime generation with progressive structure-conditional generative adversarial networks. In Proceedings of the European Conference on Computer Vision (ECCV) Workshops, Munich, Germany, 8–14 September 2018.
15. Karras, T.; Laine, S.; Aila, T. A style-based generator architecture for generative adversarial networks. In Proceedings of the IEEE/CVF Conference on Computer Vision and Pattern Recognition, Long Beach, CA, USA, 15–20 June 2019; pp. 4401–4410.
16. Wang, C.; Xu, C.; Wang, C.; Tao, D. Perceptual adversarial networks for image-to-image transformation. *IEEE Trans. Image Process.* **2018**, *27*, 4066–4079. [CrossRef] [PubMed]
17. Park, T.; Liu, M.Y.; Wang, T.C.; Zhu, J.Y. Semantic image synthesis with spatially-adaptive normalization. In Proceedings of the IEEE/CVF Conference on Computer Vision and Pattern Recognition, Long Beach, CA, USA, 15–20 June 2019; pp. 2337–2346.
18. Singh, P.; Komodakis, N. Cloud-gan: Cloud removal for sentinel-2 imagery using a cyclic consistent generative adversarial networks. In Proceedings of the IGARSS 2018—2018 IEEE International Geoscience and Remote Sensing Symposium, Valencia, Spain, 22–27 July 2018; pp. 1772–1775.
19. Wang, Z.; Jiang, K.; Yi, P.; Han, Z.; He, Z. Ultra-dense GAN for satellite imagery super-resolution. *Neurocomputing* **2020**, *398*, 328–337. [CrossRef]
20. Salvetti, F.; Mazzia, V.; Khaliq, A.; Chiaberge, M. Multi-image Super Resolution of Remotely Sensed Images using Residual Feature Attention Deep Neural Networks. *arXiv* **2020**, arXiv:2007.03107.
21. Shermeyer, J.; Hossler, T.; Van Etten, A.; Hogan, D.; Lewis, R.; Kim, D. Rareplanes: Synthetic data takes flight. In Proceedings of the IEEE/CVF Winter Conference on Applications of Computer Vision, Waikola, HI, USA, 5–9 January 2021; pp. 207–217.
22. Maggiori, E.; Tarabalka, Y.; Charpiat, G.; Alliez, P. Can Semantic Labeling Methods Generalize to Any City? The Inria Aerial Image Labeling Benchmark. In Proceedings of the IEEE International Geoscience and Remote Sensing Symposium (IGARSS), Fort Worth, Texas, USA, 23–28 July 2017.
23. Ma, J.; Wu, L.; Tang, X.; Liu, F.; Zhang, X.; Jiao, L. Building extraction of aerial images by a global and multi-scale encoder-decoder network. *Remote Sens.* **2020**, *12*, 2350. [CrossRef]
24. OpenStreetMap Contributors. Available online: https://www.openstreetmap.org (accessed on 3 October 2021).
25. Kang, J.; Körner, M.; Wang, Y.; Taubenböck, H.; Zhu, X.X. Building instance classification using street view images. *ISPRS J. Photogramm. Remote Sens.* **2018**, *145*, 44–59. [CrossRef]
26. Baier, G.; Deschemps, A.; Schmitt, M.; Yokoya, N. Synthesizing Optical and SAR Imagery From Land Cover Maps and Auxiliary Raster Data. *IEEE Trans. Geosci. Remote Sens.* **2021**. [CrossRef]
27. Vargas-Munoz, J.E.; Srivastava, S.; Tuia, D.; Falcão, A.X. OpenStreetMap: Challenges and Opportunities in Machine Learning and Remote Sensing. *IEEE Geosci. Remote Sens. Mag.* **2021**, *9*, 184–199. [CrossRef]
28. Ioffe, S.; Szegedy, C. Batch normalization: Accelerating deep network training by reducing internal covariate shift. In Proceedings of the 32nd International Conference on International Conference on Machine Learning, Lille, France, 7–9 July 2015; pp. 448–456.
29. Mao, X.; Li, Q.; Xie, H.; Lau, R.Y.; Wang, Z.; Paul Smolley, S. Least squares generative adversarial networks. In Proceedings of the IEEE International Conference on Computer Vision, Venice, Italy, 22–29 October 2017; pp. 2794–2802.
30. Miyato, T.; Kataoka, T.; Koyama, M.; Yoshida, Y. Spectral normalization for generative adversarial networks. *arXiv* **2018**, arXiv:1802.05957.
31. Ganokratanaa, T.; Aramvith, S.; Sebe, N. Unsupervised anomaly detection and localization based on deep spatiotemporal translation network. *IEEE Access* **2020**, *8*, 50312–50329. [CrossRef]
32. Huang, G.; Liu, Z.; Van Der Maaten, L.; Weinberger, K.Q. Densely connected convolutional networks. In Proceedings of the IEEE Conference on Computer Vision and Pattern Recognition, Honolulu, HI, USA, 21–26 July 2017; pp. 4700–4708.
33. Heusel, M.; Ramsauer, H.; Unterthiner, T.; Nessler, B.; Hochreiter, S. Gans trained by a two time-scale update rule converge to a local nash equilibrium. *arXiv* **2017**, arXiv:1706.08500.
34. Szegedy, C.; Vanhoucke, V.; Ioffe, S.; Shlens, J.; Wojna, Z. Rethinking the inception architecture for computer vision. In Proceedings of the IEEE Conference on Computer Vision and Pattern Recognition, Las Vegas, NV, USA, 27–30 June 2016; pp. 2818–2826.

remote sensing

MDPI

Article

Fast and High-Quality 3-D Terahertz Super-Resolution Imaging Using Lightweight SR-CNN

Lei Fan, Yang Zeng, Qi Yang *, Hongqiang Wang and Bin Deng

College of Electronic Science, National University of Defense Technology, Changsha 410073, China; fanlei15@nudt.edu.cn (L.F.); zengyang@nudt.edu.cn (Y.Z.); wanghongqiang@nudt.edu.cn (H.W.); sagitdeng@nudt.edu.cn (B.D.)

* Correspondence: yangqi08@nudt.edu.cn; Tel.: +86-731-8457-5714

Abstract: High-quality three-dimensional (3-D) radar imaging is one of the challenging problems in radar imaging enhancement. The existing sparsity regularizations are limited to the heavy computational burden and time-consuming iteration operation. Compared with the conventional sparsity regularizations, the super-resolution (SR) imaging methods based on convolution neural network (CNN) can promote imaging time and achieve more accuracy. However, they are confined to 2-D space and model training under small dataset is not competently considered. To solve these problem, a fast and high-quality 3-D terahertz radar imaging method based on lightweight super-resolution CNN (SR-CNN) is proposed in this paper. First, an original 3-D radar echo model is presented and the expected SR model is derived by the given imaging geometry. Second, the SR imaging method based on lightweight SR-CNN is proposed to improve the image quality and speed up the imaging time. Furthermore, the resolution characteristics among spectrum estimation, sparsity regularization and SR-CNN are analyzed by the point spread function (PSF). Finally, electromagnetic computation simulations are carried out to validate the effectiveness of the proposed method in terms of image quality. The robustness against noise and the stability under small are demonstrate by ablation experiments.

Keywords: three-dimensional radar imaging; convolution neural network; super-resolution; side-lobe suppression; terahertz radar

Citation: Fan, L.; Zeng, Y.; Yang, Q.; Wang, H.; Deng, B. Fast and High-Quality 3-D Terahertz Super-Resolution Imaging Using Lightweight SR-CNN. *Remote Sens.* **2021**, *13*, 3800. https://doi.org/10.3390/rs13193800

Academic Editors: Fahimeh Farahnakian, Jukka Heikkonen and Pouya Jafarzadeh

Received: 27 August 2021
Accepted: 20 September 2021
Published: 22 September 2021

1. Introduction

Three-dimension (3-D) radar imaging can prominently reflect the 3-D spatial structure of the target with respect to conventional 2-D radar imaging and serve as a significant application such as geological hazard monitoring and forewarning [1], ecological applications [2], and military reconnaissance [3]. Typical 3-D radar imaging systems encompass the interferometric synthetic aperture radar (InSAR) [4], multiple-input multiple-output inverse SAR (MIMO ISAR) [5], and tomographic SAR [6]. According to the difference in elevation dimension imaging, 3-D radar imaging systems are mainly divided into two categories. First-class imaging systems utilize the interferometry technique and equivalent geometry to retrieve target height information [7]. The interferometry imaging handles phase differences from multiple SAR/ISAR images produced by multiple receivers of different views. However, this method is limited to distinguish scatterers located at the same Range-Doppler unit. Second-class imaging systems obtain the full 3-D radar echo data, which can form the synthetic aperture in azimuth and elevation dimension. Tomographic SAR is the representative of the second class [8], which develops azimuth aperture by flying a linear trajectory in a spotlight mode while the synthetic aperture in elevation dimension is formed by multiple closely spaced tracks. However, tomographic SAR is limited to multiple equivalent flights and cannot meet real-time requirements. Different from tomographic SAR, 3-D imaging based on different configurations of antenna arrays can be an efficient and fast substitute. The matter that 3-D radar imaging based on cross-array could optimize the beam width and enhance the image quality was demonstrated theoretically [9]. The premise for array radar imaging was based on high isolation array antennas. Nevertheless, the coupling problem of antennas array cannot be ignored in real-world radar imaging [10], which is straightforwardly related to the beamforming of MIMO signal. Mutual coupling reduction method for patch arrays was designed and achieved around 22.7-dB reduction [11]. A wideband linear array antenna based on the new reflector slot-strip-foam-inverted patch antennas was validated to improve the bandwidth gain effectively [12]. The development of these antenna decoupling techniques will further promote the practical application of MIMO 3-D radar imaging.

Antenna systems are strongly associated with radar transmitting and receiving signal. In recent years, this has been further developed and has boosted radar imaging techniques for both microwave and terahertz (THz) radar [13–18]. Compared with 3-D imaging using microwave radars, THz radars take advantages of a higher carrier frequency and wider absolute bandwidth, which can form higher range resolution and reach better azimuth resolution with smaller rotating angel conspicuously [19–25]. THz radar imaging will no longer be limited to some isolated points, and attain the high-resolution image with the obvious target outline. Accordingly, it is meaningful for studying high resolution 3-D imaging in the THz band.

Since the high side-lobe degrade the image quality in high-resolution radar imaging, especially 3-D Thz radar imaging, it is necessary to research the imaging method of enhancing radar image quality and suppressing the side-lobe. The traditional imaging methods based on spectrum estimation suffer from limited resolution and high side-lobes. Because the Fourier transform (FFT) of the window function would inevitably bring Sinc function with the high side-lobe. Sparsity regularizations have been proposed to solve high side-lobe and image quality by imposing sparsity constraints on imaging processing. Cetin et al. in [26] utilized L_0 regularization to improve 3-D radar image quality during signal reconstruction process. Austin et al. in [27] further improved L_0 regularization and applied an iterative shrinkage-thresholding algorithm to avoid falling into local optima. Wang et al. used the Basis Pursuit Denoising (BPDN) method [28] to achieve side-lobe suppression effectively. BPDN transformed the imaging process into an iterative optimization process, i.e., $\hat{x} = argmin\|y - Ax\|_2^2 + \varepsilon\|x\|_1$, where x and y denotes the reflectivity of imaging area and radar echo, respectively. A denotes corresponding imaging dictionary matrix. In essence, these methods avoid falling into ill-conditioned solution by adding sparse prior and attain the high-quality images. However, sparsity regularization depends on iterative optimization process, which are computationally intensive and time-consuming. This is because it is involved in solving the inverse of the matrix. In addition, the final image quality depends on different and accurate parameters setting for different targets. Compress sensing (CS) can obtain relatively high-quality image. However, this superiority is based on the sacrifice of enormous computation and storage cost [29], especially for the dictionary matrix A in 3-D cases. For example, considering the sizes of radar echo and imaging grids are $50 \times 50 \times 50$ and $100 \times 100 \times 100$, respectively, the total memory would be as large as 1.82T [30], which poses serious requirements for memory and storage. Although many improved techniques such as slice [31], patch [32], and vectorization [33] have been proposed to improve the efficiency of CS, it is suboptimal to enhance image quality.

With the rapid development of convolution neural network (CNN), CNN has demonstrated superior performance in many fields such as SAR target recognition [34], radar imaging enhancement [35], and time-frequency analysis [36]. Radar imaging enhancement based on CNN can overcome high side-lobe of spectrum estimation and time-consuming iteration of sparse regularization. Gao et al. [37] validated the feasibility of transforming complex data into dual-channel data for CNN and proposed a simple forward complex CNN to enhance 2-D radar image quality. Qin et al. [38] further improved loss function and integrated it into generative adversarial networks (GAN), which can boost the extraction of weak scattering centers caused by the minimum square error (MSE) function. In fact, the network architecture of GAN is complex, and it is difficult to reach convergence in the case of small datasets. Zhi et al. [39] applied CNN into a 2-D MIMO virtual array radar imaging enhancement, but the phase information of the output was not adequately considered. Overall, these methods are confined to 2-D space, and the problem of network training under small datasets has not been adequately studied in the field of 3-D imaging enhancement.

To solve time-consuming iterative operation and instability under small datasets, a fast and high-quality 3-D super-resolution (SR) imaging network, namely SR-CNN, is proposed. The network architecture is designed to be lightweight to meet the demand of small datasets. The proposed method is free from manual annotation datasets and the model trained by simulated data can be utilized in real data commendably. The main contributions of this paper are as follows.

(1) A fast and high-quality 3-D SR imaging method is proposed. Compared with the method based on sparsity regularization, the imaging time is reduced by two orders of magnitude and imaging quality is improved obviously.
(2) A lightweight CNN is designed, which reduces the model parameters and computation significantly. The training model can achieve satisfactory convergence under small datasets and the accuracy can reasonably improve.
(3) The input and output of SR-CNN both are complex data. The phenomenon that the performance of dividing complex data into real part and imaginary part is better than that of amplitude and phase is found.

This paper is organized as follows. In Section 2, the data generation of input and output for SR-CNN is derived. Then, the lightweight network structure and train details are described in detail. In Section 3, resolution characteristics of different methods are compared and electromagnetic simulation data are used to validate the effectiveness of the proposed method. The discussion about advantages of the proposed method and further work is presented in Section 4. Section 5 concludes this paper.

2. Methodology

In this section, the detailed processing of 3-D SR imaging method is given. The main structure of the proposed method consists of three main parts: input and output data generation of SR-CNN, lightweight network structure, and train details. These three parts are explained in detail below.

2.1. Input and Output Data Generation of SR-CNN

The 3-D radar imaging geometry of the general spotlight mode is shown in Figure 1. The imaging geometry could be equivalent to the circular SAR and turntable ISAR with a few modifications [40]. First, it is assumed that there is an ideal point target and a reference point target in the imaging scene. Then, the azimuth angle θ and the elevation angle $90-\varphi$ with the z-axis denote the radar illumination. R_t and R_{ref} denote the range of the ideal point target and reference point away from the radar, respectively. The echo of point target s_t can be described as

$$s_t(t,t_a) = A_t \cdot rect\left(\frac{t-2R_t/c}{T_p}\right) \cdot \exp\left(j2\pi\left[f_c(t-2R_t/c)+\frac{1}{2}\gamma(t-2R_t/c)^2\right]\right) \quad (1)$$

where A_t denotes the amplitude of target signal. T_p denotes the signal time window. c denote the speed of light. t and t_a denote the fast-time and the slow-time, respectively. f_c and γ denote the carrier frequency and the frequency modulation rate, respectively.

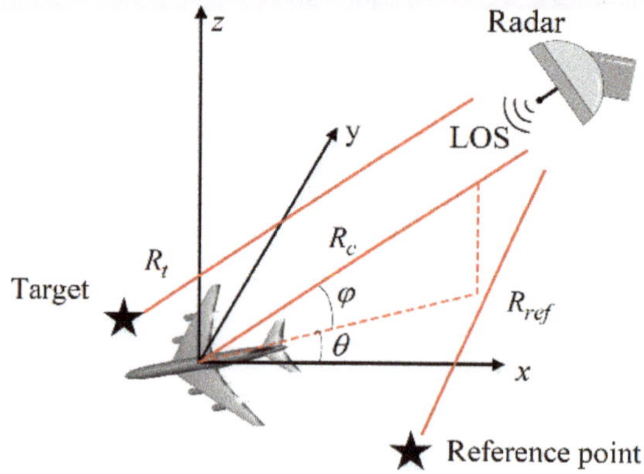

Figure 1. The imaging geometry.

The echo of the reference point is similar with (1), and it can be expressed as

$$s_{ref}(t,t_a) = A_r \cdot rect\left(\frac{t-2R_{ref}/c}{T_p}\right) \cdot \exp\left(j2\pi\left[f_c\left(t-2R_{ref}/c\right)+\frac{1}{2}\gamma(t-2R_{ref}/c)^2\right]\right) \quad (2)$$

For the convenience of derivation, we redefine the fast time as $t = t - 2R_{ref}/c$. The signal is received by de-chirp, and the expression of the de-chirp signal is

$$\begin{aligned} s(t,t_a) &= s_t(t,t_a)/s_{ref}(t,t_a) \\ &= A \cdot rect\left(\tfrac{t-2R_\Delta/c}{T_p}\right) \cdot \exp\left(-j\tfrac{4\pi}{c}\gamma\left(t-2R_{ref}/c\right)R_\Delta - j\tfrac{4\pi}{c}f_cR_\Delta + j\tfrac{4\pi\gamma}{c^2}R_\Delta^2\right) \end{aligned} \quad (3)$$

where $A = A_t/A_r$ and $R_\Delta = R_t - R_{ref}$. After the procedure of ramp phase and residual video-phase (RVP) correction, the de-chirp signal can be rewritten as

$$\begin{aligned} s(t,t_a) &= \mathrm{IFT}(\mathrm{FT}(s(t,t_a)) \cdot \exp(-j\pi f^2/\gamma)) \\ &= A \cdot rect\left(\tfrac{t}{T_p}\right) \cdot \exp(-j4\pi(f_c+\gamma t)R_\Delta/c) \end{aligned} \quad (4)$$

where FT and IFT denote Fourier transform and inverse Fourier transform, respectively. In (4), supposing the range alignment and phase correction have already been accomplished for moving target, R_Δ can be expressed with Taylor expansion under plane-wave approximation,

$$\begin{aligned} R_\Delta &\approx R_c - R_{ref} - x\cos\theta\cos\varphi - y\sin\theta\cos\varphi - z\sin\varphi \\ &\approx -x\cos\theta\cos\varphi - y\sin\theta\cos\varphi - z\sin\varphi \end{aligned} \quad (5)$$

where R_c denotes the range between the radar to the imaging center. To facilitate subsequent processing, the signal model is discretized. N, M, and L are the number of samples along frequency, azimuth, and elevation dimension, respectively. P, Q and K are the number of image grid points in range, azimuth, and elevation direction, respectively. Under the condition of far-field plane wave, the wave number along three coordinates axes can expressed as

$$\begin{cases} k_x(n,m,l) = \frac{4\pi f_n}{c}\cos\theta_m\cos\varphi_l \\ k_y(n,m,l) = \frac{4\pi f_n}{c}\sin\theta_m\cos\varphi_7 \\ k_z(n,m,I) = \frac{4\pi f_n}{c}\sin\varphi_l \end{cases} \quad (6)$$

where f_n, θ_m, and φ_l denote the discrete values of frequency, azimuth angle, and elevation angle, respectively. Based on the point spread function (PSF), the radar echo in wave number domain can be written as

$$\begin{aligned} y(k_x, k_y, k_z) &= \sum_{m=1}^{M} \sum_{n=1}^{N} \sum_{l=1}^{L} \sigma(x, y, z) \exp(-j4\pi f_n R_\Delta / c) \\ &= \sum_{m=1}^{M} \sum_{n=1}^{N} \sum_{l=1}^{L} \sigma(x, y, z) \exp(-j(k_x x + k_y y + k_z z)) \end{aligned} \tag{7}$$

where $\sigma(x, y, z)$ denotes the reflectivity of the point target.

Under the condition of small rotating angles, the 3-D imaging results can be obtained by applying 3-D IFT to radar echo $y(k_x, k_y, k_z)$ in the wave number domain:

$$I(\mathrm{p}, \mathrm{q}, \mathrm{k}) = \sum_{\mathrm{p}=1}^{P} \sum_{\mathrm{q}=1}^{Q} \sum_{\mathrm{k}=1}^{K} \left(k_x^2 + k_y^2 + k_z^2\right) \cdot \cos(\theta) \cdot y(k_x, k_y, k_z) \cdot e^{j(k_x x + k_y y + k_z z)} \tag{8}$$

where $I(\mathrm{p}, \mathrm{q}, \mathrm{k})$ denotes actually the input image of SR-CNN. According to nonparametric spectral analysis, the imaging resolutions of range (x direction), azimuth (y direction), and elevation (z direction) can be approximated as

$$R_x = \frac{c}{2B}, R_y = \frac{\lambda}{4\sin(\Delta\phi/2)}, R_z = \frac{\lambda}{4\sin(\Delta\theta/2)} \tag{9}$$

where B denotes the bandwidth. λ denotes the wavelength. $\Delta\phi$ and $\Delta\theta$ denote the rotating angles along azimuth and elevation dimension, respectively.

Based on the given imaging geometry and PSF, we extend the model in [37] into 3-D space and apply phase to output images. The expected SR output can be expressed as following:

$$O(\mathrm{p}, \mathrm{q}, \mathrm{k}) = \sum_{n=1}^{N} \sum_{m=1}^{M} \sum_{l=1}^{L} \sigma(x, y, z) \cdot \exp\left(-x^2/\sigma_x^2 - y^2/\sigma_y^2 - z^2/\sigma_z^2\right) \cdot \exp(-j(k_x x + k_y y + k_z z)) \tag{10}$$

where σ_x, σ_y, and σ_z control the width of PSF along three coordinates axes, respectively. $\exp(-j(k_x x + k_y y + k_z z))$ denotes the corresponding phase of each scattering center. According to -3 dB definition, the imaging resolution along these three dimensions for expected output images can be deduced as:

$$R_x' = 1.18\sigma_x, R_y' = 1.18\sigma_y, R_z' = 1.18\sigma_z \tag{11}$$

where R_x', R_y', R_z' denote the resolution of expected SR output along three coordinate axes, respectively.

2.2. Network Structure of SR-CNN

For lightweight CNN, computational cost and model depth are the two most important factors to be considered. The computational cost mainly concerns the number of network parameters and floating-point operations per second (FLOPs). The number of network parameters in traditional direction connection of convolution layers is enormous in the case of 3-D convolution. Due to the fact that the channel features in high dimension space are redundant, channel compression is an important way to reduce the number of network parameters. Inspired by [41], Figure 2 shows the direct connection of convolution layers and the designed convolution 'Fire' module. The mathematical formulas on the blue box and orange box represent the feature of corresponding size and convolution layers with specific kernel size, respectively. For example, $H \times W \times D \times C_1$ denotes the height, width, depth, and channel numbers of four-dimension tensor, respectively. Conv.S_1@$1 \times 1 \times 1$ denotes convolution layer of kernel size 1 and channel number S_1.

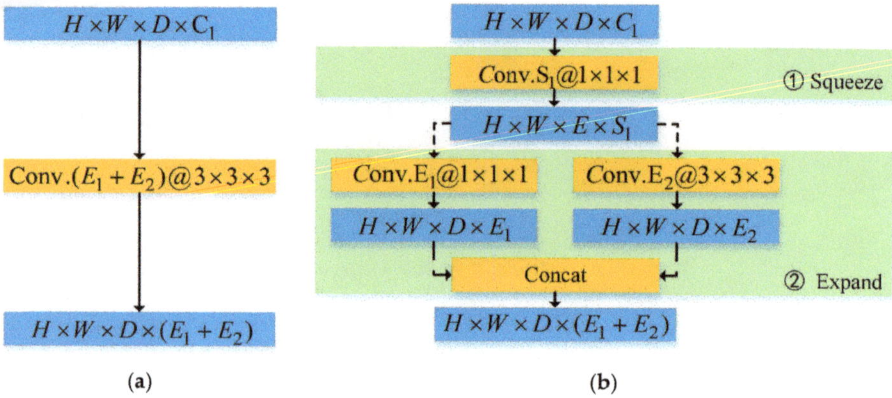

Figure 2. Schematic diagram of local network structure. (**a**) Direct connection of convolution layers (**b**) 'Fire' module.

Both connections achieve the same aim where the input feature of size $H \times W \times D \times C_1$ is transformed into the output feature of size $H \times W \times D \times (E_1 + E_2)$ by a series of convolution layers. For the traditional direct connection, the feature of size $H \times W \times D \times C_1$ is passed into one convolution layer with kernel size 3 to obtain the feature of size $H \times W \times D \times (E_1 + E_2)$. For the 'Fire' module, it contains two stages: the 'Squeeze' stage and the 'Expand' stage. For the 'Squeeze' stage, the feature of size $H \times W \times D \times C_1$ is passed into one convolution layer with kernel size 1 to obtain the feature of size $H \times W \times D \times S_1$. Thus, this feature is fed into two different convolution layers with kernel size 1 and 3 to obtain two feature of size $H \times W \times D \times E_1$ and $H \times W \times D \times E_2$, respectively. Finally, these two features are concatenated in channel dimension subsequently in the 'Expand' stage, which attain the final feature of size $H \times W \times D \times (E_1 + E_2)$.

Based on the experience of lightweight network design, the 'Fire' module needs to meet two conditions: (1) $S_1 = C_1/2$; and (2) $E_1 = E_2$. It is easy to calculate the number of parameters for the 'Fire' module are $3^3 \times E_2 \times S_1 + E_1 \times S_1 + C_1 \times S_1$, while that of the traditional direction connection is $3^3 \times C_1 \times (E_1 + E_2)$. It means that the number of local network parameters can reduce to about 1/4. Considering the feature size keeping unchanged, the Flops can also reduce to about $1/4 \approx \frac{(3^3 \times E_2 \times S_1 + E_1 \times S_1 + C_1 \times S_1) \times H \times W}{3^3 \times C_1 \times (E_1 + E_2) \times H \times W}$. These reason why the number of network parameter for the latter reduces to 1/4 is that the latter ingeniously utilizes the convolution layer with kernel size 1 to reduce parameters. In addition, the 'Fire' module can protract the depth and augment complexity of the network structure.

The whole network structure of SR-CNN is constructed as an end-to-end framework with supervised training. The specific structure adopts on the modified structure of full CNN [35], which can yield high performance with few training data set. The detailed network structure is shown in Figure 3. First, for the input and output of SR-CNN, we treat complex data as dual-channel data, which represents real and imaginary part, respectively, rather than amplitude and phase. It is because experiments have found that the latter is difficult to converge. We guess that the feature of amplitude and phase channel is huge and far from an image in conventional sense, which lead the convolution layer hardly to extract effective features. Then, the main difference between original full CNN and our modified structure is that the original direct connections of convolution layers are replaced by the 'Fire' module. In addition, the stride sizes of max pooling layers are 2 and 5 in turn, while the sizes of corresponding transpose convolution (Trans. conv) layers are reversed. Moreover, these features are concatenated in channel dimension by skip connection. The detailed size of each layer output is displayed on the top of the cubes. According to these sizes and conditions that the 'Fire' module needs to meet, it is easy to calculate the size of parameters S_1, E_1 and E_2.

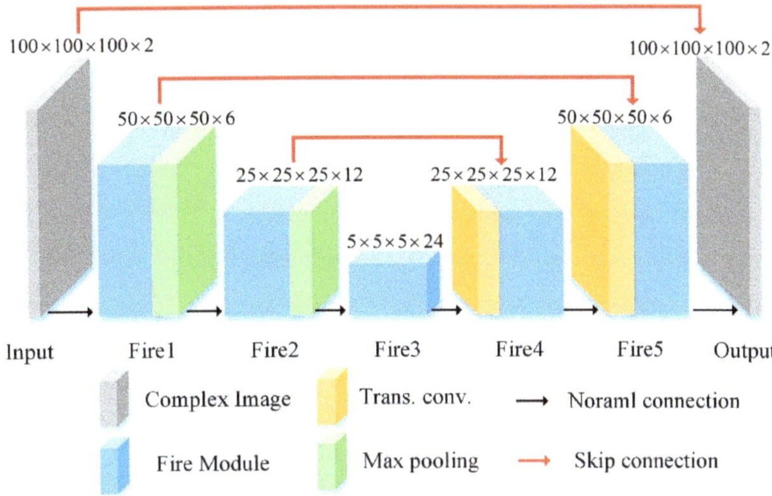

Figure 3. The whole network structure of SR-CNN.

The existing SR imaging methods based on CNN mainly consist of [37,38]. A simple forward SR imaging was designed in [37], but it did not consider the number of network parameters and multi-scale features for 3-D case. Hence, it was not optimal in terms of efficiency. Ref [38], based on GAN, argues that it is difficult to achieve 3-D SR imaging due to the limitation of small datasets. The proposed method combines full CNN and local network module 'Fire'. The 'Fire' module can reduce the network parameters significantly. Full CNN can improve the stability of the network training by multi-scale feature concatenation. A comparison about the proposed method [37] is shown in Section 3.5, which validates that the chosen architecture is best.

2.3. Simution and Training Details

The inputs and outputs of SR-CNN are generated, respectively, through (8) and (10). The imaging parameters are set as following: frequency ranges from 213.6 GHz to 226.4 GHz with 51 evenly sampling points, azimuth angle, and elevation angle both ranging from -1.68° to 1.68° with evenly 51 sampling points. It means that $N = M = L = 51$. As shown, $P = Q = K = 100$. According to (9), it is easy to calculate the resolutions in range, azimuth and elevation dimensions are 1.17 cm, 1.15 cm and 1.15 cm, respectively. For the expected SR output, σ_x, σ_y, and σ_z are set as 0.4 cm. According to (11), the resolutions in all three dimensions can be calculated as 0.47 cm, so the expected SR ratio is about 2.5 times in three directions.

Given the 3-D imaging space, the positions of hundreds of scattering centers are randomly generated according to the uniform distribution. Corresponding scattering intensities are also randomly generated and obey a complex Gaussian distribution, i.e., $N(0,1) + \mathrm{j}N(0,1)$. Although the distribution of scattering intensity varies in high frequency, the experimental performance for other distribution is not different. It is worth noting that these random operations are used to imitate the possible distribution of targets as realistically as possible. In order to intuitively understand the training and test data, we randomly select one group of input and output samples for interpretation. As shown in Figure 4, the 3-D image and 2-D image profiles of input data suffer from high side-lobe and low image quality relatively. Different from the input images, the output images own no side lobe and the ratio of the original scattering amplitude is maintained.

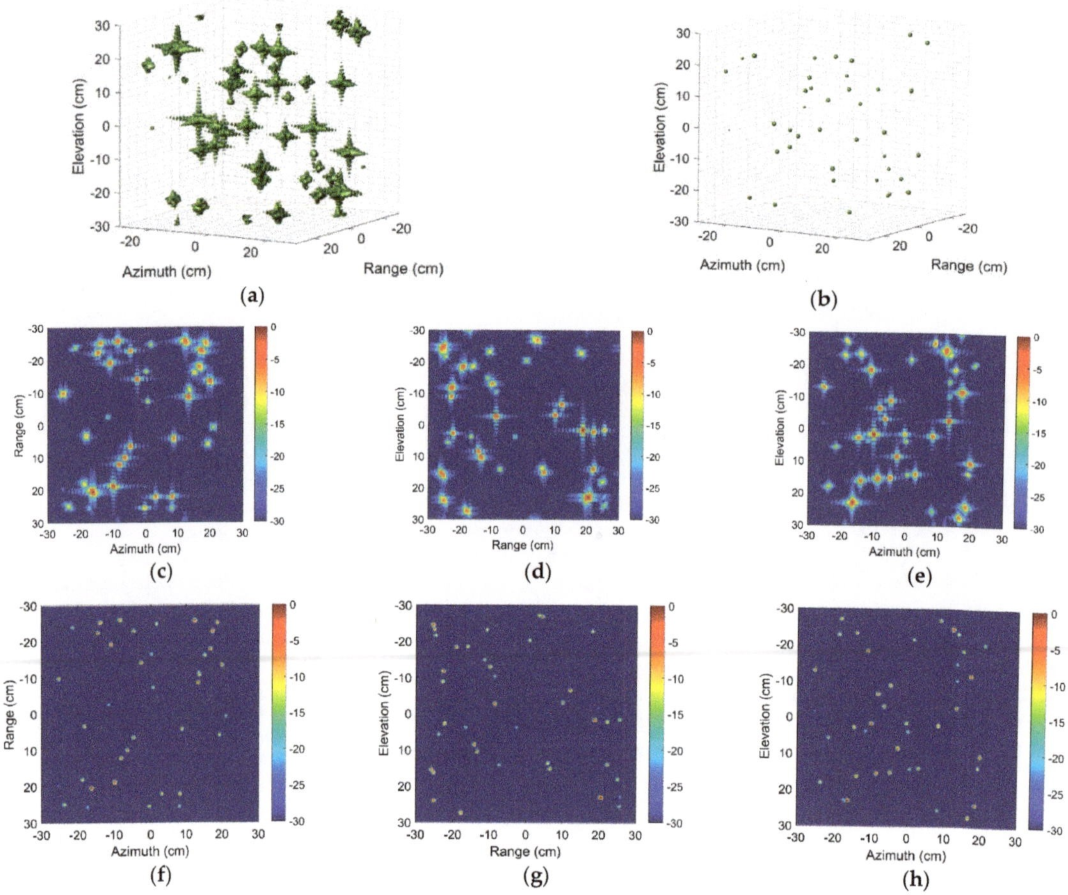

Figure 4. Three-dimensional images and two-dimensional image profiles of the input and output sample. Three-dimensional image of (**a**) Input and (**b**) Output sample. Two-dimensional image profiles of (**c**–**e**) Input and (**f**–**h**) Output sample.

For the regression problem based on supervised training, the MSE function can measure the difference between input and output. The loss function is shown in the following equation:

$$L = \frac{1}{N}\sum_{n=1}^{N}(P_n - O_n)^2 \quad (12)$$

where N denotes the total number of the train dataset and P_n denotes the predicted image under the input image I_n.

Based on the lightweight network structure, we do not need as much data as we used to need, which will be explained in Section 3.4. The total training samples are reduced to 500 and the division ratio of the training set and validation set is 9:1. The test dataset for Section 3.3 consists of an additional 100 samples. The batch size and the maximum training epochs are set to 4 and 30, respectively. Adam optimization is applied with a learning rate of 0.002.

3. Results

In this section, imaging resolutions of spectral estimation, sparsity regularization, and SR-CNN along three directions are analyzed, and 3-D imaging results of aircraft A380 are compared. Additionally, anti-noise ability and an ablation study of different network structures are provided to validate the effectiveness of the proposed method. Experiments are carried out with both MATLAB platform and Pytorch framework on a NVIDIA GeForce RTX 2080 Ti GPU card.

3.1. EXP1: Resolution Analysis of Different Methods

PSF is used to analyze resolution characteristics and side-lobe suppression. For convenience of explanation, the point target located at $(0, 0, 0)$ is selected as an example to intuitively analyze the SR performance. Figure 5a–d show the 3-D imaging results by 3D-IFFT without windowing (IFFT wo win), 3D-IFFT with windowing (IFFT w win), BPDN, and SR-CNN, in turn. These images are displayed in log magnitude and the dynamic range is 30 dB. First, Figure 5a belongs to spectrum estimation in essence. It can be found that the conventional imaging method by 3D-IFFT without windowing prompts high side-lobe compared with the ground-truth image in Figure 5e. The reasons why spectrum estimation suffers from high side-lobe is that the imaging resolution is limited by the Rayleigh criterion. These side-lobes will degrade the quality of the image. In addition, the side-lobes of strong scattering centers may tend to shelter from the weak scattering centers in the image. Then, traditional windowing is the simplest way to suppress the side-lobes, but window function will inevitably cause the expansion of the main lobe shown in Figure 5b. The window function chooses the typical Taylor window and the maximum of second side-lobe is -30 dB. From this figure, it can be found that, though the side-lobe disappears, the main-lobe will be widened obviously as expected.

Figure 5c,d show the imaging results of BPDN and SR-CNN. Both achieve image quality enhancement and a certain degree of SR compared with spectrum estimation, which are similar to the ground-truth. The detailed difference of BPDN and SR-CNN will be analyzed in details below.

To compare the difference between BPDN and SR-CNN intuitively, the contours images are chosen, which can reflect the side-lobe and fined contour structure. Figure 6 displays the azimuth–elevation contour images at range 0 m. The analysis of Figure 6a,b is consistent with the above and will not be repeated here. Comparing with Figure 6c,d, it can be found that BPDN achieves side-lobe suppression, but the contour is relatively more uneven than SR-CNN. The main reason may be that the optimization principle of BPDN is based on L1 regularization, where the stop conditions usually are met in the coordinate axes direction. In addition, since the expected output of SR-CNN owns smooth edges, the final prediction results of SR-CNN in supervised training do not exist in this problem. We also note that the prediction by SR-CNN is not completely consistent with the ground-truth, which means that SR ratio does not actually reach 2.5.

To further measure the performance of SR ratio directly, high resolution range profile (HRRP) is selected to compare -3 dB width. Figure 7 shows the HRRP at azimuth 0m and elevation 0 m. Observing the -3 dB line of the local amplification image in Figure 7b, the widths of main lobe are among these five range profiles are about 1.17 cm, 1.50 cm, 0.98 cm, 0.6 cm, and 0.47 cm, in turn. These numerical are consistent with the theoretical analysis. It can be calculated easily that adding window suffers from about 1.28 times main-lobe widening. BPDN reaches about 1.2 times SR while SR-CNN achieves 2 times. Furthermore, the EXP1 in Table 1 compares the time needs for four methods. Due to the speedup of GPU parallel and lightweight network, SR-CNN just needs less than 1 s while BPDN needs more than 130 s. This is because once the network model has already trained, the prediction process just depends on a simple forward convolution process rather than iterative optimization.

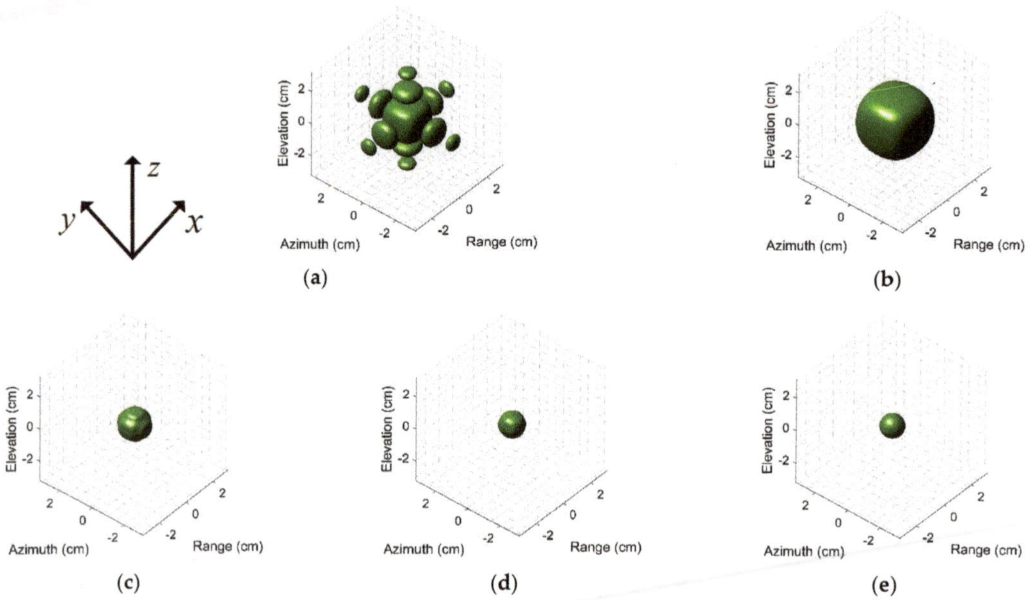

Figure 5. Three-dimensional images of point target $(0, 0, 0)$ by (**a**) 3D IFFT without windowing, (**b**) 3D IFFT with windowing, (**c**) BPDN, (**d**) SR-CNN, (**e**) Ground-truth.

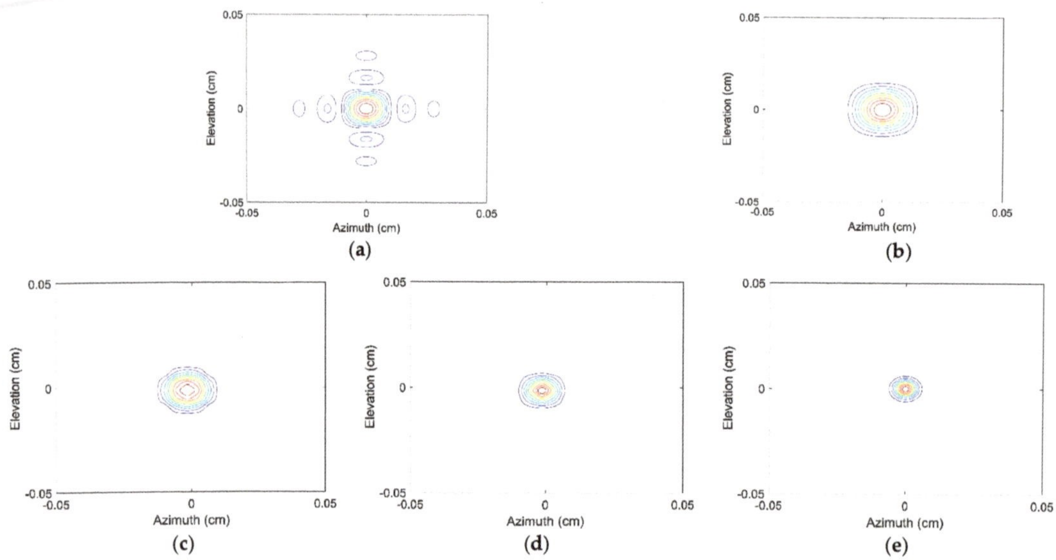

Figure 6. Azimuth-elevation images at range 0 m by (**a**) 3D IFFT without windowing, (**b**) 3D IFFT with windowing, (**c**) BPDN, (**d**) SR-CNN, (**e**) Ground-truth.

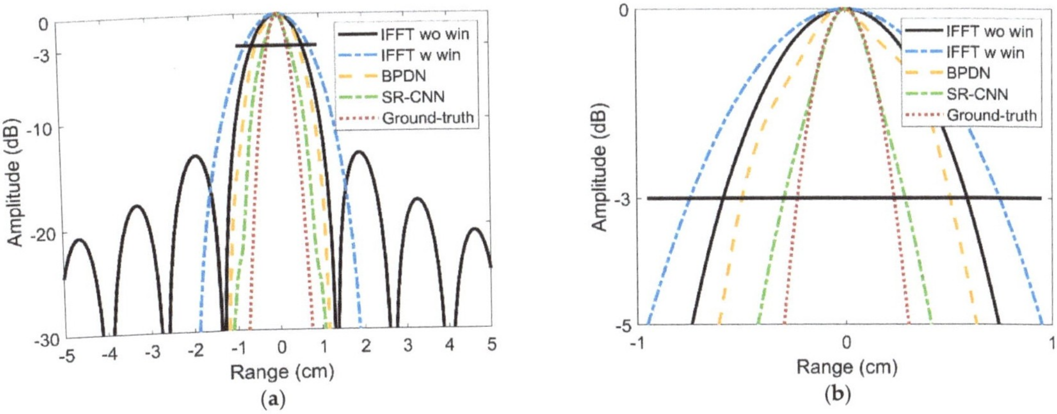

Figure 7. Range profiles of the target imaging result at azimuth 0 m and elevation 0 m. (**a**) Original results. (**b**) Local amplification.

Table 1. Comparison of time needs for different methods in two experiments.

Method	Time Needs (s)	
	EXP 1	**EXP 2**
IFFT wo win	0.064	0.084
IFFT w win	0.064	0.084
BPDN	130.144	227.142
SR-CNN	0.906	0.965

3.2. *EXP2: Electromagnetic Computation Simulation of Aircraft A380*

Electromagnetic computation simulation conducted by FEKO are used to further validate the performance of the SR-CNN on a real target. The simulation parameters of radar imaging are consistent with the above. Figure 8 shows the computer-aided design (CAD) model of aircraft A380. The material of A380 is set to the perfect electric conductor (PEC). The solver chooses large element physical optics based on full ray tracing.

Figure 8. CAD model of aircraft A380.

Figure 9 shows three-dimensional imaging results of the aircraft A380 by above four different methods. We can find that the imaging quality in Figure 9a degrades due to high side-lobe, especially the side-lobes of some strong scattering centers are even stronger than that of weak scattering centers. Figure 9b shows the imaging results of adding the

Taylor window. It is difficult to identify the target details from the image since the main lobe is widened obviously. Furthermore, some adjacent weak scattering centers may be submerged and image quality deteriorates accordingly. Both BPDN and SR-CNN enhance the resolution and suppress the side-lobes. However, by intuitively observing their visual quality, it can be found apparently that the image quality predicted by SR-CNN is superior to that of BPDN obviously. The outline of the aircraft can be clearly found in Figure 9d, which is conducive to the further refined recognition. For BPDN, there is still part of side-lobes around strong scattering centers due to L1 regularization, and it lost two scattering centers located wing edges of targets.

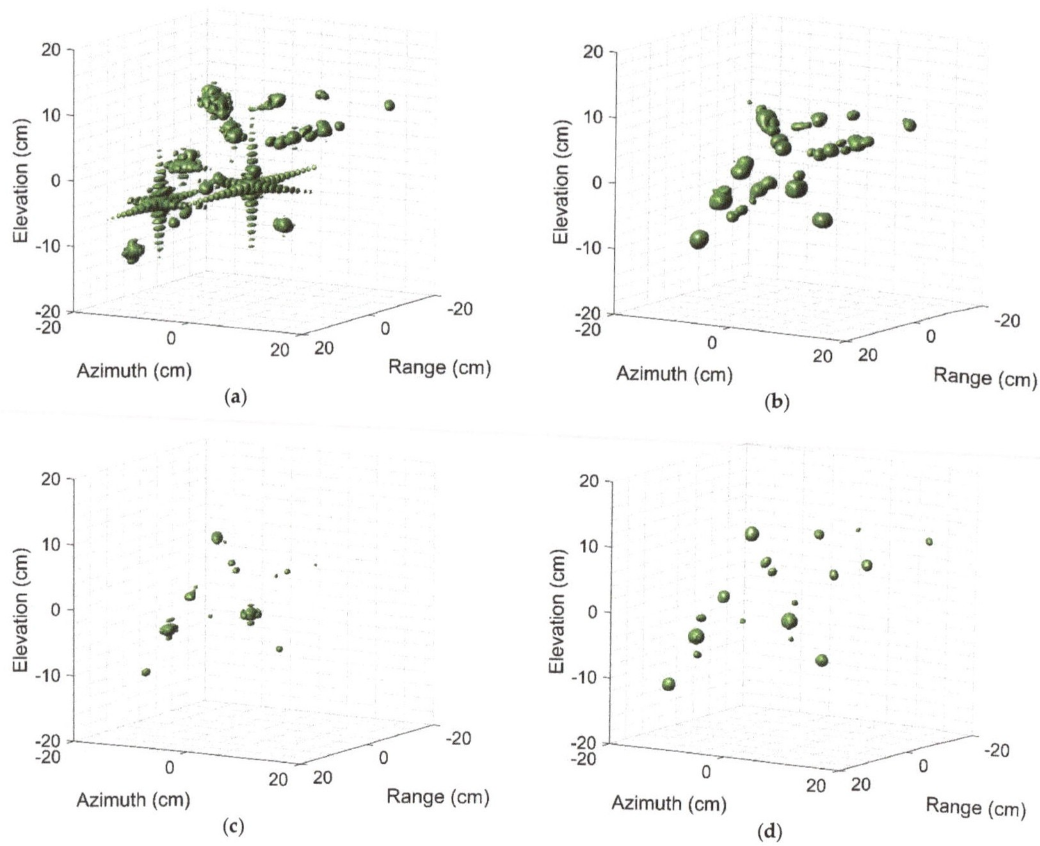

Figure 9. Three-dimensional images of aircraft A380 by (**a**) 3D IFFT without windowing, (**b**) 3D IFFT with windowing, (**c**) BPDN, (**d**) SR-CNN.

Figure 10 shows the range–azimuth profiles, range–elevation profiles, and azimuth–elevation profiles of above 3-D imaging results. These images are displayed in log magnitude and the dynamic range is 30 dB. It can be found that BPDN has a poorer ability on recovering weak scattering centers than SR-CNN. The reason is that the reconstruction of BPDN is based on the minimum of residual decomposition in the sense of orthogonal sense. The minimum loss is apt to fall into local optimum when most of the strong scattering centers are retained. Comparing the time needs of BPDN and SR-CNN in Table 1, SR-CNN can improve the imaging speed by about 230 times and the time need about prediction is stable.

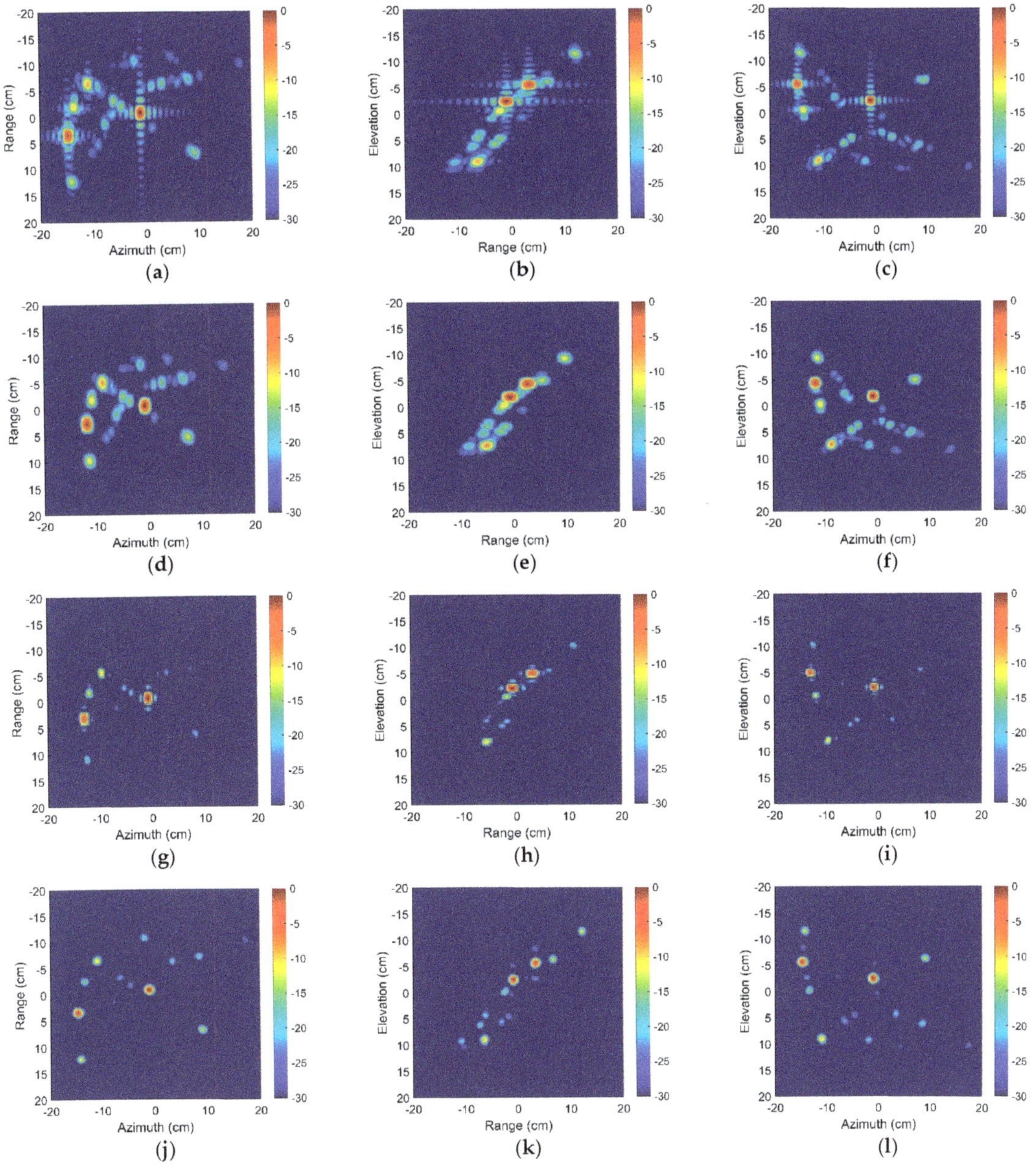

Figure 10. Two-dimension image profiles of aircraft A380 by (**a–c**) 3-D IFFT without windowing, (**d–f**) 3-D IFFT with windowing, (**g–i**) BPDN, (**j–l**) SR-CNN. First column is the range–azimuth profile. Second column is the range–elevation profile. The last column is the azimuth–elevation profile.

To understand the position of scattering centers accurately, Figure 11a shows 3-D imaging results of SR-CNN profiled on the CAD model. It can be found that the image after SR remarkably fits to the real structure of the target. Prudentially observing two-dimension profiles shows that these position mainly come from the discontinuities of the fuselage, the wings, the nose, the engines, etc. On the one hand, these strong scattering centers are caused by the specular reflection of the main components. On the other hand, the cavity represented by the engine is second main source. These facts are consistent with the reality. Since the ground truth is hard to define for real targets, qualitative comparisons can hardly be conducted and assessed. Nevertheless, recent results have shown the superiority of the proposed method in terms of image quality and imaging time.

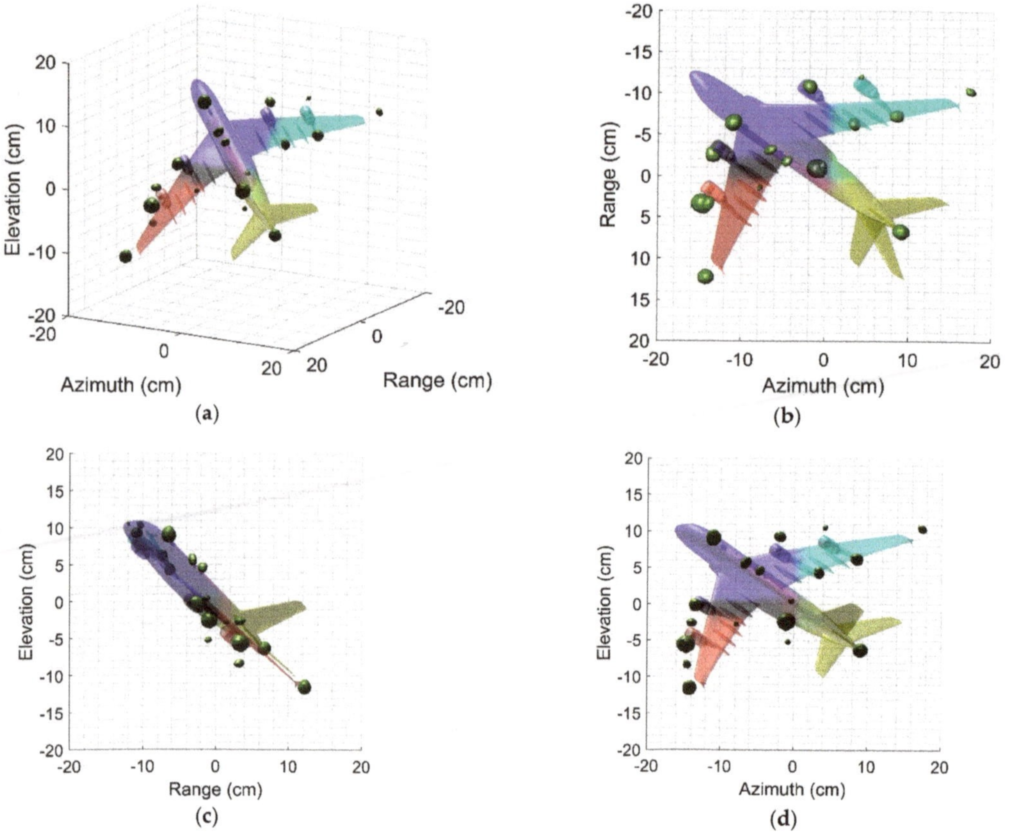

Figure 11. Three-dimension Images profiled on the CAD model. (**a**) Three-dimension image. (**b**) Range–azimuth profile. (**c**) Range–elevation profile. (**d**) Azimuth–elevation profile.

3.3. *Performance Analysis for Anti-Noise Ability and Imaging Time*

With the anti-noise ability and time needs for sparsity estimation, BPDN and SR-CNN are compared. The signal-noise-ratio (SNR) chooses −10, 0, and 10 dB. It is worth noting that 100 independent repeated tests are carried out for each SNR and each method. The root mean square error (RMSE) is used as a quantitative performance index to evaluate the accuracy. It is defined as follows:

$$RMSE = \sqrt{\frac{1}{N}\sum_{\mathrm{n}}\sum_{\mathrm{p}}\sum_{\mathrm{q}}\sum_{\mathrm{k}}(O_{\mathrm{n}}(\mathrm{p},\mathrm{q},\mathrm{k}) - P_{\mathrm{n}}(\mathrm{p},\mathrm{q},\mathrm{k}))^2} \quad (13)$$

As shown in Figure 12a, the RMSEs of SR-CNN are smallest among all methods in different SNRs. It is because spectrum estimation suffers from high-lobe or main-lobe widening and BPDN may exist weak side-lobe. We notice that RMSEs of the first method are less than that of second method. We speculate that there are two reasons mainly:

(1) The ground-truth produced by (10) encourages images to be sparse; and (2) main-lobe broadening by window function will inflate the image. Figure 12b shows the average time needs for different methods. SR-CNN is slightly larger than spectrum estimation and reduces about two orders of magnitude than sparsity-regularization BPDN. The superiority of the proposed method in terms of anti-noise ability and imaging time is further demonstrated.

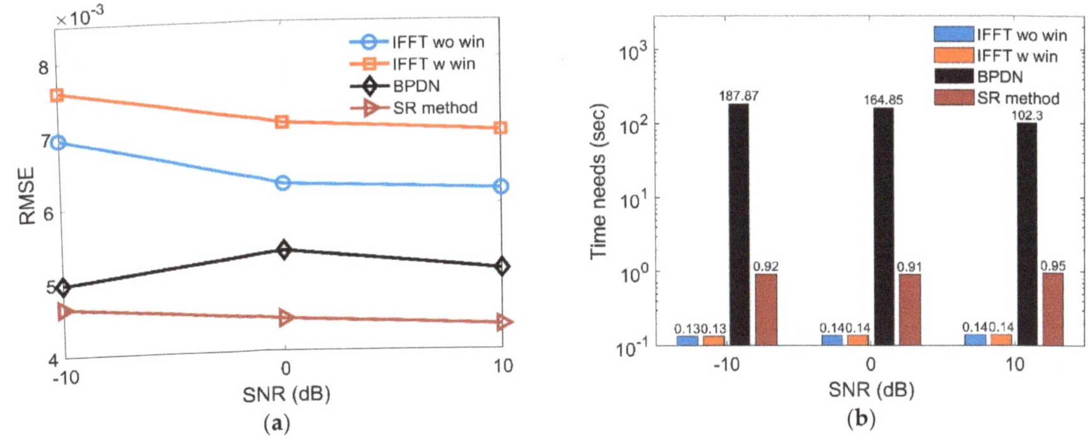

Figure 12. Comparison of (**a**) RMSE and (**b**) Average time needs among four different methods.

3.4. Ablation Experiments of Lightweight Network Structure

In order to further evaluate the effectiveness of the proposed lightweight network structure, the ablation experiments with different connection of convolution layers and dataset sizes are conducted. The details of the ablation experiments are listed in Table 2. Fire-500 represents that the structure of the proposed method utilizes the 'Fire' module and the size of dataset is 500. Direct connection represents the connection in (a).

Table 2. Structures of different networks.

Network	Connection		Dataset Size	
	Direct Connection	**Fire Module**	**500**	**2000**
Direct-500	√		√	
Fire-500		√	√	
Direct-2000	√			√
Fire-2000		√		√

Figure 13 presents the evolution of the RMSE of different network versus epochs. Comparing results of network with different connection, we can find that the networks based on the 'Fire' module can acquire higher accuracy than that of direct connection and achieve faster convergence accordingly. It is mainly because the 'Fire' module can reduce the number of network parameters and add the complexity of network. Then, we compare the performance of different dataset size for lightweight network. From the figure, we can find that, with larger datasets, the accuracy of the network is close to that of small datasets. It validates that the proposed lightweight network can reduce the required data for training while maintain high prediction performance.

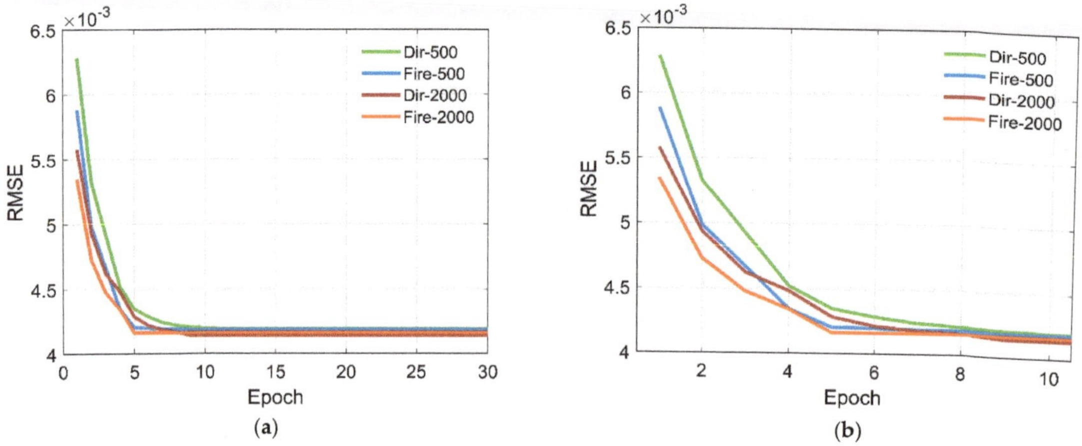

Figure 13. Evolution of the RMSE of different networks versus epoch. (**a**) Original image. (**b**) Local amplification.

3.5. Comparison with Methods Based on Neural Network

To further compare the performance of the proposed method with neural network-based methods, we select the existing method in [37], which is an efficient method to achieve SR. We convert the 2-D convolution layer to the 3-D convolution layer and use it as a comparison reference. The training parameters remain the same, and both optimal network models are selected to evaluate RMSE under SNR −10 dB, 0 dB, and 10 dB. The size of training dataset is 500. The RMSE of both methods are listed in Table 3. It can be found that the RMSE of SR-CNN is smaller than that of the method in [37]. In addition, RMSE of SR-CNN is relatively unstable under a small number of training samples. An important reason lies in the lightweight network structure. Therefore, the superiority of lightweight SR-CNN under small datasets is validated.

Table 3. Comparison with neural network-based method.

Method	RMSE × 1000 (SNR = −10 dB)	RMSE × 1000 (SNR = 0 dB)	RMSE × 1000 (SNR = 10 dB)
The method in [37]	5.46	4.81	4.48
SR-CNN	4.63	4.46	4.32

4. Discussion

A terahertz 3-D SR imaging method based on lightweight SR-CNN is proposed in this paper. First, the original 3-D radar echoes are derived based on the given imaging geometry, and corresponding expected SR images are designed using PSF. Then, training datasets are generated by randomly placing scattering centers within the given imaging region. Thus, considering the high computing demand of 3-D data and the limitation of small datasets, an effective lightweight network structure should be designed and improve the efficiency of supervised training. Using the compression of channels, we design the 'Fire' module to replace the traditional direct connection of convolution layers, which can significantly reduce the number of network parameters and FLOPs. Finally, combining the 'Fire' module with full CNN, a lightweight and efficient network structure SR-CNN is provided.

The advantages of the proposed method are as follows. (1) In terms of time needs, experimental results show that the time needs of SR-CNN can reduce two orders of magnitude compared with sparsity regularization. Because once the training of the model is completed, the prediction process of SR-CNN only consists of simple matrix addition and multiplication. However, for sparse regularization, the iteration process involves solving the inverse of the matrix, which increases the time needs drastically. (2) In terms of image quality, the proposed method achieves the best image quality compared with methods based-spectrum estimation and the methods-based sparse regularization. Since the expected output is sparse, the training final aim of the network is to become the output in the supervised training method. Therefore, the predicted result by SR-CNN is closest to the output. It needs to be pointed out that the setting of output is in line with real needs. In addition, the imaging sparsity by BPDN is dependent on effective parameter settings. (3) The proposed method has strong and stable anti-noise performance. This is because high-dimensional features extracted by SR-CNN are sparse. This sparsity is similar to the sparse sampling of CS; therefore, high-dimensional stable features of the target can be obtained accurately under different SNRs.

Future work can be considered in the following directions. (1) Considering that the current imaging parameters are known in advance, the imaging of moving targets with estimation of unknown motion parameters is an interesting direction. (2) The basis of a signal model is established with PSF, but the scattering characteristics of many structures do not satisfy PSF in reality. For example, the imaging results of a thin metal rod changes with the angle of observation. It is appealing to establish a theoretical prior model that is more in line with reality. (3) The input of the network is the complex image. Although the time needs are less than 1 s, it is worth studying whether it can directly learn from the original radar echo to reach imaging, which will observably accelerate the imaging speed in the field of 3-D radar imaging.

5. Conclusions

A fast and high-quality three-dimension SR imaging based on lightweight SR-CNN was proposed in this paper, which broke the limit of time consumption in the conventional sparsity-regularization method and outstood the SR imaging based on CNN. Based on the imaging geometry and PSF, the original 3-D echo and expected SR images were derived. By the designed lightweight network 'Fire' module and effective supervised training, the complete training framework of SR-CNN was provided in detail. In terms of resolution characteristic, the proposed method achieved at least two times SR in three dimensions compared with spectrum estimation. Additionally, the time of enhancing imaging can obtain two orders of improved magnitude compared with sparsity regularization BPDN. The effectiveness of the proposed method in terms of image quality was demonstrated by electromagnetic simulation, and the robustness against noise and the advantages of time need were verified as well. In the future, we will combine compressed sensing with neural networks, and design a fast and high-quality imaging method from the raw radar echoes promptly, which we have been already working on.

Author Contributions: Methodology, L.F. and Q.Y.; validation, L.F., Q.Y. and Y.Z.; formal analysis, L.F. and H.W.; writing—original draft preparation, L.F.; writing—review and editing, L.F. and B.D.; visualization, L.F. and Q.Y.; supervision, H.W. All authors have read and agreed to the published version of the manuscript.

Funding: This work was supported by the National Natural Science Foundation of China (No. 61871386 and No. 61971427).

Institutional Review Board Statement: No applicable.

Informed Consent Statement: Not applicable.

Data Availability Statement: No applicable.

Conflicts of Interest: The authors declare no conflict of interest.

References

1. Reigber, A.; Moreira, A. First Demonstration of Airborne SAR Tomography Using Multibaseline L-Band Data. *IEEE Trans. Geosci. Remote Sens.* **2000**, *5*, 2142–2152. [CrossRef]
2. Misezhnikov, G.S.; Shteinshleiger, V.B. SAR looks at planet Earth: On the project of a spacebased three-frequency band synthetic aperture radar (SAR) for exploring natural resources of the Earth and solving ecological problems. *IEEE Aerosp. Electr. Syst. Manag.* **1992**, *7*, 3–4. [CrossRef]
3. Pei, J.; Huang, Y.; Huo, W. SAR Automatic Target Recognition Based on Multiview Deep Learning Framework. *IEEE Trans. Geosci. Remote Sens.* **2018**, *56*, 2196–2210. [CrossRef]
4. Zhang, Y.; Yang, Q.; Deng, B.; Qin, Y.; Wang, H. Estimation of Translational Motion Parameters in Terahertz Interferometric Inverse Synthetic Aperture Radar (InISAR) Imaging Based on a Strong Scattering Centers Fusion Technique. *Remote Sens.* **2019**, *11*, 1221. [CrossRef]
5. Ma, C.; Yeo, T.S.; Tan, C.S.; Li, J.; Shang, Y. Three-Dimensional Imaging Using Colocated MIMO Radar and ISAR Technique. *IEEE Trans. Geosci. Remote Sens.* **2012**, *50*, 3189–3201. [CrossRef]
6. Zhu, X.X.; Bamler, R. Very High Resolution Spaceborne SAR Tomography in Urban Environment. *IEEE Trans. Geosci. Remote Sens.* **2010**, *48*, 4296–4308. [CrossRef]
7. Zhang, Y.; Yang, Q.; Deng, B.; Qin, Y.; Wang, H. Experimental Research on Interferometric Inverse Synthetic Aperture Radar Imaging with Multi-Channel Terahertz Radar System. *Sensors* **2019**, *19*, 2330. [CrossRef]
8. Zhou, S.; Li, Y.; Zhang, F.; Chen, L.; Bu, X. Automatic Regularization of TomoSAR Point Clouds for Buildings Using Neural Networks. *Sensors* **2019**, *19*, 3748. [CrossRef]
9. Zhang, S.; Dong, G.; Kuang, G. Superresolution Downward-Looking Linear Array Three-Dimensional SAR Imaging Based on Two-Dimensional Compressive Sensing. *IEEE J. Sel. Top. Appl. Earth Obs. Remote Sens.* **2016**, *9*, 2184–2196. [CrossRef]
10. Maleki, A.; Oskouei, H.D.; Mohammadi Shirkolaei, M. Miniaturized microstrip patch antenna with high inter-port isolation for full duplex communication system. *Int. J. RF Microw. Comput.-Aided Eng.* **2021**, *31*, e22760. [CrossRef]
11. Mohamadzade, B.; Lalbakhsh, A.; Simorangkir, R.B.V.B.; Rezaee, A.; Hashmi, R.M. Mutual Coupling Reduction in Microstrip Array Antenna by Employing Cut Side Patches and EBG Structures. *Prog. Electromagn. Res.* **2020**, *89*, 179–187. [CrossRef]
12. Mohammadi Shirkolaei, M. Wideband linear microstrip array antenna with high efficiency and low side lobe level. *Int. J. RF Microw. Comput.-Aided Eng.* **2020**, *30*. [CrossRef]
13. Afzal, M.U.; Lalbakhsh, A.; Esselle, K.P. Electromagnetic-wave beam-scanning antenna using near-field rotatable graded-dielectric plates. *J. Appl. Phys.* **2018**, *124*, 234901. [CrossRef]
14. Alibakhshikenari, M.; Virdee, B.S.; Limiti, E. Wideband planar array antenna based on SCRLH-TL for airborne synthetic aperture radar application. *J. Electromagn. Wave* **2018**, *32*, 1586–1599. [CrossRef]
15. Lalbakhsh, A.; Afzal, M.U.; Esselle, K.P.; Smith, S.L.; Zeb, B.A. Single-Dielectric Wideband Partially Reflecting Surface With Variable Reflection Components for Realization of a Compact High-Gain Resonant Cavity Antenna. *IEEE Trans. Antennas Propag.* **2019**, *67*, 1916–1921. [CrossRef]
16. Lalbakhsh, A.; Afzal, M.U.; Esselle, K.P.; Smith, S.L. A high-gain wideband ebg resonator antenna for 60 GHz unlicenced frequency band. In Proceedings of the 12th European Conference on Antennas and Propagation (EuCAP 2018), London, UK, 9–13 April 2018; pp. 1–3.
17. Alibakhshi-Kenari, M.; Naser-Moghadasi, M.; Ali Sadeghzadeh, R.; Singh Virdee, B. Metamaterial-based antennas for integration in UWB transceivers and portable microwave handsets. *Int. J. RF Microw. Comput.-Aided Eng.* **2016**, *26*, 88–96. [CrossRef]
18. Mohammadi, M.; Kashani, F.H.; Ghalibafan, J. A partially ferrite-filled rectangular waveguide with CRLH response and its application to a magnetically scannable antenna. *J. Magn. Magn. Mater.* **2019**, *491*, 165551. [CrossRef]
19. Yang, Q.; Deng, B.; Wang, H.; Qin, Y. A Doppler aliasing free micro-motion parameter estimation method in the terahertz band. *J. Wirel. Com. Netw.* **2017**, *2017*, 61. [CrossRef]
20. Li, H.; Li, C.; Wu, S.; Zheng, S.; Fang, G. Adaptive 3D Imaging for Moving Targets Based on a SIMO InISAR Imaging System in 0.2 THz Band. *Remote Sens.* **2021**, *13*, 782. [CrossRef]
21. Yang, Q.; Deng, B.; Zhang, Y.; Qin, Y.; Wang, H. Parameter estimation and imaging of rough surface rotating targets in the terahertz band. *J. Appl. Remote Sens.* **2017**, *11*, 045001. [CrossRef]
22. Liu, L.; Weng, C.; Li, S. Passive Remote Sensing of Ice Cloud Properties at Terahertz Wavelengths Based on Genetic Algorithm. *Remote Sens.* **2021**, *13*, 735. [CrossRef]
23. Li, Y.; Hu, W.; Chen, S. Spatial Resolution Matching of Microwave Radiometer Data with Convolutional Neural Network. *Remote Sens.* **2019**, *11*, 2432. [CrossRef]
24. Fan, L.; Yang, Q.; Zeng, Y.; Deng, B.; Wang, H. Multi-View HRRP Recognition Based on Denoising Features Enhancement. In Proceedings of the Global Symposium on Millimeter-Waves and Terahertz, Nanjing, China, 23–26 May 2021. [CrossRef]
25. Gao, J.; Cui, Z.; Cheng, B. Fast Three-Dimensional Image Reconstruction of a Standoff Screening System in the Terahertz Regime. *IEEE Trans. THz Sci. Technol.* **2018**, *8*, 38–51. [CrossRef]
26. Cetin, M.; Stojanovic, I.; Onhon, O. Sparsity-Driven Synthetic Aperture Radar Imaging: Reconstruction, autofocusing, moving targets, and compressed sensing. *IEEE Signal Process. Manag.* **2014**, *31*, 27–40. [CrossRef]
27. Austin, C.D.; Ertin, E.; Moses, R.L. Sparse Signal Methods for 3-D Radar Imaging. *IEEE J. Sel. Top. Signal Process.* **2011**, *5*, 408–423. [CrossRef]

28. Lu, W.; Vaswani, N. Regularized Modified BPDN for Noisy Sparse Reconstruction With Partial Erroneous Support and Signal Value Knowledge. *IEEE Trans. Signal Process.* **2011**, *60*, 182–196. [CrossRef]
29. Wang, M.; Wei, S.; Shi, J. CSR-Net: A Novel Complex-Valued Network for Fast and Precise 3-D Microwave Sparse Reconstruction. *IEEE J. Sel. Top. Appl. Earth Obser. Remote Sens.* **2020**, *13*, 4476–4492. [CrossRef]
30. Yang, D.; Ni, W.; Du, L.; Liu, H.; Wang, J. Efficient Attributed Scatter Center Extraction Based on Image-Domain Sparse Representation. *IEEE Trans. Signal Process.* **2020**, *68*, 4368–4381. [CrossRef]
31. Zhao, J.; Zhang, M.; Wang, X.; Cai, Z.; Nie, D. Three-dimensional super resolution ISAR imaging based on 2D unitary ESPRIT scattering centre extraction technique. *IET Radar Sonar Navig.* **2017**, *11*, 98–106. [CrossRef]
32. Wang, L.; Li, L.; Ding, J.; Cui, T.J. A Fast Patches-Based Imaging Algorithm for 3-D Multistatic Imaging. *IEEE Geosci. Remote Sens. Lett.* **2017**, *14*, 941–945. [CrossRef]
33. Yao, L.; Qin, C.; Chen, Q.; Wu, H. Automatic Road Marking Extraction and Vectorization from Vehicle-Borne Laser Scanning Data. *Remote Sens.* **2021**, *13*, 2612. [CrossRef]
34. Yu, J.; Zhou, G.; Zhou, S.; Yin, J. A Lightweight Fully Convolutional Neural Network for SAR Automatic Target Recognition. *Remote Sens.* **2021**, *13*, 3029. [CrossRef]
35. Hu, C.; Wang, L.; Li, Z.; Zhu, D. Inverse Synthetic Aperture Radar Imaging Using a Fully Convolutional Neural Network. *IEEE Geosci. Remote Sens. Lett.* **2020**, *17*, 1203–1207. [CrossRef]
36. Qian, J.; Huang, S.; Wang, L.; Bi, G.; Yang, X. Super-Resolution ISAR Imaging for Maneuvering Target Based on Deep-Learning-Assisted Time-Frequency Analysis. *IEEE Trans. Geosci. Remote Sens.* **2021**, 1–14. [CrossRef]
37. Gao, J.; Deng, B.; Qin, Y.; Wang, H.; Li, X. Enhanced Radar Imaging Using a Complex-Valued Convolutional Neural Network. *IEEE Geosci. Remote Sens. Lett.* **2019**, *16*, 35–39. [CrossRef]
38. Qin, D.; Gao, X. Enhancing ISAR Resolution by a Generative Adversarial Network. *IEEE Geosci. Remote Sens. Lett.* **2021**, *18*, 127–131. [CrossRef]
39. Zhao, D.; Jin, T.; Dai, Y.; Song, Y.; Su, X. A Three-Dimensional Enhanced Imaging Method on Human Body for Ultra-Wideband Multiple-Input Multiple-Output Radar. *Electronics* **2018**, *7*, 101. [CrossRef]
40. Qiu, W.; Zhou, J.; Fu, Q. Tensor Representation for Three-Dimensional Radar Target Imaging With Sparsely Sampled Data. *IEEE Trans. Comput. Imaging* **2020**, *6*, 263–275. [CrossRef]
41. Zhang, J.; Zhu, H.; Wang, P.; Ling, X. ATT Squeeze U-Net: A Lightweight Network for Forest Fire Detection and Recognition. *IEEE Access* **2021**, *9*, 10858–10870. [CrossRef]

remote sensing

MDPI

Article

Predicting Arbitrary-Oriented Objects as Points in Remote Sensing Images

Jian Wang, Le Yang and Fan Li *

School of Information and Communications Engineering, Xi'an Jiaotong University, Xi'an 710049, China; wj851329121@stu.xjtu.edu.cn (J.W.); yangle15@xjtu.edu.cn (L.Y.)
* Correspondence: lifan@mail.xjtu.edu.cn

Abstract: To detect rotated objects in remote sensing images, researchers have proposed a series of arbitrary-oriented object detection methods, which place multiple anchors with different angles, scales, and aspect ratios on the images. However, a major difference between remote sensing images and natural images is the small probability of overlap between objects in the same category, so the anchor-based design can introduce much redundancy during the detection process. In this paper, we convert the detection problem to a center point prediction problem, where the pre-defined anchors can be discarded. By directly predicting the center point, orientation, and corresponding height and width of the object, our methods can simplify the design of the model and reduce the computations related to anchors. In order to further fuse the multi-level features and get accurate object centers, a deformable feature pyramid network is proposed, to detect objects under complex backgrounds and various orientations of rotated objects. Experiments and analysis on two remote sensing datasets, DOTA and HRSC2016, demonstrate the effectiveness of our approach. Our best model, equipped with Deformable-FPN, achieved 74.75% mAP on DOTA and 96.59% on HRSC2016 with a single-stage model, single-scale training, and testing. By detecting arbitrarily oriented objects from their centers, the proposed model performs competitively against oriented anchor-based methods.

Keywords: object detection; remote sensing image; anchor free; oriented bounding boxes; deformable convolution

Citation: Wang, J.; Yang, L.; Li, F. Predicting Arbitrary-Oriented Objects as Points in Remote Sensing Images. *Remote Sens.* **2021**, *13*, 3731. https://doi.org/10.3390/rs13183731

Academic Editors: Jukka Heikkonen, Fahimeh Farahnakian and Pouya Jafarzadeh

Received: 10 August 2021
Accepted: 15 September 2021
Published: 17 September 2021

1. Introduction

With the development of modern remote sensing technology, a large number of remote sensing images with higher spatial resolution and richer content have been produced [1–4]. Object detection in remote sensing images has broad application prospects in many fields, such as environmental monitoring [5–7], disaster control [8,9], infrared detection [10,11], and the military. Benefiting from deep convolutional neural networks, considerable results have been achieved for the object detection task in natural images. However, due to the complex background, variable object scales, arbitrary orientations and shooting angles, object detection in aerial images is still a hot topic in the field of computer vision [12–16].

Compared with natural image datasets [17,18], remote sensing image detection mainly faces the following differences and challenges (Illustrated in Figure 1):

1. Low overlap and Densely arranged. Remote sensing images are usually captured by satellite, radar, and so on, from a vertical view. Unlike object detection for natural images, where overlap between objects is typically present, the rotated objects in remote sensing images have a low probability of overlapping each other, especially for objects in the same category. Furthermore, objects usually appear in densely arranged forms in some categories, such as ships and vehicles, which leads to difficulties for the detector to distinguish between adjacent objects;
2. Arbitrary orientations. Objects usually appear in the image with various directions. Compared to the widely used horizontal bounding boxes (HBBs) in natural image detection, oriented bounding boxes (OBBs) can better depict objects with arbitrary

orientations and aspect ratios than horizontal bounding boxes in remote sensing images. This not only requires the detector to correctly locate and classify the object of interest, but also to accurately predict its direction;

3. Complex background and Drastic scale changes. Compared to natural images, remote sensing images have higher resolution, with more complex and variable backgrounds. A lot of objects to be detected are easily submerged in the background, which requires the detector to be effectively focused on areas of interest. Meanwhile, the scales of objects vary drastically in remote sensing images; for example, some vehicles and bridges are only within a few pixels, while soccer fields can comprise thousands of pixels in aerial images.

Figure 1. Examples of Low overlap and Densely arranged (**Left**), Arbitrary orientations of objects (**Middle**), and Drastic scale changes (**Right**) in remote sensing images.

The above difficulties make remote sensing image detection more challenging and attractive, while requiring natural image object detection methods to be adapted to rotated objects. However, most rotated object detectors place multiple anchors per location to get a higher IoU between pre-set anchors and object bounding boxes. Dense anchors ensure the performance of the rotation detectors while having a higher computational burden. Can these anchors be discarded in the rotated object detection process, in order to improve the computational efficiency and simplify the design of the model? We find that one major difference between remote sensing images and natural images is the small probability of overlap between objects having the same category. So, the large overlap between adjacent objects per location is rare in this situation, especially when using oriented bounding boxes to represent the rotated objects. Therefore, we hope the network could directly predict the classification and regression information of the rotated object from the corresponding position, such as an object center, which can improve the overall efficiency of the detector and avoid the need for manual designs of the anchors. Meanwhile, the networks need to have robust feature extraction capabilities for objects with drastic scale changes and accurately predict the orientation of rotated objects.

To discard anchors in the detection process, we convert the rotation object detection problem into a center point prediction problem. First, we represent an oriented object by the center of its oriented bounding box. The network learns a center probability map to localize the object's center through use of a modulated focal loss. Then, inspired by [19], we use the circular smooth label to learn the object's direction, in order to accurately predict the angle of an object and avoid regression errors due to angular periodicity at the boundary. A parallel bounding-box height and width prediction branch is used to predict the object's size in a multi-task learning manner. Therefore, we can detect the oriented objects in an anchor-free way.

Further, to accurately localize the object center under drastic scale changes and various object orientations, a deformable feature pyramid network (Deformable-FPN) is proposed, in order to further fuse the multi-level features. Specifically, deformable convolution [20,21] is used to reduce the feature channels and project the features simultaneously. After mixing the adjacent-level features using an add operation, we perform another deformable convolution to reduce the aliasing effect of the add operation. By constructing the FPN in a deformable manner, the convolution kernel can be adaptively adjusted, according to the scale and direction of the object. Experiments show that our Deformable-FPN can bring significant improvements to detecting objects in remote sensing images, compared to FPN.

In summary, the main contributions of this paper are as follows:

1. We analyze that one major difference between remote sensing images and natural images is the small probability of overlap between objects with the same category and, based on the analysis, propose a center point-based arbitrary-oriented object detector without pre-set anchors;
2. We design a deformable feature pyramid network to fuse the multi-level features for rotated objects, which can get a better feature representation for accurately localizing the object center;
3. We carry out experiments on two remote sensing benchmarks—the DOTA and HRSC2016 datasets—to demonstrate the effectiveness of our approach. Specifically, our center point-based arbitrary-oriented object detector achieves 74.75% mAP on DOTA and 96.59% on HRSC2016 with a single-stage model, single-scale training, and testing.

The remainder of this paper is organized as follows. Section 2 first describes the related works. Section 3 provides a detailed description of the proposed method, including center-point based arbitrary-oriented object detector and Deformable-FPN. The experiment results and settings are provided in Section 4 and discussed in Section 5. Finally, Section 6 summarizes this paper and presents our conclusions.

2. Related Work

2.1. Object Detection in Natural Images

In recent years, horizontal object detection algorithms in natural image datasets, such as MSCOCO [17] and PASCAL VOC [18], have achieved promising progress. We classify them as follows:

Anchor-based Horizontal Object Detectors: Most region-based two-stage methods [22–26] first generate category-agnostic region proposals from the original image, then use category-specific classifiers and regressors to classify and localize the objects from the proposals. Considering their efficiency, single-stage detectors have drawn more and more attention from researchers. Single-stage methods perform bounding box (bbox) regression and classification simultaneously, such as SSD [27], YOLO [28–30], RetinaNet [31], and so on [32–35]. The above methods densely place a series of prior boxes (Anchors) with different scales and aspect ratios on the image. Multiple anchors per location are needed to cover the objects as much as possible, and classification and location refinement are performed based on these pre-set anchors.

Anchor-free Horizontal Object Detectors: Researchers have also designed some comparable detectors without complex pre-set anchors, which are inspiring to the detection process. CornerNet [36] detects an object bounding box as a pair of keypoints, demonstrating the effectiveness of anchor-free object detection. Further, CenterNet [37] models an object as a single point, then regresses the bbox parameters from this point. Based on RetinaNet [31], FCOS [38] abandoned the pre-set anchors and directly predicts the distance from a reference point to four bbox boundaries. All of these methods have achieved great performance and have avoided the use of hyper-parameters related to anchor boxes, as well as complicated calculations such as intersection over union (IoU) between bboxes during training.

2.2. Object Detection in Remote Sensing Images

Object detection also has a wide range of applications in remote sensing images. Reggiannini et al. [5] designed a sea surveillance system to detect and identify illegal maritime traffic. Almulihi et al. [7] propose a statistical framework based on gamma distributions and demonstrate the effectiveness for oil spill detection in SAR images.

Zhang et al. [8] analyze the frequency properties of motions to detect living people in disaster areas. In [10], a difference maximum loss function is used to guide the learning directions of the networks for infrared and visible image object detection.

Based on the fact that rotation detectors are needed for remote sensing images, many excellent rotated object detectors [19,39–46] have been developed from horizontal detection methods. RRPN [39] sets rotating anchors to obtain better region proposals. R-DFPN [47] propose a rotation dense feature pyramid network to solve the narrow width problems of the ship, which can effectively detect ships in different scenes. Yang et al. [19] converted an angle regression problem to a classification problem and handled the periodicity of the angle by using circular smooth label (CSL). Due to the complex background, drastic scale changes, and various object orientations problems, multi-stage rotation detectors [41–43] have been widely used.

3. Method

In this section, we first introduce the overall architecture of our proposed center point-based arbitrary-oriented object detector. Then, we detail how to localize the object's center and predict the corresponding angle and size. Finally, the detailed structure of Deformable-FPN is introduced.

3.1. Overall Architecture

The overall architecture of our methods, based on [37], is illustrated in Figure 2. ResNet [48] is used as our backbone, in order to extract multi-level feature maps (denoted as C_3, C_4, C_5). Then, these features are sent to deformable feature pyramid networks to obtain a high-resolution, strong semantic feature map, P_2, which is responsible for the following detection task. Finally, four parallel sub-networks are used to predict the relevant parameters of the oriented bounding boxes. Specifically, the Center Heatmap branch is used to predict the center probability, for localizing the object's center. A refined position of the center is obtained from the Center offset branch. The Orientation branch is responsible for predicting the object's direction by using the Circular Smooth Label, and the corresponding height and width are obtained from the Object size branch.

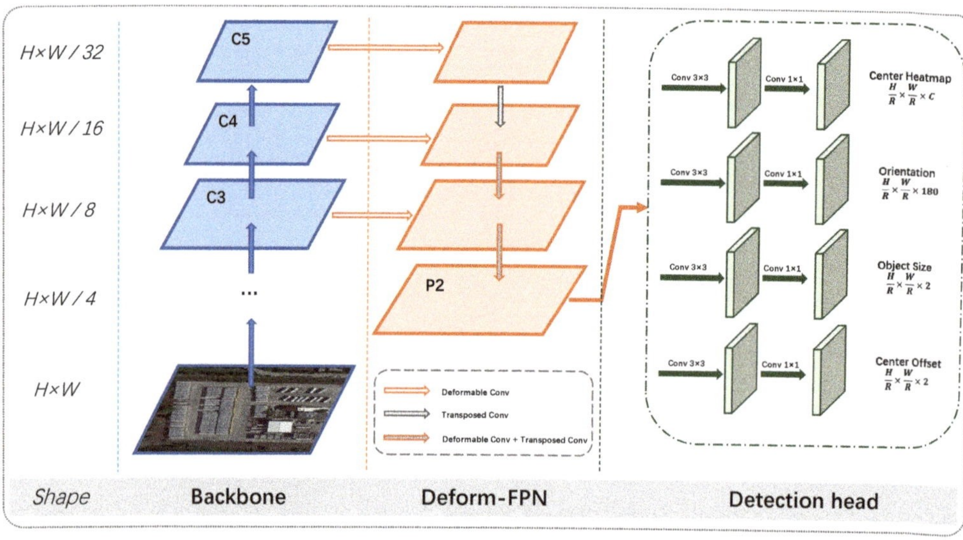

Figure 2. Overall architecture of our proposed center-point based arbitrary-oriented object detector.

3.2. Detecting Arbitrary-Oriented Object by Its Center Point

3.2.1. Center Point Localization

Let W and H be the width and height of the input image. We aim to let the network predict a category-specific center point heatmap $\hat{Y} \in [0,1]^{\frac{W}{R} \times \frac{H}{R} \times C}$, based on the features extracted from the backbone, where R is the stride between the input and feature P_2 (as shown in Figure 2), and C is the number of object categories ($C = 15$ in DOTA, 1 in HRSC2016). R was set to four, following [37]. The predicted value $\hat{Y} = 1$ denotes a detected center point of the object, while $\hat{Y} = 0$ denotes background.

We followed [36,37] to train the center prediction networks. Specifically, for each object's center (p_x, p_y) of class c, a ground-truth positive location $(\tilde{p}_x, \tilde{p}_y) = (\lfloor \frac{p_x}{R} \rfloor, \lfloor \frac{p_y}{R} \rfloor)$ is responsible for predicting it, and all other locations are negative. During training, equally penalizing negative locations can severely degrade the performance of the network; this is because, if a negative location is close to the corresponding ground-truth positive location, it can still represent the center of the object within a certain error range. Thus, simply dividing it as a negative sample will increase the difficulty of learning object centers. So, we alleviated the penalty for negative locations within a radius of the positive location. This radius, r, is determined by the object size in an adaptive manner: a pair of diagonal points within the radius can generate a bounding box exceeding a certain Intersection over Union (IoU) with the ground-truth box; the IoU threshold is set to 0.5 in this work. Finally, the ground-truth heatmap $Y \in [0,1]^{\frac{W}{R} \times \frac{H}{R} \times C}$ used to reduce the penalty is generated as follows: We split all ground truth center points into Y and pass them through the Gaussian kernel K_{xyc}:

$$K_{xyc} = exp(-\frac{(x - \tilde{p_x})^2 + (y - \tilde{p_y})^2}{2{\sigma_p}^2}) \tag{1}$$

$$\sigma_p = r/3. \tag{2}$$

We use the element-wise maximum operation if two Gaussians of the same class overlap. The loss function for center point prediction is a variant of focal loss [31], formulized as:

$$L_{center} = -\frac{1}{N} \sum_{x,y,c} \begin{cases} (1 - \hat{Y}(x,y,c))^{\alpha} log(\hat{Y}(x,y,c)) & if Y(x,y,c) = 1 \\ (1 - Y(x,y,c))^{\beta} \hat{Y}(x,y,c)^{\alpha} log(1 - \hat{Y}(x,y,c)) & otherwise, \end{cases} \tag{3}$$

where N is the total number of objects in the image, and α and β are the hyperparameters controlling the contribution of each point ($\alpha = 2$ and $\beta = 4$, by default, following [37]).

As the predicted $\hat{Y}$ has a stride of R with the input image, the center point position obtained by $\hat{Y}$ will inevitably have quantization error. Thus, a Center offset branch was introduced to eliminate this error. The model predicts $\hat{o} \in [0,1]^{\frac{W}{R} \times \frac{H}{R} \times 2}$, in order to refine the object's center. For each object's center $p = (p_x, p_y)$, smooth L1 loss [26] is used during training:

$$L_{offset} = \frac{1}{N} \sum_{p} Smooth_{L1}(\hat{o}_{\tilde{p}}, \frac{p}{R} - \lfloor \frac{p}{R} \rfloor). \tag{4}$$

Then, combining $\hat{Y}$ and $\hat{o}$, we can accurately locate the object's center.

3.2.2. Angle Prediction for Oriented Objects

In this section, we first introduce the five-parameter long side-based representation for oriented objects and analyze the angular boundary discontinuity problem. Then, we detail the circular smooth label, in order to solve the boundary discontinuity problem and predict the angles of oriented objects.

Representations for Oriented Objects. As we discussed in Section 1, the use of oriented bounding boxes can better depict objects in remote sensing images. We use five-parameter long side-based methods to represent the oriented objects. As shown in Figure 3, five parameters (C_x, C_y, h, w, θ) were used to represent an OBB, where h represents the long side of the bounding box, the other side is referred to as w, and θ is the angle between the long side and x-axis, with a 180° range. Compared to the HBB, OBB needs an extra parameter, θ, to represent the direction information.

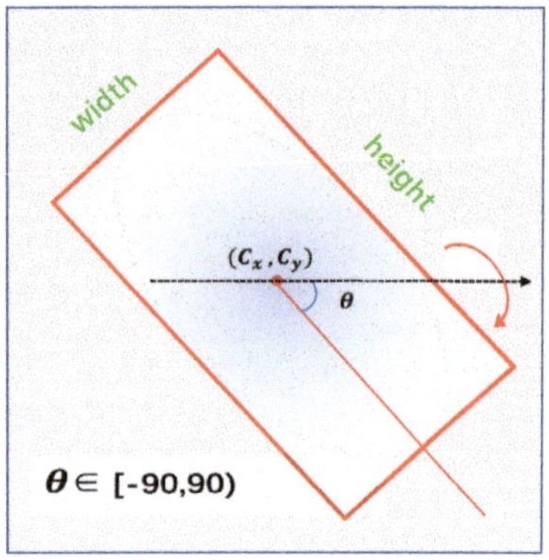

Figure 3. Five-parameter long side-based representation for oriented objects.

As there are generally various angles of an object in remote sensing images, accurately predicting the direction is important, especially for objects with large aspect ratios. Due to the periodicity of the angle, directly regressing the angle θ may lead to the boundary discontinuity problem, resulting in a large loss value during training. As illustrated in Figure 4, two oriented objects can have relatively similar directions while crossing the angular boundary, resulting in a large difference between regression values. This discontinuous boundary can interfere with the network's learning of the object direction and, thus, degrade the model's performance.

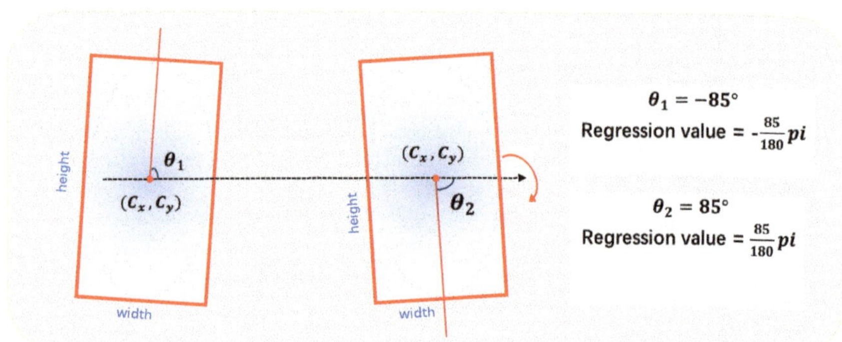

Figure 4. An example of discontinuous angular boundary based on the five-parameter long side representation.

Circular Smooth Label. Following [19], we convert the angle regression problem into a classification problem. As the five-parameter long side-based representation has 180° angle range, each 1° degree interval is referred to a category, which results in 180 categories in total. Then, the one-hot angle label passes through a periodic function, followed by a Gaussian function to smooth the label, formulized as:

$$CSL(x) = \begin{cases} g(x) & \theta - r_{csl} < x < \theta + r_{csl} \\ 0 & otherwise, \end{cases} \tag{5}$$

where $g(x)$ is the Gaussian function, which satisfies $g(x) = g(x + kT), k \in N, T = 180$; and r_{csl} is the radius of the Gaussian function, which controls the smoothing degree of the angle label. For example, when $r_{csl} = 0$, the Gaussian function becomes to pulse function and the CSL degrades into the one-hot label. We illustrate the CSL in Figure 5.

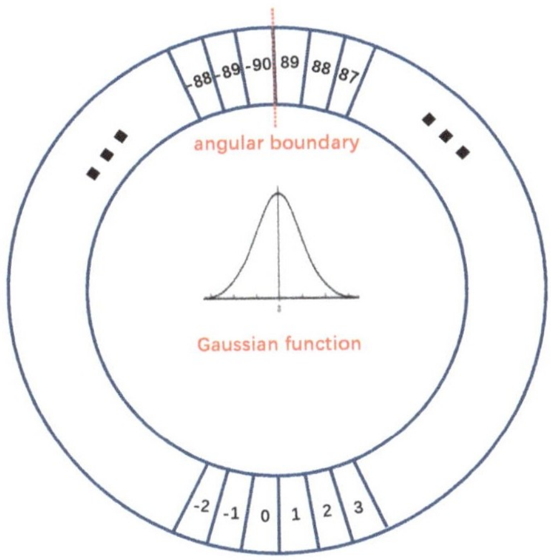

Figure 5. Visualization of the circular smooth label.

The loss function for the CSL is not the commonly used Softmax Cross-Entropy loss; as we use a smooth label, Sigmoid Binary Cross-Entropy is used to train the angle prediction network. Specifically, the model predicts $\hat{\theta} \in [0,1]^{\frac{W}{R} \times \frac{H}{R} \times 180}$ for an input image, and the loss function is:

$$L_{CSL} = \frac{1}{N} \sum_{p} BCE(\hat{\theta}_{\tilde{p}}, \theta_p), \tag{6}$$

where θ_p is the circular smooth label for object p in the image.

3.2.3. Prediction of Object Size

We have that (C_x, C_y, h, w, θ) represents the OBBs, using the center location and direction of each object obtained in Sections 3.2.1–3.2.2. The rest (i.e., the long side h and short side w) are predicted through the Object size branch shown in Figure 2. The model outputs $\hat{S} \in R^{\frac{W}{R} \times \frac{H}{R} \times 2}$ for the object size. For each object p, with corresponding size label $s_p = (h_p, w_p)$, smooth L1 loss is used:

$$L_{size} = \frac{1}{N} \sum_{p} Smooth_{L1}(\hat{S}_{\tilde{p}}, ln(\frac{s_p}{R})). \tag{7}$$

Note that the smooth L1 loss used in this paper is ($\delta = \frac{1}{9}$ by default):

$$Smooth_{L1}(x) = \begin{cases} \frac{1}{2\delta} x^2 & if |x| < \delta \\ x - \frac{\delta}{2} & otherwise. \end{cases} \tag{8}$$

The overall training objective for our arbitrary-oriented object detector is:

$$L = L_{center} + \lambda_{angle} L_{CSL} + \lambda_{size} L_{size} + \lambda_{offset} L_{offset}, \tag{9}$$

where λ_{angle}, λ_{size}, and λ_{offset} are used to balance the weighting between different tasks. In this paper, λ_{angle}, λ_{size}, and λ_{offset} are set to 0.5, 1, and 1, respectively.

3.3. Feature Enhancement by Deformable FPN

We aim to better localize the object's center and corresponding direction by building a pyramidal feature hierarchy on the network's output features. The feature maps extracted by the backbone are referred to as C_3, C_4, and C_5, shown in Figure 2. These feature maps have different spatial resolutions and large semantic gaps. Low-resolution maps have strong semantic information, which has great representational capacity for object detection, especially for large objects (e.g., Soccer fields) in aerial images, while high resolution maps have relatively low-level features but can provide more detailed information, which is very important for detecting small objects. Due to the various orientations and large scale differences of objects in remote sensing images, the standard FPN [25] used to fuse these feature maps may not work well in this situation. The standard convolution kernel appears in a regular rectangular manner, which has the characteristic of translation invariance. Meanwhile, the resolutions of these feature maps differ, and the semantic information of objects is not strictly aligned to these feature maps. Therefore, using standard convolution to project these features before the add operation may harm the representation ability of oriented objects, which is essential to accurately localize the object's center and direction. However, Deformable convolution (DConv) can learn the position of convolution kernels adaptively, which can better project the features of oriented objects in the feature pyramid network. We detail the structure of Deformable FPN in the following, and demonstrate its effectiveness in Section 4.

3.3.1. Structure of Deformable FPN

To verify the effectiveness of our method, we introduce three kinds of necks, including our Deformable FPN, to process backbone features to P_2, which are subsequently sent to the detection head. Figure 6 shows detailed architectures of the three necks, using ResNet50 [48] as a backbone. A direct Top-down pathway is constructed without building the feature pyramid structure (Figure 6) but, instead, using deformable convolutions, as originally used by [37] for ResNet. Our proposed Deformable FPN is shown in Figure 6, while a commonly used FPN structure is shown in Figure 6. We keep the same channels of features in each stage, which are 256, 128, and 64 for features with stride 16, 8, and 4, respectively.

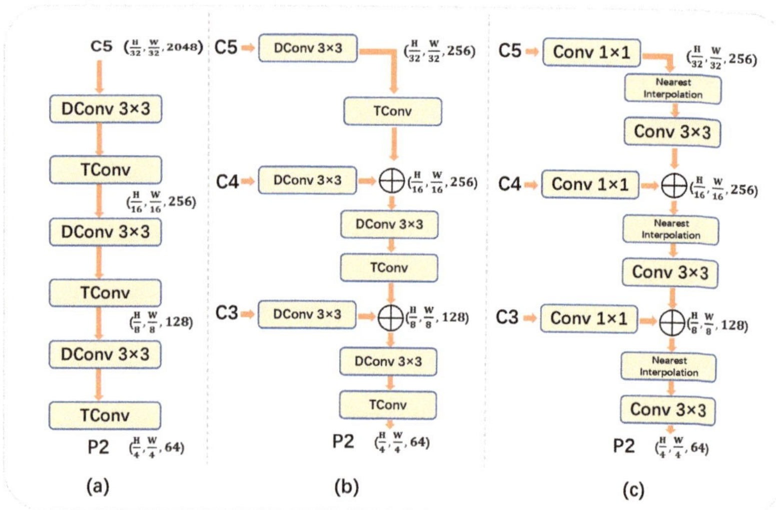

Figure 6. Different kinds of necks to process the backbone features: (**a**) A direct Top-down pathway without the feature pyramid structure; (**b**) our proposed Deformable FPN; and (**c**) standard FPN.

- **Direct Top-down pathway** As shown in Figure 6, we only use the backbone feature C5 from the last stage of ResNet to generate $P2$. A direct Top-down pathway was used, without constructing a feature pyramid structure on it. Deformable convolution is used to change the channels, and transposed convolution is used to up-sample the feature map. We refer to this Direct Top-down Structure as DTS, for simplicity.
- **Deformable FPN** Directly using $C5$ to generate $P2$ for oriented object detection may result in the loss of some detailed information, which is essential for small object detection and the accurate localization of object centers. As the feature $C5$ has a relatively large stride (of 32) and a large receptive field in the input image, we construct the Deformable FPN as follows: we use DConv 3×3 to reduce the channels and project the backbone features $C3$, $C4$, and $C5$. Transposed convolution is used to up-sample the spatial resolution of features by a factor of two. Then, the up-sampled feature map is merged with the projected feature from the backbone of same resolution, by using an element-wise add operation. After merging the features from the adjacent stage, another deformable convolution is used to further align the merged feature and reduce its channel simultaneously. We illustrate this process in Figure 6b.
- **FPN** A commonly used feature pyramid structure is shown in Figure 6c. Conv 1×1 is used to reduce the channel for $C3$, $C4$, and $C5$, and nearest neighbor interpolation is used to up-sample the spatial resolution. Note that there are two differences from [25], in order to align the architecture with our Deformable FPN. First, the feature channels are reduced along with their spatial resolution. Specifically, the channels of features in each stage are 256, 128, and 64 for features with a stride of 16, 8, and 4, respectively, while [25] consistently set the channels to 256. Second, we added an extra Conv 3×3 after the added feature map, in order to further fuse them.

Comparing our Deformable FPN with DTS, we reuse the shallow, high-resolution features of the backbone, which provide more detailed texture information to better localize the object center and detect small objects, such as vehicles and bridges, in remote sensing images. Compared with FPN, by using deformable convolution—which adaptively learns the position of convolution kernels—it can better project the features of oriented objects. Moreover, applying transposed convolution, rather than nearest neighbor interpolation, to up-sample the features can help to better localize the centers.

3.3.2. Deformable Groups

As we use deformable convolution in the feature pyramid structure, we discuss how larger Deformable groups in DConv can further enhance the representation power of the network in this section.

The deformable convolution used in this paper is DCNv2 [21]. For a convolutional kernel and K sampling locations, the deformable convolution operation can be formulized as follows:

$$y(p) = \sum_{k=1}^{K} \omega_k \cdot x(p + p_k + \Delta_{p_k}) \cdot \Delta_{m_k}, \tag{10}$$

where $x(p)$ and $y(p)$ denote the feature at location p on input feature map x and output feature map y, respectively; the pre-set convolution kernel location is denoted as p_k and ω_k is the kernel weight; and Δ_{p_k} and Δ_{m_k} are the learnable kernel offset and scalar weight based on input feature, respectively. Take a 3×3 deformable convolutional kernel as an example: there are $K = 9$ sampling locations. For each location k, a two-dimensional vector(Δ_{p_k}) is used to determine the offsets in the x- and y-axes, and a one-dimensional tensor is used for the scalar weight (Δ_{m_k}). So, the network first predicts offset maps, which have $3K$ channels based on the input features, then uses the predicted offsets to find K convolution locations at each point p. Finally, Equation (10) is used to calculate the output feature maps. We illustrate this process in Figure 7a.

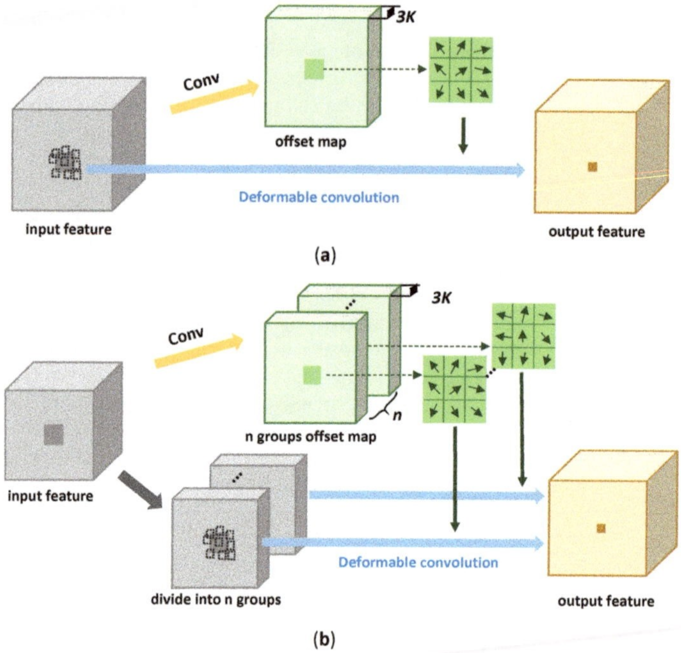

Figure 7. Illustration of 3 × 3 deformable convolution: (**a**) One deformable group; and (**b**) n deformable groups.

Note that all channels in the input feature maps share one group of offsets when the number of deformable groups is set to 1 (as shown in Figure 7a). Input features share these common offsets to perform the deformable convolution. When the number of deformable groups is n ($n > 1$), the networks first output $n \times 3K$-channel offset maps, the input feature (C channels) is divided into n groups, where each group of features has C/n channels, and the corresponding $3K$-channel offset maps are used to calculate the kernel offsets (as shown in Figure 7b). Finally, the output feature will be obtained by deformable convolution on the input feature. Different from the groups in the standard convolutional operation, each channel in the output features will be calculated on the entire input features only, with different kernel offsets. Increasing the number of deformable groups can enhance the representation ability of DConv, as different groups of input channels use different kernel offsets, and the network can generate a unique offset for each group of features, according to the characteristics of the input features.

4. Experiments

4.1. Data Sets and Evaluation Metrics

4.1.1. DOTA

DOTA is a large-scale dataset for object detection in remote sensing images. The images are collected from different sensors and platforms. There are 2806 images, with scales from 800 × 800 to 4000 × 4000 pixels. The proportions of the training set, validation set, and testing set in DOTA are $\frac{1}{2}$, $\frac{1}{6}$, and $\frac{1}{3}$, respectively. The DOTA dataset contains 15 common categories, with 188,282 instances in total. The full names (short names) for the categories are: Plane (PL), Baseball diamond (BD), Bridge (BR), Ground track field (GTF), Small vehicle (SV), Large vehicle (LV), Ship (SH), Tennis court (TC), Basketball court (BC), Storage tank (ST), Soccer-ball field (SBF), Roundabout (RA), Harbor (HA), Swimming pool (SP), and Helicopter (HC).

4.1.2. HRSC2016

HRSC2016 is a dataset for ship detection in aerial images. The HRSC2016 dataset contains images of two scenarios, including ships at sea and ships inshore at six famous harbors. There are 436, 181, and 444 images for training, validation and testing, respectively. The ground sample distances of images are between 2 m and 0.4 m, and the image resolutions range from 300 × 300 to 1500 × 900.

4.1.3. Evaluation Metrics

The Mean Average Precision (mAP) is commonly used to evaluate the performance of object detectors, where the AP is the area under the precision–recall curve for a specific category, which ranges from $[0, 1]$. It is formulized as:

$$Precision = \frac{TP}{TP + FP} \tag{11}$$

$$Recall = \frac{TP}{TP + FN} \tag{12}$$

$$mAP = \frac{1}{C}\sum_{c=1}^{C}\int P_c(R_c)dR_c, \tag{13}$$

where C is the number of categories, and TP, FP, and FN represent the numbers of correctly detected objects, incorrectly detected objects, and mis-detected objects, respectively.

4.2. Implementation Details

4.2.1. Image Pre-Processing

The images in the DOTA dataset always have a high resolution. Directly training on the original high-resolution images does not reconcile with the hardware, due to limited GPU memory. Therefore, we cropped the images into sub-images of size 1024 × 1024, with an overlap of 256 pixels, and obtained 14,560 labeled images for training. We introduce two methods for testing in this paper. In the first method, we crop the testing images using the same size as used in the training stage (1024 × 1024 pixels) and, after inference on all sub-images, the final detection results are obtained by splicing all sub-image results. This method is commonly used for inference on the test images in the DOTA dataset; however, it may generate some false results at the cutting edge, leading to poor performance especially for some categories with large sizes (e.g., Ground field track and Soccer field). The second method involves cropping the testing images with a relatively high resolution (3200 pixels i this paper) during inference. We simply padded the images if the size of the original image is smaller than the crop size. By cropping the testing images at a relatively high resolution, a large number of images will not be cut and, so, the model can detect objects based on the complete instance, thus obtaining a more accurate evaluation result. Note that the only difference between the two methods is the crop size used for testing.

For the HRSC2016 dataset, we resized the long side of images to 640 pixels and kept the same aspect ratio as the original images. Thus, the short side of each image was different and smaller than 640 pixels. Then, we uniformly padded the resized images to 640 × 640 pixels, both for training and testing.

4.2.2. Experimental Settings

All experiments were implemented in PyTorch. ImageNet [49]-pretrained ResNets were used as our default backbone. We used the Adam [50] optimizer to optimize the overall networks for 140 epochs. We set a batch size of 12 for DOTA and 32 for HRSC2016. The initial learning rates were 1.25×10^{-4} and 2×10^{-4} for DOTA and HRSC2016, with the learning rate dropped by 10 × at 100 and 130 epochs. We used a single-scale training strategy with input resolution of 1024 for DOTA and 640 for HRSC2016, as mentioned before, and the stride R was set to 4. The Gaussian radii r_{csl} for CSL were set to 4 and 6 for

DOTA and HRSC2016, respectively. Our data augmentation methods included random horizontal and vertical flipping, random graying, and random rotation. We did not use multi-scale training and testing augmentations in our experiments.

4.3. Results

4.3.1. Effectiveness of Deformable FPN

Due to the wide variety of object scales, orientations and shapes, we chose DOTA as our main dataset for validation. We implemented a standard feature pyramid network (FPN), a direct Top-down structure (DTS), and our proposed Deformable FPN (De-FPN) as necks to process features from the ResNet50 backbone.

Results are shown in Table 1. We give the average precision of each category and total mAP. HRT denotes the high resolution testing discussed in Section 4.2.1. The building detector from FPN achieved 69.68% mAP, which is already a good performance for the DOTA dataset. However, the direct Top-down structure had 1.2% higher mAP than the FPN structure. Note that the DTS does not build a feature hierarchical structure inside the network, but had a better performance than FPN, indicating that the deformable convolution can better project features for rotating objects. Furthermore, the interpolation operation used to up-sample the features may harm the representation power for predicting object centers exactly.

Our Deformable FPN achieved a remarkable improvement of 1.23% higher mAP, compared with DTS, which indicates that Deformable FPN can better fuse the multi-level features and help the detector to accurately localize the rotating objects. Compared with FPN, the advantages of building a feature hierarchical structure in our way are evident. The improvement of up to 2.43% higher mAP was obtained through use of deformable convolution and transposed convolution within the FPN structure. Further, by using original high-resolution images during testing, our detector could obtain a more accurate evaluation result. Specifically, the high-resolution test boosted the mAP by 1.79%, 2.39%, and 1.65% for FPN, DTS, and De-FPN, respectively.

Table 1. Three kinds of necks are used to build arbitrary-oriented object detectors: Feature pyramid network (FPN), direct Top-down structure (DTS), and Deformable FPN(De-FPN). HRT denotes using High-Resolution crop during Testing. All models use ImageNet-pretrained ResNet50 as a backbone.

Neck	HRT	PL	BD	BR	GTF	SV	LV	SH	TC	BC	ST	SBF	RA	HA	SP	HC	mAP
FPN		88.36	78.03	45.35	57.90	76.52	78.24	85.12	90.63	78.89	82.23	39.04	61.21	62.56	70.64	50.50	69.68
DTS		**88.79**	**82.99**	42.75	59.31	76.53	77.15	85.17	**90.76**	**79.84**	**82.27**	**49.59**	59.68	63.98	68.15	**56.28**	70.88
De-FPN		88.73	80.71	**46.36**	**67.10**	**78.16**	**80.51**	**86.32**	90.67	**79.84**	81.66	45.91	**63.66**	**66.89**	**71.22**	53.87	**72.11**
FPN	√	89.15	78.98	**47.07**	59.17	76.78	79.14	86.89	90.80	79.51	83.67	46.60	60.83	66.81	72.77	53.84	71.47
DTS	√	**89.70**	**84.72**	45.00	67.62	76.64	78.23	86.60	90.78	79.66	83.59	**54.91**	59.84	67.08	70.29	**64.32**	73.27
De-FPN	√	89.47	81.96	46.89	**70.72**	**77.01**	**81.44**	**87.32**	**90.81**	**80.06**	**83.68**	46.27	**63.55**	**73.62**	**72.91**	60.62	**73.76**

4.3.2. Results on DOTA

We compared our results with other state-of-the-art methods in the DOTA dataset. We used ResNet50, ResNet101, and ResNet152 as backbones to construct our Arbitrary-oriented anchor-free based object detector, denoted as CenterRot. The results are shown in Table 2. The DOTA dataset contains complex scenes, wherein object scales change drastically. Two-stage methods are commonly used in DOTA, in order to handle the imbalance between foregrounds and backgrounds in these complex scenes, such as ROI Transformer [42] and CAD-Net [51], which have achieved 69.59% and 69.90% mAP, respectively, when using ResNet101 as a backbone. Meanwhile, extremely large and small objects can appear in one image (as shown in Figure 1), such that multi-scale training and testing technologies are used to obtain a better performance, such as FADet [52], which obtained 73.28% mAP using ResNet101, and MFIAR-Net [53], which obtained 73.49% mAP using ResNet152 as the backbone. However, multi-scale settings need to infer one image multiple times at different sizes and merge all results after testing, which leads to a larger computational burden during inference.

Our CenterRot converts the oriented object detection problem to a center point localization problem. Based on the fact that remote sensing images have less probability of overlap between objects with the same category, directly detecting the oriented object from its center can lead to a comparable performance with oriented anchor-based methods. Specifically, CenterRot achieved 73.76% and 74.00% mAP on the OBB task of DOTA, when using ResNet50 and ResNet101 as the backbone, respectively. Due to the strong representation ability of our Deformable FPN for rotated objects , CenterRot, equipped with larger deformable groups ($n = 16$ in Deformable FPN), achieved the best performance (74.75% mAP) when using ResNet152 as the backbone, surpassing all published single-stage methods with single-scale training and testing. Detailed results for each category and method are provided in Table 2.

Table 2. State-of-the-Art comparison with other methods in the oriented object detection task in the DOTA test set. AP for each category and overall mAP on DOTA are provided (the best result is highlighted in bold), where MS denotes multi-scale training and testing and * denotes that larger deformable groups ($n = 16$ in Deformable FPN) were used.

Method	Backbone	MS	PL	BD	BR	GTF	SV	LV	SH	TC	BC	ST	SBF	RA	HA	SP	HC	mAP
SSD [27]	VGG16		41.06	24.31	4.55	17.10	15.93	7.72	13.21	39.96	12.05	46.88	9.09	30.82	1.36	3.50	0.00	17.84
YOLOv2 [29]	Darknet19		52.75	24.24	10.60	35.50	14.36	2.41	7.37	51.79	43.98	31.35	22.30	36.68	14.61	22.55	11.89	25.49
FR-H [22]	ResNet50		49.74	64.22	9.38	56.66	19.18	14.17	9.51	61.61	65.47	57.52	51.36	49.41	20.80	45.84	24.38	39.95
FR-O [1]	ResNet50		79.42	77.13	17.70	64.05	35.30	38.02	37.16	89.41	69.64	59.28	50.30	52.91	47.89	47.40	46.30	54.13
RetinaNet-R [43]	ResNet50		88.90	67.70	33.60	56.80	66.10	73.30	75.20	**90.90**	74.00	75.10	43.80	56.70	51.10	55.70	21.50	62.00
RetinaNet-H [43]	ResNet50		88.90	74.50	40.10	58.00	63.10	50.60	63.60	**90.90**	77.90	76.40	48.30	55.90	50.70	60.20	34.20	62.20
RSDet [54]	ResNet50		89.30	**82.70**	**47.70**	63.90	66.80	62.00	67.30	90.80	**85.30**	82.40	**62.30**	62.40	65.70	68.60	**64.60**	70.80
CenterRot (Ours)	ResNet50		**89.47**	81.96	46.89	**70.72**	**77.01**	**81.44**	**87.32**	90.81	80.06	**83.68**	46.27	**63.55**	**73.62**	**72.91**	60.62	**73.76**
R-FCN [24]	ResNet101		39.57	46.13	3.03	38.46	9.10	3.66	7.45	41.97	50.43	66.98	40.34	51.28	11.14	35.59	17.45	30.84
R-DFPN [47]	ResNet101		80.92	65.82	33.77	58.94	55.77	50.94	54.78	90.33	66.34	68.66	48.73	51.76	55.10	51.32	35.88	57.94
R^2CNN [55]	ResNet101		80.94	65.67	35.34	67.44	59.92	50.91	55.81	90.67	66.92	72.39	55.06	52.23	55.14	53.35	48.22	60.67
RRPN [39]	ResNet101		88.52	71.20	31.66	59.30	51.85	56.19	57.25	90.81	72.84	67.38	56.69	52.84	53.08	51.94	53.58	61.01
ICN [41]	ResNet101	√	81.40	74.30	47.70	70.30	64.90	67.80	70.00	90.80	79.10	78.20	53.60	62.90	67.00	64.20	50.20	68.20
ROI Trans [42]	ResNet101	√	88.64	78.52	43.44	75.92	68.81	73.68	83.59	90.74	77.27	81.46	58.39	53.34	62.83	58.93	47.67	69.56
CAD-Net [51]	ResNet101		87.80	82.40	49.40	73.50	71.10	63.50	76.70	**90.90**	79.20	73.30	48.40	60.90	62.00	67.00	62.20	69.90
RSDet [54]	ResNet101		89.80	82.90	48.60	65.20	69.50	70.10	70.20	90.50	85.60	83.40	62.50	63.90	65.60	67.20	68.00	72.20
BBAVectors [56]	ResNet101	√	88.35	79.96	50.69	62.18	**78.43**	78.98	**87.94**	90.85	83.58	84.35	54.13	60.24	65.22	64.28	55.70	72.32
SCRDet [57]	ResNet101	√	89.98	80.65	52.09	68.36	68.36	60.32	72.41	90.85	**87.94**	**86.86**	65.02	**66.68**	66.25	68.24	65.21	72.61
SARD [58]	ResNet101		89.93	**84.11**	**54.19**	72.04	68.41	61.18	66.00	90.82	87.79	86.59	**65.65**	64.04	66.68	68.84	**68.03**	72.95
GLS-Net [59]	ResNet101		88.65	77.40	51.20	71.03	73.30	72.16	84.68	90.87	80.43	85.38	58.33	62.27	67.58	70.69	60.42	72.96
FADet [52]	ResNet101	√	**90.21**	79.58	45.49	**76.41**	73.18	68.27	79.56	90.83	83.40	84.68	53.40	65.42	74.17	69.69	64.86	73.28
CenterRot (Ours)	ResNet101		89.74	83.57	49.53	66.45	77.07	**80.57**	86.97	90.75	81.50	84.05	54.14	64.14	**74.22**	**72.77**	54.56	**74.00**
MFIAR-Net [53]	ResNet152	√	89.62	**84.03**	52.41	**70.30**	70.13	67.64	77.81	90.85	85.40	**86.22**	63.21	64.14	68.31	70.21	62.11	73.49
R^3Det [43]	ResNet152		89.49	81.17	50.53	66.10	70.92	78.66	78.21	90.81	85.26	84.23	61.81	63.77	68.16	69.83	**67.17**	73.74
RSDet-Refine [54]	ResNet152		**90.10**	82.00	**53.80**	68.50	70.20	78.70	73.60	**91.20**	**87.10**	84.70	**64.30**	**68.20**	66.10	69.30	63.70	74.10
CenterRot * (Ours)	ResNet152		89.69	81.42	51.16	68.82	**78.77**	**81.45**	**87.23**	90.82	80.31	84.27	56.13	64.24	**75.80**	**74.68**	56.51	**74.75**

4.3.3. Results on HRSC2016

The HRSC2016 dataset has only one category—ship—where some of them have large aspect ratios and various orientations. Therefore, it is still a challenge to detect ships in this dataset. The results are shown in Table 3, from which it can be seen that our CenterRot achieved state-of-the-art performance consistently, without the use of a more complicated architecture, compared with the other methods. Specifically, CenterRot achieved 90.20% and 96.59% for mAP 07 and 12, respectively, where mAP 07 denotes using the 2007 evaluation metric, while mAP 12 denotes using the 2012 evaluation metric.

Table 3. State-of-the-art comparison of HRSC2016. mAP 07(12) means using the 2007(2012) evaluation metric.

Method	Backbone	mAP 07	mAP 12
RoI-Trans [42]	ResNet101	86.20	-
RetinaNet-R [43]	ResNet101	89.18	95.21
R^3Det [43]	ResNet101	89.26	96.01
R^3Det-DCL [60]	ResNet101	89.46	96.41
CenterRot (Ours)	ResNet50	**90.20**	**96.59**

4.3.4. Visualization

The visualization results are presented using our CenterRot. The results for DOTA are shown in Figure 8 and those for HRSC2016 are shown in Figure 9.

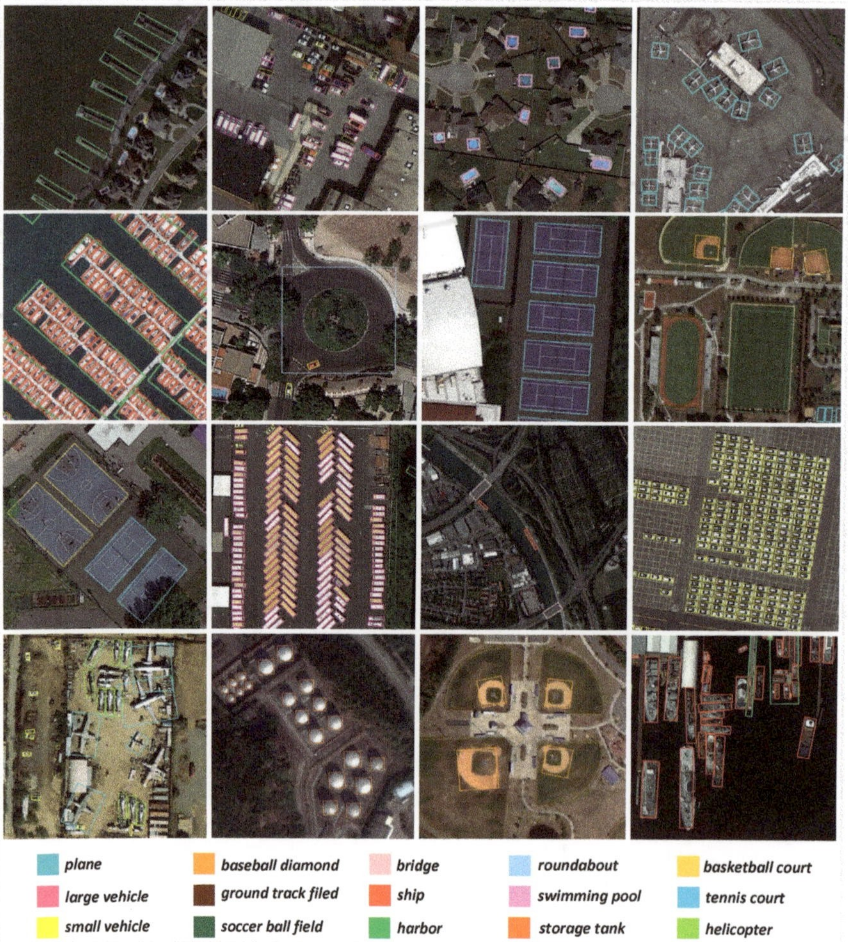

Figure 8. Visualization of detection results on DOTA.

Figure 9. Visualization of detection results on HRSC2016.

5. Discussion

The proposed CenterRot achieved prominent performance in detecting rotated objects for both of the DOTA and HRSC2016 datasets. Objects with the same category have a lower probability of overlapping each other, so directly detecting rotated objects from their center is effective and efficient. We selected several categories in order to further analyze our method. As shown in Table 4, small vehicle, large vehicle, and ship were the most common rotated objects in DOTA, which always appeared in a densely arranged manner. Anchor-based methods operate by setting anchors with different angles, scales and aspect ratios per location, in order to cover the rotated objects as much as possible. However, it is impossible to assign appropriate anchors for each object, due to the various orientations in this situation. Our methods performed well in these categories especially, due to the fact that we converted the oriented bounding box regression problem into a center point localization problem. Less overlap between objects means fewer collisions between object centers, such that the networks can learn the positions of rotated objects from their center easier. We also visualized some predicted center heatmaps, as shown in Figure 10. Moreover, since the deformable FPN can better project features for rotated objects and the use of CSL to predict the object direction, our methods still performed well for objects with large aspect ratios, such as harbors and ships in HRSC2016.

Table 4. Comparison of selected categories in DOTA. All methods use ResNet152 as a backbone.

Method	SV	LV	SH	HA	SBF	RA
MFIAR-Net	70.13	67.64	77.81	68.31	63.21	64.14
R^3Det	70.92	78.66	78.21	68.16	61.81	63.77
RSDet-Refine	70.20	78.70	73.60	66.10	**64.30**	**68.20**
CenterRot (Ours)	**78.77**	**81.45**	**87.23**	**75.80**	56.13	64.24

However, as we cut the original images, some large objects were incomplete during training, such as the soccer ball field, which may confuse our detector when localizing the exact center, resulting in relatively poor performance in these categories. Due to this, we use the five-parameter long side-based representation for oriented objects, which will create some ambiguity when representing the square-like objects (objects with small aspect ratio). So, the model will produce a large loss value when predicting the angle and size of these objects and perform poorly in these categories, such as roundabout. Other oriented representations, such as the five-parameter acute angle-based method [19], will avoid this problem while suffering EoE problems. Therefore, it is still worth studying how to better represent the rotated objects.

Future works will mainly involve improving the effectiveness and robustness of the proposed methods in real-world applications. Different from the classical benchmark datasets, the objects in input images can vary much more frequently and can be affected by other conditions, such as angle of insolation. Moreover, as cloudy weather is very common, the cloud can occlude some objects. The anchor-free rotated object detection problem in such a circumstance is also worth studying.

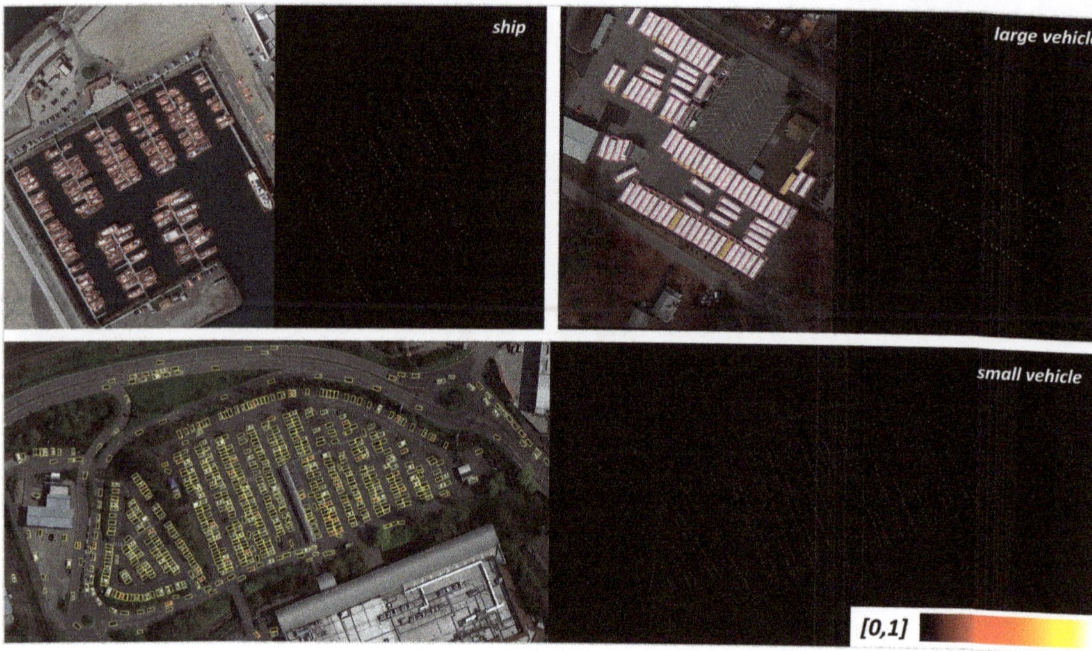

Figure 10. Visualization of predicted center heatmaps for some categories.

6. Conclusions

In this paper, we found that objects within the same category tend to have less overlap with each other in remote sensing images, and setting multiple anchors per location to detect rotated objects may not be necessary. We proposed an anchor-free based arbitrary-oriented object detector to detect the rotated objects from their centers and achieved great performance without pre-set anchors, which avoids complex computations on anchors, such as IoU. To accurately localize the object center under complex backgrounds and the arbitrary orientations of rotated objects, we proposed a deformable feature pyramid network to fuse the multi-level features and obtained a better feature representation for detecting rotated objects. Experiments on DOTA showed that our Deformable FPN can better project the features of rotated objects than standard FPN. Our CenterRot achieved a state-of-the-art performance, with 74.75% mAP on DOTA and 96.59% on HRSC2016, with a single-stage model, including single-scale training and testing. Extensive experiments demonstrated that detecting arbitrary-oriented objects from their centers is, indeed, an effective baseline choice.

Author Contributions: Conceptualization, J.W., L.Y. and F.L.; methodology, J.W.; software, J.W.; validation, J.W. and L.Y.; formal analysis, J.W., L.Y. and F.L.; investigation, J.W.; resources, F.L.; data curation, J.W.; writing—original draft preparation, J.W.; writing—review and editing, J.W., L.Y. and F.L.; visualization, J.W.; supervision, L.Y. and F.L.; project administration, F.L.; funding acquisition, F.L. All authors have read and agreed to the published version of the manuscript.

Funding: This research was funded by the National Natural Science Foundation of China grant number U1903213.

Institutional Review Board Statement: Not applicable.

Informed Consent Statement: Not applicable.

Data Availability Statement: The DOTA and HRSC2016 datasets used for this study can be accessed at https://captain-whu.github.io/DOTA/dataset.html and https://sites.google.com/site/hrsc2016/ accessed on 10 August 2021.

Conflicts of Interest: The authors declare no conflict of interest.

References

1. Xia, G.S.; Bai, X.; Ding, J.; Zhu, Z.; Belongie, S.; Luo, J.; Datcu, M.; Pelillo, M.; Zhang, L. DOTA: A large-scale dataset for object detection in aerial images. In Proceedings of the IEEE Conference on Computer Vision and Pattern Recognition (CVPR), Salt Lake City, UT, USA, 18-22 June 2018; pp. 3974-3983
2. Liu, Z.; Yuan, L.; Weng, L.; Yang, Y. A high resolution optical satellite image dataset for ship recognition and some new baselines. In Proceedings of the International Conference on Pattern Recognition Applications and Methods, SCITEPRESS, Porto, Portugal, 24–26 February 2017; Volume 2; pp. 324–331.
3. Karatzas, D.; Gomez-Bigorda, L.; Nicolaou, A.; Ghosh, S.; Bagdanov, A.; Iwamura, M.; Matas, J.; Neumann, L.; Chandrasekhar, V.R.; Lu, S.; et al. ICDAR 2015 competition on robust reading. In Proceedings of the 2015 13th International Conference on Document Analysis and Recognition (ICDAR), Nancy, France, 23–26 August 2015; pp. 1156–1160.
4. Nayef, N.; Yin, F.; Bizid, I.; Choi, H.; Feng, Y.; Karatzas, D.; Luo, Z.; Pal, U.; Rigaud, C.; Chazalon, J.; et al. Icdar2017 robust reading challenge on multi-lingual scene text detection and script identification-rrc-mlt. In Proceedings of the 2017 14th IAPR International Conference on Document analysis and Recognition (ICDAR), Kyoto, Japan, 13–15 November 2017; Volume 1, pp. 1454–1459.
5. Reggiannini, M.; Righi, M.; Tampucci, M.; Lo Duca, A.; Bacciu, C.; Bedini, L.; D'Errico, A.; Di Paola, C.; Marchetti, A.; Martinelli, M.; et al. Remote sensing for maritime prompt monitoring. *J. Mar. Sci. Eng.* **2019**, *7*, 202. [CrossRef]
6. Moroni, D.; Pieri, G.; Tampucci, M. Environmental decision support systems for monitoring small scale oil spills: Existing solutions, best practices and current challenges. *J. Mar. Sci. Eng.* **2019**, *7*, 19. [CrossRef]
7. Almulihi, A.; Alharithi, F.; Bourouis, S.; Alroobaea, R.; Pawar, Y.; Bouguila, N. Oil spill detection in SAR images using online extended variational learning of dirichlet process mixtures of gamma distributions. *Remote Sens.* **2021**, *13*, 2991. [CrossRef]
8. Zhang, L.; Yang, X.; Shen, J. Frequency variability feature for life signs detection and localization in natural disasters. *Remote Sens.* **2021**, *13*, 796. [CrossRef]
9. Zhang, T.; Zhang, X.; Shi, J.; Wei, S. Depthwise separable convolution neural network for high-speed SAR ship detection. *Remote Sens.* **2019**, *11*, 2483. [CrossRef]
10. Xiao, X.; Wang, B.; Miao, L.; Li, L.; Zhou, Z.; Ma, J.; Dong, D. Infrared and visible image object detection via focused feature enhancement and cascaded semantic extension. *Remote Sens.* **2021**, *13*, 2538. [CrossRef]
11. Tong, X.; Sun, B.; Wei, J.; Zuo, Z.; Su, S. EAAU-Net: Enhanced asymmetric attention U-Net for infrared small target detection. *Remote Sens.* **2021**, *13*, 3200. [CrossRef]
12. Zhang, Z.; Guo, W.; Zhu, S.; Yu, W. Toward arbitrary-oriented ship detection with rotated region proposal and discrimination networks. *IEEE Geosci. Remote Sens. Lett.* **2018**, *15*, 1745–1749. [CrossRef]
13. Long, Y.; Gong, Y.; Xiao, Z.; Liu, Q. Accurate object localization in remote sensing images based on convolutional neural networks. *IEEE Trans. Geosci. Remote Sens.* **2017**, *55*, 2486–2498 [CrossRef]
14. Yang, R.; Pan, Z.; Jia, X.; Zhang, L.; Deng, Y. A novel CNN-based detector for ship detection based on rotatable bounding box in SAR images. *IEEE J. Sel. Top. Appl. Earth Obs. Remote Sens.* **2021**, *14*, 1938–1958. [CrossRef]
15. Tian, L.; Cao, Y.; He, B.; Zhang, Y.; He, C.; Li, D. Image enhancement driven by object characteristics and dense feature reuse network for ship target detection in remote sensing imagery. *Remote Sens.* **2021**, *13*, 1327. [CrossRef]
16. Dong, Y.; Chen, F.; Han, S.; Liu, H. Ship object detection of remote sensing image based on visual attention. *Remote Sens.* **2021**, *13*, 3192. [CrossRef]

17. Lin, T.Y.; Maire, M.; Belongie, S.; Hays, J.; Perona, P.; Ramanan, D.; Dollár, P.; Zitnick, C.L. Microsoft coco: Common objects in context. In Proceedings of the European Conference on Computer Vision, Zurich, Switzerland, 6–12 September 2014; pp. 740–755.
18. Everingham, M.; Van Gool, L.; Williams, C.K.; Winn, J.; Zisserman, A. The pascal visual object classes (voc) challenge. *Int. J. Comput. Vis.* **2010**, *88*, 303–338. [CrossRef]
19. Yang, X.; Yan, J. Arbitrary-oriented object detection with circular smooth label. In Proceedings of the European Conference on Computer Vision, Glasgow, UK, 23–28 August 2020; pp. 677–694.
20. Dai, J.; Qi, H.; Xiong, Y.; Li, Y.; Zhang, G.; Hu, H.; Wei, Y. Deformable convolutional networks. In Proceedings of the IEEE International Conference on Computer Vision, Venice, Italy, 22–29 October 2017; pp. 764–773.
21. Zhu, X.; Hu, H.; Lin, S.; Dai, J. Deformable convnets v2: More deformable, better results. In Proceedings of the IEEE/CVF Conference on Computer Vision and Pattern Recognition, Long Beach, CA, USA, 16–20 June 2019; pp. 9308–9316.
22. Ren, S.; He, K.; Girshick, R.; Sun, J. Faster r-cnn: Towards real-time object detection with region proposal networks. *Adv. Neural Inf. Process. Syst.* **2015**, *28*, 91–99. [CrossRef] [PubMed]
23. He, K.; Gkioxari, G.; Dollár, P.; Girshick, R. Mask r-cnn. In Proceedings of the IEEE International Conference on Computer Vision, Venice, Italy, 22–29 October 2017; pp. 2961–2969.
24. Dai, J.; Li, Y.; He, K.; Sun, J. R-fcn: Object detection via region-based fully convolutional networks. In Proceedings of the Advances in Neural Information Processing Systems, Barcelona, Spain, 5–10 December 2016; pp. 379–387.
25. Lin, T.Y.; Dollár, P.; Girshick, R.; He, K.; Hariharan, B.; Belongie, S. Feature pyramid networks for object detection. In Proceedings of the IEEE Conference On Computer Vision and Pattern Recognition, Honolulu, HI, USA, 21–26 July 2017; pp. 2117–2125.
26. Girshick, R. Fast r-cnn. In Proceedings of the IEEE International Conference on Computer Vision, Santiago, Chile, 7–13 December 2015; pp. 1440–1448.
27. Liu, W.; Anguelov, D.; Erhan, D.; Szegedy, C.; Reed, S.; Fu, C.Y.; Berg, A.C. Ssd: Single shot multibox detector. In Proceedings of the European Conference on Computer Vision, Amsterdam, The Netherlands, 8–16 October 2016; pp. 21–37.
28. Redmon, J.; Divvala, S.; Girshick, R.; Farhadi, A. You only look once: Unified, real-time object detection. In Proceedings of the IEEE Conference on Computer Vision and Pattern Recognition, Las Vegas, NV, USA, 27–30 June 2016; pp. 779–788.
29. Redmon, J.; Farhadi, A. YOLO9000: Better, faster, stronger. In Proceedings of the IEEE Conference on Computer Vision and Pattern Recognition, Honolulu, HI, USA, 21–26 July 2017; pp. 7263–7271.
30. Redmon, J.; Farhadi, A. Yolov3: An incremental improvement. *arXiv* **2018**, arXiv:1804.02767.
31. Lin, T.Y.; Goyal, P.; Girshick, R.; He, K.; Dollár, P. Focal loss for dense object detection. In Proceedings of the IEEE International Conference on Computer Vision, Venice, Italy, 22–29 October 2017; pp. 2980–2988.
32. Zhang, S.; Wen, L.; Bian, X.; Lei, Z.; Li, S.Z. Single-Shot refinement neural network for object detection. In Proceedings of the IEEE Conference on Computer Vision and Pattern Recognition (CVPR), Salt Lake City, UT, USA, 18–22 June 2018; pp. 4203–4212.
33. Liu, S.; Huang, D. Receptive field block net for accurate and fast object detection. In Proceedings of the European Conference on Computer Vision (ECCV), Munich, Germany, 8–14 September 2018; pp. 385–400.
34. Zhou, X.; Zhuo, J.; Krahenbuhl, P. Bottom-up object detection by grouping extreme and center points. In Proceedings of the IEEE/CVF Conference on Computer Vision and Pattern Recognition, Long Beach, CA, USA, 16–20 June 2019; pp. 850–859.
35. Zhang, S.; Chi, C.; Yao, Y.; Lei, Z.; Li, S.Z. Bridging the gap between anchor-based and anchor-free detection via adaptive training sample selection. In Proceedings of the IEEE/CVF Conference on Computer Vision and Pattern Recognition, Seattle, WA, USA, 14–19 June 2020; pp. 9759–9768.
36. Law, H.; Deng, J. Cornernet: Detecting objects as paired keypoints. In Proceedings of the European Conference on Computer Vision (ECCV), Munich, Germany, 8–14 September 2018; pp. 734–750.
37. Zhou, X.; Wang, D.; Krähenbühl, P. Objects as points. *arXiv* **2019**, arXiv:1904.07850.
38. Tian, Z.; Shen, C.; Chen, H.; He, T. Fcos: Fully convolutional one-stage object detection. In Proceedings of the IEEE/CVF International Conference on Computer Vision, Seoul, Korea, 27 October–2 November 2019; pp. 9627–9636.
39. Ma, J.; Shao, W.; Ye, H.; Wang, L.; Wang, H.; Zheng, Y.; Xue, X. Arbitrary-oriented scene text detection via rotation proposals. *IEEE Trans. Multimed.* **2018**, *20*, 3111–3122. [CrossRef]
40. Liao, M.; Shi, B.; Bai, X. Textboxes++: A single-shot oriented scene text detector. *IEEE Trans. Image Process.* **2018**, *27*, 3676–3690. [CrossRef]
41. Azimi, S.M.; Vig, E.; Bahmanyar, R.; Körner, M.; Reinartz, P. Towards multi-class object detection in unconstrained remote sensing imagery. In Proceedings of the Asian Conference on Computer Vision, Perth, Australia, 2–6 December 2018; pp. 150–165.
42. Ding, J.; Xue, N.; Long, Y.; Xia, G.S.; Lu, Q. Learning roi transformer for oriented object detection in aerial images. In Proceedings of the IEEE/CVF Conference on Computer Vision and Pattern Recognition, Long Beach, CA, USA, 16–20 June 2019; pp. 2849–2858.
43. Yang, X.; Liu, Q.; Yan, J.; Li, A.; Zhang, Z.; Yu, G. R3det: Refined single-stage detector with feature refinement for rotating object. *arXiv* **2019**, arXiv:1908.05612.
44. Li, Y.; Mao, H.; Liu, R.; Pei, X.; Jiao, L.; Shang, R. A lightweight keypoint-based oriented object detection of remote sensing images. *Remote Sens.* **2021**, *13*, 2459. [CrossRef]
45. Ming, Q.; Miao, L.; Zhou, Z.; Song, J.; Yang, X. Sparse label assignment for oriented object detection in aerial images. *Remote Sens.* **2021**, *13*, 2664. [CrossRef]
46. Qing, Y.; Liu, W.; Feng, L.; Gao, W. Improved YOLO network for free-angle remote sensing target detection. *Remote Sens.* **2021**, *13*, 2171. [CrossRef]

47. Yang, X.; Sun, H.; Fu, K.; Yang, J.; Sun, X.; Yan, M.; Guo, Z. Automatic Ship Detection in Remote Sensing Images from Google Earth of Complex Scenes Based on Multiscale Rotation Dense Feature Pyramid Networks. *Remote Sens.* **2018**, *10*, 132. [CrossRef]
48. He, K.; Zhang, X.; Ren, S.; Sun, J. Deep residual learning for image recognition. In Proceedings of the IEEE Conference on Computer Vision and Pattern Recognition, Las Vegas, NV, USA, 27–30 June 2016; pp. 770–778.
49. Deng, J.; Dong, W.; Socher, R.; Li, L.J.; Li, K.; Fei-Fei, L. Imagenet: A large-scale hierarchical image database. In Proceedings of the 2009 IEEE Conference on Computer Vision and Pattern Recognition, Miami, FL, USA, 20–25 June 2009; pp. 248–255.
50. Kingma, D.P.; Ba, J. Adam: A method for stochastic optimization. *arXiv* **2014**, arXiv:1412.6980.
51. Zhang, G.; Lu, S.; Zhang, W. CAD-Net: A context-aware detection network for objects in remote sensing imagery. *IEEE Trans. Geosci. Remote. Sens.* **2019**, *57*, 10015–10024. [CrossRef]
52. Li, C.; Xu, C.; Cui, Z.; Wang, D.; Zhang, T.; Yang, J. Feature-attentioned object detection in remote sensing imagery. In Proceedings of the 2019 IEEE International Conference on Image Processing (ICIP), Taipei, Taiwan, 22–25 September 2019; pp. 3886–3890.
53. Yang, F.; Li, W.; Hu, H.; Li, W.; Wang, P. Multi-scale feature integrated attention-based rotation network for object detection in VHR aerial images. *Sensors* **2020**, *20*, 1686. [CrossRef]
54. Qian, W.; Yang, X.; Peng, S.; Guo, Y.; Yan, J. Learning modulated loss for rotated object detection. *arXiv* **2019**, arXiv:1911.08299.
55. Jiang, Y.; Zhu, X.; Wang, X.; Yang, S.; Li, W.; Wang, H.; Fu, P.; Luo, Z. R2cnn: Rotational region cnn for orientation robust scene text detection. *arXiv* **2017**, arXiv:1706.09579.
56. Yi, J.; Wu, P.; Liu, B.; Huang, Q.; Qu, H.; Metaxas, D. Oriented object detection in aerial images with box boundary-aware vectors. In Proceedings of the IEEE/CVF Winter Conference on Applications of Computer Vision, Waikoloa, HI, USA, 5–9 January 2021; pp. 2150–2159.
57. Yang, X.; Yang, J.; Yan, J.; Zhang, Y.; Zhang, T.; Guo, Z.; Sun, X.; Fu, K. Scrdet: Towards more robust detection for small, cluttered and rotated objects. In Proceedings of the IEEE/CVF International Conference on Computer Vision, Seoul, Korea, 27 October–2 November 2019; pp. 8232–8241.
58. Wang, Y.; Zhang, Y.; Zhang, Y.; Zhao, L.; Sun, X.; Guo, Z. SARD: Towards scale-aware rotated object detection in aerial imagery. *IEEE Access* **2019**, *7*, 173855–173865. [CrossRef]
59. Li, C.; Luo, B.; Hong, H.; Su, X.; Wang, Y.; Liu, J.; Wang, C.; Zhang, J.; Wei, L. Object Detection Based on Global-Local Saliency Constraint in Aerial Images. *Remote Sens.* **2020**, *12*, 1435. [CrossRef]
60. Yang, X.; Hou, L.; Zhou, Y.; Wang, W.; Yan, J. Dense label encoding for boundary discontinuity free rotation detection. In Proceedings of the IEEE/CVF Conference on Computer Vision and Pattern Recognition, Nashville, TN, USA, 19–25 June 2021; pp. 15819–15829.

remote sensing

MDPI

Article

Learning Rotated Inscribed Ellipse for Oriented Object Detection in Remote Sensing Images

Xu He [1], Shiping Ma [1], Linyuan He [1,2,*], Le Ru [1] and Chen Wang [1]

[1] Aeronautics Engineering College, Air Force Engineering University, Xi'an 710038, China; dvhv26@163.com (X.H.); mashiping@126.com (S.M.); ru-le@126.com (L.R.); wwangchen77@163.com (C.W.)
[2] Unbanned System Research Institute, Northwestern Polytechnical University, Xi'an 710072, China
* Correspondence: hal1983@163.com

Abstract: Oriented object detection in remote sensing images (RSIs) is a significant yet challenging Earth Vision task, as the objects in RSIs usually emerge with complicated backgrounds, arbitrary orientations, multi-scale distributions, and dramatic aspect ratio variations. Existing oriented object detectors are mostly inherited from the anchor-based paradigm. However, the prominent performance of high-precision and real-time detection with anchor-based detectors is overshadowed by the design limitations of tediously rotated anchors. By using the simplicity and efficiency of keypoint-based detection, in this work, we extend a keypoint-based detector to the task of oriented object detection in RSIs. Specifically, we first simplify the oriented bounding box (OBB) as a center-based rotated inscribed ellipse (RIE), and then employ six parameters to represent the RIE inside each OBB: the center point position of the RIE, the offsets of the long half axis, the length of the short half axis, and an orientation label. In addition, to resolve the influence of complex backgrounds and large-scale variations, a high-resolution gated aggregation network (HRGANet) is designed to identify the targets of interest from complex backgrounds and fuse multi-scale features by using a gated aggregation model (GAM). Furthermore, by analyzing the influence of eccentricity on orientation error, eccentricity-wise orientation loss (ewoLoss) is proposed to assign the penalties on the orientation loss based on the eccentricity of the RIE, which effectively improves the accuracy of the detection of oriented objects with a large aspect ratio. Extensive experimental results on the DOTA and HRSC2016 datasets demonstrate the effectiveness of the proposed method.

Keywords: oriented object detection; rotated inscribed ellipse; remote sensing images; keypoint-based detection; gated aggregation; eccentricity-wise

Citation: He, X.; Ma, S.; He, L.; Ru, L.; Wang, C. Learning Rotated Inscribed Ellipse for Oriented Object Detection in Remote Sensing Images. *Remote Sens.* **2021**, *13*, 3622. https://doi.org/10.3390/rs13183622

Academic Editors: Fahimeh Farahnakian, Jukka Heikkonen and Pouya Jafarzadeh

Received: 8 August 2021
Accepted: 7 September 2021
Published: 10 September 2021

1. Introduction

With the fast-paced development of unmanned aerial vehicles (UAVs) and remote sensing technology, the analysis of remote sensing images (RSIs) has been increasingly applied in fields such as land surveying, environmental monitoring, intelligent transportation, seabed mapping, heritage site reconstruction, and so on [1–6]. Object detection in RSIs is regarded as a high-level computer vision task with the purpose of pinpointing the targets in RSIs. Due to the characteristics of remote sensing targets, such as complex backgrounds, huge aspect ratios, multiple scales, and variations of orientations, remote sensing object detection remains a challenging and significant research issue.

In recent years, due to their outstanding learning abilities, the most advanced detection models have been developed by using deep convolutional neural networks (DCNNs). Existing natural image object detection approaches [7–18] usually leverage the horizontal detection paradigm, which has evolved into a well-established area. Nevertheless, remote sensing images are typically taken with bird's-eye views, and horizontal-detection-based methods will experience significant performance degradation when applied directly to remote sensing images, largely owing to the distinctive appearances and characteristics of remote sensing objects. For instance, compared with the detection of objects in images of natural scenes, the task of remote sensing object detection tends to encompass more challenges, such as complex backgrounds, arbitrary orientations, multi-scale distributions, and large aspect ratio variations. When we take the horizontal bounding box (HBB) in the top half of Figure 1a to represent the objects of a remote sensing image, it will introduce massive numbers of extra pixels outside of the targets, seriously damaging the accuracy of positioning. Meanwhile, the HBB used for densely arranged remote sensing oriented objects may generate a larger intersection-over-union (IoU) with adjacent boxes, which tends to introduce some missed ground-truth boxes that are restrained by non-maximum suppression (NMS), and the missed detection rate increases. To tackle these challenges, oriented object detection methods that utilize an oriented bounding box (OBB) to compactly enclose an object with orientations are preferred in RSIs.

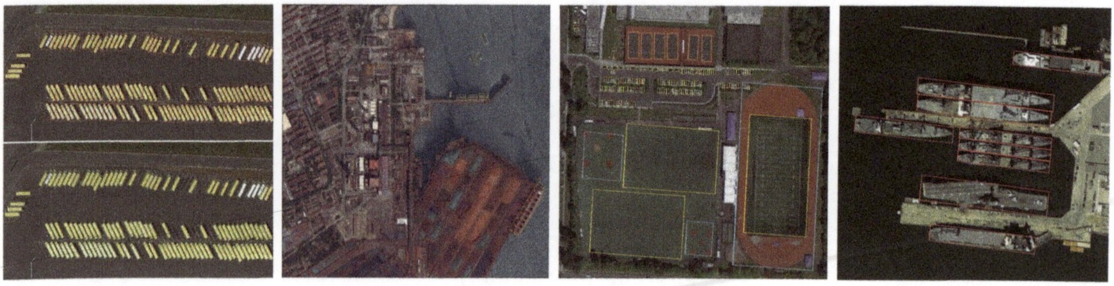

(a) (b) (c) (d)

Figure 1. Some RSIs in the DOTA and HRSC2016 datasets. (**a**) The direction of objects in RSIs is always arbitrary. The HBB (**top**) and OBB (**bottom**) are two representation methods in RSI object detection. (**b**) Remote sensing images tend to contain complex backgrounds. (**c**) The scales of objects in the same remote sensing image may also vary dramatically, such as with small vehicles and track fields on the ground. (**d**) There are many objects with large aspect ratios in RSIs, such as slender ships.

Existing oriented object detectors are mainly inherited from the anchor-based detection paradigm. Nevertheless, anchor-based oriented object detectors that rely on anchor mechanisms result in complicated computations and designs related to the rotated anchor boxes, such as those of the orientations, scales, number, and aspect ratios of the anchor boxes. Therefore, research works on anchor-free detection methods that liberate the detection model from massive computations on the anchors have drawn much attention in recent years. Specifically, as an active topic in the field of anchor-free object detection, keypoint-based methods (e.g., CornerNet [13], CenterNet [14], and ExtremeNet [15]) propose forsaking the design of anchors and directly regressing target positions by exploring the features of correlative keypoints either on the box boundary points or the center point. To the best of our knowledge, many works that were built upon the keypoint-based detection pipeline have achieved great success in the RSI object detection field. For example, P-RSDet [19] converted the task of detection of remote sensing targets into the regression of polar radii and polar angles based on the center pole point in polar coordinates. GRS-Det [20] proposed an anchor-free center-based ship detection algorithm based on a unique U-shape network design and a rotation Gaussian Mask. The VCSOP detector [21] transformed the vehicle detection task into a multitask learning problem (i.e., center, scale, orientation, and offset subtasks) via an anchor-free one-stage fully convolutional network (FCN). Due to the bird's-eye views in RSIs, center-based methods that have fewer ambiguous samples and vivid object representation are more suitable for remote sensing oriented object detection. Notably, center-based methods usually extend the CenterNet [14] to the oriented object detection task by introducing an accessional angle θ together with the width w and height h. However, due to the periodicity of the angle, angle-based approaches that represent the oriented object with the angle-oriented OBB will encounter boundary discontinuity and regression uncertainty issues [22], resulting in serious damage to the detection performance. To address this problem, our work explores an angle-free method according to the geometric characteristics of the OBB. Specifically, we describe an OBB as a center-based rotated inscribed ellipse (RIE), and then employ six parameters to describe the RIE inside each OBB: the center point position of the RIE (center point (x, y)), the offsets of the long half axis (δ_x, δ_y), the length of the short half axis b, and an orientation label ψ. In contrast to the angle-based approaches, our angle-free OBB definition guarantees the uniqueness of the representation of the OBB and effectively eliminates the boundary case, which dramatically improves the detection accuracy.

On the other hand, trapped by the complicated backgrounds and multi-scale object distribution in RSIs, as shown in Figure 1b,c, keypoint-based detectors that utilize a single-scale high-resolution feature map to make predictions may detect a large number of uninteresting objects and omit some objects with multiple scales. Therefore, it is momentous to enhance the feature extraction capability and improve the multi-scale information fusion of the backbone network. In our work, we design a high-resolution gated aggregation network (HRGANet) that better distinguishes the objects of interest from complex backgrounds and integrates the features with different scales by using a parallel multi-scale information interaction and gated aggregation information fusion mechanisms. In addition, because large aspect ratios tend to make a significant impact on the orientation error and the accuracy of the IoU, it is reasonable to assign penalties on the orientation loss based on the aspect ratio information. Taking the perspective that the eccentricity of the RIE can better reflect the aspect ratio from the side, we propose an eccentricity-wise orientation loss (ewoLoss) to penalize the orientation loss based on the eccentricity of the RIE, which effectively takes into consideration the effect of the aspect ratio on the orientation error and improves the accuracy of the detection of slender objects.

In summary, the contributions of this article are four-fold:

- We introduce a novel center-based OBB representation method called the rotated inscribed ellipse (RIE). As an angle-free OBB definition, the RIE effectively eliminates the angle periodicity and address the boundary case issues;
- We design a high-resolution gated aggregation network to capture the objects of interest from complicated backgrounds and integrate different scale features by implementing multi-scale parallel interactions and gated aggregation fusion;
- We propose an eccentricity-wise orientation loss function to fix the sensitivity of the eccentricity of the ellipse to the orientation error and effectively improve the accuracy of the detection of slender oriented objects with large aspect ratios;
- We perform extensive experiments to verify the advanced performance compared with state-of-the-art oriented object detectors on remote sensing datasets.

The rest of this article is structured as follows. Section 2 introduces the related work in detail. The detailed introduction of our method is explained in Section 3. In Section 4, we explain the extensive comparison experiments, the ablation study, and the experimental analysis at length. Finally, the conclusion is presented in Section 5.

2. Related Works

In this section, relevant works concerning deep-learning-based oriented object detection methods and anchor-free object detection methods in RSIs are briefly reviewed.

2.1. Oriented Object Detection in RSIs

Considering the rotation characteristics of remote sensing objects, it is more suitable to employ a rotated bounding box to represent objects with multiple orientations and to devise advanced oriented object detection algorithms that adapt to remote sensing scenes.

Recent advances in oriented object detection have mainly been driven by the improvements and promotion of general object detection methods that use horizontal bounding boxes to represent remote sensing objects. In general, the mainstream and classical oriented object detection algorithms in RSIs can be roughly divided into anchor-based paradigms and anchor-free object detection methods. The anchor-based detectors (e.g., YOLO [10], SSD [11], Faster-RCNN [9], and RetinaNet [12]) have dominated the field of object detection for many years. Specifically, for a remote sensing image, the anchor-based detectors first utilize many predetermined anchors with different sizes, aspect ratios, and rotation angles as a reference. Then, the detector either directly regresses the location of the object bounding box or generates region proposals on the basis of anchors and determines whether each region contains some category of an object. Inspired by this kind of ingenious anchor mechanism, a large number of oriented object detectors [22–41] have been proposed in the literature to pinpoint oriented objects in RSIs. For example, Liu et al. [23] used the Faster-RCNN framework and introduced a rotated region of interest (ROI) for the task of the detection of oriented ships in RSIs. The method in [24,25] used a rotation-invariant convolutional neural network to address the problem of inter-class similarity and intra-class diversity in multi-class RSI object detection. The RoI Transformer [30] employed a strategy of transforming from a horizontal RoI to an oriented RoI and allowed the network to obtain the OBB representation with a supervised RoI learner. With the aim of application for rotated ships, the R^2PN [31] transformed the original region proposal network (RPN) into a rotated region proposal network (R^2PN) to generate oriented proposals with orientation information. CAD-Net [32] used a local and global context network to obtain the object-level and scene contextual clues for robust oriented object detection in RSIs. The work in [33] proposed an iterative one-stage feature refinement detection network that transformed the horizontal object detection method into an oriented object detection method and effectively improved the RSI detection performance. In order to predict the angle-based OBB, SCRDet [34] applied an IoU penalty factor to the general smooth L1 loss function, which cleverly addressed the angular periodicity and boundary issues for accurate oriented object detection tasks. S^2A-Net [35] realized the effect of feature alignment between the horizontal features and oriented objects through an one-stage fully convolutional network (FCN).

In addition to effective feature extraction network designs for oriented objects mentioned above, some scholars have studied the sensitivity of angle regression errors to anchor-based detection methods and resorted to more robust angle-free OBB representations for oriented object detection in RSIs. For instance, Xu et al. [36] represented an arbitrarily oriented object by employing a gliding vertex on the four corners based on the HBB, which refrained from the regression of the angle. The work in [37] introduced a two-dimensional vector to express the rotated angle and explored a length-independent and fast IoU calculation method for the purpose of better slender object detection. Furthermore, Yang et al. [22] transformed the task of the regression of an angle into a classification task by using an ingenious circular smooth label (CSL) design, which eliminated the angle periodicity problem in the process of regression. As a continuation of the CSL work, densely coded labels (DCLs) [38] were used to further explore the defects of CSLs, and a novel coding mode that made the model more sensitive to the angular classification distance and the aspect ratios of objects was proposed. ProjBB [39] addressed the regression uncertainty issue caused by the rotation angle with a novel projection-based angle-free OBB representation approach. Not singly, but in pairs, the purpose of our work is also to explore an angle-free OBB representation for better oriented object detection in remote sensing images.

2.2. Anchor-Free Object Detection in RSIs

Recently, as an active theme in the field of remote sensing object detection, anchor-free methods have been put forward to abandon the paradigm of anchors and to regress the bounding box directly through a sequence of convolution operations. In general, anchor-free methods can be classified into per-pixel point-based detectors and keypoint-based detectors. The per-pixel point-based detectors (e.g., DenseBox [16], FoveaBox [17], and FCOS [18]) detect objects by predicting whether a pixel point is positive and the offsets from the corresponding per-pixel point to the box boundaries of the target. DenseBox [16] first attempted to employ an anchor-free pipeline to directly predict the classification confidence and bounding box localization with an FCN. FCOS [18] detected an object by predicting four distances from pixel points to four boundaries of the bounding box. Meanwhile, FCOS also introduced a weight factor, Centerness, to evaluate the importance of the positive pixel points and steer the network to distinguish discriminative features from complicated backgrounds. FoveaBox [17] located the object box by directly predicting the mapping transformation relation between center points and two corner points, and it learned the object category of confidence. Inspired by this paradigm of detection, many researchers began to explore per-pixel point-based oriented object detection approaches for RSIs. For example, based on the FCOS pipeline, IENet [42] proposed an interacting module in the detection head to bind the classification and localization branches for accurate oriented object detection in RSIs. In addition, IENet also introduced a novel OBB representation method that depicted oriented objects with an outsourcing box of the OBB. Axis Learning [43] used a per-pixel point-based detection model that detected the orientated objects by predicting the axis of an object and the width perpendicular to the axis.

Differently from per-pixel point-based methods, keypoint-based methods (e.g., CornerNet [13], CenterNet [14], and ExtremeNet [15]) pinpoint oriented objects by capturing the correlative keypoints, such as the corner point, center point, and extreme point. CornerNet is the forerunner of the keypoint-based methods; it locates the HBB of an object through heatmaps of the upper-left and bottom-right points. It groups the corner points of the box by evaluating the embedding distances. CenterNet captures an object by using a center keypoint and regressing the width, height, and offset properties of the bounding box. ExtremeNet detects an object through an extreme point (extreme points of four boundaries) and center point estimation network. In the remote sensing oriented object detection field, many works have based themselves upon the keypoint-based detection framework. For example, combining CornerNet and CenterNet, Chen et al. [44] utilized an end-to-end FCN to identity an OBB according to the corners, center, and corresponding angle of a ship. CBDA-Net [45] extracted rotated objects in RSIs by introducing a boundary region and center region attention module and used an aspect-ratio-wise angle loss for slender objects. The work in [46] proposed a pixel-wise IoU loss function that enhances the relation between the angle offset and the IoU and effectively improves the detection performance for objects with high aspect ratios. Pan et al. [47] introduced a unified dynamic refinement network to extract densely packed oriented objects according to the selected shape and orientation features. Meanwhile, there are also some works that have integrated the angle-free strategy into the keypoint-based detection pipeline for RSIs. O^2-DNet [48] utilized a center point detection network to locate the intersection point and formed an OBB representation with a pair of internal middle lines. X-LineNet [49] detected aircraft by predicting and clustering the paired vertical intersecting line segments inside each bounding box. BBAVectors [50] captured an oriented object by learning the box-boundary-aware vectors that were distributed in four independent quadrants of the Cartesian coordinate system. Continuing this angle-free thought, the method proposed in this article uses a center-based rotated inscribed ellipse to represent the OBB. At the same time, our method provides a strong feature extraction network to extract objects from complex backgrounds and implements an aspect-ratio-wise orientation loss for slender objects, which effectively boosts the performance in oriented object detection in RSIs. A more detailed introduction of the proposed method will be provided in Section 3.

3. Materials and Methods

The architecture of the proposed method is illustrated in Figure 2. The network framework mainly includes a feature extraction network—namely, a high-resolution gated aggregation network (HRGANet) and a multitask prediction head. The HRGANet is designed to tackle the problems of extensive multi-scale distributed objects and complex backgrounds in RSIs. The HRGANet can be divided into the backbone, the high-resolution network (HRNet) [51], and the gated aggregation model. The HRNet is a parallel-interaction high-resolution network that is utilized to fuse multi-resolution feature representations and render high-resolution representations for richer semantic information and more precise spatial positioning information. The gated aggregation module (GAM) is proposed to adaptively fuse different resolution feature maps for multi-scale objects through a gated aggregation mechanism. Meanwhile, there are five subnetworks for the center heatmap, center offset, long half-axis offset, eccentricity, and orientation prediction in the oriented object detection head. Finally, oriented objects are detected by predicting the inscribed ellipse with orientation information inside each OBB. In addition, we utilize ewoLoss to penalize the orientation loss based on the eccentricity of the rotated inscribed ellipse for better slender object detection. We will introduce the network from three perspectives: (1) the high-resolution gated aggregation network; (2) the rotated inscribed ellipse prediction head; (3) the eccentricity-wise orientation loss.

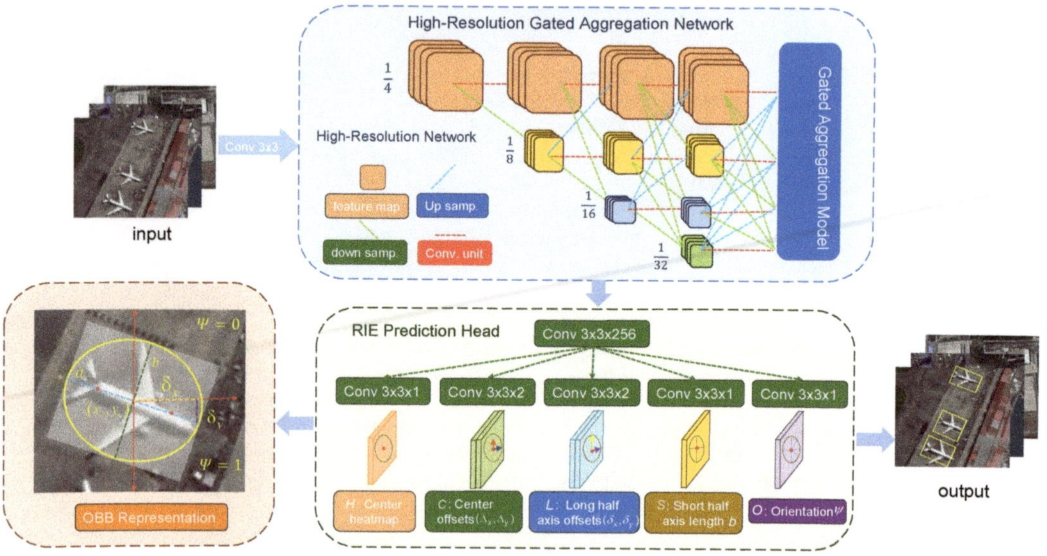

Figure 2. Framework of the our method. The backbone network, HRGANet, is followed by the RIE prediction model. The HRGANet backbone network contains HRNet and GAM. Up samp. represents a bilinear upsampling operation and a 1×1 convolution. Down samp. denotes 3×3 convolution with a stride of 2. Conv unit. is a 1×1 convolution.

3.1. High-Resolution Gated Aggregation Network

As illustrated in Section 1, the objects in remote sensing images tend to have the characteristics of large-scale variations and complicated backgrounds. Therefore, it is necessary to design an effective feature extraction network that fully exploits the multi-resolution feature representations and fuses multi-scale information for robust multi-scale feature extraction. From this point of view, we introduce a high-resolution gated aggregation network (HRGANet) to make good use of the multi-resolution feature maps. As shown in Figure 2, the HRGANet is composed of two components: the high-resolution network (HRNet) and gated aggregation model (GAM).

3.1.1. High-Resolution Network

The backbone network for feature extraction in our method uses HRNet, which performs well in keypoint detection. In contrast to the frequently used keypoint extraction networks (e.g., VGG [52], ResNet [53], and Hourglass [54]) that concatenate the multi-resolution feature maps in series, HRNet links different-resolution feature maps in parallel with repeated multi-scale fusion. The whole procedure of keypoint extraction in HRNet efficiently keeps high-resolution features while replenishing the high- and low-resolution information, which enables it to obtain abundant multi-scale feature representations. The brief sketch of HRNet is illustrated in Figure 2. First, the input image is fed into a stem, which consists of two 3 × 3 convolutions with a stride of 2. Then, the resolution is decreased to 1/4. The overall structure of HRNet has four main stages, which gradually add high-to-low resolution stages in succession. The structure of these four stages can be simplified, as indicated in the following formula:

$$\begin{array}{llll} \mathcal{S}_{11} \rightarrow \mathcal{S}_{21} \rightarrow \mathcal{S}_{31} \rightarrow \mathcal{S}_{41} \\ \qquad \searrow \mathcal{S}_{22} \rightarrow \mathcal{S}_{32} \rightarrow \mathcal{S}_{42} \\ \qquad\qquad \searrow \mathcal{S}_{33} \rightarrow \mathcal{S}_{43} \\ \qquad\qquad\qquad \searrow \mathcal{S}_{44} \end{array} \tag{1}$$

where $\{(\mathcal{S}_{ij})|i,j \in 1,2,3,4\}$ represents the ith sub-stage and $j \in \{1,2,3,4\}$ denotes that the resolution of feature maps in the corresponding sub-stage is $\frac{1}{2^{(j+1)}}$ of the original feature maps. Meanwhile, through repeated multi-resolution feature fusion and parallel high-resolution feature maintenance, HRNet can better extract multi-scale features, and then obtain richer semantic and spatial information for RSI objects. The detailed network structure of HRNet is shown in Table 1. Note that we used HRNet-W48 in our experiments.

Table 1. The structure of the backbone network of HRNet. It mainly embodies four stages. The 1st (2nd, 3rd, and 4th) stage is composed of 1 (1, 4, and 3) repeated modularized blocks. Meanwhile, each modularized block in the 1st (2nd, 3rd, and 4th) stage consists of 1 (2, 3, and 4) branch(es) belonging to a different resolution. Each branch contains four residual units and one fusion unit. In the table, each cell in the Stage box is composed of three parts: The first part ([·]) represents the residual unit, the second number denotes the iteration times of the residual units, and the third number represents the iteration times of the modularized blocks. ≡ in the Fusion column represents the fusion unit. C is the channel number of the residual unit. We set C to 48 and represent the network as HRNet-W48. Res. is the abbreviation of resolution.

Res.	Stage1	Fusion	Stage2	Fusion	Stage3	Fusion	Stage4	Fusion
$\frac{1}{4}$	$\begin{bmatrix} 1\times1,64 \\ 3\times3,64 \\ 1\times1,256 \end{bmatrix} \times 4 \times 1$	≡	$\begin{bmatrix} 3\times3,C \\ 3\times3,C \end{bmatrix} \times 4 \times 1$	≡	$\begin{bmatrix} 3\times3,C \\ 3\times3,C \end{bmatrix} \times 4 \times 4$	≡	$\begin{bmatrix} 3\times3,C \\ 3\times3,C \end{bmatrix} \times 4 \times 3$	≡
$\frac{1}{8}$		≡	$\begin{bmatrix} 3\times3,2C \\ 3\times3,2C \end{bmatrix} \times 4 \times 1$	≡	$\begin{bmatrix} 3\times3,2C \\ 3\times3,2C \end{bmatrix} \times 4 \times 4$	≡	$\begin{bmatrix} 3\times3,2C \\ 3\times3,2C \end{bmatrix} \times 4 \times 3$	≡
$\frac{1}{16}$				≡	$\begin{bmatrix} 3\times3,4C \\ 3\times3,4C \end{bmatrix} \times 4 \times 4$	≡	$\begin{bmatrix} 3\times3,4C \\ 3\times3,4C \end{bmatrix} \times 4 \times 3$	≡
$\frac{1}{32}$						≡	$\begin{bmatrix} 3\times3,8C \\ 3\times3,8C \end{bmatrix} \times 4 \times 3$	≡

3.1.2. Gated Aggregation Model

In a conventional keypoint or object detection network, the feature aggregation pattern is carried out by directly stacking or concatenating the feature maps. Nevertheless, to the best of our knowledge, feature maps with different resolutions contain serious semantic dissimilarities. In general, low-resolution feature maps provide richer semantics for object category recognition, whereas high-resolution feature maps contain more spatial information for object localization. Some works [21,50] directly up-sampled low-resolution feature maps ($\frac{1}{8}$, $\frac{1}{16}$, and $\frac{1}{32}$ feature maps) to a $\frac{1}{4}$ spatial resolution, and then fused these feature maps through a concatenation operation. This kind of fusion strategy does not consider that some of these features are meaningless or a hindrance to the inference of object identification and positioning. To enhance valuable feature representations and restrain invalid information, we designed a gated aggregation mechanism (GAM) to evaluate the availability of pixels in each feature map and effectively fuse multi-resolution feature representations. The details of the GAM are presented in Figure 3. As shown in Figure 2, the inputs $\{F_i \in \mathbf{R}^{\frac{W}{2^{i+1}} \times \frac{H}{2^{i+1}} \times 2^{i-1}C}, i \in \{1,2,3,4\}\}$ of the GAM are the output feature maps of the HRNet, where W, H, and C represent the width, height, and channel number of the feature maps, respectively. In Figure 3, F_2, F_3, and F_4 are up-sampled to the same $\frac{1}{4}$ resolution as F_1. We can obtain the feature maps of the same scale $\{X_i \in \mathbf{R}^{\frac{W}{4}\times\frac{H}{4}\times 2^{i-1}C}, i \in \{1,2,3,4\}\}$. Then, X_i is fed into a weight block to adaptively assign the weight of pixels in different feature maps and to generate the weight maps $\{W_i \in \mathbf{R}^{\frac{W}{4}\times\frac{H}{4}\times 1}, i \in \{1,2,3,4\}\}$. W_i can be defined as

$$W_i = \sigma(BN(Conv_{1\times1}(X_i))) \tag{2}$$

where $Conv_{1\times1}$ represents the 1×1 convolution operation in which the number of kernels is equal to 1, BN denotes the batch normalization operation, and σ is the ReLU activation function. These three parts compose a weight block. Then, we employ a SoftMax operation to obtain the normalized gate maps $\{G_i \in \mathbf{R}^{\frac{W}{4}\times\frac{H}{4}\times 1}, i \in \{1,2,3,4\}\}$ as:

$$G_i = \frac{e^{W_i}}{\sum_{j=1}^{4} e^{W_j}} \tag{3}$$

where $G_i \in (0,1)$ is the important gated aggregation factor. Finally, by means of these gate maps, the gated aggregation feature maps output $Y \in \mathbf{R}^{\frac{W}{4}\times\frac{H}{4}\times 15C}$ for the following prediction head, which can be calculated as:

$$Y = \sum_{i=1}^{4} G_i \otimes X_i \tag{4}$$

where the summation symbol represents the concatenation operation $\oplus$. The feature maps are concatenated along the channel direction. Note that we perform a 1×1 convolution to reconcile the final feature maps and integrate the feature maps into the 256 channels after Y. With this gated aggregation strategy, meritorious feature representations are distributed to higher gate factors, and unnecessary information will be suppressed. As a result, our feature extraction network can provide more flexible feature representations in detecting remote sensing objects of different scales from complicated backgrounds.

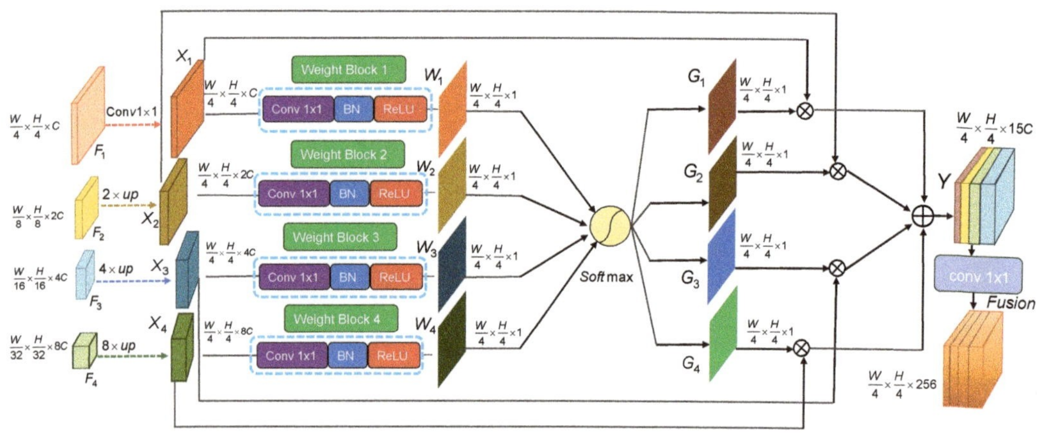

Figure 3. The network structure of the GAM. W, H, and C represent the width, height, and channel number of the feature maps, respectively. $\otimes$ represents the broadcast multiplication operation. $\oplus$ denotes the concatenation operation. $Conv1 \times 1$ is a convolution operation with 1×1 kernels, BN is a batch normalization operation, and ReLU is the ReLU activation function. A weight block is composed of a 1×1 convolution operation, a BN operation, and a ReLU operation.

3.2. Rotated Inscribed Ellipse Prediction Head

To capture the objects in RSIs, some works [21,45] described the OBB with a rotated rectangular box (RRB) representation (x, y, w, h, θ). As shown in Figure 4a, x, y, w, h, and θ represent the center abscissa, center ordinate, width, height, and angle. This representation method has some pros and cons. On the bright side, this definition method can ensure the conciseness and uniqueness of the OBB representation. Nevertheless, there still exist some problems in some extreme conditions. For example, in Figure 4a, the rotation angle θ of the RRB is defined as the angle between the horizontal axis (x-axis) corresponding to the lowest point of the RRB and the first edge encountered when it rotates counterclockwise. The first edge encountered is the width and the other is the height, which are not defined in terms of length. This angular representation has a range of angles of $[0, 90)$. However, a boundary problem emerges due to the angular periodicity when this representation encounters an angular boundary. In Figure 4a, the blue rectangle denotes the angle θ, which is equal to 0. When this rectangle is subjected to a slight jiggle, two very different conditions appear. When we rotate the blue box by a small angle $\triangle\theta$ towards the upper-right corner to reach the position of the red box, the angle of rotation is defined as $\triangle\theta$. However, when we rotate the blue box by a small angle $\triangle\theta$ towards the bottom-right corner to reach the position of the green box, the angle of rotation is defined as $(90 - \triangle\theta)$. This kind of rotation angle representation method causes a large jump in the angle's value during the rotation of the rectangular box from the top-right corner to the bottom-right corner, and the regression of the angle parameter is discontinuous and has serious jitter problems. In addition, for the red box, the long side is the width and the short side is the height. However, for the green box, the short side is the width and the long side is the height. For these two boxes in close proximity, the width and height are abruptly swapped, which makes the regression in terms of width and height less effective. In this condition, a tiny change in the rotation angle would lead to a large change in the regression target, which seriously hinders the training of the network and deteriorates the performance of the detector.

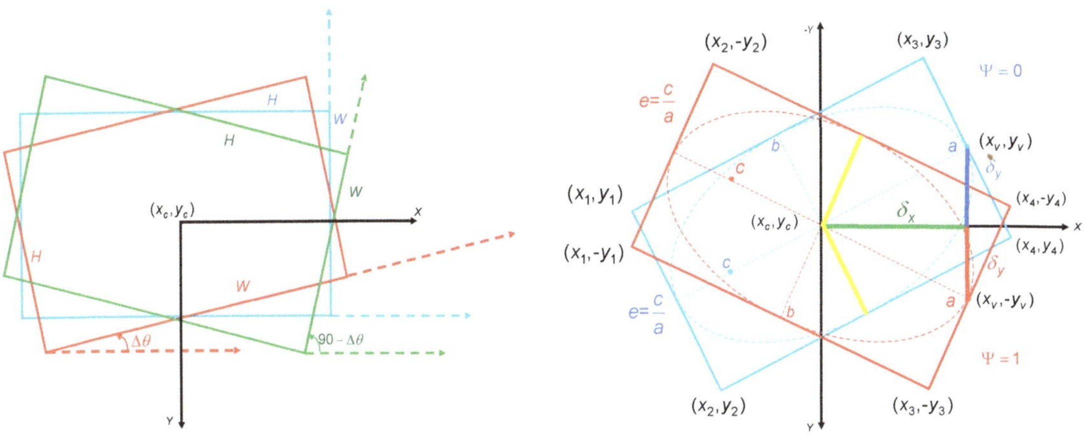

(a) (b)

Figure 4. (**a**) RRB representation (x, y, w, h, θ), where (x_c, y_c), w, h, and $\triangle\theta$ represent the center point, width, height, and small angle jitter, respectively. (**b**) RIE representation of the target used in our method. (x_c, y_c), (x_v, y_v), and $\{(x_i, y_i)|i = 1, 2, 3, 4\}$ are the center point, long half-axis vertex, and four outer rectangle vertices of the RIE. e and ψ represent the eccentricity and orientation label, respectively. Yellow lines b denote the short half axis. Red, blue, and green lines (δ_x, δ_y) represent the offsets of the long half axis a.

To address the above-mentioned problems, we propose a new angle-free OBB representation. We transform the OBB regression task into the corresponding RIE regression problem. First, as shown in Figure 4b, we represent the OBB as four vertices

$\{(x_i, y_i)|i = 1,2,3,4\}$, where the order of the four vertices is based on the values of x_i, i.e., $(x_1 \le x_2 < x_3 \le x_4)$. Then, we can calculate the coordinate of the long half-axis vertices $\{x_v = (x_3 + x_4)/2, y_v = (y_3 + y_4)/2 | x_3 < x_4\}$. When x_3 is equal to x_4, the bounding box is the HBB. The coordinates of the HBB's long half-axis vertices are defined as:

$$(x_v, y_v) = \begin{cases} (\frac{(x_3+x_4)}{2}, \frac{(y_3+y_4)}{2}), & |y_4 - yc| \le |x_4 - x_c| \\ (\frac{(x_2+x_4)}{2}, \frac{(y_2+y_4)}{2}), & |y_4 - yc| > |x_4 - x_c| \end{cases} \quad (5)$$

where $(x_c = \sum_{i=1}^{4} x_i/4, y_c = \sum_{i=1}^{4} y_i/4)$ is the coordinate of the center point. Therefore, we can obtain the long half-axis offsets $(\delta_x = |x_v - x_c|, \delta_y = |y_v - y_c|)$. By predicting the offsets between the long half-axis vertices and the center point, we can obtain the long half-axis length value $a = \sqrt{(\delta_x)^2 + (\delta_y)^2}$. Meanwhile, to obtain the complete size of the RIE, we also implement a sub-network to predict the short half-axis length b. In addition, as shown in Figure 4b, it is not well established to represent a unique RIE by predicting the center point (x, y), long half-axis offsets (δ_x, δ_y), and short half axis b because there are two obscure RIEs with mirror symmetry on the x-axis. To remove this ambiguity, we design an orientation label ψ, and the ground truth of ψ is defined as:

$$\psi = \begin{cases} 0, & (x_v = x_c \mid y_v > y_c) \\ 1, & (x_v > x_c \,\&\, y_v \le y_c) \end{cases} \quad (6)$$

When the long half-axis vertex is located in the 1st quadrant or y-axis, ψ is equal to 0. Meanwhile, when the long half-axis vertex is located in the 4th quadrant or x-axis, ψ is equal to 1. By using such a classification strategy, we can effectively ensure the uniqueness of the RIE representation and eliminate the ambiguity of the definition. Finally, the representation of the RIE can be described by a 6-D vector $(x, y, \delta_x, \delta_y, b, \psi)$. As shown in Figure 2, we introduce an RIE prediction head to obtain the parameters of the RIE. First, a $3 \times 3 \times 256$ convolutional unit is employed to reduce the channel number of the gated aggregated feature maps Y to 256. Then, five parallel 1×1 convolutional units follow to generate a center heatmap $(H \in \mathbf{R}^{\frac{W}{4} \times \frac{H}{4} \times K})$, a center offset map $(C \in \mathbf{R}^{\frac{W}{4} \times \frac{H}{4} \times 2})$, a long half-axis offset map $(L \in \mathbf{R}^{\frac{W}{4} \times \frac{H}{4} \times 2})$, a short-half axis length map $(S \in \mathbf{R}^{\frac{W}{4} \times \frac{H}{4} \times 1})$, and an orientation map $(O \in \mathbf{R}^{\frac{W}{4} \times \frac{H}{4} \times 1})$, where K is the number of categories of the corresponding datasets. Note that the output orientation map is finally processed by a sigmoid function. For the sake of brevity, we have not shown final sigmoid function in Figure 2.

3.3. *Eccentricity-Wise Orientation Loss*

In addition to the characteristics of complex backgrounds, arbitrary orientations, and multi-scale distributions, large aspect ratio variations are also salient characteristics of RSI objects. For example, the aspect ratios of a baseball diamond and a storage tank in RSIs approach 1, but the aspect ratios of long and narrow objects, such as bridges and ships, are even higher than 50. Therefore, it is worthwhile to explore the effects of objects' aspect ratios on the accuracy of the detection of rotated objects. As shown in Figure 5a, we first fix the width, height, and center point of the ground-truth rotated bounding box and the predicted rotated bounding box. Then, we record the IoU value between the ground truth and the predicted box with different angle biases and aspect ratios. We can see from the observation that the sensitivity of the IoU to the angle bias varies considerably for different aspect ratios. First, for the same aspect ratio, the IoU between two rotated boxes decreases as the angle bias increases. Meanwhile, under the same angle bias, the larger the aspect ratio is, the smaller the IoU is. That is, generally, more slender and narrow objects have greater sensitivity to angle deviations. For long and narrow objects, a small angle bias will lead to a large IoU variation. In addition, eccentricity $\{e = \frac{c}{a} \in [0, 1)\}$ is another important index that can reflect the degree of narrowness of an object. The narrower an object is, the larger the eccentricity is. As shown in Figure 5b, we also record the IoU values under different orientation offsets and the eccentricity of the RIE. Under the same orientation offsets, the larger the eccentricity is, the smaller the IoU is. To take full account of the effect of the aspect ratio on the angle prediction bias, we introduce an eccentricity-wise orientation loss (ewoLoss) that utilizes the eccentricity e of the RIE to represent the aspect ratio, and it effectively eliminates the influence of large aspect ratio variations on detection accuracy. First, we propose the utilization of the cosine similarity of the long half axis between the predicted RIE and the ground-truth RIE to calculate the orientation offset. Specifically, with the aid of the ground-truth long half-axis offsets (δ_x^*, δ_y^*) and the predicted long half-axis offsets (δ_x, δ_y), we can calculate the orientation offset $|\triangle\Theta|$ between the predicted long half axis and ground-truth long half axis:

$$\begin{aligned} |\triangle\Theta| &= arccos(\frac{\delta_x^2+\delta_y^2+(\delta_x^*)^2+(\delta_y^*)^2-(\delta_x-\delta_x^*)^2-\left(\delta_y-\delta_y^*\right)^2}{2\sqrt{\delta_x^2+\delta_y^2}*\sqrt{(\delta_x^*)^2+(\delta_y^*)^2}}) \\ &= arccos(\frac{\delta_x*\delta_x^*+\delta_y*\delta_y^*}{\sqrt{\delta_x^2+\delta_y^2}*\sqrt{(\delta_x^*)^2+(\delta_y^*)^2}}) \end{aligned} \tag{7}$$

$arccos$ denotes the inverse cosine function. $|\triangle\Theta|$ indicates the angle error between the predicted RIE and ground-truth RIE. Then, considering that the orientation offsets under different eccentricities have varying influences on the performance of rotated target detection, we hope that the orientation losses under different eccentricities are different, and the orientation losses of targets with greater eccentricities should be larger. The ewoLoss is calculated as:

$$L_{ewo} = \sum_{i=1}^{N}\{(1+exp(e_i-1))\}|\triangle\Theta| \tag{8}$$

where i is the index value of the target, and exp represents the exponential function. e_i is the eccentricity in object i, α is a constant to modulate the orientation loss, and N is the object number in one batch.

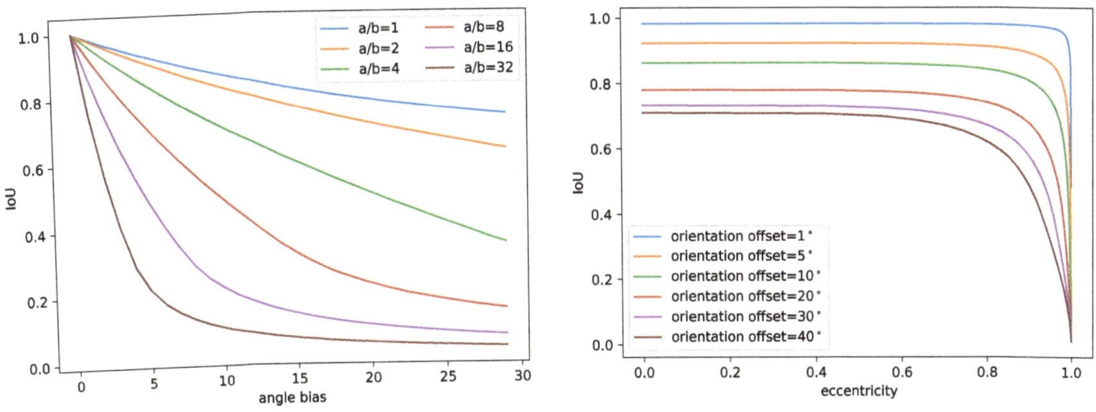

Figure 5. (**a**) IoU curves under different height–width ratios and angle biases. a/b represent the height–width ratio, i.e., the aspect ratio of the object. (**b**) IoU curves under different orientation offsets and RIE eccentricities.

3.4. Loss Functions

Our total loss function is a multi-task loss and is composed of five parts. The first part is the center heatmap loss. A heatmap is a commonly used technical tool applied to keypoint detection tasks in general images. In our work, we inherit the center point heatmap method from the CenterNet [14] network to detect the center points of oriented objects in RSIs. As described in Figure 2, the center heatmap $H \in \mathbf{R}^{\frac{W}{4} \times \frac{H}{4} \times K}$ in the RIE prediction head has K channels, with each belonging to one target category. The value of each predicted pixel point in the heatmap denotes the confidence of detection. We apply a 2-D Gaussian $exp^{(-\frac{(x_h-\tilde{x}_c)^2+(y_h-\tilde{y}_c)^2}{2s^2})}$ around the heatmap of the object's center point $(\tilde{x}_c, \tilde{y}_c)$ to form the ground-truth heatmap $H^* \in \mathbf{R}^{\frac{W}{4} \times \frac{H}{4} \times K}$, where (x_h, y_h) denotes the pixel point in heatmap H^*, and s represents the standard deviation of the adapted object size. Then, following the idea of CornerNet [13], we utilize the variant focal loss to train the regression of the center heatmap:

$$L_h = -\frac{1}{N}\sum_i \begin{cases} (1-h_i)^\gamma \log(h_i), & h_i^* = 1 \\ (1-h_i^*)^\eta h_i^\gamma \log(1-h_i), & \text{otherwise} \end{cases} \tag{9}$$

where h^* and h represent the ground-truth and predicted values of the heatmap, N is the number of targets, and i denotes the pixel location in the heatmap. The hyper-parameters γ and η are set to 2 and 4 in our method to balance the ratio of positive and negative samples. The second part of our loss is the center offset loss. Because the coordinates of the center keypoint on the heatmap are integer values, the ground-truth values of the heatmap are generated by down-sampling the input image through the HRGANet. The size of the ground-truth heatmap is reduced compared to that of the input image, and the discretization process will introduce rounding errors. Therefore, as shown in Figure 2, we introduce center offset maps $C \in \mathbf{R}^{\frac{W}{4} \times \frac{H}{4} \times 2}$ to predict the quantization loss $(\triangle x, \triangle y)$ between the integer center point coordinates and quantified center point coordinates for the mapping of the center point from the input image to the heatmap:

$$\mathbf{c} = (\triangle x, \triangle y) = \left(\frac{x_c}{4} - \left\lfloor \frac{x_c}{4} \right\rfloor, \frac{y_c}{4} - \left\lfloor \frac{y_c}{4} \right\rfloor\right) \tag{10}$$

Smooth L_1 loss is adopted to optimize the center offset as follows:

$$L_c = \frac{1}{N}\sum_{k=1}^{N} \text{Smooth}_{L_1}(\mathbf{c}_k - \mathbf{c}_k^*) \tag{11}$$

where N is the number of targets, $\mathbf{c}^*$ and $\mathbf{c}$ are the ground-truth and predicted values of the offsets, and k denotes the object index number. The smooth L_1 loss can be calculated as:

$$\text{Smooth }_{L_1}(x) = \begin{cases} 0.5x^2, & |x| < 1 \\ |x| - 0.5, & \text{otherwise} \end{cases} \tag{12}$$

The third part of our loss is the box size loss. The box size is composed of the long half-axis offsets (δ_x, δ_y) and the short half-axis length b. We describe the box size with a 3-D vector $\mathbf{B} = (\delta_x, \delta_y, b)$. We also use a smooth L_1 loss to regress the box size parameters:

$$L_b = \frac{1}{N}\sum_{k=1}^{N} \text{Smooth}_{L_1}(\mathbf{B}_k - \mathbf{B}_k^*) \tag{13}$$

where N is the number of targets, $\mathbf{B}^*$ and $\mathbf{B}$ are the ground-truth and predicted box size vectors, and k denotes the object index number. The fourth part of our loss is the orientation loss. As shown in Figure 2, we use an orientation label to determine the orientation of the RIE. We use the binary cross-entropy loss to train the orientation label loss as follows:

$$L_\psi = -\frac{1}{N}\sum_{i=1}^{N}(\psi_i^* \log(\psi_i) + (1-\psi_i^*)\log(1-\psi_i)) \tag{14}$$

where N is the number of targets, ψ^* and ψ are the ground-truth and predicted orientation labels, and i denotes object index number. The last part is the eccentricity-wise orientation loss L_{ewo}. Finally, we use the weight uncertainty loss [55] to balance the multi-task loss, and the final loss used in our method is designed as follows:

$$L = \frac{1}{\sigma_1^2}L_h + \frac{1}{\sigma_2^2}L_b + \frac{1}{\sigma_3^2}L_c + \frac{1}{\sigma_4^2}L_\psi + \frac{1}{\sigma_5^2}L_{ewo} + 2\log\sigma_1\sigma_2\sigma_3\sigma_4\sigma_5 \quad (15)$$

where σ_1, σ_2, σ_3, σ_4, and σ_5 are the learnable uncertainty indexes for balancing the weight of each loss. The uncertainty loss can automatically learn the multitask weights from training data. The detailed introduction of this multitask loss can be found in [55].

4. Experiments and Analysis of the Results

In this section, we first introduce two public remote sensing image datasets, DOTA [56] and HRSC2016 [57], as well as the evaluation metrics used in our experiments. Then, we analyze the implementation details of the network training and the inference process of our detector. Next, we analyze the experimental results on two datasets in comparison with the state-of-the-art detectors. Finally, some ablation study results and promising detection results are displayed.

4.1. Datasets

4.1.1. DOTA

DOTA [56] is composed of 2806 remote sensing images and 188,282 instances in total. Each instance is annotated with oriented bounding boxes consisting of four vertex coordinates, which are collected from multiple sensors and platforms. The images of this dataset mainly contain the following categories: storage tank (ST), plane (PL), baseball diamond (BD), tennis court (TC), swimming pool (SP), ship (SH), ground track field (GTF), harbor (HA), bridge (BR), small vehicle (SV), large vehicle (LV), roundabout (RA), helicopter (HC), soccer-ball field (SBF), and basketball court (BC). In Figure 6, we present the proportion distribution of numbers and the size distribution of the instances of each category in the DOTA dataset. We can see that this multi-class dataset contains a large number of multi-scale oriented objects in RSIs with complex backgrounds, so it is suitable for experiments. In the DOTA dataset, the splits of the training, validation, and test sets are 1/2, 1/6, and 1/3, respectively. The size of each image falls within the range of 0.8 k $\times$ 0.8 k to 4 k $\times$ 4 k pixels. The median aspect ratio of the DOTA dataset is close to 2.5, which means that the effects of various aspect ratios on the detection accuracy can be well evaluated.

4.1.2. HRSC2016

HRSC2016 [57] is a challenging dataset developed for the detection of oriented ship objects in the field of remote sensing imagery. It is composed of 1070 images and 2970 instances in various scales, orientations, and appearances. The image scales range from 300 $\times$ 300 to 1500 $\times$ 900 pixels, and all of the images were collected by Google Earth from six famous ports. The median aspect ratio of the HRSC2016 dataset approaches 5. The training, validation, and test sets contain 436, 181, and 444 images, respectively. In the experiments, both the training set and validation set were utilized for network training.

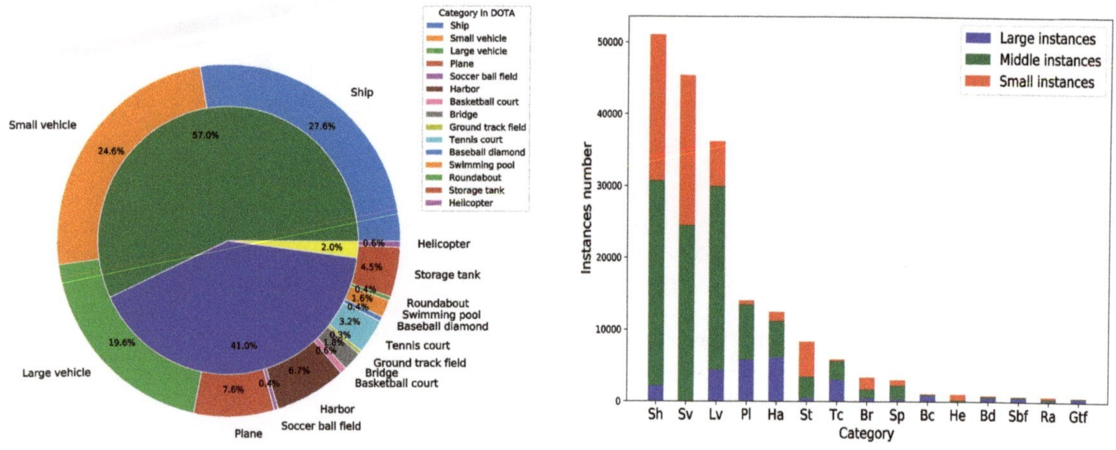

(a) (b)

Figure 6. (**a**) The proportion distribution of the numbers of instances in each category in the DOTA dataset. The outer ring represents the number distribution of 15 categories. The internal ring denotes the total distribution of small (green), middle (blue), and large instances (yellow). (**b**) The size distribution of instances in each category in the DOTA dataset. We divided all of the instances into three splits according to their OBB height: small instances for heights from 10 to 50 pixels, middle instances for heights from 50 to 300 pixels, and large instances for heights above 300 pixels.

4.2. Evaluation Metrics

In this article, three common evaluation metrics—the mean average precision (mAP), F1 score, and frames per second (FPS)—were adopted to evaluate the accuracy and speed of the oriented object detection methods. First, two fundamental evaluation metrics, precision and recall, are indispensable before calculating the metric of the mAP. The precision metric represents the ratio of true positive samples to all positive samples. The recall metric denotes the ratio of true positive samples to all predicted positive samples. They are defined as follows:

$$
\begin{aligned}
Precision &= \frac{TP}{TP+FN} \\
Recall &= \frac{TP}{TP+FP}
\end{aligned}
\tag{16}
$$

where TP, FP, and FN represent the number of predictions of true positive samples, the number of predictions of false positive samples, and the number of predictions of false negative samples. In addition to the precision and recall, we can obtain the comprehensive evaluation metric, the $F1$ score, which is used for single-category object detection.

$$F1\ score = 2/(\frac{1}{precision} + \frac{1}{recall}) = \frac{2 \times Precision \times Recall}{Precision + Recall} \tag{17}$$

Meanwhile, utilizing the precision and recall, we can calculate the corresponding average precision (AP) in each category. By calculating the AP values of all of the categories, we obtain the mean AP (i.e., mAP) value for multi-class objects as follows:

$$\text{mAP} = \frac{1}{N_c}\sum_{i=1}^{N_c}\int_0^1 P_i(R_i)dR_i \tag{18}$$

where N_c indicates the number of categories in the multi-class dataset (e.g., 15 for the DOTA dataset). P_i and R_i denote the precision and recall rates of the i-th class of predicted multi-class objects in the dataset. In addition, we use a general speed evaluation metric,

FPS, which is calculated with the number of images that can be processed per second in order to measure the speed of object detection.

4.3. Implementation Details

The experimental environment of the proposed method was implemented with the PyTorch [58] deep learning framework. For the DOTA and HRSC2016 datasets, we cropped the input image resolution to 800 × 800 pixels and 512 × 512 pixels, respectively. We applied data augmentation strategies to enrich the datasets in the network training process, which included random rotation, random flipping, color jittering, and random scaling in the range of $[0.8, 1.2]$. We trained the network on two NVIDIA GTX 1080 Ti GPUs with a batch size of 8 and utilized Adam [59] with an initial learning rate of 1.5×10^{-4} to optimize the network. In total, we trained the network for 120 epochs on the DOTA dataset and 140 epochs on the HRSC2016 dataset. The learning rate was reduced by a learning rate decay factor of 10 after the 80th and 100th epochs.

4.4. Network Inference

During network inference, the peaks in the heatmap are extracted as the center points for each class object by applying an NMS operation (3 × 3 max-pooling operation). The heatmap value is considered as the detected category confidence score. When the category confidence score is higher than 0.1, it is considered as a correct object center point. Then, we take out the predicted center offsets $\mathbf{c} = (\triangle x, \triangle y)$, long half-axis offsets (δ_x, δ_y), the short half axis b, and the orientation label ψ at the selected heatmap center point $(\bar{x}_c, \bar{y}_c)$. We first add the center offsets $(\triangle x, \triangle y)$ to adjust the heatmap center point $(\bar{x}_c, \bar{y}_c)$ and obtain the modified heatmap center point $(\hat{x}_c, \hat{y}_c) = (\bar{x}_c + \triangle x, \bar{y}_c + \triangle y)$. Finally, we can obtain the predicted rescaled center point location $(x_c, y_c) = (4\hat{x}_c, 4\hat{y}_c)$ in the input image. The coordinates $\{(v_i^x, v_i^y,)|i \in \{1,2,3,4\}\}$ for four vertexes of the predicted RIE at the center point of (x_c, y_c) can be formulated as follows:

$$
\begin{aligned}
&v_1^x = x_c + \delta_x, \quad v_1^y = y_c + \gamma \times \delta_y \\
&v_2^x = x_c - \delta_x, \quad v_2^y = y_c - \gamma \times \delta_y \\
&v_3^x = x_c + b \times \delta_y / \sqrt{(\delta_x^2 + \delta_y^2)}, \quad v_3^y = y_c - b \times \gamma \times \delta_x / \sqrt{(\delta_x^2 + \delta_y^2)} \\
&v_4^x = x_c - b \times \delta_y / \sqrt{(\delta_x^2 + \delta_y^2)}, \quad v_4^y = y_c + b \times \gamma \times \delta_x / \sqrt{(\delta_x^2 + \delta_y^2)}
\end{aligned}
\tag{19}
$$

where γ is an orientation guiding factor, and γ is defined as follows:

$$
\gamma = \begin{cases} 1, & \psi > 0.5 \\ -1, & \psi \leq 0.5 \end{cases}
\tag{20}
$$

where ψ denotes the predicted orientation label value. In addition, in the post-processing stage, there is still a large number of highly overlapping oriented boxes, which improves the false detection rate. In this situation, we employed the oriented NMS strategy from [21] to calculate the IoU between two OBBs and filter out the redundant boxes.

4.5. Comparison with State-of-the-Art Methods

In our experiments, to verify the effectiveness of our method, we compared it with state-of-the-art detectors on the task of oriented object detection in two remote sensing datasets: the DOTA [56] dataset and the HRSC2016 [57] dataset.

4.5.1. Results on DOTA

We compared our method with state-of-the-art anchor-based and anchor-free methods on the DOTA dataset. The results of the comparison of precision on the DOTA dataset are presented in Table 2. For a fair comparison, data augmentations were adopted for all of the compared methods. First, we compared the AP in fifteen categories of objects in the

DOTA dataset and the mAP values of fourteen anchor-based detectors. FR-O [56] is the official baseline method proposed in the DOTA dataset. Based on the Faster-RCNN [9] framework, R-DFPN [26] adds a parameter of angle learning and improves the accuracy of the baseline from 54.13% to 57.94%. R^2CNN [27] proposes a multi-scale regional proposal pooling layer followed by a region proposal network and boosts the accuracy to 60.67%. RRPN [28] introduces a rotating region of interest (RROI) pooling layer and realizes the detection of arbitrarily oriented objects, which improves the performance from 60.67% to 61.01%. ICN [29] designs a cascaded image network to enhance the features based on the R-DFPN [26] network and improves the performance of detection from 61.01% to 68.20%. Meanwhile, we report the detection results of nine other advanced oriented object detectors that were mentioned above, i.e., RoI Trans [30], CAD-Net [32], R^3Det [33], SCRDet [34], ProjBB [39], Gliding Vertex [36], APE [37], S^2A-Net [35], and CSL [22]. It can be noticed that our method of the RIE with the backbone of HRGANet-W48 obtained a 75.94% mAP and outperformed most of the anchor-based methods with which it was compared, except for S^2A-Net [35] (76.11%) and CSL [22] (76.17%). In comparison with the official baseline of DOTA (FR-O [56]), the improvement in accuracy was 21.81%, which demonstrates the advantage of the RIE. Meanwhile, it is worth noting that the use of the RIE under HRGANet-W48 outperformed all of the reported anchor-free methods. Specifically, the RIE outperformed IENet [42], PIoU [46], Axis Learning [43], P-RSDet [19], O^2-DNet [48], BBAVector [50], DRN [47], and CBDA-Net [45] by 18.8%, 15.44%, 9.96%, 6.12%, 4.82%, 3.62%, 2.71%, and 0.2% in terms of mAP. Moreover, the best and second-best AP values for detection in 15 categories of objects are recorded in Table 2. Our method achieved the best performance on objects with large aspect ratios, such as the large vehicle (LV) and harbor (HA), and the second-best performance on the baseball diamond (BD), bridge (BR), and ship (SH) with complicated backgrounds. In addition, we present the visualization of the detection results for the DOTA dataset in Figure 7. The detection results in Figure 7 indicate that our method can precisely capture multi-class and multi-scale objects with complex backgrounds and large aspect ratios.

Table 2. Comparison with state-of-the-art methods of oriented object detection in RSIs on the DOTA dataset. We set the IoU threshold to 0.5 when calculating the AP.

Method	Backbone	FPN	PL	BD	BR	GTF	SV	LV	SH	TC	BC	ST	SBF	RA	HA	SP	HC	mAP
Anchor-based																		
FR-O [56]	ResNet-50	-	79.42	77.13	17.70	64.05	35.30	38.02	37.16	89.41	69.64	59.28	50.30	52.91	47.89	47.40	46.30	54.13
R-DFPN [26]	ResNet-101	✓	80.92	65.82	33.77	58.94	55.77	50.94	54.78	90.33	66.34	68.66	48.73	51.76	55.10	51.32	35.88	57.94
R^2CNN [27]	ResNet-101	-	80.94	65.67	35.34	67.44	59.92	50.91	55.81	90.67	66.92	72.39	55.06	52.23	55.14	53.35	48.22	60.67
RRPN [28]	ResNet-101	-	88.52	71.20	31.66	59.30	51.85	56.19	57.25	90.81	72.84	67.38	56.69	52.84	53.08	51.94	53.58	61.01
ICN [29]	ResNet-101	✓	81.40	74.30	47.70	70.30	64.90	67.80	70.00	90.80	79.10	78.20	53.60	62.90	67.00	64.20	50.20	68.20
RoI Trans [30]	ResNet-101	✓	88.64	78.52	43.44	75.92	68.81	73.68	83.59	90.74	77.27	81.46	58.39	53.54	62.83	58.93	47.67	69.56
CAD-Net [32]	ResNet-101	✓	87.80	82.40	49.40	73.50	71.10	63.50	76.70	90.90	79.20	73.30	48.40	60.90	62.00	67.00	62.20	69.90
R^3Det [33]	ResNet-101	✓	89.54	81.99	48.46	62.52	70.48	74.29	77.54	90.80	81.39	83.54	61.97	59.82	65.44	67.46	60.05	71.69
SCRDet [34]	ResNet-101	✓	89.98	80.65	52.09	68.36	68.36	60.32	72.41	90.85	87.94	86.86	65.02	66.68	66.25	68.24	65.21	72.61
ProjBB [39]	ResNet-101	✓	88.96	79.32	53.98	70.21	60.67	76.20	89.71	90.22	78.94	76.82	60.49	63.62	73.12	71.43	61.69	73.03
Gliding Vertex [36]	ResNet-101	✓	89.64	85.00	52.26	77.34	73.01	73.14	86.82	90.74	79.02	86.81	59.55	70.91	72.94	70.86	57.32	75.02
APE [37]	ResNet-50	✓	89.96	83.62	53.42	76.03	74.01	77.16	79.45	90.83	87.15	84.51	67.72	60.33	74.61	71.84	65.55	75.75
S^2A-Net [35]	ResNet-101	✓	88.70	81.41	54.28	69.75	78.04	80.54	88.04	90.69	84.75	86.22	65.03	65.81	76.16	73.37	58.86	76.11
CSL [22]	ResNeXt101 [60]	✓	90.25	85.53	54.64	75.31	70.44	73.51	77.62	90.84	86.15	86.69	69.60	68.04	73.83	71.10	68.93	76.17
Anchor-free																		
IENet [42]	ResNet-101	✓	57.14	80.20	65.54	39.82	32.07	49.71	65.01	52.58	81.45	44.66	78.51	46.54	56.73	64.40	64.24	57.14
PIoU [46]	DLA-34 [61]	-	80.90	69.70	24.10	60.20	38.30	64.40	64.80	90.90	77.20	70.40	46.50	37.10	57.10	61.90	64.00	60.50
Axis Learning [43]	ResNet-101	✓	79.53	77.15	38.59	61.15	67.53	70.49	76.30	89.66	79.07	83.53	47.27	61.01	56.28	66.06	36.05	65.98
P-RSDet [19]	ResNet-101	✓	89.02	73.65	47.33	72.03	70.58	73.71	72.76	90.82	80.12	81.32	59.45	57.87	60.79	65.21	52.59	69.82
O^2-DNet [48]	Huorglass-104	-	89.20	76.54	48.95	67.52	71.11	75.86	78.85	90.84	78.97	78.26	61.44	60.79	59.66	63.85	64.91	71.12
BBAVectors [50]	ResNet-101	-	88.35	79.96	50.69	62.18	78.43	78.98	87.94	90.85	83.58	84.35	54.13	60.24	65.22	64.28	55.70	72.32
DRN [47]	Hourglass-104	-	89.71	82.34	47.22	64.10	76.22	74.43	85.84	90.57	86.18	84.89	57.65	61.93	69.30	69.63	58.48	73.23
CBDA-Net [45]	DLA-34 [61]	-	89.17	85.92	50.28	65.02	77.72	82.32	87.89	90.48	86.47	85.90	66.85	66.48	67.41	71.33	62.89	75.74
RIE *	HRGANet-W48	-	89.23	84.86	55.69	70.32	75.76	80.68	86.14	90.26	80.17	81.34	59.36	63.24	74.12	70.87	60.36	74.83
RIE	HRGANet-W48	-	89.85	85.68	58.81	70.56	76.66	82.47	88.09	90.56	80.89	82.27	60.46	63.67	76.63	71.56	60.89	75.94

PL: plane, BD: baseball diamond, GTF: ground track field, SV: small vehicle, LV: large vehicle, BR: bridge, TC: tennis court, ST: storage tank, SH: ship, BC: basketball court, SBF: soccer-ball field, RA: roundabout, HA: harbor, SP: swimming pool, HC: helicopter. In each column, the red and blue colors denote the best and second-best detection results. RIE * represents our method without the ewoLoss function.

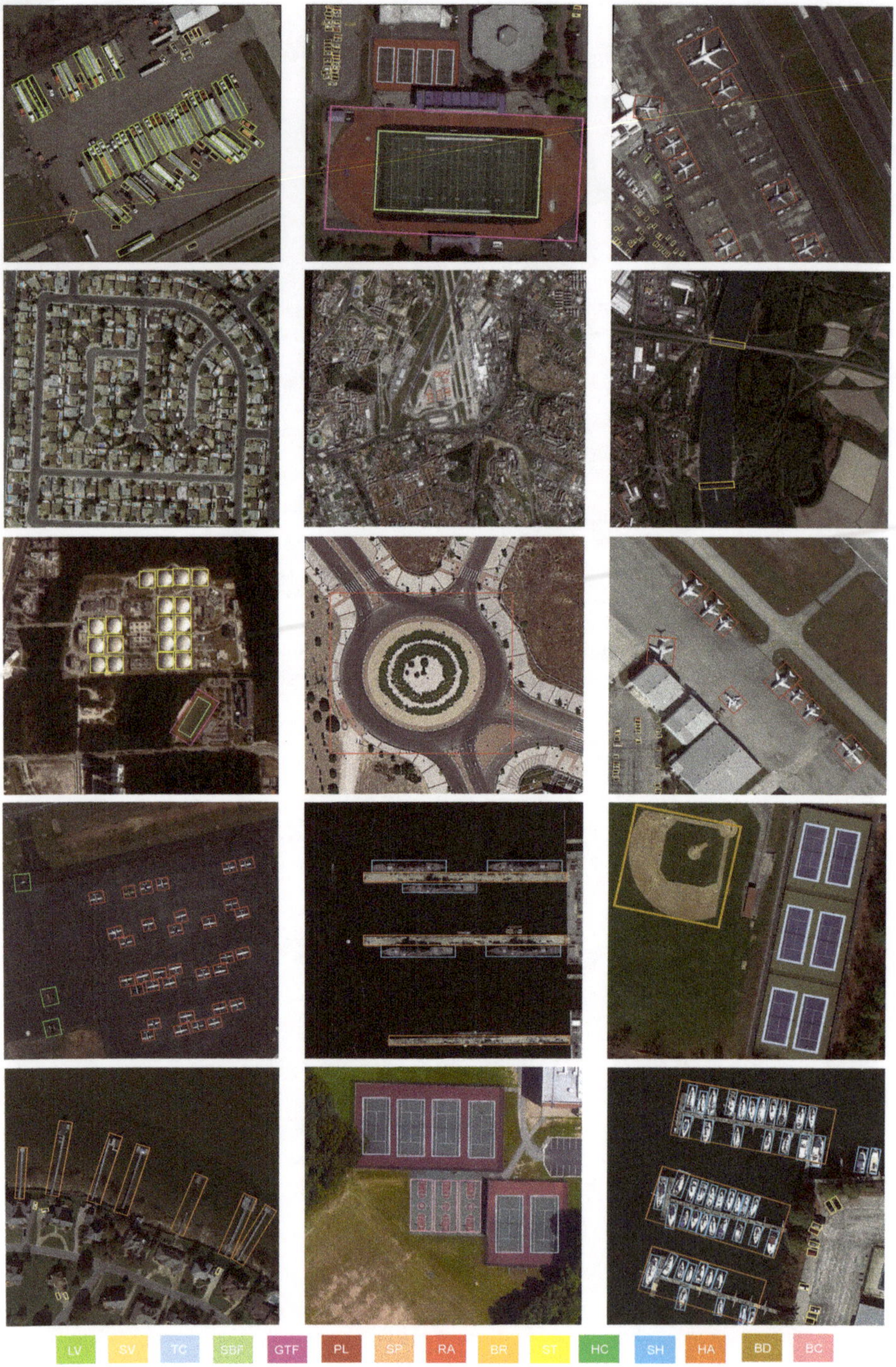

Figure 7. Visualization of the detection results of our method on the DOTA dataset.

4.5.2. Results on HRSC2016

To demonstrate the superiority of our method, we also evaluated the RIE on the HRSC2016 ship dataset and compared the RIE with sixteen other oriented object detectors, as shown in Table 3. BL2 [23], RC1, and RC2 [40] are the official baselines of the HRSC2016 dataset, achieving 69.60% AP and 75.70% AP. RRD [41] introduces an activate rotating filter (ARF) and boosts the performance to 82.89% AP. In addition to these three ship detectors, we also compared our method with six other state-of-the-art anchor-based ship detectors, which were introduced in Section 2, i.e., R^2CNN [27], RRPN [28], R^2PN [31], RoI Trans [30], R^3Det [33], and S^2A-Net [35]. It can be noticed that the RIE outperformed all of the anchor-based methods with which it was compared in terms of AP. Specifically, our method boosts the performance of ship detection from the baselines of BL2 [23] (69.60%), RC1, and RC2 [40] (75.70%) to 91.27% in terms of the AP, which indicates the remarkable performance improvement for the ship identification task. Meanwhile, compared with the state-of-the-art anchor-free methods, i.e., IENet [42], Axis Learning [43], BBAVector [50], PIoU [46], GRS-Det [20], and CBDA-Net [45], the RIE outperformed them by 16.26%, 13.12%, 2.67%, 2.07%, 1.7%, and 0.77% in terms of AP. In addition, as shown in Figure 8, the ships with a large aspect ratio and multi-scale distributions could be effectively detected, and the objects could be tightly surrounded by the predicted oriented bounding boxes. These experimental results illustrate that our method can effectively capture ships in complex sea and land backgrounds.

Table 3. Comparison of the results of accuracy and parameters on the HRSC2016 dataset.

Model	Backbone	Resolution	AP (%)	Parameters
BL2 [23]	ResNet101	-	69.60	-
R^2CNN [27]	ResNet-101	800 × 800	73.07	-
RC1&RC2 [40]	VGG-16	800 × 800	75.70	-
RRPN [28]	ResNet-101	800 × 800	79.08	181.5 MB
R^2PN [31]	VGG-16	-	79.60	-
RRD [41]	VGG-16	384 × 384	82.89	-
RoI Trans [30]	ResNet-101-FPN	512 × 800	86.20	273.8 MB
R^3Det [33]	ResNet-101-FPN	800 × 800	89.26	227.0 MB
S^2A-Net [35]	ResNet-101-FPN	512 × 800	90.17	257.0 MB
IENet [42]	ResNet-101-FPN	1024 × 1024	75.01	212.5 MB
Axis learning [43]	ResNet-101-FPN	800 × 800	78.15	-
BBAVector [50]	ResNet-101	608 × 608	88.60	276.3 MB
PIoU [46]	DLA-34 [61]	512 × 512	89.20	-
GRS-Det [20]	ResNet-101	800 × 800	89.57	200.0 MB
DRN [47]	Hourglass-104	768 × 768	92.70	-
CBDA-Net [45]	DLA-34 [61]	-	90.50	-
RIE	HRGANet-W48	800 × 800	91.27	207.5 MB

4.6. Accuracy–Speed Trade-Off

As shown in Figure 9, we plotted the results of the comparison of the accuracy and speed trade-off with our method and twelve other advanced oriented object detectors for the HRSC2016 dataset. Note that the circular sign denotes the anchor-based methods, while the triangular sign represents the anchor-free methods. The results show that the proposed method can obtain a 91.27% mAP and 20.7 FPS. For the accuracy performance, our method outperformed all of the recorded methods in Figure 9, except for the accuracy of 92.70% achieved by the DRN [47]. However, the DRN [47] under Hourglass-104 [54] runs at a slower detection speed of 5.7 FPS, while under HRGA-Net-W48, our method can run at a faster speed of 20.7 FPS. It is worth noting that our detection speed was faster than those of all other methods with which it was compared, except for the CBDA-Net [45] (50 FPS) and PIoU [46] (55 FPS), which use a more lightweight DLA-34 [61] backbone as the backbone network. At the same time, our method outperformed the two fastest detection methods, CBDA-Net [45] and PIoU [46], by 2.07% and 0.77% in terms of AP. Therefore, this confirms that our method can achieve an excellent accuracy–speed trade-off, which boosts its practical value.

Figure 8. Visualization of the detection results of our method on the HRSC2016 dataset.

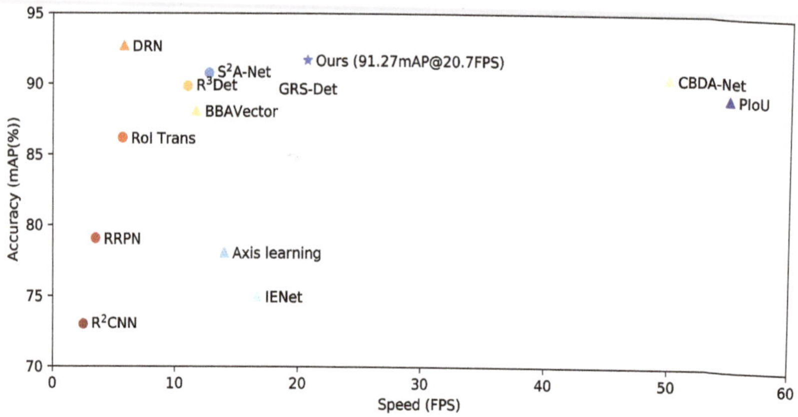

Figure 9. Accuracy versus speed on the HRSC2016 dataset.

4.7. Ablation Study

We implemented an ablation study in terms of the GAM and ewoLoss on the HRSC2016 [57] dataset, as shown in Table 4. The RIE without the GAM and ewoLoss was adopted as the baseline in the first row of Table 4. It can be seen that the baseline only achieved 81.19% and 86.15% in terms of the $F1$-score and mAP, respectively. For ewoLoss, we observed 4.29% and 2.48% increases in terms of the $F1$-score and mAP, as shown in the second row of Table 4. Furthermore, by adding the GAM, the experimental results show a 5.53% improvement in the $F1$-score and a 3.75% improvement in the mAP. It should be noticed that our method achieved a salient improvement in terms of precision while maintaining a higher recall metric. That indicates that our improved backbone,

HRGANet-W48, can capture more robust multi-scale features of the objects with the help of the GAM, while HAGANet-W48 filters the complex background interference and further improves the detection performance. When we added the GAM and ewoLoss at the same time, the *F*1-score and mAP reached 87.94% and 91.27%, which are 6.75% and 5.12% higher than the baseline. Meanwhile, as shown in Table 2, we recorded the detection results of our method on the DOTA [56] dataset both with and without ewoLoss. The results indicate that the performance of the detection of objects with large aspect ratios, such as the bridge (BR), large vehicle (LV), ship (SH), and harbor (HA), was dramatically improved by adding the ewoLoss. This indicates that the proposed ewoLoss exactly boosts the accuracy of the detection of slender oriented objects with large aspect ratios. In addition, as shown in the last column of Table 4, by adding the GAM and ewoLoss, the detection results had a maximal 4.05% improvement in the mAP. These experimental results demonstrate that the GAM and ewoLoss are both conducive to the performance of oriented object identification. When both the GAM and ewoLoss are adopted, the performance is the best.

Table 4. Ablation study of the RIE. All of the models were implemented on the HRSC2016 and DOTA datasets.

Model	GAM	ewoLoss	Recall	Precision	*F*1-Score	HRSC2016 mAP	DOTA mAP
Baseline	-	-	91.76	72.81	81.19	86.15	71.89
RIE	-	✓	93.18	78.95	85.48 (**+4.29**)	88.63 (**+2.48**)	73.71 (**+1.82**)
	✓	-	94.21	80.33	86.72 (**+5.53**)	89.90 (**+3.75**)	74.83 (**+2.94**)
	✓	✓	95.11	81.78	87.94 (**+6.75**)	91.27 (**+5.12**)	75.94 (**+4.05**)

As shown in Table 5, to further explore the impacts of different representation methods, as described in Section 3.2, we compared the RIE-based representation method $(x, y, \delta_x, \delta_y, b, \psi)$ with the angle-based representation method (x, y, w, h, θ) on the DOTA and HRSC2016 datasets. Meanwhile, we chose three backbone networks, i.e., ResNet-101, HRNet-W48, and HRGANet-W48, to strengthen the contrast and further prove the effectiveness of the GAM. Table 5 shows the results of the comparison between the angle-based and RIE-based representation methods. The bold part represents the increment of *F*1-score and mAP values. On the DOTA dataset, our RIE-based representation obtained a remarkable increase of 4.41%, 3.79%, and 4.48% in the mAP under the same implementation configuration based on the ResNet-101, HRNet-W48, and HRGANet-W48 backbone networks. At the same time, on the HRSC2016 dataset, our RIE-based representation achieved a salient improvement of 4.23%, 4.30%, and 3.80% in the mAP under the same implementation configuration based on the ResNet-101, HRNet-W48, and HRGANet-W48 backbone networks. These improvement effects under different backbone networks further prove the effectiveness and robustness of our RIE-based representation method. In addition, from Table 5, we can conclude that HRGANet-W48 with the GAM can produce more improvement for each model compared with the original HRNet-W48, which further verifies the effectiveness of the GAM.

4.8. Complexity Analysis

In our method, we designed a gated aggregation model (GAM) and ewoLoss to boost the detection accuracy. Our backbone network, HRGANet-W48, increased some additional parameters and memory compared with the original HRNet-W48, which is mainly attributed to the GAM. Therefore, as shown in Table 6, we analyzed the complexity of the GAM and presented the parameters of the GAM in detail. It can be noticed that the GAM has a total of 18,504 parameters, which spend about 0.7059 MB of memory. We can see that the lightweight and efficient GAM contributes to the performance of our method with negligible computational complexity. In addition, we also recorded the total parameters of our method and several other state-of-the-art methods, such as RRPN [28], RoI Trans [30], R^3Det [33], S^2-Net [33], IENet [42], BBAVector [50], and GRS-Det [20], in Table 3. Our method took only approximately 207.5 MB of memory for the parameters, which is lighter than all of the other reported methods, except for the RRPN [28] and GRS-Det [20].

Table 5. Results of the comparison between the angle-based and RIE-based representation methods on the DOTA and HRSC2016 datasets based on three backbone networks.

Dataset	Representation Method	Backbone	mAP (%)
DOTA	Angle-based	ResNet-101 HRNet-W48 HRGANet-W48	68.87 70.36 71.46
DOTA	RIE-based	ResNet-101 HRNet-W32 HRGANet-W48	73.28 (**+4.41**) 74.15 (**+3.79**) 75.94 (**+4.48**)
HRSC2016	Angle-based	ResNet-101 HRNet-W48 HRGANet-W48	83.40 85.60 87.47
HRSC2016	RIE-based	ResNet-101 HRNet-W48 HRGANet-W48	87.63 (**+4.23**) 89.90 (**+4.30**) 91.27 (**+3.80**)

Table 6. Statistical results of the GAM parameters.

GAM Architecture	GAM Layers	Parameters	Memory (MB)
Weight block 1	Conv1 × 1 BN ReLU	$1 \times (1 \times 1 \times 48) = 48$ 2 0	1.9×10^{-4}
Weight block 2	Conv1 × 1 BN ReLU	$1 \times (1 \times 1 \times 48 \times 2) = 96$ 2 0	3.7×10^{-4}
Weight block 3	Conv1 × 1 BN ReLU	$1 \times (1 \times 1 \times 48 \times 4) = 192$ 2 0	7.4×10^{-4}
Weight block 4	Conv1 × 1 BN ReLU	$1 \times (1 \times 1 \times 48 \times 8) = 384$ 2 0	1.47×10^{-3}
Fusion	Conv1 × 1	$256 \times (1 \times 1 \times 48 \times 15) =$ 184,320	0.7031
Softmax	softmax function	0	0
total	-	185,048	0.7059

4.9. Applications and Limitations

The method proposed in this article mainly aims at the detection of objects in remote sensing images. We only evaluated the proposed method in terms of the oriented object detection task on the DOTA and HRSC2016 remote sensing image datasets. To our knowledge, oriented object detectors can also be used in oblique text detection, synthetic aperture radar (SAR) image object detection, UAV target detection, seabed pockmark detection, and so on. The application prospects of our method are very broad. We will perform some experiments on these tasks to verify the superiority of our method in the future. Meanwhile, there are still many limitations in the proposed method. First, the detection speed of our method only approaches 20 FPS, which is far from reaching the standard of real-time detection. Therefore, reducing the parameters of the model and speeding up the calculation speed are the focus of the next study. Second, from the detection results on the DOTA dataset, we can see that the detection performance of our method on some objects with inter-class similarity (e.g., BC and TC) is not satisfactory. Meanwhile, our method cannot identify targets with intra-class diversity, such as different categories of ships. The overall category discrimination ability of this model is not strong. We will utilize the attention mechanism to boost the discrimination ability of our method in future work. Third, due to cloud occlusion during remote sensing image shooting, the detection performance of our method will be greatly affected. Therefore, the removal of cloud occlusion while detecting is an important research direction.

5. Conclusions

In this article, we designed a novel anchor-free center-based oriented object detector for remote sensing imagery. The proposed method abandons the angle-based bounding box representation paradigm and uses instead a six-parameter rotated inscribed ellipse (RIE) representation method $(x, y, \delta_x, \delta_y, b, \psi)$. By learning the RIE in each rectangular bounding box, we can address the boundary case and angular periodicity issues of angle-based methods. Moreover, aiming at the problems of complex backgrounds and large-scale variations, we propose a high-resolution gated aggregation network to eliminate background interference and reconcile features of different scales based on a high-resolution network (HRNet) and a gated aggregation model (GAM). In addition, an eccentricity-wise orientation loss function was designed to fix the sensitivity of the RIE's eccentricity to the orientation loss, which prominently improves the performance in the detection of objects with large aspect ratios. We performed extensive comparisons and ablation experiments on the DOTA and HRSC2016 datasets. The experimental results prove the effectiveness of our method for oriented object detection in remote sensing images. Meanwhile, the results also demonstrate that our method can achieve an excellent accuracy and speed trade-off. In future work, we will explore more efficient backbone networks and more ingenious bounding box representation methods to boost the performance in oriented object detection in remote sensing images.

Author Contributions: Methodology, L.H.; software, C.W.; validation, L.R.; formal analysis, L.H.; investigation, S.M.; resources, L.R.; data curation, S.M.; writing—original draft preparation, X.H.; writing—review and editing, X.H.; visualization, L.R.; supervision, S.M.; project administration, C.W.; funding acquisition, L.H. All authors have read and agreed to the published version of the manuscript.

Funding: This work was supported in part by the National Natural Science Foundation of China under Grant 61701524 and in part by the China Postdoctoral Science Foundation under Grant 2019M653742 (corresponding author: L.H.).

Institutional Review Board Statement: Not applicable.

Informed Consent Statement: Not applicable.

Data Availability Statement: The data used to support the findings of this study are available from the corresponding author upon request.

Conflicts of Interest: The authors declare no conflict of interest.

References

1. Kamusoko, C. Importance of remote sensing and land change modeling for urbanization studies. In *Urban Development in Asia and Africa*; Springer: Singapore, 2017.
2. Ahmad, K.; Pogorelov, K.; Riegler, M.; Conci, N.; Halvorsen, P. Social media and satellites. *Multimed. Tools Appl.* **2016**, *78*, 2837–2875. [CrossRef]
3. Tang, T.; Zhou, S.; Deng, Z.; Zou, H.; Lei, L. Vehicle detection in aerial images based on region convolutional neural networks and hard negative example mining. *Sensors* **2017**, *17*, 336. [CrossRef] [PubMed]
4. Janowski, L.; Wroblewski, R.; Dworniczak, J.; Kolakowski, M.; Rogowska, K.; Wojcik, M.; Gajewski, J. Offshore benthic habitat mapping based on object-based image analysis and geomorphometric approach. A case study from the Slupsk Bank, Southern Baltic Sea. *Sci. Total Environ.* **2021**, *11*, 149712. [CrossRef]
5. Madricardo, F.; Bassani, M.; D'Acunto, G.; Calandriello, A.; Foglini, F. New evidence of a Roman road in the Venice Lagoon (Italy) based on high resolution seafloor reconstruction. *Sci. Rep.* **2021**, *11*, 1–19.

6. Li, S.; Xu, Y.L.; Zhu, M.M.; Ma, S.P.; Tang, H. Remote sensing airport detection based on End-to-End deep transferable convolutional neural networks. *IEEE Trans. Geosci. Remote Sens.* **2019**, *16*, 1640–1644. [CrossRef]
7. Girshick, R. Fast r-cnn. In Proceedings of the IEEE International Conference on Computer Vision, Santiago, Chile, 7–13 December 2015; pp. 1440–1448.
8. Cai, Z.; Vasconcelos, N. Cascade r-cnn: Delving into high quality object detection. In Proceedings of the IEEE Conference on Computer Vision and Pattern Recognition, Salt Lake City, UT, USA, 18–22 June 2018; pp. 6154–6162.
9. Ren, S.; He, K.; Girshick, R.; Sun, J. Faster r-cnn: Towards real-time object detection with region proposal networks. *IEEE Trans. Pattern Anal. Mach. Intell.* **2016**, *39*, 1137–1149. [CrossRef]
10. Redmon, J.; Divvala, S.; Girshick, R.; Farhadi, A. You only look once: Unified, real-time object detection. In Proceedings of the IEEE Conference on Computer Vision and Pattern Recognition, Las Vegas, NV, USA, 27–30 June 2016; pp. 779–788.
11. Liu, W.; Anguelov, D.; Erhan, D.; Szegedy, C.; Reed, S.; Fu, C.; Berg, A.C. Ssd: Single shot multibox detector. In Proceedings of the European Conference on Computer Vision (ECCV), Amsterdam, The Netherlands, 11–14 October 2016; pp. 21–37.
12. Lin, T.Y.; Goyal, P.; Girshick, R.; He, K.; Dollár, P. Focal loss for dense object detection. In Proceedings of the IEEE International Conference on Computer Vision, Venice, Italy, 22–29 October 2017; pp. 2980–2988.
13. Law, H.; Deng, J. Cornernet: Detecting objects as paired keypoints. In Proceedings of the European Conference on Computer Vision (ECCV), Munich, Germany, 8–14 September 2018; pp. 734–750.
14. Duan, K.; Bai, S.; Xie, L.; Qi, H.; Huang, Q.; Tian, Q. Centernet: Keypoint triplets for object detection. In Proceedings of the IEEE International Conference on Computer Vision, Seoul, Korea, 27–28 October 2019; pp. 6569–6578.
15. Zhou, X.Y.; Zhuo, J.C.; Krahenbuhl, P. Bottom-up object detection by grouping extreme and center points. In Proceedings of the IEEE Conference on Computer Vision and Pattern Recognition, Long Beach, CA, USA, 16–20 June 2019; pp. 850–859.
16. Huang, L.; Yang, Y.; Deng, Y.; Yu, Y. Densebox: Unifying landmark localization with end to end object detection. *arXiv* **2015**, arXiv:1509.04874.
17. Kong, T.; Sun, F.; Liu, H.; Jiang, Y.; Li, L.; Shi, J. Foveabox: Beyound anchor-based object detection. *IEEE Trans. Image Process.* **2020**, *29*, 7389–7398. [CrossRef]
18. Tian, Z.; Shen, C.; Chen, H.; He, T. Fcos: Fully convolutional one-stage object detection. In Proceedings of the IEEE International Conference on Computer Vision, Thessaloniki, Greece, 23–25 September 2019; pp. 9627–9636.
19. Zhou, L.; Wei, H.; Li, H.; Zhao, W.; Zhang, Y.; Zhang, Y. Arbitrary-Oriented Object Detection in Remote Sensing Images Based on Polar Coordinates. *IEEE Access* **2020**, *8*, 223373–223384. [CrossRef]
20. Zhang, X.; Wang, G.; Zhu, P.; Zhang, T.; Li, C.; Jiao, L. GRS-Det: An Anchor-Free Rotation Ship Detector Based on Gaussian-Mask in Remote Sensing Images. *IEEE Trans. Geosci. Remote Sens.* **2020**, *59*, 3518–3531. [CrossRef]
21. Shi, F.; Zhang, T.; Zhang, T. Orientation-Aware Vehicle Detection in Aerial Images via an Anchor-Free Object Detection Approach. *IEEE Trans. Geosci. Remote Sens.* **2020**, *59*, 5221–5233. [CrossRef]
22. Yang, X.; Yan, J. Arbitrary-Oriented Object Detection with Circular Smooth Label. In Proceedings of the 16th European Conference on Computer Vision (ECCV), Glasgow, UK, 23–28 August 2020; pp. 677–694.
23. Liu, Z.; Hu, J.; Weng, L.; Yang, Y. Rotated region based CNN for ship detection. In Proceedings of the 2017 IEEE International Conference on Image Processing (ICIP), Beijing, China, 17–20 September 2017; pp. 900–904.
24. Cheng, G.; Zhou, P.; Han, J. Learning Rotation-Invariant Convolutional Neural Networks for Object Detection in VHR Optical Remote Sensing Images. *IEEE Trans. Geosci. Remote Sens.* **2016**, *54*, 7405–7415. [CrossRef]
25. Cheng, G.; Han, J.; Zhou, P.; Xu, D. Learning Rotation-Invariant and Fisher Discriminative Convolutional Neural Networks for Object Detection. *IEEE Trans. Geosci. Remote Sens.* **2019**, *28*, 265–278. [CrossRef] [PubMed]
26. Yang, X.; Sun, H.; Fu, K.; Yang, J.; Sun, X.; Yan, M.; Guo, Z. Automatic Ship Detection in Remote Sensing Images from Google Earth of Complex Scenes Based on Multiscale Rotation Dense Feature Pyramid Networks. *Remote Sens.* **2018**, *10*, 132. [CrossRef]
27. Jiang, Y.; Zhu, X.; Wang, X.; Yang, S.; Li, W.; Wang, H.; Fu, P.; Luo, Z. R2cnn: Rotational region cnn for orientation robust scene text detection. *arXiv* **2017**, arXiv:1706.09579.
28. Ma, J.; Shao, W.; Ye, H.; Wang, L.; Wang, H.; Zheng, Y.; Xue, X. Arbitrary-oriented scene text detection via rotation proposals. *IEEE Trans. Multimed.* **2018**, *20*, 3111–3122. [CrossRef]
29. Azimi, S.M.; Vig, E.; Bahmanyar, R. Towards multi-class object detection in unconstrained remote sensing imagery. In Proceedings of the Asian Conference on Computer Vision, Perth, WA, Australia, 2–6 December 2018; Springer: Berlin/Heidelberg, Germany, 2018; pp. 150–165.
30. Ding, J.; Xue, N.; Long, Y.; Xia, G.S.; Lu, Q. Learning RoI transformer for oriented object detection in aerial images. In Proceedings of the IEEE Conference on Computer Vision and Pattern Recognition, Long Beach, CA, USA, 16–20 June 2019; pp. 2849–2858.
31. Zhang, Z.; Guo, W.; Zhu, S.; Yu, W. Toward arbitrarily oriented ship detection with rotated region proposal and discrimination networks. *IEEE Geosci. Remote Sens. Lett.* **2018**, *15*, 1745–1749. [CrossRef]
32. Zhang, G.; Lu, S.; Zhang, W. Cad-net: A context-aware detection network for objects in remote sensing imagery. *IEEE Trans. Geosci. Remote Sens.* **2019**, *57*, 10015–10024. [CrossRef]
33. Yang, X.; Liu, Q.; Yan, J.; Li, A.; Zhang, Z.; Yu, G. R3det: Refined single-stage detector with feature refinement for rotating object. *arXiv* **2019**, arXiv:1908.05612.

34. Yang, X.; Yang, J.; Yan, J.; Zhang, Y.; Zhang, T.; Guo, Z.; Sun, X.; Fu, K. Scrdet: Towards more robust detection for small, cluttered and rotated objects. In Proceedings of the IEEE International Conference on Computer Vision, Thessaloniki, Greece, 23–25 September 2019; pp. 8232–8241.
35. Han, J.; Ding, J.; Li, J.; Xia, G.S. Align Deep Features for Oriented Object Detection. *IEEE Trans. Geosci. Remote Sens.* **2020**. [CrossRef]
36. Xu, Y.; Fu, M.; Wang, Q.; Wang, Y.; Chen, K.; Xia, G.S.; Bai, X. Gliding vertex on the horizontal bounding box for multi-oriented object detection. *IEEE Trans. Pattern Anal. Mach. Intell.* **2020**, *43*, 1452–1459. [CrossRef]
37. Zhu, Y.; Du, J.; Wu, X. Adaptive Period Embedding for Representing Oriented Objects in Aerial Images. *IEEE Trans. Geosci. Remote Sens.* **2020**, *58*, 7247–7257. [CrossRef]
38. Yang, X.; Hou, L.; Zhou, Y. Dense label encoding for boundary discontinuity free rotation detection. In Proceedings of the IEEE Conference on Computer Vision and Pattern Recognition, Las Vegas, NV, USA, 19–25 June 2016; pp. 15819–15829.
39. Wu, Q.; Xiang, W.; Tang, R.; Zhu, J. Bounding Box Projection for Regression Uncertainty in Oriented Object Detection. *IEEE Access* **2021**, *9*, 58768–58779. [CrossRef]
40. Liu, Z.; Yuan, L.; Weng, L.; Yang, Y. A high resolution optical satellite image dataset for ship recognition and some new baselines. In Proceedings of the International Conference on Pattern Recognition Applications and Methods, Porto, Portugal, 24–26 February 2017; Volume 2, pp. 324–331.
41. Liao, M.; Zhu, Z.; Shi, B.; Xia, G.S.; Bai, X. Rotation-Sensitive Regression for Oriented Scene Text Detection. In Proceedings of the IEEE Conference on Computer Vision and Pattern Recognition (CVPR), Salt Lake City, UT, USA, 18–22 June 2018.
42. Lin, Y.; Feng, P.; Guan, J. Ienet: Interacting embranchment one stage anchor free detector for orientation aerial object detection. *arXiv* **2019**, arXiv:1912.00969.
43. Xiao, Z.; Qian, L.; Shao, W.; Tan, X.; Wang, K. Axis Learning for Orientated Objects Detection in Aerial Images. *Remote Sens.* **2020**, *12*, 908. [CrossRef]
44. Chen, J.; Xie, F.; Lu, Y.; Jiang, Z. Finding Arbitrary-Oriented Ships From Remote Sensing Images Using Corner Detection. *IEEE Geosci. Remote Sens. Lett.* **2019**, *17*, 1712–1716. [CrossRef]
45. Liu, S.; Zhang, L.; Lu, H.; He, Y. Center-Boundary Dual Attention for Oriented Object Detection in Remote Sensing Images. *IEEE Trans. Geosci. Remote Sens.* **2021**. [CrossRef]
46. Chen, Z.; Chen, K.; Lin, W.; See, J.; Yu, H.; Ke, Y.; Yang, C. Piou loss: Towards accurate oriented object detection in complex environments. In Proceedings of the European Conference on Computer Vision (ECCV), Glasgow, UK, 23–28 August 2020; pp. 195–211.
47. Pan, X.; Ren, Y.; Sheng, K.; Dong, W.; Yuan, H.; Guo, X.; Xu, C. Dynamic refinement network for oriented and densely packed object detection. In Proceedings of the IEEE Conference on Computer Vision and Pattern Recognition, Seattle, WA, USA, 14–19 June 2020; pp. 11207–11216.
48. Wei, H.; Zhang, Y.; Chang, Z.; Li, H.; Wang, H.; Sun, X. Oriented objects as pairs of middle lines. *ISPRS J. Photogramm. Remote Sens.* **2020**, *169*, 268–279. [CrossRef]
49. Wei, H.; Zhang, Y.; Wang, B.; Yang, Y.; Li, H.; Wang, H. X-LineNet: Detecting Aircraft in Remote Sensing Images by a Pair of Intersecting Line Segments. *IEEE Trans. Geosci. Remote Sens.* **2021**, *59*, 1645–1659. [CrossRef]
50. Yi, J.; Wu, P.; Liu, B.; Huang, Q.; Qu, H.; Metaxas, D. Oriented Object Detection in Aerial Images with Box Boundary-Aware Vectors. In Proceedings of the IEEE/CVF Winter Conference on Applications of Computer Vision, Snowmass Village, CO, USA, 1–5 March 2020; pp. 2150–2159.
51. Wang, J. Deep High-Resolution Representation Learning for Visual Recognition. *IEEE Trans. Pattern Anal. Mach. Intell.* **2020**, *43*, 3349–3364. [CrossRef]
52. Simonyan, K.; Zisserman, A. Very deep convolutional networks for large-scale image recognition. In Proceedings of the International Conference on Learning Representations, San Diego, CA, USA, 7–9 May 2015; pp. 1–14.
53. He, K.; Zhang, X.; Ren, S.; Sun, J. Deep residual learning for image recognition. In Proceedings of the IEEE Conference on Computer Vision and Pattern Recognition, Las Vegas, NV, USA, 27–30 June 2016; pp. 770–778.
54. Newell, A.; Yang, K.; Deng, J. Stacked hourglass networks for human pose estimation. In Proceedings of the European Conference on Computer Vision (ECCV), Amsterdam, The Netherlands, 8–16 October 2016; pp. 483–499.
55. Kendall, A.; Gal, Y.; Cipolla, R. Multi-task learning using uncertainty to weigh losses for scene geometry and semantics. In Proceedings of the IEEE Conference on Computer Vision and Pattern Recognition, Salt Lake City, UT, USA, 18–22 June 2018; pp. 7482–7491.
56. Xia, G.; Bai, X.; Ding, J.; Zhu, Z.; Belongie, S.; Luo, J.; Datcu, M.; Pelillo, M.; Zhang, L. DOTA: A large-scale dataset for object detection in aerial images. In Proceedings of the IEEE Conference on Computer Vision and Pattern Recognition, Salt Lake City, UT, USA, 18–22 June 2018; pp. 3974–3983.
57. Liu, Z.; Wang, H.; Weng, L.; Yang, Y. Ship rotated bounding box space for ship extraction from high-resolution optical satellite images with complex backgrounds. *IEEE Geosci. Remote Sens. Lett.* **2016**, *13*, 1074–1078. [CrossRef]
58. Paszke, A.; Gross, S.; Massa, F.; Lerer, A.; Bradbury, J.; Chanan, G.; Killeen, T.; Lin, Z.M.; Gimelshein, N.; Antiga, L. Pytorch: An imperative style, high-performance deep learning library. *Adv. Neural Inf. Process. Syst.* **2019**, *32*, 8026–8037.
59. Kingma, D.P.; Ba, J. Adam: A method for stochastic optimization. *arXiv* **2014**, arXiv:1412.6980.

60. Xie, S.; Girshick, R.; Dollár, P.; Tu, Z.; He, K. Aggregated residual transformations for deep neural networks. In Proceedings of the IEEE Conference on Computer Vision and Pattern Recognition, Honolulu, HI, USA, 21–26 July 2017; pp. 1492–1500.
61. Yu, F.; Wang, D.; Shelhamer, E.; Darrell, T. Deep layer aggregation. In Proceedings of the IEEE Conference on Computer Vision and Pattern Recognition, Salt Lake City, UT, USA, 18–22 June 2018; pp. 2403–2412.

remote sensing

MDPI

Article

Split-Attention Networks with Self-Calibrated Convolution for Moon Impact Crater Detection from Multi-Source Data

Yutong Jia [1], Gang Wan [1,*], Lei Liu [1], Jue Wang [1], Yitian Wu [1], Naiyang Xue [2], Ying Wang [1] and Rixin Yang [1]

1 Department of Surveying and Mapping and Space Environment, Space Engineering University, Beijing 101407, China; jiayutong@st.btbu.edu.cn (Y.J.); liuleiargis@pku.edu.cn (L.L.); 3120160445@bit.edu.cn (J.W.); wuyt@radi.ac.cn (Y.W.); wang18922950496@163.com (Y.W.); naruto_young@163.com (R.Y.)

2 Department of Electronic and Optical Engineering, Space Engineering University, Beijing 101407, China; hgdxny15@163.com

* Correspondence: casper_51@163.com; Tel.: +86-131-4521-4654

Abstract: Impact craters are the most prominent features on the surface of the Moon, Mars, and Mercury. They play an essential role in constructing lunar bases, the dating of Mars and Mercury, and the surface exploration of other celestial bodies. The traditional crater detection algorithms (CDA) are mainly based on manual interpretation which is combined with classical image processing techniques. The traditional CDAs are, however, inefficient for detecting smaller or overlapped impact craters. In this paper, we propose a Split-Attention Networks with Self-Calibrated Convolution (SCNeSt) architecture, in which the channel-wise attention with multi-path representation and self-calibrated convolutions can generate more prosperous and more discriminative feature representations. The algorithm first extracts the crater feature model under the well-known target detection R-FCN network framework. The trained models are then applied to detecting the impact craters on Mercury and Mars using the transfer learning method. In the lunar impact crater detection experiment, we managed to extract a total of 157,389 impact craters with diameters between 0.6 and 860 km. Our proposed model outperforms the ResNet, ResNeXt, ScNet, and ResNeSt models in terms of recall rate and accuracy is more efficient than that other residual network models. Without training for Mars and Mercury remote sensing data, our model can also identify craters of different scales and demonstrates outstanding robustness and transferability.

Keywords: crater detection algorithm (CDA); R-FCN; self-calibrated convolution; split attention mechanism; transfer learning; remote sensing

Citation: Jia, Y.; Wan, G.; Liu, L.; Wang, J.; Wu, Y.; Xue, N.; Wang, Y.; Yang, R. Split-Attention Networks with Self-Calibrated Convolution for Moon Impact Crater Detection from Multi-Source Data. *Remote Sens.* **2021**, *13*, 3193. https://doi.org/10.3390/rs13163193

Academic Editors: Jukka Heikkonen, Fahimeh Farahnakian and Pouya Jafarzadeh

Received: 5 July 2021
Accepted: 9 August 2021
Published: 12 August 2021

1. Introduction

Impact craters are considered to be one of the most important features of the Moon, Mars, and Mercury [1]. They gradually evolve because of colliding objects, such as meteorites, satellites, or massive asteroids [2]. Most of the impact craters on the lunar surface have circular pit structures with different sizes and uneven aggregations.

The impact craters on the surface of deep space stars contain significant geological data. This is because they are the product of the meteorite's high-speed movement, impact on the surface of celestial bodies, and lava eruption inside heavenly bodies. Therefore, such data can be used to retrieve the geological age of the stars [3], analyze the tectonic history of the lead [4], and explore the existence of iced water [5]. In addition, it can be used for autonomous navigation [6], landing site selection [7], base selection, and other missions of deep space probs.

The precise and rapid discovery of impact craters has always been a priority for deep space exploration since the beginning of the Moon and Mars exploration activities. Several deep space star surface impact crater extraction algorithms have also been proposed. These algorithms are broadly classified as (i) traditional algorithms, which use image processing technology to identify impact craters, and (ii) automatic algorithms [8–11], which use deep learning models to extract impact craters [12–14].

The traditional automatic feature extraction algorithms for impact crater morphology are mainly based on classical image processing methods, including Hough transform, feature matching, curve fitting, and other recognition techniques. For example, [15] used the Hough Transform to obtain more than 75 percent of the current impact craters with a diameter greater than 10 km based on data from the Mars Orbiter Laser Altimeter (MOLA). Hough transform is the most widely used method in this area which is efficient for impact crater identification and recognition of the discontinuous edges. However, for irregular shapes, the computational complexity of such methods is very high. Further, [16] used the conic curve-fitting approach to automatically classify asteroid impact craters to aid optical navigation of the spacecraft to solve this problem. The proposed method in [15] successfully identified about 90% of impact craters with an error rate of less than 5%. Based on the Mars Orbiter Camera (MOC), Mars Orbiter Laser Altimeter (MOLA), and High-Resolution 3D Camera (HRSC)), [9] proposed a least-squares fitting method (DLS) for the identification of Mars impact craters. By comparing the recognition results of the Hough ring transform algorithm, they then showed that the conic fitting method is more reliable, but its computational complexity is higher.

The construction and matching of data quality and crater characteristics are central to traditional crater recognition algorithms. The main goals are to create a more accurate crater function model and a faster template matching algorithm. Nonetheless, the geomorphic features of impact craters are many. The impact craters in an area may also be nested and overlapped. The available data samples are also insufficient in many cases.

Artificial intelligence has developed rapidly by introducing deep learning models in recent years. Among deep learning techniques, convolutional neural networks (CNN) are shown to offer significant practical advantages for image processing. CNN have been successfully applied to many classic image processing problems, such as image denoising, super-resolution image reconstruction, image segmentation, target detection, and object classification. Crater detection and segmentation of the image data can be used to solve the problem of crater recognition.

Cohen [17] considered the classification of meteorite craters, proposing a meteorite crater identification and classification algorithm based on a genetic algorithm. Yang [3] also proposed an impact crater detection model on the lunar surface based on the target detection R-FCN model and further studied the lunar age estimation. Furthermore, [12] suggested the DeepMoon model for lunar surface impact crater identification based on the U-Net model of image semantic segmentation in deep learning. They then transferred their model to the Mercury surface impact crater recognition and achieved reasonable results. The DeepMoon model's structure was applied to the impact craters on Mars' surface in [18], and the DeepMars model was proposed to achieve rapid detection of impact craters on Mars' surface. Jia [19] also improved the model and suggested a need-attention-aware U-NET (NAU-NET) in the DEM impact crater trial and obtained Recall and Precision of 0.791 and 0.856, respectively.

Intelligent impact crater identification methods based on deep learning are more efficient than the traditional identification methods in recognizing significant differences in the radius of the impact crater and their complex morphological characteristics. However, due to the variety of deep space objects, the recognition model based on single star surface impact craters offers a poor generalization ability, especially in recognizing overlapping and small impact craters. To address this issue, in this paper, we consider the deep space star surface impact crater and combine the existing Moon image and DEM data of the Moon, Mars, and Mercury surfaces to establish a deep learning-based deep space star surface impact crater intelligent identification framework. The proposed model improves the model generalization ability through transfer learning. An improved residual network and multi-scale target extraction are introduced to accelerate the model convergence and improve the accuracy of feature extraction. In addition, a more efficient pooling operation and Soft-NMS algorithm are proposed, which effectively reduces false-negative errors of the detection model.

The main contributions of this paper are as follows:

1. We propose a SCNeSt architecture in which the channel-wise attention with multi-path representation and self-calibrated convolutions provide a higher detection and estimation accuracy for small impact craters.
2. To address the issues caused by a single data source with low resolution and insufficient impact crater features, we extract the profile and curvature of the impact crater from Chang 'e-1 DEM data, integrated it with Chang 'e-1 DOM data, and combined it with International Astronomical Union (IAU) impact crater database, and constructed the VOC data set.
3. The lunar crater model is trained, and transfer learning is used to detect the impact craters on Mercury and Mars. This is shown to increase the model's generalization ability.

The rest of this paper is organized as follows. In Section 2, we introduce the R-FCN network for target detection and SCNeSt, RPN, and ROI Pooling. The model is then applied for impact crater detection on Mercury and Mars surfaces using transfer learning. Section 3 then introduces the experimental data, evaluation indexes, and experimental conditions. Furthermore, Section 4 evaluates the lunar impact crater detection results and compares the proposed network with other existing networks. Finally, Section 5 provides our conclusions and offers insights on the direction of future work.

2. Methods

We adopted a combination of deep learning and transfer learning, as shown in Figure 1. In the first stage, CE-1 images of 4800 × 4800 pixels and 1200 × 1200 pixels were used (image fusion method referred to 3.1), achieving a recall rate of 95.82%, where almost all identified craters in the test set were recovered. In the second stage, we transferred the detection model of the first stage to the SLDEM [20] images without any training samples. The learning process in the second stage followed transfer learning, hence extracts the learning features and knowledge from the SLDEM data with a recall rate of 91.35%. We finally found 157,389 impact craters on the Moon, ranging in size from 0.6 to 860 km. The number of detected craters was almost 20 times larger than the known craters, with 91.14 percent of them smaller than 10 km in diameter.

For the meteorite craters that were in both CE-1 and SLDEM, we selected $D \geq 20$ km for CE-1 detection, and $D < 20$ km for SLDEM data detection. The average detection time of an image was 0.13 s.

2.1. SCNeSt Backbone Network

Inspired by the ResNeSt network framework and the self-calibrated convolution in the ScNet [21], in this paper, we improved the ResNeSt. To enhance the diversity of output features, self-calibrated convolution in the ScNet was substituted with the second convolution layer of the ResNeSt Block to obtain more features and more efficient classification performance. Meanwhile, in a split-attention radix group of ResNeSt, we used the method of combining MaxPooling and AvgPooling to replace the original GlobalPooling. This enabled obtaining more texture features at the same time. MaxPooling reduces useless information, and AvgPooling obtains the texture information.

The SCNeSt Block structure is shown in Figure 2. The self-calibrated Conv evenly divided the input into four parts and then performed different operations for each position. First, the input X was evenly divided into and various functions that process the input X. Then, X_1 was sent up to the first branch (self-calibrated branch) and X_2 to the second branch (conventional transform branch). Finally, the processed features were concatenated as the output.

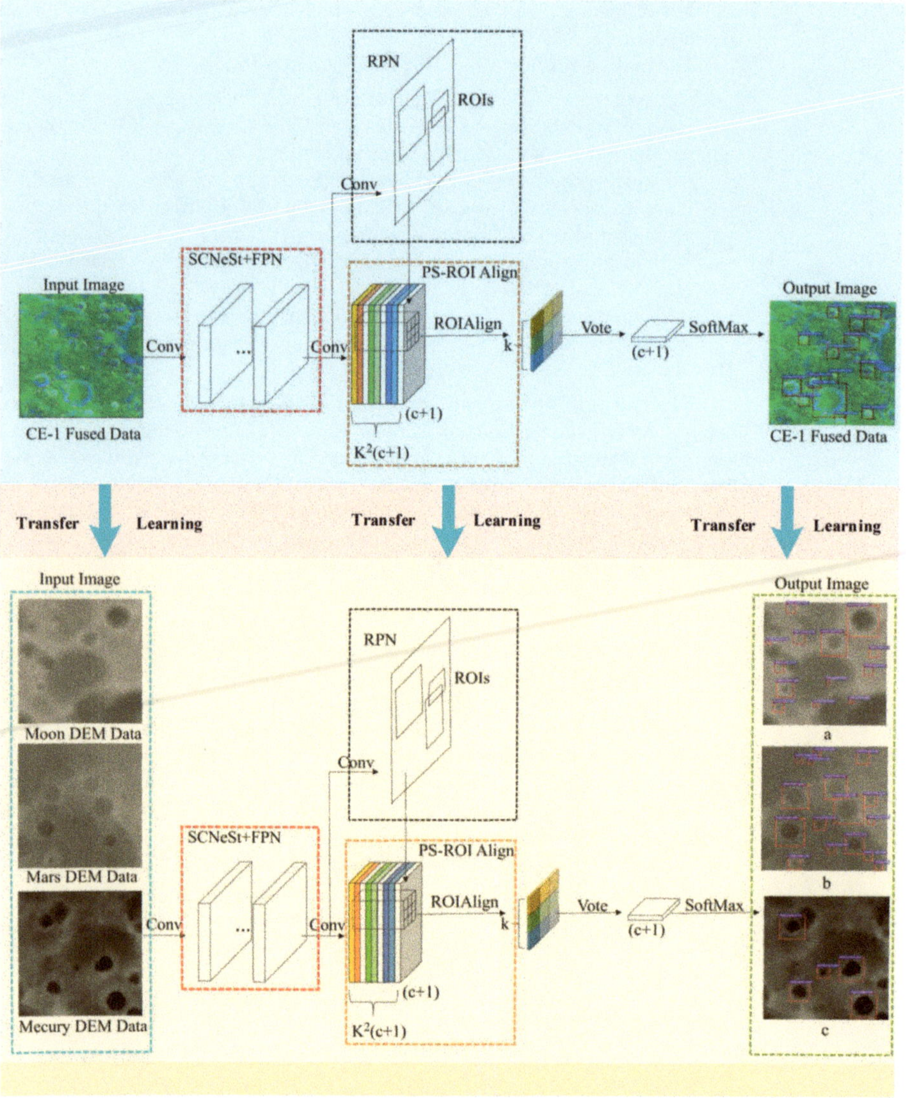

Figure 1. Deep space impact crater detection framework based on the improved R-FCN.

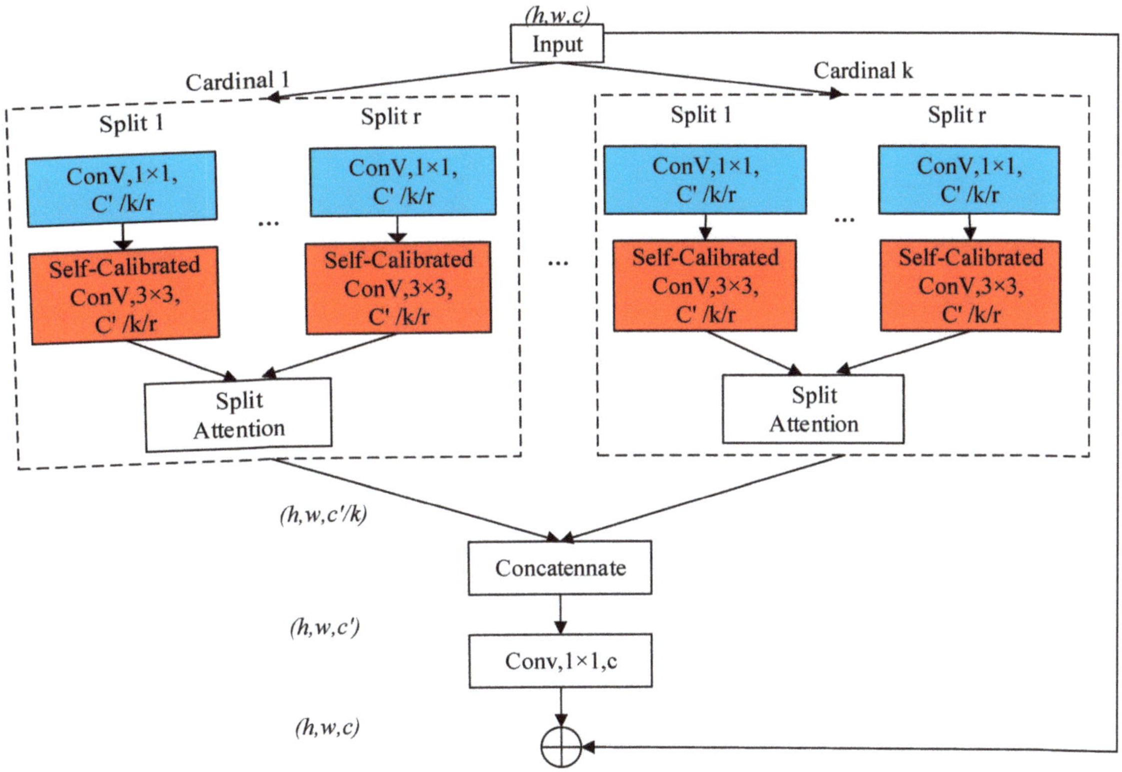

Figure 2. The SCNeSt block. The blue module represents vanilla convolutions, and the red module describes self-calibrated convolutions.

In the self-calibrated branch, for input X_1, average subsampling, convolution feature transformation, and bilinear up-sampling were performed. The input was then added to obtain the attention feature map at the spatial level. The acquired spatial attention map was fused with the transformed X_1. The process is described as:

$$\begin{cases} X_1' = Up(T_1) = Up(T_1 \times K_2) = Up(Down(X_1) \times K_2) \\ Y_1' = F_3(X_1) + \sigma(X_1 + X_1') \end{cases} \quad (1)$$

The schematic diagram of the self-calibrated Conv module is shown in Figure 3. The self-calibrated Conv proposed in this paper has the following three advantages:

(1) Self-calibrated branching significantly increases the receptive field of the output features and acquires more features.
(2) The self-calibrated branch only considers the information of the airspace position, avoiding the information of the unwanted region, hence uses resources more efficiently.
(3) Self-calibrated branching also encodes multi-scale feature information and further enriches the feature content.

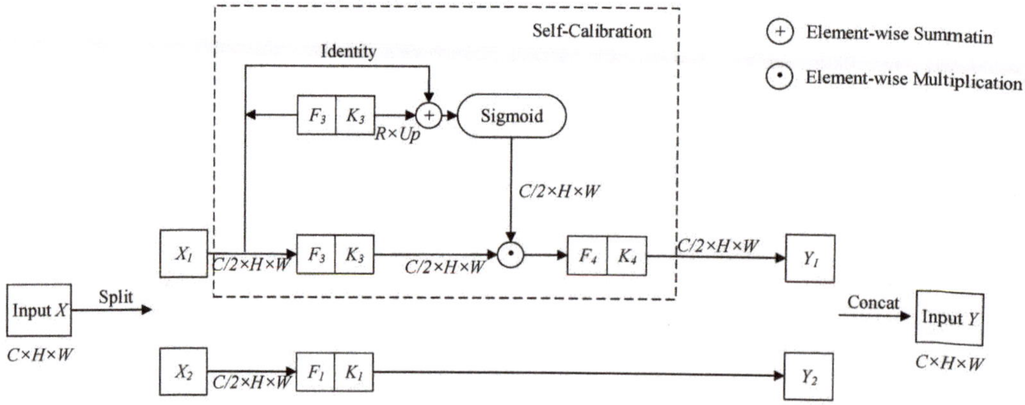

Figure 3. The schematic diagram of the self-calibrated Conv module. In self-calibrated convolutions, the original filters were separated into four portions, each in charge of different functionality. This makes self-calibrated convolutions quite different from traditional convolutions or grouped convolutions performed homogeneously.

2.2. Multi-Scale Feature Extractor

Although the external network detects small targets, the external network has weak semantics. If we only carried out the deconvolution operation without feature fusion, part of the information would be lost after repeated convolution and deconvolution. This is more harmful to detecting the small targets. To address this issue, we synchronized with the deconvolution process, and the high-level features were successively fused with the shallow elements. This preserved the semantic information and resolution of the feature layer.

The FPN [22] consisted of three parts, as shown in Figure 4d. The first part was the feature extraction using the feedforward process of the general convolutional neural network from bottom to top.

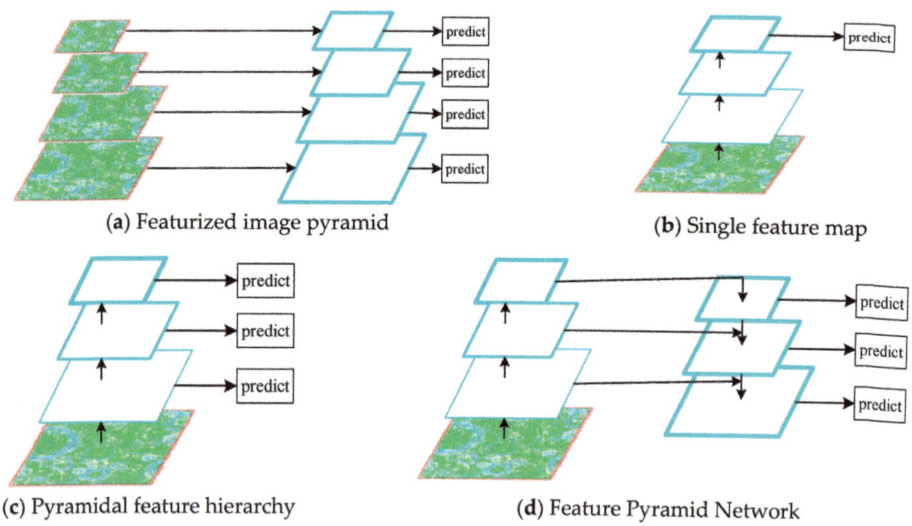

Figure 4. The multi-scale detection methods.

In the second part, we first selected the upper-level feature graphs with more vital semantic information in the feature graphs obtained in the first part. Then, they were up-sampled from top to bottom to strengthen the upper-level features. This also equalized the sizes of the feature graphs in the adjacent layers. In the third part, the feature graphs of the first two steps were combined using horizontal connections. Through these three parts, the high- and low-level features were connected to enrich the semantic information of each scale.

The whole FPN network was embedded into the RPN to generate features of different scales. These features were then fused as the input of the RPN network to improve the accuracy of the two-stage target detection algorithm, as shown in Figure 5.

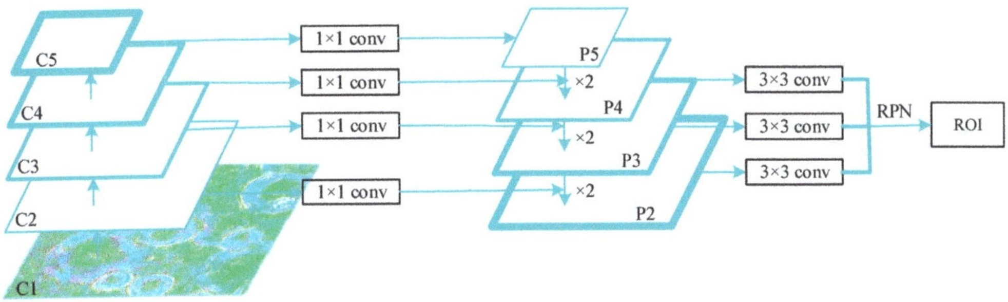

Figure 5. RPN network with FPN.

2.3. *Position-Sensitive ROI Align*

The ROI Pooling layer [23] improves the detection accuracy and speeds up the training and testing process. Nevertheless, two rounds of quantization operations were required, generating the candidate box and determining the corresponding grid position. The first step was to round up the two sampling points selected in the original ROI Pooling layer. This ensured that the generated sampling points were aligned with the standard coordinate points, and the subsequent Pooling operations would round up again. Since the feature map obtained by the CNN was 16 times smaller than that of the original image, X/16 needed to be used for the calculation in the corresponding process. Hence, there existed floating-point numbers with decimals in the calculations. The coordinate point deviation on the feature map caused by the two-step rounding operation corresponded to the pixel deviation on the original image, which was 16 times. The pixel deviation led to mismatching between the image and the feature map so that the ROI on the feature map could not correspond to the original image. This, however, had an impact on the regression positioning of the back layer.

To avoid the round-off operation of the floating-point numbers by two rounds quantization, a bilinear difference pair was introduced to improve the alignment method. A particular region of the feature map corresponding to the ROI was divided into 2×2 region blocks. Each region block was then quartered, and each small grid center was taken as the sampling point. As illustrated in Figure 6, the coordinates of the 16 sampling points in vertices A, B, C, D, and the evenly divided 2×2 region were not integers. After determining the sampling points, the bilinear difference evaluation was directly mapped to the feature map, and each sampling point was evaluated in the X and Y directions. After the difference was completed, the maximum pooling operation was carried out, and the final feature map was obtained by analogy. The whole procedure did not operate on specific coordinate values. The decimal was retained in the coordinate calculation process to avoid the discrete quantization error of the two ROI round-off operations and make the final detection box position more accurate.

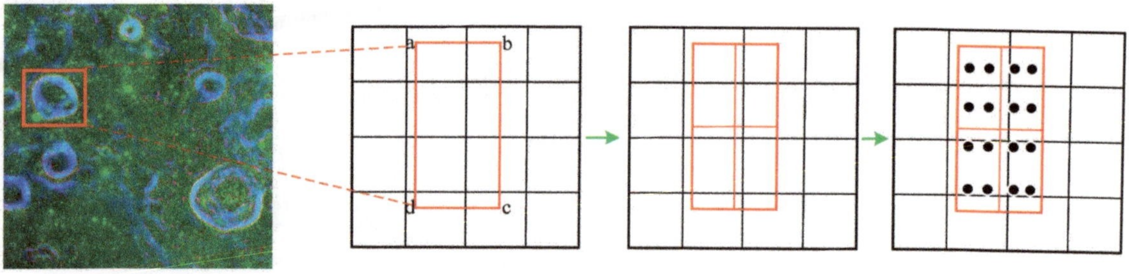

(**a**) RoI coordinates (**b**) Coordinate mapping regression (**c**) RoI Pooling (**d**) Sampling point evaluation

Figure 6. The improved RoI Pooling using bilinear interpolation.

A Position Sensitive ROI Align algorithm was implemented by porting ROI Align into PS-ROI Pooling. The PS-ROI Align improved the detection performance of the model and significantly improved the perception ability for the small objects.

2.4. Soft-NMS

After obtaining the detection box by the R-FCN model, we used the non-maximum suppression (NMS) [24] algorithm to accurately convey the best coordinates of the target and remove the repeated boundary box. For the same object, multiple detection scores were generated as the detection windows were overlapped. In such cases, the NMS kept the correct detection box (with the highest confidence). The remaining detection boxes were removed from the optimal position (with the confidence reduced to 0) to obtain the most accurate bounding box. The NMS can be expressed by the score reset function:

$$Q_i = \begin{cases} Q_i, \ iou(M, b_i) < N_t \\ 0, \ iou(M, b_i) \geq N_t \end{cases} \quad (2)$$

where Q_i is the confidence of the detection box, M is the position of the detection box with the highest confidence, b_i is the position of the detection box, N_t is the set overlap threshold, and $iou(M, b_i)$ is the overlap rate between M and b_i.

Note that non-maximum suppression may cause a critical issue by forcing the scores of adjacent detection boxes to 0. In such cases, if different impact craters appear in the overlapping area, the detection of impact craters will fail. This reduces the detection rate of the algorithm, as in Figure 7a.

Soft non-maximum suppression algorithm (Soft-NMS) [25] replaces the score reset in the NMS algorithm with:

$$Q_i \leftarrow Q_i f(iou(M, b_i)) \quad (3)$$

Noting that the impact craters were rectangular targets in the image, and considering overlapping impact craters, a linear weighted fraction resetting function was used as the following:

$$Q_i = \begin{cases} Q_i, \ iou(M, b_i) < N_t \\ Q_i(1 - iou(M, b_i)), \ iou(M, b_i) \geq N_t \end{cases} \quad (4)$$

In Figure 7b, the confidence of the dashed line detection box was changed to 1.0, indicating that Soft-NMS can effectively avoid missing the impact craters in the overlapping areas. This significantly improved the detection rate of the model.

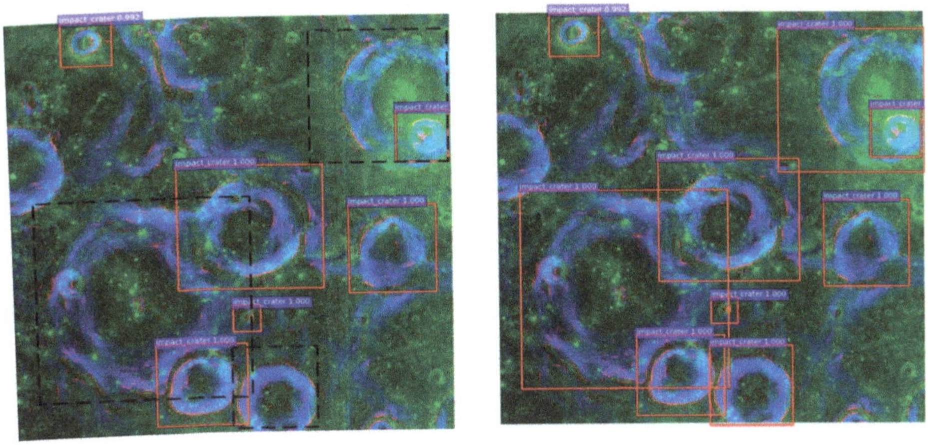

(**a**) Non-maximum suppression algorithm. (**b**) Soft non-maximum suppression algorithm.

Figure 7. Comparison of NMS and Soft-NMS algorithms.

3. Experiments

Our algorithm was divided into two parts. First, the features of impact craters were extracted under the Structure of the R-FCN network based on the SCNeSt network skeleton, and the data were DOM and DEM fusion data from CE-1. Multi-scale Feature Extractor and Position-Sensitive ROI Align could better detect impact craters of different scales. They were combined with the Soft-NMS algorithm to accurately convey the best coordinates of the target and remove the repeated boundary box. In the first stage, the craters with D > 20 km were mainly extracted. In the second stage, the trained model was applied to SLDEM data to extract small craters with D < 20 km. What is more, the trained models were then applied to detecting the impact craters on Mercury and Mars using the transfer learning method.

3.1. Dataset

The area studied on the Moon was latitude $-65°\sim65°$, longitude $-180°\sim65°$, and longitude $65°\sim180°$. The DOM and DEM data adopt equiangular cylindrical projection. During the crater exploration mission, DEM data from CE-1 was resampled to 120 m/pixel. The slop information and profile curvature were also extracted from DEM data. DOM data was integrated with DEM data. The crater in the study area was marked by using the lunar data set published by the IAU impact crater VOC dataset generated by combining with Labelimg. The CE-1 fusion data were then clipped into 1200×1200, 4800×4800 images at a 50% overlap rate, 8000, 1000, and 1000 images were randomly selected and used for training, validation, and testing, respectively. Due to the low resolution of CE-1 data, we used it to identify large impact craters ranging from 20 km to 550 km in diameter. The detailed data generation was shown in Figure 8.

The SLDEM from the Lunar Reconnaissance Orbiter (LRO) and the Kaguya merged digital elevation model had a resolution of 59 m/pixel and spans ±60 degrees latitude (and the maximum range in longitude). The Plate Carree projection was used to create this global grayscale map, which had a resolution of $184{,}320 \times 61{,}440$ pixels and a bit depth of 16 bits per pixel. We cropped it into 1000×1000-pixel images to detect small impact craters. The SLDEM data has a high resolution and has a good identification effect for small impact craters and degraded impact craters. We used it to identify impact craters with a diameter less than 20 km.

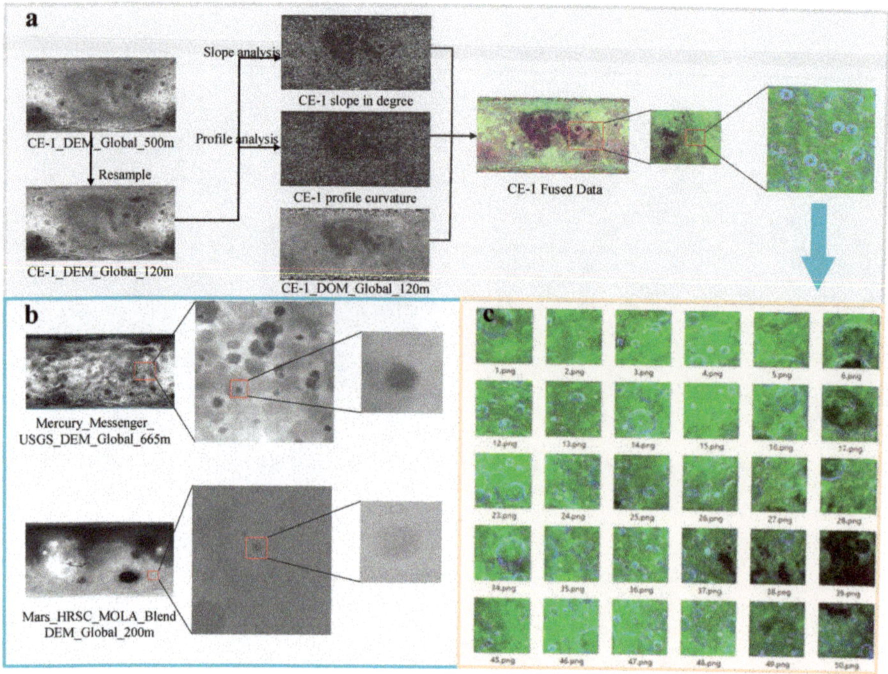

Figure 8. Deep space impact craters data: ((**a**) CE-1 data fusion process. (**b**). Mercury and Mars DEM data. (**c**). The CE-1 fusion dataset).

The Mercury MESSENGER Global DEM has a resolution of 665 m per pixel and spans ±90 degrees latitude and Longitude range from 0° to 360°, which is different from our Moon DEM in terms of image properties. This global grayscale map is an Equirectangular projection with a resolution of 23,040 × 11,520 pixels. Mercury differs from the Moon in gravitational acceleration, surface structure, terrain, and impact background.

The Mars HRSC and MOLA Blended Global DEM had a resolution of 200 m per pixel and spans ±90 degrees latitude (and the maximum range in longitude). This global grayscale map was a Simple Cylindrical projection with a resolution of 106,694 × 53,347 pixels. We also cropped it into 1000 × 1000-pixel images to detect small impact craters.

3.2. Evaluation Metrics

Computer configuration in the experiment comprised two NVIDIA GeForce 2080 Ti RTX GPUs, 64 Gb of memory, Ubuntu16.04 operating system, Cuda10.0, Cudnn7.5, and Opencv3.5.6, and used Caffe framework for training.

The Precision–Recall (P-R) curve and Average Precision (AP) values were used in this experiment to objectively test the accuracy of the target detection algorithm.

$$P = \frac{N_{tp}}{N_{tp} + N_{fp}} \tag{5}$$

where N_{tp} is the number of correctly detected crater targets in the formula, and N_{fp} is the number of miss-detected targets. The Recall in the P-R curve represents the missed detection rate of the algorithm:

$$R = \frac{N_{tp}}{N_{tp} + N_{fn}} \tag{6}$$

where N_{fn} is the missed meteorite crater target.

With Precision as the longitudinal axis and Recall as the horizontal axis, the P-R curve was then fitted by changing the threshold condition. In addition, for the target detection task, the IOU of the predicted location and the actual location of the target were considered when calculating the P-R curve. This was to reflect the accuracy of the target location prediction. In this experiment, IOU was set to 0.5.

The F_1 value is a statistical index used to measure the accuracy of the dichotomous model. This index takes into account both the accuracy and recall rate of the classification model. The F_1 value can be defined as a weighted average of model accuracy and recall rate as:

$$F_1 = 2 * \frac{PR}{P+R} \tag{7}$$

where P and R are the accuracy and recall rates, respectively.

3.3. Training Details

In training the convolutional neural network, it is necessary to set some super parameters, e.g., learning rate, training iteration volume, selection of loss function. The parameter settings are shown in Table 1.

Table 1. The model super parameters.

Parameter	Value
Learning rate	0.0001
Training batches	10,000
Training wheels	1000
Objective function	Cross-entropy and MSE

We used the Adam algorithm for optimization with the momentum of the SGD gradient descent algorithm. We used the first-moment estimation and second-order moments of the gradient vector to estimate the dynamic adjustment of each parameter. In each iteration update, the iteration vector had a specific scope to stabilize the parameter. The introduction of the near iterative gradient direction of the penalty term improved the convergence speed of the models.

The objective function was divided into classification and regression. The Mean Square Error (MSE) algorithm realized the target location by calculating the lowest square value of the predicted site and the actual location. The cross-entropy function also calculated the probability difference between the prediction confidence of the target classification and the essential target category. Furthermore, having the cross-entropy as the loss function prevented the learning rate reduction in the MSE loss function in the case of gradient descent. Therefore, we set

$$C = -\frac{1}{N}\sum_{n} y \ln a + (1-y)\ln(1-a) \tag{8}$$

to be optimized where y is the expected output, a denotes the actual output, N is the total number of training data, n represents the input sample.

4. Results and Discussion

4.1. Analysis of the Lunar Impact Crater Detection Results

In Figure 9, we compare the proposed model in this paper with the identified crater distribution. As it is seen, the number of identified lunar craters was significantly higher than that of the number of identified craters with diameters between 1 and 100 km. This indicates that the proposed model identified many craters in the small and medium diameter ranges. Despite the irregular, severely eroded, and scattered nature of the major lunar craters, the proposed model recognized 46 craters with diameters ranging from 200 to 550 km.

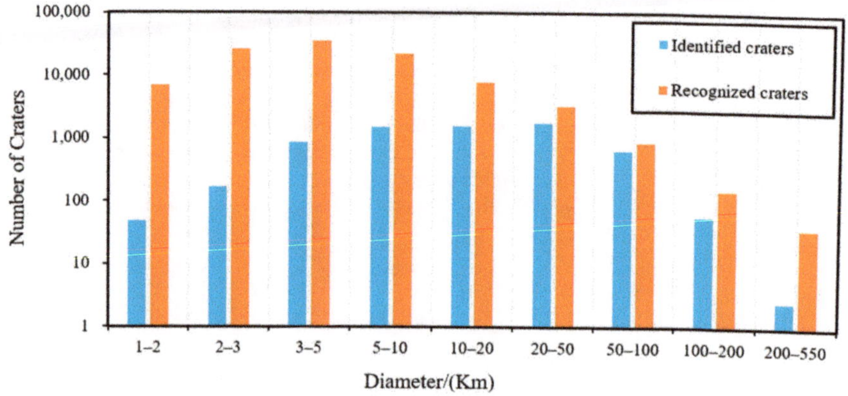

Figure 9. Comparison of the distribution of lunar craters with different diameters identified by the IAU. (The yellow column represents the number of craters recognized by the model. The blue column represents the number of identified craters.).

We also studied the detected craters to ensure their authenticity. We compared them to three databases of artificially acquired lunar craters:

(1) Head et al. [26], where a total of 5185 craters with a diameter of D $\geq$ 20 km was obtained by the Digital Terrestrial Model (DTM) of the Lunar Reconnaissance Orbiter (LRO) Lunar Orbiter Laser Altimeter (LOLA);

(2) Povilaitis et al. [27], in which the previously described database was expanded to 22,746 craters with D = 5–20 km;

(3) The Robbins database [28] holds over 2 million lunar craters, including 1.3 million with D $\geq$ 1 km. This database contains the largest number of lunar craters.

In addition, three kinds of automatic crater directories were considered:

(4) Salamunićcar et al. [29], in which LU78287GT was generated based on Hough transform;

(5) Wang et al. [30], which was based on CE-1 data, and included 106,016 impact craters with D > 500 m;

(6) Silburt et al. [12], which was based on the DEM data from CNN and LRO and generated a meteorite crater database.

(7) Yang et al. [3] adopted the CE-1 and CE-2 data and compiled 117,240 impact craters with D $\geq$ 1–2 km.

Figure 10 shows the comparison results of the number of matched craters at different scales. For manual annotation, it is seen that the matching degree of Povilaitis et al. is consistent with that obtained in our model for craters with diameters of 5–550 km. For the manually annotated Robins database, the number of craters between 1 and 2 km is close to the number identified by our model. This is because of the efficiency of the proposed model in the identification of smaller craters. However, the number of craters between 2 and 20 km is far greater than that of our model. This is because degradation of craters and other reasons leads to insufficient feature extraction. For the overall matching percentage of manually annotated data, the consistency of our recognition results reaches 88.78% for craters with diameters between 5–550 km.

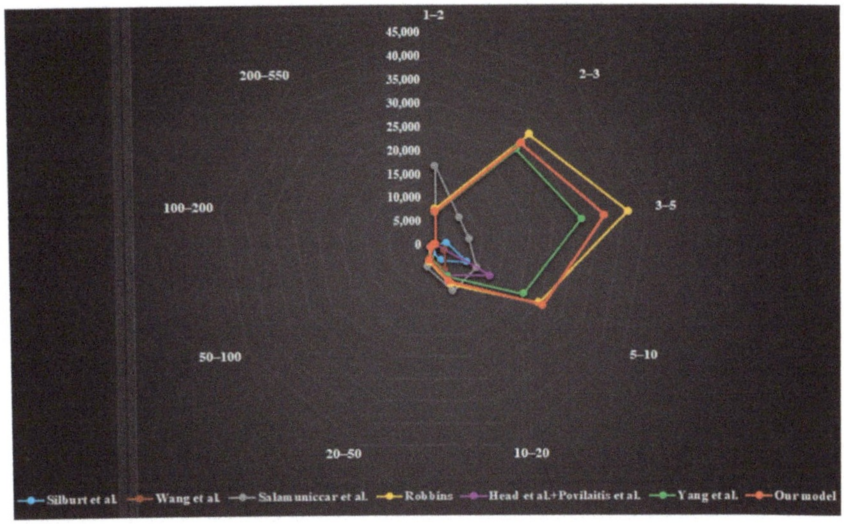

Figure 10. Comparison results of the number of the matched craters at different scales.

For the automatically labeled database and Yang's database, and the impact craters diameter D ranging from 1 to 5 km, our model outperformed the others. This is because we used CE-1 fusion data and SLDEM data, and the trained designed network had a higher identification efficiency for smaller impact craters. According to Wang et al., the number of impact craters with diameters between 1 and 5 km is less than the number of identified craters. Again, the number of impact craters with larger diameters was less than that of the identified craters. At 100 km, they almost overlap, and there is also no global correction. Wang et al.'s crater center location has a different offset from the rest of the databases. Only the craters detected in CE-1 were used for comparison, which accounted for 15% of the total number of craters seen.

According to the initial study results, the accuracy of most of the craters derived from CE-1 data was D = 10~50 km. For the Sliburt et al. Impact Crater Database, the identification number was small for $D \leq 3$ km and $D \geq 50$ km. This indicates that compared with the deep learning method, the transfer learning-based detection identified a larger number of craters in the small and large diameter ranges with fuzzy and severe degradation. Note that it is challenging to detect the secondary craters using the automated methods.

4.2. *Network Performance Comparison*

4.2.1. Comparison of Crater Detection Performance of Different Networks

We trained a total of 2 groups of 10 residual network modules in the R-FCN models, including the groups with different residual network depths of 50 and 101 layers. Using random seeds to divide data into the training set and verification set, each model operated three different sources for training. The results for each model in the validation set are shown in Table 2. The Precision, Recall, F1 Score, test time of each image, and the required memory size of the models were considered as the performance measure.

As it is seen in Table 2, for the network depth of 50 layers, the detection accuracy and recall rate increased by using various improved ResNet modules. The SCNeSt-50-FPN model achieved an accuracy rate of 89.6 and a recall rate of 81.2, which was 3% higher than that of the ResNeSt-50-FPN model. It can also be seen that adaptive convolution and different pooling methods resulted in more accurate crater contour extraction. By increasing the depth of the network, the performance of each residual network was also improved. Compared with other residual networks, the accuracy rate and recall rate of the SCNeSt-101-FPN reached 92.7 and 90.1, respectively, and its F1 total score reached 91.3, which suggests an excellent detection result. Compared with the ResNeSt, the memory requirement of our proposed model was reduced, and the time to detect a picture was about 0.125 s.

Table 2. Detection index results for different networks.

Backbone	Precision (%)	Recall (%)	F_1 Score (%)	Times (s)	Params (M)
ResNet-50-FPN	79.2	63.5	70.4	0.140	25.6
SCNet-50-FPN	80.1	75.6	77.7	0.141	25.6
ResNeXt-50-FPN	84.2	79.3	81.6	0.132	25.0
ResNeSt-50-FPN	86.3	80.1	83.1	0.141	27.5
SCNeSt -50-FPN	89.6	81.2	85.2	0.136	27.5
ResNet-101-FPN	80.2	69.8	74.6	0.134	44.5
SCNet-101-FPN	82.5	83.2	82.9	0.135	44.6
ResNeXt-101-FPN	87.9	85.3	86.5	0.121	44.2
ResNeSt-101-FPN	89.3	88.3	88.7	0.136	48.2
SCNeSt -101-FPN	92.7	90.1	91.3	0.125	48.1

The P-R curve of the training process is shown in Figure 11. The SCNeSt model achieved the highest performance on the test dataset. This is mainly due to its improvements in pooling and the self-calibrated branch, which completed the seamless fusion of multi-scale features.

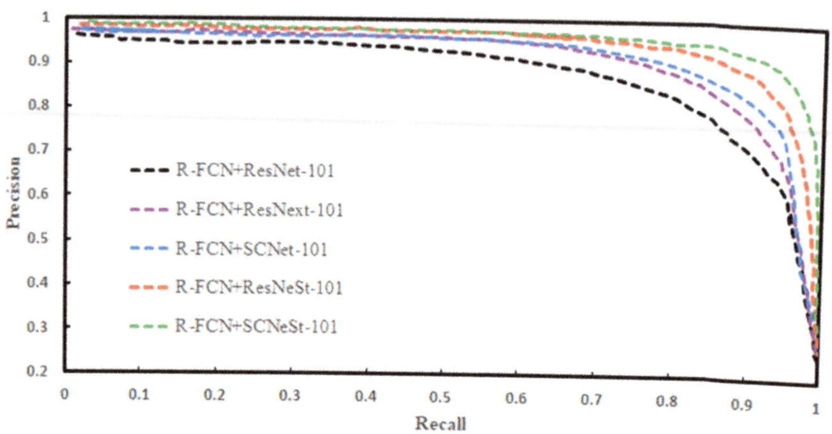

Figure 11. The P-R curves for different models.

To further demonstrate the results of each model, we chose 3 CE-1 fusion images and 2 SLDEM images in the verification set to compare the products, as shown in Figure 12.

Figure 12 shows samples of the impact crater detection. It is seen that the proposed model in this paper had a better detection effect on craters of different scales. Compared with the impact crater detection results of different models in Figure 12b, other models cannot detect small and prominent impact craters. It can also be seen in Figure 12c that ResNext can identify large impact craters, which is attributed to the Group Convolution. As shown in Figure 12d, some small impact craters could be accurately detected, which means that self-calibrated Conv can establish small space and inter-channel dependency around each spatial location. Therefore, it can help CNN generate feature expressions with more discriminant ability because it has more abundant information. Figure 12e also shows that large impact craters and some minor impact craters were efficiently detected but many small impact craters were still missed. In Figure 12f, impact craters of different scales can be effectively detected. Thanks to the combination of adaptive convolution and split attention, more features can be extracted. To further test the influence of the PS-ROI Align module and Soft-NMS on the performance of the R-FCN network, two groups of control tests were conducted. The results are presented in Tables 3 and 4.

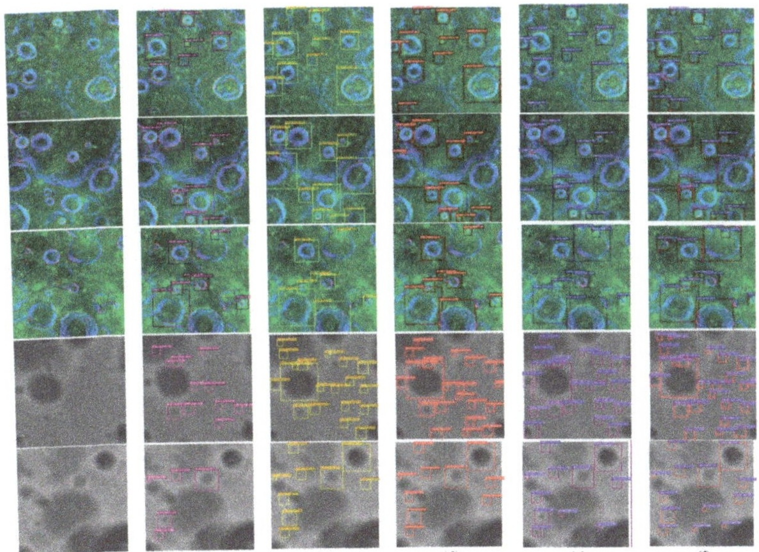

Figure 12. Comparison of the impact crater detection for different models: (**a**) Origin DEM, (**b**) ResNet, (**c**) ResNeXt, (**d**) ScNet, (**e**) ResNeSt, and (**f**) Our model.

Table 3 shows that the PS-ROI Align was superior to ROI Pooling in terms of accuracy, recall rate, and F_1 score at different network depths. This means that the ROI Align cancels the quantization operation. The pixels with floating-point coordinates in the quantization process were calculated by bilinear interpolation, which resulted in higher detection accuracy for small impact craters. Table 4 further shows the experimental results of the Soft-NMS and NMS detection boxes. It is seen that the improved Soft-NMS offered a higher detection performance than that of NMS. It is worth noting that the Soft-NMS needed no further training and was simple to implement. It is also simple to incorporate into any object detection operation.

4.2.2. Performance Comparison of Multi-Scale Impact Crater Networks

To verify the robustness and obtain the portability of the model, four lunar remote sensing data with different resolutions were selected for detection. They were SLDEM data with a resolution of 118 m/piex and 59 m/piex, LRO DEM data with a resolution of 29 m/pix, and DOM data with 7 m/pix. The test results are presented in Figure 13.

Table 3. Added ROI network parameter comparison.

Basic Net	Target Detection Network	ROI Pooling	PS-ROI Align	Recall (%)	Recall (%)	F_1
SCNeSt-50	R-FCN	1	0	85.3	79.6	82.3
		0	1	86.3	80.1	83.1
SCNeSt-101	R-FCN	1	0	90.7	87.1	88.8
		0	1	92.7	90.1	91.3

Table 4. Added Soft-NMS network parameter comparison.

Basic Net	Target Detection Network	NMS	Soft-NMS	Recall (%)	Recall (%)	F_1
SCNeSt-50	R-FCN	1	0	85.4	79.6	80.3
		0	1	86.3	80.1	83.1
SCNeSt-101	R-FCN	1	0	91.2	88.7	82.9
		0	1	92.7	90.1	91.3

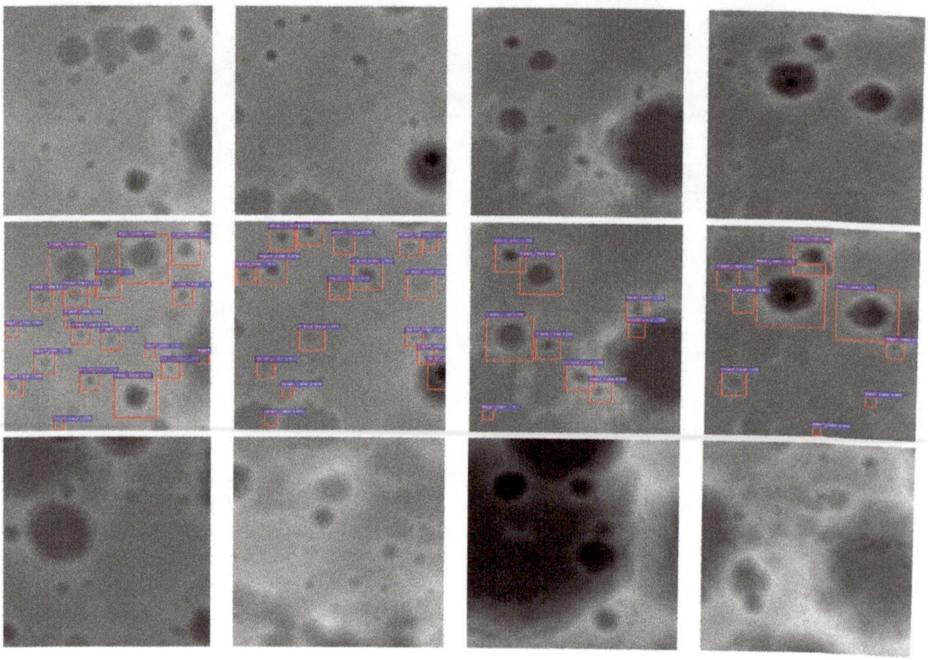

Figure 13. *Cont.*

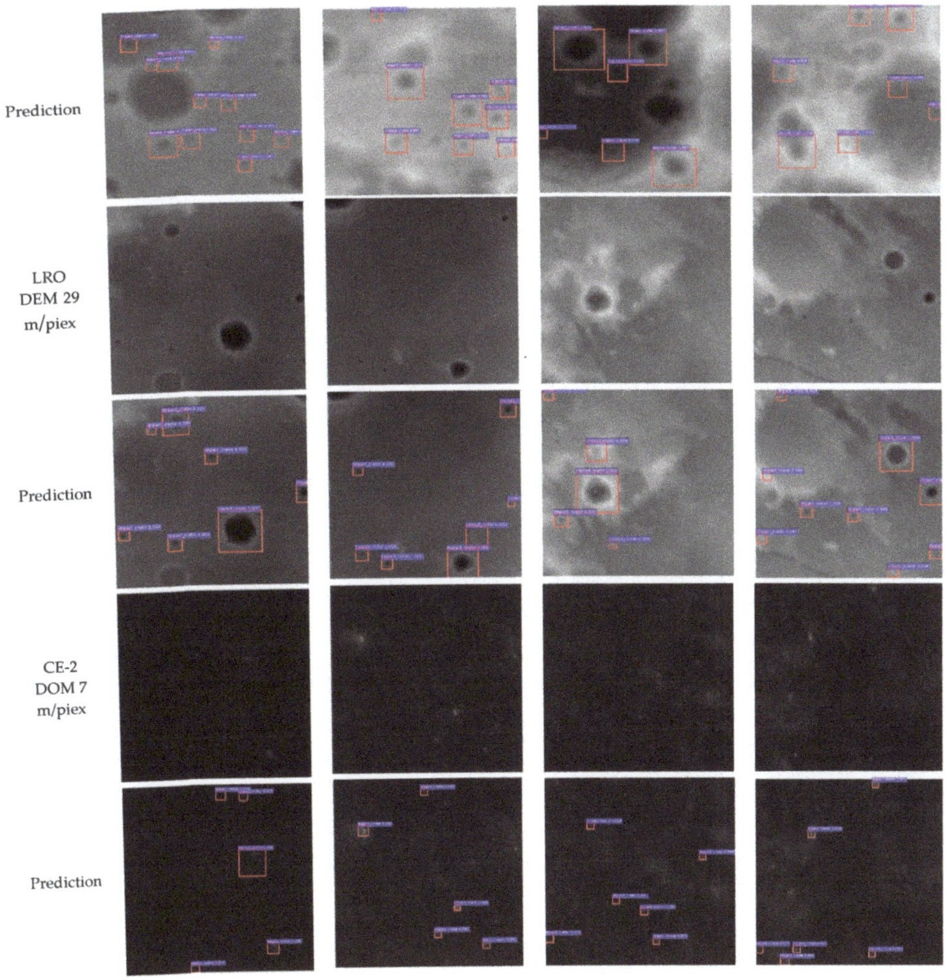

Figure 13. Crater detection results for data with different resolutions.

It is seen that the LRO DEM 29 m/pix results were more accurate in crater detection for different sensor resolutions. However, for more precise illumination data, the detection performance was rather low. Although some impact craters with high pixel points could be detected, most of them were not detected. This may be because DOM data is affected by illumination, which is not ideal for our model detection. For high-resolution DEM data, however, our model provided high detection performance.

4.3. *Transfer Learning in Mars and Mercury Impact Crater Detection Analysis*

Identifying the secondary impact craters is a critical step in the crater counting process for surface age determination. Failure to take these factors into account may result in a significant overestimation of the measured crater density, leading to incorrect model ages. We applied our model to Mars and Mercury data to examine the robustness of our model. The MARS_HRSC_MOLA_BLENDDEM_GLOBAL_200m and MERCURY_MESSENGER_USGS_DEM_GLOBAL_665m datasets were selected for Mars and Mercury, respectively. The results are shown in Figure 14.

Figure 14 shows that the detection recall rate for medium and small impact craters on Mars was 96.8, and multi-scale impact craters were detected. For Mercury, due to the resolution of the dataset and the irregular shape of the craters, some craters were miss-detected. Note that the model trained using the lunar data was applied to Mars and Mercury. In terms of the overall test results, our model achieved a high level of robustness, especially for multi-scale Mars craters.

5. Conclusions

In this study, a new deep-space crater detection network model was proposed, which was trained end-to-end for lunar, Mars, and Mercury data. The CE-1 DEM and DOM data were used as the training data. Based on the R-FCN network architecture, self-calibrated Conv and split attention mechanisms were used for feature extraction. Combined with the multi-scale RPN model, our proposed model efficiently extracted the features of the large, medium, and small impact craters. We further introduced a Position-Sensitive ROI Align network structure that can effectively remove the contour of irregular impact craters. Combined with the improved Soft-NMS framework, the overlapping craters can be efficiently detected. Our model evaluated the proposed network on four resolution lunar data and Mars and Mercury data through transfer learning, and the results demonstrated its advantages for crater-detection missions. Therefore, we will continue to look for small impact craters (D < 1 km) to lay the groundwork for lunar and Mars lander landings and navigation applications.

Author Contributions: Data curation, R.Y.; Funding acquisition, G.W.; Project administration, Y.W. (Yitian Wu); Resources, J.W.; Software, L.L.; Validation, Y.W. (Ying Wang); Visualization, N.X.; Writing—original draft, Y.J.; Writing—review & editing, Y.J. All authors have read and agreed to the published version of the manuscript.

Funding: This research received no external funding.

Data Availability Statement: In this study, using Chang-E data download address for Chinese lunar exploration data and information system, web site for https://moon.bao.ac.cn/moonGisMap.search (accessed on 4 July 2021). In addition, the use of the LRO DEM data and SLDEM data, as well as Mars and Mercury in the USGS DEM data, download website, https://planetarymaps.usgs.gov/mosaic/ (accessed on 4 July 2021). International Astronomical Union. https://planetarynames.wr.usgs.gov/Page/MOON/target (accessed on 4 July 2021).

Acknowledgments: The authors would like to thank Space Engineering University for its hardware support and NASA's Lunar digital elevation model data. In addition, the author is incredibly grateful to Zhao Haishi for his advice.

Conflicts of Interest: The authors declare no conflict of interest.

Abbreviations

The following abbreviations are used in this manuscript:

CDA	Crater detection algorithm
LRO	Lunar Reconnaissance Orbiter
MOLA	Mars Orbiter Laser Altimeter
MOC	Mars Orbiter Camera
HRSC	High Resolution Stereo Camera
CNN	Convolutional neural networks
IAU	International Astronomical Union
RPN	Region proposal network
NMS	Non-maximum suppression
RoI	Region of interest
FPN	Feature pyramid network
DEM	Digital Elevation Model
DTM	Digital Terrestrial Model
DOM	Digital Orthophoto Map

References

1. Fudali, R.F. Impact cratering: A geologic process. *J. Geol.* **1989**, *97*, 773. [CrossRef]
2. Neukum, G.; Nig, B.; Arkani-Hamed, J. A study of lunar impact crater size-distributions. *Moon* **1975**, *12*, 201–229. [CrossRef]
3. Yang, C.; Zhao, H.; Bruzzone, L.; Benediktsson, J.A.; Liang, Y.; Liu, B.; Zeng, X.; Guan, R.; Li, C.; Ouyang, Z. Lunar impact crater identification and age estimation with Chang'E data by deep and transfer learning. *Nat. Commun.* **2020**, *11*, 6358. [CrossRef] [PubMed]
4. Craddock, R.A.; Maxwell, T.A.; Howard, A.D. Crater morphometry and modification in the Sinus Sabaeus and Margaritifer Sinus regions of Mars. *J. Geo. Res.* **1997**, *102*, 13321–13340. [CrossRef]
5. Biswas, J.; Sheridan, S.; Pitcher, C.; Richter, L.; Reiss, P. Searching for potential ice-rich mining sites on the Moon with the Lunar Volatiles Scout. *Planet. Space Sci.* **2019**, *181*, 104826. [CrossRef]
6. De Rosa, D.; Bussey, B.; Cahill, J.T.; Lutz, T.; Crawford, I.A.; Hackwill, T.; van Gasselt, S.; Neukum, G.; Witte, L.; McGovern, A.; et al. Characterisation of potential landing sites for the European Space Agency's Lunar Lander project. *Planet. Space Sci.* **2012**, *74*, 224–246. [CrossRef]
7. Iqbal, W.; Hiesinger, H.; Bogert, C. Geological mapping and chronology of lunar landing sites: Apollo 11. *Icarus* **2019**, *333*, 528–547. [CrossRef]
8. Yan, W.; Gang, Y.; Lei, G. A novel sparse boosting method for crater detection in the high resolution planetary image. *Adv. Space Res.* **2015**, *56*, 982–991.
9. Kim, J.R.; Muller, J.P.; Van Gasselt, S.; Morley, J.G.; Neukum, G. Automated Crater Detection, A New Tool for Mars Cartography and Chronology. *Photogramm. Eng. Remote Sens.* **2015**, *71*, 1205–1218. [CrossRef]
10. Salamunićcar, G.; Lončarić, S.; Mazarico, E. LU60645GT and MA132843GT catalogues of Lunar and Martian impact craters developed using a Crater Shape-based interpolation crater detection algorithm for topography data. *Planet. Space Sci.* **2012**, *60*, 236–247. [CrossRef]
11. Karachevtseva, I.P.; Oberst, J.; Zubarev, A.E.; Nadezhdina, I.E.; Kokhanov, A.A.; Garov, A.S.; Uchaev, D.V.; Uchaev, D.V.; Malinnikov, V.A.; Klimkin, N.D. The Phobos information system. *Planet. Space Sci.* **2014**, *102*, 74–85. [CrossRef]
12. Silburt, A.; Ali-Dib, M.; Zhu, C.; Jackson, A.; Valencia, D.; Kissin, Y.; Tamayo, D.; Menou, K. Lunar crater identification via deep learning. *Icarus* **2019**, *317*, 27–38. [CrossRef]
13. Ali-Dib, M.; Menou, K.; Jackson, A.P.; Zhu, C.; Hammond, N. Automated crater shape retrieval using weakly-supervised deep learning. *Icarus* **2020**, *345*, 113749. [CrossRef]
14. DeLatte, D.M.; Crites, S.T.; Guttenberg, N.; Yairi, T. Automated crater detection algorithms from a machine learning perspective in the convolutional neural network era. *Adv. Space Res.* **2019**, *64*, 1615–1628. [CrossRef]
15. Michael, G.G. Coordinate registration by automated crater recognition. *Planet. Space Sci.* **2003**, *51*, 563–568. [CrossRef]
16. Cheng, Y.; Johnson, A.E.; Matthies, L.H.; Olson, C.F. Optical Landmark Detection for Spacecraft Navigation. In Proceedings of the 13th AAS/AIAA Space Flight Mechanics Meeting, Ponce, PR, USA, 24–27 March 2003; pp. 1785–1803.
17. Cohen, J.P.; Ding, W. Crater detection via genetic search methods to reduce image features. *Adv. Space Res.* **2014**, *53*, 1768–1782. [CrossRef]
18. Zheng, Z.; Zhang, S.; Yu, B.; Li, Q.; Zhang, Y. Defect Inspection in Tire Radiographic Image Using Concise Semantic Segmentation. *IEEE Access* **2020**, *8*, 112674–112687. [CrossRef]
19. Jia, Y.; Liu, L.; Zhang, C. Moon Impact Crater Detection Using Nested Attention Mechanism Based UNet++. *IEEE Access* **2021**, *9*, 44107–44116. [CrossRef]
20. Barker, M.K.; Mazarico, E.M.; Neumann, G.A.; Zuber, M.T.; Smith, D.E. A new lunar digital elevation model from the Lunar Orbiter Laser Altimeter and SELENE Terrain Camera. *Icarus* **2016**, *273*, 346–355. [CrossRef]
21. Liu, J.; Hou, Q.; Cheng, M.; Wang, C.; Feng, J. Improving Convolutional Networks With Self-Calibrated Convolutions. In Proceedings of the 2020 IEEE/CVF Conference on Computer Vision and Pattern Recognition (CVPR), Seattle, WA, USA, 13–19 June 2020; pp. 10093–10102.
22. Lin, T.Y.; Dollar, P.; Girshick, R.; He, K.; Hariharan, B.; Belongie, S. Feature Pyramid Networks for Object Detection. In Proceedings of the 2017 IEEE Conference on Computer Vision and Pattern Recognition (CVPR), Honolulu, HI, USA, 21–26 July 2017; pp. 936–944.
23. Ren, S.; He, K.; Girshick, R.; Sun, J. Faster R-CNN: Towards Real-Time Object Detection with Region Proposal Networks. *IEEE Trans. Pattern Anal. Mach. Intell.* **2016**, *39*, 91–99. [CrossRef]
24. Neubeck, A.; Gool, L. Efficient Non-Maximum Suppression. In Proceedings of the 18th International Conference on Pattern Recognition (ICPR'06), Hong Kong, China, 20–24 August 2006; pp. 850–855.
25. Bodla, N.; Singh, B.; Chellappa, R.; Davis, L.S. Soft-NMS—Improving Object Detection with One Line of Code. In Proceedings of the 2017 IEEE International Conference on Computer Vision (ICCV), Venice, Italy, 22–29 October 2017; pp. 5562–5570.
26. Head, J.W.; Fassett, C.I.; Kadish, S.J.; Smith, D.E.; Zuber, M.T.; Neumann, G.A.; Mazarico, E. Global Distribution of Large Lunar Craters: Implications for Resurfacing and Impactor Populations. *Science* **2010**, *329*, 1504–1507. [CrossRef] [PubMed]
27. Povilaitis, R.Z.; Robinson, M.S.; van der Bogert, C.H.; Hiesinger, H.; Meyer, H.M.; Ostrach, L.R. Crater density differences: Exploring regional resurfacing, secondary crater populations, and crater saturation equilibrium on the moon. *Planet. Space Sci.* **2018**, *162*, 41–51. [CrossRef]

28. Robbins, S.J. A New Global Database of Lunar Impact Craters >1–2 km: 1. Crater Locations and Sizes, Comparisons with Published Databases, and Global Analysis. *J. Geophys. Res. Planets* **2019**, *124*, 871–892. [CrossRef]
29. Salamunićcar, G.; Lončarić, S.; Grumpe, A.; Wöhler, C. Hybrid method for crater detection based on topography reconstruction from optical images and the new LU78287GT catalogue of Lunar impact craters. *Adv. Space Res.* **2014**, *53*, 1783–1797. [CrossRef]
30. Wang, J.; Cheng, W.; Zhou, C. A Chang'E-1 global catalog of lunar impact craters. *Planet. Space Sci.* **2015**, *112*, 42–45. [CrossRef]

remote sensing

MDPI

Article

Variational Generative Adversarial Network with Crossed Spatial and Spectral Interactions for Hyperspectral Image Classification

Zhongwei Li [1], Xue Zhu [2], Ziqi Xin [2], Fangming Guo [1], Xingshuai Cui [2] and Leiquan Wang [2,*]

1 College of Oceanography and Space Informatics, China University of Petroleum (East China), Qingdao 266580, China; lizhongwei@upc.edu.cn (Z.L.); guofangming@s.upc.edu.cn (F.G.)
2 College of Computer Science and Technology, China University of Petroleum (East China), Qingdao 266580, China; S19070024@s.upc.edu.cn (X.Z.); S20070012@s.upc.edu.cn (Z.X.); cuixingshuai@s.upc.edu.cn (X.C.)
* Correspondence: 20060068@upc.edu.cn

Abstract: Variational Autoencoders (VAEs) and Generative Adversarial Networks (GANs) have been widely used in hyperspectral image classification (HSIC) tasks. However, the generated HSI virtual samples by VAEs are often ambiguous, and GANs are prone to the mode collapse, which lead the poor generalization abilities ultimately. Moreover, most of these models only consider the extraction of spectral or spatial features. They fail to combine the two branches interactively and ignore the correlation between them. Consequently, the variational generative adversarial network with crossed spatial and spectral interactions (CSSVGAN) was proposed in this paper, which includes a dual-branch variational Encoder to map spectral and spatial information to different latent spaces, a crossed interactive Generator to improve the quality of generated virtual samples, and a Discriminator stuck with a classifier to enhance the classification performance. Combining these three subnetworks, the proposed CSSVGAN achieves excellent classification by ensuring the diversity and interacting spectral and spatial features in a crossed manner. The superior experimental results on three datasets verify the effectiveness of this method.

Keywords: hyperspectral image classification; variational autoencoder; generative adversarial network; crossed spatial and spectral interactions

Citation: Li, Z.; Zhu, X.; Xin, Z.; Guo, F.; Cui, X.; Wang, L. Variational Generative Adversarial Network with Crossed Spatial and Spectral Interactions for Hyperspectral Image Classification. *Remote Sens.* **2021**, *13*, 3131. https://doi.org/10.3390/rs13163131

Academic Editors: Fahimeh Farahnakian, Jukka Heikkonen and Pouya Jafarzadeh

Received: 12 July 2021
Accepted: 5 August 2021
Published: 7 August 2021

1. Introduction

Hyperspectral images (HSI) contain hundreds of continuous and diverse bands rich in spectral and spatial information, which can distinguish land-cover types more efficiently compared with ordinary remote sensing images [1,2]. In recent years, Hyperspectral images classification (HSIC) has become one of the most important tasks in the field of remote sensing with wide application in scenarios such as urban planning, geological exploration, and agricultural monitoring [3–6].

Originally, models such as support vector machines (SVM) [7], logistic regression (LR) [8] and and k-nearest neighbors algorithm (KNN) [9], have been widely used in HSI classification tasks for their intuitive outcomes. However, most of them only utilize handcrafted features, which fail to embody the distribution characteristics of different objects. To solve this problem, a series of deep discriminative models, such as convolutional neural networks (CNNs) [10–12], recurrent neural network (RNN) [13] and Deep Neural Networks (DNN) [14] have been proposed to optimize the classification results by fully utilizing and abstracting the limited data. Though having gained great progress, these methods only analyze the spectral characteristics through an end-to-end neural network without full consideration of special properties contained in HSI. Therefore, the extraction of high-level and abstract features in HSIC remains a challenging task. Meanwhile, the jointed spectral-spatial features extraction methods [15,16] have aroused wide interest in Geosciences and Remote Sensing community [17]. Du proposed a jointed network to extract spectral and spatial features with dimensionality reduction [18]. Zhao et al. proposed a hybrid spectral CNN (HybridSN) to better extract double-way features [19], which combined spectral-spatial 3D-CNN with spatial 2D-CNN to improve the classification accuracy.

Although the methods above enhance the abilities of spectral and spatial features extraction, they are still based on the discriminative model in essence, which can neither calculate prior probability nor describe the unique features of HSI data. In addition, the access to acquire HSI data is very expensive and scarce, requiring huge human resources to label the samples by field investigation. These characteristics make it impractical to obtain enough markable samples for training. Therefore, the deep generative models have emerged at the call of the time. Variational auto encoder (VAE) [20] and generative adversarial network (GAN) [21] are the representative methods of generative models.

Liu [22] and Su [23] used VAEs to ensure the diversity of the generated data that were sampled from the latent space. However, the generated HSI virtual samples are often ambiguous, which cannot guarantee similarities with the real HSI data. Therefore, GANs have also been applied for HSI generation to improve the quality of generated virtual data. GANs strengthen the ability of discriminators to distinguish the true data sources from the false by introducing "Nash equilibrium" [24–29]. For example, Zhan [30] designed a 1-D GAN (HSGAN) to generate the virtual HSI pixels similar to the real ones, thus improving the performance of the classifier. Feng [31] devised two generators to generate 2D-spatial and 1D-spectral information respectively. Zhu [32] exploited 1D-GAN and 3D-GAN architectures to enhance the classification performance. However, GANs are prone to mode collapse, resulting in poor generalization ability of HSI classification.

To overcome the limitations of VAEs and GANs, VAE-GAN jointed framework has been proposed for HSIC. Wang proposed a conditional variational autoencoder with an adversarial training process for HSIC (CVA^{2}E) [33]. In this work, GAN was spliced with VAE to realize high-quality restoration of the samples and achieve diversity. Tao et al. [34] proposed the semi-supervised variational generative adversarial networks with a collaborative relationship between the generation network and the classification network to produce meaningful samples that contribute to the final classification. To sum up, in VAE-GAN frameworks, VAE focuses on encoding the latent space, providing creativity of generated samples, while GAN concentrates on replicating the data, contributing to the high quality of virtual samples.

Spectral and spatial are two typical characteristics of HSI, both of which must be taken into account for HSIC. Nevertheless, the distributions of spectral and spatial features are not identical. Therefore, it is difficult to cope with such a complex situation for a single encoder in VAEs. Meanwhile, most of the existing generative methods use spectral and spatial features respectively for HSIC, which affects the generative model to generate realistic virtual samples. In fact, the spectral and spatial features are closely correlated, which cannot be treated separately. Interaction between spectral and spatial information should be established to refine the generated virtual samples for better classification performance.

In this paper, a variational generative adversarial network with crossed spatial and spectral interactions (CSSVGAN) was proposed for HSIC, which consists of a dual-branch variational Encoder, a crossed interactive Generator, and a Discriminator stuck together with a classifier. The dual-branch variational Encoder maps spectral and spatial information to different latent spaces. The crossed interactive Generator reconstructs the spatial and spectral samples from the latent spectral and spatial distribution in a crossed manner. Notably, the intersectional generation process promotes the consistency of learned spatial and spectral features and simulates the highly correlated spatial and spectral characteristics of true HSI. The Discriminator receives the samples from both generator and original training data to distinguish the authenticity of the data. To sum up, the variational Encoder ensures diversity, and the Generator guarantees authenticity. The two components place higher demands on the Discriminator to achieve better classification performance.

Compared with the existing literature, this paper is expected to make the following contributions:

- The dual-branch variational Encoder in the jointed VAE-GAN framework is developed to map spectral and spatial information into different latent spaces, provides discriminative spectral and spatial features, and ensures the diversity of generated virtual samples.
- The crossed interactive Generator is proposed to improve the quality of generated virtual samples, which exploits the consistency of learned spatial and spectral features to imitate the highly correlated spatial and spectral characteristics of HSI.
- The variational generative adversarial network with crossed spatial and spectral interactions is proposed for HSIC, where the diversity and authenticity of generated samples are enhanced simultaneously.
- Experimental results on the three public datasets demonstrate that the proposed CSSVGAN achieves better performance compared with other well-known models.

The remainder of this paper is arranged as follows. Section 2 introduces VAEs and GANs. Section 3 provides the details of the CSSVGAN framework and the crossed interactive module. Section 4 evaluates the performance of the proposed CSSVGAN through comparison with other methods. The results of the experiment are discussed in Section 5 and the conclusion is given in Section 6.

2. Related Work

2.1. Variational Autoencoder

Variational autoencoder is one variant of the standard AE, proposed by Kingma et al. for the first time [35]. The essence of VAE is to construct an exclusive distribution for each sample X and then sample it represented by Z. It brings Kullback–Leibler [36] divergence penalty method into the process of sampling and constrains it. Then the reconstructed data can be translated to generated simulation data through deep training. The above principle gives VAE a significant advantage in processing hyperspectral images with expensive and rare samples. VAE model adopts the posterior distribution method to verify that $\rho(Z|X)$ rather than $\rho(Z)$ obeys the normal distribution. Then it manages to find the mean μ and variance σ of $\rho(Z|X_k))$ corresponding to each X_k through the training of neural networks (where X_k represents the sample of the original data and $\rho(Z|X_k)$ represents the posterior distribution). Another particularity of VAE is that it makes all $\rho(Z|X)$ align with the standard normal distribution $N \sim (0,1)$. Taking account of the complexity of HSI data, VAE has superiority over AE in terms of noise interference [37]. It can prevent the occurrence of zero noise, increase the diversity of samples, and further ensure the generation ability of the model.

A VAE model is consists of two parts: Encoder M and Decoder N. M is an approximator for the probability function $m_\tau(z|x)$, and N is to generate the posterior's approximate value $n\theta(x,z)$. τ and θ are the parameters of the deep neural network, aiming to optimize the following objective functions jointly.

$$V(P,Q) = -KL(m_\tau(z|x)||p_\theta(z|x)) + R(x), \tag{1}$$

Among them, R is to calculate the reconstruction loss of a given sample x in the VAE model. The framework of VAE is described in Figure 1, where e_i represents the sample of standard normal distribution, corresponding with X_k one to one.

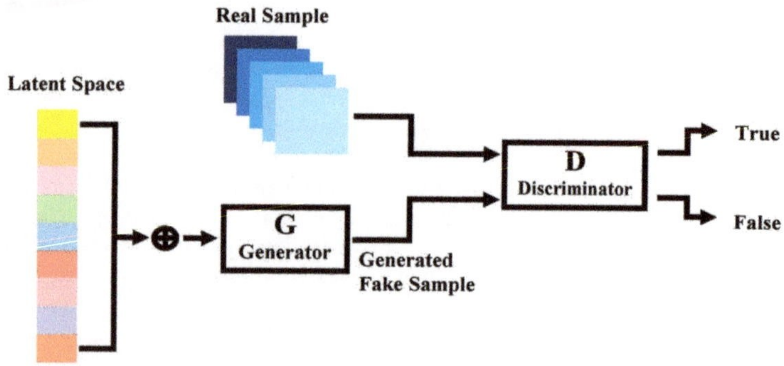

Figure 1. The framework of VAE.

2.2. *Generative Adversarial Network*

Generative adversarial network is put forward by Goodfellow et al. [24], which trains the generation model with a minimax game based on the game theory. The GAN has gained remarkable results in representing the distribution of latent variables for its special structure, which has attracted more attention from the field of visual image processing. A GAN model includes two subnets: the generator G, denoted as $G(z;\theta_g)$ and the discriminator D, denoted as $G(x;\theta_d)$, and θ_g and θ_d are defined as parameters of the deep neural networks. G shows a prominent capacity in learning the mapping of latent variables and synthesizing new similar data from mapping represented by $G(z)$. The function of D is to take the original HSI or the fake image generated by G as input and then distinguish its authenticity. The architecture of GAN is shown in Figure 2.

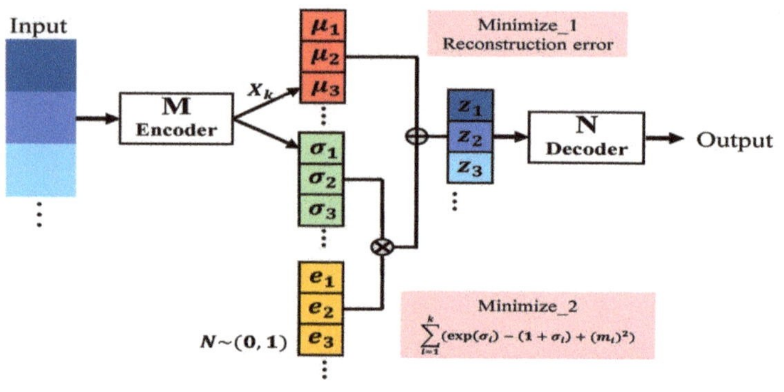

Figure 2. The architecture of GAN.

After the game training, G and D would maximize log-likelihood respectively and achieve the best generation effect by competing with each other. The expression of the above process is as follows:

$$min_G max_G V(G,D) = E_{x\sim P(x)}[\log_D(x)] + E_{x\sim P_{g(z)}}[\log(1 - D(G(z)))], \quad (2)$$

where $P_{(x)}$ represents the real data distribution and $P_{g(z)}$ means the samples' distribution generated by G. The game would reach a global equilibrium situation between the two players when $P_{(x)}$ equaling to $P_{g(z)}$ happened. In this case, the best performance of $D(x)$ can be expressed as:

$$D(x)_{max} = P_{(x)+P_{g(x)}}, \quad (3)$$

However, the over-confidence of D would cause inaccurate results of GAN's identification and make the generated data far away from the original HSI. To tackle the problem, endeavors have been made to improve the accuracy of HSIC by modifying the loss, such as WGAN [38], LSGAN [39], CycleGAN [40] and so on. Salimans [41] raised a deep convolutional generative adversarial network (DCGAN) to enhance the stability of the training and improve the quality of the results. Subsequently, Alec et al. [42] proposed a one-side label smoothing idea named improved DCGAN, which multiplied the positive sample label by alpha and the negative sample label by beta, that is, the coefficients of positive and negative samples in the objective function of D were no longer from 0 to 1, but from α to β. (β in the real application could be set to 0.9). It aimed to solve the problems described as follows:

$$D(x) = \frac{\alpha P_{(x)} + \beta P_{g(x)}}{P_{(x)} + P_{g(x)}}, \tag{4}$$

In this instance, GAN can reduce the disadvantage of overconfidence and make the generated samples more authentic.

3. Methodology

3.1. The Overall Framework of CSSVGAN

The overall framework of CSSVGAN is shown in Figure 3. In the process of data preprocessing, assuming that HSI cuboid X contains n pixels; the spectral band of each pixel is defined as p_x; and X can be expressed as $X \epsilon R^{n*p_x}$. Then HSI is divided into several patch cubes of the same size. The labeled pixels are marked as $X_1 = x_i^1 \epsilon R^{(s*s*p_x*n_1)}$, and the unlabeled pixels are marked as $X_2 = x_i^2 \epsilon R^{(s*s*p_x*n_2)}$. Among them, s, n_1 and n_2 stand for the adjacent spatial sizes of HSI cuboids, the number of labeled samples and the number of unlabeled samples respectively, and n equals to n_1 plus n_2.

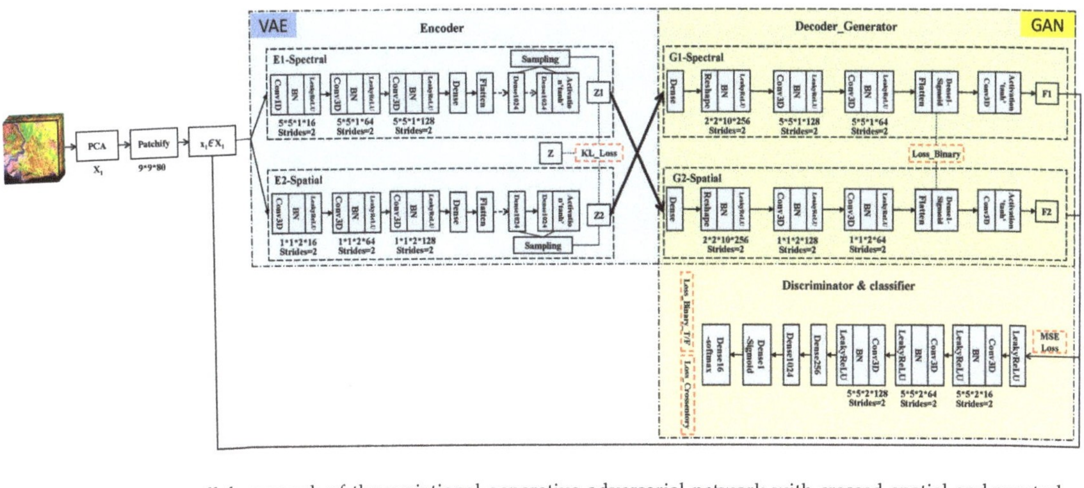

Figure 3. The overall framework of the variational generative adversarial network with crossed spatial and spectral interactions (CSSVGAN) for HSIC.

It is noteworthy that HSI classification is developed at the pixel level. Therefore, in this paper, the CSSVGAN framework uses a cube composed of patches of size $9 \times 9 \times p_x$ as the inputs of the Encoder, where p denotes the spectral bands of each pixel. Then a tensor represents the variables and outputs of each layer. Firstly, the spectral latent variable Z_1 and the spatial latent variable Z_2 are obtained by taking the above X_1 as input into the dual-branch variational Encoder. Secondly, these two inputs are taken to the crossed interactive Generator module to obtain the virtual data F_1 and F_2. Finally, the data are mixed with

X_1 into the Discriminator for adversarial training to get the predicted classification results $\hat{Y} = \hat{y}_i$ by the classifier.

3.2. The Dual-Branch Variational Encoder in CSSVGAN

In the CSSVGAN model mentioned above, the Encoder (Figure 4) is composed of a dual-branch spatial feature extraction E_1 and a spectral feature extraction E_1 to generate more diverse samples. In the E_1 module, the size of the 3D convolution kernel is $(1 \times 1 \times 2)$, the stride is $(2,2,2)$ and the spectral features are marked as Z_1. The implementation details are described in Table 1. Identically, in the E_2 module, the 3D convolution kernels, the strides and the spatial features are presented by $(5 \times 5 \times 1)$, $(2,2,2)$ and Z_2 respectively, as described in Table 2.

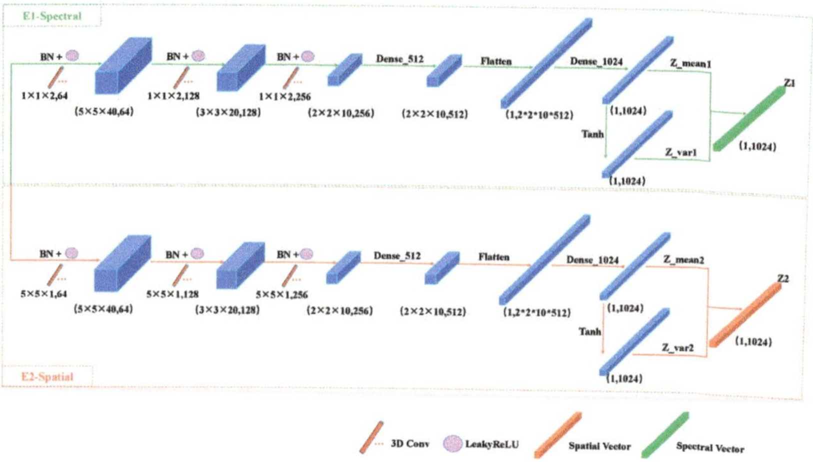

Figure 4. The dual-branch Encoder in CSSVGAN.

Table 1. The implementation details of the Spectral feature extraction E_1.

Input Size	Layer Operations	Output Size
$(9 \times 9 \times 80, 1)$	Conv3D $(1 \times 1 \times 2, 64) - BN - LeakyReLU$	$(5 \times 5 \times 40, 64)$
$(5 \times 5 \times 40, 64)$	Conv3D $(1 \times 1 \times 2, 128) - BN - LeakyReLU$	$(3 \times 3 \times 20, 128)$
$(3 \times 3 \times 20, 128)$	Conv3D $(1 \times 1 \times 2, 256) - BN - LeakyReLU$	$(2 \times 2 \times 10, 256)$
$(2 \times 2 \times 10, 256)$	Dense $(512) - BN - LeakyReLU$	$(2 \times 2 \times 10, 512)$
$(2 \times 2 \times 10, 512)$	Flatten	$(, 20, 480)$
$(, 20, 480)$	Dense (1024)	$(, 1024)$
$(, 1024)$	Dense $(1024) - Tanh$	$(, 1024)$
$(, 1024)$	Lambda $(Sampling)$	$(, 1024)$

Table 2. The implementation details of the Spatial feature extraction E_2.

Input Size	Layer Operations	Output Size
$(9 \times 9 \times 80, 1)$	Conv3D $(5 \times 5 \times 1, 64) - BN - LeakyReLU$	$(5 \times 5 \times 40, 64)$
$(5 \times 5 \times 40, 64)$	Conv3D $(5 \times 5 \times 1, 128) - BN - LeakyReLU$	$(3 \times 3 \times 20, 128)$
$(3 \times 3 \times 20, 128)$	Conv3D $(5 \times 5 \times 1, 256) - BN - LeakyReLU$	$(2 \times 2 \times 10, 256)$
$(2 \times 2 \times 10, 256)$	Dense $(512) - BN - LeakyReLU$	$(2 \times 2 \times 10, 512)$
$(2 \times 2 \times 10, 512)$	Flatten	$(, 20, 480)$
$(, 20, 480)$	Dense (1024)	$(, 1024)$
$(, 1024)$	Dense $(1024) - Tanh$	$(, 1024)$
$(, 1024)$	Lambda $(Sampling)$	$(, 1024)$

Meanwhile, to ensure the consistent distribution of samples and original data, KL divergence principle is utilized to constrain Z_1 and Z_2 separately. Assuming that the mean and variance of Zi are expressed as Z_{meani} and $Z_{vari}(i = 1,2)$, the loss function in the training process is as follows:

$$L_i(\theta,\varphi) = -KL(q_\varphi(z_i|x)\|p_\theta(z_i|x)), \tag{5}$$

where $p(z_i|x)$ is the posterior distribution of potential eigenvectors in the Encoder module, and its calculation is based on the Bayesian formula as shown below. But when the dimension of Z is too high, the calculation of $P(x)$ is not feasible. At this time, a known distribution $q(z_i|x)$ is required to approximate $p(z_i|x)$, which is given by KL divergence. By minimizing KL divergence, the approximate $p(z_i|x)$ can be obtained. θ and φ represent the parameters of distribution function p and q separately.

$$L_i(\theta,\varphi) = E_{q_\varphi(z_i,x)}[\log \tfrac{p_\theta(x,z_i)}{q_\varphi(z_i,x)}] - E_q(x)[\log_q(x)], \tag{6}$$

Formula (6) in the back is provided with a constant term $logN$, the entropy of empirical distribution $q(x)$. The advantage of it is that the optimization objective function is more explicit, that is, when $p_\theta(z_i,x)$ is equal to $q_\varphi(z_i,x)$, KL dispersion can be minimized.

3.3. The Crossed Interactive Generator in CSSVGAN

In CSSVGAN, the crossed interactive Generator module plays a role in data restoration of VAE and data expansion of GAN, which includes the spectral Generator G_1 and the spatial Generator G_2 in the crossed manner. G_1 accepts the spatial latent variables Z_2 to generate spectral virtual data F_1, and G_2 accepts the spectral latent variables Z_1 to generate spatial virtual data F_2.

As shown in Figure 5, the 3D convolution of spectral Generator G_1 is $(1\times1\times2)$ that uses $(2,2,2)$ strides to convert the spatial latent variables Z_2 to the generated samples. Similarly, the spatial Generator G_2 with $(5\times5\times1)$ convolution uses $(2,2,2)$ strides to transform the spectral latent variables Z_1 into generated samples. Therefore, the correlation between spectral and spatial features in HSI can be fully considered to further improve the quality and authenticity of the generated samples. The implementation details of G_1 and G_2 are described in Tables 3 and 4.

Table 3. The implementation details of spectral Generator G_1.

Input Size	Layer Operations	Output Size
(, 1024)	Dense $(2*2*10*256)$	$(10,240)$
(, 10, 240)	Reshape $(2\times2\times10\times256)BN-LeakyReLU$	$(2,2,10,256)$
$(2,2,10,256)$	Conv3DTranspose $(1\times1\times2,128)BN-LeakyReLU$	$(4,4,20,128)$
$(4,4,20,128)$	Conv3DTranspose $(1\times1\times2,64)BN-LeakyReLU$	$(8,8,40,64)$
$(8,8,40,64)$	Conv3DTranspose $(1\times1\times2,1)LeakyReLU-Tanh$	$(9,9,80,1)$

Table 4. The implementation details of spatial Generator G_2.

Input Size	Layer Operations	Output Size
(, 1024)	Dense $(2*2*10*256)$	(, 10, 240)
(, 10, 240)	Reshape $(2\times2\times10\times256)BN-LeakyReLU$	$(2,2,10,256)$
$(2,2,10,256)$	Conv3DTranspose $(5\times5\times1,128)BN-LeakyReLU$	$(4,4,20,128)$
$(4,4,20,128)$	Conv3DTranspose $(5\times5\times1,64)BN-LeakyReLU$	$(8,8,40,64)$
$(8,8,40,64)$	Conv3DTranspose $(5\times5\times1,1)LeakyReLU-Tanh$	$(9,9,80,1)$

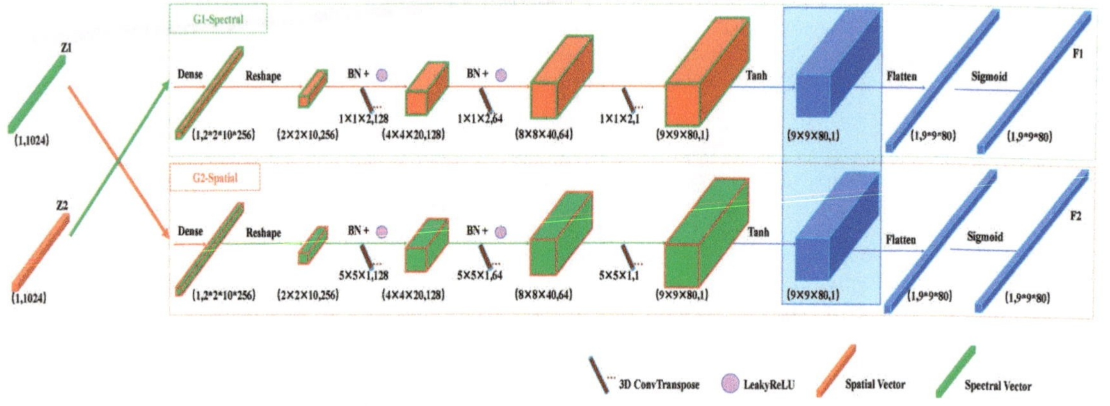

Figure 5. The Crossed Interactive Generator in CSSVGAN.

Because the mechanism of GAN is that the Generator and Discriminator are against each other before reaching the Nash equilibrium, the Generator has two target functions, as shown below.

$$MSE_{Loss\_i} = \frac{1}{n}\sum(y_{ij} - \bar{y}ij)^2, \tag{7}$$

where n is the number of samples, $i = 1, 2$, y_j means the label of virtual samples, and $\bar{y}_j$ represents the label of the original data corresponding to y_j. The above formula makes the virtual samples generated by crossed interactive Generator as similar as possible to the original data.

$$Binary_{Loss\_i} = -\frac{1}{N}\sum_{j=1}^{N} y_{ij} \cdot \log(p(y_{ij})) + (1 - y_{ij} \cdot (1 - p(y_{ij}))), \tag{8}$$

$Binary_{Loss}$ is a logarithmic loss function and can be applied to the binary classification task. Where y is the label (either true or false), and $p(y)$ is the probability that N sample points belonging to the real label. Only if y_j equals to $p(y_i)$, the total loss would be zero.

3.4. The Discriminator Stuck with a Classifier in CSSVGAN

As shown in Figure 6, the Discriminator needs to specifically identify the generated data as false and the real HSI data as true. This process can be regarded as a two-category task using one-sided label smoothing: defining the real HSI data as 0.9 and the false as zero. The loss function of it marked with $Binary_{(Loss_D)}$ is the same as the Formula (10) enumerated above. Moreover, the classifier is stuck as an interface to the output of Discriminator and the classification results are calculated directly through the SoftMax layer, where C represents the total number of labels in training data. As mentioned above, the Encoder ensures diversity and the Generator guarantees authenticity. All these contributions place higher demands on Discriminator to achieve better classification performance. Thus, the CSSVGAN framework yields a better classification result.

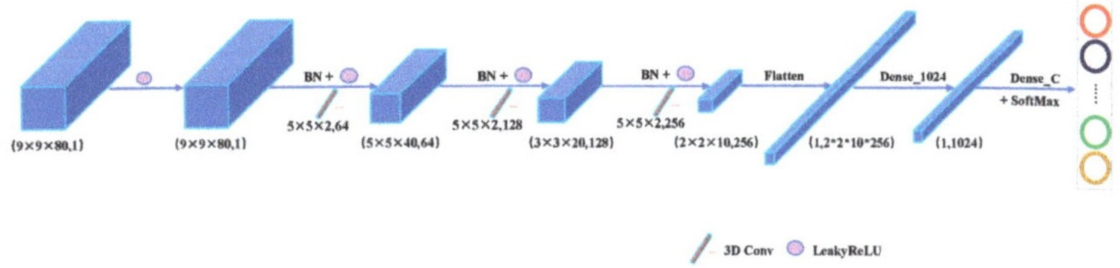

Figure 6. The Discriminator stuck with a classifier in CSSVGAN.

The implementation details of the Discriminator in CSSVGAN are described in Table 5 with the 3D convolution of $(5 \times 5 \times 2)$ and strides of $(2,2,2)$. Identifying C categories belongs to a multi-classification assignment. The SoftMax method is taken as the standard for HSIC. As shown below, the CSSVGAN method should allocate the sample x of each class c to the most likely one of the C classes to get the predicted classification results. The specific formula is as follows:

$$y_i = S(x_i) = \frac{e^{xi}}{\sum_{j=1}^{C} e^{xj}}, \tag{9}$$

Then the category of X can be expressed as the formula below:

$$class(c) = \arg\max_i (y_i = S(x_i)), \tag{10}$$

where S, C, X, Y_i signify the SoftMax function, the total number of categories, the input of SoftMax, and the probability that the prediction object belongs to class C, respectively. X_i similar with X_j is a sample of one certain category. Therefore, the following formula can be used for the loss function of objective constraint.

$$C_{Loss} = -\sum_{i=1}^{n} p(y_{i1}) \cdot \log y_{i1} + p(y_{i2}) \cdot \log(y_{i2}) + \cdots + p(y_{ic}) \cdot \log(y_{ic}), \tag{11}$$

where n means the total number of samples, C represents the total number of categories, and y denotes the single label (either true or false) with the same description as above.

Table 5. The implementation details in Discriminator.

Input Size	Layer Operations	Output Size
$(9 \times 9 \times 80, 1)$	$BN - LeakyReLU$	$(9 \times 9 \times 80, 1)$
$(9 \times 9 \times 80, 1)$	Conv3D $(5 \times 5 \times 2, 64) - BN - LeakyReLU$	$(5 \times 5 \times 40, 64)$
$(5 \times 5 \times 40, 64)$	Conv3D $(5 \times 5 \times 2, 128) - BN - LeakyReLU$	$(3 \times 3 \times 20, 128)$
$(3 \times 3 \times 20, 128)$	Conv3D $(5 \times 5 \times 2, 256) - BN - LeakyReLU$	$(2 \times 2 \times 10, 256)$
$(2 \times 2 \times 10, 256)$	Flatten	(, 10, 240)
(, 10, 240)	Dense (16)	(, 16)

3.5. The Total Loss of CSSVGAN

As illustrated in Figure 3, up till now, the final goal of the total loss of the CSSVGAN model can be divided into four parts: two KL divergence constraint losses and a mean-square error loss from the Encoder, two binary losses from the Generator, one binary loss from the Discriminator and one multi-classification loss from the multi classifier. The ensemble formula can be expressed as:

$$L_{Total} = \underbrace{\sigma_1 L_1(\theta,\varphi) + \sigma_2 L_2(\theta,\varphi) + \sigma_3 MSE_{Loss_{1\_2}}}_{Encoder\_Loss} + \underbrace{\sigma_4 Binary_{Loss_1} + \sigma_5 Binary_{Loss_2}}_{Generator\_Loss} + \underbrace{Binary\_Loss_D}_{Discrminator\_Loss} + \underbrace{C_{Loss}}_{Classifier\_Loss}, \quad (12)$$

where L_1 and L_2 represent the loss between Z_1 or Z_2 and the standard normal distribution respectively in Section 3.2. MSE_{Loss_1} and MSE_{Loss_2} signify the mean square error of y_1 and y_2 in Section 3.3 separately. $MSE_{Loss_{1\_2}}$ calculates the mean square error between y_1 and y_2. The purpose of $Binary_{Loss_1}$ and $Binary_{Loss_2}$ is to assume that the virtual data F_1 and F_2 (in Section 3.3) are true with a value of one. $Binary_{Loss_D}$ denotes that the Discriminator identifies F_1 and F_2 as false data with a value of zero. Finally, the C_{Loss} is the loss of multi classes of the classifier.

4. Experiments

4.1. Dataset Description

In this paper, three representative hyperspectral datasets recognized by the remote sensing community (i.e., Indian Pines, Pavia University and Salinas) are accepted as benchmark datasets. The details of them are as follows:

(1) Indian pines (IP): The first dataset was accepted for HSI classification imaged by Airborne Visible/Infrared Imaging Spectrometer (AVIRIS) in Northwestern Indiana in the USA. It includes 16 categories with a spatial resolution of approximately 20 m per pixel. Samples are shown in Figure 7. The spectral of AVIRIS coverage ranges from 0.4 to 2.5 μm and includes 200 bands for continuous imaging of ground objects (20 bands are influenced by noise or steam, so only 200 bands are left for research), bring about the total image size of 145 × 145 × 200. However, since it contains a complex sample distribution, the category samples of training labels were very imbalanced. As some classes have more than 2000 samples while some have less than 30 merely, it is relatively difficult to achieve a high-precision classification of IP HSI.

(2) Pavia University (PU): The second dataset was a part of the hyperspectral image data of the Pavia city in Italy, photographed by the German airborne reflective optics spectral imaging system (Rosis-03) in 2003, containing 9 categories (see Figure 8). The resolution of this spectral imager is 1.3 m, including continuously 115 wavebands in the range of 0.43–0.86 μm. Among these bands, 12 bands were eliminated due to the influence of noise. Therefore, the images with the remaining 103 spectral bands in size 610 × 340 are normally used.

(3) Salinas (SA): The third dataset recorded the image of Salinas Valley in California, USA, which was also captured by AVIRIS. Unlike the IP dataset, it has a spatial resolution of 3.7 m and consists of 224 bands. However, researchers generally utilize the image of 204 bands after excluding 20 bands affected by water reflection. Thus, the size of the Salinas is 512 × 217, and Figure 9 depicts the color composite of the image as well as the ground truth map.

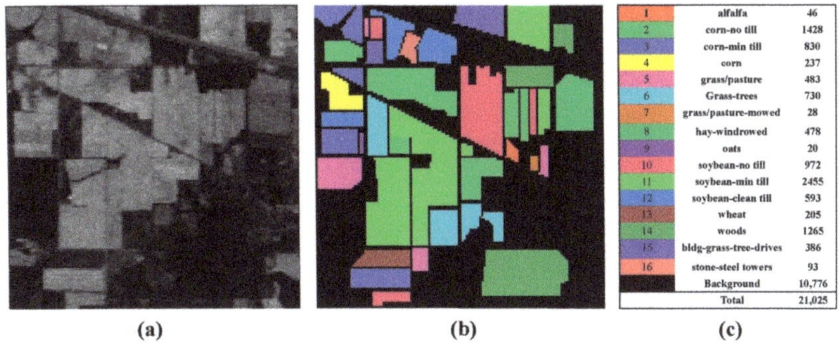

Figure 7. Indian Pines imagery: (**a**) color composite with RGB, (**b**) ground truth, and (**c**) category names with labeled samples.

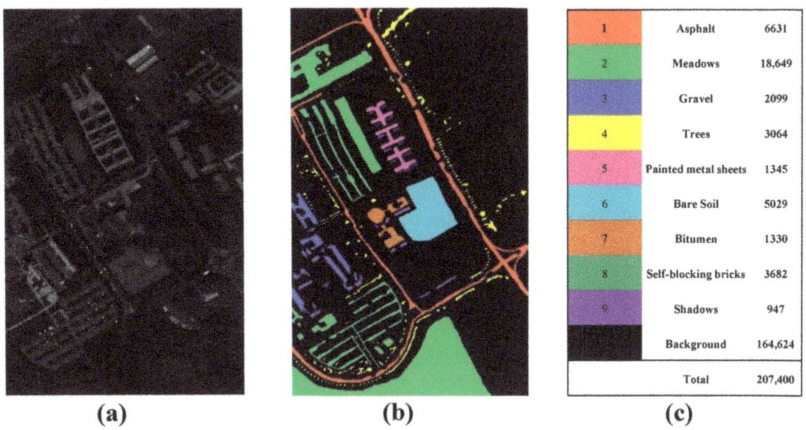

Figure 8. Pavia University imagery: (**a**) color composite with RGB, (**b**) ground truth, and (**c**) class names with available samples.

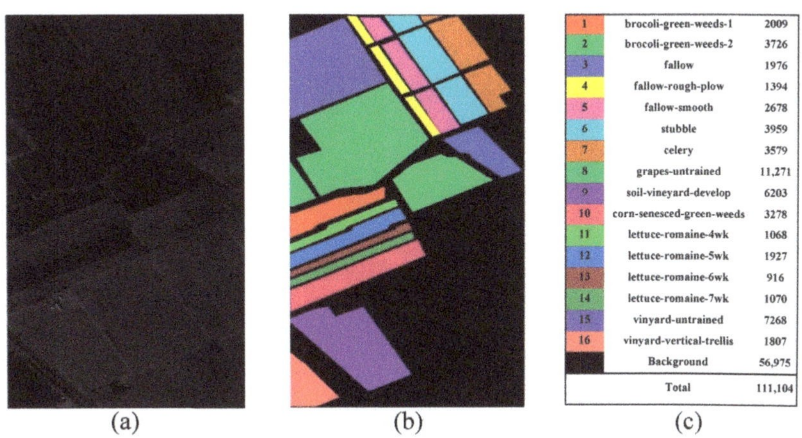

Figure 9. Salinas imagery: (**a**) color composite with RGB, (**b**) ground truth, and (**c**) class names with available samples.

4.2. Evaluation Measures

In the experiments, the available data of these datasets were randomly divided into two parts, a small part for training and the rest for testing. Whether the training samples or the testing samples were arranged according to the pixels, whose size was in $1 \times p_x$ (p_x is selected as 80 in this paper). Each pixel can be treated as a feature of a certain class, corresponding to a unique label and classified by the classifier stuck to the Discriminator. Tables 6–8 list the sample numbers for the training and testing of three datasets.

Table 6. The samples for each category of training and testing for the Indian Pines dataset.

Number	Class	Train	Test	Total
1	Alfalfa	3	43	46
2	Corn-notill	71	1357	1428
3	Corn-mintill	41	789	830
4	Corn	11	226	237
5	Grass-pasture	24	459	483
6	Grass-trees	36	694	730
7	Grass-pasture-mowed	3	25	28
8	Hay-windrowed	23	455	478
9	Oats	3	17	20
10	Soybean-notill	48	924	972
11	Soybean-mintill	122	2333	2455
12	Soybean-clean	29	564	593
13	Wheat	10	195	205
14	Woods	63	1202	1265
15	Buildings-Grass-Trees-Drives	19	367	386
16	Stone-Steel-Towers	4	89	93
	Total	510	9739	10,249

Table 7. The samples for each category of training and testing for the Pavia University dataset.

Number	Class	Train	Test	Total
1	Asphalt	66	6565	6631
2	Meadows	186	18,463	18,649
3	Gravel	20	2079	2099
4	Trees	30	3034	3064
5	Painted metal sheets	13	1333	1345
6	Bare Soil	50	4979	5029
7	Bitumen	13	1317	1330
8	Self-Blocking Bricks	36	3646	3682
9	Shadows	9	938	947
	Total	423	42,353	42,776

Table 8. The samples for each category of training and testing for the Salinas dataset.

Number	Class	Train	Test	Total
1	Broccoli_green_weeds_1	20	1989	2009
2	Broccoli_green_weeds_2	37	3689	3726
3	Fallow	19	1960	1976
4	Fallow_rough_plow	13	1381	1394
5	Fallow_smooth	26	2652	2678
6	Stubble	39	3920	3959
7	Celery	35	3544	3579
8	Grapes_untrained	112	11,159	11,271
9	Soil_vineyard_develop	62	6141	6203
10	Corn_senesced_green_weeds	32	3236	3278
11	Lettuce_romaine_4wk	10	1058	1068
12	Lettuce_romaine_5wk	19	1908	1927
13	Lettuce_romaine_6wk	9	909	916
14	Lettuce_romaine_7wk	10	1060	1070
15	Vineyard_untrained	72	7196	7268
16	Vineyard_vertical_trellis	18	1789	1807
	Total	533	53,596	54,129

Taking the phenomenon of "foreign matter of the same spectrum in surface cover" [15,43] into consideration, the average accuracy was reported to evaluate the experiment results quantitatively. Meanwhile, the proposed method was contrasted with the comparative method by three famous indexes, i.e., overall accuracy (OA), average accuracy (AA) and kappa coefficient (KA) [44], which can be denoted as below:

$$\mathrm{OA} = sum(diag(M))/sum(M), \tag{13}$$

$$\mathrm{AA} = mean((diag(M)./(sum(M,2)), \tag{14}$$

$$\mathrm{Kappa} = \frac{\mathrm{OA} - sum(M,1) \times sum(M,2)/(sum(M))^2}{1 - sum(M,2)/(sum(M))^2}, \tag{15}$$

where m represents the number of land cover categories and $M\epsilon R^{(m\times n)}$ symbolizes the confusion matrix of the classification results. Then, $diag(M)\epsilon R^{m\times 1}$ comes to be a vector of diagonal elements in M, $sum()\epsilon R^1$ proves to be the sum of all elements of matrices, where $(,1)$ means each column and $(,2)$ means each row. Finally, the $mean()\epsilon R^1$ describes the mean value of all elements along with the ./, which implies the element-wise division.

4.3. Experimental Setting

In this section, for the sake of verifying the effectiveness of CSSVGAN, several classical hyperspectral classification methods such as SVM [45], Mulit-3DCNN [46], SS3DCNN [47], SSRN [15] and certain deep generative algorithms like VAE, GAN and some jointed VAE-GAN models like the CVA^{2}E [33] and the semisupervised variational generative adversarial networks (SSVGAN) [34] were used for comparison.

To ensure the fairness of the comparative experiments, the best hyperparameter settings were adopted for each method based on their papers. All experiments were executed on the NVIDIA GeForce GTX 2070 SUPER GPU with a memory of 32 GB. Moreover, Adam [48] was used as the optimizer with an initial learning rate of 1×10^{-3} for Generator and 1×10^{-4} for Discriminator, and the training epoch was set to 200.

4.4. Experiments Results

All experiments in this paper were randomly selected train samples from the labeled pixels, and the accuracies of three datasets were reported to two decimal places in this chapter.

4.4.1. Experiments on the IP Dataset

The experimental test on IP Dataset was performed to evaluate the proposed CSSVGAN model quantitatively with other methods for HSIC. For the labeled samples, 5% of each class was randomly selected for training. The quantitative evaluation of various methods is shown in Table 9, which describes the classification accuracy of different categories in detail, as well as the indicators including OA, AA and kappa for different methods. The best value is marked in dark gray.

Table 9. The classification results for the IP dataset with 5% training samples.

Num/IP	ClassName	SVM	M3DCNN	SS3DCNN	SSRN	VAE	GAN	CVA^{2}E	SSVGAN	CSSVGAN
1	Alfalfa	58.33	0.00	0.00	100.00	100.00	60.29	67.35	90.00	50.00
2	Corn-notill	65.52	34.35	39.61	89.94	73.86	90.61	90.61	90.81	90.61
3	Corn-mintill	73.85	17.83	33.75	93.36	97.66	92.97	93.56	94.77	92.30
4	Corn	58.72	9.40	10.41	82.56	100.00	93.48	98.91	98.47	95.29
5	Grass-pasture	85.75	33.46	32.33	100.00	82.00	98.03	96.48	97.72	87.27
6	Grass-trees	83.04	90.68	82.10	95.93	91.98	93.69	95.69	90.49	97.60
7	Grass-pasture-mowed	88.00	0.00	0.00	94.73	0.00	0.00	100.00	82.76	93.33
8	Hay-windrowed	90.51	87.70	85.29	95.68	100.00	97.22	98.70	99.34	91.71
9	Oats	66.67	0.00	0.00	39.29	100.00	50.00	100.00	100.00	100.00
10	Soybean-notill	69.84	37.46	51.53	79.08	92.88	80.04	94.77	86.52	94.74
11	Soybean-mintill	67.23	57.98	64.71	88.80	92.42	94.40	88.56	98.51	95.75
12	Soybean-clean	46.11	21.08	21.26	94.43	84.48	80.84	81.30	84.03	84.48
13	Wheat	87.56	83.33	41.18	99.45	100.00	77.63	98.99	94.20	100.00
14	Woods	85.95	83.00	85.04	95.26	98.38	97.62	98.19	87.67	98.04
15	Buildings-GT-Drives	73.56	34.16	31.43	97.18	100.00	91.35	95.63	83.49	97.08
16	Stone-Steel-Towers	100.00	0.00	0.00	93.10	98.21	96.55	98.72	90.14	91.30
	OA(%)	72.82	53.54	56.23	91.04	90.07	91.01	92.48	91.99	**93.61**
	AA(%)	75.02	34.48	33.57	89.92	73.82	82.47	85.69	89.49	**91.16**
	Kappa(%)	68.57	45.73	49.46	89.75	88.61	89.77	91.40	90.91	**93.58**

First of all, although SVM achieves good exactitude, there is still a certain gap from the exact classification because of the IP dataset containing high texture spatial information, which leads to bad performance. Secondly, some conventional deep learning methods (such as M3DCNN, SS3DCNN) does not perform well in some categories due to the limitation of the number of training samples. Thirdly, the algorithms with jointed spectral-spatial feature extraction (like SSRN, etc.) show a better performance, which indicate a necessity to combine spectral information and spatial information for HSIC. Moreover, it is obvious that the generated virtual samples by VAE tend to be fuzzy and cannot guarantee similarities with the real data. While GAN lacks sampling constraints, leading to the low quality of the generated samples. Contrasted with these two deep generative models, CSSVGAN overcomes their shortcomings. Finally, compared with CVA^{2}E and SSVGAN, the two latest jointed models published in IEEE, CSSVGAN uses dual-branch feature extractions and crossed interactive method, which proves that these manners are more suitable for HSIC works. It can increase the diversity of samples and promote the generated data more similar to the original.

Among these comparative methods, CSSVGAN acquires the best accuracy in OA, AA and kappa, which improves by 2.57%, 1.24% and 3.81% respectively, at least. In addition, although all the methods have different degrees of misclassification, CSSVGAN achieves perfect accuracy in "Oats" "Wheat" and so on. The classification visualizations on the Indian Pines of comparative experiments are shown in Figure 10.

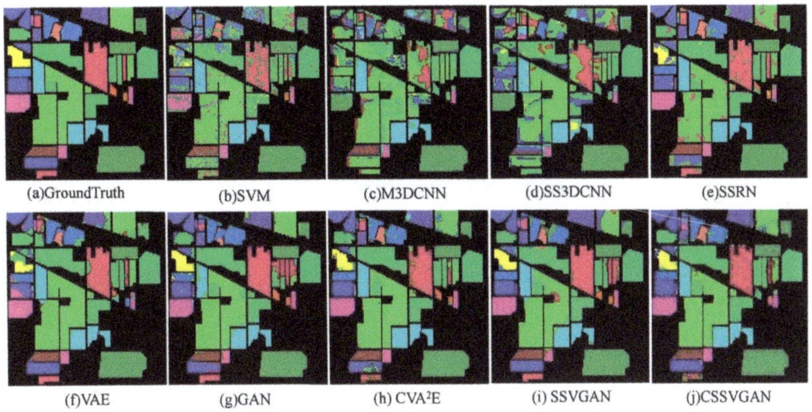

Figure 10. Classification maps for the IP dataset with 5% labeled training samples: (**a**) GroungTruth (**b**) SVM (**c**) M3DCNN (**d**) SS3DCNN (**e**) SSRN (**f**) VAE (**g**) GAN (**h**) CVA^{2}E (**i**) SSVGAN (**j**) CSSVGAN.

From Figure 10, it can be seen that CSSVGAN reduces the noisy scattering points and effectively improves the regional uniformity. That is because CSSVGAN can generate more realistic images from diverse samples.

4.4.2. Experiments on the PU Dataset

Differ from the IP dataset experiments, 1% labeled samples were selected for training and the rest for testing. Table 10 shows the quantitative evaluation of each class in comparative experiments. The best accurate value is marked in dark gray to emphasize, and the classification visualizations on the Pavia university are shown in Figure 11.

Table 10. The classification results for the PU dataset with 1% training samples.

Num/PU	ClassName	SVM	M3DCNN	SS3DCNN	SSRN	VAE	GAN	CVA^{2}E	SSVGAN	CSSVGAN
1	Asphalt	86.21	71.39	80.28	97.24	87.96	97.13	86.99	90.18	98.78
2	Meadows	90.79	82.38	86.38	83.38	86.39	96.32	96.91	94.90	99.89
3	Gravel	67.56	17.85	33.76	93.70	93.46	58.95	87.91	78.30	97.70
4	Trees	92.41	80.24	87.04	99.51	93.04	78.38	97.86	95.11	98.91
5	Painted metal sheets	95.34	99.09	99.67	99.55	99.92	93.50	96.86	96.70	99.70
6	Bare Soil	84.57	25.37	51.71	96.70	98.15	99.64	98.48	98.00	99.42
7	Bitumen	60.87	47.14	49.60	98.72	75.06	52.11	75.25	86.92	99.47
8	Self-Blocking Bricks	75.36	44.69	68.81	86.33	62.53	84.06	72.50	91.17	96.03
9	Shadows	100.00	88.35	97.80	100.00	82.86	42.57	97.13	82.53	99.14
	OA(%)	86.36	68.43	76.59	89.27	85.08	87.58	91.97	92.93	**99.11**
	AA(%)	83.68	53.00	64.14	95.01	73.45	83.58	89.32	87.83	**98.47**
	Kappa(%)	81.76	56.60	68.80	85.21	79.58	83.67	85.64	90.53	**98.83**

Table 10 shows that, as a non-deep learning algorithm, SVM has been able to improve the classification result to 86.36%, which is wonderful to some extent. VAE shows good performance in the training of the "Painted metal sheets" class but low accuracy in the "Self-blocking bricks" class, which leads to the "fuzzy" phenomenon of a single VAE network in the training of individual classes. SSRN achieves a completely correct classification in "shadows," but it lost to the CSSVGAN overall. In the index of OA results, CSSVGAN improved 12.75%, 30.68%, 22.52%, 9.83%, 14.03%, 11.53%, 7.14% and 6.18% respectively and in the index of Kappa results, CSSVGAN improved 17.07%, 42.23%, 30.03%, 13.62%, 19.25%, 15.16%, 13.19% and 8.3% respectively compared with the other eight algorithms.

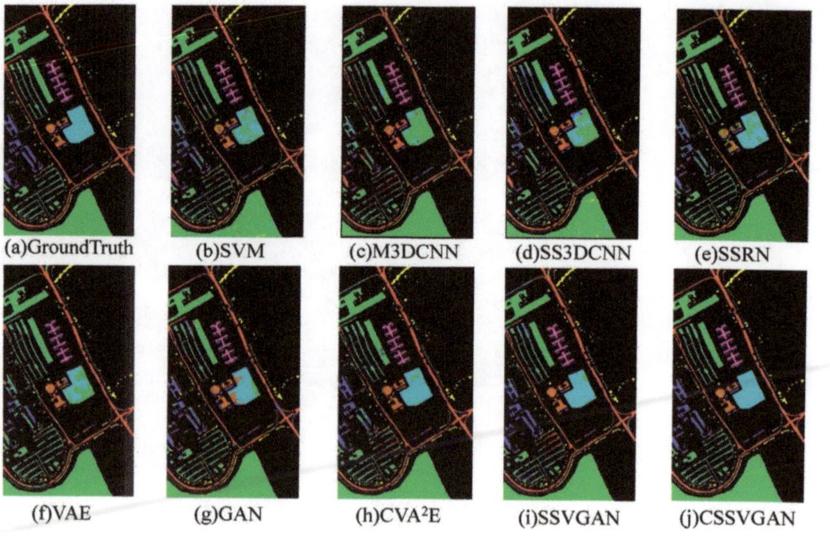

Figure 11. Classification maps for the PU dataset with 1% labeled training samples: (**a**) GroungTruth (**b**) SVM (**c**) M3DCNN (**d**) SS3DCNN (**e**) SSRN (**f**) VAE (**g**) GAN (**h**) CVA^{2}E (**i**) SSVGAN (**j**) CSSVGAN.

In Figure 11, the proposed CSSVGAN has better boundary integrity and better classification accuracy in most of the classes because the Encoder can ensure the diversity of samples, the Generator can promote the authenticity of the generated virtual data, and the Discriminator can adjust the overall framework to obtain the optimal results.

4.4.3. Experiments on the SA Dataset

The experimental setting on the Salinas dataset is the same as PU. Table 11 shows the quantitative evaluation of each class in various methods with dark gray to emphasize the best results. The classification visualization of the comparative experiments on Salinas is shown in Figure 12.

Table 11 shows that in the index of OA, AA and Kappa, CSSVGAN improved 0.57%, 1.27% and 0.62% at least compared with others. Moreover, it has a better performance in the "brocoli-green-weeds-1" and "stubble" class with a test accuracy of 100%. For the precisions of other classes, although SSRN, VAE or SSRN prevails, CSSVGAN is almost equal to them. It can be seen that CSSVGAN has smoother edges and the minimum misclassification in Figure 12, which further proves that the proposed CSSVGAN can generate more realistic virtual data according to the diversity of extracted features of samples.

Table 11. The classification results for the SA dataset with 1% training samples.

Num/SA	ClassName	SVM	M3DCNN	SS3DCNN	SSRN	VAE	GAN	CVA^{2}E	SSVGAN	CSSVGAN
1	Broccoli_green_weeds_1	99.95	94.85	56.23	100.00	97.10	100.00	100.00	100.00	100.00
2	Broccoli_green_weeds_2	98.03	65.16	81.56	98.86	97.13	62.32	99.34	97.51	99.92
3	Fallow	88.58	40.61	92.40	99.40	100.00	99.78	100.00	93.74	98.99
4	Fallow_rough_plow	99.16	97.04	95.63	96.00	98.68	93.91	99.76	91.88	99.35
5	Fallow_smooth	90.38	89.31	95.08	95.11	99.26	97.67	99.30	94.08	99.08
6	Stubble	99.64	95.64	98.78	99.69	99.24	94.36	90.53	99.31	100.00
7	Celery	98.58	75.75	98.90	99.32	97.98	98.93	99.39	99.54	99.66
8	Grapes_untrained	77.58	65.28	81.87	89.16	96.55	96.87	89.36	93.57	92.79
9	Soil_vineyard_develop	99.50	96.04	96.20	98.33	99.74	89.66	89.85	98.53	99.56
10	Corn_sg_weeds	95.01	44.82	84.13	97.67	96.79	91.71	95.71	92.44	97.81
11	Lettuce_romaine_4wk	94.00	44.66	79.64	96.02	100.00	87.95	96.82	91.62	97.76
12	Lettuce_romaine_5wk	97.40	36.69	96.19	98.45	90.89	98.73	100.00	99.42	99.32
13	Lettuce_romaine_6wk	95.93	12.17	91.50	99.76	99.87	100.00	91.97	96.78	99.67
14	Lettuce_romaine_7wk	94.86	79.53	66.83	97.72	95.83	94.14	100.00	95.85	99.71
15	Vineyard_untrained	79.87	40.93	69.11	83.74	88.09	57.33	85.41	85.17	91.75
16	Vineyard_vertical_trellis	98.76	57.78	85.09	97.07	99.61	97.32	97.00	99.11	99.66
	OA(%)	90.54	66.90	85.14	94.40	96.43	86.97	95.06	94.60	**97.00**
	AA(%)	94.20	56.78	78.89	96.65	95.87	92.17	97.08	95.50	**98.35**
	Kappa(%)	89.44	62.94	83.41	93.76	96.03	85.50	94.48	94.00	**96.65**

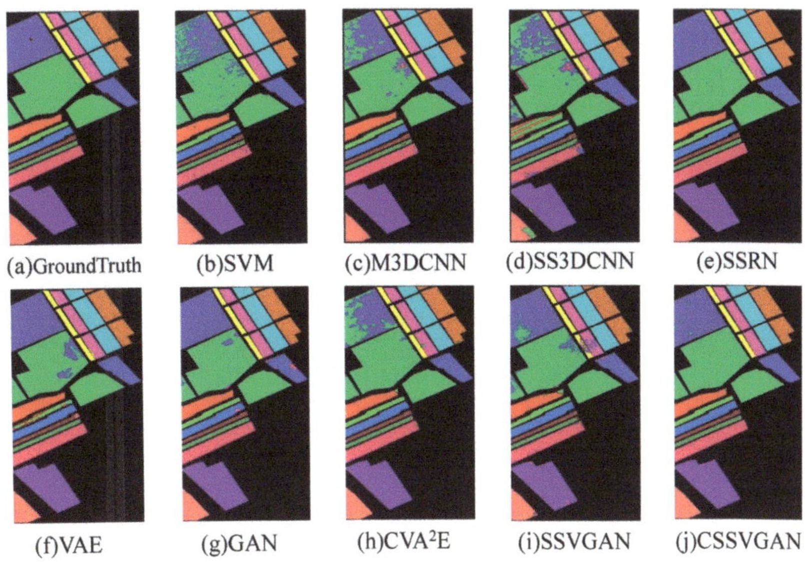

Figure 12. Classification maps for the SA dataset with 1% labeled training samples: (**a**) GroungTruth (**b**) SVM (**c**) M3DCNN (**d**) SS3DCNN (**e**) SSRN (**f**) VAE (**g**) GAN (**h**) CVA^{2}E (**i**) SSVGAN (**j**) CSSVGAN.

5. Discussions

5.1. The Ablation Experiment in CSSVGAN

Taking IP, PU and SA datasets as examples, the frameworks of ablation experiments are shown in Figure 13, including NSSNCSG, SSNCSG and SSNCDG.

As shown in Table 12, compared with NSSNCSG, the OA of CSSVGAN on IP, PU and SA datasets increased by 1.02%, 6.90% and 4.63%, respectively.

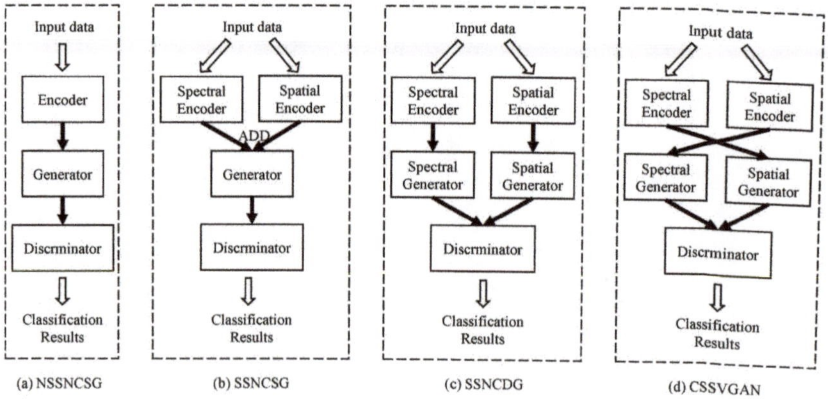

Figure 13. The frameworks of ablation experiments: (**a**) NSSNCSG (**b**) SSNCSG (**c**) SSNCDG (**d**) CSSVGAN.

Table 12. The OA(%) of Ablation experiments.

Name	Dual Branch	Crossed Interaction	Single Generator	Double Generator	IP	PU	SA
NSSNCSG	×	×	√	×	92.59	92.21	92.07
SSNCSG	√	×	√	×	92.62	98.54	96.61
SSNCDG	√	×	×	√	92.36	98.67	96.26
CSSVGAN	√	√	×	√	**93.61**	**99.11**	**97.00**

It shows that the effect of using dual-branch special-spatial feature extraction is better than not using it because the distributions of spectral and spatial features are not identical, and a single Encoder cannot handle this complex situation. Consequently, using the dual-branch variational Encoder can increase the diversity of samples. Under the constraint of KL divergence, the distribution of latent variables is more consistent with the distribution of real data.

Contrasted with SSNCSG, the OA index on IP, PU and SA datasets increase by 0.99%, 1.07% and 0.39% respectively, which means that the result of utilizing the crossed interactive method is more effective, and further influences that the crossed interactive double Generator can fully learn the spectral and spatial information and generate spatial and spectral virtual samples in higher qualities.

Finally, a comparison is made between SSNCDG and CSSVGAN, where the latter can better improve the authenticity of virtual samples by crossed manner. All these contributions of both the Encoder and the Generator put forward higher requirements to the Discriminator, optimizing Discriminator's ability to identify the true or false data and further achieve the final classification results more accurately.

5.2. Sensitivity to the Proportion of Training Samples

To verify the effectiveness of the proposed CSSVGAN, three datasets were taken as examples. The percentage of training samples was changed for each class from 1% to 9% at 4% intervals and added 10%. Figures 14–16 shows the OAs of all the comparative algorithms with various percentages of training samples.

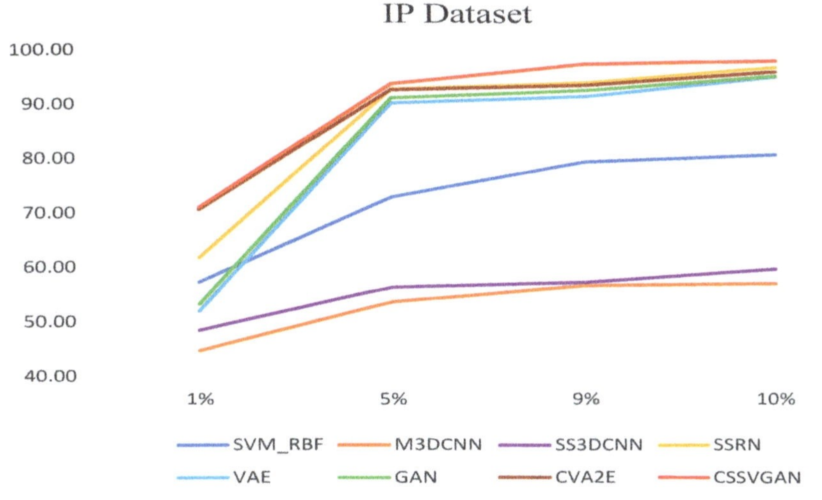

Figure 14. Sensitivity to the Proportion of Training Samples in IP dataset.

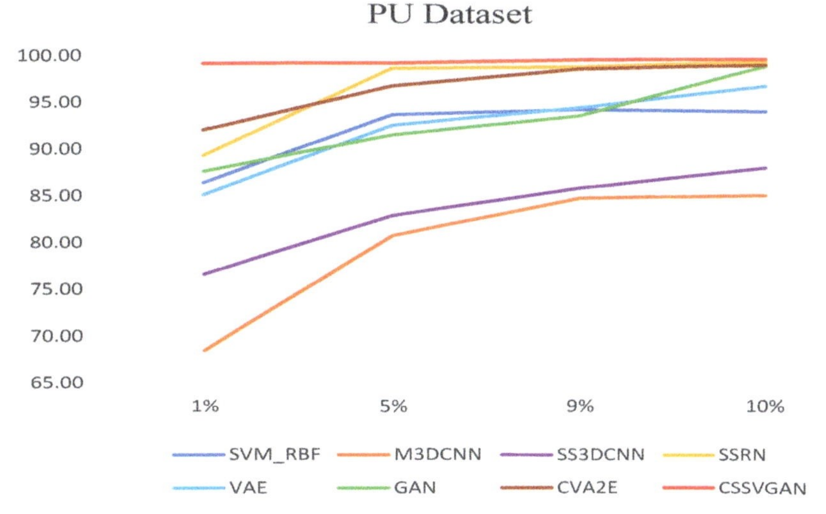

Figure 15. Sensitivity to the Proportion of Training Samples in PU dataset.

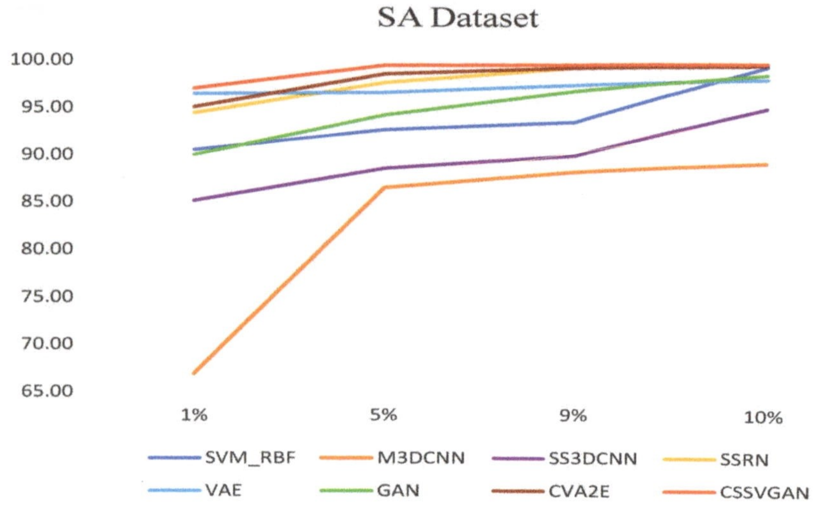

Figure 16. Sensitivity to the Proportion of Training Samples in SA dataset.

It can be seen that the CSSVGAN has the optimal effect in each proportion of training samples in three datasets because CSSVGAN can learn the extracted features interactively, ensure diverse samples and improve the quality of generated images.

5.3. *Investigation of the Proportion of Loss Function*

Taking the IP dataset as an example, the proportion σ_i $(i = 1, 2, \ldots 5)$ of loss functions and other super parameters of each module are adjusted to observe their impact on classification accuracy and the results are recorded in Table 13 (the best results are marked in dark gray). Moreover, the learning rate is also an important factor, which will not be repeated here. It can be obtained by experiments that using 1×10^{-3} for Generator and 1×10^{-4} for Discriminator are the best assignments.

Table 13. Investigation of the proportion σ_i of loss functions in IP dataset with 5% training samples.

σ_1	σ_2	σ_3	σ_4	σ_5	**IP_Result**
0.25	0.25	0.15	0.15	0.2	91.88
0.3	0.3	0.15	0.15	0.1	91.23
0.3	0.3	0.1	0.1	0.2	92.87
0.35	0.35	0.05	0.05	0.2	92.75
0.35	0.35	0.1	0.1	0.1	**93.61**

Analyzing Table 13 reveals that when $\sigma1 \sim \sigma5$ are set as 0.35, 0.35, 0.1, 0.1 and 0.1 respectively, the CSSVGAN model achieves the best performance. Under this condition, the Encoder can acquire the maximum diversity of samples. The Discriminator is able to realize the most accurate classification, and the Generator is capable of generating the images most like the original data. Moreover, the best parameter combination $\sigma1 \sim \sigma5$ on the SA dataset is similar to IP, while in the PU dataset, they are set as 0.3, 0.3, 0.1, 0.1 and 0.2.

6. Conclusions

In this paper, variational generative adversarial network with crossed spatial and spectral interactions (CSSVGAN) is proposed for HSIC. It mainly consists of three modules: a dual-branch variational Encoder, a crossed interactive Generator, and a Discriminator stuck with a classifier. From the experiment results of these three datasets, it showed that CSSVGAN can outperform the other methods in the index of OA, AA and Kappa in its abilities because of the dual-branch and the crossed interactive manners. Moreover, using the dual-branch Encoder can ensure the diversity of generated samples by mapping spectral and spatial information into different latent spaces, and utilizing the crossed interactive Generator can imitate the highly correlated spatial and spectral characteristics of HSI by exploiting the consistency of learned spectral and spatial features. All these contributions made the proposed CSSVGAN give the best performance in three datasets. In the future, we will develop towards to realize lightweight generative models and explore the application of the jointed "Transformer and GAN" model for HSIC.

Author Contributions: Conceptualization, Z.L. and X.Z.; methodology, Z.L., X.Z. and L.W.; software, Z.L., X.Z., L.W. and Z.X.; validation, Z.L., F.G. and X.C.; writing—original draft preparation, L.W. and X.Z.; writing—review and editing, Z.L., Z.X. and F.G.; project administration, Z.L. and L.W.; funding acquisition, Z.L. and L.W. All authors read and agreed to the published version of the manuscript.

Funding: This research was funded by the Joint Funds of the General Program of the National Natural Science Foundation of China, Grant Number 62071491, the National Natural Science Foundation of China, Grant Number U1906217, and the Fundamental Research Funds for the Central Universities, Grant No. 19CX05003A-11.

Informed Consent Statement: Informed consent was obtained from all subjects involved in the study.

Data Availability Statement: Publicly available datasets were analyzed in this study , which can be found here: http://www.ehu.eus/ccwintco/index.php?title=Hyperspectral_Remote_Sensing_Scenes, latest accessed on 29 July 2021.

Acknowledgments: The authors are grateful for the positive and constructive comments of editor and reviewers, which have significantly improved this work.

Conflicts of Interest: The authors declare no conflict of interest. The funders had no role in the design of the study; in the collection, analyses, or interpretation of data; in the writing of the manuscript; nor in the decision to publish the results.

References

1. Chen, P.; Jiao, L.; Liu, F.; Zhao, J.; Zhao, Z. Dimensionality reduction for hyperspectral image classification based on multiview graphs ensemble. *J. Appl. Remote Sens.* **2016**, *10*, 030501. [CrossRef]
2. Shi, G.; Luo, F.; Tang, Y.; Li, Y. Dimensionality Reduction of Hyperspectral Image Based on Local Constrained Manifold Structure Collaborative Preserving Embedding. *Remote Sens.* **2021**, *13*, 1363. [CrossRef]
3. Atzberger, C. Advances in remote sensing of agriculture: Context description, existing operational monitoring systems and major information needs. *Remote Sens.* **2013**, *5*, 949–981. [CrossRef]
4. Sun, Y.; Wang, S.; Liu, Q.; Hang, R.; Liu, G. Hypergraph embedding for spatial-spectral joint feature extraction in hyperspectral images. *Remote Sens.* **2017**, *9*, 506. [CrossRef]
5. Abbate, G.; Fiumi, L.; De Lorenzo, C.; Vintila, R. Evaluation of remote sensing data for urban planning. Applicative examples by means of multispectral and hyperspectral data. In Proceedings of the 2003 2nd GRSS/ISPRS Joint Workshop on Remote Sensing and Data Fusion over Urban Areas, Berlin, Germany, 22–23 May 2003; pp. 201–205.
6. Yuen, P.W.; Richardson, M. An introduction to hyperspectral imaging and its application for security, surveillance and target acquisition. *Imaging Sci. J.* **2010**, *58*, 241–253. [CrossRef]
7. Tan, K.; Zhang, J.; Du, Q.; Wang, X. GPU parallel implementation of support vector machines for hyperspectral image classification. *IEEE J. Sel. Top. Appl. Earth Obs. Remote Sens.* **2015**, *8*, 4647–4656. [CrossRef]
8. Li, J.; Bioucas-Dias, J.M.; Plaza, A. Semisupervised hyperspectral image classification using soft sparse multinomial logistic regression. *IEEE Geosci. Remote Sens. Lett.* **2012**, *10*, 318–322.
9. Tan, K.; Hu, J.; Li, J.; Du, P. A novel semi-supervised hyperspectral image classification approach based on spatial neighborhood information and classifier combination. *ISPRS J. Photogramm. Remote Sens.* **2015**, *105*, 19–29. [CrossRef]
10. Gao, Q.; Lim, S.; Jia, X. Hyperspectral image classification using convolutional neural networks and multiple feature learning. *Remote Sens.* **2018**, *10*, 299. [CrossRef]
11. Chen, Y.; Jiang, H.; Li, C.; Jia, X.; Ghamisi, P. Deep feature extraction and classification of hyperspectral images based on convolutional neural networks. *IEEE Trans. Geosci. Remote Sens.* **2016**, *54*, 6232–6251. [CrossRef]
12. Zhang, B.; Zhao, L.; Zhang, X. Three-dimensional convolutional neural network model for tree species classification using airborne hyperspectral images. *Remote Sens. Environ.* **2020**, *247*, 111938. [CrossRef]

13. Chen, Y.C.; Lei, T.C.; Yao, S.; Wang, H.P. PM2. 5 Prediction Model Based on Combinational Hammerstein Recurrent Neural Networks. *Mathematics* **2020**, *8*, 2178. [CrossRef]
14. Nezami, S.; Khoramshahi, E.; Nevalainen, O.; Pölönen, I.; Honkavaara, E. Tree species classification of drone hyperspectral and rgb imagery with deep learning convolutional neural networks. *Remote Sens.* **2020**, *12*, 1070. [CrossRef]
15. Zhong, Z.; Li, J.; Luo, Z.; Chapman, M. Spectral–spatial residual network for hyperspectral image classification: A 3-D deep learning framework. *IEEE Trans. Geosci. Remote Sens.* **2017**, *56*, 847–858. [CrossRef]
16. Xu, Y.; Zhang, L.; Du, B.; Zhang, F. Spectral–spatial unified networks for hyperspectral image classification. *IEEE Trans. Geosci. Remote Sens.* **2018**, *56*, 5893–5909. [CrossRef]
17. Liu, G.; Gao, L.; Qi, L. Hyperspectral Image Classification via Multieatureased Correlation Adaptive Representation. *Remote Sens.* **2021**, *13*, 1253. [CrossRef]
18. Zhao, W.; Du, S. Spectral–spatial feature extraction for hyperspectral image classification: A dimension reduction and deep learning approach. *IEEE Trans. Geosci. Remote Sens.* **2016**, *54*, 4544–4554. [CrossRef]
19. Roy, S.K.; Krishna, G.; Dubey, S.R.; Chaudhuri, B.B. HybridSN: Exploring 3-D–2-D CNN feature hierarchy for hyperspectral image classification. *IEEE Geosci. Remote Sens. Lett.* **2019**, *17*, 277–281. [CrossRef]
20. Belwalkar, A.; Nath, A.; Dikshit, O. Spectral-Spatial Classification of Hyperspectral Remote Sensing Images Using Variational Autoencoder and Convolution Neural Network. In Proceedings of the International Archives of the Photogrammetry, Remote Sensing and Spatial Information Sciences, Dehradun, India, 20–23 November 2018.
21. Goodfellow, I.; Pouget-Abadie, J.; Mirza, M.; Xu, B.; Warde-Farley, D.; Ozair, S.; Courville, A.; Bengio, Y. Generative adversarial nets. *arXiv* **2014**, arXiv:1406.2661v1.
22. Liu, X.; Gherbi, A.; Wei, Z.; Li, W.; Cheriet, M. Multispectral image reconstruction from color images using enhanced variational autoencoder and generative adversarial network. *IEEE Access* **2020**, *9*, 1666–1679. [CrossRef]
23. Su, Y.; Li, J.; Plaza, A.; Marinoni, A.; Gamba, P.; Chakravortty, S. DAEN: Deep autoencoder networks for hyperspectral unmixing. *IEEE Trans. Geosci. Remote Sens.* **2019**, *57*, 4309–4321. [CrossRef]
24. Makhzani, A.; Shlens, J.; Jaitly, N.; Goodfellow, I.; Frey, B. Adversarial autoencoders. *arXiv* **2015**, arXiv:1511.05644.
25. Bao, J.; Chen, D.; Wen, F.; Li, H.; Hua, G. CVAE-GAN: Fine-grained image generation through asymmetric training. In Proceedings of the IEEE International Conference on Computer Vision, Venice, Italy, 22–29 October 2017; pp. 2745–2754.
26. He, Z.; Liu, H.; Wang, Y.; Hu, J. Generative adversarial networks-based semi-supervised learning for hyperspectral image classification. *Remote Sens.* **2017**, *9*, 1042. [CrossRef]
27. Isola, P.; Zhu, J.Y.; Zhou, T.; Efros, A.A. Image-to-image translation with conditional adversarial networks. In Proceedings of the IEEE Conference on Computer Vision and Pattern Recognition, Honolulu, HI, USA, 21–26 July 2017; pp. 1125–1134.
28. Chen, X.; Duan, Y.; Houthooft, R.; Schulman, J.; Sutskever, I.; Abbeel, P. Infogan: Interpretable representation learning by information maximizing generative adversarial nets. In Proceedings of the 30th International Conference on Neural Information Processing Systems, Kyoto, Japan, 16–21 October, 2016; pp. 2180–2188
29. Feng, J.; Feng, X.; Chen, J.; Cao, X.; Zhang, X.; Jiao, L.; Yu, T. Generative adversarial networks based on collaborative learning and attention mechanism for hyperspectral image classification. *Remote Sens.* **2020**, *12*, 1149. [CrossRef]
30. Zhan, Y.; Hu, D.; Wang, Y.; Yu, X. Semisupervised hyperspectral image classification based on generative adversarial networks. *IEEE Geosci. Remote Sens. Lett.* **2017**, *15*, 212–216. [CrossRef]
31. Feng, J.; Yu, H.; Wang, L.; Cao, X.; Zhang, X.; Jiao, L. Classification of hyperspectral images based on multiclass spatial–spectral generative adversarial networks. *IEEE Trans. Geosci. Remote Sens.* **2019**, *57*, 5329–5343. [CrossRef]
32. Zhu, L.; Chen, Y.; Ghamisi, P.; Benediktsson, J.A. Generative adversarial networks for hyperspectral image classification. *IEEE Trans. Geosci. Remote Sens.* **2018**, *56*, 5046–5063. [CrossRef]
33. Wang, X.; Tan, K.; Du, Q.; Chen, Y.; Du, P. CVA2E: A conditional variational autoencoder with an adversarial training process for hyperspectral imagery classification. *IEEE Trans. Geosci. Remote Sens.* **2020**, *58*, 5676–5692. [CrossRef]
34. Wang, H.; Tao, C.; Qi, J.; Li, H.; Tang, Y. Semi-supervised variational generative adversarial networks for hyperspectral image classification. In Proceedings of the IGARSS 2019-2019 IEEE International Geoscience and Remote Sensing Symposium, Yokohama, Japan, 28 July–2 August 2019; pp. 9792–9794.
35. Kingma, D.P.; Welling, M. Auto-encoding variational bayes. *arXiv* **2013**, arXiv:1312.6114.
36. Kullback, S.; Leibler, R.A. On information and sufficiency. *Ann. Math. Stat.* **1951**, *22*, 79–86. [CrossRef]
37. Wu, C.; Wu, F.; Wu, S.; Yuan, Z.; Liu, J.; Huang, Y. Semi-supervised dimensional sentiment analysis with variational autoencoder. *Knowl. Based Syst.* **2019**, *165*, 30–39. [CrossRef]
38. Arjovsky, M.; Chintala, S.; Bottou, L. Wasserstein GAN. *arXiv* **2017**, arXiv:1701.07875.
39. Mao, X.; Li, Q.; Xie, H.; Lau, R.Y.; Wang, Z.; Paul Smolley, S. Least squares generative adversarial networks. In Proceedings of the IEEE International Conference on Computer Vision, Venice, Italy, 22–29 October 2017; pp. 2794–2802.
40. Zhu, J.Y.; Park, T.; Isola, P.; Efros, A.A. Unpaired image-to-image translation using cycle-consistent adversarial networks. In Proceedings of the IEEE International Conference on Computer Vision, Venice, Italy, 22–29 October 2017; pp. 2223–2232.
41. Salimans, T.; Goodfellow, I.; Zaremba, W.; Cheung, V.; Radford, A.; Chen, X. Improved techniques for training gans. *Adv. Neural Inf. Process. Syst.* **2016**, *29*, 2234–2242.
42. Radford, A.; Metz, L.; Chintala, S. Unsupervised representation learning with deep convolutional generative adversarial networks. *arXiv* **2015**, arXiv:1511.06434.

43. Imani, M.; Ghassemian, H. An overview on spectral and spatial information fusion for hyperspectral image classification: Current trends and challenges. *Inf. Fusion* **2020**, *59*, 59–83. [CrossRef]
44. Sun, H.; Zheng, X.; Lu, X.; Wu, S. Spectral–spatial attention network for hyperspectral image classification. *IEEE Trans. Geosci. Remote Sens.* **2019**, *58*, 3232–3245. [CrossRef]
45. Melgani, F.; Bruzzone, L. Classification of hyperspectral remote sensing images with support vector machines. *IEEE Trans. Geosci. Remote Sens.* **2004**, *42*, 1778–1790. [CrossRef]
46. He, M.; Li, B.; Chen, H. Multi-scale 3D deep convolutional neural network for hyperspectral image classification. In Proceedings of the 2017 IEEE International Conference on Image Processing (ICIP), Beijing, China, 17–20 September 2017; pp. 3904–3908.
47. Li, Y.; Zhang, H.; Shen, Q. Spectral–spatial classification of hyperspectral imagery with 3D convolutional neural network. *Remote Sens.* **2017**, *9*, 67. [CrossRef]
48. Loshchilov, I.; Hutter, F. Sgdr: Stochastic gradient descent with warm restarts. *arXiv* **2016**, arXiv:1608.03983.

remote sensing

MDPI

Article

An Attention-Guided Multilayer Feature Aggregation Network for Remote Sensing Image Scene Classification

Ming Li [1], Lin Lei [1,*], Yuqi Tang [2], Yuli Sun [1] and Gangyao Kuang [1]

1 The College of Electronic Science and Technology, National University of Defense Technology, Changsha 410073, China; liming17@nudt.edu.cn (M.L.); sunyuli@mail.ustc (Y.S.); kuangyeats@hotmail.com (G.K.)

2 School of Geosciences and Info-Physics, Central South University, Changsha 410083, China; yqtang@csu.edu.cn

* Correspondence: alaleilin@163.com

Abstract: Remote sensing image scene classification (RSISC) has broad application prospects, but related challenges still exist and urgently need to be addressed. One of the most important challenges is how to learn a strong discriminative scene representation. Recently, convolutional neural networks (CNNs) have shown great potential in RSISC due to their powerful feature learning ability; however, their performance may be restricted by the complexity of remote sensing images, such as spatial layout, varying scales, complex backgrounds, category diversity, etc. In this paper, we propose an attention-guided multilayer feature aggregation network (AGMFA-Net) that attempts to improve the scene classification performance by effectively aggregating features from different layers. Specifically, to reduce the discrepancies between different layers, we employed the channel–spatial attention on multiple high-level convolutional feature maps to capture more accurately semantic regions that correspond to the content of the given scene. Then, we utilized the learned semantic regions as guidance to aggregate the valuable information from multilayer convolutional features, so as to achieve stronger scene features for classification. Experimental results on three remote sensing scene datasets indicated that our approach achieved competitive classification performance in comparison to the baselines and other state-of-the-art methods.

Keywords: convolutional neural networks (CNNs); multilayer feature aggregation; attention mechanism; remote sensing image scene classification (RSISC)

Citation: Li, M.; Lei, L.; Tang, Y.; Sun, Y.; Kuang, G. An Attention-Guided Multilayer Feature Aggregation Network for Remote Sensing Image Scene Classification. *Remote Sens.* **2021**, *13*, 3113. https://doi.org/10.3390/rs13163113

Academic Editors: Fahimeh Farahnakian, Jukka Heikkonen and Pouya Jafarzadeh

Received: 21 June 2021
Accepted: 3 August 2021
Published: 6 August 2021

1. Introduction

With the rapid development of remote sensing imaging technology, a large amount of high-resolution remote sensing images, captured from space or air, can provide rich detail information, e.g., spatial layout, shape, and texture, about the Earth's surface. This information is a significant data source and has been used to many applications, such as land use classification [1,2], land use change detection and management [3,4], geospatial object detection [5], etc. As a fundamental and challenging task in remote sensing image understanding, remote sensing image scene classification (RSISC) has already become one of the hot topics in research in recent years, the main purpose being to automatically assign one or multiple predefined tags (e.g., airport, river, bridge) to a given remote sensing scene according to its semantic content. In this paper, we mainly concentrated on the single-label remote sensing image scene classification problem.

Due to the imaging characteristics of high-resolution remote sensing images, a remote sensing scene is usually composed of different land use units, and different combinations of them may generate different scene categories. As shown in Figure 1, a remote sensing scene labeled "bridge" consists of five different land cover units including vehicle, trees, ship, river, and bridge. However, to classify this scene, we only need to pay more attention to the "bridge" regions, i.e., the red-box-covered region; the other regions can be considered as interference. In addition, imaging viewpoint, spatial resolution, illumination, and scale variation also significantly influence the final classification accuracy [6]. Therefore, how to learn discriminative and robust feature representation is very crucial for improving scene classification performance.

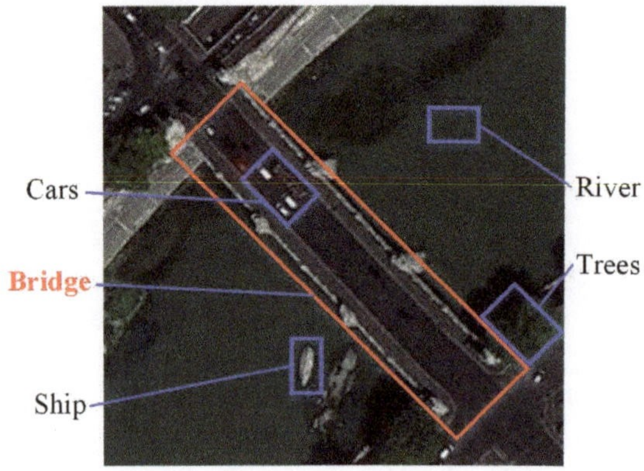

Figure 1. The characteristics of a remote sensing scene image. A remote sensing scene consists of many types of land cover units. However, to classify this scene, we only need to pay more attention to the key regions, i.e., bridge, while other regions can be regarded as interference.

To address the RSISC problem, the traditional approaches mainly rely on some hand-crafted visual features, for example the color histogram [7], texture [8], scale-invariant feature transformation [9], or the histogram of oriented gradients [10], and try to extract discriminative scene representation for the classification. However, the performance of these methods was compromised by the limited expressive capacity of the hand-crafted features, especially when dealing with some complex scenes.

Recently, deep learning techniques, especially convolutional neural networks (CNNs), have achieved state-of-the-art performance in all kinds of computer vision tasks, e.g., image classification [11,12], object detection [13], and semantic segmentation [14], due to their powerful feature learning ability. Compared with the hand-crafted features, deep features have richer semantic information, which is more suitable for describing the true content of images. Starting from the earliest convolutional neural network, i.e., AlexNet [11], many high-performance CNNs, such as VGGNet [12], ResNet [15], and DenseNet [16], have been developed and successfully employed in many other domains.

In the task of remote sensing scene classification, capturing scene representation with sufficient discriminative ability is important to improve the classification accuracy. In recent years, deep learning has also shown great potential on this task and a large number of deep-learning-based approaches [17–22] have been developed. Among them, considering the complementarity features of different layers of a convolutional neural network is an effective strategy to improve scene classification accuracy [6,23–25]. To comprehensively utilize different layers' convolutional features, the simplest way is to directly concatenate them together [25]. The other solution is to concatenate them after using a certain feature selection mechanism. However, these methods have some limitations. First, the direct concatenation strategy can simply merge the features in different layers, but it suffers from a limited ability to suppress feature redundancy and interference information, which is not conducive to highlight discriminative features. Second, some current methods generally operate under the belief that features from the last convolutional layer can best represent the semantic regions of the given scene, so they usually utilize the last convolutional features to guide the multilayer feature fusion. However, by referencing some research conclusions and convolutional feature visualization experiments, we found that the last convolutional features can only extract the most discriminative features while ignoring other crucial information that is also important for classification. In other words, only using the last convolutional features may lack semantic integrity. Third, in order to maximize the fusion feature's representation ability, the multilayer feature aggregation operation should follow certain rules, that is, for different layers' convolutional features, we should only fuse those valuable regions of different layers and selectively suppress irrelevant information. Through this adaptive selection mechanism, more powerful scene representation can finally be obtained.

Inspired by this, we propose an attention-guided multilayer feature aggregation network (AGMFA-Net). Specifically, we first extracted multiple convolutional feature maps with different spatial resolutions from the backbone network. Then, the channel–spatial attention was adopted on multiple high-level convolutional feature maps to obtain complete semantic regions that were consistent with the given scene as accurately as possible. Third, in order to integrate the valuable information from different convolutional layers and alleviate the impacts of discrepancies between them, we used the learned semantic regions to guide the multilayer feature aggregation operation. Finally, the aggregated features were fed into the classifier to perform remote sensing scene classification.

The main contributions of this paper are listed as follows:

(1) We propose an attention-guided multilayer feature aggregation network, which can capture more powerful scene representation by aggregating valuable information from different convolutional layers, as well as suppressing irrelevant interference between them;

(2) Instead of only considering discriminative features from the last convolutional feature map, we employed channel–spatial attention on multiple high-level convolutional feature maps simultaneously to make up for information loss and capture more complete semantic regions that were consistent with the given scene,. The visualization and qualitative results in the experiments demonstrated its effectiveness;

(3) We evaluated the proposed AGMFA-Net on three widely used benchmark datasets, and the experimental results showed that the proposed method can achieve better classification performance in comparison to some other state-of-the-art methods.

The rest of the paper is organized as follows. Related work is reviewed in Section 2, followed by the detailed presentation of the proposed method in Section 3. Experiments and the analysis are presented in Section 4. Section 5 is the conclusion.

2. Related Works

Over the past few years, many RSISC approaches have been proposed. Among them, deep-learning-based methods have gradually become the main stream. In this section, we mainly review the relevant deep learning methods and then briefly describe some attention methods that are related to the proposed AGMFA-Net. As for the traditional RSISC approaches based on hand-crafted features, we recommend reading the papers [17,18].

2.1. Deep-Learning-Based Remote Sensing Image Scene Classification

The advent of deep learning techniques, especially convolutional neural networks, has brought huge performance gains to remote sensing image scene classification. In comparison to the hand-crafted features, deep features contain more abstract and discriminative semantics, which can describe the given scene more precisely. In this subsection, we summarize the existing deep-learning-based scene classification methods as follows.

2.1.1. Fine-Tuning Methods

In the early stage, it is generally acknowledged that fully training a new CNN model on the target remote sensing datasets is a good strategy. However, compared with natural image datasets, e.g., ImageNet [26], the available remote sensing scene datasets are relatively insufficient, which cannot train a good model because they easily suffer from the overfitting problem. Therefore, some works [17,27] attempted to directly fine-tune the parameters of pretrained CNN models (e.g., AlexNet [11], GoogLeNet [28]) for remote sensing image scene classification. Although good performance has been witnessed, these methods commonly use the features from fully connected layers for classification, while ignoring the spatial information in remote sensing scenes, which is also crucial.

2.1.2. Deep Feature Encoding Methods

Instead of directly using the features from a pretrained CNN as the final scene representation, deep feature encoding methods regard the deep CNN as a feature extractor to capture various different levels of features, then encode these features using some unsupervised feature encoding techniques. Zhao and Du [29] utilized bag of words (BoW) [30] to encode local spatial patterns into a new scene representation. Zheng et al. [31] extracted multiscale local feature information from the last convolutional layer using the proposed multiscale pooling strategy and then generated the holistic scene representation with the Fisher vector (FV) [32]. Several methods attempt to encode multilayer convolutional features to capture more discriminative scene features due to the complementarity between them. Wang et al. [33] used the vectors of locally aggregated descriptors (VLADs) [34] to aggregate multilayer convolutional features. He et al. [35] presented a covariance pooling algorithm to integrate multilayer convolutional features and achieved great performance.

2.1.3. Multiple Feature Fusion Methods

It is generally believed that features from different scales have different representation abilities to describe the given scene. Therefore, fusing different features is a good solution to improve classification performance. According to the types of features used, existing multiple feature fusion methods can be roughly classified into two categories: the methods fusing both deep and hand-crafted features and the methods fusing different deep features. For the former, hand-crafted features have been proven to be effective in describing some special scenes; thus, some works [36,37] attempted to combine hand-crafted features with deep features to improve the feature representation ability. For example, Lu et al. [36] proposed a bidirectional adaptive fusion model to effectively fuse SIFT features and deep features together and successfully addressed the problem of scale and rotation variability. Yu et al. [37] proposed two feature-level fusion architectures, which used the mapped local binary pattern (LBP) and saliency coded networks as two auxiliary streams and then separately integrated them with the raw RGB network for further enhancing the scene representation capacity. The second category of methods have been popular in recent years, which mainly fuse multilayer deep features from a single CNN [6,23–25,38] or multilevel deep features from multiple different CNN branches [39–42] to obtain diverse features for classification.

In addition, to solve the scale variation of the objects in remote sensing imagery, Liu et al. [43] proposed a dual-branch multiscale CNN architecture. Furthermore, Zhang et al. [44] utilized the attention mechanism to extract discriminative features at different scales and then fused them for classification.

2.1.4. Other Methods

Recently, a variety of new ideas and theories have been introduced into the remote sensing image scene classification task, such as the attention mechanism [45–47], CapsNet [48], GAN [49], loss function optimization [50], deep bilinear transformation [51], neural architecture search [52], meta learning [53], etc. It should be noted that these approaches aim to solve specific issues, such as capturing discriminative scene representation, solving the problem of small training samples, searching the optimal network architecture for classification, etc.

2.2. Attention in CNNs

Inspired by the human sensing process, attention mechanisms have been studied extensively in computer vision (CV) [54–56] and natural language processing (NLP) [57]. The basic idea of attention is to construct a constraint mechanism that can selectively emphasize and reserve the key regions to extract the important features while depreciating other harmful interference information. Currently, many attention mechanisms have been proposed and successfully applied in various fields. Hu et al. [54] presented the squeeze-and-excitation network (SENet) to model correlations between different channels for capturing the importance of different feature channels. In addition, CBAM [55] considers capturing feature information from spatial and channel attention simultaneously, which significantly improves the feature representation ability. Recently, the nonlocal neural network [56] has been widely used in salient object detection [58], image superresolution [59], etc. Its main purpose is to enhance the features of the current position by aggregating contextual information from other positions and solve the problem that the receptive field of a single convolutional layer is ineffective to cover correlated regions. Compared with the typical convolution operation, the nonlocal structure can capture global receptive field information and further improve the feature discrimination. Later, some improved algorithms were proposed, such as the GCNet [60] and the CCNet [61], to address the problem of computational complexity. Recently, some studies [62,63] introduced the self-attention mechanism into remote sensing image scene classification and achieved promising results. Benefiting from the advantages of the attention mechanism, we introduced the channel and spatial attention in this paper simultaneously in order to capture more accurate semantic regions for multilayer feature aggregation.

3. The Proposed Method

In this section, we first introduce the overall architecture of the proposed AGMFA-Net in Section 3.1. Section 3.2 gives the details of the multilayer feature extraction module. Finally, the implementation of the multilayer feature aggregation module is provided in Section 3.3.

3.1. Overall Architecture

The goal of the proposed method is to learn discriminative feature representation for remote sensing image scene classification. Figure 2 illustrates the overall architecture of AGMFA-Net, which consists of three main components: feature extraction module, multilayer feature aggregation module, and classification module. Our network was built on ResNet-50 [15] as the backbone. Firstly, the input image is fed into the backbone to generate a series of convolutional feature maps that contain different levels of information about the given scene; we denote them as Res2, Res3, Res4_1, Res4_2, and Res4_3. Then, the multilayer feature aggregation module is utilized to fuse these features to generate a new feature with more powerful scene representation ability. Concretely, in order to achieve semantic regions corresponding to the given scene as accurately as possible, the channel–spatial attention module was simultaneously employed on multiple high-level feature maps, i.e., Res4_1, Res4_2, and Res4_3, and a new attention mask is generated. Then, we used this mask to guide the multilayer feature aggregation procedure. Through this process, discriminative information of different feature maps will be well fused to generate a more powerful scene representation, as well as suppress some interference or useless information caused by low-level feature maps. After that, a block operation (including convolution, ReLU, normalization) was employed to merge the information of the aggregated features among the channel. Finally, a fully connected layer and a softmax layer followed to predict the label of the input scene. In the following subsections, we introduce each component in detail.

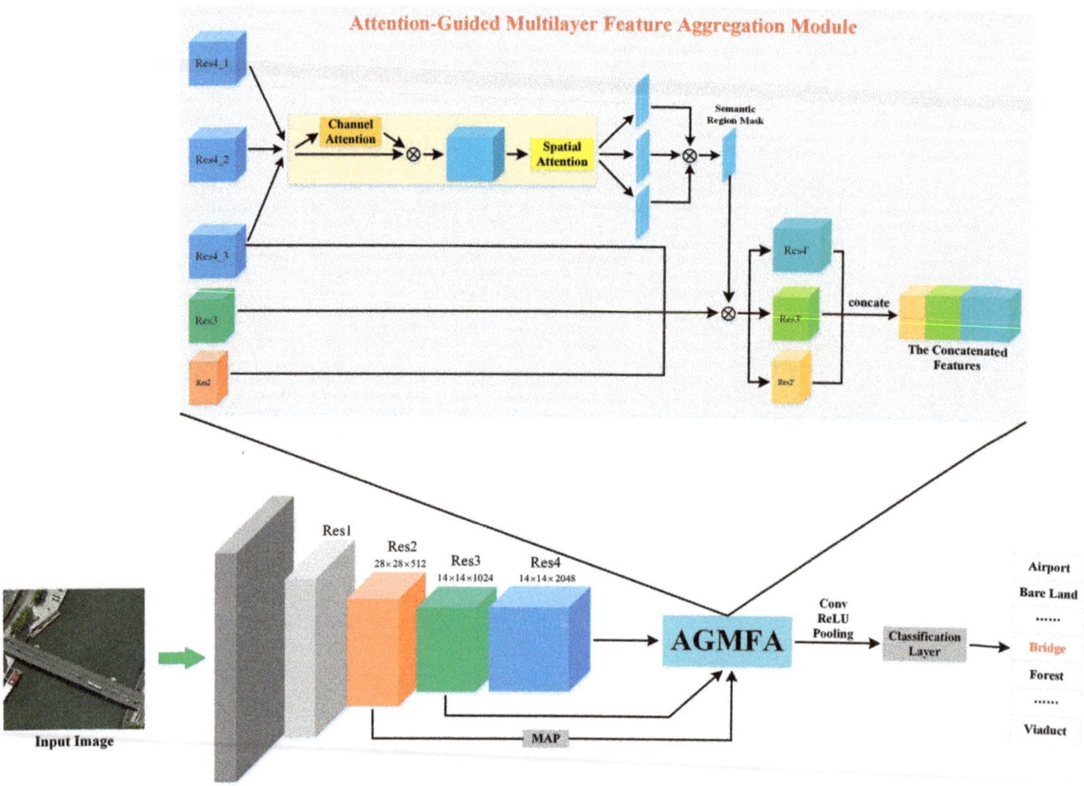

Figure 2. The overall architecture of our proposed AGMFA-Net.

3.2. Multilayer Feature Extraction

Limited by the scarcity of training samples in remote sensing images, many existing methods capture multilayer convolutional features using the pretrained CNN models. Currently, many famous CNN architectures have been developed, e.g., AlexNet, VGGNet, ResNet, etc. Considering the excellent classification performance of ResNet on ImageNet, in this paper, we used the modified ResNet-50 to extract multilayer convolutional feature maps from remote sensing scenes. For the ResNet-50 model, it starts with an initial convolutional layer with a kernel of size 7 × 7 and a stride of 2. Then, a max-pooling layer is added with a 3 × 3 window and a stride of 2. The later portion is composed of four residual blocks; we denote the outputs of each residual block as Res1, Res2, Res3, and Res4, respectively. Because we only extracted multilayer feature maps, we deleted all layers after Res4. In addition, to retain more spatial information, we changed the stride of Res4 from 2 to 1. Assuming the size of the input image is 3 × 224 × 224, the sizes of Res2, Res3, and Res4 are 512 × 28 × 28, 1024 × 14 × 14, and 2048 × 14 × 14, respectively. At the same time, the size of high-level convolutional feature maps (e.g., Res4_1, Res4_2, and Res4_3) was the same, i.e., 512 × 14 × 14. It is worth noting that Res4 and Res4_3 denote the same feature map; they both represent the output of the last residual block of ResNet-50. To ensure that the size of each feature map is consistent, we downsampled Res2 to change its size to 512 × 14 × 14 by using a max-pooling operation. The main motivation to extract multilayer convolutional features was that they can complement each other, which has been proven to be helpful for improving the remote sensing image scene classification accuracy.

3.3. *Multilayer Feature Aggregation*

In general, the features from deeper layers can describe the semantic information of the given scenes better, while the features from lower layers have rich appearance information; they are both important for classification. Thus, fusing features from different layers has become a commonly used strategy to obtain a more comprehensive scene representation. However, directly aggregating multilayer features without considering the discrepancies between them, e.g., feature redundancy, semantic ambiguity, and background interference, may result in reducing the discriminative ability. To aggregate multilayer convolutional features more effectively and obtain more valuable information of each feature map, an attention-guided multilayer feature aggregation module was designed, as shown in Figure 2. It mainly consists of two parts: semantic region extraction and multilayer feature aggregation.

To reduce the impacts of semantic interference, feature redundancy, etc., between different convolutional layers, we followed a rule that only aggregates multilayer features corresponding to the semantic regions of the given scene. Therefore, there are two key issues that need to be considered: (1) how to accurately obtain the semantic regions of the input scenes; (2) how to fuse different levels of feature maps based on the learned semantic regions?

3.3.1. Semantic Region Extraction

For the first issue, a commonly used solution is to only use the last convolutional activation as the semantic regions. However, this solution is not effective because the semantic regions are incomplete and ignore other discriminative regions, which are also important for scene classification. To address this problem, we first analyzed the activation characteristics of different high-level convolutional feature maps in the last residual block of ResNet-50, and the visualization results are shown in Figure 3 by using the gradient-weighted class activation mapping (Grad-CAM) algorithm [64]. It can be observed that a single convolutional feature map usually only activates the most discriminative regions of the given scene, while ignoring the importance of other semantic areas. In addition, the activation regions of different convolutional feature maps are different, but also overlap. Furthermore, multiple convolutional feature maps can compensate each other to achieve more complete activation regions.

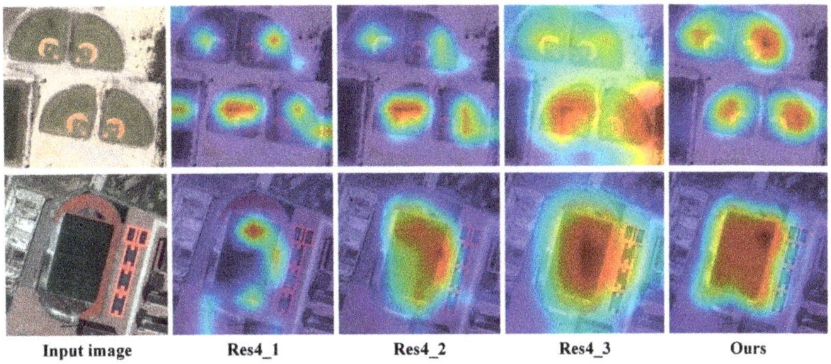

Figure 3. Grad-CAM visualization results. We compare the visualization results of our proposed channel–spatial attention with three other high-level convolutional feature maps of the last residual block of ResNet-50.

In order to capture more semantic regions of the given scene accurately, we proposed to simultaneously aggregate multiple high-level convolutional features based on the channel–spatial attention mechanism. Recently, benefiting from the human visual system, various attention mechanisms have been developed and have achieved great success in many fields, which aim to selectively concentrate on the prominent regions to extract the discriminative features from the given scene while discarding other interference information. Among them, the CBAM [55] algorithm is excellent and has been introduced in remote sensing scene classification. CBAM considers two different dimensions of the channel and spatial information simultaneously to capture important features and suppress useless features more effectively. Therefore, we employed CBAM in this paper to obtain important semantic regions from each high-level convolutional feature map.

Suppose Res4_1 $\in \mathbb{R}^{C\times H\times W}$, Res4_2 $\in \mathbb{R}^{C\times H\times W}$, and Res4_3 $\in \mathbb{R}^{C\times H\times W}$ denote three high-level convolutional feature maps from the last residual block of ResNet-50, respectively. C, H, and W represent the channel number, height, and width of each feature map. As shown in Figure 2, each high-level convolutional feature map is first separately passed to the channel–spatial attention module to generate three different attention masks, and these masks are then multiplied to obtain the final semantic regions.

Figure 4 demonstrates the detailed workflow of the channel–spatial attention operation, which consists of two components: the channel stream and the spatial stream. Let the input feature map be $\mathrm{X} \in \mathbb{R}^{C\times H\times W}$, where C, H, and W are the number of channels, height, and width, respectively. Firstly, two pooling operations, i.e., global max pooling and global average pooling, are employed to aggregate the spatial information of X and generate two $C \times 1 \times 1$ spatial contextual descriptors; we denote them as $\mathrm{X}^{C}_{\max} \in \mathbb{R}^{C\times 1\times 1}$ and $\mathrm{X}^{C}_{\mathrm{avg}} \in \mathbb{R}^{C\times 1\times 1}$, respectively. Then, two descriptors are fed into a shared network with a hidden layer and multilayer perception. To reduce the computational overhead, the activation size of the hidden layer is $\mathbb{R}^{C/r\times 1\times 1}$, where r is the reduction ratio. After that, two output features of the shared network are added after a sigmoid activation function to obtain the channel attention map $\mathrm{M_C} \in \mathbb{R}^{C\times 1\times 1}$. Finally, the refined feature X' is obtained by multiplying $\mathrm{M_C}$ with the input feature map X. In summary, the entire process of channel attention can be expressed as follows:

$$\mathrm{X}' = \mathrm{M}_c(\mathrm{X}) \otimes \mathrm{X} \tag{1}$$

where $\otimes$ represents elementwise multiplication and $\mathrm{M}_c(\mathrm{X})$ denotes the channel attention map, which can be described as:

$$\begin{aligned} \mathrm{M_C}(\mathrm{X}) &= \sigma(\mathrm{MLP}(\mathrm{AvgPool}(\mathrm{X})) + \mathrm{MLP}(\mathrm{MaxPool}(\mathrm{X}))) \\ &= \sigma(W_1(W_0(\mathrm{X}^{C}_{avg})) + W_1(W_0(\mathrm{X}^{C}_{\max}))) \end{aligned} \tag{2}$$

where σ denotes the sigmoid function, MLP represents the multi-layer perceptron, AvgPool and MaxPool denote the global average pooling and global max pooling, respectively, and $W_0 \in \mathbb{R}^{C/r\times C}$ and $W_1 \in \mathbb{R}^{C\times C/r}$ are the weights of the MLP.

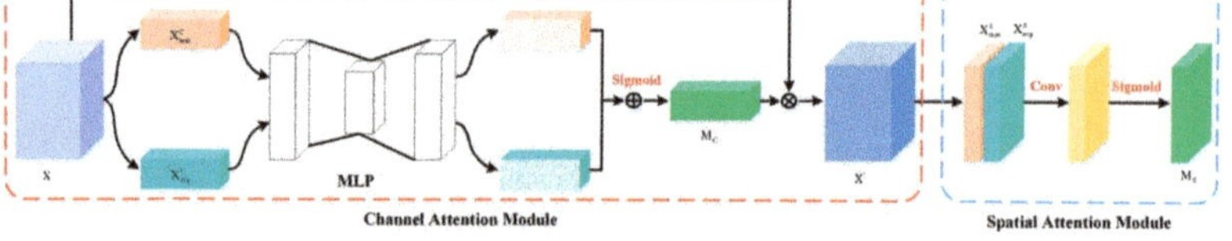

Figure 4. Diagram of the channel–spatial attention module.

Different from channel attention, spatial attention aims to utilize the interspatial relationships of features to generate a spatial attention map, which mainly focuses on the discriminative areas. To obtain the spatial attention map $\mathrm{M_S} \in \mathbb{R}^{H\times W}$, the average pooling and max pooling operations are adopted along the channel dimension at first to generate two $1 \times H \times W$ channel descriptors, which are denoted as $\mathrm{X}^{S}_{avg} \in \mathbb{R}^{1\times H\times W}$ and $\mathrm{X}^{S}_{\max} \in \mathbb{R}^{1\times H\times W}$. Then, these two channel descriptors are concatenated to generate a new descriptor. After that, a 7×7 convolution and sigmoid function are used to capture a spatial attention map $\mathrm{M_S}$, which can highlight the important regions of the given scenes while suppressing other interference regions. It should be noted that we only need to generate the spatial attention map, instead of reweighting the input feature map X' to generate a refined feature map. Therefore, the spatial attention is computed as:

$$\begin{aligned} M_S(X') &= \sigma(f^{7\times7}\text{concat}[\text{AvgPool}(X');\text{MaxPool}(X')]) \\ &= \sigma(f^{7\times7}\text{concat}[X^S_{avg};X^S_{max}]) \end{aligned} \tag{3}$$

where σ and concat denote the sigmoid function and concatenation operation, respectively, $f^{7\times7}$ represents a convolution operation with a filter size of 7×7, and AvgPool and MaxPool represent the average pooling and max pooling along the channel dimension. By referring to [55], we connected channel attention and spatial attention in a sequential arrangement manner, which can more effectively focus on important semantic regions of the given scene.

For high-level convolutional feature maps, Res4_1, Res4_2, and Res4_3, we separately pass them into the channel–spatial attention module to capture different attention masks, denoted as M4_1, M4_2, and M4_3. It is worth noting that each mask mainly concentrates on discriminative regions, but they complement each other. To obtain a more accurate semantic region mask, we conducted the matrix multiplication operation on the above three masks, and the newly generated semantic region mask is denoted as M. Compared with the discriminative mask only using the last convolutional features of ResNet-50, our method makes full use of the information from multiple high-level convolutional feature maps to obtain a more efficient and complete semantic region mask, as shown in the last column in Figure 3. The expression of this procedure can be written as follows.

$$M = M4\_1 \otimes M4\_2 \otimes M4\_3 \tag{4}$$

where $\otimes$ denotes the elementwise multiplication operation.

3.3.2. Multilayer Feature Aggregation

It is acknowledged that convolutional features extracted from different layers can describe different levels of information of the given scene; some published research [6,33,38] has also proven that fusing multiple convolutional features can significantly promote the scene classification performance. However, integrating multilayer convolutional features indiscriminately may be easily affected by the differences, e.g., semantic ambiguity, feature redundancy, and background interference, resulting in the discrimination of the learned scene representation being insufficient. To solve this problem, we designed a novel multilayer feature fusion strategy. Specifically, we first obtained semantic regions in terms of the semantic region extraction operation, then used the learned semantic regions to guide the process of multilayer feature aggregation. Compared with other fusion strategies, e.g., fusion by addition, our method not only fuses valuable feature information of each convolutional layer effectively, but also avoids the interference of unfavorable factors.

As shown in Figure 2, Res2, Res3, and Res4 are three different convolutional feature maps captured from the backbone network. M represents the semantic region mask. To aggregate multilayer convolutional features, we separately multiply Res2, Res3, and Res4 by M to generate new features; we present them as $Res2'$, $Res3'$, and $Res4'$. By this step, different convolutional layers' important information, which is consistent with M, is selected for the subsequent fusion procedure. After that, these features are concatenated along the channel dimension. Specifically, in order to reduce the feature dimension and merge the information of the concatenated features among the channels, a 1×1 convolution operation and a ReLU operation are followed; we denote the output features as Y. Therefore, we can use the formula to express this as follows:

$$\begin{aligned} &Res2' = Res2 \otimes M \\ &Res3' = Res3 \otimes M \\ &Res4' = Res4 \otimes M \\ &Y = \delta(f^{1\times1}\text{concat}[Res2';Res3';Res4']) \end{aligned} \tag{5}$$

where ⊗ denotes the elementwise operation, δ represents the ReLU function, $f^{1\times1}$ denotes a convolution operation with the filter size of 1 × 1, and concat represents the concatenation operation.

After obtaining Y, it is sent into the classifier for scene classification.

3.4. Loss Function

During training, the cross entropy loss function is used to minimize a weighted cumulation loss. Suppose that $I=\{(x_1,y_1),\ldots,(x_N,y_N)\}$ is a training batch of N images, where y_i, a one-hot vector, is the label of the i-th image x_i. p_i is a vector in which the j-th element is the probability that image x_i is classified into the j-th class. Then, the cross entropy loss can be formulated as follows:

$$L = -\frac{1}{N}\sum_{i=1}^{N}\left(y_i{}^{I}\log(p_i)\right) \tag{6}$$

4. Experiments

In this section, we conduct a series of experiments to verify the effectiveness of the proposed AGMFA-Net.

4.1. Datasets

To evaluate the performance of the proposed method, the following commonly used remote sensing scene classification datasets were employed: the UC Merced Land Use dataset [30], the more challenging large-scale Aerial Image Dataset (AID) [18], and the NWPU-RESISC45 dataset [17].

(1) UC Merced Land Use dataset (UCML): The UCML dataset is a classical benchmark for remote sensing scene classification. It consists of 21 different classes of land use images with a pixel resolution of 0.3 m. It contains a total of 2100 remote sensing images with 100 samples for each class. These samples are all annotated from a publicly available aerial image, and the size of each sample is 256 × 256 pixels. The example images of each class are shown in Figure 5.

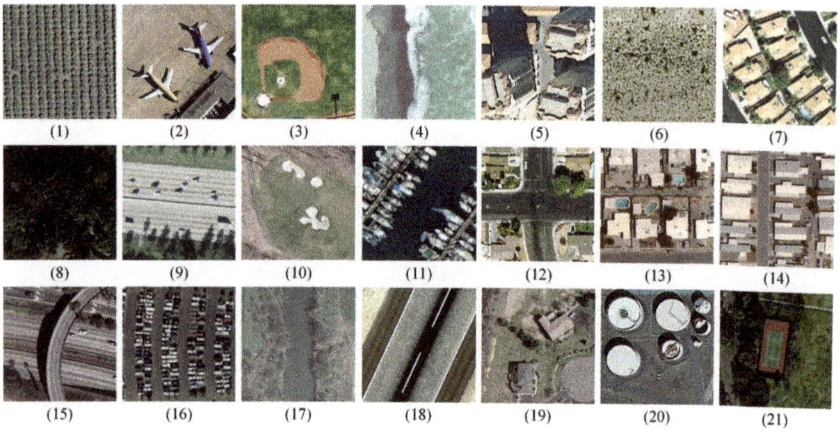

Figure 5. Examples of the UCML dataset.

(2) Aerial Image Dataset (AID): The AID dataset has 10,000 remote sensing scene images, which are divided into 30 different land cover categories. Each category's number varies from 220 to 420. The size of each image is 600 × 600 pixels, and the spatial resolution ranges from about 8 m to 0.5 m. It is noted that the AID dataset is a relatively large-scale remote sensing scene dataset and is challenging for classifying. Some examples of each category are presented in Figure 6.

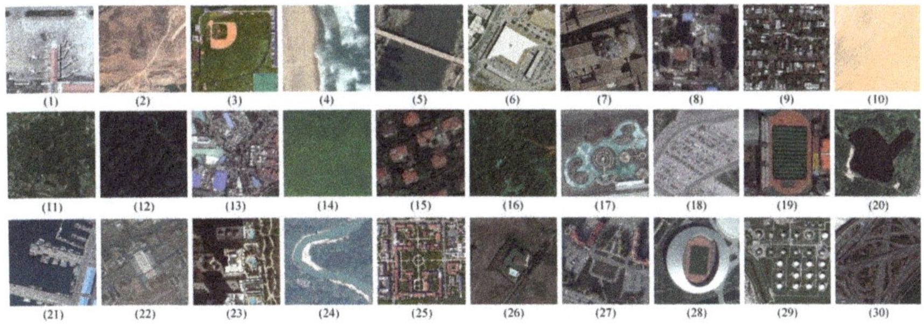

Figure 6. Examples of the AID dataset.

(3) NWPU-RESISC45 dataset: This dataset is more complex and challenging compared with the above three datasets. It contains a total of 31,500 images divided into 45 different scenes. Each scene has 700 images with an image size of 256 × 256 pixels. Because of the more diverse scenes, the spatial resolution of the images varies from 0.2 m to 30 m. Figure 7 shows some examples of this dataset.

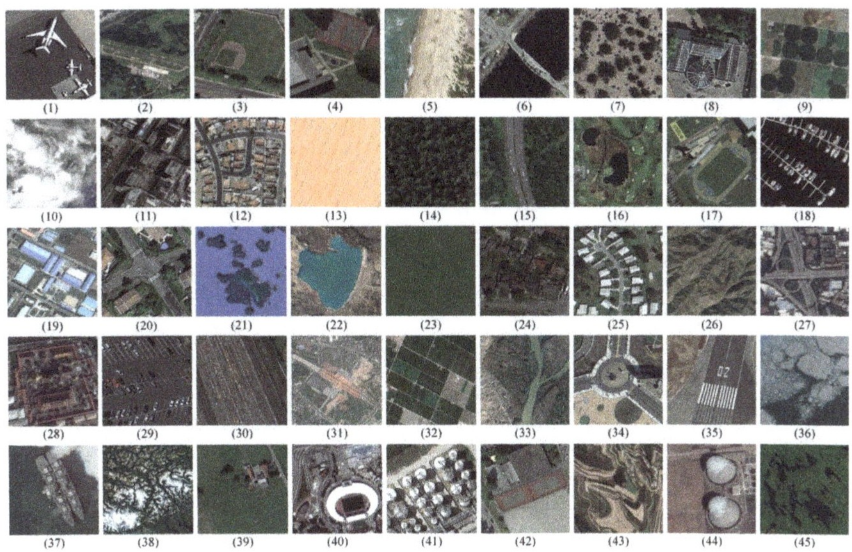

Figure 7. Examples of the NWPU-RESISC45 dataset.

To ensure a fair comparison, we employed the commonly used training ratios to divide each dataset. For the UCML dataset, we set the training ratio to 80% and the rest of the samples (20%) for testing. For the AID dataset, we set two training–testing ratios, i.e., 20–80% and 50–50%, respectively. Similarly, two training ratios, i.e., 10–90% and 20–80%, were used for the NWPU-RESISC45 dataset.

4.2. Implementation Details

All experiments were completed using the PyTorch [65] deep learning library. We employed ResNet-50 as the backbone network. To verify the scalability of the proposed method, we also conducted experiments with the VGGNet-16 network. All networks were trained using one NVIDIA GeForce RTX 2070 Super GPU. To make the network converge quickly, all the experimental networks were first pretrained on the ImageNet and then fine-tuned with the above three benchmark datasets. Our proposed network was optimized by the stochastic gradient descent (SGD) algorithm with the momentum as 0.9, the initial learning as 0.001, and the weight decay penalty as 1×10^{-5}. After every 30 epochs, the learning rate decayed by 10 times. The batch size and maximum training iterations were set to 32 and 150, respectively. In the training stage, data augmentation was adopted to improve the generalization performance. Concretely, the input images were first resized to 256×256 pixels, then randomly cropped to 224×224 pixels as the network input after random horizontal flipping.

4.3. *Evaluation Metrics*

To comprehensively evaluate the classification of the proposed method, three evaluation metrics were used in this paper. They include the overall accuracy and the confusion matrix. Each evaluation metric is explained as follows:

(1) Overall accuracy (OA): The OA is defined as the ratio between the number of correctly classified images and the total number of testing images;

(2) Confusion matrix (CM): The CM is a special matrix used to visually evaluate the performance of the algorithm. In this matrix, the column represents the ground truth and the row denotes the prediction. From it, we can observe the classification accuracy of each scene, as well as the categories that are easily confused with each other.

4.4. *Ablation Study*

In our proposed method, we mainly improved the discriminative capability of the multilayer feature aggregation from two aspects. To separately demonstrate the effectiveness of each component, we conducted ablation experiments on the AID and NWPU-RESISC45 datasets using ResNet-50 as the backbone network.

4.4.1. The Effectiveness of Semantic Region Extraction

We conducted experiments to qualitatively analyze the effectiveness of semantic region extraction. In the following, we compare the following network architectures, i.e., ResNet-50, ResNet-50+DA (direct aggregation), ResNet-50+WA (without attention), ResNet-50+SA (spatial attention), Ours (low-level features), Ours (multiple high-level features). Specifically, ResNet-50 was the baseline network. ResNet-50+DA represents directly aggregating multiple high-level convolutional feature maps indiscriminately. ResNet-50+WA denotes aggregating multiple high-level convolutional feature maps without using attention. Instead, we employed the method in [66], which captures semantic regions by utilizing multiple high-level convolutional feature maps in an unsupervised way. ResNet-50+SA represents using the spatial attention following each high-level convolutional feature map, then aggregating them to generate new semantic regions. Ours (low-level features) and Ours (high-level features) are two methods that adopt channel attention and spatial attention separately on low-level and high-level features to capture semantic regions. More intuitively, we illustrate the activation maps of the aggregated features between different compared methods using the Grad-CAM algorithm in Figure 8. It can be observed that the above six methods can activate the discriminative regions, which are consistent with the semantic label of the scenes; however, the activation regions of our proposed method are more complete and can accurately cover the overall discriminative regions.

4.4.2. The Effectiveness of Multilayer Feature Aggregation

We also conducted experiments on the AID and NWPU-RESISC45 datasets to quantitatively evaluate the performance of the proposed multilayer feature aggregation strategy, and the results are shown in Table 1. From Table 1, we can make the following conclusions: (1) For the AID and NWPU-RESISC45 datasets, the multilayer feature aggregation methods can further promote the classification accuracy when compared with the baseline. This observation verified that fusing features from different layers can indeed achieve better results. (2) The classification accuracy of ResNet-50+DA and ResNet-50+WA was similar. We considered the reason is partly that ResNet-50+WA employs an unsupervised method to obtain semantic regions, which cannot suppress the impacts of complex backgrounds, resulting in worse accuracy. (3) The methods based on attention were better than ResNet-50+DA and ResNet-50+WA, except the training ratio of the NWPU-RESISC45 dataset was 10%. We also respectively compared the classification performance when obtaining semantic regions based on low-level and high-level features in our method. (4) We found that when using low-level features, its classification performance on the AID and NWPU-RESISC45 datasets was better than the baseline, but lower than other methods. We considered the reason to be that the use of low-level revolutionary features cannot effectively reduce the interference of background noise and semantic ambiguity, resulting in the captured semantic regions being inaccurate, which further reduces the performance of multilayer feature fusion. (5) When using multiple high-level convolutional features to capture semantic regions, our method can achieve optimal classification accuracy because we used channel and spatial attention together to obtain more accurate semantic regions. Therefore, the final aggregated features have better discrimination.

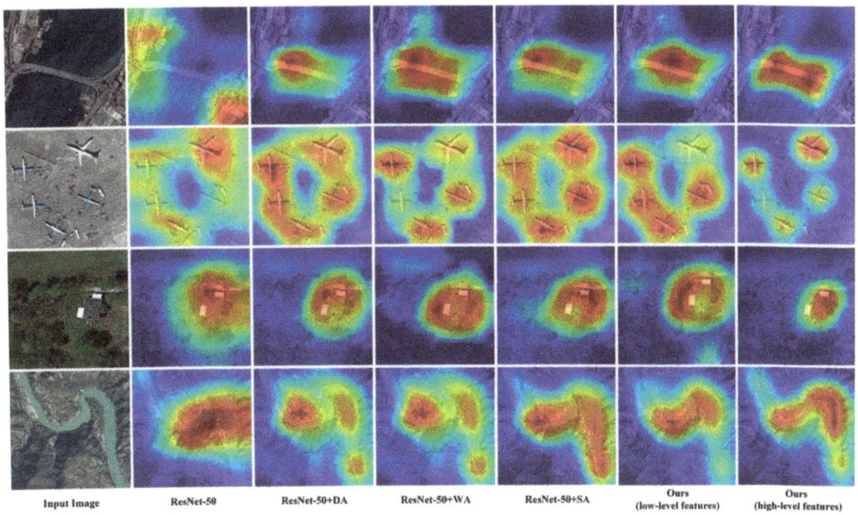

Figure 8. Grad-CAM visualization results. We compare the visualization results of the proposed AGMFA-Net (ResNet-50) with the baseline (ResNet-50) and three other multilayer feature aggregation methods. The Grad-CAM visualization is computed for the last convolutional outputs.

Table 1. Ablation experimental results on two datasets with different training ratios.

Method	AID		NWPU-RESISC45	
	20%	50%	10%	20%
ResNet-50 (Baseline)	92.93 ± 0.25	95.40 ± 0.18	89.06 ± 0.34	91.91 ± 0.09
ResNet-50+DA	93.54 ± 0.30	96.08 ± 0.34	90.26 ± 0.04	93.21 ± 0.16
ResNet-50+WA	93.66 ± 0.28	96.15 ± 0.28	90.24 ± 0.07	93.08 ± 0.04
ResNet-50+SA	93.77 ± 0.31	96.32 ± 0.18	90.13 ± 0.59	93.22 ± 0.10
Ours (low-level features)	93.51 ± 0.51	95.98 ± 0.20	89.16 ± 0.36	92.76 ± 0.11
Ours (high-level features)	94.25 ± 0.13	96.68 ± 0.21	91.01 ± 0.18	93.70 ± 0.08

4.5. *State-of-the-Art Comparison and Analysis*

4.5.1. Results on the UCML Dataset

UCML is a classical dataset for evaluating the performance of remote sensing image scene classification. To illustrate the superiority of our proposed method, we compared it with some state-of-the-art scene classification methods that are reviewed in Section 2, and the comparison results are shown in Table 2. As can be seen from Table 2, our method, which employed ResNet-50 as the backbone, achieved the optimal overall classification accuracy. In addition, when using VGGNet-16, our method also surpassed most of the methods and obtained a competitive classification performance. It is worth noting that the overall accuracy of most of the compared methods reached above 98%, but our method still showed good superiority and demonstrated its effectiveness.

Table 2. The OA (%) and STD (%) of different methods on the UCML dataset.

Methods	Accuracy
VGGNet-16 [12]	96.10 ± 0.46
ResNet-50 [15]	98.76 ± 0.20
MCNN [43]	96.66 ± 0.90
Multi-CNN [41]	99.05 ± 0.48
Fusion by Addition [25]	97.42 ± 1.79
Two-Stream Fusion [39]	98.02 ± 1.03
VGG-VD16+MSCP [35]	98.40 ± 0.34
VGG-VD16+MSCP+MRA [35]	98.40 ± 0.34
ARCNet-VGG16 [45]	99.12 ± 0.40
VGG-16-CapsNet [48]	98.81 ± 0.22
MG-CAP (Bilinear) [22]	98.60 ± 0.26
MG-CAP (Sqrt-E) [22]	99.00 ± 0.10
GBNet+global feature [38]	98.57 ± 0.48
EfficientNet-B0-aux [50]	99.04 ± 0.33
EfficientNet-B3-aux [50]	99.09 ± 0.17
IB-CNN(M) [51]	98.90 ± 0.21
TEX-TS-Net [37]	98.40 ± 0.76
SAL-TS-Net [37]	98.90 ± 0.95
ResNet-50+EAM [47]	98.98 ± 0.37
Ours (VGGNet-16)	98.71 ± 0.49
Ours (ResNet-50)	**99.33 ± 0.31**

Figure 9 shows the confusion matrix of our proposed method when the training ratio was 80%. It can be seen that almost all scenes can be accurately classified except for some easily confused categories, such as freeway and overpass, medium residential and dense residential, and forest and sparse residential. This is because some scenes are composed of multiple different land use units (e.g., sparse residential contains forest and building together) or show different spatial layout characteristics (e.g., freeway and overpass both contain road, but they have different spatial layouts). These issues make them difficult to classify.

4.5.2. Results on the AID Dataset

AID is a larger and more challenging dataset than the UCML dataset. We compared our method with other scene classification methods with two training ratios, 20% and 50%. For both training ratios, our method performed better than other competitors, as shown in Table 3. For a training ratio of 50%, our method with VGGNet-16 as the backbone surpassed almost all the compared methods that use the same backbone, such as Fusion by Addition [25], VGG-16+MSCP [35], ARCNet-VGG16 [45], MF^{2}Net [6], VGG-16-CapsNet [48], etc. Similarly, when using ResNet-50 as the backbone, our method achieved the highest classification accuracy, which exceeded other methods that use ResNet or more advanced network as the backbone. For example, our method increased by 0.06% over ResNet-50+EAM [47], 0.11 over IB-CNN (M) [51], and 0.12 over EfficientNet-B3-aux [50]. For a training ratio of 20%, our method that used VGGNet-16 showed mediocre performance; however, when using ResNet-50 as the backbone, our method performed better than all the other methods. Specifically, our method was slightly higher than EfficientNet-B3-aux and IB-CNN(M) by 0.16% and 0.02% and exceeded ResNet-50+EAM by 0.16%.

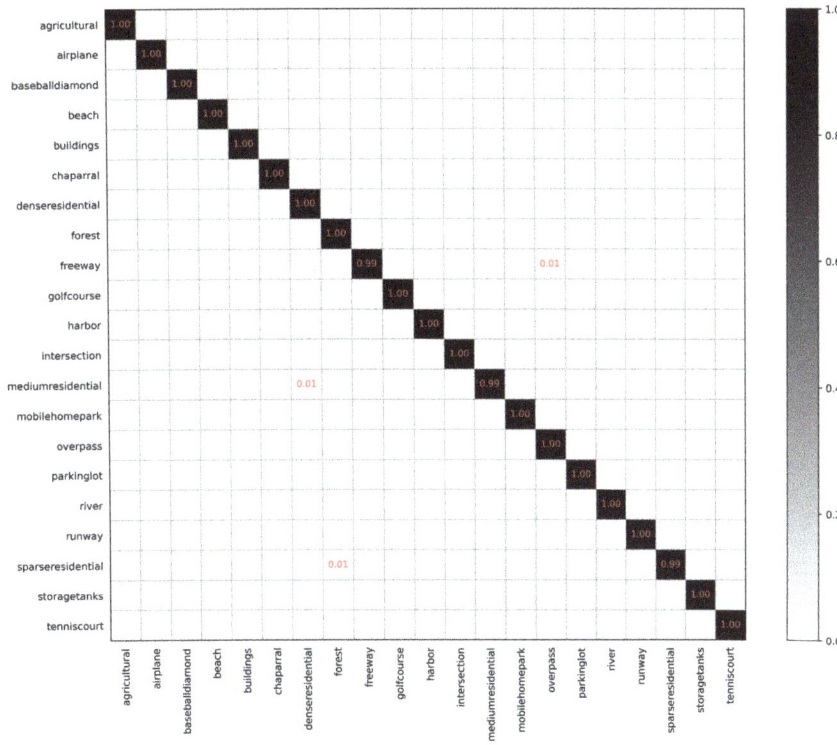

Figure 9. Confusion matrix of the proposed method on the UCML dataset with a training ratio of 80%.

The CMs of different training ratios are illustrated in Figures 10 and 11, respectively. For a training ratio of 50% in Figure 10, most of the categories achieved a classification accuracy higher than 95%, except the scenes of resort (92%) and school (93%). Specifically, the most difficult scenes to classify were resort and park, because they are composed of some similar land use units and also have the same spatial structures. In addition, school is easily confused with square and industrial. For a training ratio of 20% in Figure 11, our method can also obtain excellent classification accuracy, except for the following four scenes: center (87%), resort (79%), school (84%), and square (86%).

4.5.3. Results on the NWPU-RESISC45 Dataset

For the larger NWPU-RESISC45 dataset, the comparison results are shown in Table 4. For two training ratios, our methods obtained remarkable performance. When the training ratio was 20%, our method that used ResNet-50 as the backbone exceeded all the competitors. Specifically, in comparison to the baselines, our method separately improved by 1.79% (ResNet-50) and 2.86% (VGGNet-16) when using different networks. When using VGGNet-16 as the backbone, we surpassed other methods that use the same backbone, e.g., Two-Stream [39], VGGNet16+MSCP, MF^2Net, and VGG-16-CapsNet. In addition, our method achieved the highest classification accuracy when using ResNet-50, higher than ResNet-50+EAM by 0.19% and higher than IB-CNN (M) by 0.37%. For the training ratio of 10%, our methods can also obtain excellent classification performance.

Table 3. Overall accuracy and standard deviation (%) of different methods on the AID dataset.

Method	Training Ratio	
	20%	50%
VGGNet-16 [12]	88.81 ± 0.35	92.84 ± 0.27
ResNet-50 [15]	92.93 ± 0.25	95.40 ± 0.18
Fusion by Addition [25]	-	91.87 ± 0.36
Two-Stream Fusion [39]	80.22 ± 0.22	93.16 ± 0.18
Multilevel Fusion [40]	-	95.36 ± 0.22
VGG-16+MSCP [35]	91.52 ± 0.21	94.42 ± 0.17
ARCNet-VGG16 [45]	88.75 ± 0.40	93.10 ± 0.55
MF^{2}Net [6]	91.34 ± 0.35	94.84 ± 0.27
MSP [31]	93.90	-
MCNN [43]	-	91.80 ± 0.22
VGG-16-CapsNet [48]	91.63 ± 0.19	94.74 ± 0.17
Inception-v3-CapsNet [48]	93.79 ± 0.13	96.32 ± 0.12
MG-CAP (Bilinear) [22]	92.11 ± 0.15	95.14 ± 0.12
MG-CAP (Sqrt-E) [22]	93.34 ± 0.18	96.12 ± 0.12
EfficientNet-B0-aux [50]	93.69 ± 0.11	96.17 ± 0.16
EfficientNet-B3-aux [50]	94.19 ± 0.15	96.56 ± 0.14
IB-CNN(M) [51]	94.23 ± 0.16	96.57 ± 0.28
TEX-TS-Net [37]	93.31 ± 0.11	95.17 ± 0.21
SAL-TS-Net [37]	94.09 ± 0.34	95.99 ± 0.35
ResNet-50+EAM [47]	93.64 ± 0.25	96.62 ± 0.13
Ours (VGGNet-16)	91.09 ± 0.30	95.10 ± 0.78
Ours (ResNet-50)	**94.25 ± 0.13**	**96.68 ± 0.21**

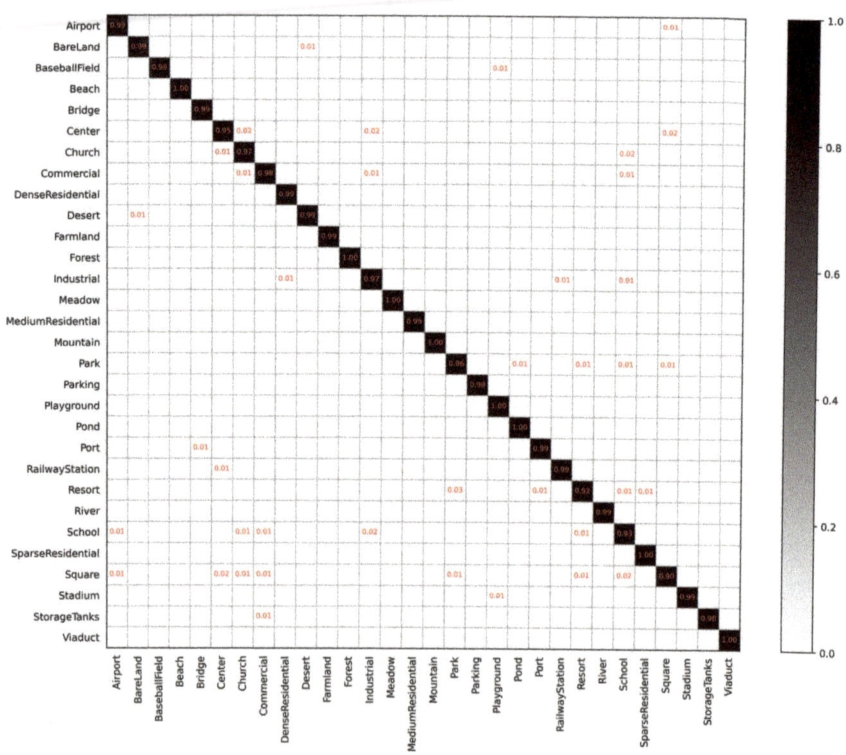

Figure 10. Confusion matrix of the proposed method on the AID dataset with a training ratio of 50%.

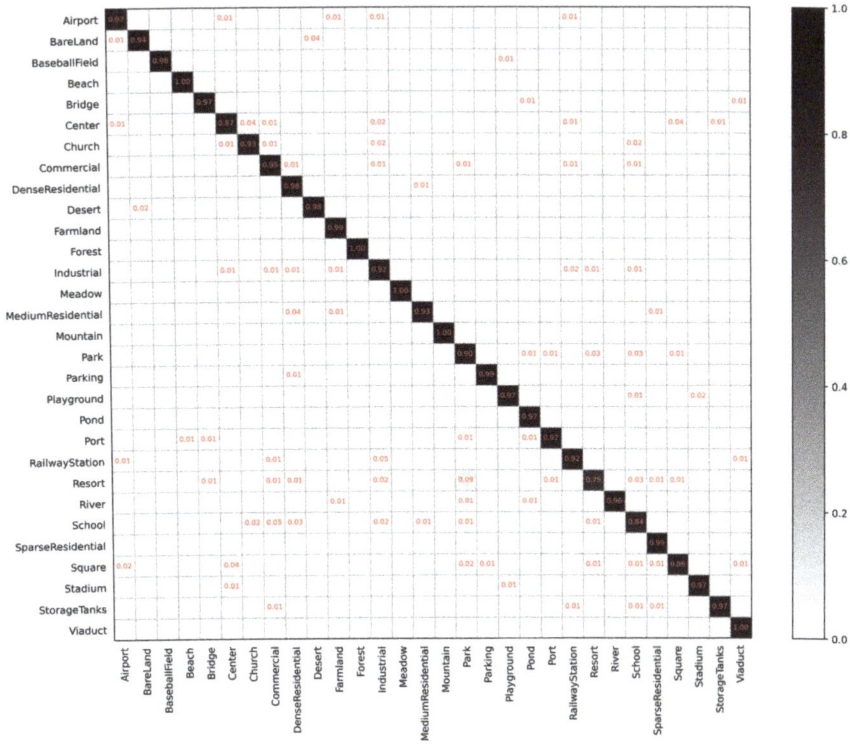

Figure 11. Confusion matrix of the proposed method on the AID dataset with a training ratio of 20%.

Table 4. Overall accuracy and standard deviation (%) of different methods on the NWPU-RESISC45 dataset.

Method	Training Ratio	
	10%	20%
VGGNet-16 [12]	81.15 ± 0.35	86.52 ± 0.21
ResNet-50 [15]	89.06 ± 0.34	91.91 ± 0.09
Two-Stream [39]	80.22 ± 0.22	83.16 ± 0.18
VGG-16+MSCP [35]	85.33 ± 0.17	88.93 ± 0.14
MF^{2}Net [6]	85.54 ± 0.36	89.76 ± 0.27
VGG-16-CapsNet [48]	85.08 ± 0.13	89.18 ± 0.14
Inception-v3-CapsNet [48]	89.03 ± 0.21	92.60 ± 0.11
MG-CAP (Bilinear) [22]	89.42 ± 0.19	91.72 ± 0.16
MG-CAP (Sqrt-E) [22]	90.83 ± 0.12	92.95 ± 0.13
EfficientNet-B0-aux [50]	89.96 ± 0.27	92.89 ± 0.16
IB-CNN(M) [51]	90.49 ± 0.17	93.33 ± 0.21
TEX-TS-Net [37]	84.77 ± 0.24	86.36 ± 0.19
SAL-TS-Net [37]	85.02 ± 0.25	87.01 ± 0.19
ResNet-50+EAM [47]	90.87 ± 0.15	93.51 ± 0.12
Ours (VGGNet-16)	86.87 ± 0.19	90.38 ± 0.16
Ours (ResNet-50)	**91.01 ± 0.18**	**93.70 ± 0.08**

Figures 12 and 13 are the confusion matrix results for the training ratios of 20% and 10%, respectively. It can be observed that when setting the training ratio to 20%, almost all the scenes can achieve above 90% classification accuracy, except two scenes, i.e., church (83%) and palace (83%), which are very easily confused with each other. In addition, for the training ratio of 10%, most of the scenes can be classified well; the scenes with the lowest classification accuracy still remain church (77%) and palace (75%).

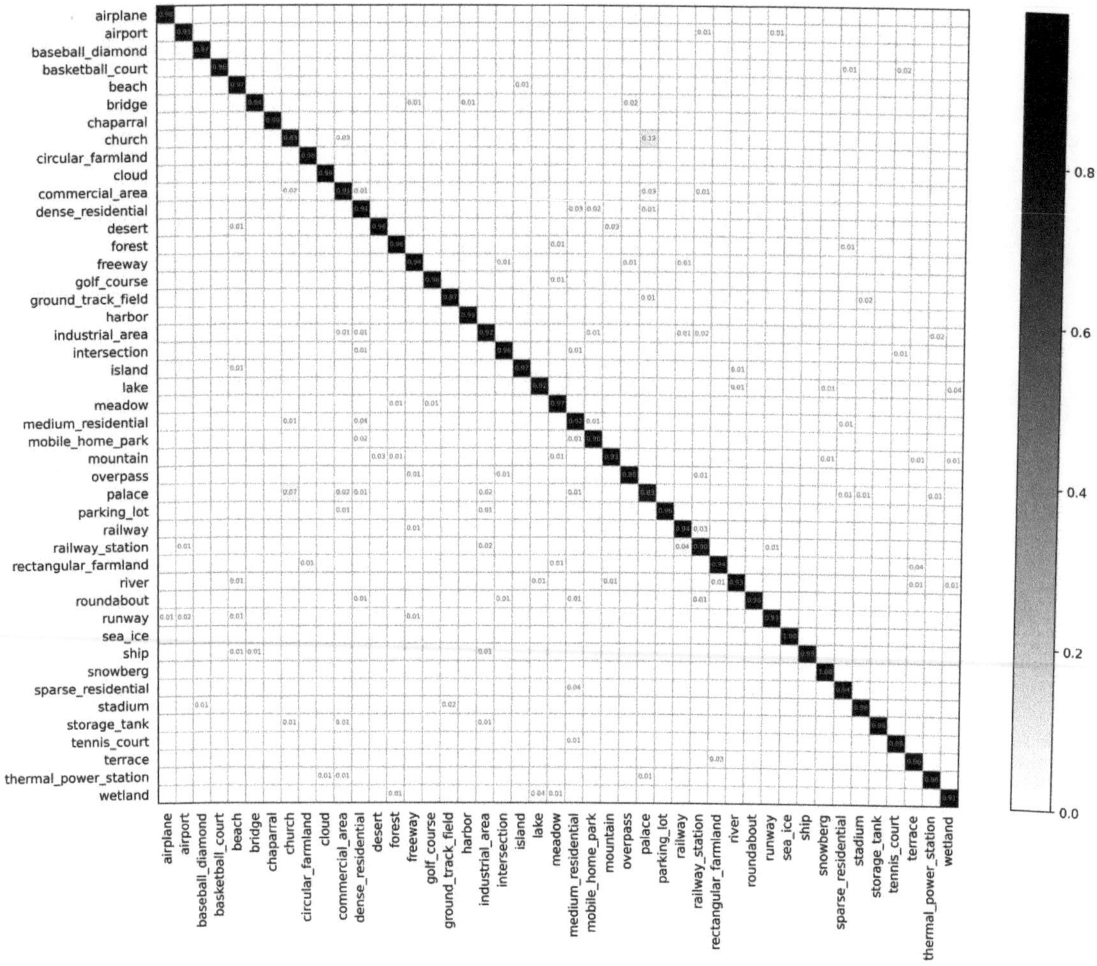

Figure 12. Confusion matrix of the proposed method on the NWPU-RESISC45 dataset with a training ratio of 20%.

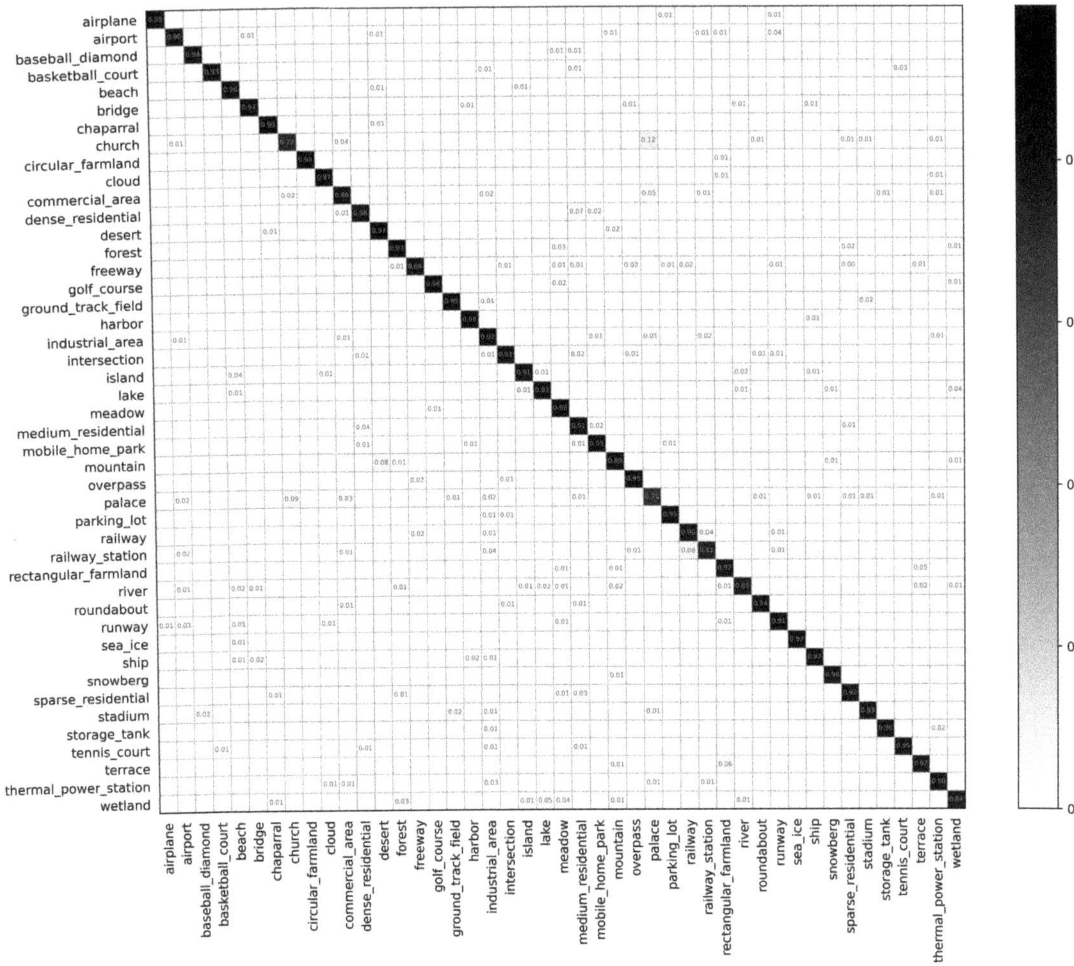

Figure 13. Confusion matrix of the proposed method on the NWPU-RESISC45 dataset with a training ratio of 10%.

5. Conclusions

One of the crucial challenges of remote sensing image scene classification is how to learn a powerful scene representation. To address this problem, we presented a novel attention-guided multilayer feature aggregation network in this paper, which consisted of three parts: the multilayer feature extraction module, the multilayer feature aggregation module, and the classification module. Concretely, we first used the backbone network to extract multiple convolutional feature maps with different spatial resolutions. Then, a semantically guided multilayer feature aggregation module was used to integrate features from different convolutional layers to reduce the interferences of useless information and at the same time improve the scene representation capacity. Specifically, to capture semantic regions that were consistent with the given scene accurately, we employed channel–spatial attention to make full use of the feature information of multiple high-level convolutional feature layers. Compared with the semantic regions captured from a single convolutional layer, our method showed better results. Finally, the aggregated features were fed into the classifier for scene classification. Experiments on three benchmark datasets were conducted, and the results demonstrated that our proposed method can achieve promising classification performance and outperform other remote sensing image scene classification methods.

Author Contributions: Conceptualization, M.L.; data curation, M.L. and L.L.; formal analysis, M.L.; methodology, M.L. and Y.S.; software, M.L.; validation, M.L. and Y.S.; writing—original draft, M.L.; writing—review and editing, L.L., Y.T. and G.K. All authors read and agreed to the published version of the manuscript.

Funding: This research received no external funding.

Institutional Review Board Statement: Not applicable.

Informed Consent Statement: Not applicable.

Data Availability Statement: The UC Merced Land Use, AID and NWPU-RESISC45 datasets used in this study are openly and freely available at http://weegee.vision.ucmerced.edu/datasets/landuse.html, https://captain-whu.github.io/AID/, and https://gcheng-nwpu.github.io/datasets#RESISC45, respectively.

Acknowledgments: We would like to thank the handling Editor and the anonymous reviewers for their careful reading and helpful suggestions.

Conflicts of Interest: The authors declare no conflict of interest.

References

1. Mou, L.; Ghamisi, P.; Zhu, X.X. Deep recurrent neural networks for hyperspectral image classification. *IEEE Trans. Geosci. Remote Sens.* **2017**, *55*, 3639–3655. [CrossRef]
2. Li, X.; Lei, L.; Sun, Y.; Li, M.; Kuang, G. Multimodal bilinear fusion network with second-order attention-based channel selection for land cover classification. *IEEE J. Sel. Top. Appl. Earth Obs. Remote Sens.* **2020**, *13*, 1011–1026. [CrossRef]
3. Wu, C.; Zhang, L.; Zhang, L. A scene change detection framework for multi-temporal very high resolution remote sensing images. *Signal Process* **2015**, *124*, 84–197. [CrossRef]
4. Hu, Q.; Wu, W.; Xia, T.; Yu, Q.; Yang, P.; Li, Z.; Song, Q. Exploring the use of Google Earth imagery and object-based methods in land use/cover mapping. *Remote Sens.* **2013**, *105*, 6026–6042. [CrossRef]
5. Wang, C.; Shi, J.; Yang, X.; Zhou, Y.; Wei, S.; Li, L.; Zhang, X. Geospatial object detection via deconvolutional region proposal network. *IEEE J. Sel. Topics Appl. Earth Observ. Remote Sens.* **2019**, *12*, 3014–3027. [CrossRef]
6. Xu, K.; Huang, H.; Li, Y.; Shi, G. Multilayer feature fusion network for scene classification in remote sensing. *IEEE Geosci. Remote Sens. Lett.* **2020**, *17*, 1894–1898. [CrossRef]
7. Swain, M.J.; Ballard, D.H. Color indexing. *Int. J. Comput. Vis.* **1991**, *7*, 11–32. [CrossRef]
8. Haralick, R.M.; Shanmugam, K.; Dinstein, I.H. Textural features for image classification. *IEEE Trans. Syst. Man Cybern.* **1973**, *SMC-3*, 610–621. [CrossRef]
9. Lowe, D.G. Distinctive image features from scale-invariant key-points. *Int. J. Comput. Vis.* **2004**, *60*, 91–110. [CrossRef]
10. Dalal, N.; Triggs, B. Histograms of oriented gradients for human detection. In Proceedings of the 2005 IEEE Computer Society Conference on Computer Vision and Pattern Recognition (CVPR'05), San Diego, CA, USA, 20–25 June 2005; Volume 1, pp. 886–893.
11. Krizhevsky, A.; Sutskever, I.; Hinton, G. Imagenet classification with deep convolutional neural networks. *Adv. Neural Inf. Process. Syst.* **2012**, 25, 1097–1105. [CrossRef]
12. Simonyan, K.; Zisserman, A. Very deep convolutional networks for large-scale image recognition. *arXiv* **2014**, arXiv:1409.1556.
13. Ren, S.; He, K.; Girshick, R.; Sun, J. Faster R-CNN: Towards real-time object detection with region proposal networks. *IEEE Trans. Pattern Anal. Mach. Intell* **2017**, *39*, 1137–1149. [CrossRef] [PubMed]
14. Chen, L.; Papandreou, G.; Kokkinos, I.; Murphy, K.; Yuille, A.L. DeepLab: Semantic image segmentation with deep convolutional nets, atrous convolution, and fully connected CRFs. *IEEE Trans. Pattern Anal. Mach. Intell.* **2018**, *40*, 834–848. [CrossRef] [PubMed]
15. He, K.; Zhang, X.; Ren, S.; Sun, J. Deep residual learning for image recognition. In Proceedings of the 2016 IEEE Conference on Computer Vision and Pattern Recognition (CVPR), Las Vegas, NV, USA, 27–30 June 2016; pp. 770–778.
16. Huang, G.; Liu, Z.; Van Der Maaten, L.; Weinberger, K.Q. Densely connected convolutional networks. In Proceedings of the IEEE Conference on Computer Vision and Pattern Recognition, Honolulu, HI, USA, 21–26 July 2017; pp. 2261–2269.
17. Cheng, G.; Han, J.; Lu, X. Remote sensing image scene classification: Benchmark and state of the art. *Proc. IEEE* **2017**, *105*, 1865–1883. [CrossRef]
18. Xia, G.-S.; Hu, J.; Hu, F.; Shi, B.; Bai, X.; Zhong, Y.; Zhang, L.; Lu, X. AID: A benchmark data set for performance evaluation of aerial scene classification. *IEEE Trans. Geosci. Remote Sens.* **2017**, *55*, 3965–3981. [CrossRef]
19. Cheng, G.; Xie, X.; Han, J.; Guo, L.; Xia, G.S. Remote sensing image scene classification meets deep learning: Challenges, methods, benchmarks, and opportunities. *IEEE J. Sel. Topics. Appl. Earth Observ. Remote Sens.* **2020**, *13*, 3735–3756. [CrossRef]
20. Nogueira, K.; Penatti, O.A.; Santos, J.A.D. Towards better exploiting convolutional neural networks for remote sensing scene classification. *Pattern Recognit* **2017**, *61*, 539–556. [CrossRef]

21. Cheng, G.; Li, Z.; Yao, X.; Guo, L.; Wei, Z. Remote sensing image scene classification using bag of convolutional features. *IEEE Geosci. Remote Sens. Lett.* **2017**, *14*, 1735–1739. [CrossRef]
22. Wang, S.; Guan, Y.; Shao, L. Multi-granularity canonical appearance pooling for remote sensing scene classification. *IEEE Trans. Image Process* **2020**, *29*, 5396–5407. [CrossRef]
23. Li, E.; Xia, J.; Du, P.; Lin, C.; Samat, A. Integrating multilayer features of convolutional neural networks for remote sensing scene classification. *IEEE Trans. Geosci. Remote Sens.* **2017**, *55*, 5653–5665. [CrossRef]
24. Lu, X.; Sun, H.; Zheng, X. A feature aggregation convolutional neural network for remote sensing scene classification. *IEEE Trans. Geosci. Remote Sens.* **2019**, *57*, 7894–7906. [CrossRef]
25. Chaib, S.; Liu, H.; Gu, Y.; Yao, H. Deep feature fusion for VHR remote sensing scene classification. *IEEE Trans. Geosci. Remote Sens.* **2017**, *55*, 4775–4784. [CrossRef]
26. Deng, J.; Dong, W.; Socher, R.; Li, L.-J.; Li, K.; Fei-Fei, L. ImageNet: A large-scale hierarchical image database. In Proceedings of the 2009 IEEE Conference on Computer Vision and Pattern Recognition, Miami, FL, USA, 20–25 June 2009; pp. 248–255.
27. Liang, Y.; Monteiro, S.T.; Saber, E.S. Transfer learning for high resolution aerial image classification. In Proceedings of the 2016 IEEE Applied Imagery Pattern Recognition Workshop (AIPR), Washington, DC, USA, 18–20 October 2016; pp. 1–8.
28. Szegedy, C.; Liu, W.; Jia, Y.; Sermanet, P.; Reed, S.; Anguelov, D.; Erhan, D; Vanhoucke, V.; Rabinovich, A. Going deeper with convolutions. In Proceedings of the 2015 IEEE Conference on Computer Vision and Pattern Recognition (CVPR), Boston, MA, USA, 7–12 June 2015; pp. 1–9.
29. Zhao, W.; Du, S. Scene classification using multi-scale deeply described visual words. *Int. J. Remote Sens.* **2016**, *37*, 4119–4131. [CrossRef]
30. Yang, Y.; Newsam, S. Bag-of-visual-words and spatial extensions for land use classification. In Proceedings of the GIS '10: 18th Sigspatial International Conference on Advances in Geographic Information Systems, San Jose, CA, USA, 2–5 November 2010; ACM: New York, NY, USA, 2010; pp. 270–279.
31. Zheng, X.; Yuan, Y.; Lu, X. A deep scene representation for aerial scene classification. *IEEE Trans. Geosci. Remote Sens.* **2019**, *57*, 4799–4809. [CrossRef]
32. Sanchez, J.; Perronnin, F.; Mensink, T.; Verbeek, J. Image classification with the fisher vector: Theory and practice. *Int. J. Comput. Vis.* **2013**, *105*, 222–245. [CrossRef]
33. Wang, G.; Fan, B.; Xiang, S.; Pan, C. Aggregating rich hierarchical features for scene classification in remote sensing imagery. *IEEE J. Sel. Topics. Appl. Earth Observ. Remote Sens.* **2017**, *10*, 4104–4115. [CrossRef]
34. Negrel, R.; Picard, D.; Gosselin, P.-H. Evaluation of second-order visual features for land use classification. In Proceedings of the 2014 12th International Workshop on Content-Based Multimedia Indexing (CBMI), Klagenfurt, Austria, 18–20 June 2014; pp. 1–5.
35. He, N.; Fang, L.; Li, S.; Plaza, A.; Plaza, J. Remote sensing scene classification using multilayer stacked covariance pooling. *IEEE Trans. Geosci. Remote Sens.* **2018**, *56*, 6899–6910. [CrossRef]
36. Lu, X.; Ji, W.; Li, X.; Zheng, X. Bidirectional adaptive feature fusion for remote sensing scene classification. *Neurocomputing* **2019**, 328, 135–146. [CrossRef]
37. Yu, Y.; Liu, F. Dense connectivity based two-stream deep feature fusion framework for aerial scene classification. *Remote Sens.* **2018**, *10*, 1158. [CrossRef]
38. Sun, H.; Li, S.; Zheng, X.; Lu, X. Remote sensing scene classification by gated bidirectional network. *IEEE Trans. Geosci. Remote Sens.* **2020**, *58*, 82–96. [CrossRef]
39. Yu, Y.; Liu, F. A two-stream deep fusion framework for high-resolution aerial scene classification. *Comput. Intell. Neurosci.* **2018**, *2018*, 8639367. [CrossRef] [PubMed]
40. Yu, Y.; Liu, F. Aerial scene classification via multilevel fusion based on deep convolutional neural networks. *IEEE Geosci. Remote Sens. Lett.* **2018**, *15*, 287–291. [CrossRef]
41. Du, P.; Li, E.; Xia, J.; Samat, A.; Bai, X. Feature and model level fusion of pretrained CNN for remote sensing scene classification. *IEEE J. Sel. Topics Appl. Earth Observ. Remote Sens.* **2019**, *12*, 2600–2611. [CrossRef]
42. Zeng, D.; Chen, S.; Chen, B.; Li, S. Improving remote sensing scene classification by integrating global-context and local-object features. *Remote Sens.* **2018**, *10*, 734. [CrossRef]
43. Liu, Y.; Zhong, Y.; Qin, Q. Scene classification based on multiscale convolutional neural network. *IEEE Trans. Geosci. Remote Sens.* **2018**, *56*, 7109–7121. [CrossRef]
44. Ji, J.; Zhang, T.; Jiang, L.; Zhong, W.; Xiong, H. Combining multilevel features for remote sensing image scene classification with attention model. *IEEE Geosci. Remote Sens. Lett.* **2020**, *17*, 1647–1651. [CrossRef]
45. Wang, Q.; Liu, S.; Chanussot, J.; Li, X. Scene classification with recurrent attention of VHR remote sensing images. *IEEE Trans. Geosci. Remote Sens.* **2019**, *57*, 1155–1167. [CrossRef]
46. Cao, R.; Fang, L.; Lu, T.; He, N. Self-attention-based deep feature fusion for remote sensing scene classification. *IEEE Geosci. Remote Sens. Lett.* **2021**, *18*, 43–47. [CrossRef]
47. Zhao, Z.; Li, J.; Luo, Z.; Li, J.; Chen, C. Remote sensing image scene classification based on an enhanced attention module. *IEEE Geosci. Remote Sens. Lett.* **2020**. [CrossRef]
48. Zhang, W.; Tang, P.; Zhao, L. Remote sensing image scene classification using CNN-CapsNet. *Remote Sens.* **2019**, *11*, 494. [CrossRef]

49. Yu, Y.; Li, X.; Liu, F. Attention GANs: Unsupervised deep feature learning for aerial scene classification. *IEEE Trans. Geosci. Remote Sens.* **2020**, *58*, 519–531. [CrossRef]
50. Bazi, Y.; Rahhal, A.; Alhichri, M.M.H.; Alajlan, N. Simple yet effective fine-tuning of deep CNNs using an auxiliary classification loss for remote sensing scene classification. *Remote Sens.* **2019**, *11*, 2908. [CrossRef]
51. Li, E.; Samat, A.; Du, P.; Liu, W.; Hu, J. Improved Bilinear CNN Model for Remote Sensing Scene Classification. *IEEE Geosci. Remote Sens. Lett.* **2020**. [CrossRef]
52. Peng, C.; Li, Y.; Jiao, L.; Shang, R. Efficient Convolutional Neural Architecture Search for Remote Sensing Image Scene Classification. *IEEE Trans. Geosci. Remote Sens.* **2021**, *59*, 6092–6105. [CrossRef]
53. Zhang, P.; Bai, Y.; Wang, D.; Bai, B.; Li, Y. Few-shot classification of aerial scene images via meta-learning. *Remote Sens.* **2021**, *13*, 108. [CrossRef]
54. Hu, J.; Shen, L.; Sun, G. Squeeze-and-excitation networks. In Proceedings of the 2018 IEEE/CVF Conference on Computer Vision and Pattern Recognition, Salt Lake City, UT, USA, 18–23 June 2018; pp. 7132–7141.
55. Woo, S.; Park, J.; Lee, J.Y. Cbam: Convolutional block attention module. In Proceedings of the European Conference on Computer Vision, Munich, Germany, 8–14 September 2018; pp. 3–19.
56. Wang, X.; Girshick, R.; Gupta, A.; He, K. Non-local neural networks. In Proceedings of the 2018 IEEE/CVF Conference on Computer Vision and Pattern Recognition Workshops (CVPRW), Salt Lake City, UT, USA, 18–22 June 2018; pp. 7794–7803.
57. Vaswani, A.; Shazeer, N.; Parmar, N.; Uszkoreit, J.; Jones, L.; Gomez, A.N.; Kaiser, L; Polosukhin, I. Attention is all you need. In Proceedings of the Advances in Neural Information Processing Systems, Long Beach, CA, USA, 4–9 December 2017; pp. 5998–6008.
58. Gu, Y.; Wang, L.; Wang, Z.; Liu, Y.; Cheng, M.-M.; Lu, S.-P. Pyramid Constrained Selfw-Attention Network for Fast Video Salient Object Detection. *Proc. AAAI Conf. Artif. Intell* **2020**, *34*, 10869–10876.
59. Zhu, F.; Fang, C.; Ma, K.-K. PNEN: Pyramid Non-Local Enhanced Networks. *IEEE Trans. Image Process.* **2020**, *29*, 8831–8841. [CrossRef]
60. Cao, Y.; Xu, J.; Lin, S.; Wei, F.; Hu, H. GCNet: Non-local networks meet squeeze-excitation networks and beyond. In Proceedings of the 2019 IEEE/CVF International Conference on Computer Vision Workshops, ICCV Workshops 2019, Seoul, Korea, 27–28 October 2019; pp. 1971–1980.
61. Huang, Z.; Wang, X.; Huang, L.; Huang, C.; Wei, Y.; Liu, W. CCNet: Criss-cross attention for semantic segmentation. In Proceedings of the 2018 IEEE/CVF Conference on Computer Vision and Pattern Recognition (CVPR 2018), Salt Lake City, UT, USA, 18–22 June 2018; pp. 603–612.
62. Zhang, D.; Li, N.; Ye, Q. Positional context aggregation network for remote sensing scene classification. *IEEE Geosci. Remote Sens. Lett.* **2020**, *17*, 943–947. [CrossRef]
63. Fu, L.; Zhang, D.; Ye, Q. Recurrent Thrifty Attention Network for Remote Sensing Scene Recognition. *IEEE Trans. Geosci. Remote Sens.* **2020**. [CrossRef]
64. Selvaraju, R.R.; Cogswell, M.; Das, A.; Vedantam, R.; Parikh, D.; Batra, D. Grad-CAM: Visual explanations from deep networks via gradient-based localization. In Proceedings of the 2017 IEEE International Conference on Computer Vision (ICCV), Venice, Italy, 22–29 October 2017; pp. 618–626.
65. Paszke, A.; Gross, S.; Massa, F.; Lerer, A.; Bradbury, J.; Chanan, G.; Killeen, T.; Lin, Z.; Gimelshein, N.; Antiga, L.; et al. PyTorch: An imperative style, high-performance deep learning library. In Proceedings of the Advances in Neural Information Processing Systems, Vancouver, BC, Canada, 8–14 December 2019; pp. 8024–8035.
66. Wei, X.; Luo, J.; Wu, J.; Zhou, Z. Selective convolution descriptor aggregation for fine-grained image retrieval. *IEEE Trans. Image Process* **2017**, *26*, 2868–2881. [CrossRef]

remote sensing

MDPI

Article

Learning the Incremental Warp for 3D Vehicle Tracking in LiDAR Point Clouds

Shengjing Tian [1], Xiuping Liu [1,*], Meng Liu [2], Yuhao Bian [1], Junbin Gao [3] and Baocai Yin [4]

1 School of Mathematical Sciences, Dalian University of Technology, Dalian 116024, China; tye@mail.dlut.edu.cn (S.T.); yhbian@mail.dlut.edu.cn (Y.B.)
2 School of Computer and Technology, Shan Dong Jianzhu University, Jinan 250101, China; mengliu.sdu@gmail.com
3 Discipline of Business Analytics, Business School, The University of Sydney, Sydney, NSW 2006, Australia; junbin.gao@sydney.edu.au
4 Faculty of Electronic Information and Electrical Engineering, Dalian University of Technology, Dalian 116024, China; ybc@dlut.edu.cn
* Correspondence: xpliu@dlut.edu.cn

Abstract: Object tracking from LiDAR point clouds, which are always incomplete, sparse, and unstructured, plays a crucial role in urban navigation. Some existing methods utilize a learned similarity network for locating the target, immensely limiting the advancements in tracking accuracy. In this study, we leveraged a powerful target discriminator and an accurate state estimator to robustly track target objects in challenging point cloud scenarios. Considering the complex nature of estimating the state, we extended the traditional Lucas and Kanade (LK) algorithm to 3D point cloud tracking. Specifically, we propose a state estimation subnetwork that aims to learn the incremental warp for updating the coarse target state. Moreover, to obtain a coarse state, we present a simple yet efficient discrimination subnetwork. It can project 3D shapes into a more discriminatory latent space by integrating the global feature into each point-wise feature. Experiments on KITTI and PandaSet datasets showed that compared with the most advanced of other methods, our proposed method can achieve significant improvements—in particular, up to 13.68% on KITTI.

Keywords: point clouds; 3D tracking; state estimation; Siamese network; deep LK

Citation: Tian, S.; Liu, X.; Liu, M.; Bian, Y.; Gao, J.; Yin, B. Learning the Incremental Warp for 3D Vehicle Tracking in LiDAR Point Clouds. *Remote Sens.* **2021**, *13*, 2770. https://doi.org/10.3390/rs13142770

Academic Editor: Fahimeh Farahnakian

Received: 24 June 2021
Accepted: 9 July 2021
Published: 14 July 2021

1. Introduction

Single object tracking in point clouds aims to localize the time-varying target represented by point clouds with the supervision of a 3D bounding box in the first frame. It is a challenging yet indispensable task in many real-world applications, such as autonomous driving [1,2] and mobile robot tracking [3,4]. Generally, object tracking encompasses two subtasks, target discrimination and state estimation, which are the fundamental steps for an agent to sense the surrounding environment and conduct motion planning [5]. Over the last few years, 2D single object tracking task has been explored extensively [6–9]. Inspired by that success, many RGB-D based methods refer to the pattern of 2D tracking to conduct 3D tracking [10–13]. Although working well in the conventional 2D domain, these methods rely heavily on the RGB modality. They hence may fail when color information is low-quality or even unavailable. In this work, we focus on the 3D vehicle tracking task using deep learning in point clouds, which is still in the development stage due to several factors, such as self-occlusion, disorder, density change, and the difficulty of the state estimation.

Recently, the high-end sensors for LiDAR (light detection and ranging) have attracted much attention, since they have high accuracy and are less sensitive to weather conditions than most other sensors. More importantly, it can capture the structure of a scene by producing plenty of 3D point clouds to provide reliable geometric information from far away. However, the source data constitute the unstructured representation, where the standard convolution operation is not applicable. This hampers the application of deep learning models in 3D object tracking. To overcome this barrier, some studies projected point clouds onto planes from a bird's-eye view (BEV), and then discretized them into 2D images [5,14,15].

Although they could conduct tracking by detecting frame by frame, the BEV loses abundant geometric information. Consequently, starting from source point clouds, Giancola et al. [2] proposed SiamTrack3D to learn a generically template matching function which is trained by the shape completion regularization. Qi et al. [16] leveraged deep Hough voting [17] to produce potential bounding boxes. It is worth mentioning that the aforementioned approaches merely focus on determining the best proposal from a set of object proposals. In other words, they thoroughly ignore the importance of comprehensively considering both the target discrimination and the state estimation [18].

To address this problem, we elaborately designed a 3D point cloud tracking framework with the purpose of bridging the gap between target discrimination and state estimation. It is mainly comprised of two components, a powerful target discriminator and an accurate target state estimator, which realize their respective functions through the Siamese network. The state estimation subnetwork (SES) is proposed to estimate a optimal warp using the template and candidates extracted from the tracked frame. This subnetwork extends the 2D Lucas and Kanade (LK) algorithm [19] to the 3D point cloud tracking problem by incorporating it into a deep network. However, it is non-trivial, since the Jacobian matrix from the first-order Taylor expansion cannot be calculated as in the RGB image, where the Jacobian matrix can be split into two partial terms using the chain rule. The reason is that the gradients in x, y, and z cannot be calculated, as connections among points are lacking in 3D point clouds. To circumvent this issue, we thoughtfully present an approximation-based solution and a learning-based solution. By integrating them into a deep network in an end-to-end manner, our state estimation subnetwork can take a pair of point clouds as inputs to predict the incremental warp parameters. Additionally, we introduce an efficient target discrimination subnetwork (TDS) to remedy the deficiency of the SES. In order to project 3D shapes into a more discriminatory latent space, we designed a new loss that takes global semantic information into consideration. During online tracking, by forcing these two components to cooperate with each other properly, our proposed model could cope with the challenging point cloud scenarios robustly.

The key contributions of our work are three-fold:

- A novel state estimation subnetwork was designed, which extends the 2D LK algorithm to 3D point cloud tracking. In particular, based on the Siamese architecture, this subnetwork can learn the incremental warp for meliorating the coarse target state.
- A simple yet powerful discrimination subnetwork architecture is introduced, which projects 3D shapes into a more discriminatory latent space by integrating the global semantic feature into each point-wise feature. More importantly, it surpasses the 3D tracker using sole shape completion regularization [2].
- An efficient framework for 3D point cloud tracking is proposed to bridge the performance difference between the state estimation component and the target discrimination component. Due to the complementarity of these two components, our method achieved a significant improvement, from 40.09%/56.17% to 53.77%/69.65% (success/precision), on the KITTI tracking dataset.

2. Related Work

2.1. 2D Object Tracking

In this paper, we focus on single object tracking problem, which can be divided into two subtasks: target discrimination and state estimation [18]. Regarding 2D visual tracking, some discrimination-based methods [7,20] have recently shown outstanding performance. In particular, the family of the correlation filter trackers [8,20] have enjoyed great popularity in the tracking research community. These methods leverage the properties of circulant matrices, which can be diagonalized by discrete Fourier transformation (DFT) to learn a classifier online. With the help of the background context and the implicit exhaustive convolution in a 2D-grid, correlation filter methods achieve impressive performance. In addition, other discrimination approaches based on deep learning have achieved competitive results on 2D tracking benchmarks [21,22]. For instance, MDNet [7] first learns general feature representation in a multi-domain way, and then it captures domain-specific information via online updating. CFNet [23] creatively integrates the correlation filter into SiamFC [24]. However, the majority of these approaches put attention into developing a powerful discriminator and simply rely on brute-force multi-scale searching to adjust the target state. There also exist several special methods such as DeepLK [19] and GONTURN [25] which are merely derived from state estimation, but these obtain merely passable performances. To sum up, the situation is that most approaches for tracking the target start only with one aspect of the two subtasks.

To mitigate this situation, Danelljan et al. [18] designed a tracker called ATOM that seamlessly combines the intersection-over-union (IoU) network [26] with the fast online classifier. Taking ATOM as baseline, Zhao et al. [27] proposed an adaptive feature fusion and obtained considerable improvements. Their method takes into account both state estimation and target discrimination, thereby achieving better accuracy and robustness. Afterwards, following the same state estimation component presented in ATOM [18], Bhat et al. [28] proposed an efficient discriminator which resorts to a target predictor employing an iterative optimization technique. Our work is motivated to bridge the gap between state estimation and target discrimination via deep network, and can be thought of as 3D counterpart of them. However, utilizing a deep network to exert the potentiality of the unstructured point cloud is still challenging in 3D tracking tasks. In this work, we present a unified framework to track the target in point cloud with a dedicated state estimation subnetwork (Section 3.2) and discrimination subnetwork (Section 3.3).

2.2. 3D Point Cloud Tracking

Point cloud is a prevalent trend for representing objects in the real 3D world. A number of advanced algorithms based on point clouds have been flourishing in object classification [29,30], detection [17,31,32], registration [19], and segmentation [33,34]. Nevertheless, point cloud tracking based on deep learning has been untapped. As we all know, many algorithms dedicated to RGB-D data have been widely studied [10,12,13,35], but most of them are mainly used to boost the 2D tracking methods with the depth channel and are not good at tackling the long-range scenario. Therefore, designing an effective pattern for tracking those partial point clouds is a very promising problem.

In the past few years, there has emerged some approaches to track the target in 3D spatial data [2,5,14,36–38]. For instance, Held et al. [36] used color-augmented search alignment algorithm to obtain the separated vehicle's velocity. Subsequently, combining shape, color, and motion information, Held et al. [37] utilized the dynamic Bayesian probabilistic model to explore the state space. However, these methods is bound to segmentation and data association algorithms. Different from them, Xiao et al. [39] simultaneously detect and track pedestrian using motion prior. All these traditional methods only obtain point segments instead of 3D orientated bounding boxes. Recently, some deep-learning-based methods infused new energy into point cloud tracking. For example, AVOD [14], FaF [5], and PIXOR [15] are designed for object detection based on BEV inputs, but can be applied tracking task in a tracking-by-detection manner. Specially for 3D tracking, Giancola et al. [2] introduced completion regularization to train a Siamese network. Subsequently, in light of the limitation of candidate box generation, Qi et al. [16] designed a point-to-box network, Zou [40] reduced redundant search space using a 3D frustum, and Fang et al. extended the region proposal network into pointNet++ [41] for 3D tracking. Nevertheless, all of above methods put more emphasis on distinguishing the target from a lot of proposals. We aimed to deal with both target discrimination and state estimation, with a dedicated Siamese network and extended LK algorithm.

2.3. *Jacobian Matrix Estimation*

Many tasks involve the estimation of the Jacobian matrix. As we know, the visual servoing field [42,43] usually relies on approximating the inverse Jacobian to control an agent favorably. In addition, for facial image alignment, Xiong et al. [44] proposed a supervised descent method (SDM), which avoids the calculation of the Jacobian and Hessian matrices with a sequence of learned descent directions based on hand-crafted feature. Lin et al. [45] proposed the conditional Lucas-Kanade algorithm to improve the SDM. Subsequently, Han et al. [46] dealt with the image-to-image alignment problem by jointly learning the feature representation for each pixel and partial derivatives. In this work, we innovatively estimate the Jacobian of extended LK algorithm in point cloud tracking.

3. Method

3.1. *Overview*

The proposed 3D point cloud tracking approach not only discriminates the target from distractors but also estimates the target state in a unified framework. Its pipeline is shown in Figure 1. Firstly, the template cropped from the reference frame and the current tracked frame are fed into the target discrimination subnetwork (TDS). It can select the best candidate in terms of the confidence score. Then, such selected candidate and the template are fed into the state estimation subnetwork (SES) to produce a incremental warp parameters $\Delta\rho$. These parameters are applied to the rough state of the best candidate, leading to a new state. Next, the warped point cloud extracted by the new state is sent into the SES again, producing $\Delta\rho$ together with the template. This procedure is implemented iteratively until the terminal condition is satisfied. We use the same feature backbone but train the TDS and SES separately. In Section 3.2, we first present our SES in detail, which extends the LK algorithm for 2D tracking to 3D point clouds. In Section 3.3, the powerful TDS is introduced. Finally, in Section 3.4, we describe an online tracking strategy that illustrates how two components cooperate with each other.

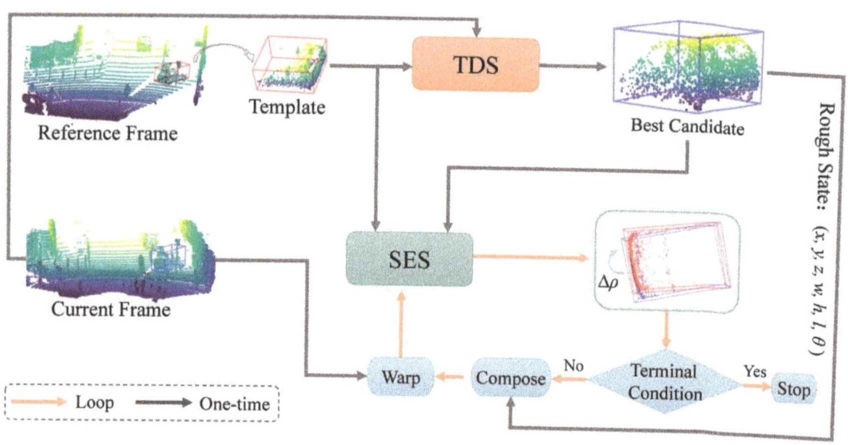

Figure 1. Overview of the proposed method for 3D point cloud tracking. During online tracking, the TDS first provides a rough state of the best candidate. Afterwards, provided with the template from the reference frame, the SES produces the incremental warp of the rough state. It is implemented iteratively until the terminal condition ($\|\Delta\rho\| < \epsilon$) is satisfied. The state estimation subnetwork (SES) and the target determination subnetwork (TDS) are separately trained using the KITTI tracking dataset.

3.2. *State Estimation Subnetwork*

Our state estimation subnetwork (SES) is designed to learn the incremental warp parameters between the template cropped from the first frame and candidate point clouds, so as to accommodate any motion variations. We took inspiration from DeepLK [19] and extended it to the 3D point cloud tracking task. To describe the state estimation subnetwork, we briefly revisit the inverse compositional (IC) LK algorithm [47] for 2D tracking.

The IC formulation is very ingenious and efficient because it avoids the repeated computation of the Jacobian on the warped source image. Given a template image T and a source image I, the essence of IC-LK is to solve the incremental warp parameters $\Delta\rho$ on T using sum-of-squared-error criterion. Therefore its objective function for one pixel $\mathbf{x}$ can be formulated as follows:

$$\min_{\Delta\rho} \| I(\mathbf{x}) - T(\mathcal{W}(\mathbf{x};\rho + \Delta\rho))\|_2^2, \tag{1}$$

where $\rho \in R^{D\times 1}$ are currently known state parameters, $\Delta\rho$ is the number of increments the state parameters are to go through, $\mathbf{x} = (x, y)^\top$ are the pixel coordinates, and $\mathcal{W}$ is the warp function. More concretely, if one considers the location shift and scale, i.e., $\rho = (\delta_x, \delta_y, \delta_s)^\top$, the warp function can be written as $\mathcal{W}(\mathbf{x};\rho) = (\delta_s x + \delta_x, \delta_s y + \delta_y)^\top \in R^{2\times 1}$. Using the first-order Taylor expansion at the identity warp ρ_0, the Equation (1) can be rewritten as

$$\min_{\Delta\rho} \left\| I(\mathbf{x}) - T(\mathcal{W}(\mathbf{x};\rho_0)) - \nabla T \frac{\partial \mathcal{W}(\mathbf{x};\rho_0)}{\partial \rho} \Delta\rho \right\|_2^2, \tag{2}$$

where $\mathcal{W}(\mathbf{x}, \rho_0) = \mathbf{x}$ is the identity mapping and $\nabla T = \left(\frac{\partial T}{\partial x}, \frac{\partial T}{\partial y}\right) \in R^{1\times 2}$ represents the image gradients. Let the Jacobian $J = \nabla T \frac{\partial \mathcal{W}(\mathbf{x};\rho_0)}{\partial \rho^\top} \in R^{1\times D}$. We hence can obtain $\Delta\rho$ by minimizing the above Equation (2); namely,

$$\Delta\rho = (J^\top J)^{-1} J^\top [I(\mathbf{x}) - T(\mathcal{W}(\mathbf{x};\rho_0))]. \tag{3}$$

Compared with 2D visual tracking, 3D point cloud tracking has an unstructured data representation and high-dimensional search space for state parameters. Let $P_T \in R^{3\times N}$ denote the template point cloud. $P_I \in R^{3\times N}$ denotes the source point cloud in the tracked frames, which is extracted by a bounding box with inaccurate center and orientation. Note that we set the quantities of both P_T and P_I to N, and when their totals of points are less than N, we repeat sampling from existing points. In this work, we treat the deep network $\phi : R^{3\times N} \mapsto R^{K\times 1}$ as a learnable "image" function. In light of this, the template point cloud P_T and the source point cloud P_I can obtain their descriptors using the network ϕ after transforming them into the canonical coordinate system. In addition, we regard the rigid transformation $G \in R^{3\times 4}$ between P_T and P_I as the "warp" function. In this way, we can apply the philosophy of IC-LK to the 3D point cloud tracking problem. In practice, the 3D bounding box is usually utilized to represent the target state which can be parametrized by $\mathcal{S} = (x, y, z, h, w, l, \theta)$ in the LiDAR system, as shown in Figure 2. Therein, (x, y, z) is the target center coordinate, (h, w, l) represents the target size, and θ is the rotation angle around the y-axis. Due to the target size remaining almost unchanged in 3D spatial space, it is sufficient to focus only on the state variations in the angle and x, y, and z axes. Consequently, the transformation G will be represented by four warping parameters $\rho = (x, y, z, \theta)^\top$. More concretely, it can be simplified as follows:

$$G(\rho) = \begin{pmatrix} \cos(\theta) & 0 & -\sin(\theta) & x \\ 0 & 1 & 0 & y \\ \sin(\theta) & 0 & \cos(\theta) & z \end{pmatrix}. \tag{4}$$

Now the state estimation problem in 3D tracking can be transformed to find $G(\rho)$ satisfying $\phi(G(\rho) \circ P_T) = \phi(P_I)$, where $(\circ)$ is the warp operation on the homogeneous coordinate with $G(\rho)$. Being analogous to the aforementioned IC-LK in Equation (2), the objective of state estimation can be written as

$$\min_{\Delta\rho} \left\| \phi(P_I) - \phi(P_T) - \frac{\partial \phi(G(\rho_0) \circ P_T)}{\partial \rho} \Delta\rho \right\|_2^2. \tag{5}$$

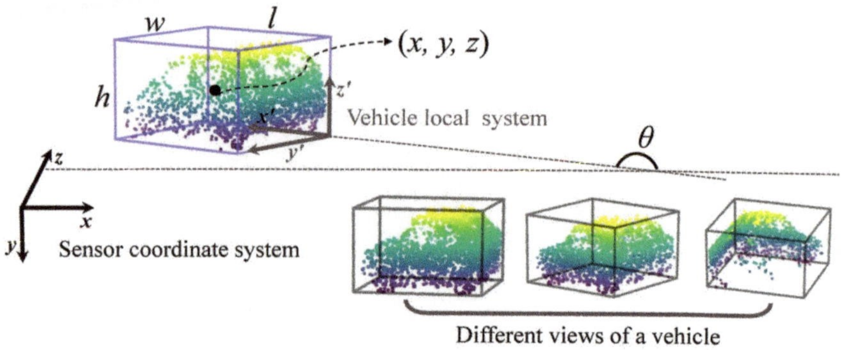

Figure 2. Object state representation in the sensor coordinate system. An object can be encompassed with a 3D bounding box (blue). Therein, (x, y, z) represents the object center location in the LiDAR coordinate system. (h, w, l) are the height, width, and length of object, respectively. θ is the radian between the motion direction and x-axis. The right-bottom also exhibits different views of object point clouds produced by LiDAR.

Similar to the Equation (2), we could solve the incremental warp $\Delta\rho = (\hat{J}^\top \hat{J})^{-1}\hat{J}^\top[\phi(P_I) - \phi(P_T)]$ with the Jacobian matrix $\hat{J} = \partial\phi(G(\rho_0) \circ P_T)/\partial\rho$. Unfortunately, this Jacobian matrix cannot be calculated like the classical image manner. The core obstacle is that the gradients in x, y, and z cannot be calculated in the scattered point clouds due to the lack of connections among points or another regular convolution structure.

We introduce two solutions to circumvent this problem. One direct solution is to approximate the Jacobian matrix through a finite difference gradient [48]. Each column of the Jacobian matrix $\hat{J}$ can be computed as

$$\hat{J}_i = \frac{\phi(G_i \circ P_T) - \phi(P_T)}{\mu_i} \tag{6}$$

where μ_i are infinitesimal perturbations of the warp parameters $\Delta\rho$, and G_i is a transformation involving only one of the warp parameters. (In other words, only the i-th warp parameter has a non-zero value μ_i. Please refer to Appendix A for details).

On the other hand, we treat the construction of $\hat{J}$ as a non-linear function $\mathcal{F}$ with respect to $\phi(P_T)$. We hence propose an alternative: to learn the Jacobian matrix using a multi-layer perceptron, which consists of three fully-connected layers and ReLU activation functions (Figure 3). In Section 4.4, we report the comparison experiments.

Based on the above extension, we can analogously solve the incremental warp $\Delta\rho$ of the 3D point cloud in terms of Equation (3) as follows:

$$\Delta\rho = J^\dagger[\phi(P_I) - \phi(P_T)], \tag{7}$$

where $J^\dagger = (\hat{J}^\top \hat{J})^{-1}\hat{J}^\top$ is a Moore–Penrose inverse of $\hat{J}$. Afterwards, the source point cloud cropped from the coming frame can adjust its state by the following formula

$$\mathcal{S}_I \leftarrow \texttt{Compose}(\mathcal{S}_I, \Delta\rho), \tag{8}$$

where `Compose` is the inverse compositional function and $\mathcal{S}_I$ is the state representation of the source point cloud P_I.

Network Architecture. Figure 3 summarizes the architecture of the state estimation subnetwork. Owing to the inherent complexity of the state estimation, it is non-trivial to train a powerful estimator on the fly under the sole supervision of the first point cloud scenario. We hence train the SES offline to learn general properties for predicting the incremental warp. It is natural that we opt to adopt a Siamese architecture for producing the incremental warp parameters between the template and candidate. In particular, our network contains two branches sharing the same feature backbone, each of which consists of two blocks. As shown in Figure 3, Block-1 first generates the global descriptor. Then Block-2 consumes the aggregation of the global descriptor and the point-wise features to generate the final K-dimensional descriptor, based on which the Jacobian matrix $\hat{J}$ can be calculated. Finally, the LK module jointly considers $\phi(P_I)$, $\phi(P_T)$, and $\hat{J}$ to predict $\Delta\rho$. It is notable that this module theoretically provides the fusion strategy, namely, $\phi(P_I) - \phi(P_T)$, between two features produced by the Siamese network. Moreover, we adopt the conditional LK loss [19] to train this subnetwork in an end-to-end manner. It is formulated as

$$L_{ses} = \frac{1}{M}\sum_{m} \mathcal{L}_1\left(J^{+(m)}[\phi(P_I^{(m)}) - \phi(P_T^{(m)})], \Delta\rho_{gt}^{(m)}\right), \tag{9}$$

where $\Delta\rho_{gt}$ is the ground-truth warp parameter, $\mathcal{L}_1$ is the smooth L_1 function [49], and M is the number of paired point clouds in a mini-batch. This loss can propagate back to update the network when the derivative of the batch inverse matrix is implemented.

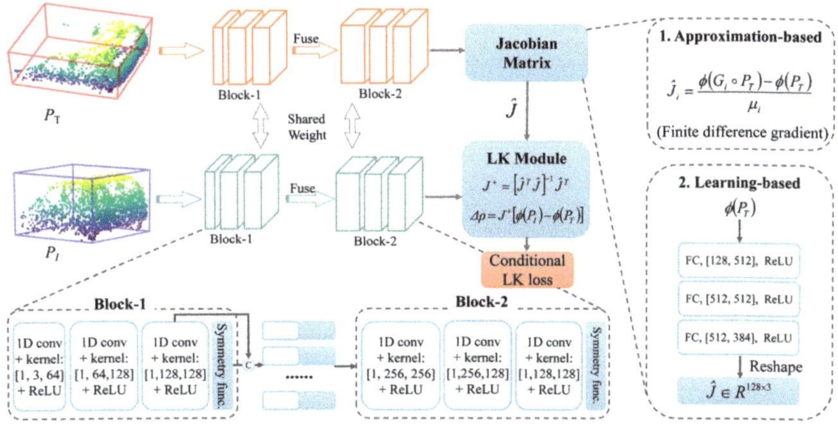

Figure 3. Illustration of the proposed state estimation subnetwork (SES). Firstly, the SES extracts the shape descriptors of the paired point clouds using the designed architecture. Its details are shown in the bottom dashed box. Subsequently, the Jacobian matrix is computed by one of the solutions: approximation-based or learning-based. Its details are shown in the right dashed box. Finally, the LK module generates the incremental warp parameters $\Delta\rho$.

3.3. *Target Discrimination Subnetwork*

In the 3D search space, how to efficiently determine the presence of the target is very critical for an agent to conduct state estimation. In this section, considering that the SES lacks discrimination ability, we present the design of a target discrimination subnetwork (TDS) to realize a strong alliance with the SES. It aims to distinguish the best candidate from distractors, thereby providing a rough target state. We leverage the matching function based method [2] to track the target. Generally, its model can be written as

$$\Psi(P_T, P_I) = g(\psi(P_T), \psi(P_I)), \tag{10}$$

where Ψ is the confidence score function, $\psi : R^{3\times N} \mapsto R^{K\times 1}$ is the feature extractor, and g is a similarity metric. Under this framework, the candidate with the highest score is selected as the target.

In this work, to equip the model with global semantic information, we incorporate the intermediate features generated by the first block to point-wise features, and then pass them to the second block, as shown in Figure 4. Consequently, our TDS could project 3D partial shapes into a more discriminatory latent space, which allows an agent to distinguish the target more accurately from distractors.

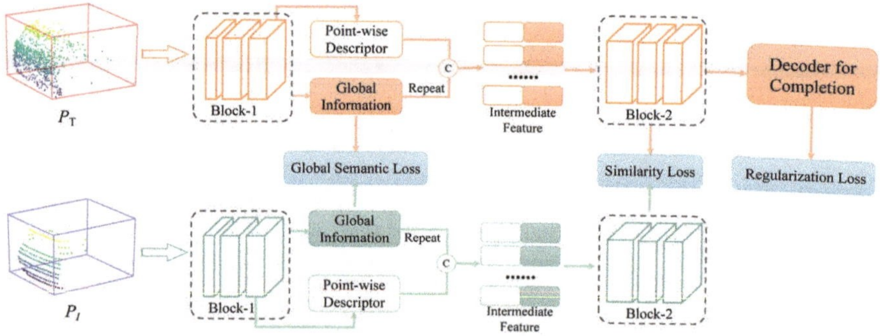

Figure 4. The scheme of our TDS. Its feature backbone is composed of two blocks as same as the SES. Particularly, the global feature generated from Block-1 is repeated and concatenated with each point-wise feature. Afterwards, the intermediate aggregation feature is further fed into Block-2. Finally, we use the combination of similarity loss, global semantic loss, and regularization completion loss to train the TDS.

Network Architecture. We trained the TDS offline from scratch with an annotated KITTI dataset. Based on the Siamese network, the TDS takes paired point clouds as inputs and directly produces their similarity scores. Specifically, its feature extractor also consists of two blocks the same as in the SES. As can be seen in Figure 4, Block-1 ψ_1 generates the global descriptor, and Block-2 ψ_2 utilizes the aggregation of the global point-wise features to generate the more discriminative descriptor. As for the similarity metric g, we conservatively utilize hand-crafted cosine function. Finally, the similarity loss, global semantic loss, and regularization completion loss are combined in order to train this subnetwork; i.e.,

$$L_{tds} = \mathcal{L}_2(g(\psi_2(P_T), \psi_2(P_I)), s) + \lambda_1 \mathcal{L}_2(g(\psi_1(P_T), \psi_1(P_T)), s) + \mathcal{L}_3(\hat{P_T}, P_T), \quad (11)$$

where $\mathcal{L}_2$ is mean square error loss, s is the ground-truth score, λ_1 is the balance factor, and $\mathcal{L}_3$ is the completion loss for regularization [2], where $\hat{P_T}$ represents each template point cloud predicted via shape completion network [2].

3.4. Online Tracking

Once trained offline, the two subnetworks can be combined for online tracking. We denote our whole framework SETD. For one scenario (frame) $F_t, t \in \{1, 2, \cdots, Q\}$, where Q is the total number of frames, we first sampled a collection of candidates C_t in terms of Kalman filter as SiamTrack3D does, and then passed them into the TDS, which provided the rough state representation $\mathcal{S}_t^{(0)}$ of the optimal candidate. Afterwards, it was adjusted iteratively by the SES until the termination condition ϵ was satisfied, and the best state $\mathcal{S}_t$ was thus determined. Finally, the template point cloud $P_T^{(0)}$ was updated by appending the selected candidate to itself. Algorithm 1 shows the whole process in detail.

Algorithm 1: SETD online tracking.

Input: Frames:$\{F_t | t = 1, \cdots, Q\}$, Template: $P_T^{(0)}$
Output: Target state: $\{\mathcal{S}_t\}$
for $t = 1, \cdots, Q$ **do**
  $\mathcal{C}_t \leftarrow \texttt{Sample}(F_t)$
  $\mathcal{S}_t^{(0)} \leftarrow \texttt{TDS}(\mathcal{C}_t, P_T^{(t-1)})$
  $P_I^{(0)} \leftarrow \texttt{Crop}(F_t, \mathcal{S}_t^{(0)})$
  for $i = 1, \cdots, N_{iter}$ **do**
    $\Delta\rho \leftarrow \texttt{SES}(P_I^{(i-1)}, P_T^{(t-1)})$
    if $\|\Delta\rho\| \leq \epsilon$ **then**
      Break
    else
      $\mathcal{S}_t^{(i)} \leftarrow \texttt{Compose}(\mathcal{S}_t^{(i-1)}, \Delta\rho)$
      $P_I^{(i)} \leftarrow \texttt{Crop}(F_t, \mathcal{S}_t^{(i)})$
    end
  end
  $\mathcal{S}_t \leftarrow \texttt{BestState}(\{\mathcal{S}_t^{(i)}\})$
  $P_T^{(t)} \leftarrow \texttt{Update}(P_T^{(t-1)}, \texttt{Crop}(F_t, \mathcal{S}_t))$
end
*Crop means getting the points inside the 3D bounding box parameterized by $\mathcal{S}$.

4. Experiments

KITTI [50] is a prevalent dataset of outdoor LiDAR point clouds. Its training set of contains 21 scenes (over 27,000 frames), and each frame has about 1.2 million points. For a fair comparison, we followed [2] to divide this dataset into a training set (scene 0–16), a validation set (scene 17–18), and a testing set (scene 19–20). In addition, to validate the effectiveness of different trackers, we also evaluated them on another large-scale point cloud dataset—PandaSet [51]. It covers complex driving scenarios, including lighting conditions at day time and night, steep hills, and dense traffic. In this dataset, more than 25 scenes were collected for testing, and the tracked instances are split into three levels (easy, middle, and hard) according to the LiDAR range.

4.1. Evaluation Metrics

To evaluate the tracking results, we adopted one-pass evaluation (OPE) [21] based on the location error and the overlap. The overlap represents the intersection-over-union (IoU) between the predicted bounding box B_P and the corresponding ground-truth bounding box B_G, i.e., $\texttt{volume}(B_T \cap B_G)/\texttt{volume}(B_T \cup B_G)$. The location error measures the Euclidean distance between the centers of B_G and B_T. In this paper, the success and precision metrics are utilized as evaluation metrics. Specifically, the success metric is defined as the area-under-curve (AUC) where the x-axis denotes the overlap threshold ranging from 0 to 1 and the y-axis refers to the ratio above the threshold. The precision metric is defined as the AUC where the x-axis represents the location error threshold ranging from 0 to 2 meters and the y-axis is the ratio below this threshold.

4.2. Implementation Details

Training. We conducted experiments with PyTorch and Python 3.7 on a PC equipped with a GTX 2080Ti, 32 GB RAM, and 4.00 GHz Intel Core i7-4790K CPU. When training our SES, we first sampled a pair of target shapes (template and source point clouds) from the same sequence. Additionally, these two point clouds were transformed into a canonical coordinate system according to the respective bounding box. Assuming that the target motion obeyed the Gaussian distribution, we randomly produced the warp parameters

$\Delta\rho_{gt}$ and applied them to the source point clouds for the supervised learning. In practice, only when the IoU between the warped bounding box and its corresponding ground truth is larger than 0.1 can this paired data be fed into the SES. The mean of Gaussian distribution was set to zero, and the covariance was a diagonal matrix $\texttt{diag}(0.5, 0.5, 5.0)$. The dimension K of shape descriptor generated by ϕ was set to 128. The network was trained from scratch using the Adam optimizer with the batch size of 32 and the initial learning rate of 1×10^{-3}.

Regarding our TDS, the input data were the paired point clouds transformed into a canonical coordinate system. The outputs were similarity scores. The ground-truth score is the soft distance obtained by the Gaussian function. The output dimensions K of ψ were set to 128. Our proposed loss (Equation (11)) was utilized to train it from scratch using an Adam optimizer. The batch size and initial learning rate were set to 32 and 1×10^{-3}, respectively. As for λ_1, we reported its performance using several metrics in Section 4.4. The learning rates of both subnetworks were reduced via multiplying by a ratio of 0.1 when the loss of the validation set reached a plateau, and the maximum number of epochs was set to 40.

Testing. During the online testing phase, the tracked vehicle instance was usually specified in the first frame. When dealing with a coming frame, we exhaustively drew a set of 3D candidate boxes $\mathcal{C}_t$ over the search space [2]. The number of $\mathcal{C}_t$ was set to 125. The number of iterations N_{iter} was set to 2, and the termination parameter ϵ was set to 1×10^{-5}. Besides, for each frame, our SETD tracker only took about 120 ms of GPU time (50 ms for the TDS and 70 ms for the SES) to determine the final state. We did not take into account the time cost of the generation and normalization of the template and candidates, which was 300 ms of CPU time, approximately.

4.3. *Performance Comparison*

We first compare the proposed SETD with the baseline in relation to several attributes, such as dynamics and occlusion. Then, we evaluate the tracking performances of recent related trackers on large-scale datasets.

4.3.1. Comparison with Baseline

SiamTrack3D [2] was the first method made to deal with this special task using the Siamese network and gives some referential insights via adequate ablation studies. It is a strong baseline for state-of-the-art tracking performance. We first show visualization comparison results for different attributes in Figures 5–7, and then report quantitative comparison results in Table 1.

Figure 5 shows a visualization of density variation. As can be seen, "Object-1" changed from dense to sparse, and "Object-2" varied from sparse to dense. For all these scenes, our method tracked the target accurately, whereas SiamTrack3D exhibited skewing. Figure 6 presents some tracking results for dynamic scenes. Generally, the scene is treated as dynamic when the center distance between consecutive frames is larger than 0.709 [2]. As shown in Figure 6, our method performed better than SiamTrack3D when the target moved quickly, which is attributed to the seamless integration of target discrimination and state estimation. In particular, even though the target was partly occluded and moving at high speed in the scenario presented in the second row, SETD obtained satisfying results, whereas SiamTrack3D produced greatly varying results. Figure 7 plots the tracking results from when the target suffered from different degrees of occlusion. As can be seen, whether the target was visible, partly occluded, or largely occluded, our SETD performed better than SiamTrack3D.

Table 1. Performance comparison with the baseline [2] in terms of several attributes.

Attribute	SiamTrack3D		SETD	
	Success (%)	Precision (%)	Success (%)	Precision (%)
Visible	37.38	55.14	53.87	68.75
Occluded	42.45	55.90	53.76	70.33
Static	38.01	53.37	54.55	70.10
Dynamic	40.78	58.42	48.46	66.34

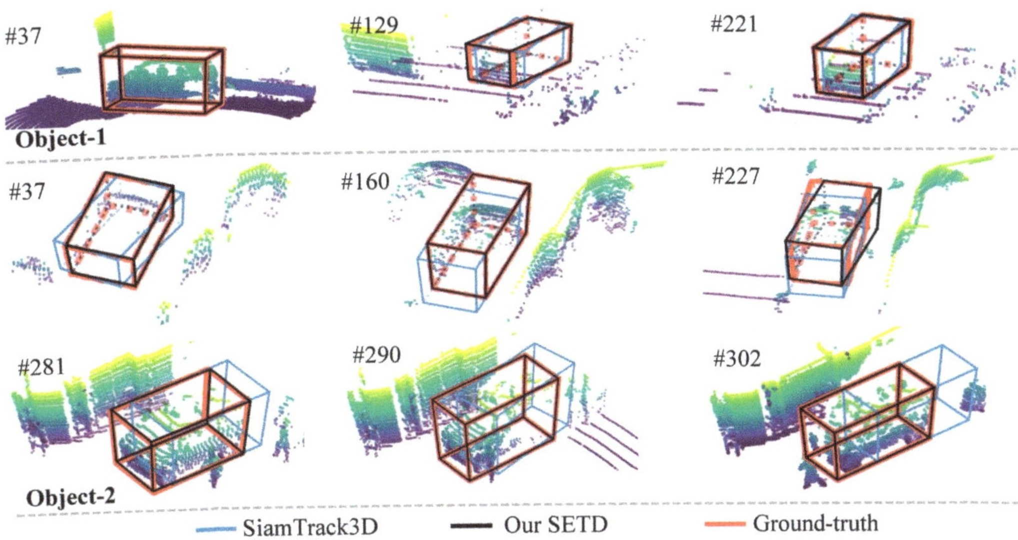

Figure 5. Visual results on density change. We exhibit some key frames of two different objects. Compared with SiamTrack3D (**blue**), our SETD (**black**) has a larger overlap with the ground truth (**red**). The number after # refers to the frame ID.

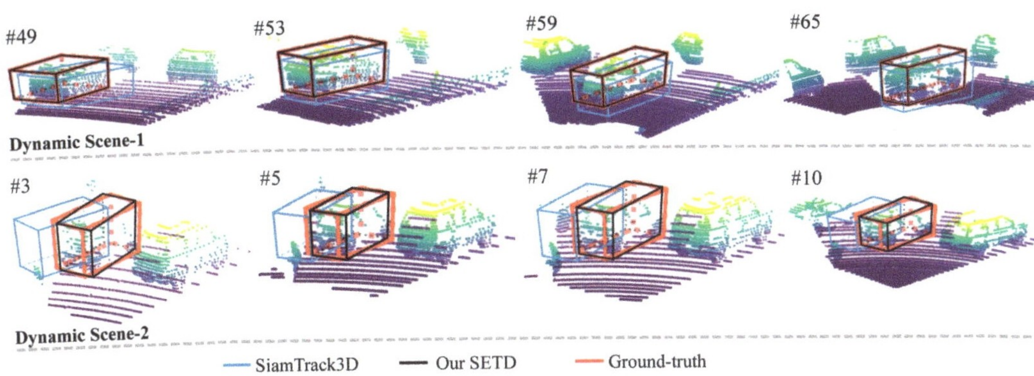

Figure 6. Visual results in terms of dynamics. We show two dynamic scenes. When the target ran at a high speed (dynamic), our SETD obtained better results, whereas SiamTrack3D resulted in significant skewing.

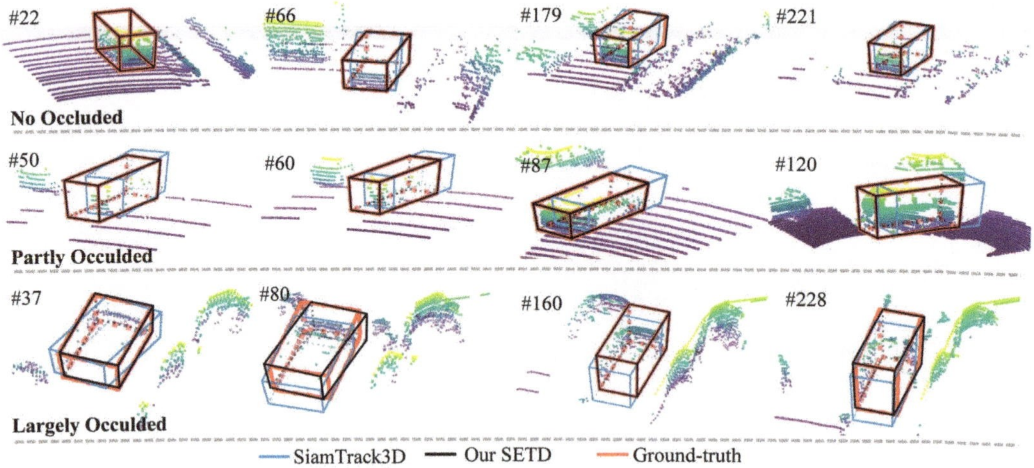

Figure 7. Visual results in terms of occlusion. The first row shows the results of a visible vehicle. The second row shows partly occluded vehicles. The last row is a largely occluded vehicle. Our SETD performed better than SiamTrack3D in all three degrees of occlusion.

In addition, we quantitatively compared their performances according to four conditions: visible, occluded, dynamic, and static. The success and precision metrics for said attributes are shown in Table 1. Overall, the proposed SETD fully outperformed SiamTrack3D. Specifically, compared with SiamTrack3D, SETD not only significantly improved detection by 16.45%/13.61% (success/precision) in visible scenes, but also 11.31%/14.43% in occluded scenes. Meanwhile, as shown in the last two rows of Table 1, SETD also achieved great improvements of 7.68%/7.92% when detecting dynamic scenes, and 16.54%/16.73% when detecting static scenes. These significant improvements on four types of scenes thoroughly and powerfully demonstrate the effectiveness of learning incremental warp for accurate 3D point cloud tracking.

4.3.2. Comparison with Recent Methods

Apart from SiamTrack3D, we compared with other methods—AVODTrack [14], P2B [16], SiamTrack3D-RPN [52], and ICP&TDS—on the testing set. AVODTrack is a tracking-by-detection method which evolved from an advanced 3D detector, AVOD [14], by equipping it with an online association algorithm. To be more precise, it consumes point cloud BEVs and RGB images to generate a 3D detection box for every frame; then the final box is the one that has the highest IoU with the previous bounding box. P2B is a new, advanced method which integrates the target feature augmentation module into deep Hough voting [17]. SiamTrack3D-RPN evolved from SiamTrack3D by jointly learning on 2D BEV images and 3D point clouds. Furthermore, to estimate the state of moving vehicles, one may intend to obtain the motion transformation using the iterative closest point (ICP) [53], and then apply this transformation to previous bounding box. In this work, we first leveraged the proposed TDS to obtain a candidate point cloud and then run the ICP algorithm between the template and this candidate in a canonical coordinate system. This method is called ICP&TDS. In SiamTrack3D, as in the Kalman filter, the ground truth of the tracked frame is also utilized to approximate dense sampling for further testing of a tracker's discrimination ability. It applies grid search centered at the tracked ground truth. We also considered this sampling strategy for comprehensive evaluation. Note that a method with the suffix "Dense" means that it adopts this dense sampling.

Table 2 summarizes the above methods' performances on KITTI. In addition to conducting OPE of the 3D bounding box, we also present the results of the 2D BEV box, which was obtained by projecting the 3D box onto a rectangle from a bird's-eye view. As shown in this table, SETD-Dense is superior to all other methods, given its high success and precision metrics. Specifically, the success and precision metrics of our SETD constituted 13.68% and 13.48% improvements compared with SiamTrack3D, and our SETD-Dense provided 5.12% and 6.77% improvements over SiamTrack3D-Dense. This demonstrates the validity of bridging the gap between state estimation and target discrimination. ICP&TDS and ICP&TDS-Dense obtained poor performances in these specific outdoor scenes. We deem that ICP lacks strength for partial scanned point clouds. This also proves that the proposed SES plays a critical role in the point cloud tracking task. In addition, even when using multiple modalities of RGB images and LiDAR point clouds, AVODTrack was inferior to the dense sampling models (SiamTrack3D-Dense and SETD-Dense) by large margins. P2B obtained better performances than SiamTrack3D and SETD, because P2B uses a learning procedure based on deep Hough voting to generate high quality candidates, whereas SiamTrack3D and SETD only use traditional Kalman filter sampling. Hence, better candidate generation is important for the following tracking process, and we recon that integrating a learning-based candidate generation strategy into SiamTrack3D and SETD will facilitate improving their accuracy.

Table 2. Performance comparison with the state-of-the-art methods on KITTI. The OPE evaluations of 3D bounding boxes and 2D BEV boxes are reported.

Method	3D Bounding Box		2D BEV Box	
	Success (%)	Precision (%)	Success (%)	Precision (%)
SiamTrack3D-RPN	36.30	51.00	-	-
AVODTrack	63.16	69.74	67.46	69.74
P2B	56.20	72.80	-	-
SiamTrack3D	40.09	56.17	48.89	60.13
SiamTrack3D-Dense	76.94	81.38	76.86	81.37
ICP&TDS	15.55	20.19	17.08	20.60
ICP&TDS-Dense	51.07	64.82	51.07	64.82
SETD	53.77	69.65	61.14	71.56
SETD-Dense	81.98	88.14	81.98	88.14

Table 3 reports the tracking results on PandaSet. We compare the proposed method with two advanced open-source trackers: P2B and SiamTrack3D. Their performances were obtained by running their official code on our PC. As shown in the Table 3, our SETD performed considerably better than P2B in all easy, middle, and hard sets, especially in obtaining the success/precision improvements of 6.77/9.34% with the middle set. When compared with SiamTrack3D, SETD also outperformed it by a large margin on easy and middle sets. Nevertheless, on the hard set, SETD was inferior to SiamTrack3D. The reasons were that: (1) there exist some extremely sparse objects in the hard set, which makes the SES product a worse warp parameter, (2) SiamTrack3D has a better prior because it is first trained on ShapeNet and then fine-tuned on KITTI, whereas SETD is trained only from scratch.

Table 3. Performance comparison with the state-of-the-art methods on PandaSet. The results on three sets of different difficulty levels are reported.

Method	Easy		Middle		Hard	
	Success (%)	Precision (%)	Success (%)	Precision (%)	Success (%)	Precision (%)
P2B	53.49	59.97	35.76	40.56	19.13	19.64
SiamTrack3D	51.61	62.09	40.55	49.73	25.09	30.11
SETD	54.34	65.12	42.53	49.90	24.39	28.60

4.4. Ablation Studies

We carried out five self-contrast experiments to demonstrate the necessity of each part.

SES and TDS. In order to prove the effectiveness of the combination of the state estimation and the target discrimination, we examined the tracking performance only using TDS or SES. TDS-only tracks the target merely via the target discrimination subnetwork, which selects a candidate bounding box with the highest confidence score. Based on the state estimation subnetwork, SES-only directly rectifies the estimated bounding box of the previous frame for tracking the target. Our SETD properly combines these two components to make up for their performance gap.

The results are shown in Table 4. According to the success metric, our SETD achieved 10.41% and 11.28% improvements in comparison with SES-only and TDS-only, respectively. As for the precision metric, SETD (69.65%) also significantly surpassed both SES-only (49.70%) and TDS-only (60.19%). These improvements of SETD highlight the importance of combining these two components. In addition, we observed that SES-only performed worse than TDS-only according to the success and precision metrics. The main reason lies in that (1) TDS-only selects the best one of many candidates generated by the Kalman filter, but SES-only directly uses the previous result while lacking discrimination; (2) the previous result used by SES often drifts due to self-occlusion and density variations, leading to far-fetched warp parameters. This also proves our observation mentioned in Section 3.3: that determining the presence of the target is crucial for an agent to conduct state estimation.

Table 4. Self-contrast experiments evaluated by the success and precision ratio. SETD achieved the best performance.

Variants	TDS	SES	Iter.	Success (%)	Precision (%)
TDS-only	√			43.36	60.19
SES-only		√		41.09	49.70
Iter-non	√	√		44.67	59.13
SETD	√	√	√	53.77	69.65

Moreover, Figure 8 presents some tracking results obtained without or with our state estimation subnetwork. As can be seen when going through the state estimation subnetwork, some inaccurate results (blue boxes), which were predicted solely via a target discrimination subnetwork, can be adjusted towards the corresponding ground truth.

Iteration or not. We also investigated the effect of iteratively adjusting the target state. Specifically, we designed a variant model named Iter-non, which does not apply the iterative online tracking strategy (Algorithm 1). In other words, it directly uses the first prediction of the SES as the final state increment. As shown in Table 4, Iter-non obtained a 44.67% success ratio and a 59.13% precision ratio on the KITTI tracking dataset. Compared with SETD (53.77%/69.65%), Iter-non fell short by 9% in success and precision metrics, which proves the effectiveness of our iterative online tracking strategy. In fact, the iteration process is an explicit cascaded regression that is more effective and verifiable for tasks solved in continuous solution spaces [19,48,54].

Jacobian Approximation or Learning. Two solutions have been provided to tackle the Jacobi matrix issue that occurred in the SES. We can approximate it via finite difference gradient or learn it using a multi-layer perceptron. To comprehensively compare these two solutions, we plotted their loss curves during the training phase in addition to reporting the success and precision metrics on the testing set. As shown in Figure 9, the learning-based solution had a far lower cost and flatter trend than the approximation-based one. Moreover, the last row of Table 5 shows the approximation-based solution achieved success/precision of 49.93%/67.15%; the learning-based solution reached 53.77%/69.65%. The reason may be that a teachable Jacobian module could be coupled with the shape descriptor $\phi(P_T)$, whereas the finite difference gradient defined by a hand-crafted formula is a hard constraint. Please refer to Appendix B for more details.

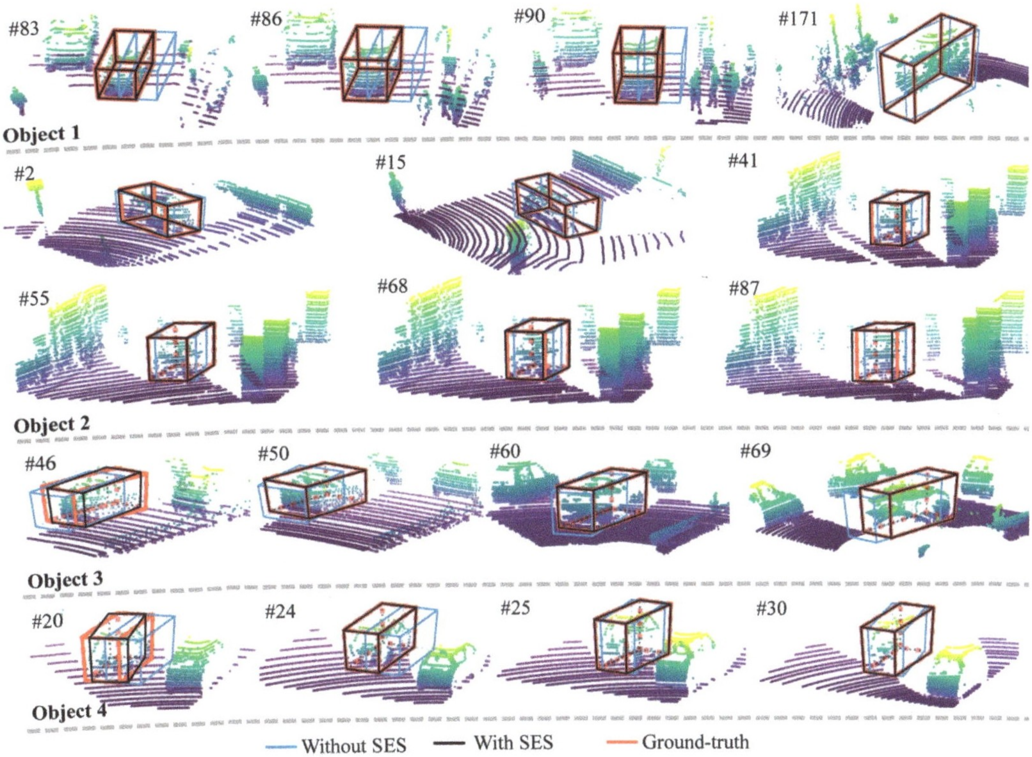

Figure 8. Tracking results with or without the state estimation subnetwork (SES). The black bounding boxes were obtained with SES, and the blue bounding boxes without SES. As we can see, with the help of SES, a rough state (blue boxes) can be favorably meliorated. The number after # is the frame ID.

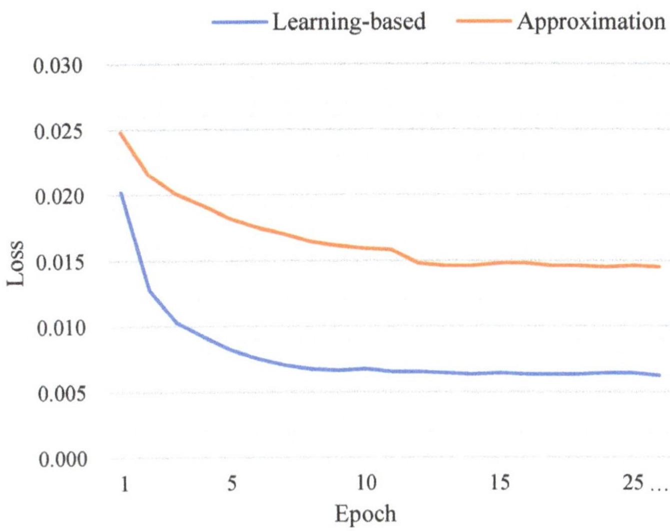

Figure 9. Loss curves during the training phase, where the blue and red curves correspond to the learning-based solution and the approximation-based solution, respectively. Obviously, the former had a low cost and fast convergence.

Descriptor Using Block-1 or Block-2. A cascaded network architecture is proposed for the 3D point cloud tracking problem. We explored the impact of using descriptors generated by different feature blocks when extending the traditional LK algorithm to the 3D point cloud tracking task. Each column of Table 5 shows that the descriptor from Block-2 is superior that from Block-1. This benefits from the novelty that we incorporate the global semantic information into point-wise features.

Table 5. Performance comparison between models using different solutions for the Jacobian problem. Each row shows results using a different feature block.

Descriptor	Learning-Based		Approximation-Based	
	Success (%)	Precision (%)	Success (%)	Precision (%)
Block-1	51.29	66.59	46.33	61.51
Block-2	53.77	69.65	49.93	67.15

Key Parameter Analysis. In Section 3.3, in order to robustly determine the presence of the target in a point cloud scenario, we proposed a new loss that combines similarity loss, global semantic loss, and regularization completion loss. Therein, the parameter λ_1 in Equation (11) plays a key role in the global information trade-off. In Figure 10, we compared different values of λ_1. As we can see, it obtained the best performance in success and precision metrics when $\lambda_1 = 1 \times 10^{-4}$.

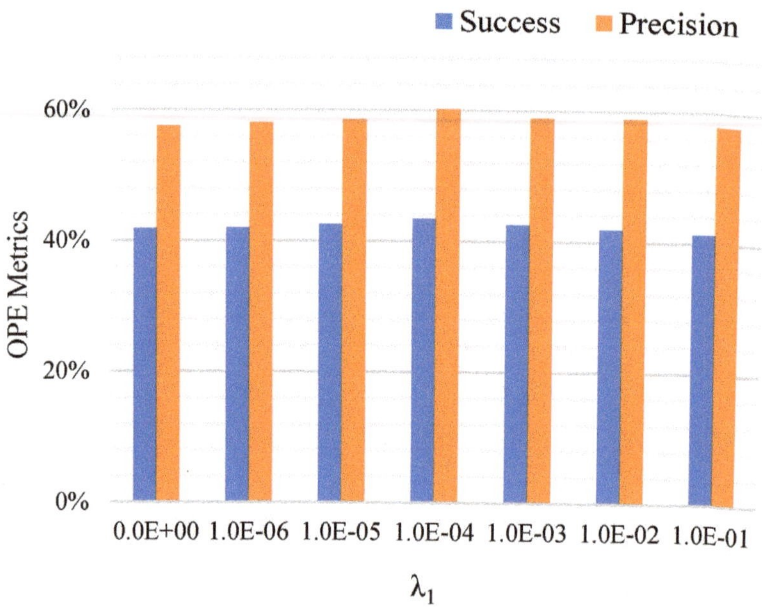

Figure 10. Influence of the parameter λ_1. The OPE success and precision metrics for different values of λ_1 are reported.

4.5. Failure Cases

Figure 11 shows some failure cases of our proposed model. In this figure, "Object 1" (the first row) and "Object 2" (the second row) could not be tracked accurately by our SETD (black). The reason is that the extremely spare points could not extract an explicit pattern to discriminate target or estimate state. "Object 3" (the last row) drifted to similar distractors surrounding the target. This is because the previous bounding box, when applied to the current frame, covered similar adjacent objects due to its very fast movement.

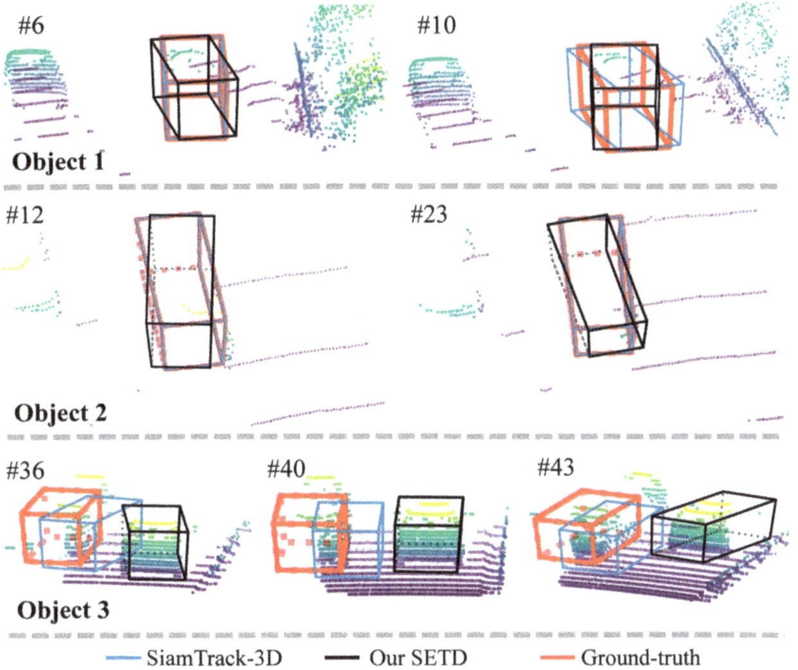

Figure 11. Failure cases of our proposed method. The first and second rows show failure to track the target due to extremely sparse point clouds. The last row shows failure due to similar distractors. The number after # is the frame ID.

5. Conclusions and Future Work

This paper presents a 3D point cloud tracking framework bridging the gap between state estimation and target discrimination subnetworks. Particularly, the traditional LK algorithm has been creatively extended to the case of 3D tracking for accurate state estimation. Meanwhile, a new loss method was proposed in the hopes of providing more powerful target discrimination. Experiments on the KITTI and PandaSet datasets have shown our method significantly outperforms others. Last but not least, the ablation studies fully demonstrated the effectiveness of each part and gave some key analyses of the descriptor, iteration strategy, and Jacobian matrix calculation.

SiamTrack3D [2] is the first point cloud tracker based on the Siamese network. This method starts with state estimation and target discrimination (inspired by 2D tracker [18,19]), and extends them to 3D point cloud tracking. Although achieving promising performance, it has huge room for improvement. For example, both SiamTrack3D and SETD are struggling with the proposal extraction issue. To be specific, they obtain proposals via Kalman filter or a dense sampling strategy. In light of this, in the future, it will be very important to explore an efficient proposal extraction algorithm. Despite the recent literature [52] providing better proposals via BEV, the joint learning on 2D BEV and 3D point cloud Siamese networks even drops the final discrimination ability. Besides, a new feature backbone [41,55] is also worth studying instead of using pointNet, which is used alone in SiamTrack3D. Last but not least, it will be important to study an end-to-end network, including the flow embedding layer [56], proposal generation, similarity metric, and state refinement.

Author Contributions: Conceptualization, S.T. and X.L.; methodology, S.T. and X.L.; validation, S.T.; formal analysis, S.T., M.L. and X.L.; investigation, S.T. and Y.B.; writing—original draft preparation, S.T. and M.L.; writing—review and editing, M.L. and J.G.; visualization, S.T. and Y.B.; supervision, X.L. and B.Y.; project administration, S.T. All authors have read and agreed to the published version of the manuscript.

Funding: This work was funded by the National Natural Science Foundation of China (grant number 61976040 and U1811463) and the National Key Research and Development Program of China (grant number 2020YFB1708902).

Institutional Review Board Statement: Not applicable.

Informed Consent Statement: Not applicable.

Data Availability Statement: The KITTI tracking dataset is available at http://www.cvlibs.net/download.php?file=data_tracking_velodyne.zip; accessed on 24 June 2021. The PandaSet dataset is available at https://scale.com/resources/download/pandaset; accessed on 24 June 2021. For the reported results, one can obtain it by request to corresponding author.

Conflicts of Interest: The authors declare no conflict of interest.

Appendix A. Details of Approximation-Based Solution

After extending the LK algoritm [19] designed for 2D visual tracking to 3D point cloud tracking task, we have the following objective

$$\min_{\Delta\rho \in R^3} \left\| \phi(P_I) - \phi(P_T) - \frac{\partial \phi(G(\rho_0) \circ P_T)}{\partial \rho} \Delta\rho \right\|_2^2. \quad \text{(A1)}$$

As the warp parameters $\rho \in R^{3\times 1}$ and $\phi(P_T) \in R^{K\times 1}$, the Jacobian matrix $\hat{J}$ in the Equation (A1) belongs to $R^{K\times 3}$. The formula of the finite difference gradient [48] is as follows:

$$\hat{J}_i = \frac{\phi(G_i \circ P_T) - \phi(P_T)}{\mu_i}. \quad \text{(A2)}$$

We use infinitesimal perturbations to approximate each column $\hat{J}_i$ of $\hat{J}$. Therein, G_i, $i = 1,2,3,4$ corresponds to the transformation that is obtained by only perturbing the i-th warp parameter, which can be formulated as

$$G_1 = \begin{pmatrix} 1 & 0 & 0 & \mu_1 \\ 0 & 1 & 0 & 0 \\ 0 & 0 & 1 & 0 \end{pmatrix}, G_2 = \begin{pmatrix} 1 & 0 & 0 & 0 \\ 0 & 1 & 0 & \mu_2 \\ 0 & 0 & 1 & 0 \end{pmatrix}, G_3 = \begin{pmatrix} 1 & 0 & 0 & 0 \\ 0 & 1 & 0 & 0 \\ 0 & 0 & 1 & \mu_3 \end{pmatrix}, G_4 = \begin{pmatrix} \cos(\mu_4) & 0 & -\sin(\mu_4) & 0 \\ 0 & 1 & 0 & 0 \\ \sin(\mu_4) & 0 & \cos(\mu_4) & 0 \end{pmatrix}. \quad \text{(A3)}$$

When training the SES by approximation-based solution, we set the infinitesimal perturbations μ_i to 0.1.

Appendix B. Analysis of the Jacobian Module

Our loss is defined as follows:

$$L_{ses} = \frac{1}{M} \sum_m \mathcal{L}_1(J^{\dagger(m)}[\phi(P_I^{(m)}) - \phi(P_T^{(m)})], \Delta\rho_{gt}^{(m)}). \quad \text{(A4)}$$

To enable the state estimation network to be trained in an end-to-end way, the differentiation of the Moore-Penrose inverse in Equation (A4) needs to be derived as [19] did. Concretely, the partial derivative of smooth L_1 function over the feature component $\phi_i(P_I)$ can be written as

$$\frac{\partial \mathcal{L}_1}{\partial \phi_i(P_I)} = \nabla \mathcal{L}_1 J^{\dagger} \delta_i, \quad \text{(A5)}$$

where $\delta_i \in \{0,1\}^{K\times 1}$ is one-hot vector and $\nabla \mathcal{L}_1$ is the derivative of the smooth L_1 loss. Besides, the partial derivative of smooth L_1 function over the $\phi_i(P_T)$ is

$$\frac{\partial \mathcal{L}_1}{\partial \phi_i(P_T)} = \nabla \mathcal{L}_1 \left(\frac{\partial J^\dagger}{\partial \phi_i(P_T)} [\phi(P_I) - \phi(P_T)] - J^\dagger \delta_i \right). \tag{A6}$$

Therein, the key step is to obtain the differentiation of $J^\dagger = (\hat{J}^\top \hat{J})^{-1} \hat{J}^\top$. By the chain rule, it can be written as

$$\frac{\partial J^\dagger}{\partial \phi_i(P_T)} = \frac{\partial (\hat{J}^\top \hat{J})^{-1}}{\partial \phi_i(P_T)} \hat{J}^\top + (\hat{J}^\top \hat{J})^{-1} \frac{\partial \hat{J}^\top}{\partial \phi_i(P_T)}, \tag{A7}$$

where

$$\frac{\partial (\hat{J}^\top \hat{J})^{-1}}{\partial \phi_i(P_T)} = -(\hat{J}^\top \hat{J})^{-1} \left(\hat{J}^\top \frac{\partial \hat{J}}{\partial \phi_i(P_T)} + \frac{\partial \hat{J}^\top}{\partial \phi_i(P_T)} \hat{J} \right) (\hat{J}^\top \hat{J})^{-1}. \tag{A8}$$

According to the above equation, the derivative of a batch inverse matrix can be implemented in PyTorch such that the SES network can be trained in an end-to-end manner.

In this work, we present two solutions for calculating the Jacobian in our paper. Here we give their back-propagation formulae to deeply compare them with each other. When using multi-layer perceptron $\mathcal{F}$ (learning-based) to calculate the Jacobian, the elements of $\hat{J}$ are related to each component of $\phi(P_T)$. We hence have

$$\frac{\partial \hat{J}}{\partial \theta_j^\phi} = \frac{\partial \hat{J}}{\partial \phi(P_T)} \frac{\partial \phi(P_T)}{\partial \theta_j^\phi}, \tag{A9}$$

where $\frac{\partial \hat{J}}{\partial \phi_i(P_T)} = \frac{\partial \mathcal{F}(\phi(P_T))}{\partial \phi_i(P_T)}$ is adaptively updated.

When using finite difference gradient (approximition-based), we have

$$\frac{\partial \hat{J}}{\partial \theta_j^\phi} = \frac{\partial \hat{J}}{\partial \phi(P_T)} \frac{\partial \phi(P_T)}{\partial \theta_j^\phi} + \sum_k \frac{\partial \hat{J}}{\partial \phi(G_k \circ P_T)} \frac{\partial \phi(G_k \circ P_T)}{\partial \theta_j^\phi}, \tag{A10}$$

where

$$\frac{\partial \hat{J}}{\partial \phi_i(P_T)} = \begin{pmatrix} 0 & \cdots & 0 \\ \vdots & & \vdots \\ -\frac{1}{\mu_1} & \cdots & -\frac{1}{\mu_4} \\ \vdots & & \vdots \\ 0 & \cdots & 0 \end{pmatrix}, \tag{A11}$$

and

$$\frac{\partial \hat{J}}{\partial \phi_i(G_k \circ P_T)} = (a_{mn}), a_{mn} = \begin{cases} \frac{1}{\mu_k}, & \text{if } m = i, n = k; \\ 0, & \text{otherwise.} \end{cases} \tag{A12}$$

As can be seen from the above formulae, $\frac{\partial \hat{J}}{\partial \phi_i(P_T)}$ in the finite difference is fixed while the learning function is updated adaptively in the multi-layer perception. Thus, the learning-based solution may be more easily coupled with the feature extractor $\phi(P_T)$ than the approximation-based one.

References

1. Ma, Y.; Anderson, J.; Crouch, S.; Shan, J. Moving Object Detection and Tracking with Doppler LiDAR. *Remote Sens.* **2019**, *11*, 1154 [CrossRef]
2. Giancola, S.; Zarzar, J.; Ghanem, B. *Leveraging Shape Completion for 3D Siamese Tracking*; CVPR: Salt Lake City, UT, USA, 2019; pp. 1359–1368.
3. Comport, A.I.; Marchand, E.; Chaumette, F. *Robust Model-Based Tracking for Robot Vision*; IROS: Prague, Czech Republic, 2004; pp. 692–697.

4. Wang, M.; Su, D.; Shi, L.; Liu, Y.; Miró, J.V. *Real-time 3D Human Tracking for Mobile Robots with Multisensors;* ICRA: Philadelphia, PA, USA, 2017; pp. 5081–5087.
5. Luo, W.; Yang, B.; Urtasun, R. *Fast and Furious: Real Time End-to-End 3D Detection, Tracking and Motion Forecasting with a Single Convolutional Net;* CVPR: Salt Lake City, UT, USA, 2018; pp. 3569–3577.
6. Schindler, K.; Ess, A.; Leibe, B.; Gool, L.V. Automatic detection and tracking of pedestrians from a moving stereo rig. *ISPRS J. Photogramm. Remote Sens.* **2010**, *65*, 523–537. [CrossRef]
7. Nam, H.; Han, B. *Learning Multi-Domain Convolutional Neural Networks for Visual Tracking;* CVPR: Salt Lake City, UT, USA, 2016; pp. 4293–4302.
8. Henriques, J.F.; Caseiro, R.; Martins, P.; Batista, J. High-Speed Tracking with Kernelized Correlation Filters. *IEEE Trans. Pattern Anal. Mach. Intell.* **2015**, *37*, 583–596. [CrossRef] [PubMed]
9. Bertinetto, L.; Valmadre, J.; Golodetz, S.; Miksik, O.; Torr, P.H.S. *Staple: Complementary Learners for Real-Time Tracking;* CVPR: Salt Lake City, UT, USA, 2016; pp. 1401–1409.
10. Liu, Y.; Jing, X.Y.; Nie, J.; Gao, H.; Liu, J.; Jiang, G.P. Context-Aware Three-Dimensional Mean-Shift With Occlusion Handling for Robust Object Tracking in RGB-D Videos. *IEEE Trans. Multimed.* **2019**, *21*, 664–676. [CrossRef]
11. Kart, U.; Kamarainen, J.K.; Matas, J. *How to Make an RGBD Tracker?* ECCV: Munich, Germany, 2018; pp. 148–161.
12. Bibi, A.; Zhang, T.; Ghanem, B. *3D Part-Based Sparse Tracker with Automatic Synchronization and Registration;* CVPR: Salt Lake City, UT, USA, 2016; pp. 1439–1448.
13. Luber, M.; Spinello, L.; Arras, K.O. *People Tracking in RGB-D Data With On-Line Boosted Target Models;* IROS: Prague, Czech Republic, 2011; pp. 3844–3849.
14. Ku, J.; Mozifian, M.; Lee, J.; Harakeh, A.; Waslander, S.L. *Joint 3D Proposal Generation and Object Detection from View;* ICRA: Philadelphia, PA, USA, 2018; pp. 5750–5757.
15. Yang, B.; Luo, W.; Urtasun, R. *PIXOR: Real-Time 3D Object Detection from Point Clouds;* CVPR: Salt Lake City, UT, USA, 2018; pp. 7652–7660.
16. Qi, H.; Feng, C.; Cao, Z.; Zhao, F.; Xiao, Y. *P2B: Point-to-Box Network for 3D Object Tracking in Point Clouds;* CVPR: Salt Lake City, UT, USA, 2020; pp. 6328–6337.
17. Qi, C.R.; Litany, O.; He, K.; Guibas, L.J. *Deep Hough Voting for 3D Object Detection in Point Clouds;* ICCV: Seoul, Korea, 2019; pp. 9276–9285.
18. Danelljan, M.; Bhat, G.; Khan, F.S.; Felsberg, M. *ATOM: Accurate Tracking by Overlap Maximization;* CVPR: Salt Lake City, UT, USA, 2019; pp. 4655–4664.
19. Wang, C.; Galoogahi, H.K.; Lin, C.H.; Lucey, S. *Deep-LK for Efficient Adaptive Object Tracking;* ICRA: Brisbane, Australia, 2018; pp. 626–634.
20. Danelljan, M.; Bhat, G.; Khan, F.S.; Felsberg, M. *ECO: Efficient Convolution Operators for Tracking;* CVPR: Salt Lake City, UT, USA, 2017; pp. 6931–6939.
21. Wu, Y.; Lim, J.; Yang, M.H. Object Tracking Benchmark. *IEEE Trans. Pattern Anal. Mach. Intell.* **2015**, *37*, 1834–1848. [CrossRef] [PubMed]
22. Kristan, M.; Leonardis, A.; Matas, J.; Felsberg, M. *The Sixth Visual Object Tracking VOT2018 Challenge Results;* ECCV: Munich, Germany, 2018; pp. 3–53.
23. Valmadre, J.; Bertinetto, L.; Henriques, J.F.; Vedaldi, A.; Torr, P.H.S. *End-to-End Representation Learning for Correlation Filter Based Tracking;* CVPR: Salt Lake City, UT, USA, 2017; pp. 5000–5008.
24. Bertinetto, L.; Valmadre, J.; Henriques, J.F.; Vedaldi, A.; Torr, P.H.S. *Fully-Convolutional Siamese Networks for Object Tracking;* ECCV: Munich, Germany, 2016; pp. 850–865.
25. Held, D.; Thrun, S.; Savarese, S. *Learning to Track at 100 FPS with Deep Regression Networks;* ECCV: Munich, Germany, 2016; pp. 749–765.
26. Jiang, B.; Luo, R.; Mao, J.; Xiao, T.; Jiang, Y. *Acquisition of Localization Confidence for Accurate Object Detection;* ECCV: Munich, Germany, 2018; pp. 816–832.
27. Zhao, S.; Xu, T.; Wu, X.J.; Zhu, X.F. Adaptive feature fusion for visual object tracking. *Pattern Recognit.* **2021**, *111*, 107679. [CrossRef]
28. Bhat, G.; Danelljan, M.; Gool, L.V.; Timofte, R. *Learning Discriminative Model Prediction for Tracking;* ICCV: Seoul, Korea, 2019; pp. 6181–6190.
29. Qi, C.R.; Su, H.; Mo, K.; Guibas, L.J. *PointNet: Deep Learning on Point Sets for 3D Classification and Segmentation;* CVPR: Salt Lake City, UT, USA, 2017; pp. 77–85.
30. Lee, J.; Cheon, S.U.; Yang, J. Connectivity-based convolutional neural network for classifying point clouds. *Pattern Recognit.* **2020**, *112*, 107708. [CrossRef]
31. Zhou, Y.; Tuzel, O. *VoxelNet: End-to-End Learning for Point Cloud Based 3D Object Detection;* CVPR: Salt Lake City, UT, USA, 2018; pp. 4490–4499.
32. Shi, S.; Wang, X.; Li, H. *PointRCNN: 3D Object Proposal Generation and Detection from Point Cloud;* CVPR: Salt Lake City, UT, USA, 2019; pp. 770–779.
33. Yi, L.; Zhao, W.; Wang, H.; Sung, M.; Guibas, L. *GSPN: Generative Shape Proposal Network for 3D Instance Segmentation in Point Cloud;* CVPR: Salt Lake City, UT, USA, 2019; pp. 3942–3951.

34. Wang, W.; Yu, R.; Huang, Q.; Neumann, U. *SGPN: Similarity Group Proposal Network for 3D Point Cloud Instance Segmentation*; CVPR: Salt Lake City, UT, USA, 2018; pp. 2569–2578.
35. Song, S.; Xiao, J. *Tracking Revisited Using RGBD Camera: Unified Benchmark and Baselines*; ICCV: Seoul, Korea, 2013; pp. 233–240.
36. Held, D.; Levinson, J.; Thrun, S. *Precision Tracking with Sparse 3D and Dense Color 2D Data*; ICRA: Karlsruhe, Germany, 2013; pp. 1138–1145.
37. Held, D.; Levinson, J.; Thrun, S.; Savarese, S. Robust real-time tracking combining 3D shape, color, and motion. *Int. J. Robot. Res.* **2016**, *35*, 30–49. [CrossRef]
38. Spinello, L.; Arras, K.O.; Triebel, R.; Siegwart, R. *A Layered Approach to People Detection in 3D Range Data*; AAAI: Palo Alto, CA, USA, 2010; pp. 1625–1630.
39. Xiao, W.; Vallet, B.; Schindler, K.; Paparoditis, N. Simultaneous detection and tracking of pedestrian from velodyne laser scanning data. *ISPRS Ann. Photogramm. Remote Sens. Spat. Inf. Sci.* **2016**, *3*, 295–302. [CrossRef]
40. Zou, H.; Cui, J.; Kong, X.; Zhang, C.; Liu, Y.; Wen, F.; Li, W. *F-Siamese Tracker: A Frustum-based Double Siamese Network for 3D Single Object Tracking*; IROS: Prague, Czech Republic, 2020; pp. 8133–8139.
41. Qi, C.R.; Yi, L.; Su, H.; Guibas, L.J. *PointNet++: Deep Hierarchical Feature Learning on Point Sets in a Metric Space*; NeurIPS: Vancouver, BC, Canada, 2017; pp. 5100–5109.
42. Chaumette, F.; Seth, H. Visual servo control, Part I: Basic approaches. *IEEE Robot. Autom. Mag.* **2006**, *13*, 82–90. [CrossRef]
43. Quentin, B.; Eric, M.; Juxi, L.; François, C.; Peter, C. Visual Servoing from Deep Neural Networks. In Proceedings of the Robotics: Science and Systems Workshop, Cambridge, MA, USA, 12–16 July 2017; pp. 1–6.
44. Xiong, X.; la Torre, F.D. *Supervised Descent Method and Its Applications to Face Alignment*; CVPR: Salt Lake City, UT, USA, 2013; pp. 532–539.
45. Lin, C.H.; Zhu, R.; Lucey, S. *The Conditional Lucas-Kanade Algorithm*; ECCV: Amsterdam, The Netherlands, 2016; pp. 793–808.
46. Han, L.; Ji, M.; Fang, L.; Nießner, M. RegNet: Learning the Optimization of Direct Image-to-Image Pose Registration. *arXiv* **2018**, arXiv:1812.10212.
47. Baker, S.; Matthews, I. Lucas-Kanade 20 years on: A unifying framework. *Int. J. Comput. Vis.* **2004**, *56*, 221–255. [CrossRef]
48. Aoki, Y.; Goforth, H.; Srivatsan, R.A.; Lucey, S. *PointNetLK: Robust & Efficient Point Cloud Registration using PointNet*; CVPR: Salt Lake City, UT, USA, 2019; pp. 7156–7165.
49. Girshick, R.B. *Fast R-CNN*; ICCV: Santiago, Chile, 2015; pp. 1440–1448.
50. Geiger, A.; Lenz, P.; Stiller, C.; Urtasun, R. Vision meets robotics: The KITTI dataset. *Int. J. Robot. Res.* **2013**, *32*, 1231–1237. [CrossRef]
51. Hesai, I.S. PandaSet by Hesai and Scale AI. Available online: https://pandaset.org/ (accessed on 24 June 2021).
52. Zarzar, J.; Giancola, S.; Ghanem, B. Efficient Bird Eye View Proposals for 3D Siamese Tracking. *arXiv* **2019**, arXiv:1903.10168.
53. Besl, P.J.; McKay, N.D. A method for registration of 3-D shapes. *IEEE Trans. Pattern Anal. Mach. Intell.* **1992**, *14*, 239–256. [CrossRef]
54. Sun, X.; Wei, Y.; Liang, S.; Tang, X.; Sun, J. *Cascaded Hand Pose Regression*; CVPR: Salt Lake City, UT, USA, 2015; pp. 824–832.
55. Wang, Y.; Sun, Y.; Liu, Z.; Sarma, S.E.; Bronstein, M.M.; Solomon, J.M. Dynamic Graph CNN for Learning on Point Clouds. *ACM Trans. Graph.* **2019**, *38*, 1–12. [CrossRef]
56. Liu, X.; Qi, C.R.; Guibas, L.J. *FlowNet3D: Learning Scene Flow in 3D Point Clouds*; CVPR: Salt Lake City, UT, USA, 2019; pp. 529–537.

remote sensing

MDPI

Article

Improved YOLO Network for Free-Angle Remote Sensing Target Detection

Yuhao Qing, Wenyi Liu *, Liuyan Feng and Wanjia Gao

School of Instrument and Electronics, North University of China, Taiyuan 030000, China; s2006262@st.nuc.edu.cn (Y.Q.); s2006261@st.nuc.edu.cn (L.F.); b1806014@st.nuc.edu.cn (W.G.)
* Correspondence: liuwenyi@nuc.edu.cn; Tel.: +86-139-3460-7107

Abstract: Despite significant progress in object detection tasks, remote sensing image target detection is still challenging owing to complex backgrounds, large differences in target sizes, and uneven distribution of rotating objects. In this study, we consider model accuracy, inference speed, and detection of objects at any angle. We also propose a RepVGG-YOLO network using an improved RepVGG model as the backbone feature extraction network, which performs the initial feature extraction from the input image and considers network training accuracy and inference speed. We use an improved feature pyramid network (FPN) and path aggregation network (PANet) to reprocess feature output by the backbone network. The FPN and PANet module integrates feature maps of different layers, combines context information on multiple scales, accumulates multiple features, and strengthens feature information extraction. Finally, to maximize the detection accuracy of objects of all sizes, we use four target detection scales at the network output to enhance feature extraction from small remote sensing target pixels. To solve the angle problem of any object, we improved the loss function for classification using circular smooth label technology, turning the angle regression problem into a classification problem, and increasing the detection accuracy of objects at any angle. We conducted experiments on two public datasets, DOTA and HRSC2016. Our results show the proposed method performs better than previous methods.

Keywords: image target detection; deep learning; multiple scales; any angle object; remote sensing of small objects

Citation: Qing, Y.; Liu, W.; Feng, L.; Gao, W. Improved YOLO Network for Free-Angle Remote Sensing Target Detection. *Remote Sens.* **2021**, *13*, 2171. https://doi.org/10.3390/rs13112171

Academic Editor: Fahimeh Farahnakian

Received: 24 April 2021
Accepted: 29 May 2021
Published: 1 June 2021

Publisher's Note: MDPI stays neutral with regard to jurisdictional claims in published maps and institutional affiliations.

1. Introduction

Target detection is a basic task in computer vision and helps estimate the category of objects in a scene and mark their locations. The rapid deployment of airborne and spaceborne sensors has made ultra-high-resolution aerial images common. However, object detection in remote sensing images remains a challenging task. Research on remote sensing images has crucial applications in the military, disaster control, environmental management, and transportation planning [1–4]. Therefore, it has attracted significant attention from researchers in recent years.

Object detection in aerial images has become a prevalent topic in computer vision [5–7]. In the past few years, machine learning methods have been successfully applied for remote sensing target detection [8–10]. David et al. [8] used the Defense Science and Technology Organization Analysts' Detection Support System, which is a system developed particularly for ship detection in remote sensing images. Wang et al. [9] proposed an intensity-space domain constant false alarm rate ship detector. Leng et al. [10] presented a highly adaptive ship detection scheme for spaceborne synthetic-aperture radar (SAR) imagery.

Although these remote sensing target detection methods based on machine learning have achieved good results, the missed detection rate remains very high in complex ground environments. Deep neural networks, particularly the convolutional neural network (CNN) class, significantly improve the detection of objects in natural images owing to the advantages in robust feature extraction using large-scale datasets. In recent years, systems employing the powerful feature learning capabilities of CNN have demonstrated remarkable success in various visual tasks such as classification [11,12], segmentation [13], tracking [14], and detection [15–17]. CNN-based target detectors can be divided into two categories: single-stage and two-stage target detection networks. Single-stage target detection networks discussed in the literature [18–21] include a you only look once (YOLO) detector optimized end-to-end, which was proposed by Joseph et al. [18,19]. Liu et al. [20] presented a method for detecting objects in images using a deep neural network single-shot detector (SSD). Lin et al. [21] designed and trained a simple dense object detector, RetinaNet, to evaluate the effectiveness of the focal loss. The works of [22–27], describing two-stage target detection networks, include the proposal by Girshick et al. [22] of a simple and scalable detection algorithm that combines the region proposal network (RPN) with a CNN (R-CNN). Subsequently, Girshick et al. [23] developed a fast region-based convolutional network (fast R-CNN) to efficiently classify targets and improve the training speed and detection accuracy of the network. Ren et al. [24] merged the convolutional features of RPN and fast R-CNN into a neural network with an attention mechanism (faster R-CNN). Dai et al. [25] proposed a region-based fully convolutional network (R-FCN), and Lin et al. [26] proposed a top-down structure, feature pyramid network (FPN), with horizontal connections, which considerably improved the accuracy of target detection.

General object detection methods, generally based on horizontal bounding boxes (HBBs), have proven quite successful in natural scenes. Recently, HBB-based methods have also been widely used for target detection in aerial images [27–31]. Li et al. [27] proposed a weakly supervised deep learning method that uses separate scene category information and mutual prompts between scene pairs to fully train deep networks. Ming et al. [28] proposed a deep learning method for remote sensing image object detection using a polarized attention module and a dynamic anchor learning strategy. Pang et al. [29] proposed a self-enhanced convolutional neural network, rotational region CNN (R^2-CNN), based on the content of remotely sensed regions. Han et al. [30] used a feature alignment module and orientation detection module to form a single-shot alignment network (S^2A-Net) for target detection in remote sensing images. Deng et al. [31] redesigned the feature extractor using cascaded rectified linear unit and inception modules, used two detection networks with different functions, and proposed a new target detection method.

Most targets in remote sensing images have the characteristics of arbitrary directionality, high aspect ratio, and dense distribution. Therefore, the HBB-based model may cause severe overlap and noise. In subsequent work, an oriented bounding box (OBB) was used to process rotating remote sensing targets [32–40], enabling more accurate target capture and introducing considerably less background noise. Feng et al. [32] proposed a robust Student's t-distribution-aided one-stage orientation detector. Ding et al. [34] proposed an RoI transformer that transforms horizontal regions of interest into rotating regions of interest. Azimi et al. [36] minimized the joint horizontal and OBB loss functions. Liu et al. [37] applied a newly defined rotatable bounding box (RBox) to develop a method to detect objects at any angle. Yang et al. [39] proposed a rotating dense feature pyramid framework (R-DFPN), and Yang et al. [40] designed a circular smooth label (CSL) technology to analyze the angle of rotating objects.

To improve feature extraction, a few studies have integrated the attention mechanism into their network model [41–43]. Chen et al. [41] proposed a multi-scale spatial and channel attention mechanism remote sensing target detector, and Cui et al. [42] proposed using a dense attention pyramid network to detect multi-sized ships in SAR images. Zhang et al. [43] used attention-modulated features and context information to develop a novel object detection network (CAD-Net).

A few studies have focused on the effect of context information in table checks, extracting different proportions of context information as well as deep low-resolution high-level and high-resolution low-level semantic features [44–49]. Zhu et al. [44] constructed a target detection problem as an inference in a Markov random field. Gidaris et al. [45] proposed an object detection system that relies on a multi-region deep CNN. Zhang et al. [46] proposed a hierarchical target detector with deep environmental characteristics. Bell et al. [47] used a spatial recurrent neural network (S-RNN) to integrate contextual information outside the region of interest, proposing an object detector that uses information both inside and outside the target. Marcu et al. [48] proposed a dual-stream deep neural network model using two independent paths to process local and global information inference. Kang et al. [49] proposed a multi-layer neural network that tends to merge based on context.

In this article, we propose the RepVGG-YOLO model to detect targets in remote sensing images. RepVGG-YOLO uses the improved RepVGG module as the backbone feature extraction network (Backbone) of the model; spatial pyramid pooling (SPP), multi-layer FPN, and path aggregation network (PANet) as the enhanced feature extraction networks; and CSL to correct the rotating angle of objects. In this model, we increased the number of target detection scales to four. The main contributions of this article are as follows:

1. We used the improved RepVGG as the backbone feature extraction module. This module employs different networks in the training and inference parts, while considering the training accuracy and inference speed. The module uses a single-channel architecture, which has high speed, high parallelism, good flexibility, and memory-saving features. It provides a research foundation for the deployment of models on hardware systems.
2. We used the combined FPN and PANet and the top-down and bottom-up feature pyramid structures to accumulate low-level and process high-level features. Simultaneously, we used the network detection scales to enhance the network's ability to detect small remote sensing targets. The pixel feature extraction portion ensures accurate detection of objects of all sizes.
3. We used CSL to determine the angle of rotating objects, thereby turning the angle regression problem into a classification problem and more accurately detecting objects at any angle.
4. Compared with seven other recent remote sensing target detection networks, the proposed RepVGG-YOLO network demonstrated the best performance on two public datasets.

The rest of this paper is arranged as follows. Section 2 introduces the proposed model for remote sensing image target detection. Section 3 describes the experimental validation and discusses the results. Section 4 summarizes the study.

2. Materials and Methods

In this section, we first introduce the proposed network framework for target detection in remote sensing images. Next, we present a formula derivation of the Backbone network and multi-scale pyramid structure (Neck) for extracting and processing target features. Then, we discuss the prediction structure of the proposed model and, finally, we detail the loss function of the model.

2.1. Overview of the Proposed Model

We first perform operations such as random scaling, random cropping, and random arrangement of the original dataset images, followed by data enhancement on the data to balance the size and target sample ratio and segmentation of the image with overlapping areas to retain the small target edge information. Simultaneously, we crop the original data of the different sized segments into pictures of 608×608 pixels, which serve as the input to the model. As shown in Figure 1, we first extract the low-level general features from the processed image through the Backbone network. To detect targets of different scales and categories, Backbone provides several combinations of receptive field size and center step length. Then, we select the corresponding feature maps from different parts of the Backbone input for Neck. Feature maps of varying sizes $\{152 \times 152, 76 \times 76, 38 \times 38, 19 \times 19\}$ are selected from the hierarchical feature maps to detect targets of different sizes. By coupling the feature maps of different receptive field sizes, Neck enhances the network expressivity and distributes the multi-scale learning tasks to multiple networks. The Backbone aligns the feature maps by width once, and directly outputs the feature maps of the same width to the head network. Finally, we integrate the feature information and convert it into detection predictions. We elaborate on these parts in the following sections.

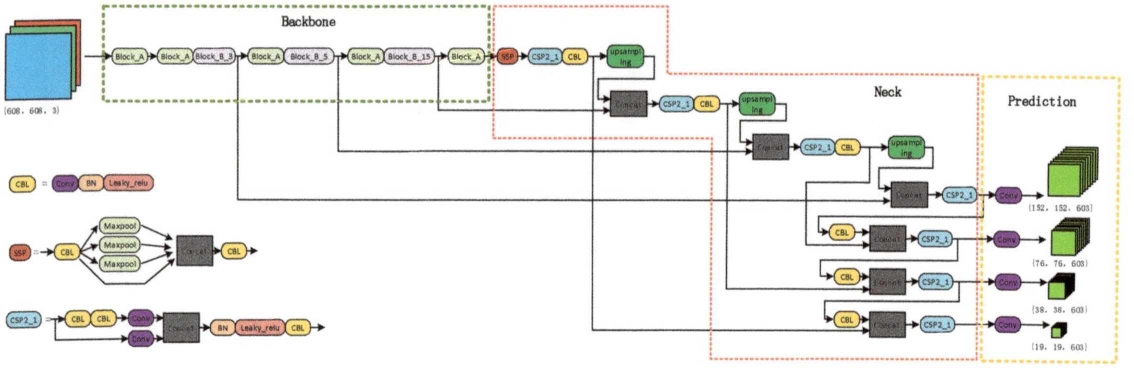

Figure 1. Overall network framework model.

2.2. Backbone Feature Extraction Network

The Backbone network is a reference network for many computer tasks, often used to extract low-level general features, such as color, shape, and texture. It can provide several combinations of receptive field size and center step length to meet the requirements of different scales and categories in target detection. ResNet and MobileNet comprise two networks often used in various computer-vision tasks. The former can realize a combination of different resolution features and extract a robust feature representation. The latter, with its faster inference speed and fewer network parameters, finds use in embedded devices with low computing power. The RepVGG [50] model has improved speed and accuracy compared with Resnet34, ResNet50, ResNet101, ResNet152, and VGG-16. While MobileNet and VGG have improved inference speed compared with models such as VGG-16, they have lower accuracy. Therefore, considering both accuracy and inference speed, we use the improved RepVGG as the backbone network in this study. The network improvements arise from VGG network enhancements: identity and residual branches are added to the VGG network block to utilize the advantages of the ResNet network. On the basis of the RepVGG-B [50] network, we add a Block_A module at the end of the network to enhance feature extraction and, at the same time, pass the feature map input of a specific shape to the subsequent network. Figure 2 shows the execution process of the backbone feature extraction network. The two-dimensional convolution in the Block_A module has a step size of 2; thus, the feature map size will be halved after the Block_A module. Similarly, because the two-dimensional convolution in the Block_B module has a step size of 1, the size of the feature map remains unchanged after the Block_B module.

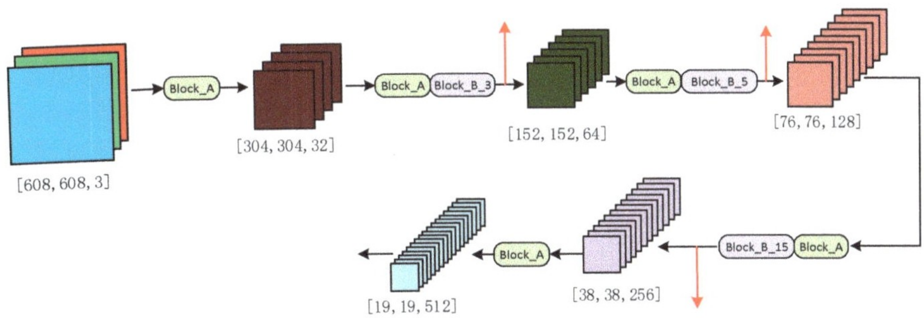

Figure 2. Backbone feature extraction network.

For the input picture size of 608 × 608, Figure 2 shows the shape of the output feature map of each layer. After each continuous Block_B module (Block_B_3, Block_B_5, Block_B_15), a branch is output, and the high-level features are passed to the subsequent network for feature fusion, thereby enhancing the feature extraction capability of the model. Finally, the feature map with the shape {19, 19, 512} is passed to strengthen the feature extraction network.

In addition, different network architectures are used in the training and inference stages while considering training accuracy and inference speed. Figure 3 shows the training and structural re-parameterization network architectures.

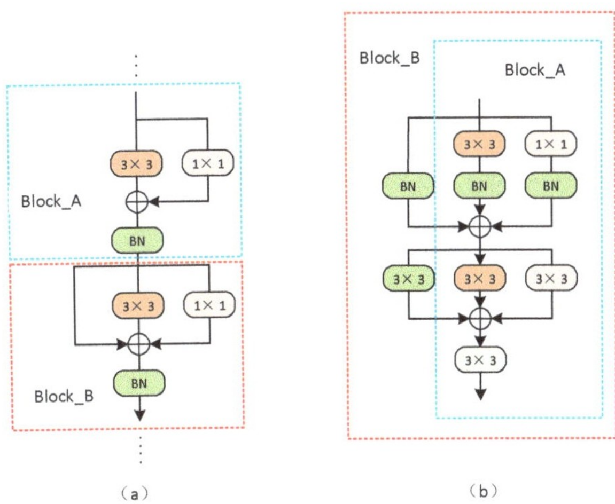

Figure 3. (**a**) Block_A and Block_B modules in the training phase; (**b**) structural re-parameterization of Block_A and Block_B.

Figure 3a shows the training network of the RepVGG. The network uses two branch structures: the residual structure that contains only Block_A of the Conv1*1 residual branch, the residual structure of Conv1*1, and the identity residual; and structure Block_B. Because the training network has multiple gradient flow paths, a deeper network model can not only handle the problem of gradient disappearance in the deep layer of the network, but also obtain a more robust feature representation in the deep layer.

Figure 3b shows that RepVGG converts the multi-channel training model to a single-channel test model. To improve the inference speed, the convolutional and batch normalization (BN) layers are merged. Equations (1) and (2) express the formulas for the convolutional and BN layers, respectively.

$$\mathrm{Conv(x)} = \mathrm{W(x)} + \mathrm{b} \tag{1}$$

$$\mathrm{BN(x)} = \gamma * \frac{(\mathrm{x} - \mathrm{mean})}{\sigma} + \beta \tag{2}$$

Replacing the argument in the BN layer equation with the convolution layer formula yields the following:

$$\begin{aligned} \mathrm{BN(Conv(x))} &= \frac{\gamma * \mathrm{W(x)}}{\sigma} + \frac{\gamma * (\mathrm{b} - \mathrm{mean})}{\sigma} + \beta \\ &= \frac{\gamma * \mathrm{W(x)}}{\sigma} + \frac{\gamma * \mu}{\sigma} + \beta \end{aligned} \tag{3}$$

Here, μ, σ, γ, and β represent the cumulative average, standard deviation, scaling factor, and deviation, respectively. We use $W^k \epsilon R^{C_2 \times C_1 \times k \times k}$ to represent the input C_1, the output C_2, and the convolution kernel of the convolution of k. With $M^1 \epsilon R^{N \times C_1 \times H_1 \times W_1}$ and $M^2 \epsilon R^{N \times C_2 \times H_2 \times W_2}$ denoting the input and output, respectively, the BN layer of the fusion convolution can be simplified to yield the following:

$$\left.\begin{aligned} W'_{i,:,:,:} &= \frac{\gamma_i}{\sigma_i} W_{i,:,:,:} \\ b'_i &= -\frac{\mu_i \gamma_i}{\sigma_i} W_{i,:,:,:} + \beta_i \\ \mathrm{BN}(\mathrm{M} * \mathrm{W}, \mu, \sigma, \gamma, \beta)_{:,\, i,\, :,:} &= (M * W')_{:,i,:,:} + b'_i \end{aligned}\right\} \tag{4}$$

where i ranges in the interval from 1 to C_2; * represents the convolution operation; and W' and b'_i the weight and bias of the convolution after fusion, respectively. Let $C_1 = C_2$, $H_1 = H_2$, and $W_1 = W_2$; then, the output can be expressed as follows:

$$\begin{aligned} M^2 &= BN\left(M^1 \times W^3, \mu^3, \sigma^3, \gamma^3, \beta^3\right) \\ &+ BN\left(M^1 \times W^1, \mu^1, \sigma^1, \gamma^1, \beta^1\right) \\ &+ BN\left(M^1, \mu^0, \sigma^0, \gamma^0, \beta^0\right) \end{aligned} \tag{5}$$

where μ^k, σ^k, γ^k, and β^k represent the BN parameters obtained after the k $\times$ k convolution and $\mu^0, \sigma^0, \gamma^0$, and β^0 represent the parameters of the identity branch. For the output of three different scales, we adopt the following strategy for fusion. We can regard the identity branch structure as a 1 $\times$ 1 convolution; for the Conv1*1 and the identity branches, the 1 $\times$ 1 convolution kernel can be filled and converted into a 3 $\times$ 3 convolution kernel; finally, we add the three 3 $\times$ 3 convolution kernels from the three output scales to obtain the final convolution kernel, and add the three deviations to obtain the final deviation. The Block_B module can be represented by Equation (5); further, because the Block_A module does not contain the identity branch structure, it can be represented by the first two items in Equation (5).

2.3. Strengthening the Feature Extraction Network (Neck)

In the target detection task, to make the model learn diverse features and improve detection performance, the Neck network can reprocess the features extracted by the Backbone, disperse the learning of different scales applied to the multiple levels of feature maps, and couple the feature maps with different receptive field sizes. In this study, we use SPP [51], improved FPN [26], and PANet [52] structure to extract the features. Figure 4 shows the detailed execution process of the model. The SPP structure uses pooling methods of different scales to perform multi-scale feature fusion, which can improve the receptive field of the model, significantly increase the receiving range of the main features, and more effectively separate the most important context features, thereby avoiding problems such as image distortion caused by cropping and zooming the image area. The computer-based learning (CBL) module comprises a two-dimensional convolution process, BN, and

Leaky_ReLU activation function. The input of the CSP2_1 module is divided into two parts. One part goes through two CBL modules and then through a two-dimensional convolution; the other part directly undergoes a two-dimensional convolution operation. Finally, the feature maps obtained from the two parts are spliced, then put through the BN layer and Leaky_ReLU activation function, and output after the CBL module.

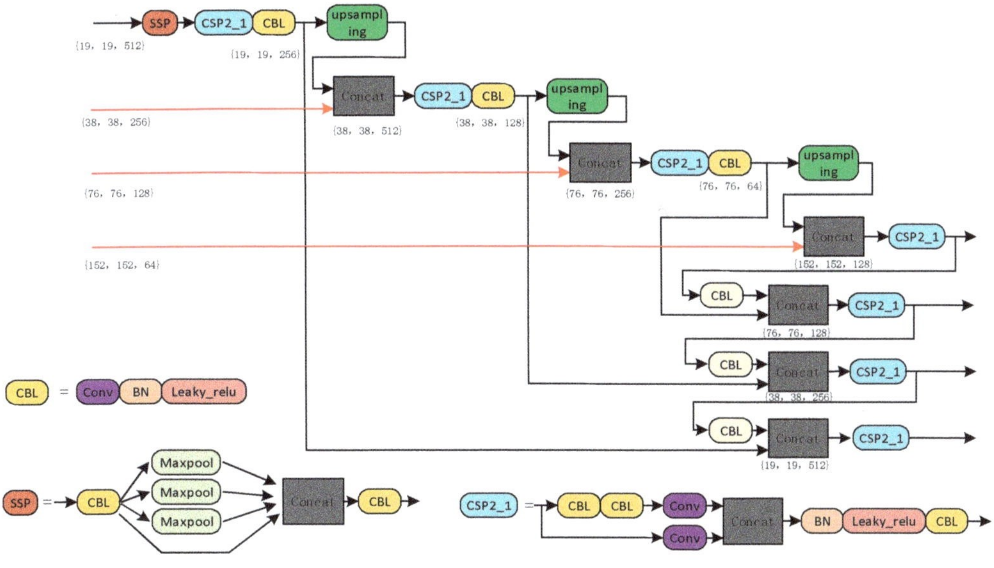

Figure 4. Strengthening the feature extraction network.

Figure 4 shows the shape of the feature map of the key parts of the entire network. Note that the light-colored CBL module (the three detection scale output parts at the bottom right) has a two-bit convolution step size of 2, whereas the other two-dimensional convolutions have a step size of 1. FPN is top-down, and transfers and integrates high-level feature information through up-sampling. FPN also transfers high-level strong semantic features to enhance the entire pyramid, but only enhances semantic information, not positioning information. We also added a bottom-up feature pyramid behind the FPN layer that accumulates low-level and processed high-level features. Because low-level features can provide more accurate location information, the additional layer creates a deeper feature pyramid, adding the ability to aggregate different detection layers from different backbone layers, which enhances the feature extraction performance of the network.

2.4. Target Boundary Processing at Any Angle

Because remote sensing images contain many complex and dense rotating targets, we need to correct these rotating objects for more accurate detection of objects at any angle. Common angle regression methods include the open source computer-vision, long edge, and ordered quadrilateral definition methods. The predictions of these methods often exceed the initial set range. Because the target parameters of learning are periodic, they can be at the boundary of periodic changes. This condition can cause a sudden increase in the loss value that increases the difficulty of learning by the network, leading to boundary problems. We use circular smooth label (CSL) [40] to handle the angle problem, as shown in Figure 5.

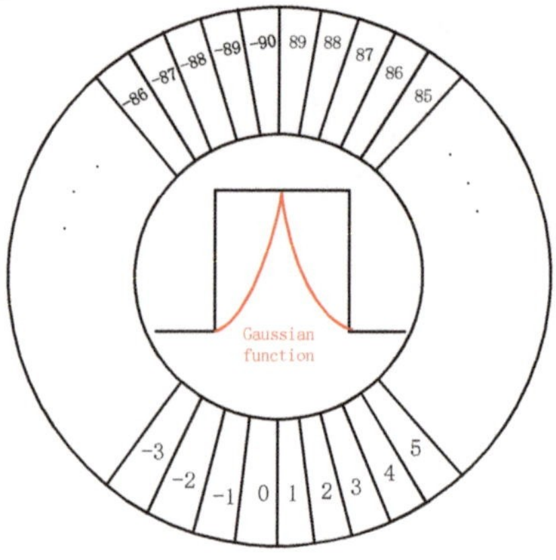

Figure 5. Circular smooth label.

Equation (6) expresses CSL, where $g(x)$ is the window function.

$$\mathrm{CSL}(x) = \begin{cases} g(x), & \theta - r < x < \theta + r \\ 0, & otherwise \end{cases} \tag{6}$$

where θ represents the angle passed by the longest side when the x-axis rotates clockwise, and r represents the window radius. We convert angle prediction from a regression problem to a classification problem and place the entire defined angle range into one category. We choose a Gaussian function for the window function to measure the angular distance between the predicted and ground truth labels. The predicted value loss becomes smaller the closer it comes to the true value within a certain range. Introducing periodicity, i.e., the two degrees, 89 and −90, become neighbors, solves the problem of angular periodicity. Using discrete rather than continuous angle predictions avoids boundary problems.

2.5. Target Prediction Network

After subjecting the image to feature extraction twice, we integrate the feature information and transform it into a prediction, as shown in Figure 6. We use the k-means clustering algorithm to generate 12 prior boxes with different scales according to the labels of the training set. Because remote sensing target detection involves detecting small targets, to enhance the feature extraction of small pixel targets, we use four detection scales with sizes of 19 × 19, 38 × 38, 76 × 76, and 152 × 152.

Taking the 19 × 19 detection scale as an example, we divide the input image into multiple 19 × 19 grids. Each grid point is preset with three boxes of corresponding scales. When these grids enclose an object, we use the corresponding grid for object detection. Finally, the shape of the feature map output by the detection feature layer is {19, 19, 603}. The third quantity implies that each of the three anchors in the corresponding grid consists of 201 dimension predictions. The width and height of the box and the coordinates of the center point (x_offset, y_offset, h, w), confidence, 16 classification results, and 180 classification angles (described in Section 2.4). Based on the set loss function (described in Section 2.6.3), iterative calculations for the backpropagation operation are performed and the position and angle of the prediction box are continually adjusted and, finally, to attain the highest confidence test results, non-maximum suppression screening is applied [53].

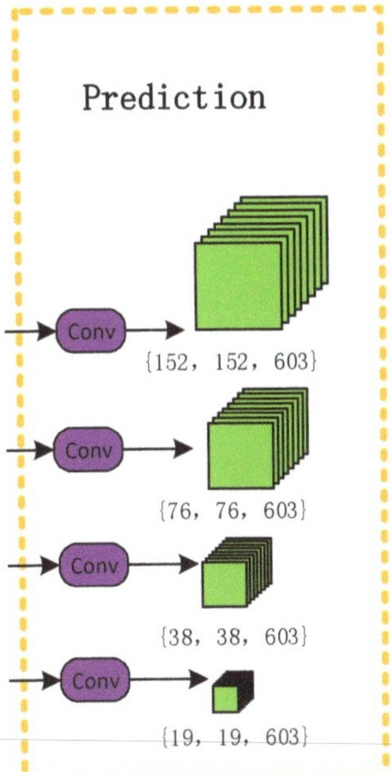

Figure 6. Target prediction network.

2.6. *Loss Function*

In this section, we describe the bounding box regression loss function, the confidence loss function with weight coefficients, and the classification loss function with increased angle calculation.

2.6.1. Bounding Box Border Regression Loss

The most commonly used indicator in target detection, often used to calculate the bounding box regression loss, the intersection over union (IoU) [54] value, is defined as the ratio of the intersection and union of the areas of two rectangular boxes. Equation (7) shows the IoU and the bounding box regression loss.

$$\left.\begin{array}{r} IoU = \frac{|B \cap B^{gt}|}{|B \cup B^{gt}|} \\ \mathrm{LOSS_{IoU}} = 1 - IoU \end{array}\right\} \tag{7}$$

where B represents the predicted bounding box, B^{gt} represents the real bounding box, $|B \cap B^{gt}|$ represents the B and B^{gt} intersection area, and $|B \cup B^{gt}|$ represents the B and B^{gt} union area. The following problems arise in calculating the loss function defined in Equation (7):

1. When B and B^{gt} do not intersect, IoU = 0, the distance between B and B^{gt} cannot be expressed, and the loss function LOSS_IoU cannot be directed or optimized.

2. When the size of B remains the same in different situations, the IoU values obtained do not change, making it impossible to distinguish different intersections of B and B^{gt}.

To overcome these problems, the generalized IoU (GIoU) [55] was proposed in 2019, with the formulation shown below:

$$\left.\begin{array}{r} GIoU = IoU - \frac{|C \setminus (B \cup B^{gt})|}{|C|} \\ \mathrm{LOSS_{GIoU}} = 1 - GIoU \end{array}\right\}, \tag{8}$$

where $|C|$ represents the area of the smallest rectangular box containing B and B^{gt}, and $|C\setminus(B\cup B^{gt})|$ represents the area of the C rectangle excluding $|B\cup B^{gt}|$. The calculation of the bounding box frame regression loss uses the GIoU. Compared with using the IoU, using the GIoU improves the measurement method of the intersection scale and alleviates the above-mentioned problems to a certain extent, but still does not consider the situation when B is inside B^{gt}. Furthermore, when the size of B remains the same and the position changes, the GIoU value also remains the same, and the model cannot be optimized.

In response to this situation, distance-IoU (DIoU) [56] was proposed in 2020. Based on IoU and GIoU, and incorporating the center point of the bounding box, DIoU can be expressed as follows:

$$\left.\begin{array}{r} DIoU = 1 - IoU + \frac{\rho^2(B,\, B^{gt})}{c^2} \\ LOSS_{DIoU} = 1 - DIoU \end{array}\right\}, \tag{9}$$

where $\rho^2(B, B^{gt})$ represents the Euclidean distance between the center points of B and B^{gt}, and c represents the diagonal distance of the smallest rectangle that can cover B and B^{gt} simultaneously. $LOSS_{DIoU}$ can be minimized by calculating the distance between B and B^{gt} and using the distance between the center points of B and B^{gt} as a penalty term, which improves the convergence speed.

Using both GIoU and DIoU, recalculating the aspect ratio of B and B^{gt}, and increasing the impact factor av, the complete IoU (CIoU) [56] was proposed, as expressed below:

$$\left.\begin{array}{r} CIoU = IoU - \frac{\rho^2(B,\, B^{gt})}{c^2} - av \\ a = \frac{v}{1-IOU+v} \\ v = \frac{4}{\pi^2}\left(arc\tan\frac{w^{gt}}{h^{gt}} - arc\tan\frac{w}{h}\right)^2 \\ LOSS_{CIoU} = 1 - IoU + \frac{\rho^2(B,\, B^{gt})}{c^2} + av \end{array}\right\} \tag{10}$$

where h^{gt} and w^{gt} are the length and width of B^{gt}, respectively; h and w are the length and width of B, respectively; a is the weight coefficient; and v is the distance between the aspect ratios of B and B^{gt}. We use $LOSS_{CIoU}$ as the bounding box border regression loss function, which brings the predicted bounding box more in line with the real bounding box, and improves the model convergence speed, regression accuracy, and detection performance.

2.6.2. Confidence Loss Function

We use cross-entropy to calculate the object confidence loss. Regardless of whether there is an object to be detected in the grid, the confidence error must be calculated. Because only a small part of the input image may contain objects to be detected, we add a weight coefficient (λ_{no}) to constrain the confidence loss for the image area that does not contain the target object, thereby reducing the number of negative samples. The object confidence loss can be expressed as follows:

$$\begin{array}{rl} LOSS_{Conf} = & -\sum_{i=0}^{S^2}\sum_{j=0}^{B} I_{ij}\left(\hat{C}_i^j \log C_i^j + (1-\hat{C}_i^j)\log(1-C_i^j)\right)R_{Iou} \\ & +(1-I_{ij})\left(\hat{C}_i \log C_i + (1-\hat{C}_i^j)\log(1-C_i)\right)\lambda_{no}. \end{array} \tag{11}$$

where S is the number of grids in the network output layer and B is the number of anchors. I_i^j indicates whether the j-th anchor in the i-th grid can detect this object (the detected value is 1 and the undetected value is 0), and the value of $\hat{C}_i^j$ is determined by whether the bounding box of the grid is responsible for predicting an object (if it is responsible for prediction, the value of $\hat{C}_i^j$ is 1, otherwise it is 0). C_i^j is the predicted value after parameter normalization (the value lies between 0 and 1). R_{Iou} represents the IoU of the rotating bounding box.

The complete decoupling of the correlation between the prediction angle and the prediction confidence means the confidence loss is not only related to the frame parameters, but also to the rotation angle. Table 1 summarizes the recalculation of the IoU [35] of the rotating bounding box as the confidence loss coefficient, along with its pseudocode.

Table 1. Rotating intersection over union (IoU) calculation pseudocode.

Algorithm 1 RIoU computation
1: **Input**: Rectangles *R*1; *R*2; :::; *RN*
2: **Output**: RIoU between rectangle pairs *RIoU*
3: **for** each pair <*Ri*; *Rj*> ($i < j$) **do**
4: Point set *PSet* φ
5: Add intersection points of *Ri* and *Rj* to *PSet*
6: Add the vertices of *Ri* inside *Rj* to *PSet*
7: Add the vertices of *Rj* inside *Ri* to *PSet*
8: Sort *PSet* into anticlockwise order
9: Compute intersection *I* of *PSet* by triangulation
10: $RIoU[i;j] \frac{\text{Area(I)}}{Area(R_i)+Area(R_j)-Area(I)}$
11: **end for**

Figure 7 shows the geometric principle of rotating IoU calculations. We divide the overlapping part into multiple triangles with the same vertex, calculate the area of each triangle separately, and finally add the calculated areas to obtain the area of the overlapping polygons. The detailed calculation principle is as follows. Given a set of rotating rectangles R1, R2, . . . , RN, calculate the RIoU of each pair of <Ri, Rj>. First, the intersection set, PSet, of Ri and Rj (the intersection of two rectangles and the vertices of one rectangle in the other rectangle form a set, PSet, corresponding to rows 4–7 of Table 1); then, calculate the intersection area, I, of PSet and, finally, calculate the RIoU according to the formula in row 10 of Table 1 (combine the points generated by the PSet into a polygon, divide the polygon into multiple triangles, calculate the sum of the area of the multiple triangles as the polygon area, and finally calculate the polygon area and remove the rotation of the polygon area; corresponding to rows 8–10 of Table 1).

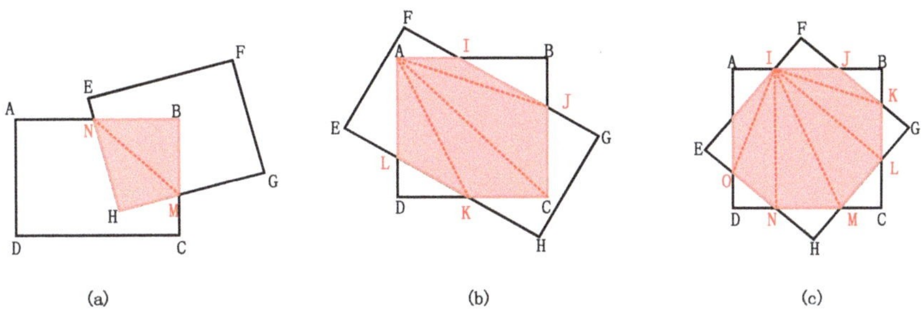

Figure 7. Intersection over union (IoU) calculation for rotating intersecting rectangles: (**a**) intersecting graph is a quadrilateral, (**b**) intersecting graph is a hexagon, and (**c**) intersecting graph is an octagon.

2.6.3. Classification Loss Function

Because we converted the angle calculation from a regression problem into a classification problem, we calculate both the category and angle loss when calculating the classification loss function. Here, we use the cross-entropy loss function for the calculation. When the j-th anchor box of the i-th grid is responsible for a real target, we calculate the classification loss function for the bounding box generated by this anchor box, using Equation (12).

$$\begin{aligned} \mathrm{LOSS}_{\mathrm{Class}} = & -\sum_{i=0}^{S^2}\sum_{j=0}^{B} I_{ij} \sum_{c\in Class,\ \theta\in(0,180]} (\hat{P}_i(c+\theta)\log P_i(c+\theta) \\ & +(1-\hat{P}_i(c+\theta))\log(1-P_i(c+\theta))) \end{aligned} \tag{12}$$

where c belongs to the target classification category; θ belongs to the angle processed by the CSL [40] algorithm; S is the number of grids in the network output layer; B is the number of anchors; and I_i^j indicates whether the j-th anchor in the i-th grid can detect this object (the detected value is 1 and the undetected value is 0).

The final total loss function equals the sum of the three loss functions, as shown in Equation (13). Furthermore, the three loss functions have the same effect on the total loss function; that is, the reduction of any one of the loss functions will lead to the optimization of the total loss function.

$$LOSS = \mathrm{LOSS}_{\mathrm{CIoU}} + \mathrm{LOSS}_{\mathrm{Conf}} + \mathrm{LOSS}_{\mathrm{Class}} \tag{13}$$

3. Experiments, Results, and Discussion

3.1. Introduction to DOTA and HRSC2016 Datasets

3.1.1. DOTA Dataset

The DOTA dataset [57] comprises 2806 aerial images obtained from different sensors and platforms, including 15 classification categories: plane (PL), baseball diamond (BD), bridge (BR), ground track (GTF), small vehicle (SV), large vehicle (LV), ship (SH), tennis court (TC), basketball court (BC), oil storage tank (ST), football field (SBF), roundabout (RA), airport and helipad (HA), swimming pool (SP), and helicopter (HC). The image data can be divided into 1411 training sets, 937 test sets, and 458 verification sets. The image size ranges between 800 × 800 and 4000 × 4000 pixels. Dataset labeling consisted of a horizontal and a directional bounding box for a total of 188,282 instances.

3.1.2. HRSC2016 Dataset

The HRSC2016 dataset [58] comes from six different ports, with a total of 1061 remote sensing pictures. Examples of detection objects include ships on the sea and ships docked on the shore. The images can be divided into 436 training sets (1207 labeled examples in total), 444 test sets (1228 labeled examples in total), and 181 validation sets (541 labeled examples in total). The image size ranges from 300 × 300 to 1500 × 900 pixels.

3.2. Image Preprocessing and Parameter Optimization

In this section, we describe image preprocessing, experimental parameter settings, and experimental evaluation standards.

3.2.1. Image Preprocessing

Owing to the complex background of remote sensing target detection [59], large changes in the target scale [60], special viewing angle [61–63], unbalanced categories [31], and so on, we preprocess the original data. Directly processing the original high-resolution remote sensing images not only increases equipment requirements, but also significantly reduces detection accuracy. We cut the entire picture and send it to the proposed model training module. During the test, we cut the test pictures into pictures of the same size as those in the training set, and after the test, we splice the predicted results one by one to obtain the total result. To ensure the loss of small target information at the cutting edge during the cutting process, we allow the cut image to have a certain proportion of overlap area (in this study, we set the overlap area to 30%). If the size of the original image is smaller than the size of the cut image, we perform an edge pixel filling operation on the original image to make its size reach the training size. In the remote sensing dataset (e.g., DOTA), the sample target size changes drastically, and small targets can be densely distributed and large and small targets can be considerably unevenly distributed (the number of small targets is much larger than the number of large targets). In this regard, we use the Mosaic data enhancement method to splice the pictures in random zooming, cropping, and arrangement, which substantially enriches the dataset and makes the distribution of targets of different sizes more uniform. Mixed multiple images can have different semantics. Enhanced network robustness occurs when the picture information allows the detector to detect targets beyond the conventional context.

3.2.2. Experimental Parameter Settings

We evaluated the performance of the proposed model on two NVIDIA GeForce RTX 2080 Ti GPUs with 11 GB of RAM. We used the PyTorch 1.7 deep learning framework and Python 3.7 compiler run on Windows 10. To optimize the network, we used stochastic gradient descent with momentum, setting the learning rate momentum and weight decay coefficients to 0.857 and 0.00005, respectively; the iterative learning rate for the first 50 K to 0.001; and the later iterative learning rate to 0.0001. The CIoU loss and classification loss coefficients were set to 0.0337 and 0.313, respectively. The weight coefficient, λ_{no}, of the confidence loss function was set to 0.4. The batch size was set to eight, and the epoch was set to 500.

3.2.3. Evaluation Criteria

To verify the performance of the proposed method, two broad criteria were used to evaluate the test results [64]: precision and recall. The accuracy rate indicates the detection rate of the predicted true-positive samples, and the recall rate indicates the rate of correctly identified true-positive samples. Accuracy and recall can be expressed as follows.

$$\text{Precision} = \frac{\text{TP}}{TP + FP} \tag{14}$$

$$\text{Recall} = \frac{\text{TP}}{TP + FN} \tag{15}$$

TP represents a real positive sample, TN represents a real negative sample, FP is a false positive sample, and FN is a false negative sample. This study adopts the mean average precision (mAP) [45–47] to evaluate all methods, which can be expressed as follows:

$$\text{mAP} = \frac{\sum_{i=1}^{N_{class}} \int \text{P}_i(R_i)dR_i}{N_{class}} \tag{16}$$

where P_i and R_i represent the accuracy and recall rate of the i-th class of classified objects, respectively. N_{class} represents the total number of detected objects in the dataset.

3.3. *Experimental Results*

Figure 8 shows the precision–recall curve of the DOTA detection object category. We focus on the interval between 0.6 and 0.9, where the recall rate is concentrated. Except for BR, when the recall value is greater than 0.6, the decline in the curves of the other types of objects increases. The BD, PL, and TC curves all drop sharply when the recall value is greater than 0.8. The results show that the overall performance of the proposed method is stable and has good detection effectiveness.

To prove that the proposed method has better performance, we compared the proposed method (RepVGG-YOLO NET) to seven other recent methods: SSD [20], joint training method for target detection and classification (YOLOV2) [19], rotation dense feature pyramid network (R-DFPN) [39], toward real-time object detection with RPN (FR-C) [25], joint image cascade and functional pyramid network and multi-size convolution kernel to extract multi-scale strong and weak semantic feature framework (ICN) [36], fine FPN and multi-layer attention network (RADET) [65], and end-to-end refined single-stage rotation detector (R3Det) [66]. Table 2 summarizes the quantitative comparison results of the eight methods on the DOTA dataset. The table indicates that the proposed model has achieved the most advanced results, achieving relatively stable detection results in all categories, with an mAP of 74.13%. SSD and YOLOV2 networks have poor detection effectiveness and relatively low detection effectiveness on small targets; their poor feature extraction network performance needs improvement. The FR-C, ICN, and RADET network models achieved good detection results.

Compared with other methods, owing to the increased processing of targets at any angle and the use of four target detection scales, the proposed model achieved good classification results for small objects with complex backgrounds and dense distributions (for example, SV and SH achieved 71.02% and 78.41% mAP values). Compared with the suboptimal method (i.e., R3Det), the suggested method achieved a 1.32% better mAP value. In addition, using the FPN and PANet structures to accumulate high-level and low-level features helped the improvement in the detection of categories with large differences in the target scale of the same image (for example, BR and LV on the same image), with BR and LV achieving classification results of 52.34% and 76.27%, respectively. We also obtained relatively stable mAP values in single-category detection (PL, BR, SV, LV, TC, BC, SBF, RA, SP, and HC achieved the highest mAP values).

Table 3 summarizes the proposed model and five other methods (i.e., rotation-sensitive regression for oriented scene text detection (RRD) [67], rotated region-based CNN for ship detection (BL2 and RC2) [68], refined single-stage detector with feature refinement for rotating object (R3 DET) [66], and rotated region proposal and discrimination networks (R2PN) [69]). Table 3 summarizes quantitative comparison results on the HRSC2016 dataset. The results demonstrate that the proposed method achieves an mAP detection result of 91.54, which is better than the other methods evaluated on this dataset. Compared with the suboptimal method (R3Det), the mAP for the proposed model was better by 2.21%. Good results were achieved for the detection of ship instances with large aspect ratios and rotation directions. The proposed method achieved 22 frames per second (FPS), which is more than that achieved by the suboptimal method (R3Det).

Figure 9 shows the partial visualization results of the proposed method on the DOTA and HRSC2016 datasets. The first three rows are the visualization results of the DOTA dataset, and the last row shows the visualization results of the HRSC2016 dataset. Figure 9 shows that the proposed model handles well the noise problem in a complex environment, and has a better detection effectiveness on densely distributed small objects. Good test results were also obtained for some samples with drastic size changes and special viewing angles.

Table 2. Comparison of the results with the other seven latest methods on the DOTA dataset (highest performance is in boldface).

Method	PL	BD	BR	GTF	SV	LV	SH	TC	BC	ST	SBF	RA	HA	SP	HC	mAP (%)
SSD	57.85	32.79	16.14	18.67	0.05	36.93	24.74	81.16	25.10	47.47	11.22	31.53	14.12	9.09	0.00	29.86
YOLOV2	76.90	33.87	22.73	34.88	38.73	32.02	52.37	61.65	48.54	33.91	29.27	36.83	36.44	38.26	11.61	39.20
R-DFPN	80.92	65.82	33.77	58.94	55.77	50.94	54.78	90.33	66.34	68.66	48.73	51.76	55.1	51.32	35.88	57.94
FR-C	80.2	77.55	32.86	68.13	53.66	52.49	50.04	90.41	75.05	59.59	57.00	49.81	61.69	56.46	41.85	60.46
ICN	81.36	74.3	47.7	70.32	64.89	67.82	69.98	90.76	79.06	78.20	53.64	62.90	67.02	64.17	50.23	68.16
RADET	79.45	76.99	48.05	**65.83**	65.46	74.40	68.86	89.70	78.14	74.97	49.92	**64.63**	66.14	**71.58**	**62.16**	69.09
R^3Det	89.24	**80.81**	51.11	65.62	70.67	76.03	**78.32**	90.83	84.89	**84.42**	65.10	57.18	**68.1**	68.98	60.88	72.81
proposed	**90.27**	79.34	**52.34**	64.35	**71.02**	**76.27**	77.41	**91.04**	**86.21**	84.17	**66.82**	63.07	67.23	69.75	62.07	**74.13**

Table 3. Comparison of the results with five other recent methods on the HRSC2016 dataset.

Method	mAP (%)	FPS
BL2	69.6	–
RC2	75.7	–
R^2PN	79.6	–
RRD	84.3	–
R^3Det	89.33	10
proposed	91.54	22

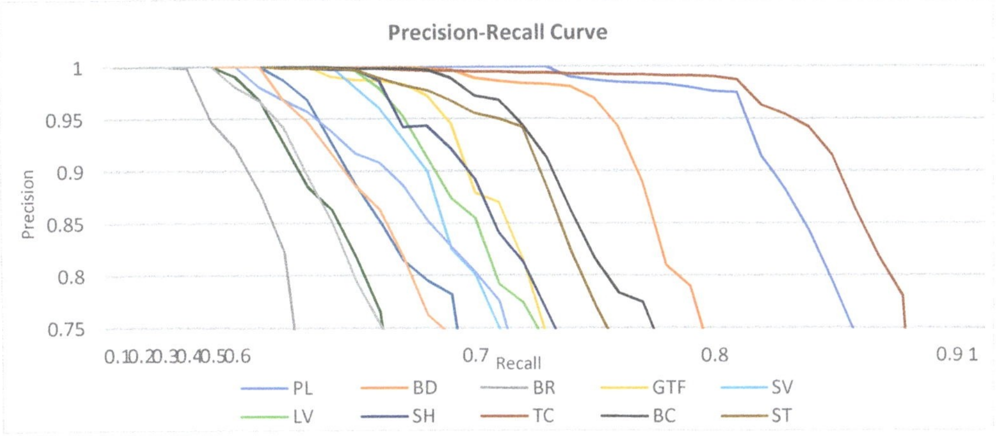

Figure 8. Precision-recall curve of the DOTA dataset.

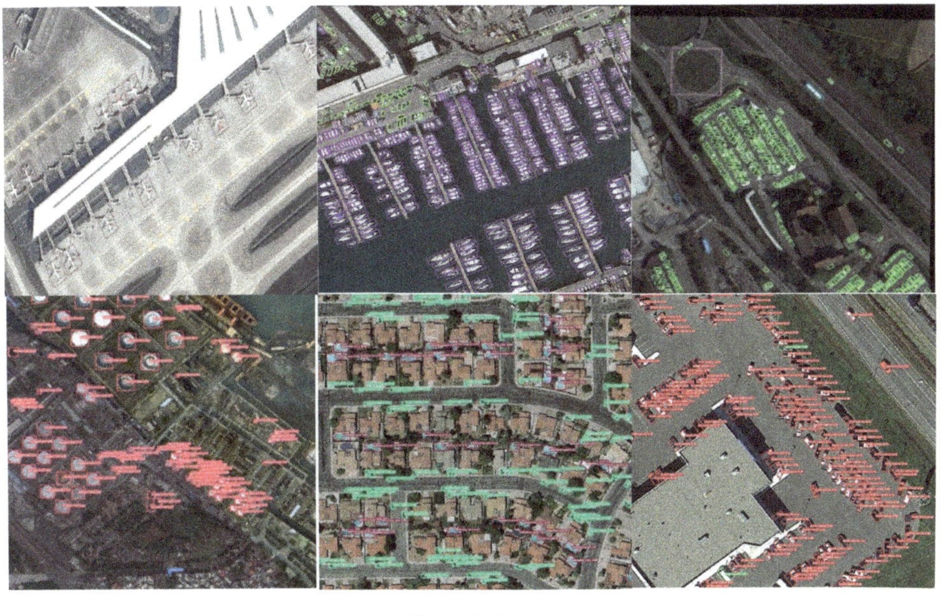

Figure 9. *Cont.*

Figure 9. Visualization results of the DOTA dataset and HRSC2016 dataset. The first three groupings of images are part of the test results of the DOTA dataset, whereas the last grouping is part of the test results of the HRSC2016 dataset.

3.4. Ablation Study

We conducted a series of comparative experiments on the DOTA data set, as shown in Table 4. We considered the influence of different combinations of the five factors of backbone network, bounding box border regression loss (BBRL), data enhancement (DE), multi-scale settings, and CSL on the final experimental results. We used mAP and FPS as evaluation criteria to verify the effectiveness of our method.

Table 4. Ablation study on components on the DOTA dataset.

N	Proposed	Backbone	BBRL	DE	Multi Scale	CSL	mAP	FPS
1	✓	RepVGG-A	DIou				66.98	25
2	✓	RepVGG-A	CIou				67.19	25
3	✓	RepVGG-B	DIou				68.03	23
4	✓	RepVGG-B	CIou				69.98	23
5	✓	RepVGG-B	CIou	✓			71.03	23
6	✓	RepVGG-B	CIou	✓	✓		72.25	22
7	✓	RepVGG-B	CIou	✓	✓	✓	74.13	22

From Table 4, the first row is the baseline, the improved RepVGG-A is used as the backbone, and the DIou is used as the BBRL. The backbone network is a reference network for many computer tasks. We set the first and third groups, and the second combination and the fourth group of experiments to verify the backbone network. The results show that RepVGG-B has more complex network parameters and is deeper than RepVGG-A. Consequently, using the improved RepVGG-B as the backbone (groups 3 and 4), mAP increased by 1.05% and 2.79%, respectively. Choosing an appropriate loss function can improve the convergence speed and prediction accuracy of the model. Here, we set the first group, the second group, and the third combination and the fourth group of experiments to analyze the BBRL. Because CIou recalculated the predicted bounding box, the aspect ratio of the bounding box and the real bounding box increased, and the influence factor increased to align the predicted bounding box with the actual box. Under the same conditions, better results were obtained when CIou was used as the BBRL. The objective of DE is to increase the number and diversity of samples, which can significantly improve the problem of sample imbalance. According to the experimental results of the fourth and fifth groups, mAP increased by 1.06% after the image was processed by cropping, zooming, and random arrangement. Because different detection scales have different sensitivities to objects of different scales, there are many detection targets with large differences in size in remote sensing images. We can observe from the experimental results of the fifth and sixth groups that mAP improved by 1.21% when four detection scales were used. The increased number of detection scales enhances the detection of small target objects. Because there are many dense rotating targets in remote sensing images, we assume that the bounding box can be predicted more accurately. Next, we set up the sixth and seventh groups of experiments. The results show that, after using CSL, we can change the angle prediction from a regression problem into a classification problem, and the periodicity problem of the angle was solved. mAP improved by 1.88% to 74.13%. We finally chose the improved RepVGG-B model as the backbone network with CIou as the BBRL loss function, using DE, Multi scale, and CSL simultaneously, and finally obtaining RepVGG-YOLO NET.

4. Conclusions

In this article, we introduce a method for detecting targets from arbitrary-angle geographic remote sensing. A RepVGG-YOLO model is proposed, which uses an improved RepVGG module as the backbone feature extraction network (Backbone) of the model, and uses SPP, feature pyramid network (FPN), and path aggregation network (PANet) as the enhanced feature extraction networks. The model combines context information on multiple scales, accumulates multi-layer features, and strengthens feature information extraction. In addition, we use four target detection scales to enhance the feature extraction of remote sensing small target pixels and the CSL method to increase the detection accuracy of objects at any angle. We redefine the classification loss function and add the angle problem to the loss calculation. The proposed model achieved the best detection performance among the eight methods evaluated. The proposed model obtained an mAP of 74.13% and 22 FPS on the DOTA dataset, wherein the mAP value exceeded that of the suboptimal method (R3Det) by 1.32%. The proposed model obtained an mAP of 91.54% on the HRSC2016 dataset. The mAP value and the FPS exceeded that of the suboptimal method (R3Det) by 2.21% and 13, respectively. We expect to conduct further research on the detection of blurred, dense small objects and obscured objects.

Author Contributions: Conceptualization, Y.Q. and W.L.; methodology, Y.Q.; software, Y.Q. and W.L.; validation, Y.Q., L.F. and W.G.; formal analysis, Y.Q. and L.F.; writing—original draft preparation, Y.Q., W.L. and L.F.; writing—review and editing, Y.Q. and W.L.; visualization, Y.Q. and W.L. All authors have read and agreed to the published version of the manuscript.

Funding: This research received no external funding.

Data Availability Statement: Not applicable.

Acknowledgments: The authors would like to thank Guigan Qing and Chaoxiu Li for their support, secondly, thanks to Lianshu Qing and Niuniu Feng for their support.

Conflicts of Interest: The authors declare no conflict of interest.

References

1. Zhang, F.; Du, B.; Zhang, L.; Xu, M. Weakly supervised learning based on coupled convolutional neural networks for aircraft detection. *IEEE Trans. Geosci. Remote Sens.* **2016**, *54*, 5553–5563. [CrossRef]
2. Kamusoko, C. Importance of remote sensing and land change modeling for urbanization studies. In *Urban Development in Asia and Africa*; Springer: Singapore, 2017.
3. Ahmad, K.; Pogorelov, K.; Riegler, M.; Conci, N.; Halvorsen, P. Social media and satellites. *Multimed. Tools Appl.* **2019**, *78*, 2837–2875. [CrossRef]
4. Tang, T.; Zhou, S.; Deng, Z.; Zou, H.; Lei, L. Vehicle detection in aerial images based on region convolutional neural networks and hard negative example mining. *Sensors* **2017**, *17*, 336. [CrossRef]

5. Cheng, G.; Zhou, P.; Han, J. RIFD-CNN: Rotation-invariant and fisher discriminative convolutional neural networks for object detection. In Proceedings of the IEEE Conference on Computer Vision and Pattern Recognition (CVPR), Las Vegas, NV, USA, 26 June–1 July 2016; pp. 2884–2893.
6. Deng, Z.; Sun, H.; Zhou, S.; Zhao, J.; Zou, H. Toward fast and accurate vehicle detection in aerial images using coupled region-based convolutional neural networks. *J-STARS* **2017**, *10*, 3652–3664. [CrossRef]
7. Long, Y.; Gong, Y.; Xiao, Z.; Liu, Q. Accurate object localization in remote sensing images based on convolutional neural networks. *IEEE Trans. Geosci. Remote Sens.* **2017**, *55*, 2486–2498. [CrossRef]
8. Crisp, D.J. A ship detection system for RADARSAT-2 dual-pol multi-look imagery implemented in the ADSS. In Proceedings of the 2013 IEEE International Conference on Radar, Adelaide, Australia, 9–12 September 2013; pp. 318–323.
9. Wang, C.; Bi, F.; Zhang, W.; Chen, L. An intensity-space domain CFAR method for ship detection in HR SAR images. *IEEE Geosci. Remote Sens. Lett.* **2017**, *14*, 529–533. [CrossRef]
10. Leng, X.; Ji, K.; Zhou, S.; Zou, H. An adaptive ship detection scheme for spaceborne SAR imagery. *Sensors* **2016**, *16*, 1345. [CrossRef] [PubMed]
11. Krizhevsky, A.; Sutskever, I.; Hinton, G.E. Imagenet classification with deep convolutional neural networks. *NIPS* **2012**, *25*, 1097–1105. [CrossRef]
12. Xie, S.; Girshick, R.; Dollár, P.; Tu, Z.; He, K. Aggregated residual transformations for deep neural networks. In Proceedings of the IEEE Conference on Computer Vision and Pattern Recognition (CVPR), Honolulu, HI, USA, 21–26 July 2017; pp. 1492–1500.
13. Chen, K.; Pang, J.; Wang, J.; Xiong, Y.; Li, X.; Sun, S.; Feng, W.; Liu, Z.; Shi, J.; Ouyang, W.; et al. Hybrid task cascade for instance segmentation. *arXiv* **2019**, arXiv:1901.07518.
14. Li, B.; Yan, J.; Wu, W.; Zhu, Z.; Hu, X. High performance visual tracking with Siamese region proposal network. In Proceedings of the IEEE Conference on Computer Vision and Pattern Recognition (CVPR), Salt Lake City, UT, USA, 18–22 June 2018; pp. 8971–8980.
15. Tian, L.; Cao, Y.; He, B.; Zhang, Y.; He, C.; Li, D. Image Enhancement Driven by Object Characteristics and Dense Feature Reuse Network for Ship Target Detection in Remote Sensing Imagery. *Remote Sens.* **2021**, *13*, 1327. [CrossRef]
16. Li, Y.; Li, X.; Zhang, C.; Lou, Z.; Zhu, Y.; Ding, Z.; Qin, T. Infrared Maritime Dim Small Target Detection Based on Spatiotemporal Cues and Directional Morphological Filtering. *Infrared Phys. Technol.* **2021**, *115*, 103657. [CrossRef]
17. Yao, Z.; Wang, L. ERBANet: Enhancing Region and Boundary Awareness for Salient Object Detection. *Neurocomputing* **2021**, *448*, 152–167. [CrossRef]
18. Redmon, J.; Divvala, S.; Girshick, R.; Farhadi, A. You only look once: Unified, real-time object detection. In Proceedings of the IEEE Conference on Computer Vision and Pattern Recognition (CVPR), Las Vegas, NV, USA, 26 June–1 July 2016; pp. 779–788.
19. Redmon, J.; Farhadi, A. YOLO9000: Better, faster, stronger. In Proceedings of the IEEE Conference on Computer Vision and Pattern Recognition (CVPR), Honolulu, HI, USA, 21–26 July 2017; pp. 7263–7271.
20. Liu, W.; Anguelov, D.; Erhan, D.; Szegedy, S.; Reed, S.; Fu, C.-Y.; Berg, A.C. SSD: Single shot multibox detector. In Proceedings of the European Conference on Computer Vision, Amsterdam, The Netherlands, 11–14 October 2016; Springer: Cham, Switzerland, 2016; pp. 21–37.
21. Lin, T.-Y.; Goyal, P.; Girshick, R.; He, K.; Dollár, P. Focal loss for dense object detection. In Proceedings of the IEEE International Conference on Computer Vision (ICCV), Venice, Italy, 22–29 October 2017; pp. 2980–2988.
22. Girshick, R.; Donahue, J.; Darrell, T.; Malik, J. Region-based convolutional networks for accurate object detection and segmentation. *IEEE Trans. Pattern Anal. Mach. Intell.* **2016**, *38*, 142–158. [CrossRef]
23. Girshick, R. Fast R-CNN. In Proceedings of the IEEE International Conference on Computer Vision (ICCV), Araucano Park, Las Condes, Chile, 11–18 December 2015; pp. 1440–1448.
24. Ren, S.; He, K.; Girshick, R.; Sun, J. Faster R-CNN: Towards real-time object detection with region proposal networks. *IEEE Trans. Pattern Anal. Mach. Intell.* **2017**, *39*, 1137–1149. [CrossRef] [PubMed]
25. Dai, J.; Li, Y.; He, K.; Sun, J. R-FCN: Object detection via region-based fully convolutional networks. *NIPS* **2016**, *29*, 379–387.
26. Lin, T.-Y.; Dollár, P.; Girshick, R.; He, K.; Hariharan, B.; Belongie, S. Feature pyramid networks for object detection. In Proceedings of the IEEE Conference on Computer Vision and Pattern Recognition (CVPR), Honolulu, HI, USA, 21–26 July 2017; pp. 2117–2125.
27. Li, Y.; Zhang, Y.; Huang, X.; Yuille, A.L. Deep networks under scene-level supervision for multi-class geospatial object detection from remote sensing images. *ISPRS J. Photogramm. Remote Sens.* **2018**, *146*, 182–196. [CrossRef]
28. Ming, Q.; Miao, L.; Zhou, Z.; Dong, Y. CFC-Net: A critical feature capturing network for arbitrary-oriented object detection in remote sensing images. *arXiv* **2021**, arXiv:2101.06849.
29. Pang, J.; Li, C.; Shi, J.; Xu, Z.; Feng, H. R2-CNN: Fast tiny object detection in large-scale remote sensing images. *IEEE Trans. Geosci. Remote Sens.* **2019**, *57*, 5512–5524. [CrossRef]
30. Han, J.; Ding, J.; Li, J.; Xia, G.S. Align deep features for oriented object detection. *IEEE Trans. Geosci. Remote Sens.* **2021**, 1–11.
31. Deng, Z.; Sun, H.; Zhou, S.; Zhao, J.; Lei, L.; Zou, H. Multi-scale object detection in remote sensing imagery with convolutional neural networks. *ISPRS J. Photogramm. Remote Sens.* **2018**, *145*, 3–22. [CrossRef]
32. Feng, P.; Lin, Y.; Guan, J.; He, G.; Shi, H.; Chambers, J. TOSO: Student's-T distribution aided one-stage orientation target detection in remote sensing images. In Proceedings of the 2020 IEEE International Conference on Acoustics, Speech and Signal Processing (ICASSP), Barcelona, Spain, 4–8 May 2020; pp. 4057–4061.

33. Xu, Y.; Fu, M.; Wang, Q.; Wang, Y.; Chen, K.; Xia, G.; Bai, X. Gliding vertex on the horizontal bounding box for multi-oriented object detection. *IEEE Trans. Pattern Anal. Mach. Intell.* **2021**, *43*, 1452–1459. [CrossRef]
34. Ding, J.; Xue, N.; Long, Y.; Xia, G.S.; Lu, Q. Learning RoI Transformer for Detecting Oriented Objects in Aerial Images. In Proceedings of the IEEE/CVF Conference on Computer Vision and Pattern Recognition, Los Angeles, CA, USA, 16–19 June 2019.
35. Xia, G.S.; Bai, X.; Ding, J.; Zhu, Z.; Belongie, S.; Luo, J.; Datcu, M.; Pelillo, M.; Zhang, L. DOTA: A Large-Scale Dataset for Object Detection in Aerial Images. In Proceedings of the IEEE Conference on Computer Vision and Pattern Recognition, Salt Lake City, UT, USA, 18–23 June 2018; pp. 3974–3983.
36. Azimi, S.M.; Vig, E.; Bahmanyar, R.; Körner, M.; Reinartz, P. Towards multi-class object detection in unconstrained remote sensing imagery. *arXiv* **2018**, arXiv:1807.02700.
37. Liu, L.; Pan, Z.; Lei, B. Learning a rotation invariant detector with rotatable bounding box. *arXiv* **2017**, arXiv:1711.09405.
38. Wang, J.; Ding, J.; Guo, H.; Cheng, W.; Pan, T.; Yang, W. Mask OBB: A Semantic Attention-Based Mask Oriented Bounding Box Representation for Multi-Category Object Detection in Aerial Images. *Remote Sens.* **2019**, *11*, 2930. [CrossRef]
39. Yang, X.; Sun, H.; Fu, K.; Yang, J.; Sun, X.; Yan, M.; Guo, Z. Automatic ship detection in remote sensing images from Google Earth of complex scenes based on multiscale rotation dense feature pyramid networks. *Remote Sens.* **2018**, *10*, 132. [CrossRef]
40. Yang, X.; Yan, J. Arbitrary-oriented object detection with circular smooth label. In Proceedings of the 16th European Conference on Computer Vision, Glasgow, UK, 23–28 August 2020; Springer: Cham, Switzerland, 2020; pp. 677–694.
41. Chen, J.; Wan, L.; Zhu, J.; Xu, G.; Deng, M. Multi-scale spatial and channel-wise attention for improving object detection in remote sensing imagery. *IEEE Geosci. Remote Sens. Lett.* **2020**, *17*, 681–685. [CrossRef]
42. Cui, Z.; Li, Q.; Cao, Z.; Liu, N. Dense attention pyramid networks for multi-scale ship detection in SAR images. *IEEE Trans. Geosci. Remote Sens.* **2019**, *57*, 8983–8997. [CrossRef]
43. Zhang, G.; Lu, S.; Zhang, W. CAD-net: A context-aware detection network for objects in remote sensing imagery. *IEEE Trans. Geosci. Remote Sens.* **2019**, *57*, 10015–10024. [CrossRef]
44. Zhu, Y.; Urtasun, R.; Salakhutdinov, R.; Fidler, S. segDeepM: Exploiting segmentation and context in deep neural networks for object detection. In Proceedings of the IEEE Conference on Computer Vision and Pattern Recognition (CVPR), Boston, MA, USA, 7–12 June 2015; pp. 4703–4711.
45. Gidaris, S.; Komodakis, N. Object detection via a multi-region and semantic segmentation-aware CNN model. In Proceedings of the IEEE International Conference on Computer Vision (ICCV), Araucano Park, Las Condes, Chile, 11–18 December 2015; pp. 1134–1142.
46. Zhang, L.; Shi, Z.; Wu, J. A hierarchical oil tank detector with deep surrounding features for high-resolution optical satellite imagery. *IEEE J. Sel. Top. Appl. Earth Observ. Remote Sens.* **2015**, *8*, 4895–4909. [CrossRef]
47. Bell, S.; Zitnick, C.L.; Bala, K.; Girshick, R. Inside-outside net: Detecting objects in context with skip pooling and recurrent neural networks. In Proceedings of the IEEE Conference on Computer Vision and Pattern Recognition (CVPR), Las Vegas, NV, USA, 26 June–1 July 2016; pp. 2874–2883.
48. Marcu, A.; Leordeanu, M. Dual local-global contextual pathways for recognition in aerial imagery. *arXiv* **2016**, arXiv:1605.05462.
49. Kang, M.; Ji, K.; Leng, X.; Lin, Z. Contextual region-based convolutional neural network with multilayer fusion for SAR ship detection. *Remote Sens.* **2017**, *9*, 860. [CrossRef]
50. Ding, X.; Zhang, X.; Ma, N.; Han, J.; Ding, G.; Sun, J. RepVGG: Making VGG-style ConvNets Great Again. *arXiv* **2021**, arXiv:2101.03697v3.
51. He, K.; Zhang, X.; Ren, S.; Sun, J. Spatial pyramid pooling in deep convolutional networks for visual recognition. *IEEE Trans. Pattern Anal. Mach. Intell.* **2015**, *37*, 1904–1916. [CrossRef] [PubMed]
52. Liu, S.; Qi, L.; Qin, H.; Shi, J.; Jia, J. Path aggregation network for instance segmentation. In Proceedings of the IEEE Conference on Computer Vision and Pattern Recognition (CVPR), Salt Lake City, UT, USA, 18–22 June 2018; pp. 8759–8768.
53. Bai, J.; Zhu, J.; Zhao, R.; Gu, F.; Wang, J. Area-based non-maximum suppression algorithm for multi-object fault detection. *Front. Optoelectron.* **2020**, *13*, 425–432. [CrossRef]
54. Rezatofighi, H.; Tsoi, N.; Gwak, J.Y.; Sadeghian, A.; Reid, I.; Savarese, S. Generalized intersection over union: A metric and a loss for bounding box regression. In Proceedings of the 2019 IEEE/CVF Conference on Computer Vision and Pattern Recognition (CVPR), Long Beach, CA, USA, 15–20 June 2019; pp. 658–666. [CrossRef]
55. Zheng, Z.; Wang, P.; Liu, W.; Li, J.; Ye, R.; Ren, D. Distance-IoU loss: Faster and better learning for bounding box regression. In Proceedings of the AAAI Conference on Artificial Intelligence, New York, NY, USA, 7–12 February 2020; Volume 34, pp. 12993–13000. [CrossRef]
56. Ma, J.; Shao, W.; Ye, H.; Wang, L.; Wang, H.; Zheng, Y.; Xue, X. Arbitrary-oriented scene text detection via rotation proposals. *IEEE Trans. Multimed.* **2018**, *20*, 3111–3122. [CrossRef]
57. Liu, Z.; Yuan, L.; Weng, L.; Yang, Y. A high resolution optical satellite image dataset for ship recognition and some new baselines. In Proceedings of the 6th International Conference on Pattern Recognition Applications and Methods (ICPRAM), Porto, Portugal, 24–26 February 2017; pp. 324–331.
58. Wang, C.; Bai, X.; Wang, S.; Zhou, J.; Ren, P. Multiscale visual attention networks for object detection in VHR remote sensing images. *IEEE Geosci. Remote Sens. Lett.* **2018**, *16*, 310–314. [CrossRef]
59. Zhang, Y.; Yuan, Y.; Feng, Y.; Liu, X. Hierarchical and robust convolutional neural network for very high-resolution remote sensing object detection. *IEEE Trans. Geosci. Remote Sens.* **2019**, *57*, 5535–5548. [CrossRef]

60. Cheng, G.; Zhou, P.; Han, J. Learning rotation-invariant convolutional neural networks for object detection in VHR optical remote sensing images. *IEEE Trans. Geosci. Remote Sens.* **2016**, *54*, 7405–7415. [CrossRef]
61. Li, K.; Cheng, G.; Bu, S.; You, X. Rotation-insensitive and context-augmented object detection in remote sensing images. *IEEE Trans. Geosci. Remote Sens.* **2017**, *56*, 2337–2348. [CrossRef]
62. Wu, X.; Hong, D.; Tian, J.; Chanussot, J.; Li, W.; Tao, R. ORSIm detector: A novel object detection framework in optical remote sensing imagery using spatial-frequency channel features. *IEEE Trans. Geosci. Remote Sens.* **2019**, *57*, 5146–5158. [CrossRef]
63. Zou, Z.; Shi, Z. Random access memories: A new paradigm for target detection in high resolution aerial remote sensing images. *IEEE Trans. Image Process.* **2017**, *27*, 1100–1111. [CrossRef] [PubMed]
64. Guo, W.; Yang, W.; Zhang, H.; Hua, G. Geospatial object detection in high resolution satellite images based on multi-scale convolutional neural network. *Remote Sens.* **2018**, *10*, 131. [CrossRef]
65. Li, Y.; Huang, Q.; Pei, X.; Jiao, L.; Shang, R. RADet: Refine feature pyramid network and multi-layer attention network for arbitrary-oriented object detection of remote sensing images. *Remote Sens.* **2020**, *12*, 389. [CrossRef]
66. Yang, X.; Liu, Q.; Yan, J.; Li, A.; Zhang, Z.; Yu, G. R3det: Refined single-stage detector with feature refinement for rotating object. *arXiv* **2019**, arXiv:1908.05612.
67. Liao, M.; Zhu, Z.; Shi, B.; Xia, G.S.; Bai, X. Rotation-sensitive regression for oriented scene text detection. In Proceedings of the IEEE Conference on Computer Vision and Pattern Recognition (CVPR), Salt Lake City, UT, USA, 18–22 June 2018; pp. 5909–5918.
68. Liu, Z.; Hu, J.; Weng, L.; Yang, Y. Rotated region based CNN for ship detection. In Proceedings of the IEEE International Conference on Image Processing, Beijing, China, 17–20 September 2017; pp. 900–904.
69. Zhang, Z.; Guo, W.; Zhu, S.; Yu, W. Toward arbitrary-oriented ship detection with rotated region proposal and discrimination networks. *IEEE Geosci. Remote Sens. Lett.* **2018**, *15*, 1745–1749. [CrossRef]

remote sensing

Technical Note

NDFTC: A New Detection Framework of Tropical Cyclones from Meteorological Satellite Images with Deep Transfer Learning

Shanchen Pang [1], Pengfei Xie [1], Danya Xu [2], Fan Meng [3], Xixi Tao [1], Bowen Li [4], Ying Li [1] and Tao Song [1,*]

[1] College of Computer Science and Technology, China University of Petroleum, Qingdao 266580, China; pangsc@upc.edu.cn (S.P.); s19070028@s.upc.edu.cn (P.X.); s19070020@s.upc.edu.cn (X.T.); s19070042@s.upc.edu.cn (Y.L.)
[2] Southern Marine Science and Engineering Guangdong Laboratory (Zhuhai), Zhuhai 519080, China; xudanya@sml-zhuhai.cn
[3] School of Geosciences, China University of Petroleum, Qingdao 266580, China; B19010078@s.upc.edu.cn
[4] School of Computer Science and Engineering, South China University of Technology, Guangzhou 510006, China; csbooven@mail.scut.edu.cn
* Correspondence: tsong@upc.edu.cn

Abstract: Accurate detection of tropical cyclones (TCs) is important to prevent and mitigate natural disasters associated with TCs. Deep transfer learning methods have advantages in detection tasks, because they can further improve the stability and accuracy of the detection model. Therefore, on the basis of deep transfer learning, we propose a new detection framework of tropical cyclones (NDFTC) from meteorological satellite images by combining the deep convolutional generative adversarial networks (DCGAN) and You Only Look Once (YOLO) v3 model. The algorithm process of NDFTC consists of three major steps: data augmentation, a pre-training phase, and transfer learning. First, to improve the utilization of finite data, DCGAN is used as the data augmentation method to generate images simulated to TCs. Second, to extract the salient characteristics of TCs, the generated images obtained from DCGAN are inputted into the detection model YOLOv3 in the pre-training phase. Furthermore, based on the network-based deep transfer learning method, we train the detection model with real images of TCs and its initial weights are transferred from the YOLOv3 trained with generated images. Training with real images helps to extract universal characteristics of TCs and using transferred weights as initial weights can improve the stability and accuracy of the model. The experimental results show that the NDFTC has a better performance, with an accuracy (ACC) of 97.78% and average precision (AP) of 81.39%, in comparison to the YOLOv3, with an ACC of 93.96% and AP of 80.64%.

Keywords: tropical cyclone detection; meteorological satellite images; deep learning; deep transfer learning; generative adversarial networks

Citation: Pang, S.; Xie, P.; Xu, D.; Meng, F.; Tao, X.; Li, B.; Li, Y.; Song, T. NDFTC: A New Detection Framework of Tropical Cyclones from Meteorological Satellite Images with Deep Transfer Learning. *Remote Sens.* **2021**, *13*, 1860. https://doi.org/10.3390/rs13091860

Academic Editors: Jihwan Choi and Fahimeh Farahnakian

Received: 29 March 2021
Accepted: 6 May 2021
Published: 10 May 2021

1. Introduction

A tropical cyclone (TC) is a kind of catastrophic weather system with enormous destructive force [1,2]. TCs encompass hurricanes, typhoons, and cyclone equivalents, and they pose a serious threat to the safety of people's lives and property and cause huge losses to agricultural production and transportation [3–7]. Therefore, accurate detection of TCs is the key to reducing the hazards [8,9].

Traditionally, the mainstream detection methods for TCs are numerical weather prediction (NWP) models, which have done a great deal of work in the development of a forecast system to provide guidance for TC prediction based on physics parameterizations and modeling techniques [10,11]. For example, the Met Office has been objectively providing real-time guidance for TC prediction and detection using its global numerical weather forecast model in recent years [12]. However, the predicted error increases because of the initial value dependency if numerical dynamical models try to simulate farther into the future [13].

The significant advantage of machine learning (ML) methods over traditional detection methods based on NWP is that ML methods do not require any assumption [14]. Decision trees (DT) are trained to classify different levels of TCs and the accuracy of TC prediction prior to 24 h was about 84.6% [15]. In addition, a convective initiation algorithm was developed from the Communication, Ocean, and Meteorological Satellite Meteorological Imager based on the DT, random forest (RF), and support vector machines (SVM) [16,17].

Recently, deep learning models, as a subset of ML methods, have had good performance in detection tasks [18–21]. For the detection task in images, object detection models based on deep learning are mainly divided into two streams based on different processing stages, which are one-stage detection models and two-stage detection models. YOLO series [22–24], SSD [25], and RetinaNet [26] are typical one-stage detection models, and R-CNN [27], Fast R-CNN [28], and Faster R-CNN [29] are classic two-stage detection models. Broadly speaking, two-stage detection models obtain high accuracy by region proposal with large-scale computing resources, whereas one-stage detection models have better performance with finite computing resources.

Additionally, deep learning models have been introduced in TC detection as well, for example, the use of deep neural networks (DNN) for existing TC detection [30], precursor detection of TCs [31], tropical and extratropical cyclone detection [32], TC track forecasting [33], and TC precursor detection by a cloud-resolving global nonhydrostatic atmospheric model [34]. However, deep learning models usually require a large number of training samples, because it is difficult to achieve high accuracy in case of finite training samples in computer vision and other fields [35–37]. At this time, transfer learning can effectively alleviate this problem by transferring the knowledge from the source domain to the target domain, and further improve the accuracy of deep learning models [38–41].

Deep transfer learning studies how to make use of knowledge transferred from other fields by DNN [42]. On the basis of different kinds of transfer techniques, there are four main categories: instance-based deep transfer learning, mapping-based deep transfer learning, network-based deep transfer learning, and adversarial-based deep transfer learning [42–46]. Instance-based deep transfer learning refers to selecting partial instances from the source domain to the training set in the target domain [43]. Mapping-based deep transfer learning refers to mapping partial instances from the source domain and target domain into a new data space [44]. Network-based deep transfer learning refers to reusing the partial network and connection parameters in the source domain and transferring it to be a part of DNN used in the target domain [45]. Adversarial-based deep transfer learning refers to introducing adversarial technologies such as generative adversarial nets (GAN) to find transferable formulations that apply to both the source domain and the target domain [46]. It is also worth noting that GAN has advantages in image processing and few-shot learning [47–49].

In order to improve the accuracy of a TC detection model in case of finite training samples, on the basis of deep transfer learning, we propose a new detection framework of tropical cyclones (NDFTC) from meteorological satellite images by combining the deep convolutional generative adversarial networks (DCGAN) and You Only Look Once (YOLO) v3 model.

The main contributions of this paper are as follows:

(1) In view of the finite data volume and complex backgrounds encountered in meteorological satellite images, a new detection framework of tropical cyclones (NDFTC) is proposed for accurate TC detection. The algorithm process of NDFTC consists of three major steps: data augmentation, a pre-training phase, and transfer learning, which ensures the effectiveness of detecting different kinds of TCs in complex backgrounds with finite data volume.

(2) We used DCGAN as the data augmentation method instead of traditional data augmentation methods such as flip and crop. DCGAN can generate images simulated to

TCs by learning the salient characteristics of TCs, which improves the utilization of finite data.

(3) We used the YOLOv3 model as the detection model in the pre-training phase. The detection model is trained with the generated images obtained from DCGAN, which can help the model to learn the salient characteristics of TCs.

(4) In the transfer learning phase, YOLOv3 is still the detection model, and it is trained with real TC images. Most importantly, the initial weights of the model are weights transferred from the model trained with generated images, which is a typically network-based deep transfer learning method. After that, the detection model can extract universal characteristics from real images of TCs and obtain a high accuracy.

2. Materials and Methods

The flowchart of the NDFTC in this paper is illustrated in Figure 1. The framework can be summarized in the following steps: (1) a dataset based on meteorological satellite images of TCs is created; (2) the dataset is divided into three sub-datasets, which are training dataset 1, training dataset 2, and test dataset; (3) DCGAN is used as the data augmentation method to generate images simulated to TCs; (4) the generated images obtained from DCGAN are inputted into the detection model YOLOv3 in the pre-training phase; and (5) the detection model is trained with real images of TCs and its initial weights are transferred from the YOLOv3 trained with generated images.

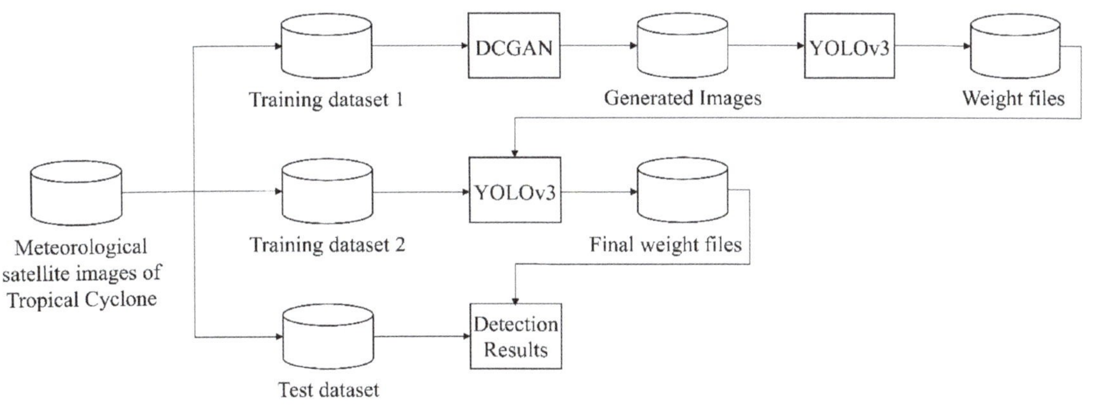

Figure 1. Overview of the proposed new detection framework of tropical cyclones (NDFTC).

2.1. Deep Convolutional Generative Adversarial Networks

As one of the research hotspots of artificial intelligence, generative adversarial networks (GAN) have developed rapidly in recent years and are widely used in image generation [50], image repair [51], visual prediction of typhoon clouds [52], and other fields.

GAN contains a generator and a discriminator [50]. The purpose of the generator is to make the discriminator unable to distinguish between the real images and generated images, whereas the purpose of the discriminator is to distinguish between real and generated images as much as possible. For the generator, an n-dimensional vector is required for input and the output is an image. The generator can be any model that can produce images, such as the simple fully connected neural network. For the discriminator, the input is a picture, and the output is the label of the picture. Similarly, the discriminator structure is similar to the generator structure, such as a network that contains convolution, and so on.

Deep convolutional generative adversarial networks (DCGANs) are an improvement on the original GAN [53]. The improvement does not include strict mathematical proof and the main contents of the improvement are as follows. Both the generator and discriminator use convolutional neural networks (CNN). Batch normalization is used in both generators and discriminators. Neither the generator nor the discriminator uses the pooling layer. The generator uses ReLU as the activation function except tanh for the output layer. The discriminator retains the structure of CNN, and the generator replaces the convolution layer with fractionally strided convolution. All layers of the discriminator use Leaky ReLU as the activation function.

2.2. *You Only Look Once (YOLO) v3 Model*

The detection model of NDFTC is the YOLOv3 model [24]. The reason why YOLOv3 is used as the detection model is that the detection speed of YOLOv3 is at least 2 times faster than SSD, RetinaNet, and Faster R-CNN [24], which can realize real-time detection of TCs and provide guarantee for disaster prevention and mitigation of TCs. In addition, YOLOv3 refers to the idea of feature pyramid networks and it ensures accurate detection of both large-size and small-size objects.

The base network of the YOLOv3 is Darknet-53. Darknet-53 uses successive 3 × 3 and 1 × 1 convolutional layers. It has 53 convolutional layers in total, as shown in Figure 1, which is why it is called Darknet-53. In addition, a large number of residual blocks are added to Darknet-53 to prevent the exploding gradient problem from network layer deepening. In the model, batch normalization is placed before the activation function Leaky ReLU, which alleviates the gradient disappearance problem. It should be noted that the concat is not the numerical addition operation for different feature graphs, but rather a direct concatenation. This means that the feature map is concatenated directly according to the channel dimension.

As for the change in image size during TC detection, the input meteorological satellite images has a size of 512 × 512 pixels. The model outputs feature maps of three sizes. The first feature map is obtained by down-sampling 32 times, and the size is 16 × 16 pixels. The second feature map is obtained by down-sampling 16 times, and the size is 32 × 32 pixels. The third feature map is obtained by down-sampling 8 times, and the size is 64 × 64 pixels. The above down-sampling is done under the guidance of YOLOv3 model by Redmon et al., which is a uniform operation of YOLOv3 and aims to obtain TC features at different scales and thus improve the detection accuracy of different kinds of TCs. Besides, the third dimension of these three feature maps is 18. Because there are three anchor boxes and each box has 1-dimensional confidence values, 4-dimensional prediction values (x^p, y^p, w^p, h^p), and 1-dimensional object class numbers, the final calculation formula is (3 × (4 + 1 + 1)) and the result is 18.

It is important to note that once the number of anchor boxes is determined, confidence values, prediction values, and object class numbers are also determined [23]. In general, an anchor box has 1-dimensional confidence values, because it is the IOU of the bounding box and the prediction box, reflecting the detection effect of this anchor box [22]. An anchor box has 4-dimensional prediction values, reflecting the coordinate information of the anchor box [22]. An anchor box has only 1-dimensional object class numbers, because our study only detects TC and not other objects.

2.3. *Loss Function*

The loss function is the error between the predicted value and the real value, which is one of the important parameters to determine the detection performance. The loss of the NDFTC includes the loss of DCGAN and the loss of YOLOv3.

2.3.1. Loss Function of DCGAN

The loss function of DCGAN includes the loss function of generator G and the loss function of discriminator D. When the generator is trained, parameters of the discriminator are fixed. When training the discriminator, parameters of the generator are fixed.

The purpose of the generator is to make the discriminator unable to distinguish between the real TC images and the generated TC images. First, the adversarial loss is introduced. $G(X)$ represents the TC images generated by the generator, Y represents the real images corresponding to it, and $D(\cdot)$ represents the discriminant probability of the generated images. The adversarial loss is as follows:

$$L_G^{adv} = \log(1 - D(G(X)) \tag{1}$$

By minimizing Formula (1), the generator can fool the discriminator, which means that the discriminator cannot distinguish between real images and generated images. Next, the L_1 loss function is introduced to measure the distance between generated images and real images.

$$L_1 = \sum_{i=1}^{P_w} \sum_{j=1}^{P_h} ||G(X)(i,j) - Y(i,j)||_1 \tag{2}$$

where (i,j) represents pixel coordinates, and P_w and P_h are the width and height of TC images, respectively.

The generator's total loss function is as follows:

$$L_G = \lambda_1 L_G^{adv} + \lambda_2 L_1 \tag{3}$$

where λ_1 and λ_2 are empirical weight parameters. The generator can generate high-quality images of TCs by minimizing Formula (3).

The purpose of the discriminator D is to distinguish between the real TC images and the generated TC images. To achieve this goal, the adversarial loss function of the discriminator is as follows:

$$L_D^{adv} = -\log(D(Y)) - \log(1 - D(G(X)) \tag{4}$$

For Equation (4), if the real image is wrongly judged as the generated image, or the generated image is wrongly judged as the real image, then an infinite situation will appear in Formula (4), which means that the discriminator should still be optimized. If the value of Formula (4) decreases gradually, it means that the discriminator is trained better and better.

2.3.2. Loss Function of YOLOv3

The loss function of YOLOv3 includes boundary box loss, confidence loss, and classification loss. The smaller the loss value, the better the performance of the model. The parameters involved in the loss function are introduced below.

The model divides the input image into an $S \times S$ grid. Each grid cell is responsible for detecting TCs if the center of a TC falls into a grid cell. The grid cell predicts B bounding boxes and confidence scores. These scores reflect how confident the model is that the box contains an object.

The first part of the total loss function is the boundary box loss, which is used to measure the difference between the real box and the predicted box, as follows:

$$L_{box} = \sum_{i=1}^{s^2 \times B} [\left(x_i^p - x_i^g\right)^2 + \left(y_i^p - y_i^g\right)^2 + \left(w_i^p - w_i^g\right)^2 + \left(h_i^p - h_i^g\right)^2] \tag{5}$$

where i is the number of bounding boxes, and $\left(x_i^p, y_i^p, w_i^p, h_i^p\right)$ is the positional parameter of the predicted box. x^p and y^p represent the center point coordinates of the predicted box, and w^p and h^p represent the width and height of the predicted box, respectively. Similarly, $\left(x_i^g, y_i^g, w_i^g, h_i^g\right)$ is the parameter of the true box.

The second part of the total loss function is the confidence loss, which reflects how confident the model is that the box contains an object. The confidence loss is as follows:

$$L_{conf} = -\sum_{i=1}^{s^2 \times B} [h_i \times lnc_i + (1 - h_i) \times \ln(1 - c_i)] \tag{6}$$

where c_i represents the probability of the object in the anchor box i. $h_i \in \{0, 1\}$ represents whether the object is present in the anchor box i, in which 1 means yes and 0 means no.

The third part of the total loss function is the classification loss as follows:

$$L_{class} = -\sum_{i=1}^{s^2 \times B} \sum_{k \in classes} [h_{ik} \times lnc_{ik}] \tag{7}$$

where c_{ik} represents the probability of the object of class k in the anchor box i. $h_{ik} \in \{0, 1\}$ represents whether the object of class k is present in the anchor box i, in which 1 means yes and 0 means no. In this paper, there is only one kind of object, so $k = 1$.

To sum up, the total loss function of the YOLOv3 model is as follows:

$$L_{total} = \lambda_1 L_{box} + \lambda_2 L_{conf} + \lambda_3 L_{class} \tag{8}$$

where λ_1, λ_2, and λ_3 are empirical weight parameters, and $\lambda_1 = \lambda_2 = \lambda_3 = 1$ in this paper.

2.4. Algorithm Process

According to the above description, the specific algorithm process is shown as follows.

Algorithm 1 The algorithm process of NDFTC.

Start
Input: 2400 meteorological satellite images of TCs; the images were collected from 1979 to 2019 in the South West Pacific Area.
A. Data Augmentation
(1) A total of 600 meteorological satellite images are input into the DCGAN model. The selection rule for these images is to randomly select 18 images from the TCs that occur every year (1979–2010), which contains the common characteristics of TCs over these years.
(2) A total of 1440 generated images with TC characteristics are obtained in the DCGAN model. These generated images are only used as training samples in the pre-training phase.
B. Pre-Training Phase
(3) The generated images obtained from step (2) are inputted into the YOLOv3 model.
(4) Feature extraction and preliminary detection of the generated images are completed.
(5) The weight trained to 10,000 times in step (4) is reserved in this phase.
C. Transfer Learning
(6) A total of 1800 meteorological satellite images are still available after step (1). A total of 80% of these data are used as the training samples in this phase. In other words, 1440 meteorological satellite images from 1979 to 2011 are used as training samples.
(7) The model starts to train with training samples of step (6) and weights of step (5) are initial weights in this phase, which is a typically network-based deep transfer learning method.
(8) A total of 360 meteorological satellite images from 2011 to 2019 are used as the testing samples. Then, the test is completed.
Output: detection results, accuracy, average precision.
End

3. Experimental Results

3.1. Data Set

The data set we used includes meteorological satellite observation images in the Southwest Pacific area from 1979 to 2019. These images, provided by the National Institute of Informatics, are meteorological satellite images with a size of 512 × 512 pixels. For more details on the meteorological satellite images we used in this study [54], see the website: http://agora.ex.nii.ac.jp/digital-typhoon/search_date.html.en#id2 (accessed on 29 March 2021).

In this paper, a total of 2400 real TC images were used. Among them, 600 real images were input into DCGAN model to produce 1440 generated images for training the detection model in the pre-training phase. Additionally, 80% of the remaining 1800 real TC images, which were from 1979 to 2011, were used to train the model. A total of 20% of the remaining 1800 real TC images, which were from 2011 to 2019, were used to test the model.

In other words, in the transfer learning phase, the selection rule for training and test data was based on the time when the TC was captured by the meteorological satellite. A total of 80% of the data used for training was historical data occurring from 1979 to 2011, whereas 20% of the data used for testing was recent data occurring from 2011 to 2019. Such a data selection method of training with historical data and testing with recent data is effective in the application of deep learning in meteorology [55], and thus we also adopted this data selection method.

3.2. Experiment Setup

In order to show the superiority of NDFTC in the training process and detection results, a TC detection model for comparison was also trained, which was only based on YOLOv3 and did not use NDFTC. In order to train and test this TC detection model for comparison, we still used 2400 real TC images, 80% of which were used for training and 20% for testing.

For the sake of fairness, the total number of training times for both NDFTC and YOLOv3 was 50,000. For the NDFTC, it used generated TC images to train 10,000 times, and then it used real TC images to train 40,000 times. For the detection model only based on YOLOv3, it was trained 50,000 times using real TC images. In the training process, the change of loss function values of NDFTC and detection model only based on YOLOv3 are shown in Figure 2.

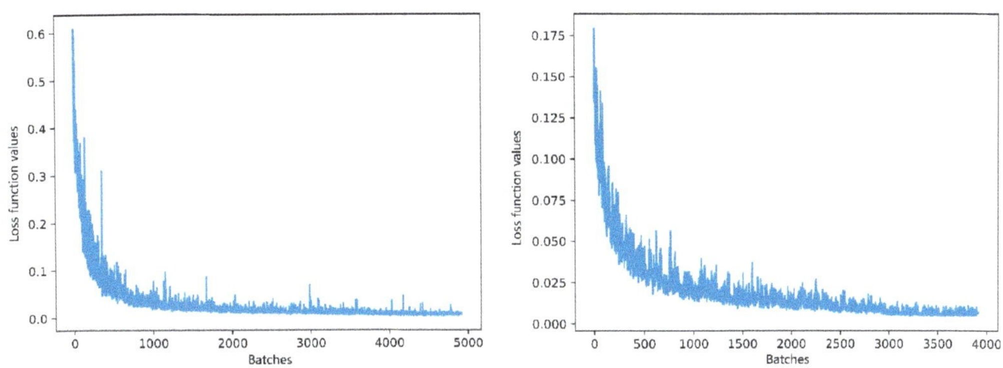

(a) (b)

Figure 2. (**a**) The change of loss function values of YOLOv3 to train real TC images; (**b**) the change of loss function values of NDFTC to train real TC images.

Figure 2 visualizes the change of loss function values of YOLOv3 and NDFTC in the training process. Compared with the TC detection model only including YOLOv3, the NDFTC proposed in this paper had smaller loss function values and a more stable training process.

In order to show the stability of NDFTC during the training process from another perspective, the changes of region average IOU are also visualized in Figure 3. Region average IOU is the intersection over union (IOU) between the predicted box and the ground truth [22]. It is one of the most important indicators to measure the stability of models in the training process, and is commonly found in deep learning models such as YOLOv1 [22], YOLOv2 [23], YOLOv3 [24], and YOLOv4 [56]. In general, the closer it is to 1, the better the model is trained.

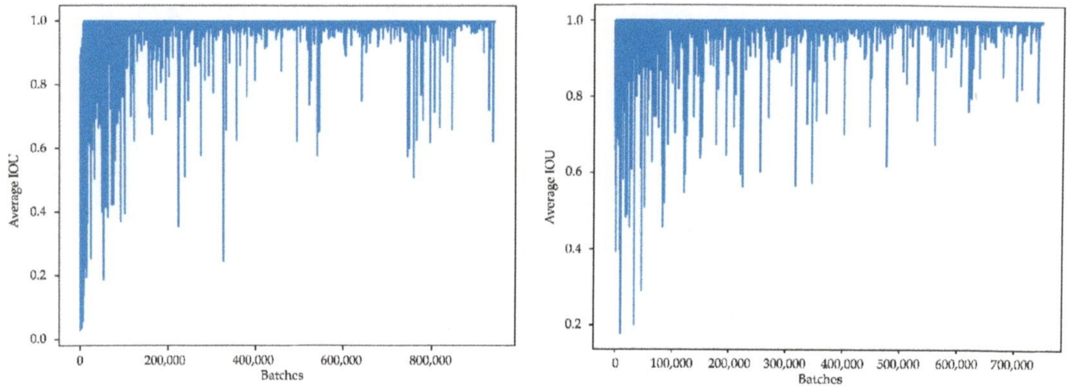

(**a**) (**b**)

Figure 3. (**a**) The change in region average IOU of YOLOv3 to train real TC images; (**b**) the change in region average IOU of NDFTC to train real TC images.

In Figure 3, the region average IOU of the models in the training process was generally decreasing. However, the region average IOU of YOLOv3 oscillated more sharply when the training reached a later stage. Compared with the TC detection model only including YOLOv3, the NDFTC oscillated less in the whole training process. This means that the NDFTC converged faster and was more stable in the training process.

3.3. Results and Discussion

In order to evaluate the detection effect of the NDFTC proposed in this paper, ACC and AP were used as evaluation indexes.

ACC refers to accuracy, which means the proportion of TCs detected correctly by the model in all images. The definition of ACC is as follows:

$$Accuracy = \frac{TP}{ALL} \tag{9}$$

where *TP* refers to the number of TC images detected correctly by the model, and *ALL* refers to the number of all images.

AP refers to average precision, which takes into account cases such as detection error and detection omission phenomenon, and it is a common index for evaluating YOLO series models such as YOLOv1, YOLOv2, and YOLOv3 by Redmon et al. [22–24]. AP is defined by precision and recall:

$$Precision = \frac{TP}{TP + FP} \tag{10}$$

$$Recall = \frac{TP}{TP + FN} \tag{11}$$

where *TP* refers to the number of TCs correctly recognized as TCs by the detection model, *FP* refers to the number of other objects recognized as TCs by the detection model, and *FN* refers to the number of TCs recognized as other objects by the detection model [57,58]. Then the P–R curve can be obtained by using the recall of TCs as the x-coordinate and the precision of TCs as the y-coordinate [59], and the area under the curve is AP, which is the index that evaluates the detection effectiveness of the NFDTC.

Figure 4 shows the ACC and AP of NDFTC and other models in the test set when the training times were 10,000, 20,000, 30,000, 40,000, and 50,000. Apparently, Figure 4 reflects that NDFTC performed better than YOLOv3 and other models with the same training times. Finally, the experimental results show that the NDFTC had better performance, with an ACC of 97.78% and AP of 81.39%, in comparison to the YOLOv3, with an ACC of 93.96% and AP of 80.64%.

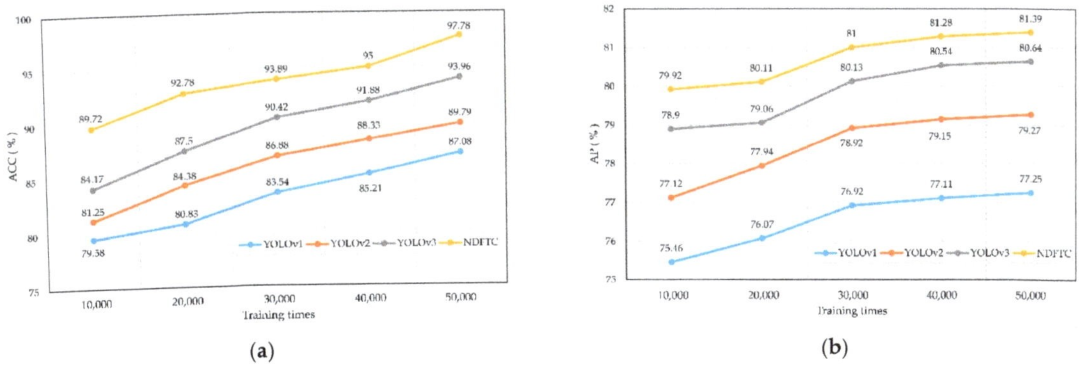

Figure 4. Performance of NDFTC and other models with ACC and AP: (**a**) ACC of NDFTC and other models; (**b**) AP of NDFTC and other models.

In order to evaluate the detection effect on different kinds of TCs, all TCs in the test set were divided into five categories. According to the National Standard for Tropical Cyclone Grade (GB/T 19201-2006), TC intensity includes tropical storm (TS), severe tropical storm (STS), typhoon (TY), severe typhoon (STY), and super typhoon (SuperTY). The ACC performance of the NDFTC and other models on the test set is shown in Table 1. It shows that the NDFTC generally had a higher ACC. The best result was from NDFTC for SuperTY detection, and at that time the ACC reached 98.59%.

Table 1. ACC performance of the NDFTC and other models on the test set for five kinds of TCs.

Model	Typhoon Types	10,000 Times	20,000 Times	30,000 Times	40,000 Times	50,000 Times
YOLOv3	TS	71.21	80.30	87.88	90.91	92.42
	STS	83.46	86.47	89.47	90.98	94.74
	TY	85.59	88.29	90.09	91.89	92.79
	STY	88.75	90.00	91.25	92.50	95.00
	SuperTY	88.89	91.11	93.33	93.33	94.44
NDFTC	TS	87.50	92.50	92.50	95.00	97.50
	STS	88.46	91.35	92.31	93.27	98.07
	TY	89.41	92.94	94.12	95.29	96.47
	STY	91.67	93.33	95.00	96.67	98.33
	SuperTY	91.55	94.37	95.77	97.18	98.59

Next, the AP performance of the NDFTC and other models on the test set is shown in Table 2. It can be found that the NDFTC basically had a higher AP. The best result was from NDFTC for STY detection, which was 91.34%.

Table 2. AP performance of the NDFTC and other models on the test set for five kinds of TCs.

Model	Typhoon Types	10,000 Times	20,000 Times	30,000 Times	40,000 Times	50,000 Times
YOLOv3	TS	60.91	61.24	63.96	68.26	66.85
	STS	80.77	83.46	83.59	82.42	86.84
	TY	79.16	76.93	79.91	80.90	78.11
	STY	88.66	89.12	87.12	87.60	88.63
	SuperTY	82.82	81.14	83.23	81.43	79.81
NDFTC	TS	67.16	69.12	63.55	67.96	63.89
	STS	78.13	74.64	84.15	81.40	82.22
	TY	79.76	83.60	81.57	86.70	83.04
	STY	89.23	86.97	89.79	84.89	91.34
	SuperTY	84.03	85.20	79.89	80.50	82.52

Last but not least, an example of TC detection results is shown in Figure 5, which is the super typhoon Marcus in 2018. It can be found that the NDFTC had a more detailed detection result, because the prediction box of NDFTC fit Marcus better. More importantly, compared with the TC detection model only including YOLOv3, the detection result of NDFTC was more consistent with the physical characteristics of TCs, because the spiral rainbands at the bottom of Marcus were also included in the detection box of NDFTC.

(a)

(b)

Figure 5. An example of TC detection results, which is the super typhoon Marcus in 2018. (**a**) The detection result of YOLOv3; (**b**) the detection result of NDFTC.

4. Discussion

To begin with, the complexity of NDFTC is explained here. Compared to the complex network architecture and huge number of parameters of YOLOv3, the complexity of DCGAN, which is a relatively simple network, could be negligible [60]. Therefore, the complexity of the NDFTC in this paper was approximately equal to that of the YOLOv3 model, conditional on a finite data set and the same scale of computing resources. More importantly, compared with the YOLOv3 model, NDFTC further improved the detection accuracy of TCs with almost no increase in complexity, which proves that NDFTC ensures generalization performance.

Then, the way in which the generated and real images are used in different phases needs to be emphasized again. In 2020, Maryam Hammami et al. proposed a CycleGAN and YOLO combined model for data augmentation and used generated data and real data to train a YOLO detector, in which generated data and real data are simultaneously input into YOLO for training [61]. In our study, the detector was trained using only generated images in the pre-training phase and only real images in the transfer learning phase, which is a typically network-based deep transfer learning method. Additionally, the average IOU and loss function values during the training process are plotted in this paper to reflect the stability of NDFTC.

Furthermore, it is necessary to explain the proportion of the data set allocated. In NDFTC, the initial dataset is composed of meteorological satellite images of TCs, and when it is divided into training dataset 1, training dataset 2, and test dataset according to Algorithm 1, then training datasets 1 and 2 must include the real images of TC. This means that training datasets 1 and 2 must contain TC features at the same time, which is a prerequisite for the adoption of NDFTC.

Finally, we need to explain the reason why 80% of the real images of TC were used for training and the rest for testing. In general, for finite datasets that are not very large, such a training and testing ratio is a common method in the field of deep learning [62,63]. It is generally believed that when the total number of images in the dataset reaches tens of thousands or even hundreds of thousands, the proportion of the training set can exceed 90% [63]. Of course, considering that the dataset of TCs used in this paper has only thousands of images, 80% was acceptable. More importantly, for object detection tasks with finite datasets, setting a smaller training dataset usually leads to lower accuracy, so we chose the common ratio of 80% over others.

5. Conclusions

In this paper, on the basis of deep transfer learning, we propose a new detection framework of tropical cyclones (NDFTC) from meteorological satellite images by combining the DCGAN and YOLOv3. The algorithm process of NDFTC consists of three major steps: data augmentation, a pre-training phase, and transfer learning, which ensures the effectiveness of detecting different kinds of TCs in complex backgrounds with finite data volume. We used DCGAN as the data augmentation method instead of traditional data augmentation methods because DCGAN can generate images simulated to TCs by learning the salient characteristics of TCs, which improves the utilization of finite data. In the pre-training phase, we used YOLOv3 as the detection model and it was trained with the generated images obtained from DCGAN, which helped the model learn the salient characteristics of TCs. In the transfer learning phase, we trained the detection model with real images of TCs and its initial weights were transferred from the YOLOv3 trained with generated images, which is a typically network-based deep transfer learning method and can improve the stability and accuracy of the model. The experimental results show that the NDFTC had better performance, with an ACC of 97.78% and AP of 81.39%, in comparison to the YOLOv3, with an ACC of 93.96% and AP of 80.64%. On the basis of the above conclusions, we think that our NDFTC with high accuracy has promising potential for detecting different kinds of TCs and we believe that NDFTC could benefit current TC-detection tasks and similar detection tasks, especially for those tasks with finite data volume.

Author Contributions: Conceptualization, T.S. and P.X.; data curation, P.X. and Y.L.; formal analysis, P.X., F.M., X.T. and B.L.; funding acquisition, S.P., T.S. and D.X.; methodology, T.S. and P.X.; project administration, S.P., D.X., T.S. and F.M.; validation, P.X.; writing—original draft, P.X. All authors have read and agreed to the published version of the manuscript.

Funding: This work was supported by the National Key Research and Development Program (no. 2018YFC1406201) and the Natural Science Foundation of China (grant: U1811464). The project was supported by the Innovation Group Project of the Southern Marine Science and Engineering Guangdong Laboratory (Zhuhai) (no. 311020008), the Natural Science Foundation of Shandong Province (grant no. ZR2019MF012), and the Taishan Scholars Fund (grant no. ZX20190157).

Data Availability Statement: The data used in this study are openly available at the National Institute of Informatics (NII) at http://agora.ex.nii.ac.jp/digital-typhoon/search_date.html.en#id2 (accessed on 29 March 2021).

Conflicts of Interest: The authors declare no conflict of interest.

Abbreviations

The following abbreviations are used in this manuscript:

TC	Tropical cyclone
TCs	Tropical cyclones
NDFTC	New detection framework of tropical cyclones
GAN	Generative adversarial nets
DCGAN	Deep convolutional generative adversarial networks
YOLO	You Only Look Once
NWP	Numerical weather prediction
ML	Machine learning
DT	Decision trees
RF	Random forest
SVM	Support vector machines
DNN	Deep neural networks
ReLU	Rectified linear unit
TP	True positive
TN	True negative
FP	False positive
FN	False negative
ACC	Accuracy
AP	Average precision
IOU	Intersection over union

References

1. Khalil, G.M. Cyclones and storm surges in Bangladesh: Some mitigative measures. *Nat. Hazards* **1992**, *6*, 11–24. [CrossRef]
2. Hunter, L.M. Migration and Environmental Hazards. *Popul. Environ.* **2005**, *26*, 273–302. [CrossRef] [PubMed]
3. Mabry, C.M.; Hamburg, S.P.; Lin, T.-C.; Horng, F.-W.; King, H.-B.; Hsia, Y.-J. Typhoon Disturbance and Stand-level Damage Patterns at a Subtropical Forest in Taiwan[1]. *Biotropica* **1998**, *30*, 238–250. [CrossRef]
4. Dale, V.H.; Joyce, L.A.; McNulty, S.; Neilson, R.P.; Ayres, M.P.; Flannigan, M.D.; Hanson, P.J.; Irland, L.C.; Lugo, A.E.; Peterson, C.J.; et al. Climate Change and Forest Disturbances. *Bioscience* **2001**, *51*, 723. [CrossRef]
5. Pielke, R.A., Jr.; Gratz, J.; Landsea, C.W.; Collins, D.; Saunders, M.A.; Musulin, R. Normalized hurricane damage in the united states: 1900–2005. *Nat. Hazards Rev.* **2008**, *9*, 29–42. [CrossRef]
6. Zhang, Q.; Liu, Q.; Wu, L. Tropical Cyclone Damages in China 1983–2006. *Am. Meteorol. Soc.* **2009**, *90*, 489–496. [CrossRef]
7. Lian, Y.; Liu, Y.; Dong, X. Strategies for controlling false online information during natural disasters: The case of Typhoon Mangkhut in China. *Technol. Soc.* **2020**, *62*, 101265. [CrossRef]
8. Kang, H.Y.; Kim, J.S.; Kim, S.Y.; Moon, Y.I. Changes in High- and Low-Flow Regimes: A Diagnostic Analysis of Tropical Cyclones in the Western North Pacific. *Water Resour. Manag.* **2017**, *31*, 3939–3951. [CrossRef]
9. Kim, J.S.; Jain, S.; Kang, H.Y.; Moon, Y.I.; Lee, J.H. Inflow into Korea's Soyang Dam: Hydrologic variability and links to typhoon impacts. *J. Hydro Environ. Res.* **2019**, *22*, 50–56. [CrossRef]
10. Burton, D.; Bernardet, L.; Faure, G.; Herndon, D.; Knaff, J.; Li, Y.; Mayers, J.; Radjab, F.; Sampson, C.; Waqaicelua, A. Structure and intensity change: Operational guidance. In Proceedings of the 7th International Workshop on Tropical Cyclones, La Réunion, France, 15–20 November 2010.
11. Halperin, D.J.; Fuelberg, H.E.; Hart, R.E.; Cossuth, J.H.; Sura, P.; Pasch, R.J. An Evaluation of Tropical Cyclone Genesis Forecasts from Global Numerical Models. *Weather Forecast.* **2013**, *28*, 1423–1445. [CrossRef]
12. Heming, J.T. Tropical cyclone tracking and verification techniques for Met Office numerical weather prediction models. *Meteorol. Appl.* **2017**, *26*, 1–8. [CrossRef]
13. Park, M.-S.; Elsberry, R.L. Latent Heating and Cooling Rates in Developing and Nondeveloping Tropical Disturbances during TCS-08: TRMM PR versus ELDORA Retrievals*. *J. Atmos. Sci.* **2013**, *70*, 15–35. [CrossRef]

14. Rhee, J.; Im, J.; Carbone, G.J.; Jensen, J.R. Delineation of climate regions using in-situ and remotely-sensed data for the Carolinas. *Remote Sens. Environ.* **2008**, *112*, 3099–3111. [CrossRef]
15. Zhang, W.; Fu, B.; Peng, M.S.; Li, T. Discriminating Developing versus Nondeveloping Tropical Disturbances in the Western North Pacific through Decision Tree Analysis. *Weather Forecast.* **2015**, *30*, 446–454. [CrossRef]
16. Han, H.; Lee, S.; Im, J.; Kim, M.; Lee, M.-I.; Ahn, M.H.; Chung, S.-R. Detection of Convective Initiation Using Meteorological Imager Onboard Communication, Ocean, and Meteorological Satellite Based on Machine Learning Approaches. *Remote Sens.* **2015**, *7*, 9184–9204. [CrossRef]
17. Kim, D.H.; Ahn, M.H. Introduction of the in-orbit test and its performance for the first meteorological imager of the Communication, Ocean, and Meteorological Satellite. *Atmos. Meas. Tech.* **2014**, *7*, 2471–2485. [CrossRef]
18. Xu, Y.; Meng, X.; Li, Y.; Xu, X. Research on privacy disclosure detection method in social networks based on multi-dimensional deep learning. *Comput. Mater. Contin.* **2020**, *62*, 137–155. [CrossRef]
19. Peng, H.; Li, Q. Research on the automatic extraction method of web data objects based on deep learning. *Intell. Autom. Soft Comput.* **2020**, *26*, 609–616. [CrossRef]
20. He, S.; Li, Z.; Tang, Y.; Liao, Z.; Li, F.; Lim, S.-J. Parameters compressing in deep learning. *Comput. Mater. Contin.* **2020**, *62*, 321–336. [CrossRef]
21. Courtrai, L.; Pham, M.-T.; Lefèvre, S. Small Object Detection in Remote Sensing Images Based on Super-Resolution with Auxiliary Generative Adversarial Networks. *Remote Sens.* **2020**, *12*, 3152. [CrossRef]
22. Redmon, J.; Divvala, S.; Girshick, R.; Farhadi, A. You only look once: Unified, real-time object detection. In Proceedings of the IEEE Conference on Computer Vision and Pattern Recognition, Las Vegas, NV, USA, 27–30 June 2016; pp. 779–788.
23. Redmon, J.; Farhadi, A. YOLO9000: Better, faster, stronger. In Proceedings of the IEEE Conference on Computer Vision and Pattern Recognition, Honolulu, HI, USA, 21–26 July 2017; pp. 7263–7271.
24. Redmon, J.; Farhadi, A. Yolov3: An incremental improvement. *arXiv* **2018**, arXiv:1804.02767.
25. Liu, W.; Anguelov, D.; Erhan, D.; Szegedy, C.; Reed, S.; Fu, C.Y.; Berg, A.C. Ssd: Single shot multibox detector. In *European Conference on Computer Vision*; Springer: Amsterdam, The Netherlands, 2016; pp. 21–37.
26. Lin, T.Y.; Goyal, P.; Girshick, R.; He, K.; Dollár, P. Focal loss for dense object detection. In Proceedings of the IEEE International Conference on Computer Vision, Venice, Italy, 22–29 October 2017; pp. 2980–2988.
27. Girshick, R.; Donahue, J.; Darrell, T.; Malik, J. Rich Feature Hierarchies for Accurate Object Detection and Semantic Segmentation. In Proceedings of the CVPR, Columbus, OH, USA, 24–27 June 2014.
28. Girshick, R. Fast r-cnn. In Proceedings of the IEEE International Conference on Computer Vision, Santiago, Chile, 11–18 December 2015; pp. 1440–1448.
29. Ren, S.; He, K.; Girshick, R.; Sun, J. Faster r-cnn: Towards real-time object detection with region proposal networks. In Proceedings of the Advances in Neural Information Processing Systems 28 (NIPS 2015), Montreal, QC, Canada, 7–12 December 2015; pp. 91–99.
30. Liu, Y.; Racah, E.; Correa, J. Application of deep convolutional neural networks for detecting extreme weather in climate datasets. *arXiv* **2016**, arXiv:1605.01156.
31. Nakano, D.M.; Sugiyama, D. Detecting Precursors of Tropical Cyclone using Deep Neural Networks. In Proceedings of the 7th International Workshop on Climate Informatics, Boulder, CO, USA, 20–22 September 2017.
32. Kumler-Bonfanti, C.; Stewart, J.; Hall, D. Tropical and Extratropical Cyclone Detection Using Deep Learning. *J. Appl. Meteorol. Climatol.* **2020**, *59*, 1971–1985. [CrossRef]
33. Giffard-Roisin, S.; Yang, M.; Charpiat, G. Tropical cyclone track forecasting using fused deep learning from aligned reanalysis data. *Front. Big Data* **2020**, *3*, 1. [CrossRef] [PubMed]
34. Matsuoka, D.; Nakano, M.; Sugiyama, D. Deep learning approach for detecting tropical cyclones and their precursors in the simulation by a cloud-resolving global nonhydrostatic atmospheric model. *Prog. Earth Planet. Sci.* **2018**, *5*, 1–16. [CrossRef]
35. Cao, J.; Chen, Z.; Wang, B. Deep Convolutional networks with superpixel segmentation for hyperspectral image classification. In Proceedings of the 2015 IEEE International Geoscience and Remote Sensing Symposium (IGARSS), Beijing, China, 10–15 July 2016; pp. 3310–3313.
36. Li, Z.; Guo, F.; Li, Q.; Ren, G.; Wang, L. An Encoder–Decoder Convolution Network with Fine-Grained Spatial Information for Hyperspectral Images Classification. *IEEE Access* **2020**, *8*, 33600. [CrossRef]
37. Gorban, A.; Mirkes, E.; Tukin, I. How deep should be the depth of convolutional neural networks: A backyard dog case study. *Cogn. Comput.* **2020**, *12*, 388. [CrossRef]
38. Pan, S.J.; Tsang, I.W.; Kwok, J.T.; Yang, Q. Domain Adaptation via Transfer Component Analysis. *IEEE Trans. Neural Netw.* **2011**, *22*, 199–210. [CrossRef] [PubMed]
39. Yang, J.; Zhao, Y.; Chan, J. Learning and transferring deep joint spectral–spatial features for hyperspectral classification. *IEEE Trans. Geosci. Remote Sens.* **2017**, *55*, 4729–4742. [CrossRef]
40. Liu, X.; Sun, Q.; Meng, Y.; Fu, M.; Bourennane, S. Hyperspectral image classification based on parameter-optimized 3D-CNNs combined with transfer learning and virtual samples. *Remote Sens.* **2018**, *10*, 1425. [CrossRef]
41. Jiang, Y.; Li, Y.; Zhang, H. Hyperspectral image classification based on 3-D separable ResNet and transfer learning. *IEEE Geosci. Remote Sens. Lett.* **2019**, *16*, 1949–1953. [CrossRef]
42. Tan, C.; Sun, F.; Kong, T.; Zhang, W.; Yang, C.; Liu, C. A Survey on Deep Transfer Learning. *arXiv* **2018**, arXiv:1808.01974.

43. Liu, X.; Liu, Z.; Wang, G.; Cai, Z.; Zhang, H. Ensemble transfer learning algorithm. *IEEE Access* **2018**, *6*, 2389–2396. [CrossRef]
44. Tzeng, E.; Hoffman, J.; Zhang, N.; Saenko, K.; Darrell, T. Deep domain confusion: Maximizing for domain invariance. *arXiv* **2014**, arXiv:1412.3474.
45. Yosinski, J.; Clune, J.; Bengio, Y.; Lipson, H. How transferable are features in deep neural networks? *arXiv* **2014**, arXiv:1411.1792.
46. Long, M.; Cao, Z.; Wang, J.; Jordan, M.I. Domain adaptation with randomized multilinear adversarial networks. *arXiv* **2017**, arXiv:1705.10667.
47. Zhao, M.; Liu, X.; Yao, X. Better Visual Image Super-Resolution with Laplacian Pyramid of Generative Adversarial Networks. *CMC Comput. Mater. Contin.* **2020**, *64*, 1601–1614. [CrossRef]
48. Fu, K.; Peng, J.; Zhang, H. Image super-resolution based on generative adversarial networks: A brief review. *Comput. Mater. Contin.* **2020**, *64*, 1977–1997. [CrossRef]
49. Li, X.; Liang, Y.; Zhao, M. Few-shot learning with generative adversarial networks based on WOA13 data. *Comput. Mater. Contin.* **2019**, *60*, 1073–1085. [CrossRef]
50. Goodfellow, I.; Pouget-Abadie, J.; Mirza, M.; Xu, B.; Warde-Farley, D.; Ozair, S.; Courville, A.; Bengio, Y. Generative adversarial nets. *Adv. Neural Inf. Process. Syst.* **2014**, *27*, 2672–2680.
51. Denton, E.; Gross, S.; Fergus, R. Semi-supervised learning with context-conditional generative adversarial networks. *arXiv* **2016**, arXiv:1611.06430.
52. Li, H.; Gao, S.; Liu, G.; Guo, D.L.; Grecos, C.; Ren, P. Visual Prediction of Typhoon Clouds With Hierarchical Generative Adversarial Networks. *IEEE Geosci. Remote Sens. Lett.* **2020**, *17*, 1478–1482. [CrossRef]
53. Radford, A.; Metz, L.; Chintala, S. Unsupervised representation learning with deep convolutional generative adversarial networks. *arXiv* **2015**, arXiv:1511.06434.
54. National Institute of Informatics. Digital Typhoon. 2009. Available online: http://agora.ex.nii.ac.jp/digital-typhoon/search_date.html.en#id2 (accessed on 29 March 2021).
55. Ham, Y.; Kim, J.; Luo, J. Deep learning for multi-year ENSO forecasts. *Nature* **2019**, *573*, 568–572. [CrossRef]
56. Bochkovskiy, A.; Wang, C.Y.; Liao, H.Y.M. YOLOv4: Optimal speed and accuracy of object detection. *arXiv* **2020**, arXiv:2004.10934.
57. Rafael Padilla. Object Detection Metrics. 2018. Available online: https://github.com/rafaelpadilla/Object-Detection-Metrics (accessed on 22 June 2018).
58. Everingham, M.; Van Gool, L.; Williams, C. The pascal visual object classes (voc) challenge. *Int. J. Comput. Vis.* **2010**, *88*, 303–338. [CrossRef]
59. Davis, J.; Goadrich, M. The relationship between Precision-Recall and ROC curves. In Proceedings of the 23rd International Conference on Machine Learning, Pittsburgh, PA, USA, 25–29 June 2006; pp. 233–240.
60. Neyshabur, B.; Bhojanapalli, S.; McAllester, D.; Srebro, N. Exploring generalization in deep learning. *arXiv* **2017**, arXiv:1706.08947.
61. Hammami, M.; Friboulet, D.; Kechichian, R. Cycle GAN-Based Data Augmentation for Multi-Organ Detection in CT Images Via Yolo. In Proceedings of the IEEE International Conference on Image Processing (ICIP), Anchorage, AK, USA, 25–28 October 2020; pp. 390–393.
62. Song, T.; Jiang, J.; Li, W. A deep learning method with merged LSTM Neural Networks for SSHA Prediction. *IEEE J. Sel. Top. Appl. Earth Obs. Remote Sens.* **2020**, *13*, 2853–2860. [CrossRef]
63. Song, T.; Wang, Z.; Xie, P. A novel dual path gated recurrent unit model for sea surface salinity prediction. *J. Atmos. Ocean. Technol.* **2020**, *37*, 317–325. [CrossRef]